Lecture Notes in Physics

Springer-Verlag Berlin Heidelberg GmbH

The Editorial Policy for Proceedings

The series Lecture Notes in Physics reports new developments in physical research and teaching – quickly, informally, and at a high level. The proceedings to be considered for publication in this series should be limited to only a few areas of research, and these should be closely related to each other. The contributions should be of a high standard and should avoid lengthy redraftings of papers already published or about to be published elsewhere. As a whole, the proceedings should aim for a balanced presentation of the theme of the conference including a description of the techniques used and enough motivation for a broad readership. It should not be assumed that the published proceedings must reflect the conference in its entirety. (A listing or abstracts of papers presented at the meeting but not included in the proceedings could be added as an appendix.)
When applying for publication in the series Lecture Notes in Physics the volume's editor(s) should submit sufficient material to enable the series editors and their referees to make a fairly accurate evaluation (e.g. a complete list of speakers and titles of papers to be presented and abstracts). If, based on this information, the proceedings are (tentatively) accepted, the volume's editor(s), whose name(s) will appear on the title pages, should select the papers suitable for publication and have them refereed (as for a journal) when appropriate. As a rule discussions will not be accepted. The series editors and Springer-Verlag will normally not interfere with the detailed editing except in fairly obvious cases or on technical matters.
Final acceptance is expressed by the series editor in charge, in consultation with Springer-Verlag only after receiving the complete manuscript. It might help to send a copy of the authors' manuscripts in advance to the editor in charge to discuss possible revisions with him. As a general rule, the series editor will confirm his tentative acceptance if the final manuscript corresponds to the original concept discussed, if the quality of the contribution meets the requirements of the series, and if the final size of the manuscript does not greatly exceed the number of pages originally agreed upon. The manuscript should be forwarded to Springer-Verlag shortly after the meeting. In cases of extreme delay (more than six months after the conference) the series editors will check once more the timeliness of the papers. Therefore, the volume's editor(s) should establish strict deadlines, or collect the articles during the conference and have them revised on the spot. If a delay is unavoidable, one should encourage the authors to update their contributions if appropriate. The editors of proceedings are strongly advised to inform contributors about these points at an early stage.
The final manuscript should contain a table of contents and an informative introduction accessible also to readers not particularly familiar with the topic of the conference. The contributions should be in English. The volume's editor(s) should check the contributions for the correct use of language. At Springer-Verlag only the prefaces will be checked by a copy-editor for language and style. Grave linguistic or technical shortcomings may lead to the rejection of contributions by the series editors. A conference report should not exceed a total of 500 pages. Keeping the size within this bound should be achieved by a stricter selection of articles and not by imposing an upper limit to the length of the individual papers. Editors receive jointly 30 complimentary copies of their book. They are entitled to purchase further copies of their book at a reduced rate. As a rule no reprints of individual contributions can be supplied. No royalty is paid on Lecture Notes in Physics volumes. Commitment to publish is made by letter of interest rather than by signing a formal contract. Springer-Verlag secures the copyright for each volume.

The Production Process

The books are hardbound, and the publisher will select quality paper appropriate to the needs of the author(s). Publication time is about ten weeks. More than twenty years of experience guarantee authors the best possible service. To reach the goal of rapid publication at a low price the technique of photographic reproduction from a camera-ready manuscript was chosen. This process shifts the main responsibility for the technical quality considerably from the publisher to the authors. We therefore urge all authors and editors of proceedings to observe very carefully the essentials for the preparation of camera-ready manuscripts, which we will supply on request. This applies especially to the quality of figures and halftones submitted for publication. In addition, it might be useful to look at some of the volumes already published. As a special service, we offer free of charge LaTeX and TeX macro packages to format the text according to Springer-Verlag's quality requirements. We strongly recommend that you make use of this offer, since the result will be a book of considerably improved technical quality. To avoid mistakes and time-consuming correspondence during the production period the conference editors should request special instructions from the publisher well before the beginning of the conference. Manuscripts not meeting the technical standard of the series will have to be returned for improvement.

For further information please contact Springer-Verlag, Physics Editorial Department II, Tiergartenstrasse 17, D-69121 Heidelberg, Germany

D. Breitschwerdt M. J. Freyberg
J. Trümper (Eds.)

The Local Bubble and Beyond

Lyman-Spitzer-Colloquium

Proceedings of the IAU Colloquium No. 166
Held in Garching, Germany, 21–25 April 1997

Springer

Editors

Dieter Breitschwerdt
Michael Freyberg
Joachim Trümper
Max-Planck-Institut für extraterrestrische Physik
D-85740 Garching, Germany

Scientific Organising Committee:

J. Boulanger (France), D. Breitschwerdt (Germany), P. Frisch (USA),
J. Lequeux (France), C. McKee (USA), U. Mebold (Germany),
Y. Tanaka (Japan), J. Trümper (Germany, Chair), J. Vallerga (USA)

Local Organising Committee:

D. Breitschwerdt (Chair), W. Brinkmann, R. Egger, L. Falke, M. J. Freyberg, P. Nass,
L. Falke (Conference Secretary)

Cataloging-in-Publication Data applied for.

Die Deutsche Bibliothek - CIP-Einheitsaufnahme

International Astronomical Union:
Proceedings of the IAU colloquium. - Dordrecht; Boston; London: Kluwer Acad. Publ.
Teilw. im Verl.
- Früher Schriftenreihe. - Früher angezeigt u.d.T.: International Astronomical Union:
Colloquium No. 166. The local bubble and beyond. - 1998
The local bubble and beyond: held in Garching, 21–25 April 1997 / D. Breitschwerdt... (ed.).
(Proceedings of the IAU colloquium; No. 166) (Lecture notes in physics; Vol. 506)
ISBN 978-3-662-14180-9 ISBN 978-3-540-69726-8 (eBook)
DOI 10.1007/978-3-540-69726-8

ISSN 0075-8450
ISBN 978-3-662-14180-9

Originally published by Springer-Verlag Berlin Heidelberg New York in 1998
Softcover reprint of the hardcover 1st edition 1998

Typesetting: Camera-ready by the authors/editors
Cover design: *design & production* GmbH, Heidelberg
SPIN: 10644123 55/3144-543210 - Printed on acid-free paper

Preface

The last International Astronomical Union (IAU) Colloquium that was solely dedicated to the physics of the local interstellar medium (LISM) took place in Madison, Wisconsin 4 – 6 June, 1984. The editors of the proceedings (F. Bruhweiler, Y. Kondo, and B. Savage) justified the timeliness of their conference, because "... there has recently been substantial improvement in our knowledge of the local interstellar medium [...] due to the availability of X-ray and ultraviolet observations from space."

Thirteen years later we are in a similar situation, since now the sensitivity, low background and high spatial resolution of ROSAT and the high spectral resolution capabilities of DXS and EUVE have enabled us to enter a new era of observational and theoretical study of our local environment. Further satellite missions that have revealed new insights and challenged our previous concepts of the LISM are HST, ISO and ULYSSES, as well as a number of small payloads of EUV spectrometers. Also the availability of HIPPARCOS data on distances and proper motions of stellar sources has led to a considerable improvement in absorption line studies. As to ground-based observations, the new Leiden/Dwingeloo Survey has provided an excellent data basis for the absorbing HI column in the northern hemisphere in order to give a more precise interpretation of soft X-ray background observations. It combines good spatial resolution with a large coverage of velocity space ranging from −450 to +400 km/s.

Shadowing experiments, which became possible with the ROSAT PSPC instrument, have taught us that the Local Bubble as the largest coherent structure embedded in the local interstellar medium is not the only contributor to soft X-ray emission in the 1/4 keV band. Instead, there can be a sizeable fraction from outside. Thus it seemed quite straightforward to extend our view to what happens "beyond". In this context we also felt that we had to restate the "Copernican" problem: is the LISM a typical and common place in the Galaxy or is it rather unique? If the former were true, then we might be in the fortunate position of having a laboratory for the study of the interstellar medium just in front of our doorstep. We think that many of us felt like that after the conference, although it goes without saying, that we still need more and better information before we can give a conclusive answer. Nevertheless, we are quite pleased to state that we have made substantial progress during the past few years.

The idea for this conference came from a meeting on "The Heliosphere in the Local Interstellar Medium" that was organized by the International Space Science Institute (ISSI) at Bern in the fall of 1995. Priscilla Frisch and one of us (DB) were co-chairing a working group on the Local Bubble. We had excellent contributions and lively discussions which were, however, heavily controversial, mainly due to the lack of information on various subjects. Thus, during one of the lunch breaks, together with James Lequeux and John Vallerga, we came to the conclusion that it was time for a full-fledged conference addressing a much wider range of topics. Subsequently, we applied to the IAU and received full support from the commissions 34 (Interstellar Matter), 40 (Radio Astronomy) and 44 (Space and High Energy Astrophysics).

The conference was extremely well attended with about 150 scientists and students from 16 different countries from all over the world. The scientific programme consisted of 24 invited papers, which were divided into review (8) and highlight (14) talks, supplemented by an opening lecture and a conference summary. There were 81 contributed papers, comprising both oral and poster presentations. The latter were displayed for the whole length of the conference and each advertised by one transparency during the meeting. The policy to treat poster and oral contributions as equally as possible was very well received. Hence we do not make any distinction in the proceedings either.

The plenary presentations were grouped into eight sessions with Wednesday afternoon being reserved for our "Discover Munich tour", which even attracted quite a few locals and continued into the "Bavarian evening" in the Hofbräuhaus. No losses or casualties were reported the next day and most of the participants were present at the conference dinner, which took place in the candle-lit vaults of the traditional 15th century Bavarian restaurant "Altes Hackerhaus". Despite the free drinks and many after-dinner speeches, all participants reassembled the next morning for the last session.

For the organizers of the conference some of the most remarkable and rewarding experiences were the lively and stimulating discussions throughout the meeting. This enthusiasm and the unbroken high spirits of the participants manifested themselves quite impressively in the spontaneous organization of a "Workshop for Continued Discussion of the Local Bubble" on Friday afternoon, which was convened by Wilt Sanders and attended by more than 50 scientists. A summary by Wilt Sanders and Priscilla Frisch is included in the proceedings. All scientific, social and cultural activities are well documented by photographs and a selection of them can be found scattered throughout the book at the beginning of each part. This was only possible because of the efforts of Petra Nass who was always on the spot to take good pictures.

Shortly before the conference, Lyman Spitzer, who was invited to give the opening lecture, died. There was unanimous agreement between all the members of the Scientific Organizing Committee, that this conference on the local and general interstellar medium, for the development of which he had been a key figure, should be dedicated to him. We have received the official permission of the president of the IAU, L. Woltjer, to henceforth subtitle IAU Colloquium No. 166 as the *Lyman Spitzer Colloquium*. Lyman Spitzer's outstanding achievements and his warm and friendly personality are commemorated in a short contribution by F.D. Kahn[1] in this volume.

That we live in a hot environment was impressively demonstrated by a fire which broke out shortly before the conference. Despite the high sound speed in such an environment, the information of a complete switch-off of the electricity in the building that hosts the lecture theatre at the Institut für Plasmaphysik (IPP) on Monday morning did not reach the organizers. After some hectic activities, we managed to move to the seminar room at ESO with only half an hour delay. We would like to thank the staff at ESO for their efforts and also the IPP, which otherwise ensured a smooth run of the conference. Our apologies to the participants, who always remained calm and civilized throughout the incident.

Financial support for the meeting was provided by the IAU, whose Assistant General Secretary J. Andersen we would like to thank for his many valuable suggestions and his unbureaucratic help in preparing the meeting. We further acknowledge support by the DFG (Deutsche Forschungsgemeinschaft), the DARA (Deutsche Agentur für Raumfahrt-Angelegenheiten) and the Max-Planck-Institut für extraterrestrische Physik (MPE).

Space does not allow us to thank all the individuals who directly or indirectly contributed to the success of the conference. We would like to express our sincere gratitude to all the staff at MPE, and also to Bernie Tucker and Renate Lange who assisted at the conference desk, and to Dietmar Heise who did all the printing. We are also grateful to Lin Kayser from outside MPE, who spent many hours in designing the conference poster. If everything ran smoothly and efficiently then this was certainly due to the tireless efforts of the Local Organizing Committee. Specifically, we would like to thank Roland Egger, Lynn Falke, and Petra Nass for the many problems they solved so elegantly prior to and during the conference.

We also thank our publisher, Springer-Verlag, for their help and advice during the publication process and, last but not least, all the authors who have, apart from the inevitable exception to the rule, handed in their manuscripts in time and in the desired format. All contributions are *fully refereed* for which we are indebted to the numerous referees who took their job very seriously, as can be judged by the high quality of all the articles. We are also proud to state that we have received all contributions, except for a handful of short papers which were included in a review and a contributed talk.

---

[1] While these proceedings were in press, the sad news of Franz Kahn's sudden death had reached us.

We hope that the contributions, which are listed in chronological order as they were presented at the conference, will serve as a standard reference on the subject of the local and general interstellar medium for the next couple of years.

Garching, February 1998

D. Breitschwerdt
M.J. Freyberg
J.E. Trümper

Contents

Part I Introduction

Part II UV and Soft X-ray Observations of the LISM

Part VII Gas in Superbubbles and in the Galactic Halo

Part VIII High-Velocity Clouds, Galactic Halo Models, Observations of the LMC

Part IX External Galaxies

Part X Conference Summary

Part XI Workshop

Part XII Appendix

1 Roland Egger, 2 Frédéric Zagury, 3 Edith Falgarone, 4 Marc-Antoine Miville-Deschênes, 5 François Boulanger, 6 ?, 7 Nikolai Bochkarev, 8 ?, 9 Ramen Parui, 10 ?, 11 Aylin Yar, 12 Bulent Uyanıker, 13 Klaas de Boer, 14 Daniel Wang, 15 Dieter Breitschwerdt, 16 David Burrows, 17 Olivier Dupin, 18 Carl Heiles, 19 Cécile Gry, 20 Wilton Sanders, 21 John Vallerga, 22 Petra Nass, 23 Bryan Penprase, 24 Sangwook Park, 25 Stuart Bowyer, 26 Elly Berkhuijsen, 27 Michael Freyberg, 28 Harald Lesch, 29 Thomas Berghöfer, 30 Guido Birk, 31 Rubens Freire Ferrero, 32 Lynn Falke, 33 Steve Snowden, 34 Donald Cox, 35 Kip Kuntz, 36 Douglas Finkbeiner, 37 Randall Smith, 38 Markus Landgraf, 39 Chris McKee, 40 Ulrich Mebold, 41 James Green, 42 Jan Palouš, 43 Mark Hurwitz, 44 Daniel Welty, 45 Rosine Lallement, 46 Van Dixon, 47 Jay Holberg, 48 Maria Cioni, 49 Stefano Cortiglioni, 50 Sunil Sidher, 51 Priscilla Frisch, 52 Ramesh Bhat, 53 Rodrigo Supper, 54 Sona Ehlerová, 55 Andrea Ferrara, 56 Jonathan Slavin, 57 Fotis Mavromatakis, 58 Mordecai-Mark Mac Low, 59 Robin Shelton, 60 Kurtis Williams, 61 Ralf-Jürgen Dettmar, 62 Jakob Englhauser, 63 Andreas Vogler, 64 Edward Jenkins, 65 Uli Zimmermann, 66 Ginevra Trinchieri, 67 Anamparambu Ramaprakash, 68 Matthias Hünsch, 69 Konrad Dennerl, 70 Giorgio Palumbo, 71 Nimesh Patel, 72 Anatoli Iyudin, 73 Yasuo Tanaka, 74 Hans Böhringer, 75 Jürgen Knödlseder, 76 Markus Schwalb, 77 Gregor Morfill, 78 Uwe Oberlack, 79 Kai Kulp, 80 Roland Diehl, 81 Gerhard Hensler, 82 Yasushi Ikebe, 83 Ricardo Génova, 84 Peter Predehl, 85 Hugo van Woerden, 86 John Beckman, 87 Jerry Edelstein, 88 Felix Lockman, 89 Joaquin Trapero, 90 Anthony Minter, 91 Joachim Trümper, 92 José Gómez, 93 Satoru Ikeuchi, 94 Christian Theis, 95 Jochen Pietz, 96 Robert Warwick, 97 Vasilii Gvaramadze, 98 Jürgen Kerp, 99 Barry Welsh, 100 Sergej Moiseenko, 101 Martin Blondiau, 102 Ian Hutchinson, 103 Matthias Wrigge, 104 Martin Barstow, 105 Peter Kalberla, 106 Olivia Puyoo, 107 Uwe Herbstmeier, 108 Wagner Corradi Barbosa, 109 Arthur Chernin, 110 Guennadi Bisnovatyi-Kogan, 111 Vladimir Dogiel, 112 Natalija Bystrova, 113 ?, 114 Sungeun Kim, 115 Alexandre Wennmacher.

List of Participants

Arnoldt Cudell, Anabela `aarnoldt@ix.urz.uni-heidelberg.de`
Universität Heidelberg, D-69001 Heidelberg, Germany

Aschenbach, Bernd `bra@xray.mpe.mpg.de`
MPI für extraterrestrische Physik, D-85740 Garching, Germany

Avillez, Miguel `mavillez@fis.ce.uevora.pt`
Universidade de Évora, 7000 Évora, Portugal

Balser, Dana `dbalser@nrao.edu`
NRAO Greenbank, West Virginia 24944, USA

Banday, Anthony `banday@mpa-garching.mpg.de`
MPI für Astrophysik, D-85740 Garching, Germany

Barstow, Martin `mab@star.le.ac.uk`
University of Leicester, Leicester LE1 7RH, United Kingdom

Becker, Werner `web@xray.mpe.mpg.de`
MPI für extraterrestrische Physik, D-85740 Garching, Germany

Beckman, John `jeb@ll.iac.es`
Instituto de Astrofísica de Canarias, La Laguna, Tenerife 38200, Spain

Berghöfer, Thomas `thb@cea.berkeley.edu`
Center for EUV Astrophysics, Berkeley, CA 94720, USA

Berkhuijsen, Elly `eberkhuijsen@mpifr-bonn.mpg.de`
MPI für Radioastronomie, D-53121 Bonn, Germany

Berry, David `dberry@evunix.uevora.pt`
Universidade de Évora, 7000 Évora, Portugal

Bertoldi, Frank `fkb@mpe.mpg.de`
MPI für extraterrestrische Physik, D-85740 Garching, Germany

Beuermann, Klaus `beuermann@neptun.uni-sw.gwdg.de`
Universitätssternwarte, D-37083 Göttingen, Germany

Bhat, Ramesh `bhatnd@gmrt.ernet.in`
National Centre for Radio Astrophysics, Pune 411007, India

Birk, Guido `birk@usm.uni-muenchen.de`
Institut für Astronomie und Astrophysik, LMU, D-81679 München, Germany

Bisnovatyi-Kogan, Guennadi `gkogan@mx.iki.rssi.ru`
Space Research Institute, Moscow 117810, Russia

Blondiau, Martin `blondiau@astro.uni-bonn.de`
Radioastronomisches Institut, Universität Bonn, D-53121 Bonn, Germany

Bochkarev, Nikolai `boch@astronomy.msk.su`
Sternberg Astronomical Institute, Moscow 119899, Russia

Böhringer, Hans hxb@mpe.mpg.de
MPI für extraterrestrische Physik, D-85740 Garching, Germany
Böse, Günther gub@xray.mpe.mpg.de
MPI für extraterrestrische Physik, D-85740 Garching, Germany
Bomans, Dominik bomans@astro.uiuc.edu
University of Illinois, Urbana, IL 61801, USA
Boulanger, François boulanger@ias.fr
Institut d'Astrophysique Spatiale, F-91405 Orsay, France
Bowyer, Stuart bowyer@cea.berkeley.edu
University of California, Berkeley, CA 94720-5030, USA
Breitschwerdt, Dieter breitsch@xray.mpe.mpg.de
MPI für extraterrestrische Physik, D-85740 Garching, Germany
Briel, Ulrich ugb@xray.mpe.mpg.de
MPI für extraterrestrische Physik, D-85740 Garching, Germany
Brinkmann, Wolfgang wpb@mpe.mpg.de
MPI für extraterrestrische Physik, D-85740 Garching, Germany
Burrows, David burrows@astro.psu.edu
Penn State University, University Park, PA 16802, USA
Bystrova, Natalija bnv@fsao.spb.su
St.Petersburg Branch of the Sp. Aph. Obs., St. Petersburg 196140, Russia
Chernin, Arthur chernin@sai.msu.su
Moscow University, Moscow 119899, Russia
Cioni, Maria cioni@tesre.bo.cnr.it
I.Te.S.R.E./Consiglio Nazionale Ricerche, I-40129 Bologna, Italy
Corradi Barbosa, Wagner wag@fisica.ufmg.br
Universidade Federal de Minas Gerais, Belo Horizonte 30161-970, Brazil
Cortiglioni, Stefano cortiglioni@tesre.bo.cnr.it
I.Te.S.R.E./Consiglio Nazionale Ricerche, I-40129 Bologna, Italy
Cox, Donald cox@wisp.physics.wisc.edu
University of Wisconsin, Madison, WI 53706, USA
de Boer, Klaas deboer@astro.uni-bonn.de
Universität Bonn, D-53121 Bonn, Germany
Dennerl, Konrad kod@xray.mpe.mpg.de
MPI für extraterrestrische Physik, D-85740 Garching, Germany
Dettmar, Ralf-Jürgen dettmar@astro.ruhr-uni-bochum.de
Ruhr-Universität Bochum, D-44780 Bochum, Germany
Diehl, Roland rod@mpe.mpg.de
MPI für extraterrestrische Physik, D-85740 Garching, Germany
Dixon, Van vand@ssl.berkeley.edu
University of California, Berkeley, CA 94720-7450, USA
Dogiel, Vladimir dogiel@lpi.ac.ru
P.N. Lebedev Institute, Moscow 117924, Russia
Dumke, Michael mdumke@mpifr-bonn.mpg.de
MPI für Radioastronomie, D-53121 Bonn, Germany

Dupin, Olivier dupin@astrsp-mrs.fr
Laboratoire d'Astronomie Spatiale, F-13376 Marseille Cedex 12, France

Edelstein, Jerry jerrye@ssl.berkeley.edu
University of California, Berkeley, CA 94720-7450, USA

Egger, Roland roe@mpe.mpg.de
MPI für extraterrestrische Physik, D-85740 Garching, Germany

Ehlerová, Sona sona@ig.cas.cz
Astronomical Institute, 141 31 Prague, Czech Republic

Englhauser, Jakob jer@xray.mpe.mpg.de
MPI für extraterrestrische Physik, D-85740 Garching, Germany

Falgarone, Edith edith@ensapb.ens.fr
Ecole Normale Supérieure, F-75005 Paris, France

Falke, Lynn falke@mpe.mpg.de
MPI für extraterrestrische Physik, D-85740 Garching, Germany

Ferrara, Andrea ferrara@arcetri.astro.it
Osservatorio Astrofisico Arcetri, I-50125 Firenze, Italy

Finkbeiner, Douglas dfink@astro.berkeley.edu
University of California, Berkeley, CA 94720, USA

Freire Ferrero, Rubens freire@cdsxb6.u-strasbg.fr
Observatoire Astronomique de Strasbourg, F-67000 Strasbourg, France

Freyberg, Michael mjf@mpe.mpg.de
MPI für extraterrestrische Physik, D-85740 Garching, Germany

Frisch, Priscilla frisch@oddjob.uchicago.edu
University of Chicago, Chicago, IL 60637, USA

Génova, Ricardo rgg@ll.iac.es
Instituto de Astrofísica de Canarias, La Laguna, Tenerife 38200, Spain

Gölz, Michael goelz@astro.uni-tuebingen.de
Universität Tübingen, D-72076 Tübingen, Germany

Gómez, José jfg@laeff.esa.es
Instituto Nacional de Tecnica Aeroespacial, E-28080 Madrid, Spain

Green, James jgreen@casa.colorado.edu
University of Colorado, Boulder, CO 80303, USA

Gry, Cécile cgry@iso.vilspa.esa.es
ISO Science Operation Center of ESA, E-28080 Madrid, Spain

Gunderson, Kurt gunderso@casa.colorado.edu
University of Colorado, Boulder, CO 80303, USA

Gvaramadze, Vasilii vgvaram@mx.iki.rssi.ru
Abastumani Astrophysical Observatory, Tbilisi, Georgia

Heiles, Carl heiles@vermi.berkeley.edu
University of California, Berkeley, CA 94720, USA

Hensler, Gerhard supas030@astrophysik.uni-kiel.de
Insitut für Astronomie und Astrophysik, Universität Kiel, D-24098 Kiel, Germany

Herbstmeier, Uwe uherbst@mpia-hd.mpg.de
MPI für Astronomie, D-69117 Heidelberg, Germany
Holberg, Jay holberg@argus.lpl.arizona.edu
University of Arizona, Tucson, AZ 85721, USA
Hünsch, Matthias supas106@astrophysik.uni-kiel.de
Institut für Astronomie und Astrophysik, Universität Kiel, D-24098 Kiel, Germany
Hurwitz, Mark markh@ssl.berkeley.edu
University of California, Berkeley, CA 94720-7450, USA
Hutchinson, Ian ibh@star.le.ac.uk
University of Leicester, Leicester LE1 7RH, United Kingdom
Ikebe, Yasushi ikebe@xray.mpe.mpg.de
MPI für extraterrestrische Physik, D-85740 Garching, Germany
Ikeuchi, Satoru ikeuchi@vega.ess.sci.osaka-u.ac.jp
Osaka University, Toyonaka 560, Japan
Immler, Stefan simmler@mpe.mpg.de
MPI für extraterrestrische Physik, D-85740 Garching, Germany
Iyudin, Anatoli ani@mpe.mpg.de
MPI für extraterrestrische Physik, D-85740 Garching, Germany
Jenkins, Edward ebj@astro.princeton.edu
Princeton University, Princeton, NJ 08544-1001, USA
Kahn, Franz fk@ast.man.ac.uk
Manchester University, Manchester M13 9PL, United Kingdom
Kalberla, Peter pkalberla@astro.uni-bonn.de
Radioastronomisches Institut, Universität Bonn, D-53121 Bonn, Germany
Kappelmann, Norbert kappelmannn@ait.physik.uni-tuebingen.de
Universität Tübingen, D-72076 Tübingen, Germany
Kerp, Jürgen jkerp@astro.uni-bonn.de
Radioastronomisches Institut, Universität Bonn, D-53121 Bonn, Germany
Kim, Sungeun sek@mso.anu.edu.au
Mount Stromlo and Siding Spring Observatories, Canberra, ACT2611, Australia
Kneissl, Rüdiger ruk@mpa-garching.mpg.de
MPI für Astrophysik, D-85740 Garching, Germany
Knödlseder, Jürgen jurgen@Sigma-0.cesr.cnes.fr
CESR, UPS-CNRS, F-31028 Toulouse Cedex, France
Krämer, Gerhard kraemer@ait.physik.uni-tuebingen.de
Universität Tübingen, D-72076 Tübingen, Germany
Kulp, Kai kulp@xray.mpe.mpg.de
MPI für extraterrestrische Physik, D-85740 Garching, Germany
Kuntz, Kip kuntz@halo.gsfc.nasa.gov
University of Maryland, College Park, MD 20742, USA
Lallement, Rosine rosine.lallement@aerov.jussieu.fr
Service d'Aéronomie du CNRS, F-91371 Verrieres-Le-Buisson, France

Landgraf, Markus landgraf@galileo.mpi-hd.mpg.de
MPI für Kernphysik, D-69117 Heidelberg, Germany

Lesch, Harald lesch@usm.uni-muenchen.de
Institut für Astronomie und Astrophysik, LMU, D-81679 München, Germany

Lockman, Felix jlockman@nrao.edu
NRAO Greenbank, West Virginia 24944, USA

Mac Low, Mordecai-Mark mordecai@mpia-hd.mpg.de
MPI für Astronomie, D-69117 Heidelberg, Germany

Mavromatakis, Fotis fotis@physics.uch.gr
University of Crete, Heraklion 71003, Greece

McKee, Chris cmckee@astro.berkeley.edu
University of California, Berkeley, CA 94720,USA

McLean, Ryan mclean@casa.colorado.edu
University of Colorado, Boulder, CO 80303, USA

Mebold, Ulrich mebold@astro.uni-bonn.de
Radioastronomisches Institut, Universität Bonn, D-53121 Bonn, Germany

Minter, Anthony tminter@nrao.edu
NRAO Greenbank, West Virginia 24944, USA

Miville-Deschênes, Marc-Antoine mamd@ias.fr
Institut d'Astrophysique Spatiale, F-91405 Orsay, France

Moiseenko, Sergej moiseenko@mx.iki.rssi.ru
Space Research Institute, Moscow 117810, Russia

Morales, Carmen morales@laeff.esa.es
Instituto Nacional de Tecnica Aeroespacial, E-28080 Madrid, Spain

Morfill, Gregor gem@mpe.mpg.de
MPI für extraterrestrische Physik, D-85740 Garching, Germany

Nass, Petra pnass@mpe.mpg.de
MPI für extraterrestrische Physik, D-85740 Garching, Germany

Neuhäuser, Ralph rne@xray.mpe.mpg.de
MPI für extraterrestrische Physik, D-85740 Garching, Germany

Oberlack, Uwe ugo@mpe.mpg.de
MPI für extraterrestrische Physik, D-85740 Garching, Germany

Palouš, Jan palous@ig.cas.cz
Astronomical Institute, 141 31 Prague, Czech Republic

Palumbo, Giorgio ggcpalumbo@astbo3.bo.astro.it
University of Bologna, I-40126 Bologna, Italy

Park, Sangwook parksan@purds1.physics.purdue.edu
Purdue University, West Lafayette, 47907-1396, USA

Parui, Ramen
Aero Communication Station, Imphal 795140, India

Patel, Nimesh npatel@cfa.harvard.edu
Smithsonian Astrophysical Observatory, Cambridge, MA 02138, USA

Pavlov, George gpavlov@xray.mpe.mpg.de
MPI für extraterrestrische Physik, D-85740 Garching, Germany

Penprase, Bryan penprase@shanti.pomona.edu
Pomona College, Claremont, CA 91711, USA

Pietsch, Wolfgang wnp@xray.mpe.mpg.de
MPI für extraterrestrische Physik, D-85740 Garching, Germany

Pietz, Jochen jpietz@astro.uni-bonn.de
Radioastronomisches Institut, Universität Bonn, D-53121 Bonn, Germany

Predehl, Peter prp@xray.mpe.mpg.de
MPI für extraterrestrische Physik, D-85740 Garching, Germany

Puyoo, Olivia puyoo@iap.fr
Institut d'Astrophysique, F-75014 Paris, France

Quirrenbach, Andreas quirrenbach@mpe.mpg.de
MPI für extraterrestrische Physik, D-85740 Garching, Germany

Ramaprakash, Anamparambu anr@iucaa.ernet.in
Inter-University Center for Astronomy and Astrophysics, Pune 411 007, India

Rauchhaupt von, Ulf ulf@mpe.mpg.de
MPI für extraterrestrische Physik, D-85740 Garching, Germany

Reach, William reach@ias.fr
Université Paris XI, F-91405 Paris, France

Sanders, Wilton sanders@dxs.ssec.wisc.edu
University of Wisconsin, Madison, WI 53706, USA

Schwalb, Markus schwalb@xray.mpe.mpg.de
MPI für extraterrestrische Physik, D-85740 Garching, Germany

Schwentker, Ortwin osc@mpe.mpg.de
MPI für extraterrestrische Physik, D-85740 Garching, Germany

Shelton, Robin shelton@spots.gsfc.nasa.gov
NASA/Goddard Space Flight Center, Greenbelt, MD 20771, USA

Sidher, Sunil s.d.sidher@rl.ac.uk
Imperial College, London SW7 2BZ, United Kingdom

Siebert, Joachim jos@xray.mpe.mpg.de
MPI für extraterrestrische Physik, D-85740 Garching, Germany

Slavin, Jonathan slavin@dusty.arc.nasa.gov
NASA/Ames Research Center, CA 94035-1000, USA

Smith, Randall rsmith@stars.gsfc.nasa.gov
NASA Goddard Space Flight Center, MD 20771, USA

Snowden, Steve snowden@lheavx.gsfc.nasa.gov
NASA/GSFC/USRA, Greenbelt, MD 20771, USA

Sterzik, Michael mfs@mpe.mpg.de
MPI für extraterrestrische Physik, D-85740 Garching, Germany

Sunyaev, Rashid sunyaev@mpa-garching.mpg.de
MPI für Astrophysik, D-85740 Garching, Germany

Supper, Rodrigo ros@xray.mpe.mpg.de
MPI für extraterrestrische Physik, D-85740 Garching, Germany

Tanaka, Yasuo Y.Tanaka@sron.ruu.nl
University of Amsterdam, NL-1098 SJ, Netherlands

Theis, Christian theis@astrophysik.uni-kiel.de
Institut für Astronomie und Astrophysik, Universität Kiel, D-24098 Kiel, Germany

Thomas, Hans-Christoph hcthomas@mpa-garching.mpg.de
MPI für Astrophysik, D-85740 Garching, Germany

Timochin, Andrej timochin@mpa-garching.mpg.de
MPI für Astrophysik, D-85740 Garching, Germany

Trapero, Joaquin trapero@laeff.esa.es
Lab. de Astrofísica Espacial y Física Fundamental, E-28080 Madrid, Spain

Trinchieri, Ginevra ginevra@brera.mi.astro.it
Osservatorio Astronomico Brera, I-20121 Milano, Italy

Trümper, Joachim jtrumper@mpe.mpg.de
MPI für extraterrestrische Physik, D-85740 Garching, Germany

Uyanıker, Bulent uyaniker@mpifr-bonn.mpg.de
MPI für Radioastronomie Bonn, D-53121 Bonn, Germany

Vallerga, John eureka@netcom.com
Eureka Scientific, Oakland, CA 94602-3017, USA

van Woerden, Hugo hugo@astro.rug.nl
Groningen University, Groningen, 9700 AV, Netherlands

Varendorff, Martin mgv@mpe.mpg.de
MPI für extraterrestrische Physik, D-85740 Garching, Germany

Voges, Wolfgang whv@xray.mpe.mpg.de
MPI für extraterrestrische Physik, D-85740 Garching, Germany

Vogler, Andreas ajv@mpe.mpg.de
MPI für extraterrestrische Physik, D-85740 Garching, Germany

Wang, Daniel wqd@nwu.edu
Northwestern University, Evanston, IL 60208-2900, USA

Warwick, Robert rsw@star.le.ac.uk
University of Leicester, Leicester LE1 7RH, United Kingdom

Welsh, Barry bwelsh@ssl.berkeley.edu
Space Sciences Lab., Berkeley, CA 94720, USA

Welty, Daniel welty@oddjob.uchicago.edu
University of Chicago, Chicago IL 60637, USA

Wennmacher, Alexandre wennmach@geo.uni-koeln.de
Universität Köln, D-50923 Köln, Germany

Wessolowski, Ulf uuw@mpe.mpg.de
MPI für extraterrestrische Physik, D-85740 Garching, Germany

Williams, Kurtis williams@xray.mpe.mpg.de
MPI für extraterrestrische Physik, D-85740 Garching, Germany

Wrigge, Matthias mwrigge@hs.uni-hamburg.de
Universität Hamburg, D-21029 Hamburg, Germany
Yar, Aylin ayar@astro.uni-bonn.de
Radioastronomisches Institut, Universität Bonn, D-53121 Bonn, Germany
Zagury, Frédéric zagury@ias.fr
Institut d'Astrophysique Spatiale, F-91405 Orsay, France
Zavlin, Vyatcheslav zavlin@xray.mpe.mpg.de
MPI für extraterrestrische Physik, D-85740 Garching, Germany
Zimmermann, Uli zim@xray.mpe.mpg.de
MPI für extraterrestrische Physik, D-85740 Garching, Germany

Part I

Introduction

Sunday evening reception: the participants ...

... and the LOC in good spirits.

Lyman Spitzer 1914 – 1997

Some fifty years ago, aged 33, Lyman Spitzer was appointed to succeed the legendary Henry Norris Russell as Director of the University Observatory at Princeton. Lyman was undaunted, he knew what needed to be done and he set about doing it. There can have been few selection committees in the academic world that ever made a more appropriate choice.

Much later Lyman wrote an autobiographical sketch "Dreams, Stars and Electrons", published in the 1989 issue of Annual Reviews of Astronomy and Astrophysics. Here he explains what drove him on and how he came to achieve the many diverse successes of his long career.

He had long been interested in the study of interstellar matter. Soon after going to Princeton he also became deeply involved in more general plasma physics by starting the "Project Matterhorn". As the name implies Lyman was a keen mountaineer, but the project has the very serious aim to find a means of confining a very hot plasma with a view to generating power from controlled thermonuclear reactions.

For anyone else this would have been more than a full-time occupation. Not so for Lyman. Together with Martin Schwarzschild he began plans for astronomical observations to be made well away from the Earth's surface. In Martin's case the aim was to get clear pictures unaffected by atmospheric twinkling, and he was content with using high altitude balloons. Lyman wanted to observe in the ultraviolet and he insisted that he needed to place instruments into a satellite right outside the Earth's atmosphere.

His advocacy and planning led to the launch of the Copernicus satellite in 1972. It was an immediate success. Molecular hydrogen was detected in absorption against the background of early type stars. This was highly significant and confirmed the presence of H_2, in addition to that of atomic hydrogen (H) that had long been known. But even more important was the detection of a low density interstellar component containing some highly ionized species, like O^{+5}; this discovery revealed that large volumes of interstellar space are intensely hot, at 10^6 K. There was a question that Lyman had long been asking but that no one had asnwered. What is it that keeps the HI clouds confined to a relatively small filling factor in interstellar space? The existence of the hot intercloud medium provided the answer. In this very real sense Lyman had long anticipated the problems that have been discussed at IAU Colloquium 166 on the Local Bubble.

The organisers of this conference felt, quite naturally, that Lyman should be asked to give the opening talk, and he had accepted the invitation, but it was not to be. It is fitting that the colloquium has now been dedicated to his memory. The enthusiasm of the participants and their many splendid contributions show that Lyman's insights are as fruitful as ever, and will long continue to be so.

In St. Paul's Cathedral in London there is a plaque in honour of the architect Christopher Wren, who built it and much else besides. Incidentally Wren was also professor of astronomy at Oxford. The inscription reads "Si moenumentum requiris, circumspice", which might loosely be translated "If you need a memorial to Lyman, read on". I like to think that he would have been pleased to see what we have done with his ideas.

Manchester, August 1997
Franz Kahn

Introductory Lecture: The Local and General Interstellar Medium

Dieter Breitschwerdt[1,2]

[1] Max–Planck–Institut für extraterrestrische Physik, Giessenbachstrasse, D-85740 Garching, Germany
[2] Heisenberg Fellow

Abstract. Important progress in the field of the Interstellar Medium (ISM) in general, and the Local Interstellar Medium (LISM) in particular, has been largely due to the recent launch of orbiting satellite missions and earth-bound observations with unprecedented sensitivity and high spectral or spatial resolution. However, the interpretation of a wealth of new data is still based to a great extent on models that were designed for the previous generation of observations. Thus there remains a clear gap between our theoretical understanding of the basic physics of the ISM and the detailed information available. In such a situation it seems appropriate to focus on some key observations and discuss their implications for modelling.

1 Introduction

This opening lecture was originally supposed to be given by Lyman Spitzer. On behalf of the Scientific Organizing Committee J. Trümper and myself wrote an invitation letter to him, which remained unanswered. A couple of weeks later, we learned of his sudden death, which means a great loss for his many friends in the astronomical community and especially among students of the ISM. It was clear that just a few weeks before the conference, we would not be able to find an adequate replacement.

In this article I will content myself to give a short overview of observational and theoretical progress with no attempt of completeness and doing justice to all the people who have been so heavily involved in this field over the past years. For brevity, I will confine myself to some key observations and models. In a couple of years time it may well appear in hindsight, that other observations and models will be more relevant to our understanding. But then progress rarely evolves in straight paths, and therefore any observation that is in conflict with our theoretical understanding *now* deserves to be discussed in this context.

2 Brief Overview of the General Interstellar Medium

The subject of the ISM is nowadays a classical field of astronomy. Nevertheless, as will be discussed in more detail below, there have always been surprises even until to date, that keep the field young and prevent researchers from falling into dogmatism.

Historically, stationary absorption lines of ionized calcium in a spectroscopic binary have been studied already at the beginning of the century (Hartmann, 1904), but it was not until the early 30's when it became clear that these lines were of interstellar and not circumstellar origin (Plaskett and Pearce, 1933). Previous work on extinction of star light lead to the concept of a pervasive gaseous and dusty medium concentrated in the disk of the Galaxy. However, kinematical information on absorption lines showed that the radial velocities differed substantially in different regions of the Galaxy (Adams, 1948), and therefore the distribution of gas must be far from uniform. A global mapping of the gas became available by 21 cm observations, following an earlier theoretical prediction of the hyperfine structure transition of HI by van de Hulst (1945). It was clear that a possible source of energy for the interstellar gas were embedded stars. Strömgren (1939) showed that early-type stars should have rather sharp HI/HII boundaries, called ionization fronts. Since the interior is kept ionized by the stellar photon field, the "Strömgren sphere" will undergo dynamical expansion due to its overpressure and drive a shock wave into the ISM. Although there is an energy input of more than $10^{48}\,\mathrm{s}^{-1}$ ionizing photons from an O9 star, say, the energy conversion factor ϵ of Lyman continuum photons into kinetic energy of the ISM can be estimated to be very small: $\epsilon \leq 10^{-2}$; most of the energy is radiated away and other sources of energy input had to be sought. The first attempts to explain the nonuniform distribution of interstellar matter into distinct "phases" that would be stable against pressure perturbations were undertaken by Pikel'ner (1967) and Field, Goldsmith and Habing (1969). The latter envisaged ionization of the ISM by low energy cosmic rays to be the main source of energy input, whereas energy losses were due to collisional excitation and subsequent line cooling. This so-called 2-phase model explained the coexistence of a cold neutral phase ($T \sim 30\,\mathrm{K}$, $n \sim 100\,\mathrm{cm}^{-3}$) with a warm and weakly ionized phase ($T \sim 10^4\,\mathrm{K}$, $n \sim 0.3\,\mathrm{cm}^{-3}$).

A major thrust in our understanding of the ISM came with the launch of COPERNICUS in 1972, which was inspired and promoted by Lyman Spitzer. The detection of the ubiquitous OVI absorption line and a careful kinematic analysis, that separated the absorbing regions from the circumstellar medium, lead to the establishment of a widespread hot interstellar medium (Jenkins and Meloy, 1974; York, 1974) with $5.4 \leq \log \mathrm{T} \leq 6.3$. Already a few years before, in a pioneering rocket experiment Bowyer et al. (1968) had revealed the existence of diffuse soft X-ray emission in the energy range below 1 keV. It could be decomposed into an extragalactic and what they called at the time an "anomalous" component, which they speculated might be of Galactic origin (see also Bowyer, this volume). Such an emission source had been suggested long ago by Spitzer (1956), who argued for a hot ($T \sim 10^6\,\mathrm{K}$) Galactic corona, in order to provide a pressure confining medium for the then observed high velocity clouds. It became clear however, that the hot medium observed with COPERNICUS could not reproduce the spectral characteristics of the soft X-ray background (SXRB), so that it seemed as if both observations were sampling different gas.

Since the prime energy sources in the ISM were known by then to be Type II supernovae (SNe), the question to be answered remained how to create a pervasive hot medium. Cox and Smith (1974) argued that according to Monte-Carlo simulations of randomly occurring SNe in a homogeneous disk, which they followed until the dispersal of the remnants in the ISM, a tunnel network of hot bubbles ($T \sim 10^6$ K, $n \leq 10^{-2}\mathrm{cm}^{-3}$) with a volume filling factor of $f_{\mathrm{HIM}}^{\mathrm{V}} \sim 0.5$ could be achieved for a Galactic SN rate of $1/50\,\mathrm{yr}^{-1}$ and an average explosion energy of $\mathrm{E_{SN}} \sim 4\times10^{50}$ erg in an ambient medium of density $n_{\mathrm{ISM}} \sim 1\,\mathrm{cm}^{-3}$. In a seminal paper, McKee and Ostriker (1977) undertook the task to combine all that was known at the time into a consistent picture. It was a bold attempt and a brilliant piece of work to devise a theory for the ISM, which owing to its complexity has mostly been and still is dissected into more or less isolated problems which can then be tackled. They were lead by the compelling idea that SN explosions in an inhomogeneous cloudy ISM should result in a time-asymptotic picture as it is observed presently. The net result is a three-component ISM in the Galactic disk, consisting of a cold neutral medium (CNM) coated by warm neutral clouds (WNM), with their ionized surfaces representing the warm ionized medium (WIM). Most of the volume is filled with hot intercloud medium (HIM) with a filling factor as large as $f_{\mathrm{HIM}}^{\mathrm{V}} \approx 0.7 - 0.8$. Global mass and energy balance between cloud evaporation and dense shell formation and SN heating and radiative cooling, respectively, provides the observed density and temperature structure (see also for more details McKee, this volume).

While this model has certainly outlived all its competitors and is probably more often quoted than any other ISM paper, we have to ask ourselves, whether it is still correct in general and with respect to the Local Bubble in particular. My answer with respect to the general ISM would be yes and no. The basic ingredients I believe are correct: SNe are the main energy source of the ISM and they are responsible for producing hot and X-ray emitting plasma all over the disk and halo by interaction with an inhomogeneous substrate; however, the high value of $f_{\mathrm{HIM}}^{\mathrm{V}}$ has been challenged by observations in the 80's. H I observations in the Milky Way (Heiles, 1980) and in nearby galaxies (Brinks and Shane, 1984) have shown the existence of supershells and H I holes, respectively, which point towards a much smaller filling factor of $f_{\mathrm{HIM}}^{\mathrm{V}} \sim 0.2$, assuming an anticorrelation between hot and neutral gas. Only 100 pc away is the Loop I superbubble, which has a radius of about 150 pc and which has been filled with hot X-ray emitting plasma, as observed by ROSAT (Egger and Aschenbach, 1995), generated by about 40 SNe in concert. At present it seems that probably half of the SNe occur in a temporally and spatially correlated fashion. Thus highly overpressured regions will be expanding preferentially perpendicular to the galactic plane and deposit mass, momentum and energy into the lower halo. This may result in a "clustered" galactic fountain type flow as described in the so-called "chimney model" (Ikeuchi, 1988; Norman and Ikeuchi, 1989; see also Ikeuchi, this volume) circulating matter between the disk and the halo and/or a galactic wind (Breitschwerdt et al., 1991; 1993). In any case, $f^{\mathrm{V}}_{\mathrm{HIM}}$ in the disk will be lowered. Global pressure equilibrium between phases is not satisfied, at least locally, because a large pressure imbalance of a factor of eight between the hot component and the Local Cloud has been found from EUV cloud shadowing (Berghöfer et al., this volume). Further, pulsar dispersion measurements have revealed the existence of diffuse ionized gas (DIG), occupying a volume fraction of 0.2 with an exponential vertical scale height of $z \sim 1\,\mathrm{kpc}$ (Reynolds, 1991). No current ISM model can explain the distribution of the DIG, its characteristic line ratios, and above all how it is kept ionized.

Although the volume filling factor of the HIM has been reduced to some extent, it still appears that half of the ISM is filled by a hot and tenuous plasma. Thus it is not unusual that our solar system is also located in such a medium. It is suggestive to assume that our local environment will therefore be an ideal laboratory for investigating the hot phase of the ISM. In the remainder of this paper, I will address this problem in more detail. Before closing this section, I would like to remind the reader that the ISM is not merely the *matter* between the stars. In physical terms, it is more adequately described by the components that dominate its *energetics*. It is well-known that, at least locally, a global energy equipartition does exist between the thermal and kinetic energy density of the plasma and the energy density of the magnetic field and the cosmic rays, which are all about $1\,\mathrm{eV/cm^3}$. Rather than being a fortuitous result, this indicates that a strong dynamical coupling between all of these ISM components is present. From a theorist's point of view this is not surprising, because in the ISM the magnetic field is well frozen into the plasma, and the cosmic rays can strongly interact by scattering off the random component of the field, thus imparting momentum and energy to the plasma while propagating through the ISM. Any theory of the ISM should therefore include the momentum and energy exchange between plasma, magnetic field and cosmic rays in the disk and the galactic halo (Breitschwerdt, 1994).

3 The Local Interstellar Medium (LISM)

The LISM has been the subject of a number of reviews (Cox and Reynolds, 1987; McCammon and Sanders, 1990; Breitschwerdt, 1996) and I will therefore restrict myself to outline some key observations and models, which I consider as important for future progress.

Historically, the term Local Hot Bubble (LHB) resulted from the observational fact that an anticorrelation between soft X-ray emitting plasma and

H I exists on large angular scales. The constancy in intensity of different ultrasoft X-ray bands ($E \leq 0.3$ keV) with largely different absorption cross sections (unity optical depth corresponding to a column density as small as $N(\mathrm{H}) \sim 10^{19}\,\mathrm{cm}^{-2}$) could only be understood, if the region surrounding the solar system was essentially devoid of H I out to distances of 100 pc with considerable variation in galactic latitude. Owing to the high $N(\mathrm{H})$ in the Galactic plane (at $b^{II} = 0°$) it was evident that there had to be a local emission component. Thus the term LHB was "in the air" and appeared almost simultaneously in the literature by a Japanese and the Wisconsin group (Tanaka and Bleeker, 1977; Sanders et al., 1977). Such a displacement of the cold gas by X-ray emitting hot gas was suspected by McKee and Ostriker (1977) to be the result of a local SN remnant (SNR). Following this idea, Cox and Anderson (1982) were modelling the expansion of a single blast wave of energy 5×10^{50} erg propagating into a low density ($n_0 = 4 \times 10^{-3}\,\mathrm{cm}^{-3}$) ambient medium with non-negligible pressure. Due to the fixed radial scale, such a model is heavily constrained, with the explosion having occurred about 10^5 yrs ago. Alternatively, it has been argued that the LHB is the relic of an old superbubble (Innes and Hartquist, 1984) which has been excavated 4×10^6 yrs ago, by a collective SN energy of 10^{52} erg, where the ambient medium density is $n_0 = 1\,\mathrm{cm}^{-3}$. Since the bubble has been subject to heavy radiative losses, the spectrum becomes progressively harder with time, so that M-bands emission (0.5 - 1 keV) can be explained as well in contrast to the young SNR model. Although not conclusive, cosmic ray data from meteorites and Antarctic ice cores show that the cosmic ray flux seems to have been remarkably constant over the last 10^9 yrs with two peaks 3.5×10^4 and 7×10^4 years ago (Sonett et al., 1987; see also Morfill, this volume).

Observations tell us, that emission in the M-bands is more isotropic than in the softer bands and that roughly 15 % can be accounted for by stellar disk sources (Schmitt and Snowden, 1990). It is still unclear at present how much of the M-bands emission is produced in the Local Bubble and what fraction comes from the galactic halo and from extragalactic sources. An upper limit at present seems to be about 20% for a local contribution (Snowden et al., 1993). This value is hard to achieve in the LHB model (e.g. Snowden et al., 1990), in which originally *all* of the emission below 0.3 keV was produced in an asymmetric cavity filled with a hot ($T = 10^6$ K) and tenuous ($n_b = 4.7 \times 10^{-3}\,\mathrm{cm}^{-3}$) plasma radiating in collisional ionization equilibrium (CIE). A direct measurement of foreground M-bands emission has become possible with the ROSAT PSPC (Trümper, 1983) owing to its much higher sensitivity and angular resolution than any previous instrument. In so-called shadowing experiments, the SXRB is used to probe absorbing material between the source and the observer. Taking high column density ($N(\mathrm{H}) \geq 10^{21}\,\mathrm{cm}^{-2}$) absorbers like the molecular cloud MBM12, which is located inside the Local Bubble (distance $D \sim 60 - 70$ pc; cf. Snowden et al., 1993), M-bands emission from behind can be efficiently blocked off. An upper limit of 0.3 for the ratio of on- to off-source intensity has been derived (Snowden et al., 1993). Unfortunately, the number of such clouds inside the Local Bubble is too small to draw any far reaching conclusions. For absorbers at larger distances it is much more complicated to disentangle the various contributions of emission and absorption. Nonetheless, a careful systematic investigation would be most valuable.

At this point, it should be emphasized, that we have no *direct* evidence for the physical state of the local plasma. In other words, we cannot measure its density, temperature and pressure. Instead, within the LHB model, the temperature is commonly inferred by a broad band spectral fitting procedure using a Raymond and Smith (1977) type plasma code (for details see Böhringer, this volume) for a gas in CIE at energies below 0.3 keV. Since only the emission measure is fixed by the spectral intensity, the density is derived once the pathlength of the emitting volume has been chosen. This is usually done under the assumption that the hot gas fills the volume devoid of HI completely.

A serious problem in the framework of the LHB model is the discrepancy between the inferred pressure of $P/k_B \approx 10000\,\mathrm{cm}^{-3}\,\mathrm{K}$, where k_B is Boltzmann's constant, and that of the Local Cloud, for which $P/k_B \approx 1700 - 2600\,\mathrm{cm}^{-3}\,\mathrm{K}$ (Lallement, this volume). It is hard to imagine how such a cloud can survive being crushed within a relatively short time scale. Moreover, there is observational evidence of other HI clouds and filaments within the Local Bubble, e.g. the "finger" filament (Kerp et al., 1993).

The most promising way, we can reasonably well define the plasma state in the future seems to be the following:

- We need high resolution spectroscopy in soft X-rays and EUV in order to identify individual lines.
- Pulsars with extremely low dispersion measures and known parallaxes can provide the average electron density $\langle n_e \rangle$ in the Local Bubble, independently of spectral fitting.

Since we get only information on $\langle n_e \rangle$, the distribution of ionized matter in the Local Bubble may be far from homogeneous. It is conceivable that an intermixture of hot plasma and clouds surrounded by WIM, like in the McKee and Ostriker (1977) model, provides the bulk of the dispersion. However, taking their mean cloud radius of 2.1 pc (from their Table 1), whith the outermost 0.5 pc in radius corresponding to the WIM, we also sample on average $N(\mathrm{H}) \sim 3 \times 10^{20}\,\mathrm{cm}^{-2}$ of CNM column density per cloud. Given a volume filling factor of $f^{\mathrm{V}}_{\mathrm{WIM}} = 0.23$ one expects to intercept about 30 clouds within any line of sight extending out to a radius of 100 pc. Therefore an average $N(\mathrm{H}) \sim 9 \times 10^{21}\,\mathrm{cm}^{-2}$ should be detectable in soft X-rays. Moreover, $N(\mathrm{H})$ should increase almost linearly with distance. But observations (Fruscione et al., 1994) show that this is not the case for most sight lines; instead a shallow gradient is seen in most directions, followed by a steep increase at about 100 pc. Alternatively, making the clouds smaller and thus reducing their HI content but retaining their HII will also decrease their life time with respect to destructive mechanisms like evaporation. Hence, although we do not constrain the distribution of the medium by pulsar dispersion measures, we are restricted by the HI content of intermixed clouds and their average life time.

An intrinsic problem with nearby pulsar dispersion measures is statistics. There are only very few pulsars in the solar vicinity. PSR 0950+08 has a parallactic distance of 125 pc (Gwinn et al., 1986). It is located in a direction where the Local Bubble extension varies considerably, i.e. between 37 and 347 pc (Fruscione et al., 1994); alternatively, the large path length could also be due to a void of HI outside the Local Bubble. PSR J0108-1431 has the lowest known dispersion measure $DM = 1.83\,\mathrm{cm}^{-3}\mathrm{pc}$ (Tauris et al., 1994). The predicted distance, using a model for the Galactic electron distribution (Taylor and Cordes, 1993), is about 90 pc, corresponding to $\langle n_e \rangle \approx 2.2 \times 10^{-2}\,\mathrm{cm}^{-3}$. Unfortunately the pulsar is located in a direction where the ROSAT count rates are fairly low and the Local Bubble boundary is closer than average to the Sun. Some time ago, we have predicted the distance of the binary millisecond pulsar J0437-4715 ($DM = 2.65\,\mathrm{cm}^{-3}\mathrm{pc}$) to be about 110 pc on the basis of a non-CIE Local Bubble model with an average density of $n \sim 2.4 \times 10^{-2}\,\mathrm{cm}^{-3}$ assuming that it is located inside the Local Bubble (Breitschwerdt and Schmutzler, 1994). Using the model of Taylor and Cordes (1993), the inferred distance is 140 pc. Due to the fact that we do not know whether these pulsars are inside the Local Bubble, there is no conclusive answer to the average density yet. Also the peculiar line of sight towards β CMa, which is largely filled with HII gas reaching out to a distance of 200 pc, with $\langle n_e \rangle \approx 2 \times 10^{-2}\,\mathrm{cm}^{-3}$ and a very low temperature $T \leq 5 \times 10^4\,\mathrm{K}$ adds to the mystery of the plasma state in the Local Bubble.

With high spectral resolution in the soft X-ray bands, we will be able to fix a unique temperature to the plasma, if it is in CIE. Alternatively, if it is not in CIE, we need more sophisticated models in which the dynamical and thermal evolution is coupled (e.g. Breitschwerdt and Schmutzler, 1994) .

Both the DXS and EUVE experiments (see Sanders and Vallerga, respectively, this volume) were a major step in this direction. EUVE was not designed for observing diffuse emission in the energy range 70 – 760 Å and DXS ($0.15 \leq E \leq 0.284\,\mathrm{keV}$) has faced some problems during data collection and calibration. Nevertheless, some tentative conclusions can be drawn. Due to their low mean free path EUV photons are an excellent probe for the very LISM. Jelinsky et al. (1995) have given upper limits on the diffuse EUV flux. Bearing in mind that a CIE plasma at $10^6\,\mathrm{K}$ like in the LHB model should exhibit a plethora of EUV emission lines, it is remarkable that the only lines detected were HeI and HeII, with intensities that are consistent with geocoronal and/or interplanetary scattering of solar radiation. The plasma emission measure is a factor of 5 to 10 below of what is expected from a CIE plasma in the range $10^{5.7} \leq T \leq 10^{6.4}\,\mathrm{K}$. The spectral resolution of the DXS Bragg crystal spectrometers is $10 - 25\,\mathrm{eV}$ while the spatial resolution is only $\sim 15°$.

One of the most important results has been the detection of emission lines, which is a direct evidence for diffuse emission of a thermal plasma (Sanders, 1993). Again, it is not possible to fit the spectrum by a CIE plasma with normal or depleted abundances.

Dropping the assumption of CIE, gives rise to a new class of models that show some promising features. There are essentially two extreme possibilities. The first is fast heating (e.g. by a shock wave) of interstellar plasma, which leads to an underionization of the gas; in this case delayed *ionization* causes the heavy species to have lower ionization stages than a plasma at the same temperature in CIE. Such a case has been considered e.g. by Cox and Anderson (1982), who followed the expansion of a young SNR into an ambient medium with non-negligible pressure. The second possibility is fast adiabatic expansion of a plasma that had been heated up during its previous history to high temperatures ($T \geq 10^6$ K), at which the assumption of CIE is fairly good; in this case delayed *recombination* essentially retains the high ionization stages of the expanding gas and the emission spectrum is similar to the original CIE plasma. We have considered this process in detail in connexion with a galactic outflow expanding away from the disk, and also for the origin of the Local Bubble (Breitschwerdt and Schmutzler, 1994; 1997). In the latter case, we assumed that the Local Bubble is the relic of an old superbubble, which was created by about 20 early type stars injecting about 10^{52} erg into a dense molecular cloud ($n_0 \sim 10^4\,\mathrm{cm}^{-3}$). After about 10^7 yr, the superbubble breaks out of the cloud and undergoes fast adiabatic expansion, with the last SN having occurred roughly 1.5×10^6 yr ago. The characteristic features of this model are:

- The emission spectrum is characterized by recombination and line emission and is similar to a 10^6 K CIE spectrum; however it is deficient in EUV lines due to the low kinetic temperature and is therefore consistent with the EUVE observations.
- There is a local contribution to the M-bands (although quantitatively this depends on the initial model)
- The plasma kinetic temperature at present is only about 4.2×10^4K and the density is $2.4 \times 10^{-2}\,\mathrm{cm}^{-3}$. Thus the thermal pressure is $P/k_B \approx 2000\,\mathrm{cm}^{-3}\,\mathrm{K}$ and therefore very close to that of the Local Cloud. The density is in agreement with the one inferred from the dispersion measure of the pulsar PSR 0950+08.

It is important to emphasize that the dynamical and thermal evolution of the plasma are intrinsically coupled and hence a self-consistent solution has to be calculated. There is no modified cooling function that can be plugged into the dynamical equations, since the dynamics changes the thermal path of the gas and vice versa.

The model presented above is by no means unique; it can explain all present observations, but might fail for future ones. The natural approach for non-CIE plasma evolution is to systematically narrow down the number of possible *initial conditions* by observations, and thus eventually find a model that comes reasonably close to reality.

4 Beyond the Local Bubble

One of the key experiments concerning the Local Bubble and the SXRB are the shadowing experiments already mentioned in the previous section. Among the first targets was a deep pointed observation of the Draco nebula (Burrows and Mendenhall, 1991; Snowden et al., 1991), that revealed that about half of the C-band emission originates behind Draco (minimum distance $D \sim 300\,\mathrm{pc}$), i.e. way beyond the Local Bubble. Although the protagonists of the LHB or displacement models have argued that this does not make any serious difference (e.g. Snowden, 1996; see also Cox, this volume), because it would only change the shape of the LHB somewhat, there remains the fact to be explained how C-band emission outside the LHB is generated in the absence of any distinct superbubbles. Even shadows much further away ($D \geq 1.5\,\mathrm{kpc}$), cast by Complex M of high velocity clouds, reveals SXR emission (Herbstmeier et al., 1995). Recently, the HI information of the northern sky has been considerably improved by the new Leiden-Dwingeloo Survey.

Also C-band emission from the halo of the edge-on spiral NGC 4631 has been reported (Wang et al., 1995), and there does not seem to be much doubt of soft X-ray emission also from the Galactic halo. How does it come about? While the HIM envisaged by McKee and Ostriker (1977) with a temperature of $T \sim 10^{5.7}$ is too cool to produce significant C-band emission, their Galactic corona with $T \geq 10^6\,\mathrm{K}$ and a pressure scale height of $h = 5(T/10^6\,\mathrm{K})\,\mathrm{kpc}$ (due to the low density of cooling clouds and energy injection from halo SNe) should do so. However, such a halo is difficult to keep in hydrostatic equilibrium for various reasons. Most simply, the cooling time τ_{cool} must be much larger than the dynamical relaxation time, i.e. $P_h^{3/2}/(q\rho_h^{5/2}) \gg c_s/g_z$ or $c_s^2 \gg \rho_h q/g_z$ where Kahn's (1976) $T^{-1/2}$-cooling-law has been applied; here P_h, ρ_h, c_s and g_z denote the halo pressure and density, the speed of sound, the z-component of the gravitational acceleration, respectively and $q = 4 \times 10^{32}\,\mathrm{cm^6\,g^{-1}\,s^{-4}}$ is a parameter (see Kahn, this volume). With $\rho_h = 2 \times 10^{-27}\,\mathrm{g/cm^3}$, $c_s = 130\,\mathrm{km/s}$ and $g_z = 10^{-8}\,\mathrm{cm/s^2}$, the above condition is hard to fulfill.

Moreover, it is very likely that cosmic rays with a comparable energy density but an essentially infinite pressure scale height will couple to the plasma through the resonant generation of MHD waves and thus drive a fraction of the plasma out to infinity in the form of a supersonic wind (Breitschwerdt et al., 1991; 1993). The rest has to rain down in a fountain (Bregman, 1980; Kahn 1981; Avillez et al., this volume). The outflowing plasma is an example of fast adiabatic expansion, as described in the previous section. The resulting emission spectrum is a superposition of delayed recombination *and* line emission. I emphasize this conjunction, because on the basis of an energy argument the significance of this process has been challenged by Don Cox (see this volume). It is true that the contribution of recombination can be low in the case of a galactic wind with a disk averaged mass loss rate of $0.4\,M_{\odot}/yr$ (Breitschwerdt and Schmutzler, 1994), although depending strongly on the initial conditions chosen. However, looking along the axis of the flow, e.g. in the direction of the North Galactic Pole from beneath the disk, we sample line emission from gas at *different* "temperatures", which is always out of equilibrium. This is fundamentally different from all hydrostatic, isothermal halo models. There is circumstantial evidence for this to have happened in the case of the soft X-ray emission in the halo of the edge-on galaxy NGC 4631. Wang et al. (1995) have reported that it is not possible to fit the halo spectrum by a single temperature of a plasma in CIE.

Another important key observation which may help us in our understanding of the Local Bubble concerns other superbubbles and in particular our nearest neighbour Loop I or North Polar Spur. On the basis of ROSAT PSPC observations, it has been shown (Egger, 1993; Egger and Aschenbach, 1995) that Loop I is an active superbubble with a present temperature of $4 \times 10^6\,K$ and a density of $2.5 \times 10^{-3}\,cm^{-3}$. The dense wall at a distance of $40 \pm 25\,pc$ as inferred from optical and UV spectral analysis of stars is most likely the interaction zone between Loop I and the Local Bubble. It can be shown (Breitschwerdt and Egger, 1997) that due to the overpressure of the plasma in Loop I, it is possible to excite a hydromagnetic Rayleigh-Taylor instability even in the presence of a magnetic field parallel to the interaction shell (by the way, this was also pointed out by Franz Kahn in the plenary discussion). The typical growth time of the most unstable mode is $\tau_g \approx \sqrt{8\pi/n_{sh}}\, N_{sh} B_0 \cos\alpha/\Delta P \approx 2.3 \times 10^6\,yr$, where $n_{sh} = 15\,cm^{-3}$, B_0 and $\Delta P = 1.4 \times 10^{-12}\,g/(cm^2\,s)$ are the density of the shell with $N_{sh} = \int_0^L n_{sh}\,dl = 10^{20}\,cm^{-2}$ being its column density (and L is the distance to the wall plus the shell thickness), the magnetic field strength and the pressure difference between Loop I and the Local Bubble, respectively; α is the angle between the direction of the magnetic field and the linear perturbation wave vector, with $B_0 \cdot \cos\alpha$ taken to be $1\mu G$. It turns out that the most unstable wavelength is about 3 pc, i.e. similar to the shell thickness. After the full fledged instability has developed and the shell starts to fragment, blobs of this size will travel ballistically through the tenuous medium of the Local Bubble. A simple calculation shows that the velocity is given to be $v_{bl} = 3\Delta P \tau_g/(\bar{m} N_{sh}) \approx 22\,km/s$. This value is fairly close to the observed velocity vector of 26 km/s of the LIC cloud (see Lallement this volume).

Far from being a direct proof, I am nevertheless tempted to conclude that the Local Bubble is a separate and rather common bubble that undergoes interaction with an other bubble; such a process might be more frequent than one would commonly think of, because star forming regions are not homogeneously distributed over the galactic disk. The presence of clouds in bubbles would then be simply a consequence of an inherent dynamic instability.

5 Open Questions

Rather than giving any conclusions here, I would like to end with a list of questions about the LISM and its relation to the general ISM, which I had written down when I was organizing the scientific part of this conference.

1. What is the geometry of the Local Bubble?
2. What is the state of the plasma inside?
3. How do clouds and HI filaments in the Local Bubble survive despite their much lower apparent thermal pressure?
4. Is the Local Bubble magnetized and if so, what is the topology of the magnetic field?
5. What is the origin of the Local Bubble?
6. What do cosmic rays tell us about nearby explosive events?
7. How does the Local Bubble connect to its environment?
8. How does the Local Bubble compare to other bubbles/superbubbles?
9. How does the LISM relate to the general ISM?
10. What is the origin of soft X-ray emission outside bubbles/superbubbles in the disk and halo?
11. What can we learn from the study of the ISM in other galaxies?
12. What are the future prospects for observation and theory?

In the previous sections I have sketched my view on some possible solutions, but now it is time to let the experts give their observational facts and interpretations on the "Local Bubble and Beyond".

Acknowledgements. I thank all participants for excellent contributions and lively and stimulating discussions during the meeting. I benefitted from many useful talks with my colleagues Roland Egger and Michael Freyberg. I thank Stu Bowyer and Chris McKee for helpful comments on the manuscript. Financial support from the DFG by a Heisenberg Fellowship is gratefully acknowledged.

References

Adams, W.S. (1948): ApJ 109, 354
Bowyer, C.S., Field, G.B., Mack, J.E. (1968): Nature 217, 3
Bregman, J.N. (1980): ApJ 236, 577
Breitschwerdt, D. (1996): Sp. Sci. Rev. 78, 173
Breitschwerdt, D. (1994): Habilitationsschrift, Universität Heidelberg, 158p.
Breitschwerdt, D., Egger, R.J. (1997): A&A (in preparation)
Breitschwerdt, D., Schmutzler, T. (1997): A&A (to be submitted)
Breitschwerdt, D., Schmutzler, T. (1994): Nature 371, 774
Breitschwerdt, D., McKenzie, J.F., Völk, H.J. (1993): A&A 269, 54
Breitschwerdt, D., McKenzie, J.F., Völk, H.J. (1991): A&A 245, 79
Brinks, E., Shane, W.W. (1984): ApJS 55, 179
Burrows, D.N., Mendenhall, J.A. (1991): Nature 351, 629

Cox, D. P., Anderson, P.R. (1982): ApJ 253, 268
Cox, D.P., Reynolds, R.J. (1987): ARA&A 25, 303
Cox, D.P., Smith, B.W. (1974): ApJ 189, L105
Egger, R.J. (1993): PhD Thesis, TU München, MPE Report 249, 111p.
Egger, R.J., Aschenbach, B. (1995): A&A 294, L25
Field, G.B., Goldsmith, D.W., Habing, H.J. (1969): ApJ 155, L14
Fruscione, A., Hawkins, I., Jelinsky, P., Wiecigroch, A. (1994): ApJS 94, 127
Gwinn, C.R., Taylor, J.H., Weisberg, J.M., Rawley, L.A. (1986): AJ 91, 338
Hartman, J. (1904): in *Sitzb. Kgl. Akad. Wiss. 527*, translated in ApJ 19, 268
Heiles, C. (1980): ApJ 235, 833
Herbstmeier, U., et al. (1995): A&A 298, 606
Ikeuchi, S. (1988): Fund. Cosmic Phys. 12, 255
Innes, D.E., Hartquist, T.W. (1984): MNRAS 209, 7
Jelinsky, P., Vallerga, J.V., Edelstein, J. (1995): ApJ 442, 653
Jenkins, E.B., Meloy, D.A. (1974): ApJ 193, L12
Kahn, F.D. (1976): A&A 50, 145
Kahn, F.D., (1981): in *Investigating the Universe,* ed. F.D. Kahn, Reidel, 1
Kerp, J., Herbstmeier, U., Mebold, U. (1993): A&A 268, L2
Linsky, J. (1996): Sp. Sci. Rev. 78, 157
McCammon, D., Sanders, W.T. (1990): ARA&A 28, 657
McKee, C.F., Ostriker, J.P. (1977): ApJ 218, 148
Norman, C.A., Ikeuchi, S. (1989): ApJ 345, 372
Pikel'ner, S. (1967): Astr. Zu. 44, 915
Plaskett, J.S., Pearce, J.A. (1933): Publ. Dom. Astr. Obs. Victoria, B.C. 5, 167
Raymond, J.C., Smith, B.W.: (1977) ApJS 35, 419
Reynolds, R. (1991): in *The interstellar disk-halo connection in galaxies,* ed. H. Bloemen, Reidel, 67
Sanders, W.T. (1993): Adv. Space Res. 13, (12)77
Sanders, W.T., Kraushaar, W.L., Nousek, J.A., Fried, P.M. (1977): ApJ 217, L87
Schmitt, J.H.M.M., Snowden, S.L. (1990): ApJ 361, 207
Snowden, S.L. (1996): MPE-Report 263, 299
Snowden, S.L., Cox, D.P., McCammon, D., Sanders, W.T. (1990): ApJ 354, 211
Snowden, S.L., et al. (1991): Science 252, 1529
Snowden, S.L., McCammon, D., Verter, F. (1993): ApJ 409, L21
Sonett, C.P., Morfill, G.E., Jokipii, J.R. (1987): Nature 330, 458
Spitzer, L. jr. (1956): ApJ 124, 20
Strömgren, B. (1939): ApJ 89, 526
Tanaka, Y., Bleeker, J.A.M. (1977): Sp. Sci. Rev. 20, 815
Tauris, T.M., et al. (1994): ApJ 428, L53
Taylor, J.H., Cordes, J.M. (1993): ApJ 411, 674
Trümper, J. (1983): Adv. Sp. Res. 2(4), 241
van de Hulst, H.C. (1945): Ned. Tijdschr. Natuurk. 11, 201
Wang, Q.D., et al. (1995): ApJ 439, 176
York, D.G. (1974): ApJ 193, L127

Part II

UV and Soft X-ray Observations of the LISM

Despite the fire incident: entertaining talks ...

... critical remarks and controversial discussions.

Observations of the Local Interstellar Cloud

R. Lallement

Service d' Aéronomie du CNRS, BP 3, 91371 Verrieres-le-Buisson, France

Abstract. Constraints on the ambient (circumsolar) interstellar medium can be derived from observations of interstellar neutrals or their derivatives in the heliosphere. New results have been recently obtained which, when combined with optical and UV observations of the nearby stars (HST-GHRS, EUVE), remove long-standing contradictions and allow us to infer for our local cloud, pressure, ionization and limits on the magnetic field. The electron density in the circumsolar gas is found to be between 0.04 and 0.15 cm^{-3}. Its total thermal pressure is within the interval $1700 - 2600\,\mathrm{cm}^{-3}\,\mathrm{K}$. If the local magnetic field is nearly perpendicular to the interstellar wind flow, which is likely, then its intensity is smaller than 3.6 μG. Our Sun is located very close to the edge of the local cloud (the volume of gas which has the same physical properties as the circumsolar gas), while there are at least 5 other cloudlets within 10 pc. Abundances vary from cloudlet to cloudlet. How they are located and whether they are separated by tenuous gas or shock discontinuities is not clear yet. Semi-hot (10^5 K) gas has been detected in absorption towards nearby stars (Wood et al, 1996), which probably originates in the "H walls" surrounding our heliosphere and other asterospheres. This shows that semi-hot gas is not necessarily linked with cloud interfaces with the hot gas of the Local Bubble.

1 Our Heliosphere: the "Solar Wind Bubble"

Our Sun is travelling through an interstellar diffuse cloud (herafter Local Interstellar Cloud, or LIC), one of the moderately warm ($T = 5,000-10,000\,\mathrm{K}$) cloudlets which have been detected in the solar vicinity (Fig.1 (top) and Fig.3). By definition, here the LIC is the very small mass of gas moving at the same velocity as the interstellar helium flow in the heliosphere, at variance with the larger volume of gas within 30 pc whose mean motion has been detected by Génova et al (1990), which includes the LIC.

The Solar Wind is blowing a cavity in the LIC, whose size and shape are determined by pressure balance between the radially expanding solar plasma and the surrounding ISM. This cavity, the heliosphere, is certainly asymmetric, because in the direction ot the relative motion of the Sun with respect to the ambient medium, the dynamic pressure associated with this motion (at $\simeq$ 20 km/s) adds up to the other interstellar pressure forces such as the thermal pressure, the magnetic tension, and the low energy cosmic ray pressure (those which do not penetrate the heliosphere) (Fig.1 (middle)). Interstellar charged particles are diverted around the heliopause, the contact-discontinuity surface which, in principle, separates the solar wind and the ISM. The solar wind is probably stopped in its radial expansion and also diverted along the inner

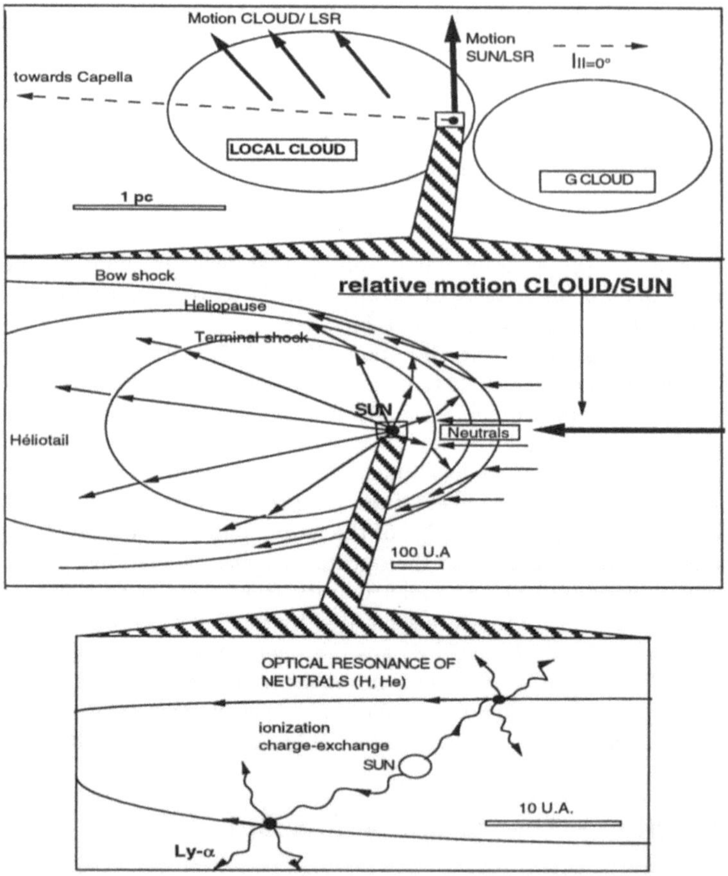

Fig. 1. The heliosphere in the Local Interstellar Cloud, at three different scales

part of the heliopause to finally merge with the ISM flow. The supersonic-subsonic transition of the solar wind is supposed to occur at the so-called "termination shock", the first discontinuity the outer probes as the Voyager and Pioneer 11 should encounter. Interstellar neutrals are the only "thermal" species to enter the heliosphere, where they can be observed through resonance scattering of solar light (Fig.1 (bottom)), but there are filtration processes at the interface due to coupling with the plasma through charge-exchange collisions. As a result, an additional and non-negligible pressure is associated to those neutrals which interact with the plasma, in particular the neutral hydrogen.

By nature, those parameters which have the strongest influence on the heliospheric size, i.e. the plasma density and the magnetic field intensity and orientation are not directly measurable in the heliosphere, and one has to rely on indirect determinations, or on interstellar measurements.

2 Constraints on the Ambient Gas from Heliospheric Observations

2.1 From Helium and Hydrogen

There has been a long-standing debate about neutral He and neutral H properties. The models predict deceleration, heating, and filtration of H at the entrance in the heliosphere while He, being not coupled to the plasma, enters the heliosphere without any perturbation. One then expects the helium flow properties to be the same in the local cloud and inside the heliosphere (before the direct interaction with the Sun). However, resonance scattering of the solar H and He Ly-alpha radiation by the inflowing H and He gives $T_{\mathrm{H}}^{\mathrm{perp}} = 8000\,\mathrm{K}$ (perpendicularly to the flow) and $T_{\mathrm{He}} = 15000 - 17000\,\mathrm{K}$. This contradiction has disappeared now, since the Ulysses in-situ detection of the neutral He (Witte et al, 1993, 1996) shows that $T_{\mathrm{He}} = 6700\,\mathrm{K}$, and $V_{\mathrm{He}} = 25.5\,\mathrm{km/s}$ from $\lambda = 254°$ and $\beta = 5°$, in perfect agreement with the properties of the LIC (see section 3). When compared with this "zero perturbation" state, H appears to be decelerated by a few km/s when entering the heliosphere (Lallement et al, 1993), providing evidence for a non-negligible plasma density outside the heliosphere, namely $n_{\mathrm{e}} = 0.05 - 0.1\,\mathrm{cm}^{-3}$ when using the Baranov and Malama (1993) filtration model.

Until very recently it was also generally admitted that helium is predominantly neutral in the local clouds if T≤10,000K. As a result, any significant ionization of the ISM implies that the neutral H density inside the heliosphere is significantly smaller than 10% of the neutral He density (for a cosmological ratio H/He= 10), due to both the ionization of the LIC itself and filtration processes. Recently, the H density in the inner heliosphere has been revised upward (e.g. Quémerais et al, 1995). The preferred values for n_{H} range from 0.10 (compiled UV glow data, see Quémerais et al, 1994) or 0.11 (H^+ pick-up ions measurements by Ulysses, Gloeckler et al, 1996) to $0.16\,\mathrm{cm}^{-3}$ from radiative transfer models. Due to the existence of the H wall (see section 4) the density found at large distance is somewhat larger than the density at 20 AU before close interaction with the sun and for this reason we will retain $0.15\,\mathrm{cm}^{-3}$ as the upper limit for the inner heliosphere density. For helium, three independent experiments have provided determinations of the neutral helium density which agree well at $n_{\mathrm{He}} = 0.015 - 0.017\,\mathrm{cm}^{-3}$ (Witte et al, 1996, Gloeckler, 1996, Möbius, 1996).

These ranges for n_{H} and n_{He} include H/He = 10 and do not require a filtration of H. However, according to the EUVE results this conclusion is now modified. As a matter of fact, helium is shown to be more ionized than hydrogen (Dupuis et al, 1995), i.e. $n_{\mathrm{H}} > 10 n_{\mathrm{He}}$ in the ISM. This leaves room for the exclusion from the heliosphere of a fraction of the neutral H, in agreement with the deceleration.

The combination of the H and He heliospheric data, the EUVE results, and the H filtration as a function of the electron density taken from a model,

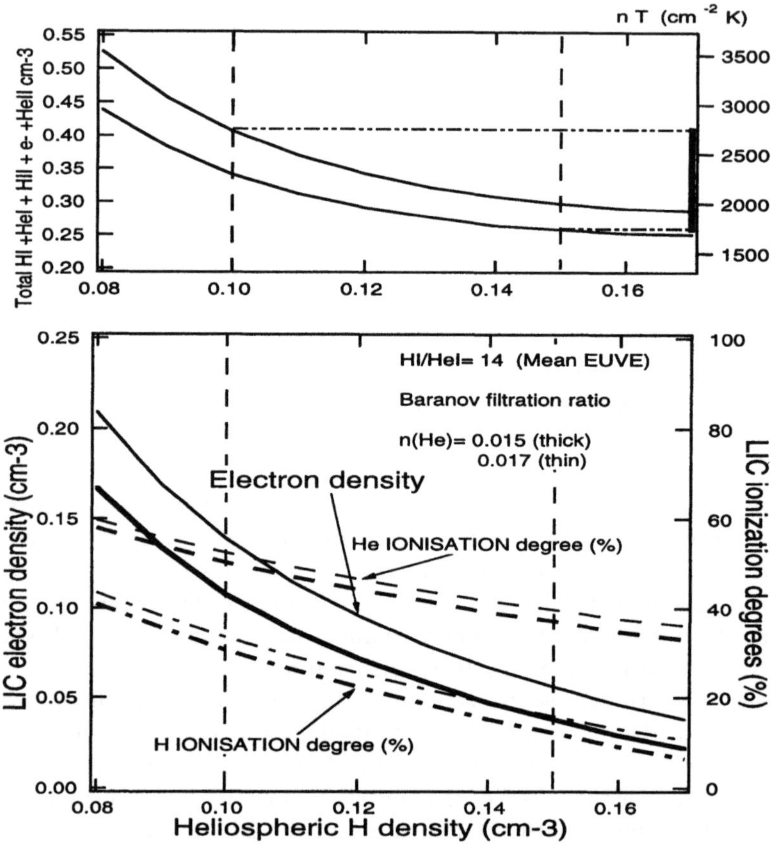

Fig. 2. Circumsolar pressure

allows us to constrain N_e. To do so, Lallement (1996) has used the Baranov and Malama (1993) model (a plasma/neutrals self-consistent model). The most appropriate value for the LIC HI/HeI ratio is 14, as measured towards G191B2B by Dupuis et al (1995). As a matter of fact, most of the neutral gas towards this star belongs to the LIC (Lallement et al, 1995, Lemoine et al, 1996), and corresponds to what is also detected towards the angularly close Capella (12.5 pc) (Linsky et al, 1995).

According to the total range for n_H, one infers for the ambient ISM $N_e = 0.04 - 0.15\,\mathrm{cm}^{-3}$ (Fig.2). This range is compatible with the deceleration of H. The corresponding neutral H density before the filtration and the total gas pressure in the LIC are shown on the top graph in Fig 2. It can be seen that the total thermal pressure is found to be $1700 - 2600\,\mathrm{cm}^{-3}\,\mathrm{K}$. This interval is much smaller than what would have been quoted three years ago.

2.2 From Oxygen Filtration

Oxygen atoms suffer significant filtration processes on entering the heliosphere, just as does neutral hydrogen, through charge-exchange with the protons. It had been concluded from studies neglecting the inverse reaction $O^+ + H \rightarrow O + H^+$ that oxygen is almost totally excluded from the heliosphere for ISM plasma densities as small as $0.01\,cm^{-3}$ (Fahr, 1991), leading to the belief that the ambient medium has to be neutral, because neutral oxygen is not underabundant in the heliosphere (e.g. Gloeckler, 1996). Recently, Izmodenov et al (1997) have used Ulysses measurements of O^+ pick-up ions (interstellar ionized oxygen), the recent precise determination of the neutral oxygen abundance in the local cloud from Capella GHRS data (Linsky et al, 1995) and a refined model of the O and O^+ flows, and shown that there is no contradiction between the presence of oxygen in the heliosphere and a moderate ionization of the ambient ISM. On the contrary, the data allow a filtration by a factor of two of this species, i.e. an ambient plasma density of up to $0.1 - 0.2\,e^- \,cm^{-3}$. Another approach has been used by Frisch (1995), using Anomalous Cosmic Rays (ACR) Voyager data. ACR are those pick-up ions which have been accelerated to high energies at the interplanetary shocks and at the heliospheric shock. This study also allows for large electron densities ($0.2\ cm^{-3}$)

2.3 From Voyager Distance and Radio Observations

If R_s is the distance to the heliospheric shock, the solar wind pressure at the heliopause is: $N_{sw}\, m\, V_{sw}^2 \,/\, K\, R_s^2$, where N_{sw}, V_{sw} are the solar wind density and velocity at 1 AU, m is the proton mass, and K is a corrective factor which takes into account the pressure increase between the solar wind shock and the heliopause (K is close to 1.1). This "inner" pressure is balanced by the "outer" pressure terms:

- the thermal and dynamic pressure of the ISM plasma: $(N_i + N_e)\, kT + 2N_i\, mV^2$, where N_i, N_e are the ion and electron densities
- the pressure of the fraction of the neutrals coupled to the plasma by charge-exchange: $\epsilon\, N(\mathrm{HI})\, m\, V^2$, where ϵ is a function of $N(\mathrm{H})$, $N(\mathrm{HII})$
- the pressure of the low-energy cosmic rays reflected by the heliosphere
- the ISM magnetic pressure: $\alpha B^2 \,/\, 8\pi$

where α depends on the field orientation and is maximum when the magnetic field is perpendicular to the relative flow. In this case the magnetic field lines are draped around the heliopause and compressed close to the stagnation line, and the magnetic tension adds to the plasma pressure. In the very improbable case of a magnetic field being by chance parallel to the flow, the situation is totally different and there is no additional pressure. In intermediate cases, one expects a partial draping and a non-zero pressure. There is evidence that the local magnetic field B is almost perpendicular to the relative flow (see Frisch, 1995) and in this case α should be close to 2 (2.2 if one uses Parker's calculations for a solar wind expanding in a pure magnetic field). The fraction of the energetic particles which are entering the heliosphere, and the neutrals which are not interacting with the plasma are not contributing to the confinement of the heliosphere.

Limits from Voyager's Present Location A straightforward way to get an upper limit on the pressure acting on the heliosphere, is by simply considering that the distance to the solar wind shock is larger than the distance already reached by the outer probes. As a matter of fact, plasma instruments on board the Voyager 1 indicate that the spacecraft are still cruising in the supersonic solar wind.

The total interstellar pressure required to confine the solar wind shock within 70 AU (Voyager 1's distance will be 69 AU at the end of 1997), is then $\Pi_\mathrm{i} < 3.0\,10^{-12}$ dynes cm^{-2}. In case of a confinement by the plasma alone the maximum density compatible with this pressure is $N_\mathrm{e} = 0.17\,\mathrm{cm}^{-3}$ (for $N(\mathrm{H}) = 0.2\,\mathrm{cm}^{-3}$). In case of pure magnetic confinement, then $\mathrm{B}_\mathrm{perp} <$ 6μG. The maximum plasma density is compatible with the range of densities derived above (but not with the 0.25 cm^{-3} derived from ACR data). As we will see, it is also compatible with some of the diagnostics derived from stellar spectroscopy, but not all of them.

Limits from the Interpretation of Voyager Radio Emission A very exciting new measurement from the voyagers is the time delay between strong solar wind events and the reception by the spacecraft of the $2-3$ kHz radio emissions believed to be generated at the heliopause by shocks and discontinuities associated to these events (Gurnett and Kurth, 1996). According to the delay of about 400 days, the distance to the heliopause should be 110-160 AU, which roughly corresponds to R_s between 75 and 95 AU. For the most probable value in this interval Π_i is about $2.6\,10^{-12}$ dynes cm^{-2}, i.e. a lower limit than what has been inferred above. This corresponds to $N_\mathrm{e} < 0.15\,\mathrm{cm}^{-3}$ in case of a pure plasma confinement. Now, what can be said about the relative contributions to confinement of the plasma and the magnetic field? Here we will distinguish between two cases which are based on different diagnostics:

i) The Voyager plasma experiment has detected a 2 kHz emission which, contrary to the 3 kHz radiation, is not stronger along the wind axis. This emission has a very precise low cutoff at 1.8 kHz. There are no satisfying explanations for this emission yet, but the presence of such a marked cutoff, which corresponds to the plasma frequency of a medium with $N_\mathrm{e} = 0.04\,\mathrm{cm}^{-3}$, suggests a phenomenon linked with the ambient ISM. If this is true, then we have an extremely precise determination of the circumsolar electron density. In this case, one can infer the plasma pressure $\Pi_i = 0.5\ 10^{-12}$ dynes cm^{-2}, and estimate the additional pressure connected with the neutrals

linked to the coupling with the plasma. A rough estimate from comparison with the Baranov and Malama models with and without neutrals gives $\Pi_{\text{neutrals}} = 0.4\,10^{-12}$ dynes cm^{-2}. Subtracting those terms from the total pressure results in a pressure associated with the magnetic field and the cosmic rays $\Pi_{\text{magnetic+CR}} < 1.7\,10^{-12}$ dynes cm^{-2}, and $\Pi_{\text{magnetic}} < 1.2\,10^{-12}$ dynes cm^{-2} if $\Pi_{\text{CR}} = 0.5\,10^{-12}$ dynes cm^{-2} (Ip and Axford, 1986). This corresponds to $B_{\text{perp}} < 4.4\mu$G if $\Pi_{\text{CR}} = 0$ and $B_{\text{perp}} < 3.7\mu$G if $\Pi_{\text{CR}} = 0.5\,10^{-12}$ dynes cm^{-2}.
ii) If the 1.8kHz cutoff is not the ISM plasma frequency, then the most likely value for N_e is of the order of $0.1\,\text{cm}^{-3}$ according to the deceleration of H (section 2.1), the filtration of H (idem), the CII/CII* ratio (see section 5). Replacing 0.04 by 0.1 in the above estimates results in $B_{\text{perp}} < 2.4\mu$G if $\Pi_{\text{CR}} = 0$ and $B_{\text{perp}} < 0$ (!!) μG if $\Pi_{\text{CR}} = 0.5\,10^{-12}$ dynes cm^{-2}.
These rough estimates should be refined in the future: (i) thanks to a better understanding of the radio emissions (ii) from better constraints on the filtration and the shape of the heliosphere (iii) by simply waiting: the progression of the spacecraft continuously changes the upper limit on the pressure, and there is hope that the heliospheric shock will be reached some time between 2010 and 2015, assuming that the solar wind is not decelerated to the point where no shock, as such, is in fact formed.

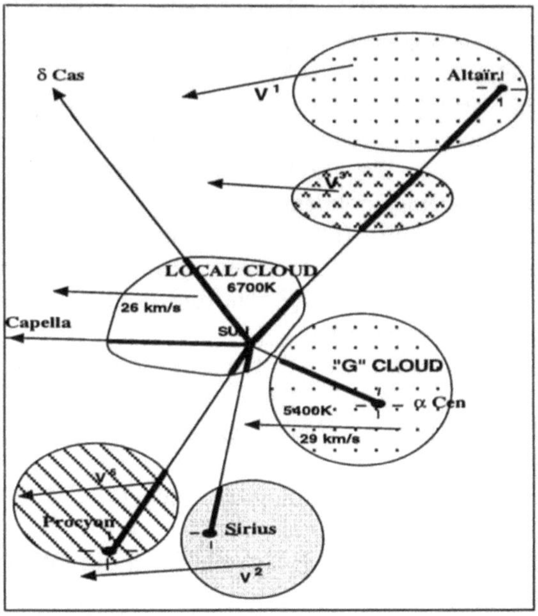

Fig. 3. Schematic view of the local cloudlets

3 LIC Characteristics from Stellar Spectroscopy

The identification of the local masses of gas relies upon the study of the Doppler shifts and the depths of the absorption lines detected in the spectra of the nearby stars. If the LIC is moving like a solid body, then for each target star the Doppler shift of the absorption line which is actually associated to this cloud is simply the projection of its velocity vector onto the line-of-sight. In other words, if one can find a set of coherent Doppler shifts in a large enough number of directions covering the sky, for a single velocity vector, then this motion is probably that of the mass of gas in which the Sun is embedded,as seen via the Doppler triangulation method.Towards α Aql, for example, three different motions are detected. One has a Doppler component compatible with the Local cloud vector (Fig 3). In fact, two masses of gas were detected in this way, in two opposite regions of the sky (see Lallement et al, 1995). The GAS instrument on board Ulysses measured 26 km/s from $\lambda = 254°$ and $\beta = 7°$ (Witte et al, 1993, 1996) in agreement with the cloud which was seen towards the largest fraction of the sky. Later the temperature measured by Ulysses (6500 K) was found to be comparable to the temperature of this cloud (6700 K) measured from D and H lines by the GHRS towards Capella (Linsky et al, 1995). The temperature of the second cloud, called the G cloud, has been determined by Linsky and Wood (1996), from α Cen spectra, and found to be 5400 K.

Once the LIC is identified, one can use all the absorption lines at the LIC Doppler shift and combine them to infer some of its physical properties. Up to now, attempts have been made with the help of the Mg^+/Mg^o , Na^+/Na^o and C^+/C^{+*} ratios (C^{+*} is the excited state). In the first two cases, N_e is deduced from the assumed equilibrium between photoionization and recombination. Those estimates are then dependent on the radiation field. The C^+/C^{+*} method has the great advantage of being independent of the field because this ratio depends only on the temperature (otherwise measured from the line widths of different species), and the electron density. As a result, this is very likely the most reliable determination. The resulting ranges for the electron density from the different methods are:

a) MgII/ MgI equilibrium with the HST-GHRS: towards Sirius (2.7 pc) (Lallement et al, 1994) $N_e = 0.3\,\mathrm{cm}^{-3}$, towards ϵ CMa (200 pc) (Gry et al, 1995) $N_e = 0.2\,\mathrm{cm}^{-3}$, towards δ Cas (22 pc) (Lallement and Ferlet, 1997) $N_e = 0.25\,\mathrm{cm}^{-3}$ assuming $T = 7000\,\mathrm{K}$.

b) Na ionization (ground): towards δ Cas (Lallement and Ferlet, 1997) $N_e = 0.05\mathrm{cm}^{-3}$

c) CII/CII* equilibrium (HST-GHRS): towards Capella (12.5 pc) (Wood and Linsky, 1997) $N_e = 0.11\mathrm{cm}^{-3}$

The results show rather strong discrepancies between the quantities derived from magnesium and those from carbon or sodium. It may be a sign of an absence of ionization equilibrium. This certainly deserves further studies because such a lack of equilibrium points in favor of a recent ionizing event in the Local Bubble. It may also simply reflect ionization gradients in the clouds (e.g. Vallerga, 1996).

4 Boundaries Between Cloudlets and Semi-hot Gas?

4.1 Existence of Shocks

Any collision between two masses of ionized gas, or partially ionized gas should produce propagating shocks in the colliding clouds. The similarity between the velocities and temperatures of the two local masses of gas (the LIC and the G cloud) may be due to the existence of a traveling weak shock dividing a single cloud into two media with slightly different properties. In other words, there would be a unique cloud and a shock front separating it into the LIC (the post-shock warmer gas) and the G (the pre-shock gas). Grzedzielski and Lallement (1996) have calculated the magnetic field intensity compatible with this scenario, assuming the shock is a perpendicular shock (those shocks are found to have the longest lifetime). The resulting values of the order of 1.5-2 μG were encouraging. However, it was recently demonstrated by Linsky and Wood (1996), that the abundances of FeII and MgII in the G cloud with respect to deuterium are a factor of 4 larger than in the LIC. There is no explanation for such a difference in terms of propagating shocks, and the existence of two very different histories for the two masses of gas is now much more likely. If these two clouds are independent, then one can argue that the faster one, the G cloud at 29 km/s, has probably not yet caught the LIC. As a matter of fact, if it had caught the LIC, two reverse shocks would have been produced, and there would be a contact discontinuity with no velocity jump at the location of the encounter. The LIC-G transition is not such a discontinuity, (the velocities are different), and unless by chance the collision has occured extremely recently, and the three surfaces (the two shocks and the contact discontinuity) are very near from each other, it is probable that the two masses of gas are not in contact. This has some importance with respect to the Local Bubble pressure, since in this case, if the clouds are embedded in hot gas, there should be hot gas between the two clouds and two hot-warm conductive interfaces close to the Sun.

4.2 Semi-hot Gas from H Walls

Semi-hot gas has been recently detected by the GHRS in the local ISM from its Ly-α absorption lines (Bertin et al, 1995, Gry et al, 1995). The existence of gas at temperatures of order of 10^5 K is of crucial importance since this gas is supposed to separate the warm clouds from the ambient hot gas of the LB. However it is clear now that there are other sources of semi-hot gas. Linsky and Wood (1996) detected an extra absorption by neutral H at 20,000 K towards α Cen and argued convincingly that this column of gas corresponds to the "H wall", a region of decelerated and compressed interstellar gas in front of the heliosphere. Spectra of the stars λ And and ϵ Ind also showed the existence of semi-hot gas, which Wood et al (1996) attributed to H walls around those stars. Moreover, Williams et al (1996) have suggested the existence of semi-hot gas in the tail of the heliosphere, due to the mixing of the main flow and neutral H resulting from charge-exchange with the fast solar wind protons. This shows that one should be careful in interpreting the amount of semi-hot gas in terms of interfaces with the ambient hot gas only.

Acknowledgements. If this paper is readable, it is due to the careful work of the native english referee from the La Palma group who corrected all errors. Many thanks to him.

References

Axford W.I., Ip W.H., 1986, Adv. Space Res. 6 (2), 27
Baranov V.B., Malama Yu.G., 1993, JGR 98, 15157
Bertin P., Vidal-Madjar A., Lallement R., et al, 1995, A&A 302, 889
Cummings A. C., Stone E.C., 1990, 21st Int. Cosmic Ray Conference 6, 202
Dupuis J., Vennes S., Bowyer S., 1995, ApJ 455, 574
Fahr H.J., 1991, A&A 241, 251
Fahr H.J., Osterbart O. and Rucinski D., 1995, A&A 294, 584
Frisch P.C., 1995, Space Sci. Rev. 72, 499
Génova R., Beckman J.E., Molaro P., and Vladilo G., 1990, ApJ 355, 150
Gloeckler G., 1996, Space Sci. Rev. 78, 335
Gry C., Lemonon L., Vidal-Madjar A., Lemoine A., Ferlet R., 1995, A&A 302, 497
Grzedzielski S., Lallement R., 1996, Space Sci. Rev. 78, 247
Gurnett D.A., Kurth W.S., 1996, Space Sci. Rev. 78, 53
Izmodenov V., Malama Yu., & Lallement R., 1997, A&A 317, 193
Lallement R., Bertaux J.L., Clarke J.T., 1993, Science 260, 1095
Lallement R., Bertin P., Ferlet R., et al,1994, A&A 286, 898
Lallement R., Ferlet R., Lagrange A.M., et al, 1995, A&A 304, 461
Lallement R., 1996, Space Sci. Rev., 78, 361
Lallement R., Ferlet R., 1997, A&A, in press
Lemoine M.,Vidal-Madjar A., Ferlet R., et al, 1996, A&A 308, 601
Linsky J.L., Brown A., Gayley K., et al, 1993, ApJ 402, 694
Linsky J.L., Diplas A., Wood B.E., et al, B.D., 1995, ApJ 451, 335
Linsky J.L., Wood B.E., 1996, ApJ 463, 254
Möbius E., 1996, Space Sci. Rev. 78, 375
Quémerais E., Bertaux J.L., Sandel B.R., Lallement R., 1994, A&A 290, 941
Quémerais E., Sandel B.R., Lallement R., Bertaux J.L., 1995, A&A 299, 249
Vallerga J., 1996, Space Sci. Rev. 78, 277
Witte M., Rosenbauer H., Banaszkiewicz M., et al, 1993, Adv. Sp. Res. 13-6,121
Witte M., Banaskiewicz M., Rosenbauer H., 1996, Space Sci. Rev. 78, 289
Williams L.L., Hall D.T., Pauls H.L., Zank G.P., 1996, ApJ 476, 366
Wood B.E., Alexander W.R., Linsky J.L., 1996, ApJ 470, 1157
Wood B.E., Linsky J.L., 1997, ApJ 474, 39

The Intrinsic Properties of the Local Interstellar Medium

Olivia Puyoo and Lotfi Ben Jaffel

Institut d'Astrophysique de Paris, 98 bis Blvd Arago, 75014 Paris, France

Abstract. We propose a new method to constrain the actual state of the interstellar cloud that surrounds the solar system. Using Voyager UVS Lyman-α sky maps and the powerful principle of invariance, we derive the H distribution all along the spacecraft path. Provided current models of the heliopause interface between the solar and the interstellar winds, we extrapolate this distribution to farther distances from the Sun and infer in a self consistent way key parameters of the local cloud. Our findings are a high interstellar hydrogen density of $\sim 0.24\,\mathrm{cm}^{-3}$ and a weak ionization $\frac{\mathrm{n(H^+)}}{\mathrm{n(H^+)+n(H)}} \simeq 14\%$.

1 Introduction

It is widely believed that the local interstellar medium represents a point sample of the more extended interstellar medium (ISM), and that its studies through its interplanetary signature is of particular interest to understand the origin and the evolution of the ISM. Because of the relative motion of the Sun with respect to the local interstellar cloud (LIC), the neutral component of the interstellar wind penetrates deeply inside the solar system, where it becomes accessible to in situ detections. After its discovery in the early 70's in the Earth neighborhood, interpretation of backscattered H Lyman-α and HeI 58.4 nm solar radiations has been extensively employed to derive H and He abundances in the local cloud. As classical interpretations of UV sky maps usually require the modeling of the H distribution in the heliosphere with the problem of filtration at the heliopause and the calculation of the radiation field including multi-scattering effects, the derivation of neutral abundances proved to be difficult, and the values found were consequently poorly constrained.

The aim of this paper is to present a new technique for deriving the intrinsic properties of the LIC by interpreting differently the UV backscattered radiation measurements. This method consists in constraining the H distribution stage by stage, starting from the Earth neighborhood up to the far unperturbed LIC. In a first step, we determine the H distribution along the Voyager trajectory by applying the principle of invariance to UVS Lyman-α sky maps recorded for several positions of the spacecraft (Puyoo et al 1997). Then we connect this distribution to the LIC one through a model of interaction between the solar and the interstellar flows. Lastly, assuming the LIC in a local steady state, we derive the intrinsic properties of the local cloud, i.e. the densities of both neutral and ionized hydrogen and helium in a self consistent way.

2 Method

Our technique is a continuation of a method recently proposed by Puyoo et al (1997) to derive in a self-consistent way the H neutral density in the inner heliosphere. This method is based on the invariance principle, which expresses, for a slab, the direct relationship between the radiation field incident on its boundaries and its optical properties, i.e. the reflexion and transmission coefficients and the contribution of inner sources. As the coefficients are dependent on the scattering gas distribution, the H distribution within the layer is therefore directly accessed. With that in view, we apply our technique to VOYAGER/UVS Lyman-α sky maps taken at different locations of the Spacecraft from 4 to 35 AU, which provides the H density distribution as shown in Fig. 1. It is worth to note that by using the principle of invariance, we avoided the modeling of the whole complex medium as it was previously required by classical techniques to interpret Voyager data. However, due to the local nature of our method, a limitation immediately appears as we can only constrain the density for regions visited by the Voyager spacecraft and for which Ly-α sky maps exist.

To reach larger distances from the Sun, we extrapolate the distribution shown in Fig. 1, using the sophisticated model of interaction between the solar wind and the supersonic flow as developed by Baranov & Malama (1993). In such models, the filtration effects due to charge exchanges between interstellar hydrogen atoms and protons at the heliopause depend mainly on two parameters: the Mach number and the proton density of the LIC. In the following, we fix the Mach number at its commonly assumed value M$\simeq$ 2 (Baranov & Malama 1993) and consider the LIC proton density as a free parameter. We then extrapolate Voyager H distribution shown in Fig. 1 by varying both the LIC proton (n_{H+}) and the H neutral ($n_{H\infty}$) densities, which provides a first relation between the two parameters (see Fig. 2). A second relation is however necessary to simultaneously constrain them. For that end, we assume the unperturbed LIC to be in a local statistical equilibrium, and we solve the charge balance equations (e.g. Vallerga & Welsh 1993) controlled by the ionization due to cosmic rays, known EUV stellar sources and the estimated contribution of the conductive interface between the local cloud and the surrounding local bubble (Slavin 1989). With the additional assumption of a cosmic ratio of H/He=10, we obtain a second relation between proton and H densities.

As shown in Fig. 2, this relation, combined with the one deduced through the Voyager distribution and the Baranov model, allows to derive in a self

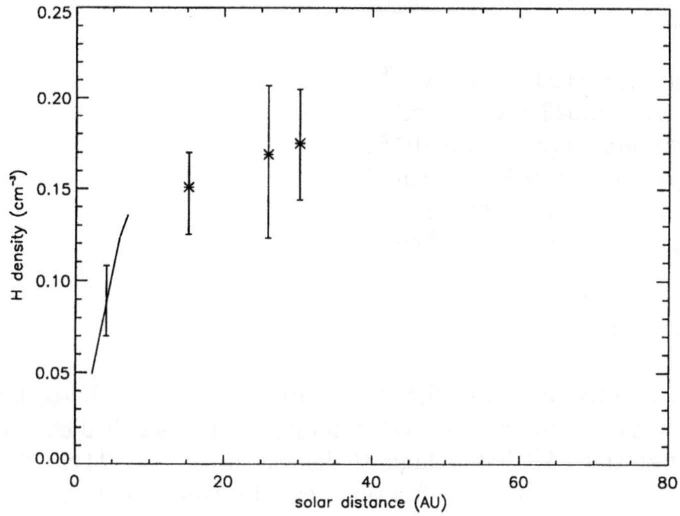

Fig. 1. H density distribution along Voyager 2 path derived from Lyman-α sky maps using the invariance principle.

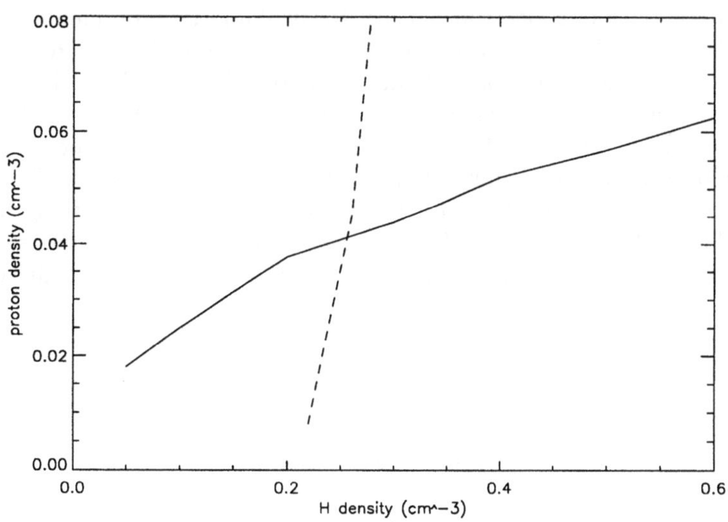

Fig. 2. Relations between the LIC proton and H neutral densities derived from the connection of the LIC to the Voyager H distribution (see Fig. 1) through Baranov model (dashed curve) and from the assumption of a local ionization equilibrium, taking $\Gamma_{\rm H} \simeq 2.5\,10^{-15}\,s^{-1}$ and $\Gamma_{\rm He} \simeq 6.9\,10^{-15}\,s^{-1}$ (solid line). The intersection gives the actual H and proton densities, $n_{\rm H\infty} \simeq 0.24\,{\rm cm}^{-3}$ $n_{\rm H+} \simeq 0.043\,{\rm cm}^{-3}$ respectively, in the local cloud.

consistent way the following set of properties of the local cloud at ~1000 AU from the Sun:

- H neutral: $n_H = 0.24 \pm 0.05\,cm^{-3}$
- proton: $n_{H^+} = 0.043 \pm 0.005\,cm^{-3}$
- He neutral: $n_{He} = (2.7 \pm 0.5)10^{-2}\,cm^{-3}$
- He^+: $n_{He^+} = (1.5 \pm 0.3)\,10^{-3}\,cm^{-3}$
- electron: $n_e = 0.044 \pm 0.005\,cm^{-3}$
- He^{++}: $n_{He^{2+}} = (1.0 \pm .2)\,10^{-5}\,cm^{-3}$

3 Conclusion

After extraction of the hydrogen distribution from the Voyager Ly-α sky maps inside the heliosphere (see Fig. 1) and assuming local statistical equilibrium, we derived that the LIC has a high H density $n_{H\infty} \simeq 0.24\,cm^{-3}$ and is weakly ionized with an ionization fraction near the Sun of ~ 15% and ~ 6% for H and He respectively. The self-consistency of the method used here, makes the derived set of LIC parameters unique for the ionization sources considered here, which could however not explain the enhanced He ionization with respect to H found by Dupuis et al (1995).
In order to check the validity of the weak ionization inferred for H, it is interesting to compare our electron density to available measurements. We apply a simple model that calculates the photoionization of H and He from the Sun out to the edge of the local cloud to different lines of sight. Toward ϵ CMa or Capella, we obtain a mean electron density $\langle n_e \rangle \simeq 0.08\,cm^{-3}$, which is in good agreement with recent HST measurements made by respectively Gry et al (1995) and Wood & Linsky (1995). As far as the ionization of H is the main pool for electron production, this result seems to confirm that this process is well described by our model. Nevertheless, an additional source that preferentially ionizes He atoms is required to explain EUVE results (Vallerga 1996).

References

Baranov, V.B, Malama, Y.G. (1993): JGR, 98 A9, 15157
Dupuis, J., Vennes, S., Bowyer, S., Pradhan, A., Thejll, P. (1995): ApJ, 455, 574
Gry, C., Lemonon, L., Vidal-Madjar, A., Lemoine, M., Ferlet, R. (1995): A&A, 302, 497
Puyoo, O., Ben Jaffel, L., Emerich, C. (1997): ApJ, 480, 262
Slavin, J.D. (1989): ApJ, 346, 718
Vallerga, J. (1996): Space Sci. Rev., 78, 277
Vallerga, J., Welsh, B.Y. (1995): ApJ, 444, 702
Wood, B., Linsky, J. (1997): ApJ, 474, 39

A Thermal Pressure Inside the Local Bubble, as Revealed by C I Fine-Structure Excitation

Edward B. Jenkins

Princeton University Observatory, Princeton, NJ 08544 USA

Abstract. Ultraviolet absorption lines from carbon atoms within a neutral cloud inside or near the edge of the Local Bubble can reveal the thermal pressure of the gas, since the relative populations of the fine-structure levels are determined by a balance between collisional excitation and radiative decay. Features from a C I multiplet appearing in the uv spectrum of the star δ Cyg ($l = 79°$, $b = +10°$, $d = 52$ pc) were observed with the echelle spectrograph of the Goddard High Resolution Spectrograph on HST. An interpretation of absorptions detected from the two lowest levels gives a thermal pressure range $10^{2.7} < p/k < 10^{3.7}\text{cm}^{-3}\text{K}$.

1 Introductory Remarks

Observations of the interstellar medium in radio, optical, uv and x-ray spectral regions highlight the existence of tremendous contrasts in density and temperature. We know that even the product of these two quantities, the thermal pressure of the gas, can exhibit significant changes from one region to the next (Jenkins, Jura, & Loewenstein 1983), indicating that processes that can lead to changes in pressure happen more rapidly than the time needed for the pressures of adjacent, large gas complexes to equalize. Thermal pressures can also be influenced by an interplay with other forms of pressure, such as turbulence, cosmic rays and magnetic fields (Kulkarni & Heiles 1988). Against this chaotic background, it is clear that we must explicitly measure pressures in specific contexts of interest and not take overall averages determined from elsewhere. Thus, in understanding the fundamental nature of the Local Bubble, our interpretations must rest upon the properties of gas that resides within its perimeter, or perhaps just beyond it. While the thermal pressure nkT is only one of several possible forms of pressure, it is a measurable physical property that must be reconciled with various plausible theoretical alternatives for the evolution and present state of the Local Bubble.

2 Method of Measurement

For any cold, mostly neutral cloud in front of an early-type star, we can observe absorption features from neutral carbon, even though the dominant ionization stage is the singly ionized form. The ground state of C I is split into three fine-structure levels, representing different J quantum states. The upper two levels have energy separations E/k from the ground level that are comparable to typical interstellar kinetic temperatures, and they are populated by collisional excitations by other gas constituents (atoms and electrons).[1] Their relative occupations are governed by an equilibrium between these collisions that scale with local density and a constant rate of depopulation via radiative decays in the infrared.

A formalism introduced by Jenkins & Shaya (1979) for expressing the relative populations,

$$f_1 = \frac{N(\mathrm{C\ I}^*)}{N(\mathrm{C\ I_{total}})} \quad \text{and} \quad f_2 = \frac{N(\mathrm{C\ I}^{**})}{N(\mathrm{C\ I_{total}})} \tag{1}$$

can be used to interpret the results in terms of various temperatures and pressures through a diagram presented in their Figure 6. $N(\mathrm{C\ I}^*)$ is the column density of carbon atoms in the first excited level ($^3P_1$), $N(\mathrm{C\ I}^{**})$ that of the second excited level ($^3P_2$), and $N(\mathrm{C\ I_{total}})$ applies to the sum of all three levels. (A particular advantage of this representation is that one can use a simple geometrical interpretation to understand the result from a superposition of two regions with different conditions, as explained by Jenkins & Shaya, but this complication will not apply to the case studied here.)

3 Observation

The investigation of pressures for the Local Bubble using C I absorption features was the objective of a Cycle 6 Guest Observer program with HST. We consider here the results for one star in this small survey, δ Cyg, a B9.5III star with a V magnitude of 2.92 and an apparent rotation velocity $v \sin i =$ 149 km s^{-1} (Hoffleit & Jaschek 1982). It is at a distance of 52 pc (Hipparcos parallax) and is located in the direction $l = 79°$, $b = +10°$. In this general direction, appreciable concentrations of H I are found at far greater distances (Paresce 1984), but some stars in the general vicinity of δ Cyg do show some Ca II absorption (Vallerga, et al. 1993). Fig. 1 shows the spectrum that was recorded by the echelle spectrograph in GHRS ($\lambda/\Delta\lambda = 80,000$), using the large, 2″ aperture with COSTAR in front.

Since C I was detected only in this one multiplet, we can not derive $N(\mathrm{C\ I})$ from a curve of growth, and therefore we must determine the line saturations indirectly. Absorption lines of Na I recorded at visible wavelengths should be good analogues to those of C I, since they too originate from a very mildly depleted element in an ionization stage below the preferred one. Very high resolution spectra of δ Cyg indicate that $b(\mathrm{Na\ I}) = 0.43$ km s^{-1} (Blades,

[1] If the atoms are near a star that emits a large flux of uv photons, optical pumping of the levels can also be important. See Fig. 6b of Jenkins & Shaya (1979).

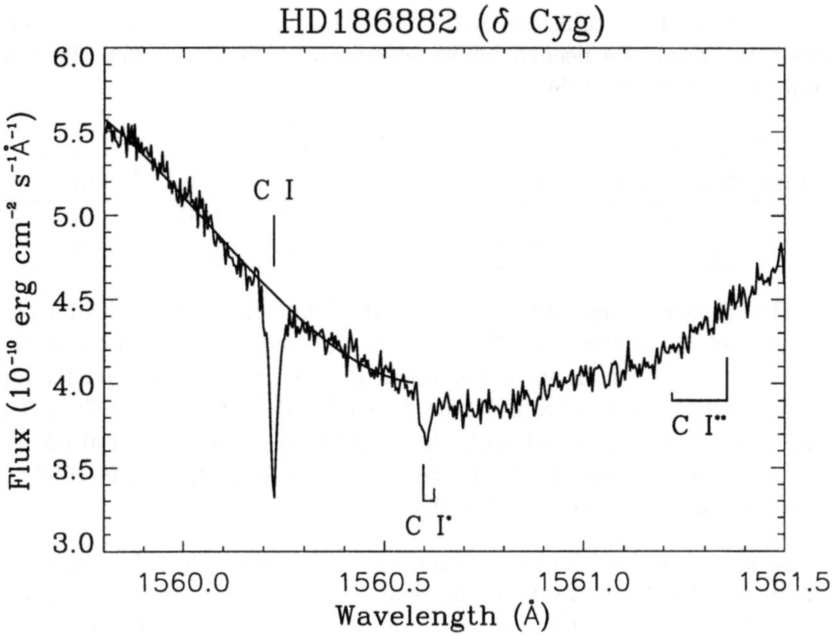

Fig. 1. The spectrum of δ Cyg covering the 1560Å multiplet of C I. The positions of absorptions out of the 3 fine-structure levels are shown with labels for C I, C I*, and C I**. The lengths of the vertical markers are proportional to the lines' f-values. Where there are two transitions from the same level, the vertical markers are joined by a horizontal line. The smooth line shows the adopted continuum level for the strong C I line.

Wynne-Jones, & Wayte 1980; Welty, Hobbs, & Kulkarni 1994), so we expect that b(C I) is somewhere between 0.43 and 0.60 km s^{-1}, depending on whether the broadening is primarily turbulent or thermal.[2] If it is turbulent, the temperature could be arbitrarily low, while for pure thermal broadening b(Na I) leads to $T = 250$ K. The table below summarizes the outcomes for these two extremes for the permissible ranges of f_1. Upper limits for f_2 do not provide useful constraints. For the physical state of the cloud in front of δ Cyg (i.e., mostly neutral, cool gas) the fine structure levels are predom-

[2] From a determination that b(Ca II) $= 0.47$ km s^{-1} toward δ Cyg (Welty, Morton, & Hobbs 1996), we might be tempted to conclude that the broadening must be mostly turbulent and the temperature is very low, since Ca atoms are almost twice as heavy as Na atoms. Unfortunately, we could be fooled by a general tendency for Ca II lines to appear broader than lines of Na I (see Fig. 7 of Welty et al.), an effect that could cancel the $m^{-1/2}$ scaling for pure thermal Doppler broadening. Our HST measurement does not have enough resolution to determine b(C I) from the apparent profile shape (Robinson, et al. 1995).

inantly excited by collisions with H I; a fractional ionization greater than about 1% would be needed before electrons could have much effect on the value of f_1 (Keenan 1989).

Source of Line Broadening	b(C I) (km s^{-1})	T (K)	f_1 l.l.	f_1 u.l.	$\log p/k$ l.l.	$\log p/k$ u.l.
pure thermal	0.60	250	0.121	0.185	3.45	3.70
pure turbulent	0.43	40	0.065	0.128	2.70	3.10

The allowed range $10^{2.7} < p/k < 10^{3.7}\mathrm{cm}^{-3}\mathrm{K}$ for the thermal pressure of the cold cloud toward δ Cyg is well centered on the estimated thermal pressure of another cloud in the Local Bubble, namely, the partly ionized, warm medium that surrounds the Sun (Wood & Linsky 1997; Lallement, 1997). However, it is significantly lower than the general estimate of $1-2\times 10^4\mathrm{cm}^{-3}\mathrm{K}$ for hot gas in the Local Bubble (Cox & Reynolds 1987) or the measurement of Bowyer, et al. (1995).

Acknowledgements. Support for this work was provided by NASA through grant number GO-06415.01-95A from the Space Telescope Science Institute, which is operated by the Association of Universities for Research in Astronomy, Inc., under NASA contract NAS5-26555.

References

Blades, J. C., Wynne-Jones, I., & Wayte, R. C. 1980: MNRAS, 193, 849
Bowyer, S., Lieu, R., Sidher, S. D., Lampton, M., & Knude, J. 1995: Nature, 375, 212
Cox, D. P., & Reynolds, R. J. 1987: Ann. Rev. A&A, 25, 303
Hoffleit, D., & Jaschek, C. 1982: The Bright Star Catalogue, (New Haven: Yale U. Obs.)
Jenkins, E. B., Jura, M., & Loewenstein, M. 1983: ApJ, 270, 88
Jenkins, E. B., & Shaya, E. J. 1979: ApJ, 231, 55
Keenan, F. P. 1989: ApJ, 339, 591
Kulkarni, S. R., & Heiles, C. 1988: in Galactic and Extragalactic Radio Astronomy, ed. G. L. Verschuur & K. I. Kellermann (Berlin: Springer), p. 95
Lallement, R. 1997: these proceedings
Paresce, F. 1984: Astron. J., 89, 1022
Robinson, R. D. 1995: in Calibrating Hubble Space Telescope: Post Servicing Mission, ed. A. Koratkar & C. Leitherer (Baltimore: Space Telescope Science Inst.), p. 193
Vallerga, J. V., Vedder, P. W., Craig, N., and Welsh, B. Y. 1993: ApJ, 411, 729
Welty, D. E., Hobbs, L. M., & Kulkarni, V. P. 1994: ApJ, 436, 152
Welty, D. E., Morton, D. C., & Hobbs, L. M. 1996: ApJS, 106, 533
Wood, B. E. & Linsky, J. L. 1997: ApJL, 474, L39

Ultra-High Resolution Ca II Observations of the LISM

Barry Y. Welsh[1], Rosine Lallement[2], and Ian Crawford[3]

[1] Eureka Scientific, Oakland, CA 94602, USA
[2] Service d'Aeronomie du CNRS, Verrieres le Buisson, France
[3] University College London, London WC1E 6BT, England

Abstract. We report on the status of a long-term program of mapping the velocity structure of the local ISM using ultra-high resolution observations of CaII K-line absorption towards nearby B and A-type stars. Absorption components common to the local interstellar cloud (LIC) have been detected in many (but not all) lines-of-sight. Our preliminary results indicate that the local ISM has a complex velocity pattern of inhomogeneous absorption, indicative of a shocked filamentary nature.

1 Introduction

The Ultra-High Resolution Facility (UHRF) at the Anglo-Australian 3.9m Telescope is an echelle spectrograph capable of a resolving power of $R = \lambda / \delta\lambda = 1{,}000{,}000$ (0.3 km s^{-1}), and is described in detail in Diego et al (1995). Such high spectral resolution enables discrete, but closely spaced, absorption components to be not only resolved, but also located with a precise velocity (0.1 km s^{-1}). This instrument is thus ideal for absorption studies of the diffuse gas "cloudlets" known to lie within 40pc of the Sun, in which previous lower resolution studies by Lallement et al (1986) and Vallerga et al (1993) have shown a complex pattern of absorption components even for lines-of-sight towards stars within 5pc. Using a Doppler triangulation method on the observed projected velocities of absorption components detected towards nearby stars, Lallement and Bertin (1992) identified two common absorption features (separated by only a few km s^{-1}) due to the local interstellar cloud (LIC) and the local "G" cloud. Both of these very nearby cloudlets may well completely surround the Sun, and hence should be detected in line-of-sight absorption studies towards all stars.

The temperature, T_c, of such a diffuse gas cloudlet can be directly determined from the Doppler broadening parameter (b-value) of the observed absorption lines, if turbulence has a small effect. For the very nearest of these clouds (the LIC) ultraviolet observations indicate the rms turbulent velocity to be < 1 km s^{-1} (Linsky and Wood 1995), and hence for an assumed LIC kinetic temperature of 7000K, the CaII K-line Doppler broadening parameter is 2.4 km s^{-1}. This observed line width is far greater than the intrinsic UHRF instrumental width, and hence we are able to gain an unambiguous direct measurement of the temperature of clouds in the local ISM.

In this paper we report on the present status of our long-term program of mapping the velocity structure of the local ISM using UHRF spectroscopy of the CaII K-line at 3333 Å observed in absorption towards nearby B and A-type stars. These observations, when used in conjunction with HST GHRS and STIS ultraviolet data, can provide better insights into (a) the physical distribution and number of diffuse interstellar clouds in the local ISM, (b) the ionization state and balance of the local ISM, (c) the density, temperature and pressure of local gas clouds, and (d) the kinematics of the local gas clouds.

2 Observations

The galactic distribution of CaII absorption column density, N(CaII), in line-of-sight towards 30 stars within 50pc has been presented by Vallerga et al (1993). For most lines of sight, the level of CaII absorption is weak (i.e. N(CaII) < 11.0 cm^{-2}). However, 2 of the 3 lines-of-sight with an appreciably high level of CaII absorption all lie in the general direction of Ophiuchus at a distance of $\sim$ 20pc. Our recent observations of 51 Oph also indicate high levels of CaII absorption in this direction (Crawford, Craig and Welsh 1997). The local distribution of neutral sodium column density, N(NaI), follows a similar pattern in that there is a distinct paucity of cold, neutral gas within 50pc (Welsh et al 1994). Bertin et al (1993) have noted that the average value of N(NaI) / N(CaII) is < 0.2 for lines-of-sight within 30pc, indicative of the presence of warm (7000K) LISM gas in which Ca is depleted by a factor of 40.

UHRF CaII spectra have already been presented for local lines-of-sight (d $<$ 40pc) towards α Oph, α Gru and α Eri (Crawford and Dunkin 1995), α Cen (Crawford 1994), and γ Aqr, ι Cen, ϵ Gru, α Hyi, γ Oph and 51 Oph (Crawford, Craig and Welsh 1997). Absorption components due to the LIC and/or the closely associated "G Cloud" have been identified in most lines-of-sight, but only in the case of the star 51 Oph are both clouds simultaneously detected. In addition, the lines-of-sight to γ Oph and α Eri reveal two and four absorption components respectively, but none of these component velocities can be associated with either the LIC or "G Cloud". Furthermore, the nearest star to the Sun, α Cen, has no detectable CaII absorption to the low level of N(CaII) $< 5 \times 10^9$ cm^{-2}.

Most of the CaII clouds detected in the LISM possess velocity dispersions of b value $\sim$ 2 km s^{-1}, suggesting physical conditions that are similar to those found for the LIC (i.e. T $\sim$ 7000K, n $\sim$ 0.1 cm^{-3}). However, several anomalous clouds have been found which do not fit into this category. For example, the cloud detected towards 51 Oph at a velocity of -20.3 km s^{-1} has an observed b value of only 0.6 km s^{-1} which corresponds to a cloud temperature of $<$ 870K. Also, the single cloud observed towards γ Aqr (d = 23pc) at a velocity corresponding the LIC has an inferred temperature of only 3500K (assuming low turbulence).

In figure 1 we show the preliminary reduction of the CaII spectrum of the B2IV star α Pav (d = 57pc) taken in November 1996 with the UHRF at the AAT. Four distinct absorption components are clearly seen, the weakest coincident with the projected velocity of the LIC component at V = -15.1 km s^{-1}. A cloud component with a similar velocity has also been observed towards α Gru (d = 18pc) and ϵ Gru (d = 23pc), which are two stars lying within 10 degrees from the line-of-sight towards α Pav. One particularly interesting aspect of the velocity structure of this line-of-sight is the narrow absorption component seen at -19.8 km s^{-1}, whose b-value would indicate an upper limit to the temperature for this cloud of only 2100K. Note also the two broad components in the red wings of the photospheric profile which may be stellar line components, due to the binary nature of α Pav.

3 Discussion

Thus far, our program of UHRF observations of absorption in the LISM clearly shows that the velocity structure of diffuse cloudlets within 40pc is far from being simple. The absorption characteristics of the nearest cloud, the LIC, indicate that the Sun can only be partially embedded within it, and that at least in the direction of α Cen the LIC is either highly asymmetric or of very low column density. There seems to be a tendency for the LIC to be observed either on its own or not at all in lines-of-sight with high negative galactic latitudes (Crawford and Dunkin 1995). Thus far, the best interpretation of all the presently available CaII absorption data point to the fact that we are not observing classical cloud-like absorption features at all, but in fact we are detecting shocked filamentary sheet-like structures in the LISM. These "interstellar wisps" may well be remnants of the proposed supernova explosion that created the Local Bubble several million years ago.

The three key remaining problems to be answered about these LISM cloudlets are (1) how long can these wisps survive if they are embedded in the alleged hot million degree gas of the Local Bubble? (2) Are these wisps in pressure or ionization equilibrium, and (3) what happens at the interface between the 7000K LIC gas and the hot LISM gas? Although our ground based UHRF data is providing important insights into the complex nature of the velocity structure of these local cloudlets, these three key questions concerning the LISM can only be answered by observations at other wavelengths, particularly using absorption spectroscopy in the ultraviolet regime, where a wide range of ionization states from many elements can be probed. Clearly the HST STIS instrument will play a key role in addressing many of these problems, and the upcoming FUSE satellite's observations of local OVI (1036Å) absorption will provide information on the relationship between both hot and cold local gas. Clearly much work has yet to be done in this field, and it is our hope to complete this UHRF survey within the next few years such that the kinematics of the LISM cloudlets can be well mapped and provide an important diagnostic database for understanding the sub-structure of the interstellar medium in regions of very low density and high ionization.

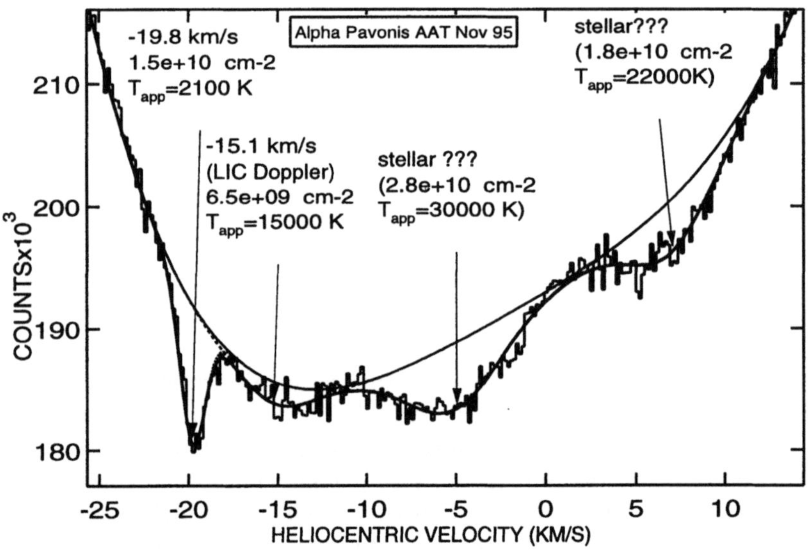

Fig. 1. UHRF fitted spectrum of the CaII absorption line towards Alpha Pav

References

Bertin, P., Lallement, R., Ferlet, R., and Vidal-Madjar, A. (1993): *Astr. & Astrophys.*, **278**, 549–560

Crawford, I. (1994): *The Observatory* **114**, 288–293

Crawford, I. and Dunkin, S. (1995): *M.N.R.A.S.*, **273**, 219–224

Crawford, I., Craig, N. and Welsh, B. (1997): *Astr. & Astrophys.*, **317**, 889–897

Diego, F. et al. (1995) : *M.N.R.A.S.*, **272**, 323–332

Lallement, R. and Bertin, P. (1992): *Astr. & Astrophys.*, **266**, 479–485

Lallement, R., Vidal-Madjar, A. and Ferlet, R. (1986): *Astr. & Astrophys.*, **168**, 225–236

Linsky, J. and Wood, B. (1995): *Astrophys. Journal*, **463**, 323–332

Vallerga, J., Vedder, P., Craig, N. and Welsh, B. (1993): *Astrophys. Journal*, **411**, 729–749

Welsh, B., Craig, N., Vedder, P. and Vallerga, J. (1994): *Astrophys. Journal*, **437**, 638–657

EURD: An Extreme Ultraviolet Spectrograph to Probe the Hot Interstellar Medium

C. Morales[1], J. Trapero[1,2], J.F. Gómez[1], S. Bowyer[3], J. Edelstein[3], and M. Lampton[3]

[1] Laboratorio de Astrofísica Espacial y Física Fundamental, INTA, Apdo. Correos 50727, 28080 Madrid, Spain
[2] Instituto de Astrofísica de Andalucía, CSIC, Apdo. Correos 3004, 18080 Granada, Spain
[3] Space Sciences Laboratory, University of California, Berkeley, CA 94720–7304, USA

Abstract. The EURD instrument has been designed to measure diffuse emission in the extreme ultraviolet (350–1100 Å). This new design provides an unprecedented 4–5 Å spectral resolution and 200 photons/sec/cm^2/sr sensitivity after only 100 hours of observations. One of the goals of this project is to search for spectral lines of highly ionized species from the high temperature component ($10^5 - 10^6$ K) of the interstellar medium that fills the Local Bubble. It is expected that EURD will detect lines due to a thermalized hot component of the interstellar medium, and it will also provide critical diagnostics of the physical properties of this gas. With EURD data we could also detect a spectral line due to the decay of massive neutrinos as well as study oxygen lines from the upper atmosphere airglow. EURD is on board the Spanish MINISAT–01 satellite.

1 Overview

EURD[1] (Espectrógrafo Ultravioleta extremo para la observación de la Radiación Difusa, i.e., Extreme Ultraviolet Spectrograph for the Observation of Diffuse Radiation) is an instrument composed of two spectrographs specially designed to detect diffuse line emission in the 350 – 1100 Å band. It is on board the first mission of the Spanish satellite MINISAT, launched on April 21, 1997. EURD is a joint project of CEA (U.C. Berkeley, USA) and INTA (Spain). A summary of the key instrument parameters are given in Table 1.

[1] Based on the development and utilization of the Espectrógrafo Ultravioleta de Radiación Difusa, a collaboration of the Spanish Instituto Nacional de Técnica Aeroespacial and the Center for EUV Astrophysics, University of California, Berkeley.

Table 1. Key Instrument Parameters

Bandpass	350-800 Å (long wave spectrometer) 500-1100Å (short wave spectrometer)
Field of View	$25.6^\circ \times 8.4^\circ$
Filters	Open, Opaque, 1000 Å Al, MgF_2
Slit	0.15×60 mm
Grating	8 cm diameter 17.6 cm focal length holographically ruled 2460 lines mm^{-1}
Grating substrate	Electroless nickel on aluminum
Grating figure	Ellipse of rotation Semi-major axis 242.87 mm, parallel to ruling Semi-minor axis 176.17 mm
Grating overcoating	Silicon carbide (long wave spectrometer) Boron carbide (short wave spectrometer)
Detector	Low-noise microchannel plate with anticoincidence guard
Detector photocathode	Chemical treatment (long wave spectrometer) MgF_2 (short wave spectrometer)
Size (each spectrograph)	$40 \times 40 \times 13$ cm
Weight (each spectrograph)	10 kg

2 Scientific Goals

2.1 Hot Phase of the ISM

There is growing evidence that the Sun is embedded in ionized, hot ($\sim 10^6$ K) gas. Whether this hot gas is a pervasive component of the interstellar medium, or is a local characteristic around the sun (a local bubble) is still a matter of discussion.

A major unknown is the temperature of this hot gas. Estimates range from 10^6 K (e.g., Bowyer, Field, & Mack 1968; Davidsen et al. 1972; McCammon et al. 1983) to 4×10^4 K (Breitschwerdt & Schmutzler 1995). To solve the current undetermination of the temperature of the hot ISM, we need to obtain spectral line data, since the lines from this hot phase will be strongly dependent upon the temperature and thermal history of this material (Breitschwerdt & Schmutzler 1995).

If the detected soft X-ray background (Bowyer, Field, & Mack 1968) is due to emission of hot gas, line emission from highly ionized atoms will be especially intense at extreme ultraviolet wavelengths. EURD, with its high sensitivity and high spectral resolution could give us first, a conclusive proof for the presence of hot gas. Second, we hope to obtain a reliable estimate of the temperature of this hot gas, and therefore provide information about its origin, and present structure and dynamics. Even the absence of lines in

EURD data will pose important constraints on our current models of the interstellar medium. An estimate of the lines produced by a steady–state collisionally ionized plasma at 10^6 K within EURD detectability can be seen in Fig. 1.

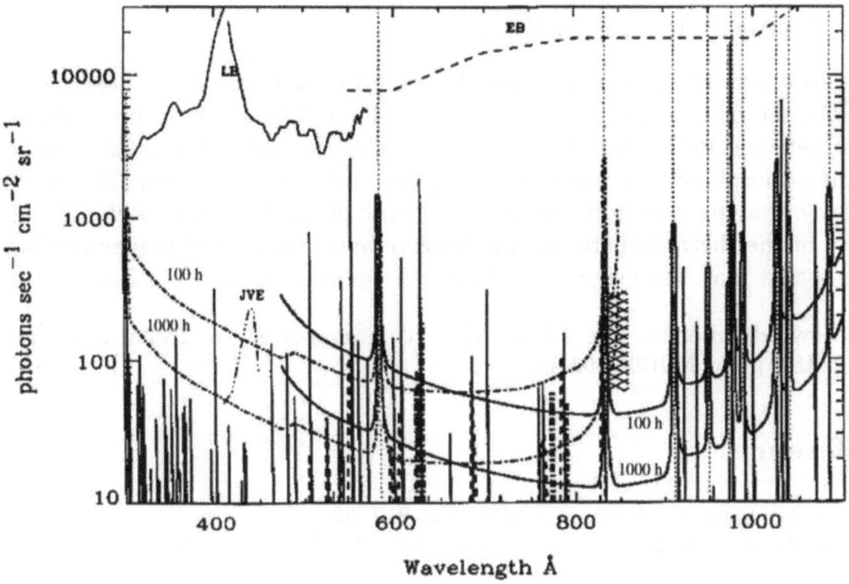

Fig. 1. EURD minimum measurable flux and existing upper limits to the diffuse EUV background. The solid vertical lines are the expected ISM emission from a steady–state collisionally ionized plasma. The dashed vertical lines are the intensities from the delayed recombination model of Breitschwerdt & Schmutzler (1994). The rectangle near 840 Å is the range of the emission signature by decaying neutrinos predicted by Sciama (1994). The dotted vertical lines show the expected nighttime airglow. The dash–dot curves are the flux limits for both short wavelength and long wavelength spectrographs, 100 and 1000 hours of observation. The solid curves are the upper limits of Labov & Bowyer (1991), labeled LB, Edelstein & Bowyer (1997), labeled EB, and Jelinsky et al. (1995), labeled JVE.

2.2 Decay of Massive Neutrinos

Some recent theories suggest that part of the dark matter in the Universe consists of massive neutrinos. Their mass and lifetime are constrained by astrophysical arguments to $\sim 29\,\mathrm{eV/c^2}$ and 10^{24} sec, respectively (Sciama 1990a,b). Their decay would produce a monochromatic photon at ~ 850 Å, thus within the EURD spectral range. The detection of these neutrinos would solve several outstanding issues, like the ionization observed in the ISM, and the classical dark matter problems: the flat rotation curves of galaxies and the flatness (and future) of the Universe.

2.3 Upper Atmosphere Nightglow

The spectrum of the upper atmosphere nightglow that will be observed by EURD, should contain several O I and O II lines. These oxygen lines have all been detected in laboratory, but only one (O II at 834 Å) has been observed in the atmosphere (Bowyer et al. 1981, Chakrabarti et al. 1984). The opportunity of having simultaneous measurements of all these lines will allow the study of the thermospheric atomic oxygen, and of solar and magnetospheric events that may leave signatures in the O II plasma around the Earth.

Acknowledgements. The EURD project is supported by INTA grant IIE- and NASA grant NGR05-003-450.

References

Bowyer, S., Field, G., Mack, J. (1968): Nat 217, 32
Bowyer, S., Kimble, R., Paresce, F., Lampton, M., Penegor, G. (1981): Appl. Opt. 20, 477
Breitschwerdt, D., Schmutzler, T. (1994): Nat 371, 774
Breitschwerdt, D., Schmutzler, T. (1995): in New Light on Galaxy Evolution, IAU Symp 171, Ed. Bender & Davies (Kluwer)
Chakrabarti, S., Kimble, R., Bowyer, S. (1984): Geophys. Res. Lett. 89, 5660
Davidsen, A., Shulman, S., Fritz, G., Meekins, J. F., Henry, R. C., Friedman, H. (1972): ApJ 177, 629
Edelstein, J., Bowyer, S. (1993): Adv. Space Res. 13, 307
Edelstein, J., Bowyer, S. (1997): ApJ, submitted
Jelinsky, P., Vallerga, J.V., Edelstein, J. (1995): ApJ 442, 653
Labov, S. E., Bowyer, S. (1991): ApJ 371, 810
McCammon, D., Burrows, D. N., Sanders, W. T., Kraushaar, W. L. (1983): ApJ 269, 107
Sciama, D.W. (1990a): Ph. Rev. Lett. 65, 2839
Sciama, D.W. (1990b): ApJ 364, 549
Sciama, D.W. (1994): in Modern Cosmology and the Dark Matter Problem (New York: Cambridge Univ. Press)

New Insights on the Local ISM from EUV Observations

Stuart Bowyer

Department of Astronomy and Space Sciences Laboratory University of California at Berkeley, California 94720-7450, USA

Abstract. New results from *EUVE* are reviewed, with emphasis on the ionization state of the ISM and the question of thermal pressure balance between warm clouds and the hot intercloud medium. A description of what may be expected from the next generation of experiments, some to be launched shortly, will also be presented.

1 Historical Comments

This conference was organized to allow us to focus on the character and origin of the ISM. It is especially concerned with the hot ($\sim 10^6$ K) and warm ($\sim 10^3$ K) components. A substantial amount of new data has been obtained since the last conference on this topic, including results from satellites including *ROSAT, ULYSSES, HST,* and *EUVE*, and from high-resolution ground-based optical studies. I will summarize some of the results on this topic obtained from the *EUVE* observatory, but first it may be worthwhile to pause and reflect upon some of the progress that we have made in this field. Let me start with reminders of a few highlights. I realize I will be slighting by omission almost everyone who has been working in this field, and I apologize for my limited choice of examples. However, a full recounting of all the effort that has been expended on this topic would overfill the Conference Proceedings all by itself.

It wasn't until the late 1960s that the hot component of the ISM was detected by Bowyer et al (1968), confirming an earlier prediction by Spitzer (1956). This initial observation showed the soft X-ray 1/4 keV flux was anti-correlated with the Galactic hydrogen column, as derived from 21 cm radio observations, thereby demonstrating that at least the majority of the flux had something to do with our Galaxy. However, the flux scaled as if the absorbing ISM cross-section was one-third of its true value, which was a truly puzzling finding.

Ten years later a number of high-resolution spectroscopy observations of hot stars in the 900 - 1200 Å band were carried out with the Copernicus satellite. These showed the O VI 1032 Å line in absorption. A perplexing fact was that this line was narrow, which is best generated by a plasma at a temperature of a few$\times 10^5$ degrees. Hence the material producing this absorption seemed to be independent of the soft X-ray producing higher temperature gas.

In the 1980s, a number of soft X-ray all-sky maps were obtained by Johns Hopkins (Davidsen et al, 1972), MIT (Marshall & Clark, 1984), and Wisconsin (McCammon et al, 1983), and the Wisconsin group carried out a number of rocket flights which provided information on the flux in bands centered at roughly 90 and 120 Å (Bloch et al, 1986). The 1/4 keV maps provided exquisite detail on the anti-correlation of this flux with hydrogen column, and confirmed that its intensity varied with an ISM cross-section of approximately one-third of its expected value. The flux in the lowest energy bands did not track with hydrogen column, which added a new dimension to the puzzle.

A variety of theorists rose to the challenge of weaving the experimental data into a coherent whole, led perhaps by the hope that this would be a simple problem, easily solved. George Field and colleagues (1969) developed a theory with thermal pressure equilibrium between warm (10^4 degree) neutral clouds and hot material. This model was dealt a decisive blow by the discovery of O VI in the ISM. McKee and Ostriker (1977) developed an overall model of the ISM with warm cloud / hot gas interfaces, which provided among other things an elegant explanation of the observations of O VI.

The overall morphology of the local high temperature gas was explored in detail by the Wisconsin group in the 1970s and 1980s, who developed the displacement model as an explanation for the soft X-ray background (Cox & Reynolds, 1987). This model reached its peak expression in the depiction of the local bubble. This term is now incorporated into the astronomical lexicon and is the title of our conference.

Jakobsen and Kahn (1986) revived the conceptual model of absorbing clouds interspersed in a hot medium, and showed that this model could also explain most of the data. While both the Jakobsen and Kahn model and the displacement model fit the observations reasonably well, each had its own set of shortcomings. Unfortunately, the available data was not sufficient to provide an overwhelming case for either model, though the interstellar clouds needed for the Jakobsen and Kahn model did not appear to be present.

Heiles and Kulkarni (1988) reviewed a substantial amount of observational material and concluded that the pervading warm neutral material is typically in sheets, rather than in clouds like those usually associated with the ISM.

The major theoretical advance of the 1990s, in my view, was the work of Breitschwerdt and Schmutzler (1994), who showed that the ionization state of the ISM was strongly dependent upon the thermal history of the material, and that in extreme cases the thermal temperature of the electrons could be vastly different from the photon distribution. Whether this concept is one that Nature has, in fact, provided, is yet to be determined.

In this sweeping overview, let me show a few illustrative examples of where we have been and where we are now. In Figure 1, I show one of the cartoons depicting the local bubble as developed by the Wisconsin group (Snowden et al, 1990). This bubble was hypothesized to have been produced

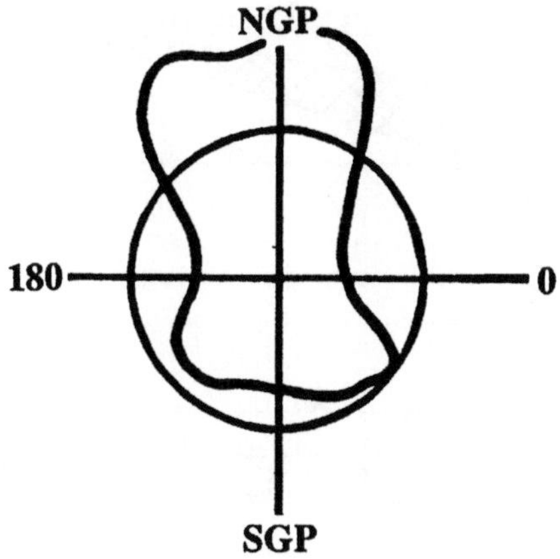

Fig. 1. A classic depiction of the "Local Bubble" as envisioned in the late 1980s. (From Snowden et al, 1990.)

by a supernova that occurred nearby which displaced the local material into a boundary. The morphology of the bubble was tuned to provide the different intensities between the poles and the plane of the Galaxy, which is roughly a two-to-one effect. Some emission was hypothesized to come from the hot material within the cavity, but since the emission measure in this cavity was not sufficient to produce all of the flux, some flux had to be produced at the boundary walls.

With the advent of *ROSAT*, a variety of new information has been obtained. Without doubt the most spectacular was the observation of soft X-ray shadows (Snowden et al, 1991; Burrows & Mendenhall, 1991) which showed that the "local bubble" was not, in fact, local, but that much of the 1/4 keV emission in the general direction of the Galactic poles was coming from distant regions. In principle, one can still invoke a wall boundary in the Galactic plane, (see, for example, Snowden, this volume) but the supporting data are quite thin and may be more easily explained in other ways.

Of substantial note is that an explanation has now been found for the curious behavior of the soft X-ray absorbing characteristics of the ISM. A number of investigations (Sidher et al, 1996, Snowden et al, 1997, several papers in this volume) have shown that several components contribute to the overall soft X-ray flux, including (among others) local emission, Galactic halo emission, and an extragalactic component. It is the summed effect of these components that produce a result that mimics a constant flux with an unnatural ISM absorbing cross section.

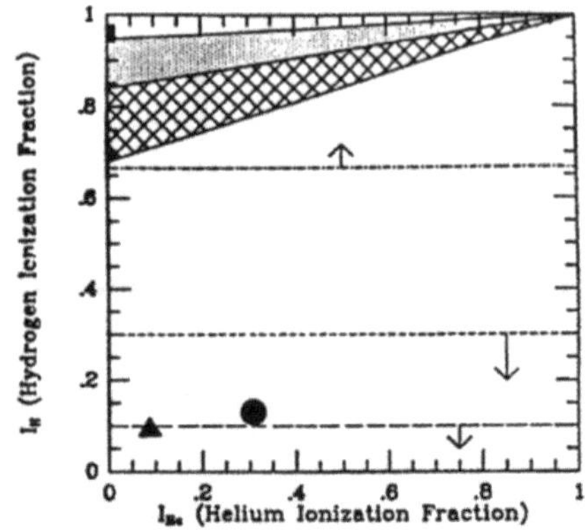

Fig. 2. The ionization state of the ISM as derived by several authors before the advent of spectroscopic data from EUV and HUT. (Results from several articles in the refereed literature whose authors need not be identified at this point.)

So much for a broad-brush overview. At this point I would like to summarize some of the work that has been carried out with the Extreme Ultraviolet Explorer, which provides us with new information on the character of the ISM.

2 Recent Results from *EUVE*

2.1 The Ionization State of the Interstellar Medium

The ionization state of the ISM was generously stated to be "uncertain" before the advent of *EUVE* observations. In Figure 2, I show a summary of most of the results available in the late 1980s. In general, authors attempted to derive a single, global ionization state of the ISM; these were obtained by rather indirect means. The wide variety of results obtained indicate the large degree of uncertainty regarding this parameter at that time.

With the advent of EUV spectroscopy, a substantial number of objects were observed which showed the effects of the intervening ISM. For studies of the ISM, the best choice of a background continuum source is a hot white dwarf. Many studies have been carried out with EUVE on hot white dwarfs, not only for the intrinsic information concerning the hot white dwarfs themselves, but also to obtain information on the ISM. The ISM H I and He I absorption in the spectra of six hot white dwarfs were derived from data obtained with EUVE. These data showed that the neutral helium to neutral hydrogen ratio was reasonably fit by a value 0.07, which is substantially

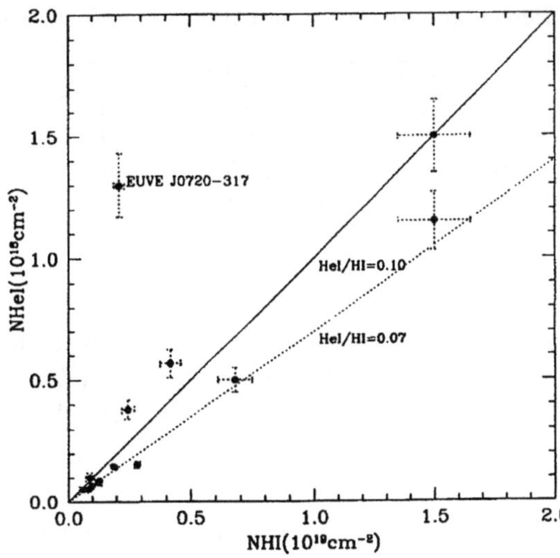

Fig. 3. Ratios of He I/H I for 12 stars within 200 pc.

less than the cosmic abundance of helium to hydrogen of 0.1 (Dupuis et al, 1995). This suggests that some of the ISM helium is ionized. There was also a suggestion of a variation in this parameter with differing look directions, but this effect was not strongly supported by the data.

The results of a study of ISM absorption to more distant stars are shown in Figure 3 where the data are combined with the earlier results cited above (Dupuis et al, in preparation). The data on the more distant stars, because of their relative faintness, have larger error bars than the data on the nearby stars; nonetheless, this extended data set clearly indicates that there are substantial variations in the ionization state of the ISM in different view directions.

We can use the He II absorption edge at 228 Å to obtain the fraction of singly ionized helium in the ISM. This requires a careful fitting of the absorption edge because of the auto-ionization effect present in the ISM, as described by Rumph et al (1994). With this information, we can then determine the total amount of ionized helium, since Heiles et al (1996) have shown that essentially no He III exists in the Galaxy. The sum of the neutral and ionized helium gives us the total helium in the line of sight, and with the assumption that the total hydrogen to total helium ratio is 10, we can obtain the total amount of hydrogen in this line of sight. A comparison of this number with the column of neutral hydrogen in this line of sight yields the column of H II, giving us a complete picture of the ionization state of the ISM.

Table 1. Percentage Ionized H and He for Four View Directions

Star	Ionized H	Ionized He
HZ43	<40	<50
GD153	<50	<60
EUVE J0720-317	~ 88	~ 28
EUVE J0723-277	~ 65	~ 62

I show in Table 1 results obtained in the direction of four stars. These examples are extreme cases which emphasize the fact that the ionization state of the ISM can vary substantially in different view directions. More work on this topic is in progress.

2.2 Is There Thermal Pressure Equilibrium Between the Warm and Hot Components of the ISM?

Almost all models of the warm and hot phases of the ISM assume thermal pressure equilibrium between these two phases, beginning with the original suggestion of Spitzer, and continuing with the two phase model of the ISM by Field and co-workers, and the McKee and Ostriker model. Other models have been suggested which invoke magnetic fields as one of the components providing pressure balance, but these have not achieved wide currency.

What is the evidence in favor of a thermal pressure balance? A wide variety of values have been obtained for the thermal pressure of the hot phase of the ISM. A number of these values have been obtained by measuring the thermal pressure of cool clouds and then assuming that the hot gas is at the same pressure. This *assumes* thermal pressure equilibrium rather than *establishes* thermal pressure equilibrium.

The best direct estimates of the thermal pressure of the hot gas come from soft X-ray measurements and EUV measurements (Bowyer et al. 1997). A problem with soft X-ray measurements is that the penetrating power of soft X-rays is relatively large, and as a consequence the flux seen by an observer may be some combination of flux in the line of sight to a cloud, plus flux from beyond the cloud that has been partially attenuated by the cloud. Or, alternatively, a cloud may be so dense that it is totally opaque to the soft X-rays, but the flux observed is partially attenuated by a smaller inconspicuous cloud in the line of sight to the totally obscuring cloud. This situation is demonstrated in Figure 4. In Figure 4 I also show the situation for EUV emission. EUV radiation is sufficiently soft that it is absorbed by even small clouds in the local ISM, and one can be certain that the flux is not coming from regions behind the cloud.

We have measured the EUV flux in the direction of a shadow cast by a cloud in the overall background, and have investigated the absorbing material

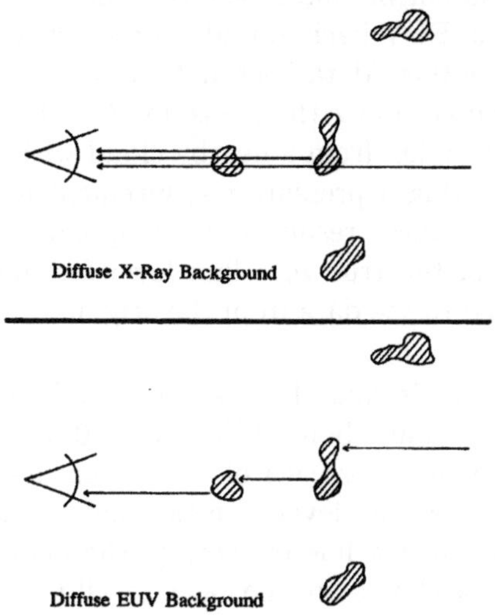

Fig. 4. A cartoon depicting the intrinsic difference between the measurement of flux in the soft X-rays and the EUV. The relatively large penetrating power of soft X-rays renders it difficult, if not impossible, to identify flux which is not the product of several emitting columns.

in the line of sight to this cloud using optical techniques (Bowyer et al. 1997). These data showed that the EUV flux measured directly in the line of sight to this absorbing cloud (at a distance of $\leq$ 40 pc) is due to unabsorbed emission between us and the cloud. A pressure of $P/K \sim 19{,}000$ was obtained for the hot component of the ISM in this direction (Bowyer et al, 1995). This pressure can be compared with the pressure of the local cloud surrounding the Sun, which has been independently determined by several techniques. The canonical value for the pressure of the local cloud is $P/K \sim 700$; a more recent determination obtained by Vallerga (1996) suggests $P/K \sim 2000$. A comparison of the pressure of the hot component with either of these two numbers shows that there is not thermal pressure equilibrium between the local cloud and the nearby hot ISM at least in this direction.

An escape which will allow us to retain the concept of thermal pressure equilibrium is provided by the fact that the pressure will certainly be different as an interstellar shock wave passes through the region. Although there is no evidence for a local shock within 40 parsecs of the Sun, this is still a hypothetical possibility. A second potential escape from this dilemma is provided by the fact that the zero-level calibration for the EUVE data employed in this study was not definitively established, and by assuming a gross miscalculation of this zero level, a lower pressure for the hot ISM could be obtained.

New EUVE measurements have been obtained for two more clouds producing shadows in the EUV background. These are at widely different lines of sight. An analysis of these data (Berghöfer et al, this volume) incorporates a new approach in determining the pressure. A differential measurement is made by comparing the flux from a nearby cloud with that from a more distant cloud, thereby giving a pressure measurement for that segment of the emitting region. When these results are extrapolated to zero distance, one obtains an estimate of the true zero flux level for the detector. This technique has been applied to the data from the original cloud shadow and to the two new shadows.

The surprising result is that the pressure in all three of these disparate view directions is the same, about $P/K \sim 16{,}000$, and the extrapolation of the results to zero distance for each of these measurements gives the same zero level flux. Using this zero level, one can make a revised estimate of the nearby pressure in the direct line of sight to the nearby clouds. Again, the pressures to the nearby clouds are the same in all three directions, and they are the same as the pressures obtained for the segments between the nearby clouds and the more distant clouds. It seems extraordinarily implausible that all of these results could give the same answer by chance. We conclude we have obtained a true measurement of the pressure in our local region.

We also note that the option of a different pressure between us and a nearby cloud which could be produced by a passing shock wave is no longer viable because of the disparate view directions sampled. We conclude that thermal pressure equilibrium between the local warm cloud and the hot component of the ISM does not exist, and hence this cloud, and by implication, other clouds, must be constrained by some other mechanism than thermal pressure.

2.3 Future Prospects

The majority of our knowledge of the hot component of the ISM has come from studies of the morphological characteristics of the soft X-ray background. Several papers in this conference have utilized the broad energy discrimination capabilities of the *ROSAT* proportional counters to obtain new insights as to the character and distribution of the emitting material, foreshadowing the potential of higher resolution spectral studies. Spectroscopy of the diffuse soft X-ray and EUV background is intrinsically quite difficult and is rooted in the fact that the intensity of the radiation from any small region is quite small. One is forced to a trade between instruments which have large flux-collecting capabilities and lower energy resolution capabilities, such as proportional counters, CCDs, and microcalorimeters, and classic slit-grating combinations which have higher spectral resolution but low flux collecting capabilities.

Table 2. Instruments to Study the Diffuse Soft X-Ray / EUV Background

Experiment	Bandpass (Å)	Resolution ($E/\Delta E$)
ROSAT (Pfeffermann et al. 1986) (lowest energy channel)	44–100	5
EUVE (Jelinsky et al. 1995)	190–250	10
	400–460	10
Alexis (Bloch & Smith 1991)	130–190	10
DXS (Sanders & Edgar 1996)	40–80	20
CUBIC (Burrows et al. 1995)	10–50	40–60
UCB/EURD	350–1100	70–200

I show in Table 2 the energy resolution capabilities of a variety of instruments which have been used to study the diffuse X-ray background, and some that are expected to be deployed in the future.

The *ROSAT PSPC* energy resolution (Pfeffermann et al, 1986) is such that the majority of the work performed with this instrument has involved morphological studies. These have been quite productive. However, it is not clear that more data of this type will be especially useful.

Jelinsky et al (1995) used the EUVE satellite to look for lines expected to be present in the hot ISM. Although the energy resolution was quite low (the instrument was not designed to study diffuse objects, but was designed to study point sources), it was nevertheless found that several lines were not present which should have been detected if the ISM were in collisional ionization equilibrium. A possible solution that was suggested was that depletion had taken place in the ISM, but a factor of 10 depletion was necessary for the results to be compatible with the upper limits obtained. Recent observations by Fitzpatrick (1996) have shown that this depletion is typically only a factor of 3 for most elements. This is certainly a strong suggestion that something in our assumptions about the character of the ISM is incorrect.

The *Alexis* instrumentation (Bloch & Smith, 1991) has not yet produced useful results pertaining to the ISM, and it now appears unlikely that it will be capable of doing so. The DXS instrument (Sanders & Edgar 1996) has somewhat higher resolution than *Alexis*, but the data produced are difficult to interpret. If one assumes that the instrument functioned correctly, the most that one can say is that line emission is present, and that higher energy resolution will certainly be required to obtain useful information on the ISM.

The *CUBIC* instrument (Burrows et al, 1995) has substantially higher spectral resolution and a large grasp. Most of its response is at higher energies, where the extragalactic diffuse X-ray background overwhelms the local ISM component. But sufficient capabilities exist at lower energies that this instrument, if successfully flown, will surely add to our knowledge of the ISM.

We have developed a slit spectrometer for the study of diffuse EUV radiation which provides much higher spectral resolution, but with a corresponding loss in grasp. However, the grasp of the instrument is almost an order of magnitude larger than that of a classic slit spectrometer. This improvement has been obtained by employing a new grating which is especially designed for the study of diffuse flux. This instrument will be capable of detecting individual lines from the thermal plasma producing the soft X-ray / EUV emission with reasonable integrations. Long integration times are feasible with this instrument because of a variety of techniques which have been employed to reduce the instrumental background to extremely low levels.

We have fabricated and calibrated three of these instruments (Bowyer et al, 1997; Morales et al, this volume). All are very similar; the most substantial difference is that one of the three employs a slightly different grating tilt that emphasizes shorter EUV wavelengths. This spectrograph employs different coatings to enhance the shorter wavelength response.

One of these spectrometers (the Ultraviolet Cosmic Background, *UCB*) (Bowyer et al. 1997), covering the 500-1100 Å band, is part of the payload of NASA's Small Spacecraft Technology Initiative satellite, *Lewis*. It is scheduled for launch in the second half of 1997. Two of these spectrometers, one covering the 500-1100 Å band and the other covering the 350-800 Å band, constitute the instrument Espectrografo Ultravioleta extremo para la observacion de la Radiacion Difusa (*EURD*). It is described in some detail in the paper by Morales et al (these proceedings). *EURD* is part of the instrument complement of the Spanish MINISAT 01 satellite. *EURD* is a collaboration of the Spanish Instituto Nacional de Technica Aeroespacial (INTA) and the Center for EUV Astrophysics, University of California at Berkeley. The reduction and analysis of *EURD* data will be carried out as a collaboration between the University of California at Berkeley and an INTA team lead by C. Morales. The MINISAT 01 satellite was successfully launched in April 1997.

An example of the lines expected in the EUV and the sensitivity of these instruments are shown in Figure 5, along with current upper limits to the diffuse EUV background. The resolution of these instruments is 4-5 Å; this is smaller than the line widths shown in the figure.

3 Coda

Several speakers have commented on open questions regarding the character of the soft X-ray / EUV flux produced in our Galaxy. In the following, I provide my own personal list of questions which I hope will be answered in the foreseeable future.

- Is the local bubble a "Local Bubble" or a generalized region produced by multiple events?

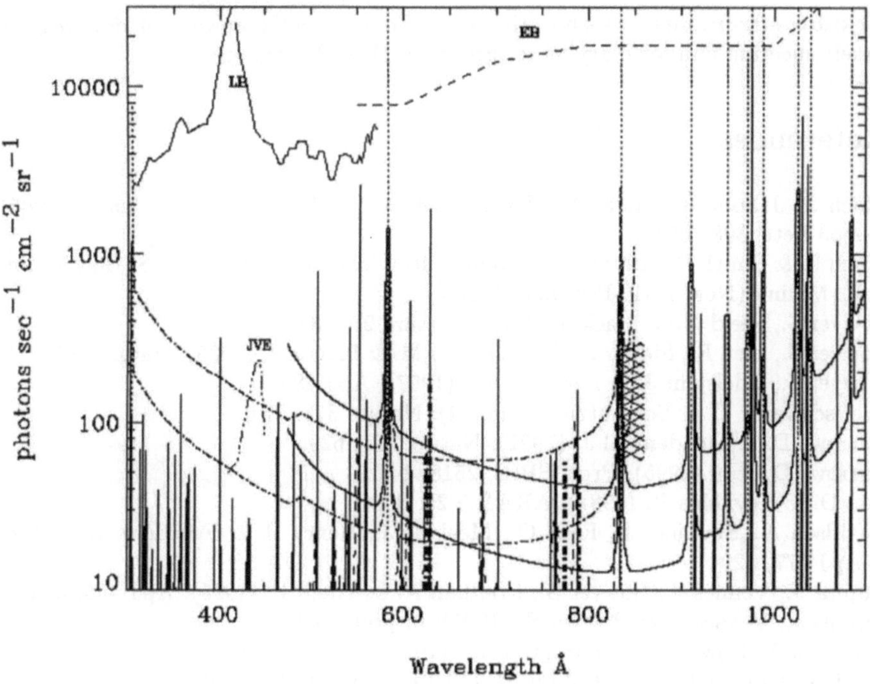

Fig. 5. The minimum measurable flux for the *UCB* and *EURD* spectrometers and existing upper limits to the diffuse EUV cosmic background. The curve marked LB is the 15Å resolution upper limit of Labov and Bowyer (1991). The curve marked EV is the 30Å resolution upper limit derived from *Voyager* data by Edelstein, Bowyer, and Lampton (1997). The curve marked JVE is the upper limit of Jelinsky et al (1995) using *EUVE* data. The curved solid lines are the flux limits provided by 200 and 1000 hr. of observation with the long wavelength *UCB* spectrometer. The dash-dotted curves are the corresponding limits for the short wavelength *EURD* spectrometer. The solid vertical lines are expected ISM emission lines from a steady-state collisionally ionized plasma. The heavy vertical bars are the intensities from the delayed recombination model of Breitschwerdt and Schmutzler (1994) and Breitschwerdt (1997, private communication). The dotted vertical lines are the expected nighttime airglow. The cross-hatched region shows the range of the emission signature predicted by Sciama (1994) for a Galactic halo of radiatively decaying neutrinos.

- Is the plasma in collisional ionization equilibrium, or is it in substantial or even extreme non-equilibrium? What are the "temperatures" of the components?
- Are clouds or sheets the primary structures for the warm neutral component?
- Are "clouds" pressure or magnetically confined?

Acknowledgements. I wish to thank the Conference Organizers for devising this timely meeting and for carrying it through with such elegance.

References

Bloch J., Jahoda K., Juda M., McCammon D., Sanders W. & Snowden S. (1986): ApJ Lett, 308, L59
Bloch J. & Smith B. (1991): in Extreme Ultraviolet Astronomy, ed. S. Bowyer and R. Malina (New York: Pergamon), 289
Bowyer S., Field G. & Mack J. (1968): Nature 217, 32
Bowyer S., Lieu R., Sidher S. D., Lampton M. & Knude J. (1995): Nature, 375, 212
Bowyer S., Edelstein J. & Lampton M. (1997): ApJ 485
Breitschwerdt D. & Schmutzler T. (1994): Nature 371, 774
Burrows D. & Mendenhall J. (1991): Nature 351, 629
Burrows D. et al.(1995): Proc. SPIE, 2518, 141
Cox D. & Reynolds R. (1987): ARA&A 25, 303
Davidsen A., Shulman J., Fritz G., Meekins J., Henry R. & Friedman H. (1972): ApJ 177, 629
Dupuis J., Vennes S., Bowyer S., Pradhan A. & Thejll P. (1995): ApJ 455, 574
Dupuis J., Vennes S. & Bowyer S. (1997): in preparation
Edelstein J., Bowyer S. & Lampton, M. (1997): submitted to ApJ
Field G., Goldsmith D. & Habing H. (1969): ApJ Lett 155, L49
Fitzpatrick E. (1996): ApJ Lett 473, L55
Heiles C. & Kulkarni S. (1988): in Galactic and Extragalactic Radio Astronomy, Second Edition, ed. Verschuur G., Kellermann K., Springer
Heiles C., Coo B.-C., Levenson N. & Reach W. (1996): ApJ 462, 326
Jakobsen P. & Kahn S. (1986): ApJ 309, 682
Jelinsky P., Vallerga J. & Edelstein J. (1995): ApJ 442, 653
Labov S. & Bowyer S. (1991): ApJ 371, 810
Marshall F. & Clark G. (1984): ApJ 287, 633
McCammon D., Burrows D., Sanders W. & Kraushaar W. (1983): ApJ 269, 107
McKee C. & Ostriker J. (1977): ApJ 218, 148
Pfeffermann E. et al.(1986): Proc. SPIE 733, 519
Rumph T., Bowyer S. & Vennes S. (1994): AJ 107, 2108
Sanders W. & Edgar R. (1996): in Astrophysics in the EUV, ed. S. Bowyer and R. Malina (Dordrecht: Kluwer), 269
Sciama D. (1994): Modern Cosmology and the Dark Matter Problem (New York: Cambridge University Press)
Sidher S., Sumner T., Quenby J. & Gambhir M. (1996): A&A 305, 308
Slavin J. & Cox D. (1992): ApJ 392, 131
Snowden S., Cox D., McCammon D. & Sanders W. (1990): ApJ 354, 211
Snowden S., Mebold U., Hirth W., Herbstmeier U. & Schmitt J. (1991): Science 252, 1529
Snowden S., Egger R., Freyberg M., McCammon D., Plucinsky P., Sanders W., Schmitt J., Trümper J. & Voges W. (1997): ApJ 485, 125
Spitzer L. (1956): ApJ 124, 20
Vallerga J. (1996): Space Sci. Rev. 78, 277

Spectral Observations of the Diffuse FUV Background with DUVE (the Diffuse UV Experiment)

Eric J. Korpela and Stuart Bowyer

Space Sciences Laboratory, University of California, Berkeley, CA 94720

Abstract. We present results from a search for FUV emission from the diffuse ISM conducted with an orbital FUV spectrometer, DUVE, which was launched in July, 1992. The DUVE spectrometer, which covers the band from 950 Å to 1080 Å with 3.2 Å resolution, observed a region of low neutral hydrogen column density near the south galactic pole for a total effective integration time of 1583 seconds. the only emission line detected was a geocoronal hydrogen line at 1025 Å . We were able to place upper limits to several emission features that provide constraints to interstellar plasma parameters. We were also able to place continuum limits in this band. We use these upper limits to place constraints upon the emission measure vs. temperature distribution of this gas using an isothermal Landini and Fossi model.

1 Introduction

Since the prediction of the existence of hot gas in the interstellar medium by Spitzer (1956) and its subsequent detection by Bowyer, Field and Mack (1968) in soft X-rays, a variety of models have been developed which attempt to explain the source of this gas, and how it evolves over time. As more data has become available, especially observations of absorption from high stage ions, these models have evolved to fit the data (Slavin and Cox, 1993). Observations of FUV emission are key to differentiating between these models.

2 The Instrument and Calibration

We have designed an instrument capable of measuring the important O VI $\lambda\lambda$ 1032,1038 emission from the diffuse interstellar medium. The instrument's small bandpass (150 Å) around the O VI lines also includes the potentially important C III λ 977, N III λ 991, and C II λ 1037 lines. The instrument, designated DUVE, The Diffuse Ultraviolet Experiment, is based on a spectrometer developed by Edelstein and Bowyer (1993). That instrument's capability for studying diffuse radiation was substantially improved. The basic design is a two stage spectrometer. A wire-grid collimator with a field of view of $2° \times 7°$ in conjunction with a diffraction grating in Wadsworth configuration and an exit slit serve to limit the instrument bandpass from 950 to 1080 Å . The second stage of the spectrometer is a dispersion stage. Light entering through the slit strikes a holographically corrected diffraction grating in a Rowland circle configuration. The second order diffracted light from the grating is focused onto a microchannel plate detector. The high spectral resolution of the dispersion stage (3.5 Å) allows the O VI doublet to be separated from the much brighter Ly β λ 1026 airglow line. Extensive calibration of the DUVE instrument was carried out using the EUV/FUV calibration facilities at the Space Sciences Laboratory (Welsh et al. 1989). The DUVE instrument was launched as a secondary payload attached to the second stage of a Delta II 7925 vehicle on 24 July, 1992. A morning launch resulted in a near midnight apogee of 1460 km (750 nmi). The mission continued for 3×10^4 seconds, until the second stage telemetry battery was depleted.

3 Observational Results

The successful mission resulted in 19 images containing 4151 seconds of observation time. Half of the observation time was made with the shutter closed, for accurate calibrations of the instrument background. The shutter opened images were examined for stellar contamination, uncorrected data dropouts and instrument anomalies and the affected portions were masked off or otherwise corrected. Where no correction was possible, the images were discarded.

The sum of these images was corrected for image distortions and shutter open/closed time, and the shutter-closed background images were subtracted from the data. The resulting image was histogrammed and convolved with a line spread function that was determined during calibration.

The only spectral line that exceeds the 3σ significance level is the geocoronal Ly β λ 1025 line. We have used the DUVE data to put upper limits to line emission in the band covered by the spectrometer assuming line widths of less than 3.5 Å . These line emission limits were determined by calculating the upper 90% statistical confidence level of the measured spectrum at each wavelength. These limits are shown in Table 1. The best determinations of continuum level were $-1.07 \pm 0.69 \times 10^3$ ph s^{-1} cm^{-2} str^{-1} Å^{-1} between 977 and 1020 Å and $-3.2 \pm 3.8 \times 10^2$ ph s^{-1} cm^{-2} str^{-1} Å^{-1} between 1028 and 1057 Å . These measurements allow us to place 2σ upper limits in these ranges of 310 and 440 ph s^{-1} cm^{-2} str^{-1} Å^{-1} respectively.

4 Isothermal Models

The simplest (and most often used) model of emission from the warm and hot phases of the interstellar medium is that of an optically thin, isothermal, collisionally excited plasma. Although collisional ionization equilibrium (CIE) is unlikely to occur in interstellar gas, virtually all observational results are expressed in terms of CIE. We do the same for purposes of comparison. For the purposes of this determination we have used the CIE code of Landini

Table 1. Upper limits to line emission placed by the DUVE data

Species	λ(Å)	I(ph s^{-1} cm^{-2} str^{-1})	I(erg s^{-1} cm^{-2} str^{-1})
H I [1]	972	$\leq 7.4\times10^4$	$\leq 1.5\times10^{-6}$
C III [2]	977	$\leq 4.0\times10^4$	$\leq 8.1\times10^{-7}$
O I [1]	989	$\leq 6.1\times10^3$	$\leq 1.2\times10^{-7}$
N III [2]	991	$\leq 5.5\times10^3$	$\leq 1.1\times10^{-7}$
Si II	992	$\leq 5.7\times10^3$	$\leq 1.1\times10^{-7}$
Si III	996	$\leq 1.6\times10^4$	$\leq 3.2\times10^{-7}$
Ne VI	1006	$\leq 1.3\times10^4$	$\leq 2.6\times10^{-7}$
Ar VI	1008	$\leq 1.0\times10^4$	$\leq 2.0\times10^{-7}$
H I [1,3]	1025	$2.26\pm0.26\times10^5$	$4.38\pm0.49\times10^{-6}$
O VI [2,4]	1032, 1038	$\leq 7.6\times10^3$	$\leq 1.4\times10^{-7}$
C II [2]	1037	$\leq 3.9\times10^3$	$\leq 7.4\times10^{-8}$
Ar I [1]	1050	$\leq 6.4\times10^3$	$\leq 1.2\times10^{-7}$
Si IV	1067	$\leq 2.9\times10^4$	$\leq 5.4\times10^{-7}$
S IV	1070	$\leq 5.1\times10^4$	$\leq 9.5\times10^{-7}$

[1] Anticipated airglow line
[2] Constrains state of hot interstellar medium
[3] Detected at 37σ.
[4] This limit is total emission from the doublet based upon joint statistics by assuming $\frac{I(1032)}{I(1038)} = 2$. Upper limits for the individual components of the doublet are $I(1032) \leq 5400$ ph s^{-1} cm^{-2} str^{-1} and $I(1038) \leq 4400$ ph s^{-1} cm^{-2} str^{-1} .

and Monsignori-Fossi (1990). Using the upper limits obtained by the DUVE instrument, we can place upper limits to the emission measure of isothermal plasma models for the interstellar medium (Fig.1). The emission measure of the interstellar medium is constrained by five emission lines in this band:

1. Mg II λ 1027, which peaks at about 1.5×10^4 K, constrains the emission measure at temperatures below 2.0×10^4 K. Because of the nearby interference from Ly β 1025 emission, the observational limit obtained is very high.
2. C II λ 1037, which peaks at a temperature of about 5×10^4 K, provides the best constraint for temperatures between 2.0×10^4 and 5.1×10^4 K. Because our simple model does not include photoionization, this may understate the constraint, as C II is expected to be produced by photoionization due to FUV radiation from stars.
3. C III λ 977, which peaks at 8.4×10^4 K, provides the best constraint between 5.1×10^4 and 1.0×10^5 K.
4. N III λ 991, which peaks at 1×10^5 K, constrains the emission measure between 1.0×10^5 and 1.8×10^5 K.
5. O VI $\lambda\lambda$ 1032,1038, which peaks at 2.8×10^5 K, provides upper limits to the emission measure above 1.8×10^5 K.

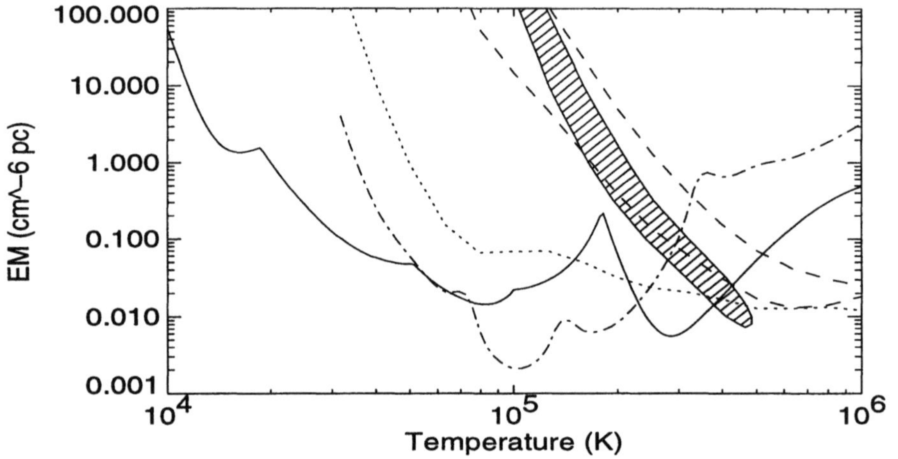

Fig. 1. Upper limit local emission measures derived from this work are shown as a solid line. The dashed lines represent X-ray emission measure limits derived for Wisconsin B and C band rocket borne observations in this direction (McCammon et al.1983). The dot-dashed line represents emission measure limits determined from UVX (1400 to 1800 Å) observations by Martin and Bowyer (1990). The dotted line represents upper limits determined from EUVE observations by Jelinsky *et al* (1995). The hatched area is the parameter space cited by Paresce and Stern (1981) as being the allowed region to create observed broadband EUV and Soft X-ray emission.

A more detailed discussion of these results may be found in Korpela et al. (1997).

References

Bowyer, C. S., Field, G. B., & Mack, J. E., Nature, 1968, **351**, 32.
Dixon, W. V., Davidsen, A. F. & Ferguson, H. C., 1996, ApJ, **465**, 288.
Edelstein, J. & Bowyer, S., 1993, AdSpR, **v13**, n12, 307.
Korpela, E., Bowyer, S. and Edelstein, J., 1997, Submitted to ApJ.
Jelinsky, P., Vallerga, J. V. & Edelstein, J., 1995, ApJ, **442**, 653.
Landini, M., & Monsignori Fossi, B. C., 1990, A&AS, **82**, 229.
Martin, C. & Bowyer, S., 1990, ApJ, **350**, 242.
McCammon, D., Burrows, D. N., Sanders, W. T., Kraushaar, W. L., 1983, ApJ, **269**, 107.
Paresce, F. & Stern, R., 1981, ApJ, **247**, 89.
Slavin, J. and Cox, D., ApJ, **417**, 187, 1993.
Spitzer, L., 1956, ApJ, **124**, 20.
Welsh, B., Vallerga, J. V., Jelinsky, P., Vedder, P. W., Bowyer, S., & Malina R. F., 1989, Proc. SPIE, **1160**, 554.

The Thermal Pressure of the Interstellar Medium Derived from Cloud Shadows in the Extreme Ultraviolet

Thomas W. Berghöfer[1], Stuart Bowyer[1], Richard Lieu[2], and Jens Knude[3]

[1] Space Sciences Laboratory, University of California, Berkeley, CA 94720, USA
[2] Department of Physics, University of Alabama, Huntsville, USA
[3] Niels Bohr Institute for Astronomy, Physics, and Geophysics, University Observatory, Juliane Mariesvej 30, DK-2100 Copenhagen OE, Denmark

Abstract. We have used the Deep Survey telescope of EUVE to investigate shadows in the diffuse EUV/Soft X-ray background cast by clouds in the interstellar medium. We confirm the existence of a shadow previously reported, and provide evidence for two new shadows. We used IRAS data to identify the clouds producing these shadows and to determine their optical depth to EUV radiation. The EUV-absorbing clouds are optically thick in the EUV, and all EUV emission detected in the direction of these shadows must be produced from material in front of the clouds. We obtained new optical data to determine the distance to these clouds. We use a new differential cloud technique to obtain the pressure of the interstellar medium. These results do not depend on any zero level calibration of the data. Our results provide evidence that the pressure of the hot interstellar gas is the same in three different directions in the local interstellar medium, and is at least 8 times higher than derived for the local cloud surrounding our Sun. This provides new evidence for large thermal pressure imbalances in the local ISM, and directly contradicts the basic assumption of thermal pressure equilibrium used in almost all present models of the interstellar medium.

1 Cloud Shadow Observations with EUVE

Prior to this paper, Bowyer et al. (1995) reported on the first detection of a spatial absorption feature in the diffuse EUV background; this feature discovered in the EUVE Deep Survey (DS) was positionally coincident with an IRAS cirrus cloud at a distance of $\lesssim 40\,\mathrm{pc}$. By attributing the background subtracted flux in front of the cloud to the hot gas in the ISM, Bowyer et al. (1995) derived a pressure (P/k) of 19,000 $\mathrm{cm}^{-3}\,\mathrm{K}$. Here we discuss a re-observation of this initial shadow and we present data for two new cloud shadows discovered in the EUV background. These shadows are located near ecliptic longitudes $\lambda = 52°$ (initial shadow), $\lambda = 309°$, and $\lambda = 214.5°$; according to the galactic coordinates of these cloud locations they are denoted as lb165-32, lb27-31, and lb329+46 in this paper. A recent re-observation of the initially discovered cloud shadow lb165-27 confirmed the existence of this cloud shadow. In both the original DS scan as well as in the re-observation, the ($3\,\sigma$) absorption feature in the EUV background is positionally coincident with an enhanced emission feature in the IRAS 100μm data. Exemplary for the three cloud shadows, Fig. 1 provides the EUVE DS observations of lb27-31.

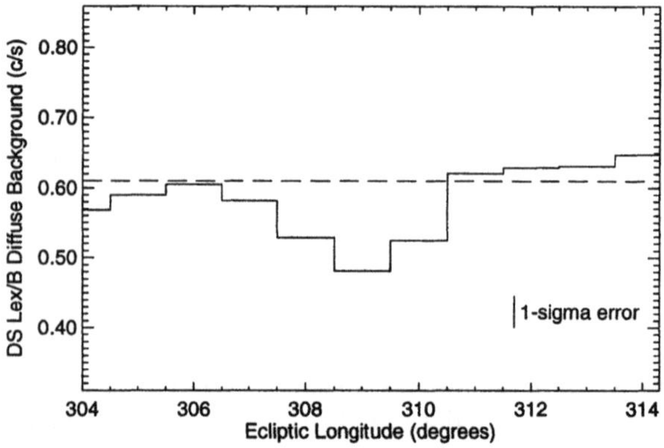

Fig. 1. EUV diffuse background count rate in the Lex/B filter as a function of ecliptic longitude

We have used detailed Strömgren photometry of stars in the cloud shadow regions to determine the distance to the EUV-absorbing clouds. The distances of the EUV-shadowing clouds have been constrained by finding the most distant unreddened star in front of a cloud and the closest reddened star behind this cloud. This method was also applied to determine the distance where optical depth reaches unity in the regions immediately adjacent to the shadowing cloud. For all three cloud shadows our optical data provide evidence for a second larger cloud ("background cloud") located behind the nearby EUV-absorbing cloud.

In Fig. 2 we plot the DS Lex/B filter count rates versus the respective distances for all three cloud shadows. For all shadows Fig. 2 shows the count rates and distances obtained at the position of the shadow as well as in front of the background cloud. For lb329+46 a third data is included that originates from a scan region showing a much larger emission column along the line of sight. The different cloud shadow regions are marked by different symbols, the distance upper limits are indicated by left arrows. As is evident from Figure 2 almost all the data points are reasonably fit with a straight line. The formal results of a linear regression analysis of the data are $0.002 \pm 0.0003\,\mathrm{cts/s\,pc^{-1}}$ for the slope and $0.331 \pm 0.045\,\mathrm{cts/s}$ for the intercept of the regression line. This line is plotted in Figure 2. The slope of the regression line gives the diffuse EUV flux per emitting length which we use to derive the pressure of the plasma in the ISM.

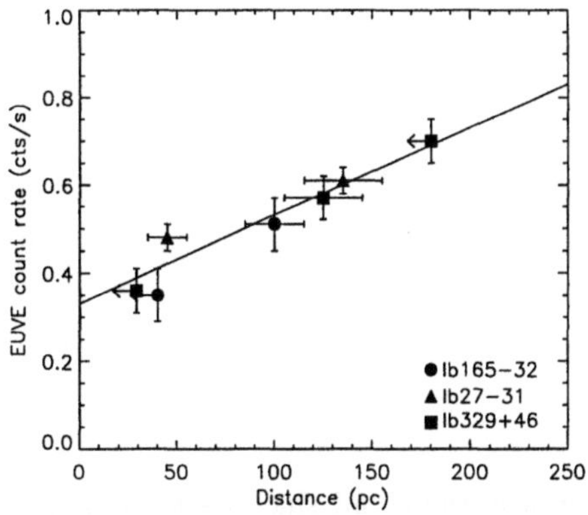

Fig. 2. Plot of all observed count rates versus the distance of the respective clouds; distance upper limits are indicated by left arrows. The solid straight line provides the results of a linear regression analysis.

2 The Thermal Pressure of the ISM

We now utilize the results of our cloud shadow observations to compute the pressure of the local ISM from the ideal gas law. For the conversion of the observed EUVE DS count rates per emitting length into EM/L we use the typically assumed model spectrum of a plasma in collisional ionization equilibrium at a temperature of 10^6 K. To account for the absorption of the local cloud around our Sun we include a foreground ISM absorption of $N_H = 5 \times 10^{18}\,cm^{-2}$ in our model. With the plasma emissivity code provided by Landini & Monsignori-Fossi we derive $EM/L = 7.3 \times 10^{-5} cm^{-6}$. The pressure of the hot ISM is thus $P/k \approx 16500\,cm^{-3}\,K$. We mention that for a Raymond & Smith model of the same parameters we obtain $EM/L = 5.1 \times 10^{-5} cm^{-6}$ and $P/k \approx 13500\,cm^{-3}\,K$. This value is close to pressure values for the "Local Bubble" published by other authors in these proceedings.

The canonical value obtained for the local cloud is $P/k = 730 \pm 30\,cm^{-3}\,K$ (e.g., Frisch 1995). Based on line measurements with EUVE, Vallerga (1996) derived a pressure for the local cloud in the range $P/k = 1700 - 2300\,cm^{-3}\,K$. A comparison of either of these results with the pressure we obtained for the hot ISM shows the pressure of the local ISM exceeds the pressure of the local cloud surrounding our Sun by a factor of ≥ 8. The tight correlation between observed diffuse EUV flux and emitting length implies the same pressure in three different cloud directions (cf. Fig. 3). The original observation (Bowyer et al. 1995) of a pressure imbalance between the local solar cloud and the surrounding ISM could have been the result of a nearby shock wave which had not yet reached the Sun. The isotropy in the results for all these directions is not consistent with this hypothesis. The zero level calibration for our cloud shadow observations with the EUVE DS instrument

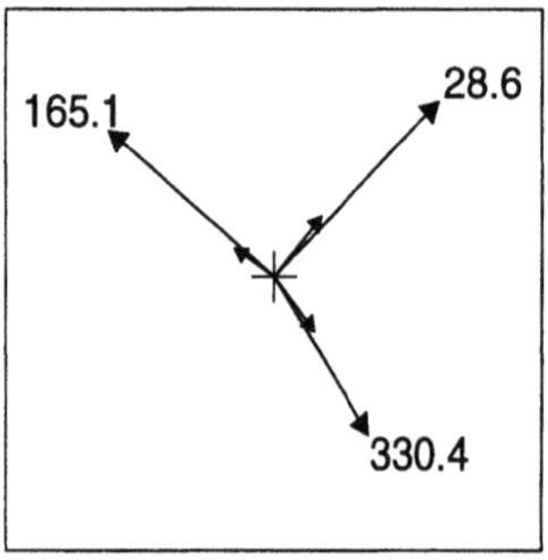

Fig. 3. Projection of the sky directions towards the cloud shadows as seen from above the Galactic plane; numbers give the Galactic longitude, the length of the arrows scales with the projected distance of the clouds.

allows us to measure absolute fluxes for the diffuse EUV background. These zero level corrected values also give the same pressure for the material in the foreground of the three shadowing clouds.

In summary, our observations provide evidence for a large pressure imbalance in the local ISM compared to the local cloud surrounding our Sun. This contradicts the basic assumption of almost all available models for the local ISM of a pressure equilibrium (e.g., McKee & Ostriker 1977). It is unclear what physical mechanism can maintain such a large pressure imbalance. We emphasize that these models as well as the calculations in the here presented work assume a hot plasma in collisional ionization equilibrium. A completely different approach is provided by Breitschwerdt & Schmutzler (1994) . Their model calculations are based on an adiabatically cooling non-equilibrium plasma. The advantage of this model is that it can at least qualitatively explain many of the observed features of the local ISM and it does not require pressure equilibrium in the local ISM.

Acknowledgements. We thank M. Lampton and J. Vallerga for useful discussions. T.W.B. acknowledges the support from the Alexander-von-Humboldt Stiftung (AvH) by a Feodor-Lynen Fellowship. This work has been supported by NASA contract NAS 5-30180. R.L. is supported by NASA grant 5-34378 awarded to UAH. J.K. thanks the European Southern Observatory for observing time and travel support.

References

Bowyer, S., Lieu, R., Sidher, S. D., Lampton, M., Knude, J. 1995, Nature, 375, 212
Breitschwerdt, D., Schmutzler, T., 1994, Nature, 371, 774
Frisch P. C., 1995, Space Sci. Rev., 72, 499
McKee, C. F., Ostriker, J. P., 1977, ApJ, 218, 148
Vallerga, J., 1996, Space Sci. Rev., 78, 277

The Structure of the Local Hot Bubble Toward $l = 165^\circ, b = 0^\circ$ Using *IUE* and *ROSAT* Data

K.S. de Boer[1] and J. Kerp[2]

[1] Sternwarte, Universität Bonn, Auf dem Hügel 71, D-53121 Bonn
[2] Radioastronomisches Institut, Univ. Bonn, Auf dem Hügel 71, D-53121 Bonn

Abstract. The first X-ray-opaque cloud in the direction of Auriga lies at 180 pc. A tenuous cloud, whose distance is derived from *IUE* absorption line studies as well as from an analysis of *ROSAT* data, is *embedded* at 55 pc. The X-ray emitting gas has a temperature of $10^{5.9-6.0}$ K. The pressure is 1.6 10^4 K cm^{-3}.

1 Distance to the Nearest Cloud from *IUE*

The distance to a cloud in the Milky Way is best determined with the help of interstellar absorption lines in the spectra of stars. Stars are selected in the general direction of the cloud, preferably well distributed in distance. The stars in front of the cloud will not show the absorption at the velocity of that cloud, whereas stars behind the cloud will show interstellar absorption at the particular velocity. Alternatively, just the sequence in column density derived from the absorption lines tells about the column density of gas in front of the stars. Knowing the distances of the stars, the cloud distance follows from the absence or presence of the said absorption lines.

IUE spectra were obtained to 6 stars in the direction of Auriga-Perseus and the absorption lines of Fe II are used to determine the column density of absorbing material. Using the depletion factor [Fe/H]$= -6.4$, the equivalent column density $N_{\rm HI}$ can be calculated. Adding data from the literature for other stars in the general direction of $l = 165^\circ$ and $b = 0^\circ$, Lilienthal & de Boer (1991) were able to derive a distance of $\simeq 55$ pc to the first cloud. Taking further stars from the compilation of Fruscione et al. (1994) shows that as of 180 pc $N_{\rm HI} > 1.6 \cdot 10^{21}$ cm^{-2} . The column density versus distance data in this direction are displayed in Figure 1.

2 X-ray Opacities and X-ray Colours

The interstellar medium (ISM) is a very efficient absorber of soft X-ray photons. The photoelectric absorption cross section of the neutral interstellar medium is roughly $\sigma \sim E^{-3}$. Within the *ROSAT* 1/4 keV energy band the neutral ISM reaches opacity $\tau \simeq 1$ at $N_{\rm HI} = 1 \cdot 10^{20}\,{\rm cm}^{-2}$. Because of the strong energy dependence the 1/4 keV energy band is ideal to search for soft X-ray absorption features caused by thin neutral clouds.

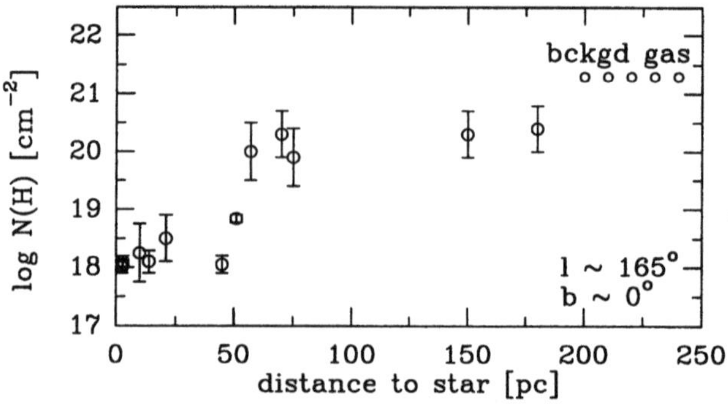

Fig. 1. The column densities $N_{\rm HI}$ toward stars in a small field in the direction of $l = 165°, b = 0°$ are plotted against the distance of the stars. The foreground gas has $N_{\rm HI} \sim 10^{18}$ cm^{-2}. At about 55 pc there is a cloud with $N_{\rm HI} \simeq 10^{20}$ cm^{-2}, then there is little gas beyond until the background gas shows up with $N_{\rm HI} > 2 \cdot 10^{21}$ cm^{-2} at $d \simeq 200$ pc. Data are from Lilienthal & de Boer (1991), Fruscione et al. (1994), and Hartman & Burton (1997)

We can split the *ROSAT* 1/4 keV band into two bands, R1 and R2, to use the maximum energy resolution of the PSPC detector. The R1 ($E \in [0.11; 0.2]$ keV) and R2 ($E \in [0.2; 0.28]$ keV) bands have a significant overlap in their energy response but the difference of the photoelectric absorption cross sections ($\sigma(\mathrm{R1}) \simeq 1.4 \cdot \sigma(\mathrm{R2})$) is sufficient to disentangle the composition of the soft X-ray emitting and absorbing matter within the local interstellar medium.

3 Combining X-ray and H I 21-cm Data

We analysed the H I 21-cm line data of the new Leiden/Dwingeloo survey (Hartmann & Burton 1997) as well as a *ROSAT* PSPC observation aimed at α Auriga (HD 34029). In front of this star the absorption line column density is $N_{\rm HI} = 1.5$-$2.0\ 10^{18}$ cm^{-2}. The background column density is, as can be seen in the H I 21-cm data, $N_{\rm HI} > 10^{21}$ cm^{-2}, for distances > 200 pc.

The Local Bubble's extent is in general assumed to be of the order of 100 pc. However, the nearest cloud on this line of sight, having low column density, lies at a distance of about 55 pc. Consequently it is likely that the cloud is located inside the Local Bubble and is embedded within the local X-ray emitting plasma.

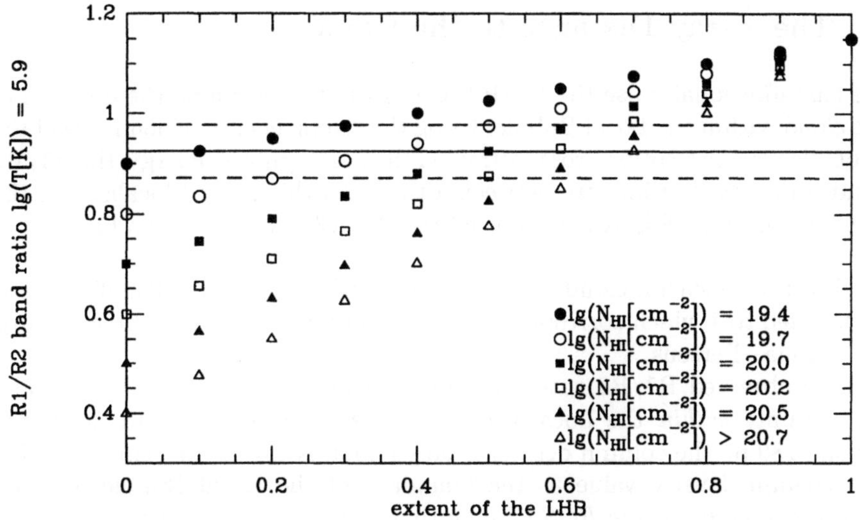

Fig. 2. A plot of the R1/R2 band ratio against the extent of the local bubble (normalised to 1) for a volume of gas at $T = 10^{5.9}$ K in relation to varying amounts of shadowing gas. For a cloud with given column density the shadowing is either strong when the cloud is nearby which leads to a small R1/R2 ratio, or is weak when it is far away resulting in a large R1/R2 ratio. The cloud with $N_{\rm HI} \simeq 1 \cdot 10^{20}$ cm^{-2} (see Fig. 1) fits with the observed R1/R2 ratio for a distance fraction of 0.3–0.4. The solid line marks the mean value of the observed R1/R2 ratio while the dashed lines indicate the 1-σ uncertainty. Since this cloud is at 55 pc (Figure 1) the bubble has an extent of $\simeq$ 180 pc, equal to the distance of the next large column density cloud

4 The Plasma Temperature of the Local Bubble

Within the *ROSAT* PSPC field of view of the pointing in the direction of α Aur we investigated the R1/R2 band ratio at several positions to determine the temperature of the X-ray emitting gas. The observed R1/R2 band ratio is well represented by emission from a plasma of $T = 10^{5.9-6.0}$ K. This temperature is in line with the values derived in different publications about the local X-ray gas.

However, the emission model used assumes *a priori* that there is no soft-X-ray-absorbing interstellar matter distributed along the line-of-sight. The amount of soft X-ray absorbing matter in front of HD 32630 is $N_{\rm HI}$=0.7 to $1.0 \cdot 10^{19}$ cm^{-2} (Fruscione et al. 1994). Correcting for this foreground absorption gives a larger R1/R2 band ratio, because the R1 band photons are attenuated more than the more energetic R2 band photons. In consequence, we have to revise the locally derived plasma temperature to $T = 10^{5.9}$ K.

5 The X-ray Distance to the Cloud

We can additionally use the R1/R2 band ratio to determine the distance of the cloud within the Local Bubble. Using the strong energy dependence of the photoelectric absorption cross section on the X-ray photon energy, the R1/R2 band ratio can discriminate between a thin or a thick cloud localized at the same distance. In Figure 2 we plotted the R1/R2 band ratio as a function of the

distance to such a cloud with a given H I column density. The distance is normalized to 1 which corresponds to the extent of the Local Bubble towards the Auriga – Perseus.

The observed R1/R2 ratio indicates the presence of a thin cloud close by, certainly *within* the Local Bubble. The absorption line data show that beyond 180 pc the column density is larger than $N_{\rm HI}= 1 \cdot 10^{21}\,{\rm cm}^{-2}$. Taking this column density value as the boundary of the Local Bubble, we find *from the analysis of the ROSAT-data* that the cloud is at a distance of 54 – 72 pc. This is in quantitative agreement with the interstellar absorption line measurements (see Fig. 1).

6 Conclusions

The quantitative analysis of the X-ray colours shows that they contain information about the 3-D structure of X-ray emitting and absorbing matter. Furthermore, we demonstrated the usefulness of combined analysis of *IUE* and other absorption line data with *ROSAT* PSPC data.

We could derive physical parameters for the Local Bubble gas such as:
1) an extent of hot gas of about 180 pc in the Perseus – Auriga direction,
2) a local X-ray gas emission measure of $EM \simeq 0.003\,{\rm cm}^{-6}\,{\rm pc}$,
3) a bubble electron density over 180 pc of $n_{\rm e} \simeq 0.004\,{\rm cm}^{-3}$,
4) a pressure within the local X-ray plasma $P_{\rm X-ray-gas} \simeq 16500\,{\rm K\,cm}^{-3}$,
5) an embedded cloud with $N_{\rm HI}= 0.5$ to $1.0 \cdot 10^{20}\,{\rm cm}^{-2}$ at ~ 55 pc,
6) which (with a thickness of 5 pc) has a density of $n({\rm HI}) \sim 5\,{\rm cm}^{-3}$,
7) at (in pressure balance with the X-ray gas) a temperature of ~ 3000 K.

References

Fruscione A., Hawkins I., Jelinsky P., Wiercigroch A., 1994, ApJS 94, 127
Hartmann D., Burton W.B., 1997, 'An Atlas of Galactic Neutral Hydrogen'. Cambride University Press
Lilienthal D., de Boer K.S., 1991, A&AS 87, 471

Part III

Observations and Modeling of the Local Bubble and the SXRB

The audience wide awake even during long sessions.

The conference desk: always on the spot.

The Interstellar and Circumstellar Environment of White Dwarfs

M.A. Barstow[1], P.D. Dobbie[1], and J.B. Holberg[2]

[1] Department of Physics and Astronomy, University of Leicester, UK
[2] Lunar and Planetary Laboratory, University of Arizona, Tucson, USA

Abstract. We have studied the EUV spectra of 13 DA white dwarfs, observed by the *EUVE* satellite, paying attention to the possible sources of absorbing material along the lines-of-sight in both the local interstellar medium and in the photospheres of the stars themselves. The range of interstellar column densities seen are consistent with our previous understanding of the local distribution of material. Absorption from interstellar He II is found in the direction of five stars, allowing us to measure directly the He ionization fraction and estimate, indirectly, that of H. The weighted mean ionization fractions along these lines-of-sight are 0.27 ± 0.04 and 0.35 ± 0.1 respectively. Where He II is directly detected, the observed ionization fractions are not correlated with direction or with the volume/column density of material along the line-of-sight. Furthermore, the limits on the amount of He II established in all other directions completely encompass the range of observed values. Indeed, all the data can be consistent with more or less constant He and H ionization fractions throughout the local ISM. However, observation of very hot DA stars, indicating higher He II columns, might contradict this picture if the material is not photospheric or circumstellar.

1 Introduction

It is generally accepted that the solar system resides inside a relatively dense local interstellar cloud a few parsecs across with a mean neutral hydrogen density of $\approx 0.1\text{cm}^{-3}$ (e.g. Frisch 1994, Gry et al. 1995). This cloud, the so-called local cloud or surrounding interstellar cloud (SIC) itself lies inside a region of much lower density, often referred to as the local bubble. The general picture that has built up is one where this bubble has been created by the shock wave from a past supernova explosion, which would also have ionized the local cloud. The current ionization state of the local cloud is then expected to depend on the recombination history of the ionized material i.e. the length of time since the shock wave passed through. However, if the flux of ionizing photons from hot stellar sources is significant, the net recombination rate may be reduced (Cheng & Bruhweiler 1990; Lyu & Bruhweiler 1996).

Prior to the launch of *EUVE*, few spectroscopic observations were available with which to study H and He in the local interstellar medium (LISM). With the ability to detect features arising from interstellar He I and He II, *EUVE* has presented us with an important and unique opportunity of providing direct tests of model predictions.

2 EUV Spectroscopy

We present here a detailed study of a sample of 13 DA white dwarfs observed by *EUVE*. The details of our data reduction and analysis techniques can all be found in our recent papers (e.g. Barstow et al. 1997a). Briefly, we compared the predictions of synthetic white dwarf spectra, folded through a model of the interstellar opacity, with the observations to determine the physical parameters describing the star and the local ISM. The combination is specified by 7 independent parameters – effective temperature (T_{eff}), log surface gravity, helium abundance (or H layer mass for a stratified model) a distance/radius related normalisation constant, H I column density, He I column density and He II column density. However, from Balmer line fits to the optical data, T_{eff} and log g are known within narrow bounds. Furthermore, a measurement of V magnitude, from optical photometry, can be used to determine the normalisation. Hence, only four parameters are completely unknown.

3 The Ionization of the Local ISM

Photoionized He II can be clearly detected in 5 of the white dwarfs studied – GD659, REJ1032, PG1123, REJ2156 and GD246 (see e.g. figure 1), from which the He ionization fraction (f_{He}) can be calculated and the H ionization fraction (f_H) inferred (assuming a cosmic He/H ratio, see Barstow et al. 1997a). In other cases, only limits to the ionization fraction can be obtained. Values of f_H ($[N_H - N_{H\,I}]/N_H$) and f_{He} ($N_{He\,II}/[N_{He\,I} + N_{He\,II}]$)calculated from the best fit column densities are listed in table 1. There appears to be no particular relationship between direction and the degree of ionization measured. The weighted mean ionization fractions of He and H in those five directions where He II is detected are 0.27 ± 0.04 and 0.35 ± 0.1 respectively. In low column density directions, the absence of detected He II features could easily be a threshold effect because the edge is simply too weak to detect. It is clear that the level of the upper limits, the errors on the measured values and the range of uncertainty associated with the values implied from the model comparisons all overlap substantially. Hence, there is no evidence that the He and H ionization fractions are not close to the weighted mean values in any of the directions probed with these observations.

4 Evidence for Helium in the Spectra of Very Hot DA Stars

Our conclusion that helium could be uniformly ionized might be contradicted by recent studies of the very hot (≈ 60000K) white dwarfs G191−B2B and REJ0457−281 (Lanz et al. 1996; Barstow et al. 1997b). To explain the EUV

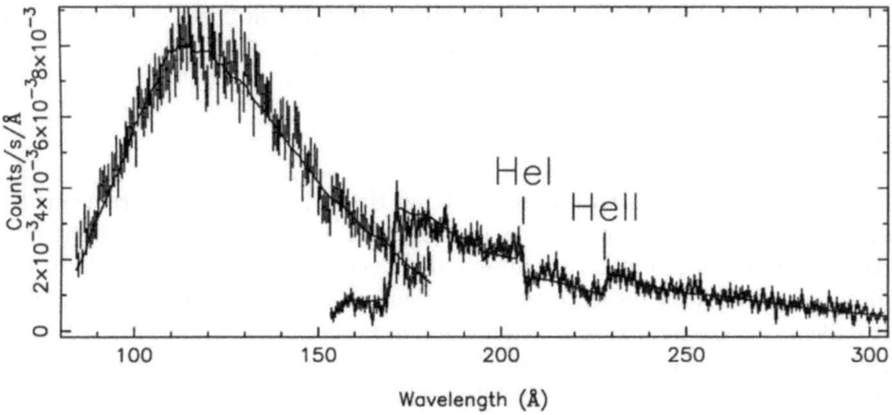

Fig. 1. EUV count spectra of REJ2156−546 compared to the best fit ISM/photospheric model, showing the interstellar absorption edges at 228Å (He II) and 206Å (He I). The apparent feature at 170Å is an instrumental artifact produced by aluminium in a thin film filter.

Table 1. The ionization fractions of H and He in the ISM.

Star	f_H			f_{He}		
	Value	-1σ	$+1\sigma$	Value	-1σ	$+1\sigma$
GD659	0.16	0.00	0.50	0.42	0.29	0.58
GD71	0.00	0.00	0.28	0.04	0.00	0.32
REJ0715−705	0.00	0.00	0.79	1.00	0.00	1.00
REJ1032+535	0.46	0.30	0.56	0.27	0.20	0.34
PG1123+189	0.43	0.13	0.65	0.20	0.15	0.26
GD153	0.00	0.00	0.28	0.00	0.00	0.36
HZ43	0.19	0.00	0.34	0.40	0.29	0.50
CoD−38 10980	0.00	0.00	1.00	0.00	0.00	1.00
BPM93487	0.17	0.00	1.00	1.00	0.00	1.00
REJ2009−605	0.00	0.00	0.63	0.04	0.00	1.00
REJ2156−546	0.34	0.00	0.57	0.39	0.28	0.46
GD246	0.20	0.04	0.45	0.23	0.17	0.30
REJ2324−547	0.57	0.00	0.74	0.13	0.00	0.35

spectra of both these stars it is necessary to include significant levels of He II opacity in the photospheric/interstellar models. If completely interstellar, the He II ionization fraction is $\approx 80\%$. Preliminary studies of several other stars near this temperature yield similar results (Barstow et al. 1997c). However, since any possible He II features must be blended with large numbers of heavy element lines in these objects, the presence of He II is only implied rather than directly detected. Furthermore, even if it is not an artifact of limitations in the stellar model atmospheres, it is impossible to decide whether the He II resides in the photosphere, in the circumstellar environment, or in the interstellar medium. Hence, while, as a group, these stars do not yet contradict the current results we must remain somewhat cautious.

5 Discussion

The value and apparent uniformity of the He and H ionization fractions we see are important in considering the physical mechanisms that might give rise to the levels of ionization observed. Recent measurements, made in the direction of ϵ CMa, indicate that the electron density in the local cloud is $\approx 0.09\text{cm}^{-3}$, giving a H ionization fraction no greater than 50% (Gry et al. 1995), in keeping with our result. Our observations are also consistent with the predictions of Lyu and Bruhweiler's (1996) time dependent ionization calculations, considering the recombination of the LISM following shock heating from a nearby supernova. For our mean ionization fractions, we estimate an elapsed time of 3-4 million years since the onset of the recombination phase.

In the light of this apparent agreement, we should consider whether we are sampling just material in local cloud or also in the surrounding medium. The latter is more tenuous but the effects are integrated over greater distances. Our data indicate that a significant fraction of the interstellar absorption observed lies outside the cloud in most of the lines-of-sight we have examined. Importantly, this is so for all those directions in which we detect ionized He. Hence, the ionization fractions reported here may not represent the conditions in the local cloud and the agreement with the measurements of Gry et al. (1995) and predictions of Lyu and Bruhweiler (1996) could be purely fortuitous.

References

Barstow M.A., Dobbie P.D., Holberg J.B., Hubeny, I, Lanz T., (1997a): MNRAS, 286, 58

Barstow M.A., Holberg J.B., Hubeny, I, Lanz T., (1997b): in White Dwarfs, eds. J. Isern, M. Hernanz and E. Garcia-Berro, Kluwer, 237

Barstow M.A., Holberg J.B., Hubeny, I, Lanz T., (1997c): in preparation.

Cheng K.-P., Bruhweiler F.C., (1990): ApJ, 364, 573

Frisch P.C., (1994): Science, 265, 1423

Gry C., Lemonon, L., Vidal-Madjar A., Lemoine M., Ferlet R., (1995): A& A, 302, 497

Lanz T., Barstow M.A., Hubeny I., Holberg J.B., (1996): ApJ, 473, 1089

Lyu C.-H., Bruhweiler F.C., (1996): ApJ, 459, 216

D/H-ratio: Observations with ORFEUS II

M. Gölz[1], N. Kappelmann[1], I. Appenzeller[2], J. Barnstedt[1], A. Fromm[1], M. Grewing[3], W. Gringel[1], C. Haas[1], W. Hopfensitz[1], G. Krämer[1], J. Krautter[2], A. Lindenberger[1], H. Mandel[2], K. Werner[1], and H. Widmann[1]

[1] Institut für Astronomie und Astrophysik, Abt. Astronomie, Waldhäuserstraße 64, D-72076 Tübingen, Germany
[2] Landessternwarte Heidelberg, Königsstuhl, D-69117 Heidelberg, Germany
[3] Institut de Radio Astronomie Millimétrique (IRAM), 300 Rue de la Piscine, F-38406 Saint Marint d'Hères, France

Abstract. During the second flight of the *ORFEUS-SPAS* satellite (Nov./Dec. 96) high resolution ($\frac{\lambda}{\Delta\lambda} = 10,000$) Echelle-spectra of BD +28° 4211 in the wavelength regime 912 - 1400 Å have been taken. Deuterium can be clearly identified in the *ORFEUS II* Echelle-spectra of this star. For the first time it was possible to take spectra of faint, not redshifted objects in the far ultraviolet with a sufficient spectral resolution to study the deuterium column density in the whole spectral range of the Lyman-series down to the Lyman-limit. We obtained a value of $\log(N_D) = 14.7\,(\pm 0.3)$ towards BD +28° 4211. The hydrogen column density has been determined using *ORFEUS* Echelle- and *IUE*-spectra of Ly-α ($\log(N_H) = 19.8\,(\pm 0.2)$). Thus a value of 8×10^{-6} can be obtained for the D/H-ratio on the line-of-sight towards BD +28° 4211.

1 Introduction

The interstellar abundance of deuterium is an indicator for the density of baryonic matter in the universe. McCullough (1992) has derived a unique interstellar D/H value for the local ISM of $1.5(\pm 0.2) \times 10^{-5}$ by reevaluating Lyman absorption-line measurements made by the *Copernicus* and the *IUE* satellite. The interstellar D/H ratio can be determined from the relative strength of the deuterium and hydrogen Lyman absorption lines, which are separated by $82\,\mathrm{km\,s^{-1}}$. The *ORFEUS* Echelle-spectrograph is the first instrument, which can access the whole Lyman-series down to the Lyman-limit with a sufficient spectral resolution to resolve the deuterium and hydrogen lines. Up to now only very few investigations of the deuterium column density have been made in the southern Milky Way in the vicinity of BD +28° 4211, a non-variable late-type object.

2 Data

The hot subdwarf BD +28° 4211 ($T_{\rm eff} = 82{,}000$ K and $\log g = 6.2$, Napiwotzki 1993) is located at a distance of about 100 pc (Perryman, priv.com.). Due to the preliminary data reduction we used only one out of several spectra of BD +28° 4211 taken during the *ORFEUS II*-mission from Nov. 19 to Dec. 04, 1996 using the *ORFEUS* Echelle-spectrograph in the wavelength-regime $\lambda\lambda = 912 - 1400$ Å with a spectral resolution of $\frac{\lambda}{\Delta\lambda} = 10{,}000$ (Krämer et al. 1990).

The spectrum of this hydrogen-rich subdwarf is dominated by stellar and interstellar Lyman lines of atomic hydrogen. The strongest line (Ly-α) has been used for the determination of the hydrogen column density. The interstellar deuterium lines Ly-β through Ly-η ($\lambda\lambda = 1025.44$ - 925.97 Å) are clearly identified in the spectrum. The Ly-α deuterium-line ($\lambda = 1215.34$ Å) is not visible due to strong geocoronal Ly-α emission (Fig. 1 (b)). We have used the deuterium lines Ly-β through Ly-λ for the determination of $\log(N_D)$.

The N I-triplet at $\lambda \approx 1134.5$ Å, which is fully resolved in the *ORFEUS II* Echelle-spectrum, has been used to check for clouds with different velocities in the ISM on the line-of-sight (Fig. 1 (a)). The profile of the N I-triplet indicates that there are no different cloud components. Several absorption lines of N I in the *ORFEUS* Echelle-spectrum have been analysed to estimate the turbulence velocity within the ISM along the line-of-sight towards BD +28° 4211.

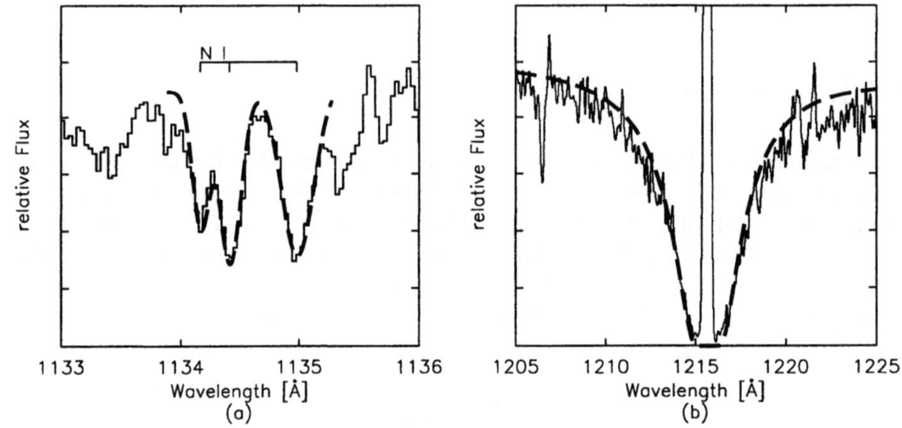

Fig. 1. (a) The N I-triplet at $\lambda \approx 1134.5$ Å in the *ORFEUS II* spectrum and a fit of the three components is shown (dashed curve). (b) The Ly-α line of hydrogen in the *ORFEUS II* spectrum of BD +28° 4211 and the theoretical line profile calculated for the value $\log(N_H) = 19.8$ is presented (dashed curve). In the center of the Ly-α absorption the strong geocoronal emission of Ly-α is visible.

3 Analysis and Results

3.1 Derivation of log(N_H)

In order to derive the hydrogen column density in the line-of-sight towards BD +28° 4211, both the Ly-α line of hydrogen and molecular hydrogen transitions have been analysed. The spectrum shows some evidence for a small amount of H_2, which seems to be neglectable for the further analysis. The *ORFEUS II* Echelle-spectrum of BD +28° 4211, as well as a high resolution *IUE*-spectrum, have been analysed to determine the atomic hydrogen column density. In order to separate the stellar and the interstellar component, recent NLTE model-spectra with $T_{eff} = 80{,}000$ K and $\log g = 6.25$ with an atmospheric composition of $n_{He}/n_H = 0.1$ have been used (Napiwotzki priv.com.). Natural damping-profiles with different interstellar column-densities have been applied to the stellar absorption lines and the resulting profiles have been compared with the observed ones (Fig. 1 (b)). We obtained a value of $\log(N_H) = 19.8\,(\pm\,0.2)$ for the interstellar hydrogen column density.

3.2 Derivation of log(N_D)

We analysed thirteen interstellar absorption lines of neutral nitrogen with the standard curve of growth technique. A turbulence velocity of $b = 6\,\mathrm{km\,s^{-1}}$ in the ISM fits the data quite well. Based on this result we determined the theoretical equivalent widths of deuterium Ly-β through Ly-λ for different hydrogen column-densities using a standard curve of growth. For each absorption we fitted the nearby hydrogen-line together with a fixed set of parameters of the deuterium absorption to the observed profiles. The continuum has been corrected for stellar contribution using a recent atmospheric model of BD +28° 4211 (Napiwotzki, priv.com.).

The investigations showed, that the weak transitions of deuterium Ly-ϵ ($\lambda = 937.5$ Å) through Ly-λ ($\lambda = 917.9$ Å) are most suitable for the estimation of log(N_D). Those lines are weak enough, that the line-profiles depend only on the column density and are independent from the turbulence velocity.

The strongest observable lines (Ly-β - Ly-δ) are dominated by the strong hydrogen absorption, which depletes the continuum, therefore the S/N- ratio of the deuterium absorption is reduced. Additionally those features are more sensitive to the background subtraction, which is only preliminary at this point. Furthermore, the equivalent widths of those strong deuterium lines depend already significantly on the turbulence velocity. Therefore we mainly stressed upon Ly-ϵ through Ly-λ for the determination of log(N_D). No further effort has been taken to derive a more precise value of b. A value of $\log(N_D) = 14.7\,(\pm\,0.3)$ has been determined from the analysed data. Although there is some evidence for H_2 in the spectrum, no absorption of HD has been found. The N_D/N_H-ratio on the line-of-sight towards BD +28° 4211 can be given as $N_D/N_H = 8\,(^{+7}_{-4}) \times 10^{-6}$. McCullough (1992) derived the mean value for the local ISM as $N_D/N_H = 1.5(\pm\,0.2) \times 10^{-5}$, which is compatible with the value derived in this publication.

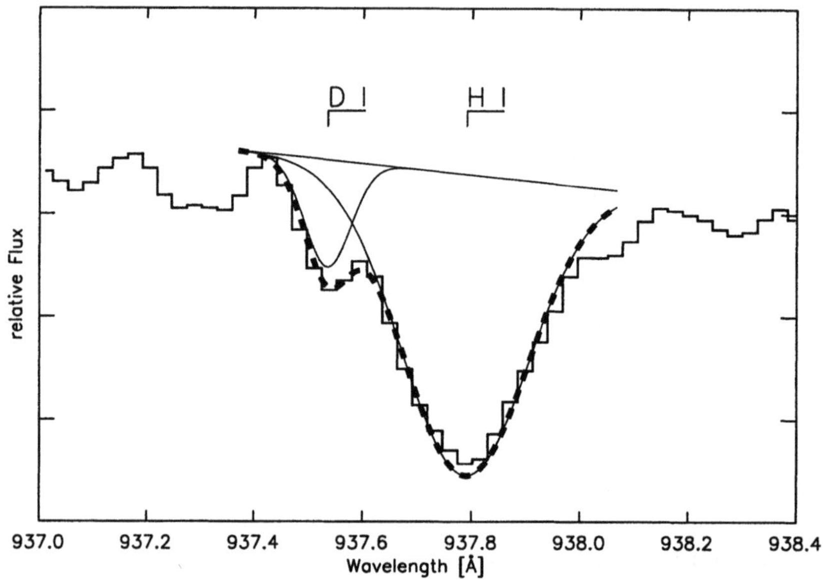

Fig. 2. The *ORFEUS II* spectrum (histogram mode) with a fit of deuterium and hydrogen Ly-ϵ (dashed curve) assuming a value of $\log(N_D) = 14.7$. Both components and the centers of the absorption lines are indicated. The spectrum has been corrected for the stellar contribution.

Acknowledgements. We wish to thank Dr. Napiwotzki for calculating the atmosphere models for BD +28° 4211 and making them available to us and Dr. Perryman for supplying us the parallax of BD +28° 4211.
Special thanks to all American and German colleagues and collaborators, who helped to realize the successful *ORFEUS*-mission. The *ORFEUS*-program was supported by DARA grants WE3 OS 8501 and WE2-QV 9304 and NASA grant NAG5-696.

References

Krämer, G. et al. (1990): *Observatories in Earth Orbit and Beyond*, ed. Y. Kondo (Kluwer Academic Publishers), 177
McCullough, P.R. (1992): ApJ 390, 213
Napiwotzki (1993): Acta Astron. 43, 343

The Diffuse Extreme Ultraviolet Background

John Vallerga and Jonathan Slavin

Eureka Scientific Inc., 2452 Delmer St. Suite 100, Oakland, CA 94602-3017, USA

Abstract. Observations of the diffuse EUV background towards 138 different directions using the spectrometers aboard the Extreme Ultraviolet Explorer satellite (EUVE) have been combined into a spectrum from 150Å to 730Å and represent an effective exposure of 18 million seconds. There is no significant evidence for any line flux from any source other than the geocorona. These results are inconsistent with the Wisconsin C and B broad-band surveys assuming the source is a logT = 5.8 - 6.1 hot plasma in ionization equilibrium with solar abundances, confirming the previous result of Jelinsky, Vallerga & Edelstein (1995) (hereafter Paper I) using an observation along the ecliptic with the same instrument. To make these results consistent with the previous broad–band surveys, the plasma responsible for the emission must either be depleted in Fe by a factor of $\sim$ 6, be behind an absorbing slab of neutral H with a column of 2×10^{19}cm^{-2}, or not be in collisional ionization equilibrium (CIE). One particular non-CIE model (Breitschwerdt & Schmutzler, 1994) that explains the soft x-ray results is also inconsistent with this EUV data.

1 Why Investigate the Diffuse EUV Background?

If the local bubble is filled with a solar abundance hot plasma (log T = 6) in equilibrium, then its dominant cooling mechanism is line emission in the EUV, specifically the Fe line complex near 170–190Å (72–64 eV). If detected, this emission must be reasonably local, as one optical depth at 170Å corresponds to a neutral H column of 10^{19}cm^{-2}. Some non-CIE models of the Local Bubble, such as the adiabatic expansion of very hot ionized plasma (Breitschwerdt & Schmutzler, 1994) can reproduce similar soft x-ray broad band count rates but differ markedly with CIE models in the EUV, since the effective temperature of the electrons is only 40,000K. The EUV flux can therefore strongly constrain CIE and non-CIE models, as well as Fe abundances in the hot gas and local absorption effects.

Though not designed as a diffuse instrument, the spectrometers aboard EUVE have a small but finite area-solid angle product ($A\Omega$=.00005 cm^{-2}sr), especially at the wavelengths of the Fe VIII–X complex near 170Å where the emission from a million degree plasma is the strongest. The spectrometers observe this background in parallel with the point source observations of EUVE Guest Observers, resulting in a combined data set of 18 million seconds of effective exposure during the first two years of the EUVE mission.

2 Data Reduction, Results and Discussion

Each of the 248 pointings of the EUVE spectrometers were reduced identically. The technique for converting the two dimensional spectral image into a one dimensional count rate spectrum with errors can be found in Paper I. A subsection of the combined spectrum of 18 million seconds exposure can be seen in Fig. 1. We have concentrated our attention on this spectral region for this paper as it is far from the He I and HeII geocoronal lines, is less affected by interstellar absorption than the longer wavelengths, and is most sensitive to a log T $\sim$ 6 plasma, purported to fill the Local Bubble.

The EUVE Medium (140Å– 380Å) and Long (280Å– 730Å) wavelength spectrometers both have a wire grid collimator to spectrally limit the geocoronal background flux, thus providing a triangular spectral signature to any diffuse monochromatic radiation. Intrinsic detector background was removed from this spectrum by determining the background under a filter support structure that shades the detector. An additional non-cosmic background is clearly present in the data which we model as a quadratic function of wavelength. Spectral models of the diffuse background were convolved through the collimator and telescope/detector response functions, added to a quadratic background and compared to the one dimensional count rate spectra.

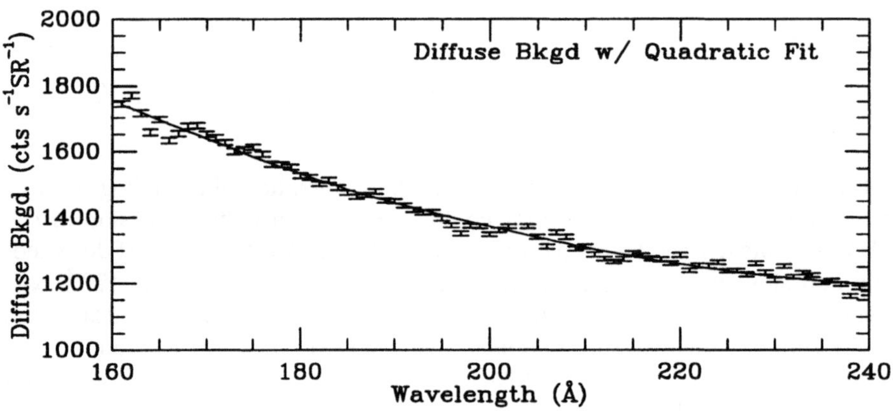

Fig. 1. The 18 million sec. spectrum of the diffuse EUV background between 160–240Å. The sloping continuum has been fit with a quadratic function. A monochromatic emission line would appear as a triangular enhancement with a FWHM of 17Å (bins).

To model the diffuse EUV flux from the Local Bubble, we have used a chi-squared fitting procedure using either a Raymond and Smith (R&S) thermal plasma code parameterized by temperature and emission measure (EM) or a single case of a Breitschwerdt & Schmutzler (B&S) non-CIE model with log T = 4.6 (electrons), log n_e = -1.6 and the flux linearly parameterized by distance through the emitting gas where a distance of 100 pc fits the soft x-ray results. In all cases we have assumed a foreground slab of neutral hydrogen with a column of 10^{18}cm^{-2}. We also let the three quadratic background parameters vary in each fit. Fig. 2 shows the predicted EUV spectra of each model folded through the instrumental response, as well as the residual of the measured spectrum after a simple quadratic fit has been subtracted. The model parameters were chosen to fit the average Wisconsin B band flux: an EM of .0036, Log T= 6.0 for the R&S model and 100 pc pathlength for the B&S model. Contrasting with the strong Fe emission lines of the CIE plasma codes, the B&S model is dominated by the HeII recombination continuum emission shortward of the edge at 228Å. Though the instrumental resolution is poor, it is evident that the parameters that fit the x-ray results for both models strongly overpredict the EUV flux. The formal fit to the EUVE data give an EM of .00057 ± .00014 cm^{-6} pc (log T = 6.0) for a R&S model and a pathlength of 21 ± 4.6 pc for the B&S model. Rather than claiming a marginal detection at this level, which requires a detailed understanding and quantification of our systematic errors, we quote a 2σ upper limit of EM =.00085 cm^{-6} pc for the R&S model or a maximum pathlength of 30 pc for this particular B&S model.

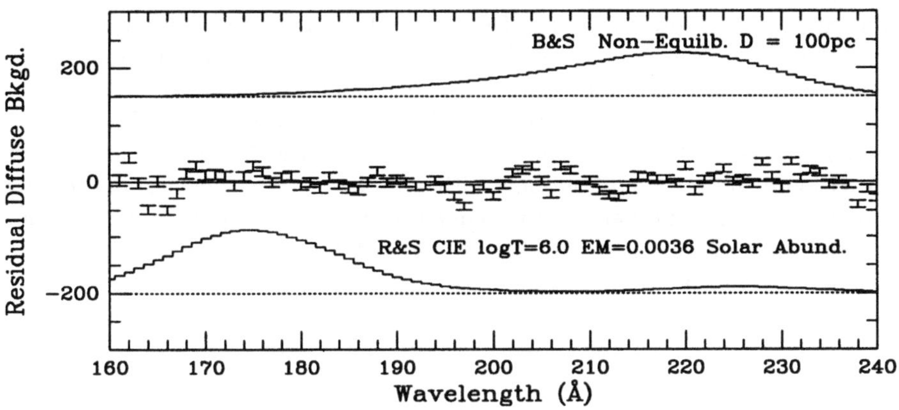

Fig. 2. A residual plot of the data shown in Fig. 1 after subtraction of the quadratic fit. Also shown (offset in Y) are two models with parameters that fit the Wisconsin soft x-ray results, both inconsistent with the residual spectrum.

When the 138 directions on the sky are analyzed separately and fit with a log T = 6.0 R&S model, none of the observations detect an EM greater than a 3 sigma confidence limit. However, there is a hint that the data favor flux from Fe lines rather than HeII recombination. If we separate the 138 directions by their B count rate, those above 75 counts s^{-1} have a weighted mean EM of .0011 ± .00032 while those directions with a count rate below 75 counts s^{-1} have a weighted mean of .00036 ± .00016, a significant difference indicating a correlation with the B band. Also, the four observations with the highest positive significance are in directions of a high B band count rate. For the non-CIE model of B&S, there was no correlation of the results with B band count rates.

The thermal plasma models used to explain the soft x-ray rocket and satellite data predict EUV flux assuming low absorbing column, solar abundances and equilibrium conditions and all three could be wrong. We chose to assume a low foreground column based on the knowledge that except in a few directions toward the galactic center, the neutral hydrogen column density measured is consistent with that in the Local Cloud of 1–2 $\times 10^{18} cm^{-2}$. To make the non-detection of EUV flux consistent with the C and B band soft x-rays using slab absorption requires a neutral column a factor of 10 higher, placing the emission beyond the local void, which is also inconsistent with the softer Be band results.

The DXS results (Sanders, this volume) detect diffuse emission in the 40Å–80Å range. However, no equilibrium models with or without elemental depletion fit the data well nor do non-equilibrium models tried to date. Our non-detection supports the view that solar abundance equilibrium models do not work, but also place a strong limit on the amount of Fe in the gas phase of a hot gas in equilibrium. A recent compilation of depletion measurements in neutral clouds towards many stars in the ISM indicate a minimum depletion of Fe of 0.5 dex, interpreted as either a component of dust grains are indestructible in the cloud environment or that the abundance of Fe in the local ISM is not solar (Fitzpatrick, 1996). The lack of Fe in emission implied by our results (0.78 dex to make the EUV derived EM limits consistent with the B band limits) would support either view. But the dust grains in this case are surviving a different environment of high T, low density sputtering.

The B&S models were the first to predict the observed absence of the strong EUV flux at the Fe line complex. Yet the one B&S model we tried overpredicts the longer wavelength HeII recombination continuum flux. We used that particular model to demonstrate the usefulness of the EUV band to differentiate between CIE and non-CIE models and to constrain both. Not much more can be gleaned from the EUVE data set – 18,000,000 seconds is probably the longest astronomical integration in history! What is needed is an instrument with a large solid angle–effective area product combined with spectral resolution to decouple the abundance and emission model effects.

References

Breitschwerdt, D. & Schmutzler, T. (1994): Nature, **371**, 774

Fitzpatrick, E.L. (1996): ApJ, **473**, L55–L58

Jelinsky, P., Vallerga, J. and Edelstein, J.(1994): ApJ, **442**, pp. 653–661 (Paper I)

The Soft X-ray Background Spectrum from DXS

W.T. Sanders[1], R.J. Edgar[2], D.A. Liedahl[3], and J.P. Morgenthaler[1]

[1] Department of Physics, University of Wisconsin–Madison, 1150 University Avenue, Madison, WI 53706, USA
[2] Center for Astrophysics, 60 Garden Street, Cambridge, MA 02138, USA
[3] V-Division, Lawrence Livermore National Lab., P. O. Box 808, L–41, Livermore, CA 94550, USA

Abstract. The Diffuse X-ray Spectrometer (DXS) obtained spectra of the low energy X-ray (44 – 83 Å) diffuse background near the galactic plane from galactic longitudes $150° \lesssim l \lesssim 300°$ with $\lesssim 3$ Å spectral resolution and $\sim 15°$ angular resolution. Thus, DXS measured X-ray spectra that arise almost entirely from within the Local Bubble. The DXS spectra show emission lines and emission-line blends, indicating that the source of the X-ray emission is thermal – hot plasma in the Local Bubble. The measured spectra are not consistent with those predicted by standard coronal models, either with solar abundances or depleted abundances, over the temperature range $10^5 - 10^7$ K. The measured spectra are also inconsistent with the predictions of various non-equilibrium models. A nearly acceptable fit to DXS spectra can be achieved using a hybrid model that combines the Raymond & Smith ionization balance calculation with recently calculated (by DAL) ionic emission lines.

1 The Diffuse X-ray Spectrometer (DXS)

The Diffuse X-ray Spectrometer (DXS) consists of a pair of identical Bragg-crystal spectrometers that were built by our research group at the University of Wisconsin-Madison and flown by NASA as an attached Shuttle payload in January 1993. It is designed to obtain spectra of the soft X-ray background in the 150 – 284 eV (83 – 44 Å) range with ~ 2.7 Å spectral resolution and 15° (FWHM) angular resolution. Each DXS instrument, one on the Port side of the shuttle and one on the Starboard side, uses a curved crystal panel to Bragg-reflect incident soft X-rays towards a collimated position-sensitive proportional counter. The collimator allows each position along the counter to "see" a restricted piece of the curved crystal panel, and each position sees its piece of the crystal panel at a different angle. Thus, each position along the counter sees a different wavelength x-ray from a different direction on the sky than does its neighboring position. By rotating the counter about an axis perpendicular to the dispersion direction of the crystals through an angle of $\sim 60°$, X-rays from one direction are successively reflected across the face of the proportional counter, so that a complete spectrum over the 150 – 284 eV range from that direction is obtained. The DXS instruments are described more fully by Sanders et al. (1992).

2 DXS Calibration

The DXS post-flight laboratory calibrations measured the instrument area-solid angle product using three fluorescent lines: the carbon K-alpha line (277 eV), the boron K-alpha line (183 eV) and the zirconium M-zeta line (151 eV). Figure 1 shows the almost-final DXS area-solid angle product as a function of energy that was used for the fits presented here. The three post-flight calibration points are indicated.

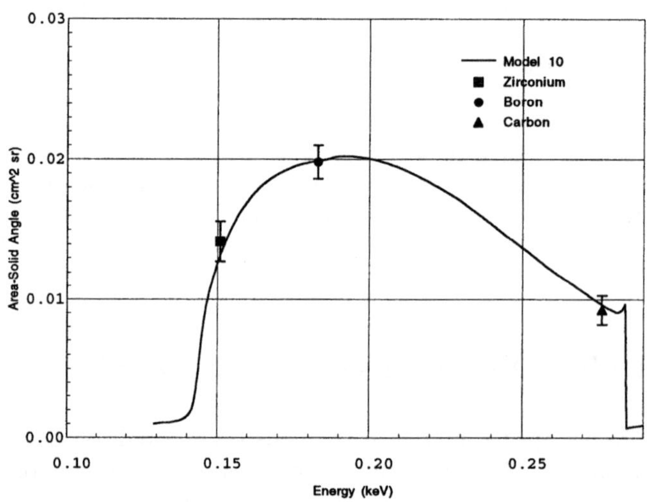

Fig. 1. Effective area - solid angle product of the DXS response function vs. energy shown with three measured calibration points.

The same laboratory calibration data were also used to determine the DXS energy scale. Bragg crystal rocking curves measured at the University of Wisconsin-Madison Synchrotron Radiation Center were used to determine the shapes of the DXS response to narrow lines. Figure 2 shows the combined laboratory data, and the DXS response function predicted output to an input model that consists of three zero-width lines at the energies indicated in parentheses. The energy scale appears to be good to ± 0.5 eV, and the widths of the laboratory fluorescent lines are greater than zero.

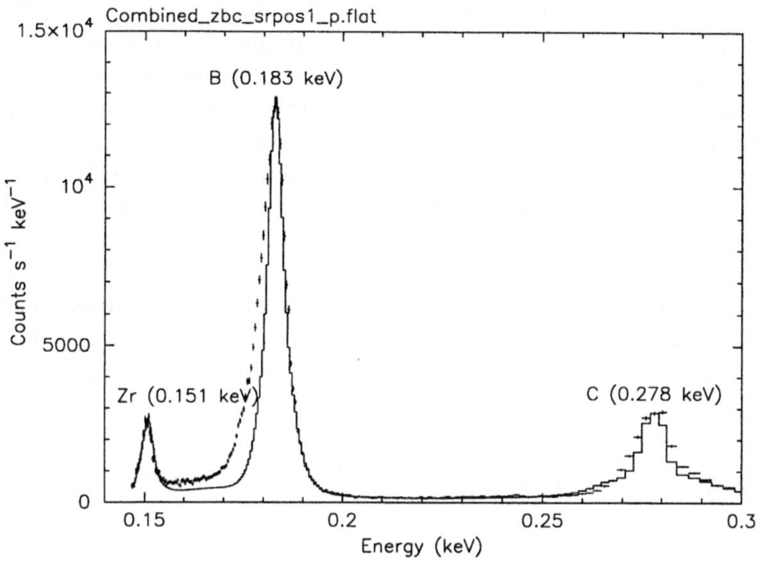

Fig. 2. Spectrum of the DXS calibration data. The solid line gives the calculated output of the DXS response function to a model of three zero-width lines at the energies in parentheses.

3 Uniqueness of the DXS Data

What is unique about the DXS data is its good spectral resolution in the 0.15 – 0.28 keV band. DXS is sensitive only in the 0.15 - 0.28 keV band, but this band, is where the Local Bubble X-ray emission is primarily seen. The DXS spectral resolution is roughly constant, ~ 2.7 Å, across its 43 - 83 Å bandpass, giving it fifteen independent spectral resolution elements across the 1/4 keV band, compared to about one for a proportional counter. The DXS energy resolution ranges from $\delta E = 5$ eV at 150 eV to $\delta E = 18$ eV at 284 eV. The drawback to the DXS approach is that the efficiency of Bragg reflection low, only 1-2%, so that a long integration time or a large field of view is required. The DXS field of view was large to get reasonable statistical precision on a shuttle flight of a few days.

4 The DXS Sky Coverage

During the 80 orbit nights of DXS data collection time during the shuttle mission, the Orbiter was oriented such that the DXS detectors repeatedly scanned the same arc on the sky – an arc within 10° of the Galactic plane from longitudes 150° - 300°. The sky covered by this arc divides into five distinct regions based on count rate: Auriga ($174° \lesssim l \lesssim 189°$), MonoGem

($189° \lesssim l \lesssim 212°$), Puppis ($212° \lesssim l \lesssim 252°$), Vela ($252° \lesssim l \lesssim 277°$), and Crux ($277° \lesssim l \lesssim 300°$). Towards Vela, half the measured counts are due to the Vela supernova remnant, and toward the MonoGem region, a quarter of the counts are due to the MonoGem Bubble, so only toward the Auriga, Puppis, and Crux directions are we seeing featureless diffuse background from the Local Hot Bubble. The DXS Auriga spectrum is from the direction of the Taurus-Auriga dark cloud complex, a high-column density interstellar cloud whose distance is ~100 pc, whereas the DXS Puppis spectrum is from the direction without significant amounts of interstellar matter for a distance of several hundred pc. The fits presented here are to data from the Puppis direction. The Auriga and Crux spectra are similar to the Puppis spectrum, but their similarities and differences are discussed elsewhere (Morgenthaler 1997).

5 DXS Spectral Fitting

Figure 3 shows the DXS Puppis spectrum in photon units as a function of energy. Subsequent spectral fits here present the data in units of counts s^{-1} $Å^{-1}$ as a function of wavelength, because those are more natural units for these data. They are plotted with decreasing wavelength for easier comparison to the energy scale plot. Although typically we would use a 3-component spectral model to fit the diffuse background spectrum at high latitudes over the spectral range 0.1 - 1 keV, only one of those three components is useful for fitting spectra from low latitudes over the spectral range 0.15 - 0.28 keV – the unabsorbed million-degree thermal component that represents emission from the Local Bubble.

First we compare the Puppis spectrum to the model by Raymond & Smith (R&S; 1977, and subsequent updates) of a solar-abundance plasma in ionization equilibrium. The temperature and the emission measure are free parameters. As we have demonstrated previously (e.g., Sanders, Edgar, & Liedahl 1996), the best-fit parameters are similar to those found by other experiments – a temperature of $T = 1.14 \times 10^6$ K and an emission measure of $n_e^2 d = 0.0026$ cm^{-6} pc. But the value of the reduced chi-squared statistic is too large, ~ 7.3 (57 degrees of freedom, hereafter, d.o.f.), so the fit is not acceptable. We also compared the Puppis spectrum to a similar model by Mewe, Kaastra, & Liedahl (MeKaL; Mewe et al. 1985, Kaastra 1992) with similar results. The best-fit temperature and emission measure are $T = 1.2 \times 10^6$ K and $n_e^2 d = 0.0029$ cm^{-6} pc, but the reduced chi-squared value is 6.0 (d.o.f. = 57), also unacceptably large. This fit is shown in Figure 4, where the solid line shows the model predictions.

There are problems with using these models to fit these data. Neither model was designed to be used to fit data with this spectral resolution, and they employ calculational approximations that may not be appropriate here. Another problem is in the atomic physics that are incorporated in these

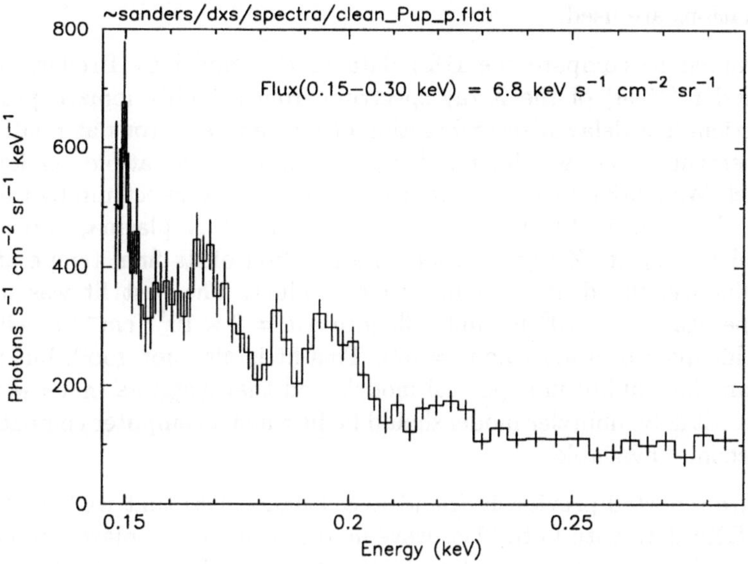

Fig. 3. DXS spectrum from the Puppis region in photon units.

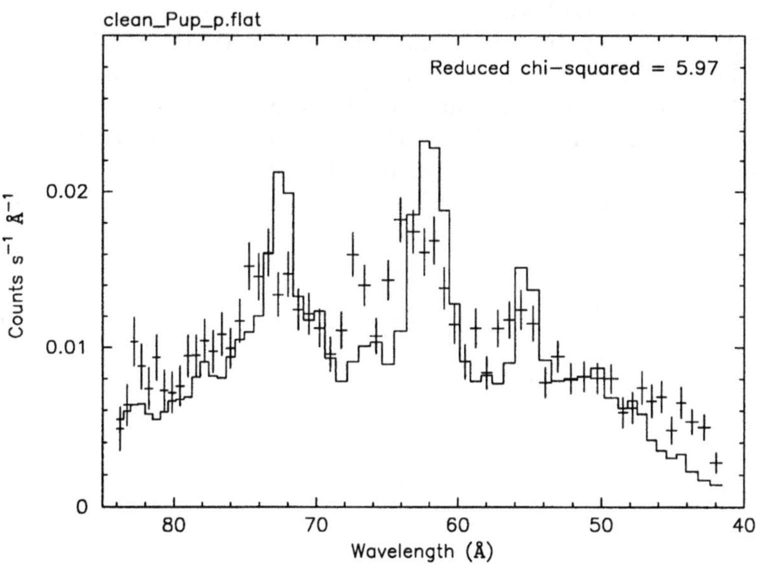

Fig. 4. Best-fit single-temperature MeKaL model to DXS Puppis data.

models. The ionic energy levels and rates are not very well known, and again approximations are used.

We wanted to compare the DXS data to the model by Breitschwerdt & Schmutzler (1994) of the X-ray spectrum from a highly ionized plasma that experiences a delayed recombination of ions and electrons at relatively low temperatures, but we did not have a computer-compatible version of that model. We (RJE) created an approximation to that spectrum by taking a plasma that was in the ionization states of a 10^7 K plasma, and as it cooled, calculating its X-ray emission as a function of its final temperature and the fluence, the density-cooling time product. The best fit was for a final temperature $T = 10^6$ K, and a fluence $n_e t = 2 \times 10^{11}$ cm^{-3} s, with a reduced chi-squared of 4.4 (d.o.f. = 57). This fit is also not good, but it is better than the equilibrium spectral models and that suggests that the real Breitschwerdt & Schmutzler model should be fit when a computer compatible version becomes available.

We have reported previously (Sanders & Edgar 1995) that we have tried to fit the DXS data with both blast wave models and cooling plasma models, both with solar abundances and with depleted abundances, and have not found acceptable fits.

The next model we tried to fit was a "hybrid" model in which we (RJE) modified the Raymond & Smith code by replacing some of the emission line parameters in that code with updated ones that we (DAL) calculated at Livermore using the HULLAC code. The model is simply a solar-abundance single-temperature equilibrium plasma with improved atomic physics data. In particular, we (DAL) calculated improved emission line spectra for 27 ions that have strong emission lines in the DXS bandpass for a range of temperatures. Those ions are Fe IX-XVI, Mg VI-X, Si VI-XII, S VII-XII, and Ne VIII. For this model, the best-fit temperature and emission measure are $T = 1.12 \times 10^6$ K and $n_e^2 d = 0.0023$ cm^{-6} pc, with 3.2 for the reduced chi-squared value (d.o.f. = 57). Figure 5 shows this model compared to the data.

The final model that we fit was the hybrid model with the elemental abundances allowed to be non-solar. For this fit, the abundances of the elements Ne, Mg, Si, S, Ar, Ca, Fe, and Ni were free parameters, as well as the temperature and the emission measure. The reduced chi-squared statistic was 1.97 (d.o.f. = 49), with a best-fit temperature and emission measure of $T = 1.3 \times 10^6$ K and $n_e^2 d = 0.0037$ cm^{-6} pc. Figure 6 shows this fit. The best-fit abundances of only three elements were significantly different from solar: Si $\sim 0.1 \pm 0.3 \times$ solar, S $\sim 0.2 \pm 0.4 \times$ solar, and Fe $\sim 0.0 \pm 0.3 \times$ solar. Using the F-statistic, we find that these three parameters are the only ones that cause the reduction in chi-squared, so if only these three elements are allowed to vary, we get the same abundance reductions with a reduced chi-squared statistic of 1.79 (d.o.f = 54).

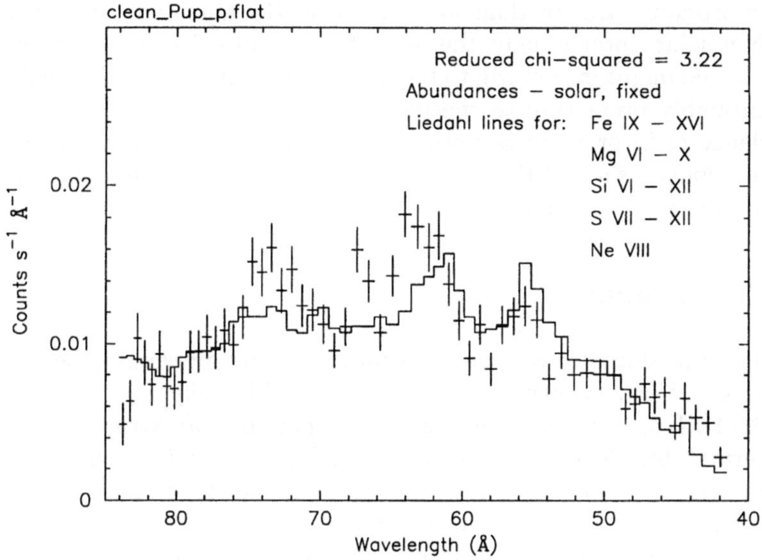

Fig. 5. Hybrid model (R & S plus Liedahl lines) fit to DXS Puppis data. Elemental abundances are solar.

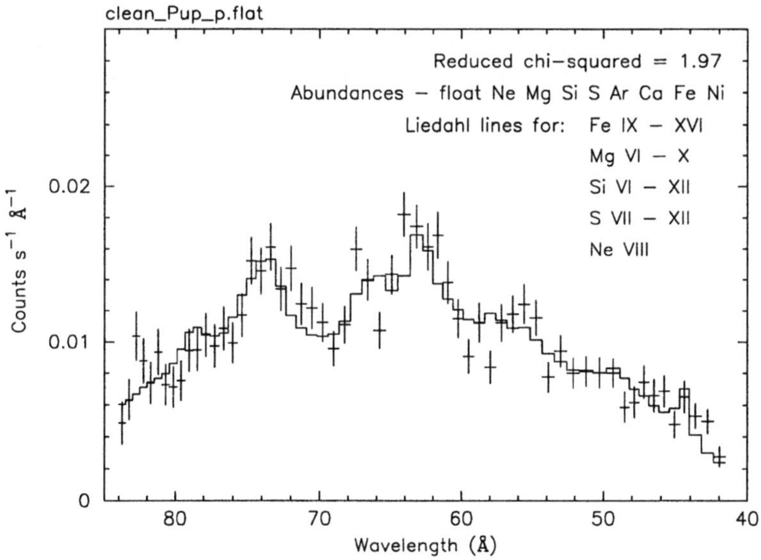

Fig. 6. Hybrid model (R & S plus Liedahl lines) fit to DXS Puppis data. Elemental abundances were allowed to float.

If this model were a good fit, we would conclude that the abundances of Si and Fe probably are less than solar, and that the abundance of S is possibly less than solar. Interestingly, the best fit abundance for Mg is very near 1.0, whereas one might expect Mg to be also depleted, if Si and Fe were depleted. It is probably premature to speculate on the significance of a possible low abundance of Si or Fe or S, until we are more comfortable with its reality. We have concerns about the uniqueness of this solution, and about where the true minimum of chi-squared is.

6 Conclusions

1. The DXS data are consistent in intensity with previous observations of the soft X-ray background from this part of the sky.
2. The DXS spectra show clear evidence that the soft X-ray background is thermal, but "good" fits to standard models are not obtained.
3. For one particular model, the Raymond & Smith coronal plasma model, substituting the recently calculated improved ionic line spectra (by DAL) allows us to obtain significantly better fits to the DXS data.
4. Allowing the elemental abundances to deviate from solar also allows better fits to the data, but the reality and reasonableness of these "depleted" abundances is still being investigated.
5. It would be desirable to fit other standard models, such as an underionized plasma or an overionized plasma, to the DXS data, incorporating the recently calculated improved ionic line spectra.
6. Fitting the DXS data may require a model with X-ray emitting plasma at more than one temperature, as well as with non-solar elemental abundances.

References

Breitschwerdt, D., Schmutzler, T. (1994): Nature 371, 774

Kaastra, J S. (1992): An X-Ray Spectral Code for Optically Thin Plasmas (Internal SRON-Leiden Report, updated version 2.0)

Mewe, R., Gronenschild, E.H.B.M., & van den Oord, G.H.J. (1985): A&AS, 62, 197

Morgenthaler, J. P. (1997): PhD Thesis, University of Wisconsin-Madison.

Raymond, J. C. & Smith, B. W. (1977): ApJS, 35, 419

Sanders, W. T., & Edgar, R. J. (1995): in Astrophysics in the Ultraviolet, eds. S. Bowyer & R. F. Malina (Dordrecht: Kluwer), 269

Sanders, W. T., Edgar, R. J., Juda, M., Kraushaar, W. L., McCammon, D., Snowden, S. L., Zhang, J., & Skinner, M. A. (1992): Proc. of the SPIE, 1743, 60

Sanders, W. T., Edgar, R. J., & Liedahl, D. A. (1996): in Röntgenstrahlung from the Universe, eds. H. U. Zimmermann, J. E. Trümper, & H. Yorke, MPE Report 263, 339

GRADES: A New Instrument for Spectroscopy of the Local Hot Gas

M. Hurwitz[1], C. McKee[1], J. Edelstein[1], J. Vallerga[1], P. Jelinsky[1], M. Freyberg[2], and D. Breitschwerdt[2]

[1] Space Sciences Laboratory, University of California, Berkeley, California 94720-7450, U.S.A.
[2] Max–Planck–Institut für extraterrestrische Physik, D-85740 Garching, Germany

Abstract. Competing models for the origin of the local component of the diffuse X-ray background span more than an order of magnitude in temperature. We propose to test these models with a new instrument, the GRating Array Diffuse EUV Spectrometer *(GRADES)*, which is sensitive to diffuse emission in the 90 - 260 Å bandpass, and offers a peak spectral resolution of about $\lambda/127$.

1 Introduction

Million-degree gas is often referred to as "X-ray gas". This misnomer belies the uncomfortable reality that extrapolation of X-ray observational limits to the total plasma luminosity is critically dependent on the details of cooling processes that are not well understood. In collisional ionization equilibrium (CIE) near 10^6 K, the total plasma luminosity divided by the luminosity at wavelengths below 100 Å is a factor of 5. Million-degree gas is extreme ultraviolet or "EUV gas." Even after foreground absorption from the local cloud at about 2×10^{18} cm^{-2} (Vallerga et al. 1993), the most energetically favorable band to study a CIE plasma at 10^6 K is near 200 Å. The band contains strong emission lines from various ionization stages of iron; if the gas is in CIE, their ratios determine the temperature independent of abundance uncertainties. The strength of the Fe lines relative to those of low-depletion elements can determine the relative CIE abundances. If underionized nonequilibrium prevails (as in conduction [Slavin 1989]), the EUV may carry a higher fraction of the radiated power; line ratios can distinguish underionized emission from the CIE case. Because the absolute strength of the collisionally excited EUV emission lines depend sensitively on the electron temperature, uncharacteristically low values (e.g. overionized nonequilibrium) dramatically alter the EUV spectrum (Breitschwerdt and Schmutzler 1994). For reference: with a mean emission measure of 0.0038 cm^{-6} pc (Snowden et al. 1997), and a temperature of 10^6 K, the strongest CIE emission line is expected to have a flux of about 120 photons cm^{-2} s^{-1} sr^{-1} (hereafter Line Units or LU).

Observational limits have been set at the 1000 LU and 500 LU level (Labov and Bowyer 1991 and Smith et al. 1996, resp.). A deep spectrum taken with *EUVE*, the Extreme Ultraviolet Explorer (Bowyer 1995), suggests that the emission measure near 10^6 K may be a factor of about four below the value from Snowden et al. (Jelinsky et al. 1993). This result, however, relies on subtraction of a very large background signal. Also using *EUVE*, Vallerga et al. find a few marginally significant detections in directions where the 1/4 keV emission is bright. Spectrographs for diffuse emission between 350 and 1100 Å have recently been developed (Bowyer et al. 1997). Observations have not begun (as of 1 June 1997). If the foreground neutral absorption is near or above 2×10^{18} cm^{-2}, the brightest CIE plasma emission features in the 350 – 912 Å band will fall below the instrument sensitivity.

2 The Proposed *GRADES* Instrument

Consider first a single slit plus cylindrical diffraction grating placed at grazing incidence with respect to the central ray (hereafter "channel"). Baffles restrict the field of view (FOV) to a cone about 10° in diameter. Diffuse sky glow within the FOV will strike the grating and be focused within the plane of dispersion onto a planar photon-counting microchannel plate detector. We vary the spacing of the grooves to provide aberration control and flatten the focal surface (Harada et al. 1991). *GRADES* comprises 16 spectrograph channels, illustrated within the payload envelope of the Spartan Lite (a potential spacecraft carrier) in Figure 1. The off-axis channels are identical to the central one save for a rotation about an axis passing through the detector surface. Outside the entrance slit, each off-axis channel carries a small pick-off mirror to coalign the FOV with that of the central channel within the dispersion plane. Perpendicular to that plane, the FOV is about 20°. *GRADES* provides no angular resolution within its FOV. The integrated flux from point sources is far below the expected diffuse flux in this band. The continuum from the brightest white dwarf in the sky (HZ 43) is comparable to the detector background. The well known flare star AU Mic in outburst (Katsova et al. 1996) produces lines corresponding to < 0.5 LU in diffuse flux when observed with *GRADES*. We summarize the top-level parameters in Table 1.

Without filters, scattering of bright emission lines of geocoronal and interplanetary origin (especially Lyman α, with a typical flux of 3500 Rayleighs [Chakrabarti et al. 1984]) would exceed the detector background. Page limits do not permit a full discussion of our rationale for filter material selection. The filter for 210 – 260 Åwill be 1500 Å thick Aluminum; for 90 – 150 Å, the filter will be 1000 Å thick Lexan plus 300 Å of Boron. Intermediate wavelengths will be spanned by both materials, each covering half of the detector area. We show the system resolution and 2σ sensitivity curve for an integration time of 26,000 seconds (1 day at 30% observing efficiency) in Figure 2.

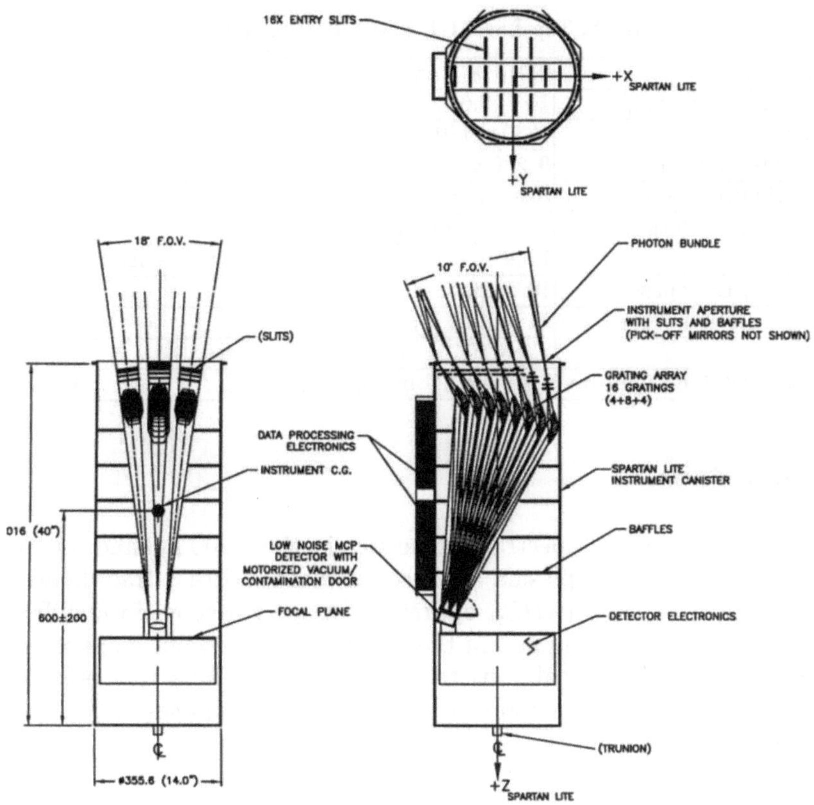

Fig. 1. Full 16-channel *GRADES* illustrated in Spartan Lite envelope.

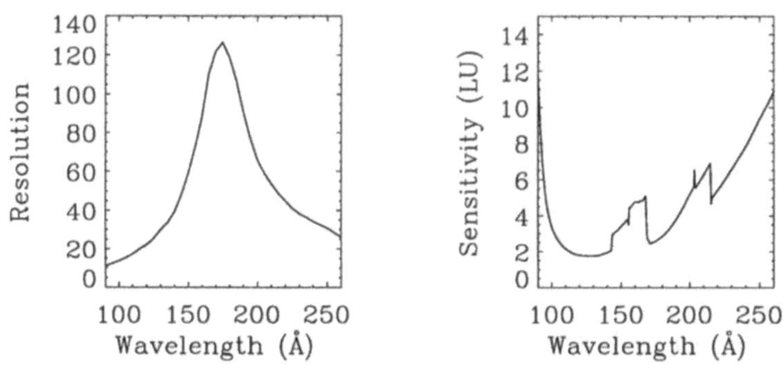

Fig. 2. *GRADES* spectral resolution $\lambda/\delta\lambda$ (76% energy width) and 2σ sensitivity to emission lines (LU) for 1 day in low earth orbit at 30% observing efficiency. Discontinuities are associated with changes in instrument throughput caused by filter support bars or filter transmission edges.

Table 1. Top-Level GRADES Parameters

Entrance slits	0.013 × 6 cm
Grating graze, blaze angle	14 °, 2.6 °
Groove density	1330 – 1800 mm^{-1}
Grating dimensions, R_{curve}	9 × 6 cm, 96 cm
Coatings, Photocathode	Rhodium, NaBr
Detector area, resolution	5.5 × 6 cm, 400 × 400 resels
Filters	1000 Å Lex. + 300 Å Boron, 1500 Å Alum.
Field of View, Bandpass	10 × 20 °, 90 – 260 Å
Spectral Resolution (76% EW)	11 – 127 (λ / $\delta\lambda$)
$A_{eff} \times \Omega$	$2.5 \times 10^{-4} - 1.6 \times 10^{-3}$ cm^2 ster
2 σ Min. Detectable Line Flux	1.8 – 12 LU in 26,000 sec

3 Conclusions

Despite the importance of spectral diagnostics in the ~200 Å band, observations of the local diffuse plasma at these wavelengths are extraordinarily scant. A new instrument *GRADES*, optimized for spectroscopy of the diffuse emission from 90 – 260 Å, will permit the first true emission line diagnostics. Data from *GRADES* will enable us to constrain the plasma temperature, emission measure, gas phase abundance of iron, and two classes of nonequilibrium effects independently.

References

Bowyer, S. (1995): in "X-ray and EUV/FUV Spectroscopy and Polarimetry", ed. S. Fineschi, *Proc. SPIE* 2517, 97
Bowyer, S., Edelstein, J., Lampton, M. (1997): accepted for *ApJ*
Breitschwerdt, D., Schmutzler, T. (1994): *Nature* 371, 774
Chakrabarti, S., Kimble, R., Bowyer, S. (1984): *JGR* 89, No. A7, 5660
Harada, T., Kita, T., Bowyer, S., Hurwitz, M. (1991): in "International Conference on the Application and Theory of Periodic Structures", ed. J. M. Lerner & W. R. McKinney, *Proc. SPIE* 1545, 2
Jelinsky, P., Vallerga, J., Edelstein, J. (1993): *ApJ Lett* 417, L41
Katsova, M., Drake, J., Livshits, M. (1996): in "Astrophysics in the EUV", ed. S. Bowyer and R. F. Malina, Kluwer Academic, Dordrecht, 175
Labov, S., Bowyer, S. (1991): *ApJ* 371, 810
Slavin, J. (1989): *ApJ* 346, 718
Smith, B., Pfafman, T., Bloch, J., Edwards, B. (1996): in "Astrophysics in the EUV", ed. S. Bowyer and R. F. Malina, Kluwer Academic, Dordrecht, 283
Snowden, S., Egger, R., Finkbeiner, D., Freyberg, M., Plucinsky, P. (1997): submitted to *ApJ*
Vallerga, J., Vedder, P., Craig, N., Welsh, B. (1993): *ApJ* 411, 729

Planned Observations of Hot Gas in the LISM

R. McLean, J.C. Green, and K.S. Gunderson

Center for Astrophysics and Space Astronomy, University of Colorado, Boulder, U.S.A.

Abstract. We plan to observe hot gas in the local interstellar medium (LISM) at high resolution using a rocket-borne spectrograph that can simultaneously observe OVI, NV and CIV at resolutions of $1 - 2\,\mathrm{km\,s^{-1}}$. Data from a single flight will have sufficient signal to noise ratio to detect a broad $50\,\mathrm{km\,s^{-1}}$ component down to a column density of $N \geq 0.5 \times 10^{13}\,\mathrm{cm^{-2}}$. High-velocity components as fast as $300\,\mathrm{km\,s^{-1}}$ can be detected. New high resolution observations of hot gas absorption line profiles will probe the kinematics of the local bubble, and comparison between the carbon, nitrogen and oxygen profiles will allow discrimination between the various production models.

1 Background

High stage ions trace hot gas in the interstellar medium. Measurements of OVI absorption are primarily from observations by Copernicus along 72 sight lines (Jenkins (1978)). The original analysis of these data found that OVI was clumped in columns of $10^{13}\,\mathrm{cm^{-2}}$ with an average separation of 165 pc. However, this analysis did not allow for a common component along all sight lines. Adding a local component of $1.6 \times 10^{13}\,\mathrm{cm^{-2}}$ with an rms velocity of $12\,\mathrm{km\,s^{-1}}$ improves the fit to the data (Shelton and Cox (1994)). Detailed profile analysis of this component is not possible with the $14\,\mathrm{km\,s^{-1}}$ resolution of Copernicus.

By adding in data from other high stage ions, it is possible to constrain the production mechanism for the hot gas. Production via radiative cooling, conductive heating and turbulent mixing layers can be differentiated by comparing the ratio of the column densities of pairs of ions (Shelton (1996)). Also possible is a mechanism where high stage ions are frozen in at cooler temperatures (Breitschwerdt and Schmutzler (1994)). Higher resolution data that make a direct comparison between OVI, NV and CIV profiles can further constrain models.

2 Objectives

We want to look at the kinematics of the hot gas in the local bubble by making simultaneous observations of OVI, NV and CIV in absorption at resolutions of $1 - 2\,\mathrm{km\,s^{-1}}$ with a signal to noise ratio of 25. In addition to

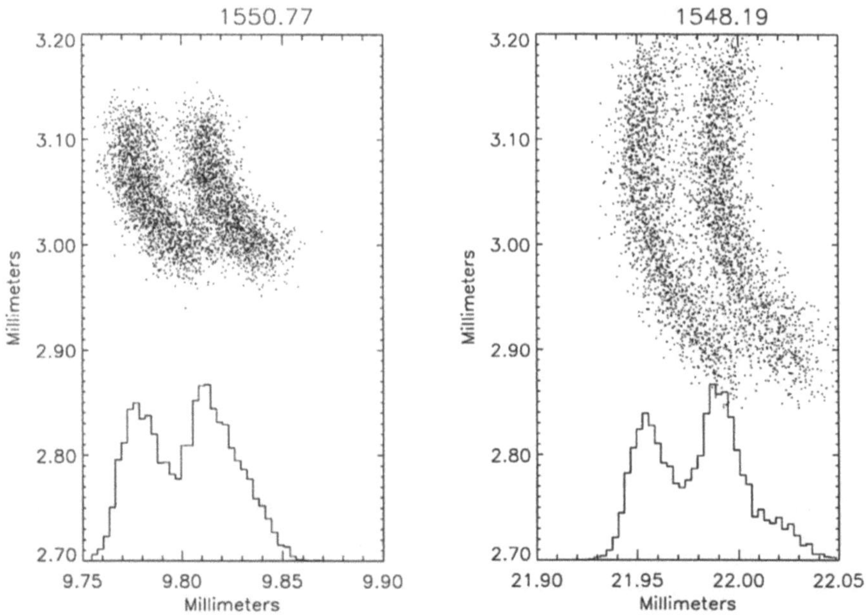

Fig. 1. Raytraces of the C IV doublet as seen by the HotCONES instrument. The two spots at each wavelength are separated by a resolution of 200,000. The histograms show line profiles in the x dimension only.

profile fitting, we will be able to detect a broad $50\,\mathrm{km\,s^{-1}}$ component with 3σ uncertainty down to a limiting column density of $0.5 \times 10^{13}\,\mathrm{cm^{-2}}$ as well as detect components as fast as $300\,\mathrm{km\,s^{-1}}$.

3 Instrument

The Hot Carbon, Oxygen, Nitrogen Echelle Spectrograph (HotCONES) uses a low order echelle as an objective grating with a cross-dispersing and focusing parabolic secondary to take simultaneous spectra at the widely separated transitions of CIV, OVI and NV. The low order echelle is ruled at a line density of 2,880 lines per millimeter with a blaze angle 63 degrees. Ruled in this way, fourth, fifth, and sixth orders have the high dispersion of an echelle without having to work with many overlapping orders. This concentrates the limited bandwidth of the detector around the high stage ion absorption lines without covering all of the intermediate wavelengths.

Raytraces of each of the transitions are shown at a separation of λ / $\delta\lambda$ of 200,000. The spectrograph will use a 25 micron resolution microchannel plate detector. The blur of the detector has been added to the raytraces shown.

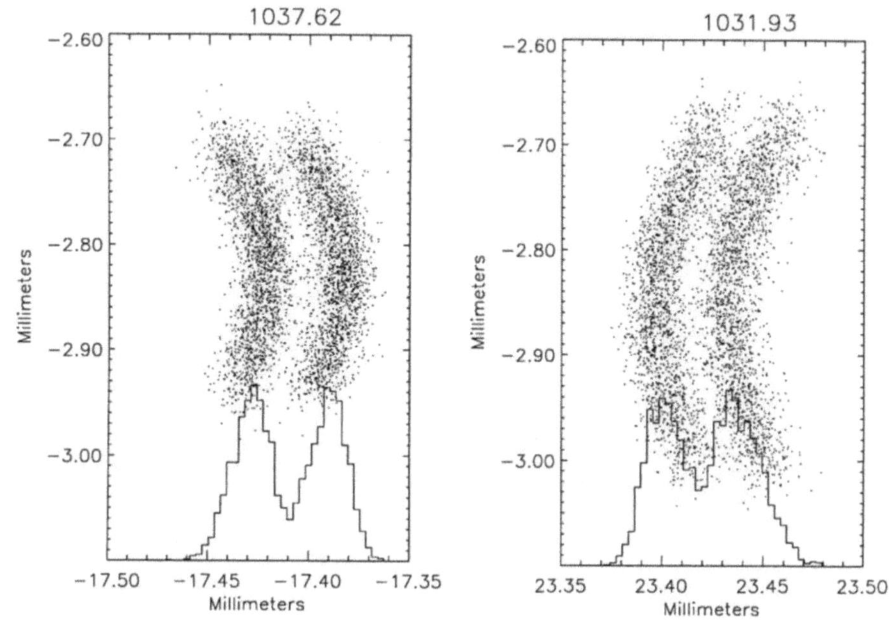

Fig. 2. Raytraces of the O VI doublet as seen by the HotCONES instrument. The two spots at each wavelength are separated by a resolution of 200,000. The histograms show line profiles in the x dimension only.

As a conservative estimate, the HotCONES experiment should have an effective area of at least $1.5\,\mathrm{cm}^2$. A sketch of the spectrograph packaged into a 22 inch diameter sounding rocket is shown in Fig. 4. In the roughly 300 seconds available from this sub-orbital platform, HotCONES can only observe bright stars to get a signal to noise ratio of 25. In the first two flights we plan to observe α Vir and δ Ori.

References

Breitschwerdt D., & Schmutzler T. (1994): Nature 371, 774
Jenkins E.B. (1978): ApJ 219, 845
Shelton R.L., & Cox D.P. (1994): ApJ 434, 599
Spitzer, L. (1996) ApJ 458, L29

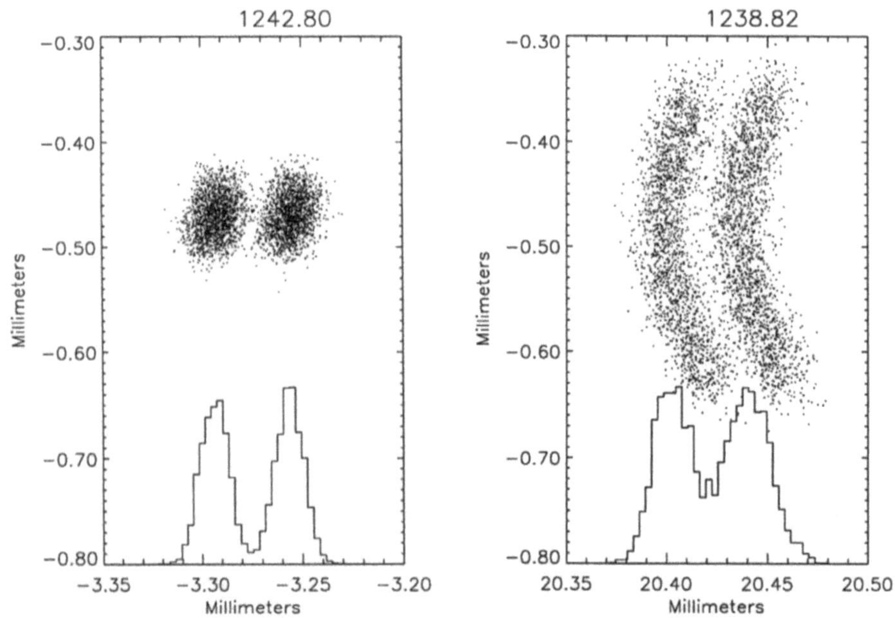

Fig. 3. Raytraces of the N V doublet as seen by the HotCONES instrument. The two spots at each wavelength are separated by a resolution of 200,000. The histograms show line profiles in the x dimension only.

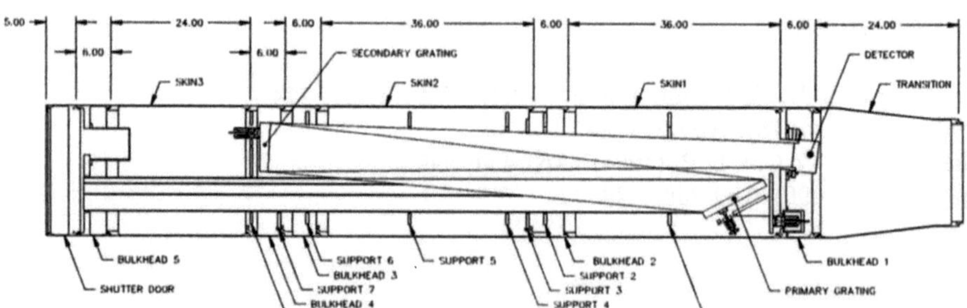

Fig. 4. Section view of the HotCONES instrument.

Interstellar C IV and Si IV in the Direction of Loop I

Rubens Freire Ferrero

Observatoire Astronomique, 11 rue de l' Université, 67000 - Strasbourg, France

Abstract. Nearby late-B and early-A stars show clear IS absorptions in Mg II, C II and other lines (IUE spectra). Some of them, in the Sco Cen direction, also show absorption features at C IV and Si IV line wavelengths. We discuss the possible formation of these lines in an intermediate temperature gas region between the dense and cold H I wall and the hot ISM, a region that could be the collision interaction between the Local Bubble and the Loop I supershell. Some individual stellar spectra are presented and discussed.

1 The Observations

A sample of late-B and early-A stars in the direction of the Loop I were observed at high resolution by IUE some years ago (Freire Ferrero 1988). The stars of the sample are not known as been shell, Ae/Be or peculiar ones, and their spectral type is suitable to search for IS high ionized species because they have neither chromosphere nor corona (Freire Ferrero 1986 and references therein), and have in general high rotational velocities allowing to distinguish between broad stellar lines (enlarged by rotation) and sharp IS lines. Nevertheless, a very careful analysis of the UV stellar spectra must be done because metallic lines produce strong stellar spectral blends at the C IV and Si IV wavelengths (Kurucz 1974, Hubený et al. 1986, Freire Ferrero 1986). Typical spectra in the C IV and Si IV spectral regions are shown in Fig. 1 for HD 119921 and HD 119361 : we see large spectral features due to photospheric metallic blends and a very narrow component displaced a little from the line wavelenth of C IV and Si IV doublets at the stellar reference frame. The displacements for both lines of the doublets are more or less the same. As we can see, the displayed spectra for both stars are practically the same but the stars are at different distances from the Sun.

From the measures of the residual absorption we can estimate column densities for C IV and Si IV species :

$N(\text{C IV}) \approx 1.3\ 10^{13} \pm 0.3\ 10^{13}\ \text{cm}^{-2}$ or $\log N(\text{C IV}) \approx 13.1 \pm 0.1$

$N(\text{Si IV}) \approx 7.5\ 10^{12} \pm 2.5\ 10^{12}\ \text{cm}^{-2}$ or $\log N(\text{Si IV}) \approx 12.85 \pm 0.15$

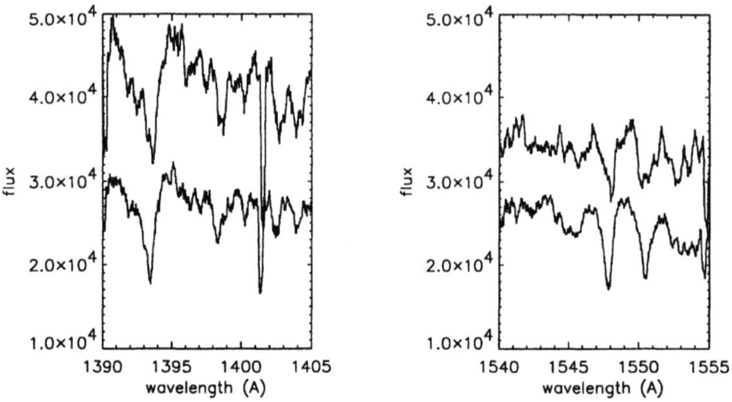

Fig. 1. Spectral similarities of HD 119921 and HD 119361

2 Discussion and Conclusions

Classical interpretation of similar spectra (for exemple from Be stars, Bruhweiler et al. 1989) attribute these absorption features to shells around these single stars. Nevertheless, it becomes difficult to support that shells give spectral signatures only in UV Si IV and C IV and not in the visible (H α emission or absorption, metallic lines; Jaschek et al. 1988).

Indeed, the fact to consider the observed asymmetric features as entirely due to C IV and Si IV lines can mislead us because they can be interpreted as produced by stellar winds, which could be used as an argument for the presence of some kind of circumstellar (CS) matter around the stars. In fact, the broad features are simply photospheric metallic line blends (without any or very little C IV and Si IV contributions) as we mentioned before, and only the residual narrow absorption can have a non-stellar origin.

Deriving abundances from the broad features assuming that are only due to C IV and Si IV leads evidently to overabundances relative to normal IS abundances. Instead, values from the residual narrow absorptions lead to reasonable values.

Some differences were observed for some stars, between two IUE spectra taken at different epochs. but superposition of these spectra show clearly that the variations are of the same order as the IUE noise. Variability is sometimes interpreted as a qualitative characteristic of shell spectra even if now it could also be admitted for IS absorptions (Frisch 1995).

Even though IS and CS lines could be difficult to disentangle, the above discussion favors the IS interpretation for the narrow C IV and Si IV features observed. In addition, plotting the observed stars in galactic coordinates and taking into account the best estimated parallaxes, we discover that the observed IS absorptions are not randomly distributed. In fact (Fig. 2) IS absorptions are concentrated into the direction of the Loop I structure. Recent

Table 1. Data on observed stars

.... Star	.. l_{II}	.... b_{II}	d (pc) ..	... Sp T ...
HD 109573	299.73	+22.91	118.	A 0 V
HD 110073	300.53	+22.83	202.	B 8 pII/III
HD 111597	302.75	+28.87	55.6	A 0 IV
HD 111774	303.04	+23.19	201.	B 7/B 8 V
HD 113852	306.46	+26.91	151.	A 0 III/IV
HD 115892	309.42	+25.79	16.8	A 2 V
HD 119361	313.20	+19.76	292.	B 8 III
HD 119921	315.28	+25.29	58.	A 0 V
HD 121847	321.66	+35.45	126.	B 8 V sh
HD 125248	330.06	+39.52	120.	A 0 pCrEu
HD 125473	321.73	+21.71	67.	A 0 IV
HD 125745	323.33	+24.46	152.	B 8 V
HD 142165	347.51	+22.15	196.	B 6 IV n
HD 147152	334.01	+0.13	220.	B 6 IV
HD 149485	327.24	-9.37	233.	B 7 V n
HD 152614	28.73	+30.66	36.	B 8 V
HD 209952	350.00	-52.47	28.	B 7 IV

work on the IS medium (Frisch 1995) confirms that the Sun is embedded into a low H I density medium and that there is a neutral gas wall at a distance of 40 ± 25 pc in the direction of Loop I (N_H goes from $\approx 10^{18}$ cm^{-2} to $\approx 10^{21}$ cm^{-2}; Cox and Reynolds 1987, Centurión and Vladilo 1991, Egger and Aschenbach 1996).

The nearby star showing C IV and Si IV IS absorption is HD 119921 at a distance corresponding to the presence of the H I wall (Centurión and Vladilo 1991). The more natural interpretation is that these ions are the result of an evaporative and conductive interface between the cold (H I wall) and the hot ISM (McKee and Ostriker 1977) or gas swept up by radiative shocks from SN at the origin of Loop I and interacting with the Local Bubble (Egger and Aschenbach 1996). This second explanation is supported by the observed X-ray shadow (0.1 - 2.0 keV) by ROSAT, corresponding to an annular enhanced H I feature attributed to the collision between the Local Bubble and the Loop I supershell (Egger and Aschenbach 1996). Colliding very hot gas forms dense colder structures than the original gas, and probably intermediate temperature transition regions develop also between the coldest and the hottest ones. The IS C IV and Si IV can be produced at T $\approx$ 5. 10^4 K in this intermediate temperature zones. We will complete this work with the new parallax data from HIPPARCOS and the reviewed IUE observations from the IUE Final Archive, when delivered.

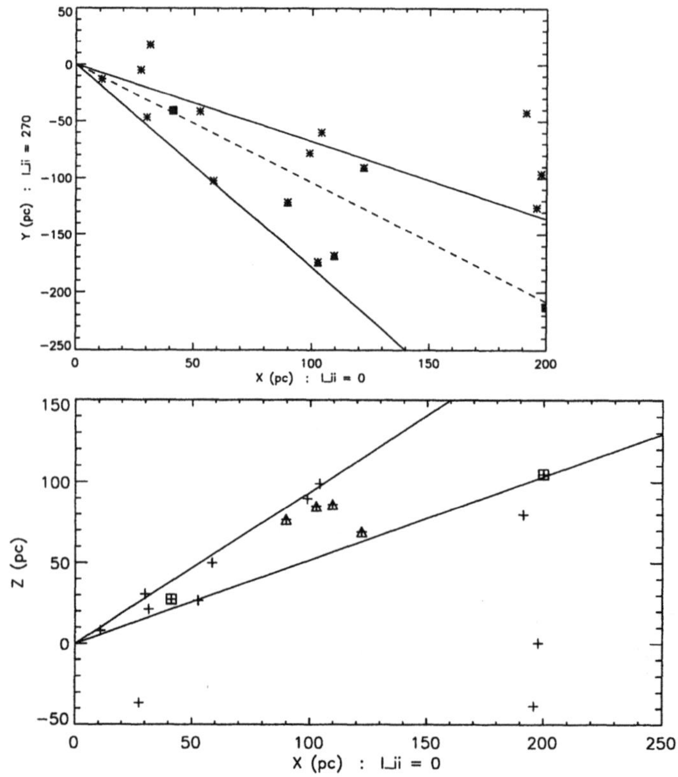

Fig. 2. The stars observed are displayed in galactic coordinates X,Y,Z. The squares are HD 119361 and HD 119921 showing the higher IS absorption and the triangles a mid absorption. The other stars do not show this IS absorption from IUE spectra.

References

Bruhweiler, F. C., Grady, C. A., Chiu, W. A. (1989): ApJ **340**, 1038–1048

Cox, D. P., Reynolds, R. J. (1987): Ann Rev Astron Astrophys **25**, 303–344

Centurión, M., Vladilo G. (1991): ApJ **372**, 494–592

Egger, R., Aschenbach, B. (1996): *Röntgenstrahlung from the Universe*, H. U. Zimmermann, J. E. Trümper and H. Yorke *eds.*, Garching, Germany, MPE Report 263, 249–250

Freire Ferrero, R. (1986): A&A **159**, 209–216

Freire Ferrero, R. (1988): *A Decade of UV Astronomy with IUE, IUE Conf.* ESA SP-281, 227–229

Frisch, P. C. (1995): *Characteristics of Nearby IS Matter*, Sp Sc Rev **72**, 499–592

Hubený, I., Harmanec, P., Stefl, S. (1986): Bull. Astron. Inst. Czechosl. **37**, 370–382

Jaschek, M., Jaschek, C., Andrillat, Y. (1988): *A Survey of Ae and A-type Shell Stars in the Photographic Region* A&AS **72**, 505–513

Kurucz, R. L. (1974): ApJ Lett. **188**, L21–L22

McKee, C., Ostriker, J. P. (1977): ApJ **218**, 148–169

A New View of the LHB and $\frac{1}{4}$ keV X-ray Halo

S.L. Snowden

NASA Goddard Space Flight Center, Code 662, Greenbelt, MD 20771, USA

Abstract. The X-ray sky at $\frac{1}{4}$ keV is completely dominated by diffuse emission. It has become clear that it originates as at least three separate components: local emission within the nearest ~ 100 pc from the Sun, halo emission from beyond most of the neutral material of the Galactic disk, and the superposition of unresolved extragalactic sources. The only way to determine the temperatures and relative emission measures of the hot plasma responsible for the Galactic components is to use the X-ray intensity variations due to column density variations in the intervening H I to separate the components. "Shadowing" studies have been pursued for individual objects using *ROSAT* data from both pointed observations and the all-sky survey with considerable success.

This paper presents the results of an all-sky analysis of the $\frac{1}{4}$ keV background from the *ROSAT* survey. A Local Hot Bubble is found consistent with, although somewhat smaller than, previous models. It has a temperature of $10^{6.1}$ K and an emission measure which varies by a factor of ~ 3.3 over large angles. The halo emission has a temperature near $10^{6.0}$ K with an emission measure which varies from near zero to more than five times that of the local emission.

1 Introduction

The source of the $\frac{1}{4}$ keV soft X-ray diffuse background (SXRB) has proved elusive since its discovery in the late 1960s (e.g., Bowyer, Field, & Mack 1968). The general negative correlation between the SXRB and the column density of Galactic neutral hydrogen (dominated by a Galactic plane-to-pole variation) was strongly suggestive of an "absorption" model, where the angular structure of the SXRB is produced by the absorption of a flux of distant origin (e.g., Marshall & Clark 1984). The absence of the expected detailed negative correlation (some specific H I features showed little indication of shadowing, e.g., Burrows et al. 1984), the shallowness of the apparent absorption, and the weakness of the energy dependence of the apparent absorption encouraged the investigation of other possibilities.

By the end of the 1980s, the displacement model (e.g., Sanders et al. 1977; Snowden et al. 1990, hereafter SCMS) proved to be the most consistent with the available data. The displacement model suggests that the irregularly shaped cavity which surrounds the Sun (known independently from interstellar absorption line studies: e.g., Frisch & York 1983, and N_{H} studies: e.g., Knapp 1975) contains an X-ray-emitting plasma, and that the cavity is extended more toward the Galactic poles than within the Galactic plane. The greater the extent of the plasma-filled cavity in a given direction, the greater the emission measure of the plasma and the lower the measured column density of H I (at constant Galactic latitudes), which produces the general negative correlation between the two. The displacement model provided a natural explanation for the weakness and energy independence of the apparent absorption and allowed for H I features which had no apparent effect on the X-ray flux.

With the launch of the *Röntgensatellit* (*ROSAT*, Trümper 1983), the situation became more complicated. One of the first unique discoveries of *ROSAT* was the observation of deep shadows in the general SXRB cast by the Draco Nebula ($l, b \sim 90°, 40°$, Burrows & Mendenhall 1991; Snowden et al. 1991). These observations unequivocally proved that there is significant $\frac{1}{4}$ keV X-ray emission in at least the lower Galactic halo: roughly 50% of the *observed* flux in that direction originates $\gtrsim 300$ pc from the Sun, $\gtrsim 200$ pc above the Galactic plane. Additional studies using both pointed observations and all-sky survey data demonstrated that the shadows were not unique to Draco but common features of the $\frac{1}{4}$ keV SXRB.

2 Data and Analysis

X-ray Data: This analysis uses the R1 and R2 band data (PSPC PI channels 8 – 19 and 20 – 41, respectively; Snowden et al. 1994) from the *ROSAT* all-sky survey (Snowden & Schmitt 1990) as presented in Snowden et al. (1997). The $\frac{1}{4}$ keV band is appropriate for the study of the LHB, as the LHB has a temperature near 10^6 K. Thermal equilibrium spectra near 10^6 K produce only $\sim 2\%$ as many counts in the $\frac{3}{4}$ keV band as in $\frac{1}{4}$ keV band. The results of this analysis show that the temperature of much of the halo emission is near 10^6 K as well.

Measure of Galactic Absorption: Corrected *IRAS* 100 μm maps (Schlegel, Finkbeiner, & Davis 1997 and Finkbeiner, this volume) are used for Galactic absorption. The low spatial frequencies in the *IRAS* 100 μm data have been replaced by those from the well-calibrated *DIRBE* data. The *DIRBE* 100 μm, 140 μm, and 240 μm data have been used to determine temperature variations and to correct the 100 μm map to a constant reference temperature. In addition, the Leiden-Dwingeloo 21-cm survey data (Hartmann & Burton, 1997) in the velocity range from -74 km s^{-1} to 24 km s^{-1} in relatively low column density directions were used to determine a global normalization. Finally, the data were scaled to a column density of hydrogen.

In order to determine the foreground and distant (relative to the Galactic absorbing gas) contributions to the observed $\frac{1}{4}$ keV SXRB, the negative correlation (shadowing) between the $\frac{1}{4}$ keV background and the column density of absorbing material along the line of sight is examined. The sky is divided into $4.°8 \times 4.°8$ regions with $24' \times 24'$ pixels, and the negative correlation between X-ray intensity and H I column density as measured by the corrected

IRAS 100 μm data is fit for each band and region with a two-component absorption model. The spacing of the regions is $3\overset{\circ}{.}2 \times 3\overset{\circ}{.}2$, providing some angular overlap. The size of the region was minimized to sample as fine an angular structure as feasible and to reduce the effect of any intrinsic larger-scale variation. The region could not be made smaller without losing the needed dynamic range in the absorbing column density and affecting the robustness of the results. The overlap simply smooths the results.

The supposition is that on small angular scales no reasonable geometry will give a significant negative correlation due to displacement of H I by X-ray emitting gas, as clearly happens on large angular scales (e.g., SCMS). This is the critical difference between this work and previous studies. With the high angular resolution and statistical significance provided by the *ROSAT* data, the negative correlation of absorption on small angular scales can be distinguished from the presumed smoother, larger-scale variations of displacement.

With the X-ray intensity and effective $N_{\rm H}$ for each $24' \times 24'$ pixel, a two-component absorption model (foreground plus absorbed halo emission) plus a fixed extragalactic contribution, which is also absorbed,

$$I_X = I_0 + I_1 \times \exp[-\sigma(N_{\rm H}, T) \times N_{\rm H}] + I_{eg} \times \exp[-\sigma(N_{\rm H}, \alpha) \times N_{\rm H}],$$

was fit to all pixels in each $4\overset{\circ}{.}8 \times 4\overset{\circ}{.}8$ region independently for the R1 and R2 bands. I_X is the observed X-ray intensity, I_0 is the fitted foreground component, I_1 is the fitted halo component which is absorbed by the column density $N_{\rm H}$, $\sigma(N_{\rm H}, T)$ is the theoretical band-averaged absorption cross section (Morrison & McCammon 1983) which is a function both of $N_{\rm H}$ and the temperature, T, of the thermal equilibrium emission model (Raymond & Smith 1977; Raymond 1991, see Snowden et al. 1994), I_{eg} is the fixed extragalactic contribution, and $\sigma(N_{\rm H}, \alpha)$ is the theoretical band-averaged absorption cross section which is a function both of $N_{\rm H}$ and power-law index, α. Temperature and emission measure results quoted here use the above thermal equilibrium emission models with solar abundances. The data were refit after the initial fitting produced an estimate of the halo temperature in order to use more appropriate absorption cross sections (a relatively insignificant effect).

I_{eg} was fixed using the results of recent observations of shadowing by extragalactic objects (a R1+R2 band flux of $\sim 400 \times 10^{-6}$ counts s^{-1} arcmin^{-2}, e.g., M101, Snowden & Pietsch 1995) and the extragalactic spectrum determined by Hasinger et al. (1993), $I(E) \sim E^{-1.96}$. The R2/R1 band ratio of 0.96 for the power law spectrum places 204×10^{-6} counts s^{-1} arcmin^{-2} in the R1 band and 196×10^{-6} counts s^{-1} arcmin^{-2} in the R2 band.

The existence of any additional emission in the region not considered by the model (e.g., if the region contains an SNR or cluster of galaxies) will increase the χ^2 statistic of the fit and affect the fitted parameters. Therefore, regions affected by a limited number of bright, small-scale extended sources such as the Cygnus Loop and Vela SNRs and the Virgo and Coma clusters of galaxies have been excluded. The existence of any clouds of sufficient column density (e.g., $> 10^{19}$ cm^{-2}) floating in the LHB will also affect the results. However, since there is no indication of a large number of such clouds, this will only be to a limited extent and not affect the general analysis.

2.1 Foreground (LHB) Emission

Fitted and smoothed values for the foreground intensity (Fig. 1a) span a factor of ~ 3.3 from ~ 250 to $\sim 820 \times 10^{-6}$ counts s^{-1} arcmin^{-2}, implying emission measures of $0.0018 - 0.0058$ cm^{-6} pc. The general appearance is similar to that of the total background, with most of the same major features present but at only about half of the contrast: higher Galactic latitudes tend to have higher intensities and the lower intensities are near the Galactic plane. The largest region of highest intensity in the south is in the Galactic center hemisphere while in the north (other than Loop I) the major enhancements are mostly in the range $90° < l < 210°$.

Figure 2a shows a histogram of the R2/R1 band ratio (data from the atypical Loop I, Monogem, and Eridanus regions are excluded). The distribution is well peaked at a value of ~ 1.12 with half-maximum values of 1.02 and 1.23. These ratios can be interpreted as model emission temperatures of $10^{6.07}$ K, $10^{6.02}$ K, and $10^{6.12}$ K, respectively, assuming no foreground absorption. The R2/R1 band ratio over the sky for the LHB emission is relatively consistent with a single-temperature plasma.

2.2 Distant (Halo and Extragalactic) Emission

Assuming that the foreground emission, I_0, has been accurately determined, it is a simple matter to determine the magnitude and effective temperature of the distant component, $I_{dist} = I_{eg} + I_1$. Instead of using the fitted values for I_1, which have errors which are highly correlated with the errors of I_0 but are greater in magnitude, the smoothed I_0 map is used to evaluate the inverted equation:

$$I_{dist} = I_{eg} + I_1 = I_{eg} + (I_X - I_0 - I_{eg} \exp[-\sigma(N_H, \alpha) \times N_H])/$$
$$\exp[-\sigma(N_H, T) \times N_H].$$

In order to examine regions where the distant flux can be reasonably evaluated, only parts of the sky where $N_H < 4.5 \times 10^{20}$ cm^{-2} were used. Figures 1b,c shows the results.

The distribution of the fitted R2/R1 band ratio in the polar regions differs slightly between north and south (see Fig. 2b, again excluding the regions contaminated by the Loop I, Monogem, and Eridanus enhancements). While both peak at R2/R1 ~ 1.01, the northern distribution is narrower than that in the south. Combining the two distributions, the half maxima are at 0.86 and 1.17 (temperatures of $10^{6.02}$ K, $10^{5.94}$ K, and $10^{6.10}$ K, respectively). The width of the curves and the fact that the extreme high-ratio wings lie outside

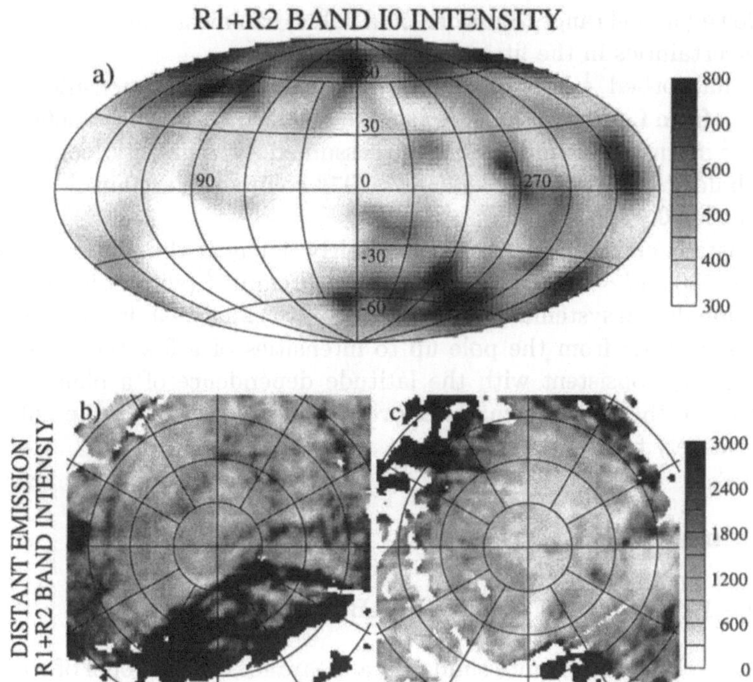

Fig. 1. a) Aitoff-Hammer projection in zero-centered Galactic coordinates of the I_0 fitted and smoothed intensity. b) North Galactic polar projection ($l = 0°$ is down, l increases clockwise) of the fitted and smoothed I_1 intensity. c) South Galactic polar projection ($l = 0°$ is up, l increases counterclockwise) of the fitted and smoothed I_1 intensity. Units are 10^{-6} counts s^{-1} $arcmin^{-2}$.

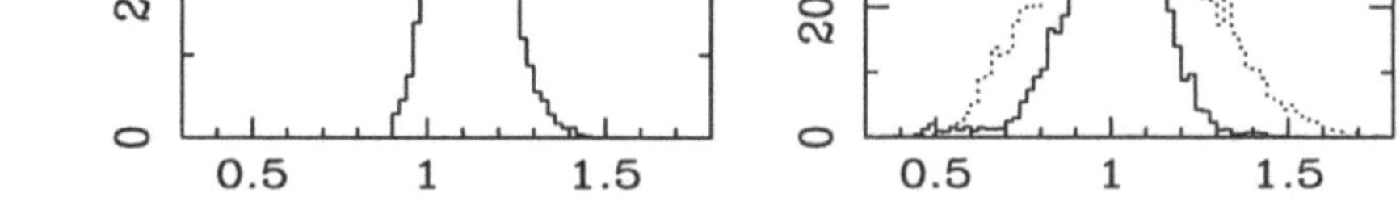

Fig. 2. Histograms of R2/R1 band ratios. a) Fitted foreground component, I_0. Typical uncertainties in the ratio are $\sim 0.05 - 0.1$. b) Fitted distant component, $I_{eg} + I_1$. typical uncertainties in the ratio are $\sim 0.1 - 0.2$. The solid line in b) shows the northern-hemisphere data while the dashed line shows the southern-hemisphere data. The vertical scaling is arbitrary as the plots oversample the data.

of the allowed model range (Snowden et al. 1994) can most likely be attributed to the uncertainties in the fitting procedure.

The unabsorbed $\frac{1}{4}$ keV band halo plus extragalactic intensities in the north away from Loop I have values ranging between ~ 400 and $\sim 3000 \times 10^{-6}$ counts s^{-1} $arcmin^{-2}$. Subtracting our assumed extragalactic contribution leaves a halo intensity of $\sim 0 - 2600 \times 10^{-6}$ counts s^{-1} $arcmin^{-2}$, emission measures of $\sim 0 - 0.017$ cm^{-6} pc.

The distant emission in the south is fairly flat near the pole with values between ~ 500 and $\sim 1200 \times 10^{-6}$ counts s^{-1} $arcmin^{-2}$ (with at least some of the variation due to systematic uncertainties in the analysis). It increases with decreasing latitude from the pole up to intensities of a few thousand. This is qualitatively consistent with the latitude dependence of a plane-parallel distribution of the emitting material over the visible region of the halo, but is quantitatively too great an increase.

3 Discussion

3.1 The LHB and the $\frac{1}{4}$ keV SXRB

Although the LHB looks quite similar in appearance to the model of a decade ago (e.g., Cox & Reynolds 1987; SCMS), this similarity is in spite of the revolution in our understanding of the $\frac{1}{4}$ keV SXRB brought about by observations of significant diffuse X-ray emission in the Galactic halo.

Following the LHB model, the I_0 intensity map can be interpreted as a map of emission from the LHB, which in turn can be interpreted as a map of the radial extent of the LHB. The MBM12 shadowing results of Snowden, McCammon, & Verter (1993) can be used to fix a distance scale of 0.156 pc (10^{-6} counts s^{-1} $arcmin^{-2})^{-1}$ to the radial extent of the LHB as a function of LHB emission. With this scale factor, the LHB extends from ~ 40 to ~ 130 pc depending on direction. Cuts of the LHB are shown in Figures 3. This is $\sim 30\%$ smaller than the less reliable minimum scaling of SCMS.

It must be noted, however, that the extent of the LHB is not necessarily the extent of the local cavity in the H I of the Galactic disk. In some directions, most notably in the direction of β CMa ($l, b = 226°, -14°, d = 203$ pc) in the third quadrant, the cavity extends $\gtrsim 200$ pc (e.g., Frisch & York 1983; Welsh et al. 1994) while the LHB, using the above scaling of the I_0 X-ray intensity, extends only to $\sim 60 - 95$ pc. The results of Gry, York, & Vidal-Madjar (1985) indicate that HII gas fills in at least some of the otherwise vacant line-of-sight. To further investigate this issue, the NaI ISM absorption line data of Welsh et al. (1994) have been used to compare the size of the cavity with the extent of the LHB. Figure 4 shows an Aitoff-Hammer projection of the positions of the Welsh et al. (1994) stars with different symbols depending on whether the data and model agree (plus signs, 54 stars, the stars have either 1) low column densities and distances and I_0 intensities which place

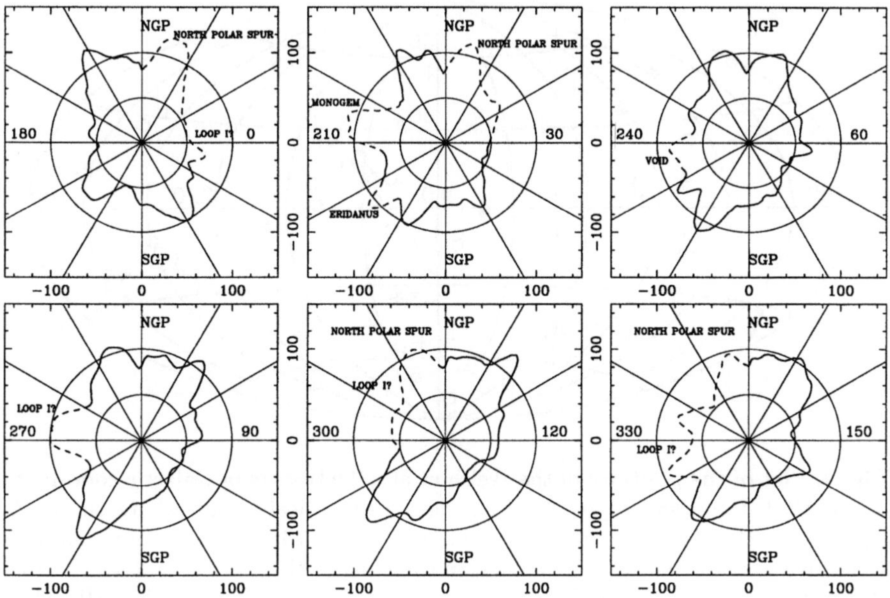

Fig. 3. Plots of the LHB radius in great-circle cuts through the Galactic poles along the labeled longitudes. The radii are derived from the fitted I_0 values: $R_{LHB} = 0.156 \times I_0$. The scale circles have radii of 50 pc and 100 pc. Regions which are contaminated by discrete features or are of other interest are labeled and the curves dashed.

them inside of the LHB or 2) high column densities and distances and I_0 intensities which place them outside of the LHB) or disagree (the stars have either 3) low column densities and distances and I_0 intensities which place them outside of the LHB or 4) high column densities and distances and I_0 intensities which place them inside of the LHB). Except for a limited number of stars (14, that appear relatively randomly distributed on the sky) which have foreground column densities placing them outside the cavity but $\frac{1}{4}$ keV I_0 intensities and distances placing them within the LHB (filled circles), the agreement appears reasonable. There are, however, a significantly greater number of stars (38) that the data imply are within the cavity but outside the LHB (open circles). About two-thirds of the latter group are found in two clusters, one centered near $(l, b \sim 340°, 0°)$ and the other centered near the third-quadrant void $(l, b \sim 240°, -15°)$, implying that the LHB may fill most if not all of the H I cavity in most other directions.

3.2 $\frac{1}{4}$ keV Emission in the Galactic Halo

Except for a limited number of regions such as Draco, the North Polar Spur, and southern latitudes with $b > -45°$, the halo emission of $\frac{1}{4}$ keV X-rays

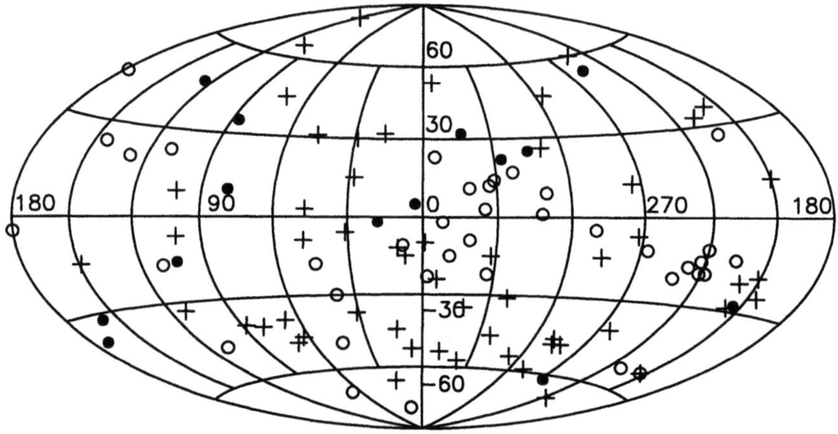

Fig. 4. Plot of the positions of the Welsh et al. (1994) stars in Galactic coordinates in an Aitoff-Hammer projection. Plus signs (+) indicate stars where the model agrees with the data, i.e., either the star has $< 5 \times 10^{18}$ cm^{-2} of foreground absorption and a distance which places it in the LHB or the star has $> 2 \times 10^{19}$ cm^{-2} of foreground absorption and a distance which places it outside of the LHB. Open circles indicate stars with distances placing them outside of the LHB but have foreground columns $< 5 \times 10^{18}$ cm^{-2}. Filled circles indicate stars with distances placing them inside of the LHB but have foreground columns $> 2 \times 10^{19}$ cm^{-2}. Errors in the stellar distances and measured column densities will obviously affect the plot.

varies over the range $\sim 200 - 1500 \times 10^{-6}$ counts s^{-1} $arcmin^{-2}$ and is relatively isothermal (at least some of the temperature variation is likely due to limitations in the analysis technique and statistical errors). The Draco region, however, shows a strong intensity enhancement over a relatively limited solid angle ($\sim$ 0.03 sr). This enhancement, the first unambiguous Galactic halo emission discovered (Burrows & Mendenhall 1991; Snowden et al. 1991), has led to a biased view of the clumpiness of the halo emission as well as to an overestimate of the mean halo surface brightness.

The halo emission is brighter in the region of the north Galactic pole than the south while halo emission in the south increases more strongly toward lower latitudes. The longitude dependence at lower latitudes is considerably different in appearance between the northern and southern hemispheres as well, although this is dominated by distinct features such as the North Polar Spur in the north and Eridanus in the south.

The average values for the distant emission for latitudes $|b| > 65°$ are 1545×10^{-6} counts s^{-1} $arcmin^{-2}$ for the north and 810×10^{-6} counts s^{-1} $arcmin^{-2}$ for the south. After subtraction of the extragalactic contribution, they are reduced to 1145 and 410×10^{-6} counts s^{-1} $arcmin^{-2}$. The southern value lies well within the range of LHB emission while the northern value is only a factor of $\sim$ 1.4 larger than peak LHB values.

The average values for the halo emission in the polar regions are roughly a factor of two lower than the $\frac{1}{4}$ keV surface brightness of the face-on spiral galaxy M101 at the equivalent position of the solar circle: $\sim 1570 \times 10^{-6}$ counts s^{-1} $arcmin^{-2}$ (Snowden & Pietsch 1995). The sum of the average halo emission from $|b| > 65°$ corresponds to a luminosity of $\sim 6.6 \times 10^{36}$ ergs s^{-1} kpc^{-2}. Using values for Type I and Type II supernova rates and the average energy dumped into the interstellar medium by their explosions (and precursor winds for Type II) listed in Ferriére (1995), an average value of $\sim 8 \times 10^{38}$ ergs s^{-1} kpc^{-2} can be derived for near the solar circle. Thus the local halo emission at $\sim 10^6$ K constitutes $\sim 1\%$ of the average supernova power at the solar circle.

The surface brightness of the distant emission in a few regions both in the north and south is particularly low, $\lesssim 600 \times 10^{-6}$ counts s^{-1} $arcmin^{-2}$. If the presumed extragalactic background is subtracted, very little is left ($\lesssim 200 \times 10^{-6}$ counts s^{-1} $arcmin^{-2}$) that can be of halo origin, which again underscores the variation of the halo emission.

4 Conclusion

ROSAT all-sky survey data and *DIRBE*-corrected *IRAS* 100 μm data have been used to study the distribution of emission responsible for the observed $\frac{1}{4}$ keV diffuse background. Emission from the Local Hot Bubble dominates the background at low Galactic latitudes and contributes much of the observed flux at high latitudes as well. The emission temperature of the LHB is in the range $10^{6.07\pm0.05}$ K and the intensity varies between ~ 250 and 820×10^{-6} counts s^{-1} $arcmin^{-2}$. The intensity of the LHB emission can be used to estimate its radial extent from the Sun, which varies between 40 pc and 130 pc. Since the intensities are generally greater at higher Galactic latitudes, they are strongly anticorrelated with total H I column density, presumably due to displacement effects as discussed in Sanders et al. (1977), McCammon & Sanders (1990), and SCMS. The LHB extends further toward the Galactic poles than in the plane, although the regions of greatest extent are not at the poles themselves. No strong enhancement is found in the direction of the third-quadrant void, implying that it is not filled with 10^6 K plasma in thermal pressure equilibrium with the LHB.

Consistent with previous shadowing studies, extensive $\frac{1}{4}$ keV emission is found in the Galactic halo with emission temperatures in the range $10^{6.02\pm0.08}$ K and the derived intensity outside the absorbing layer varies from near zero to $> 3000 \times 10^{-6}$ counts s^{-1} $arcmin^{-2}$ in some regions at lower Galactic latitude. Also consistent with previous studies, the halo emission can vary by a factor of two over angular scales of $\sim 20°$. The average halo unabsorbed intensity in the south polar region is $\sim 410 \times 10^{-6}$ counts s^{-1} $arcmin^{-2}$, which is comparable to typical values for the LHB emission. The average value for the north polar region is $\sim 1140 \times 10^{-6}$ counts s^{-1} $arcmin^{-2}$, which is about

1.4 times the peak values for the LHB emission. Thus while there are some halo regions which are a factor of four or so brighter than the brightest LHB regions (e.g., Draco), more typical high-latitude values are similar to those in the LHB.

References

Bowyer, C. S., Field, G. B., & Mack, J. E. 1968, Nature, 217, 32
Burrows, D. N. 1989, ApJ, 340, 775
Burrows, D. N., Kraushaar, W. L., McCammon, D., & Sanders, W. T. 1984, ApJ, 287, 208
Burrows, D. N., & Mendenhall, J. A. 1991, Nature, 351, 629
Cox, D. P., & Reynolds, R. J. 1987, ARA&A, 25, 303
Ferriére, K. M. 1995, ApJ, 441, 281
Frisch, P. C., & York, D. G. 1983, ApJL, 271, L59
Gry, C., York, D. G., & Vidal-Madjar, A. 1985, ApJ, 296, 593
Hasinger, G., et al. 1993, A&A, 275, 1
Hartman, D. & Burton, W. B. 1997, Atlas of Galactic Neutral Hydrogen, (Cambridge University Press:Cambridge)
Jakobsen, P., & Kahn, S. M. 1986, ApJ, 309, 682
Knapp, G. R. 1975, AJ, 80, 111
Marshall, F. J., & Clark, G. W. 1984, ApJ, 287, 633
McKee, C. F., & Ostriker, J. P. 1977, ApJ, 218, 148
Morrison, R., & McCammon, D. 1983, ApJ, 270, 119
Raymond, J. C. 1991, private communication, computer code update
Raymond, J. C., & Smith, B. W. 1977, ApJS, 35, 419
Sanders, W. T., Kraushaar, W. L., Nousek, J. A., & Fried, P. M. 1977, ApJL, 217, L87
Schlegel, D. J., Finkbeiner, D. P., Davis, M. 1997, in preparation
Snowden, S. L., Cox, D. P., McCammon, D., & Sanders, W. T. 1990, ApJ, 354, 211, SCMS
Snowden S.L., Egger R., Freyberg M.J., McCammon D., Plucinsky P.P., Sanders W.T., Schmitt J.H.M.M., Trümper J., Voges W., 1997, ApJ, 485, 125
Snowden, S. L., McCammon, D., Burrows, D. N., & Mendenhall, J. A. 1994, ApJ, 424, 714
Snowden, S. L., McCammon, D., & Verter, F. 1993, ApJL, 409, L21
Snowden, S. L., Mebold, U., Hirth, W., Herbstmeier, U., & Schmitt, J. H. M. M. 1991, Science, 252, 1529
Snowden, S. L., & Pietsch, W. 1995, ApJ, 452, 627
Snowden, S. L., & Schmitt, J. H. M. M. 1990, Ap&SS, 171, 207
Trümper, J. 1983, Adv. Space Res., 2(4), 241
Welsh, B. Y., Craig, N., Vedder, P. W., & Vallerga, J. V. 1994, ApJ, 437, 638

On the Zero-Level of the Soft X-ray Background

M.J. Freyberg

Max–Planck–Institut für extraterrestrische Physik, D-85740 Garching, Germany

Abstract. The soft X-ray background (SXRB, $0.1 - 2.0$ keV) is generally believed to be comprised of a local component, the Local (Hot) Bubble (LHB), which is only marginally absorbed, and of distant emission (galactic halo, extragalactic background). It is vital to know the foreground emission (spectrum, intensity) if one wants to disentangle and determine the 3-dimensional structure of the X-ray emitting regions. During the ROSAT PSPC All-Sky Survey non-cosmic enhancements of the measured X-ray count rate have been observed. These lasted typically for up to 8 hours (i.e. a couple of orbits). Using PSPC data it has been shown that these so-called "long-term enhancements" (LXE) are related to geomagnetic storms and variations of the solar wind. Moreover, during an observation of the comet Hyakutake indications for a correlation of the X-ray background with the X-ray brightness of the comet have been found. This can be interpreted in terms of an X-ray emitting region around the Earth. Although the variable component of these enhancements could be subtracted from the sky maps an uncertainty remained concerning an undetected and relatively constant contribution which might be present in soft X-ray observations. As a consequence this could explain or contribute to the high PSPC count rate observed from the dark side of the Moon.

1 Introduction

During the ROSAT PSPC All-Sky Survey non-cosmic enhancements of the measured X-ray count rate have been observed. They lasted typically for up to 8 hours (i.e. a couple of orbits) and were called "long-term X-ray enhancements" (LXE). As the sky was scanned in great circles (1 orbit = 1 circle ~ 96 min) the enhancements affect the whole strips of the sky ($\sim 360° \times 2°$) during the period. According to Snowden et al (1992) the LXE are a "non-local effect". Variable and/or high background values in previous survey missions (e.g., Wisconsin, HEAO 1 A2, EUVE) are also noted by, e.g., McCammon et al (1983), Garmire et al (1992), Jelinsky, Vallerga & Edelstein (1995). Using PSPC data from the Survey as well as from the Pointing Programme it has been shown (Freyberg 1994) that the LXE are related to geomagnetic storms and variations of the solar wind. Events from an LXE appear to be vignetted, i.e. have passed the X-ray optics. The X-ray spectrum is soft and hardens if the LXE is stronger. Within the PSPC resolution acceptable fits can be obtained with a thermal spectrum (Raymond-Smith plasma with solar abundances), however, significant foreground absorption ($\sim 10^{19}\,\mathrm{cm}^2$) were required, which left some doubt on the physical relevance of the fit. There are significant variations in intensity and spectrum within an individual orbit.

LXEs are also present at low geomagnetic latitudes. Accompanied is an additional component at high geographic (geomagnetic) latitudes which seems to be related to geomagnetic substorms. Although the variable component of these enhancements could be subtracted from the sky maps (Snowden et al 1995) an uncertainty remained concerning an undetected and relatively constant contribution which might be present in all ROSAT X-ray observations.

2 Does the Earth Act as an X-ray Bright Comet?

In 1996 a new field in X-ray astronomy was born with the detection of soft X-rays from Comet C/Hyakutake 1996 B2 (Lisse et al 1996). The emission was found to be unexpectedly bright in X-rays and offset from the comet nucleus. In a recent model, Cravens (1996) suggested that the X-rays are generated by charge transfer collisions of high charge state solar wind heavy ions with cometary neutrals. Almost the full field-of-view of the ROSAT HRI was filled with the comet image. Therefore the total extent cannot be accurately determined. A ROSAT WFC image indicates a size of more than 1° (Lisse et al 1996). During the 9 observation intervals variations of the X-ray count rate have been observed ("outburst" by a factor of $2-5$). However, further detections of comets in the RASS have shown that besides outbursts a general (zero-level) emission is present. The question is whether the Earth is surrounded by such an X-ray emitting region as well. One way to answer this is to look for background intensity variations that seem to be correlated with the X-rays from the comet. During the comet observations in March 1996 a calibration of the HRI background was performed. Observations with typical cosmic X-ray background values ($3-4 \times 10^3$ cts/Ms/arcmin^2) have been selected (e.g. omitting the LMC). Periods of excessively high background occuring at high geographic latitudes of the orbit were excluded; point-like as well as extended sources had been cut out by standard procedures. Finally, only observation intervals were chosen with more than 600 seconds to have sufficient signal-to-noise ratio.

3 HRI Background Excess and Zero-Level

In Fig.1 we present the ROSAT HRI X-ray light-curves of Hyakutake (Lisse et al 1996) as well as of the X-ray background (for more details see Freyberg, Dennerl & Trümper 1997). The times of comet observations are indicated by dashed lines. The data sets of the background indicate an increase of the count rate close to the outburst. This suggests that variations in the solar wind not only affect the comet's X-ray emission but also the one of the Earth. This excess is of the order of ~ 500 cts/Ms/arcmin^2 (i.e. $\sim 15\%$). Assuming that this variation can be scaled to the comet's outburst variation (i.e. if the outburst is mainly due to solar wind variation and not a cometary

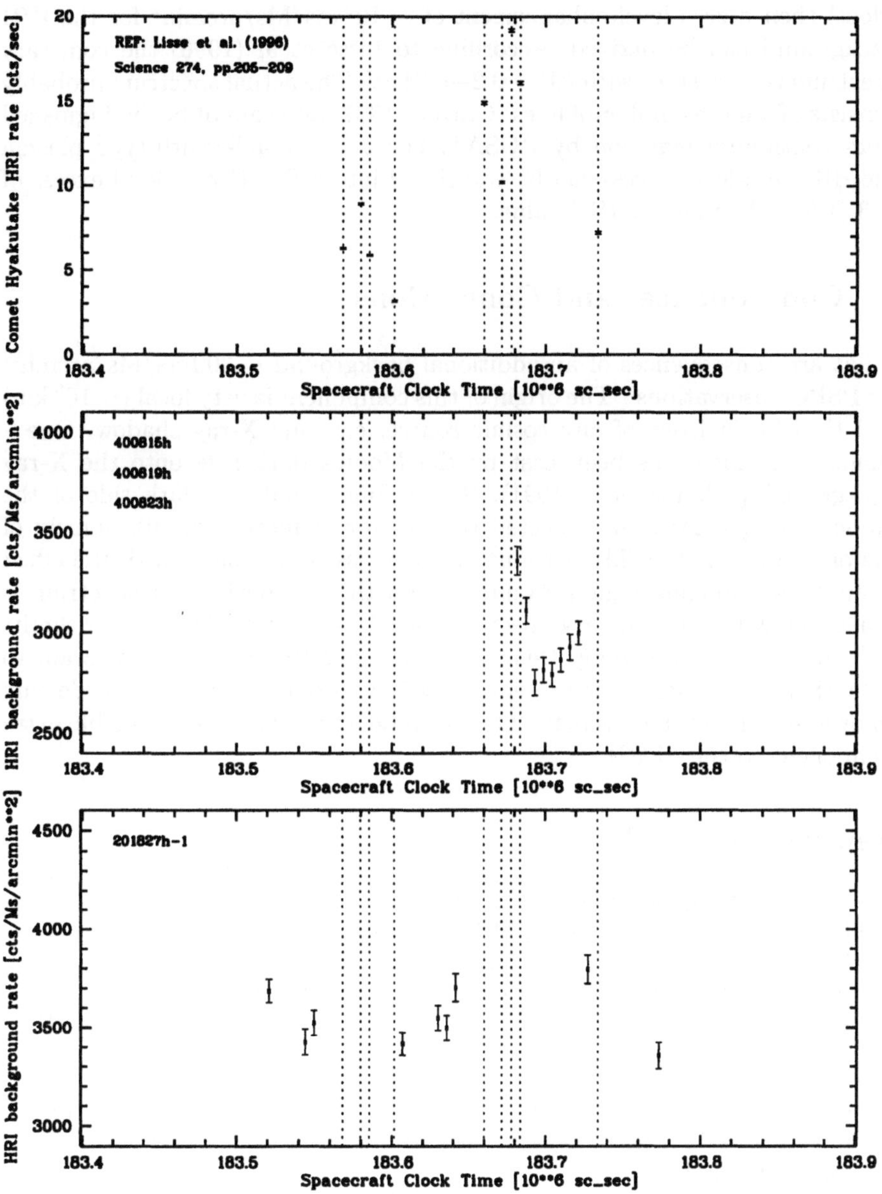

Fig. 1. *Top:* HRI count rate of Comet C/Hyakutake 1996 B2 according to Tab.1 of Lisse et al (1996). Dashed lines indicate the times of the individual observations of the comet. An outburst is visible in the $5^{th} - 8^{th}$ observation. *Middle:* HRI background count rate of 3 observations towards $(l, b) \sim (30, 0)$, with exclusion of sources, excessively high background at high geographic latitudes, short intervals. Variations of several 10^{-4} counts/s/arcmin^2 are visible. *Bottom:* HRI background count rate of one observation towards $(l, b) \sim (30, 17)$, similar to above.

effect) then a zero-level enhancement of $\sim 10^2$ cts/Ms/arcmin^2 for the HRI background can be derived. According to Lisse et al (1996) the cometary spectrum is consistent with $kT \sim 0.2-0.7$ keV. The actual spectrum probably consists of a large number of lines (Cravens 1997) and cannot be distinguished from continuum emission by ROSAT. For a Raymond-Smith-type plasma the HRI zero-level excess can be translated into a PSPC zero-level excess of ~ 3 $(1.5-5)$ times the HRI value.

4 Consequences and Conclusions

What are consequences of an additional background ~ 300 cts/Ms/arcmin^2 for PSPC observations? The origin of this component is very local ($\sim 10^5$ km) and thus be in front of any cosmic source, e.g. any X-ray shadow. One – famous – shadow has been cast by the Moon's dark side onto the X-ray background (Schmitt et al 1991). It was found that the dark side of the Moon was brighter than expected by a factor of more than 10, namely of the order of ~ 300 cts/Ms/arcmin^2. This excess is consistent with the effect of an X-ray emitting region around the Earth. The dark lunar spectrum is consistent with an cometary spectrum of $kT \sim 0.2$ keV (Freyberg, Schmitt & Snowden 1997; Freyberg, Dennerl & Trümper 1997). Finally, we want to note, that the results of the present analysis are preliminary and provide only an order-of-magnitude estimate. One purpose was to propose a possible extra component to the SXRB.

References

Cravens T.E.: 1996, Geophys.Res.Lett. 24, 105
Freyberg M.J.: 1994, PhD thesis, LMU München
Freyberg M.J., Schmitt J.H.M.M., Snowden S.L.: 1997, in preparation
Freyberg M.J., Dennerl K., Trümper J.: 1997, in preparation
Garmire G.P., Nousek J.A., Apparao K.M.V., Burrows D.N., Fink R.L., Kraft R.P.: 1992, ApJ 399, 694
Jelinsky P., Vallerga J., Edelstein J.: 1995, ApJ 442, 653
Lisse C.M., Dennerl K., Englhauser J., Harden M., Marshall F.E., Mumma M.J., Petre R., Pye J.P., Ricketts M.J., Schmitt J.H.M.M., Trümper J., West R.G.: 1996, Science 274, 205
McCammon D., Burrows D.N., Sanders W.T., Kraushaar W.L., 1983, ApJ 269, 107
Schmitt J.H.M.M., Snowden, S.L., Aschenbach B., Hasinger G., Pfeffermann E., Predehl P., Trümper J.: 1991, Nature 349, 583
Snowden S.L., Plucinsky P.P., Briel U. Hasinger G., Pfeffermann E.: 1992, ApJ 393, 819
Snowden S.L., Freyberg M.J., Plucinsky P.P., Schmitt J.H.M.M., Trümper J., Voges W., Edgar R.J., McCammon D., Sanders W.T.: 1995, ApJ 454, 643

ISOPHOT Observations of a Cold Filament in the Local Hot Bubble

Uwe Herbstmeier[1] and Alexandre Wennmacher[2]

[1] Max–Planck–Institut für Astronomie, Königstuhl 17, D-69117 Heidelberg, Germany, uherbst@mpia-hd.mpg.de
[2] Institut für Geophysik und Meteorologie der Universität, Albertus-Magnus-Platz, D-50923 Köln, Germany, wennmach@geo.uni-koeln.de

Abstract. We report on FIR measurements of the dust component towards LVC 88+36–2, a dense neutral filament in the Local Hot Bubble. The measurements were performed with ISOPHOT, the photometer on-board ESA's Infrared Space Observatory *ISO* [1]. The dust temperature derived is about 18 K and stays constant across the filament. Limb brightening at 60μm is measured in agreement with cirrus clouds not located in the Local Hot Bubble. No evidence for an increased UV radiation field along the boundaries of the cloud is found.

1 Introduction

Absorption line results towards nearby stars (e.g. Fruscione et al.1994) indicate that the sun's neighbourhood out to about 100 pc is almost void of neutral clouds. Observations with *ROSAT* have demonstrated, however, that several dense clouds are located within this bubble filled with hot gas (Local Hot Bubble, LHB) overlooked by previous measurements. These structures (e.g. Snowden et al.1993, Wang & Yu 1995) cast shadows in the soft X-ray emission of the LHB. The first example for this population was LVC 88+36–2 (Kerp et al.1993), a very elongated gas and dust filament. Sodium absorption lines in the spectra of nearby stars show that the cloud is closer than 60 pc (Wennmacher et al.1992). From HI 21 cm spectral lines Lilienthal & Wennmacher (1990) and Wennmacher (1994) derived an upper limit for the gas temperature of about 150 K and densities up to 100 cm^{-3}.

The cloud-LHB interface is presumably associated with an enhanced local UV radiation field. Conductive boundaries (e.g. Slavin, these proceedings) and/or shock fronts along the edges of the filament (Wennmacher 1994, Draine & McKee 1993) can produce intensities up to an order of magnitude brighter in the EUV than the LISRF. Here, we report on first steps to search for signatures of such a UV field. We present results from far-infrared observations of the dust emission towards LVC 88+36–2 using ISOPHOT

[1] *ISO* is an ESA project with instruments funded by ESA Member States (especially the PI countries: France, Germany, the Netherlands and the United Kingdom) with the participation of ISAS and NASA.

(Lemke et al.1996) on-board ESA's Infrared Space Observatory *ISO* (Kessler et al.1996).

2 Observations and Data Reduction

All measurements with ISOPHOT were performed along a line crossing the filament perpendicularly. Figure 1 shows this line overlayed on the *IRAS* 100 μm image of the filament.

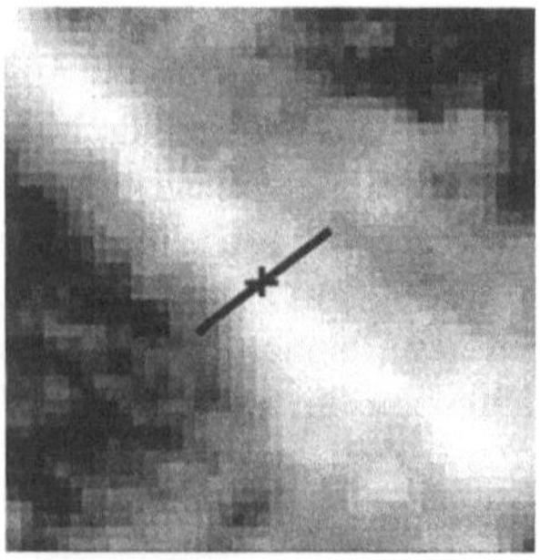

Fig. 1. The *IRAS* 100μm image of LVC 88+36–2 . The line represents the location of the ISOPHOT scans. The reference position, $(\mathrm{RA,DEC})_{\mathrm{J2000}} = (17^{\mathrm{h}}10^{\mathrm{m}}09.4^{\mathrm{s}}, +58°46'30'')$, is marked by a cross. The field extent, oriented along $(\mathrm{RA,DEC})_{\mathrm{J2000}}$, is 1.1° × 1.1°.

The FIR measurements were performed at 60, 100, and 170μm using the C100 and C200 detectors. Seven raster points were observed with a step size of 138″ (60 and 100 μm) and 180″ (170 μm). Each scan was performed a second time in opposite direction shifting the line by one pixel width to determine memory effects. The total measurement times per point were 64, 32, and 32 s for 60, 100, and 170μm respectively.

All data were analysed using the ISOPHOT Interactive Analysis PIA[2] V6.1 and associated calibration files. ISOPHOT measurements comprise of integration ramps. These are read-out values of a capacitor voltage integrating the photocurrent of the detector in regular time intervals. The data reduction starts with a correction for non-linearities of this integration process due to the electronics. It includes corrections for glitches on the ramps caused by cosmic ray particle hits. Signals are derived from the slopes of lines fitted to the ramps. A dark signal is subtracted. Transient drifts induced by flux changes were analysed. Only those values were taken for the average signal

[2] PIA is a joint development by the ESA Astrophysics Division and the ISOPHOT Consortium led by the Max-Planck-Institut für Astronomie (MPIA), Heidelberg. Contributing institutes are DIAS, RAL, AIP, MPIK, and MPIA.

per raster point which do not show large signal changes. The signals were calibrated using the associated measurements of one of the Fine Calibration Sources (FCSs), which monitor long-term responsivity drifts. To increase the signal-to-noise ratio the surface brightness values of the pixels were averaged per raster point.

3 Results

In Fig. 2a we plot the resulting surface brightness distributions. The background values derived from the minima were subtracted. The minima were found at a position off the filament towards larger declinations. The brightness peaks are attained at the same raster point with values of 0.42, 1.45, and 3.00 MJy ster^{-1} at 60, 100, and 170μm respectively. The B_{100}/B_{170} brightness ratio there gives a dust temperature of about 18 K assuming a modified (emissivity $\propto \nu^2$) blackbody spectrum.

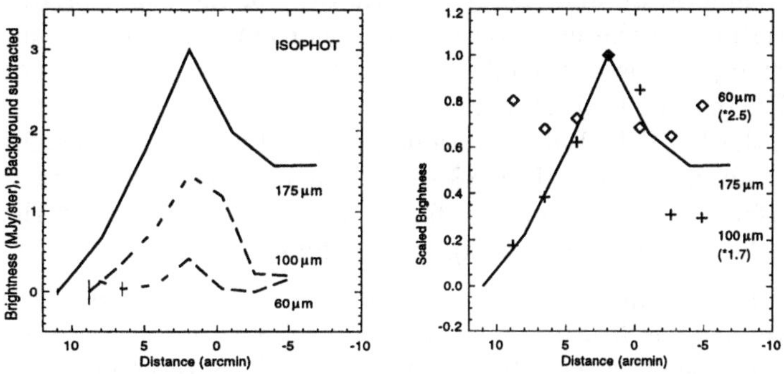

Fig. 2. a) Left: 60, 100, and 170μm surface brightness distribution measured along the scan with respect to the reference position given in Sect. 2, with the same direction as the RA-axis. Typical error bars are indicated by thin lines. **b)** Right: Same as **a)** but for scaled surface brightness. The B_λ/B_{170} ratio at the maximum is taken as the reference ratio. The height of the plotting symbols corresponds to the full width of the error bar.

To look for colour variations across the filament we applied a method developed by Laureijs et al.(1991) similar as in Bogun et al.(1996). The brightness at $\lambda = 60$ and 100μm were scaled with the B_{170}/B_λ ratio derived for the peak. These scaled values were then overplotted on the normalised 170μm scan in Fig. 2b. The distributions should coincide if there were no variations of the infrared colours along the scan. This is the case for the 100μm points towards positive scan offsets. With the exception of one point the B_{100}/B_{170} ratio on the opposite side is lower. The B_{60}/B_{170} ratios are constant in the central part but raise towards both edges.

4 Discussion

Model spectra (e.g. Désert et al.1990) predict that the B_{100}/B_{175} ratio is dominated by the modified blackbody spectrum of the big dust grain component and serves therefore as a good temperature indicator. LVC 88+36–2 is well confined towards larger equatorial coordinates. Here, no variations of the B_{100}/B_{175} ratio (i.e. temperature) are found across the filament edge. On the opposite side the filament merges with a diffuse extended halo. We extract the core's emission from the halo by subtracting a linear brightness gradient as an approximation of the halo. Then the 100/170 colours of the core with the exception of the point at the reference position agree very well with the one derived for the maximum brightness. We like to note without further discussion that the deviating ratio is found at the transition between the core and the halo emission. Also after separation between halo and core emission the B_{60}/B_{175} ratios raise towards the edges of the filament. In addition to the modified blackbody spectrum of the big grains the emission from transiently heated smaller grains contributes to the brightness at 60μm . The constancy of the big grain temperature and the systematic raise of the 60μm brightness indicate variations in the emission characteristics of the small grains. This limb brightening is, however, similarly found for other cirrus clouds not located in the LHB (e.g. Laureijs et al.1991).

Summarising, the ISOPHOT data obtained so far show no evidence for an increased UV radiation field at LVC 88+36–2 . No significant influence of the hot-cold interface or presumed shock fronts along the boundary on the characteristics of the filament's dust component is found.

Acknowledgements. We appreciate discussions with the members of the ISOPHOT Data Centre and the ISOPHOT Instrument Dedicated Team. The ISOPHOT instrument and Data Centre are funded by the Deutsche Agentur für Raumfahrtangelegenheiten DARA and the Max-Planck-Gesellschaft.

References

Bogun S., Lemke D., Klaas U., et al.(1996): A&A 315, L71
Désert F.X., Boulanger F., Puget J.L. (1990): A&A 237, 215
Draine B.T., McKee C.F. (1993): ARA&A 31, 373
Fruscione A., Hawking I., Jelinsky P., & Wiercigroch A. (1994): ApJS 94, 127
Kerp J., Herbstmeier U., & Mebold U. (1993): A&A 268, L21
Kessler M.F., Steinz J.A., Anderegg M.E., et al.(1996): A&A 315, L27
Laureijs R.J., Clark F.O., Prusti T. (1991): ApJ 372, 185
Lemke D., Klaas U., Abolins J., et al (1996): A&A 315, L64
Lilienthal D., & Wennmacher (1990): A&A 235, L13
Snowden S.L., McCammon D., Verter F. (1993): ApJ 409, L21
Wang Q.D., Yu K.C. (1995): AJ 109, 698
Wennmacher A. (1994): PhD Thesis, Universität Bonn
Wennmacher A., Lilienthal D., Herbstmeier U. (1992): A&A 261, L9

Modeling the Local Bubble

Donald P. Cox

Department of Physics, University of Wisconsin-Madison
1150 University Ave., Madison, WI 53706 USA
cox@wisp.physics.wisc.edu

Abstract. Modeling the Local Bubble is one of those activities fraught with danger. It is very easy to be too naive, to fail to consider the dependence of the model on assumptions about the nearby ambient state, or the likelihood of such a structure. It is similarly easy to become so caught up in the details of the vicinity that it is unclear where to begin a necessarily idealized modeling effort. And finally, it is important to remember that the data we have may in some cases be lying to us, and that we have not yet learned to read their facial expressions quite carefully enough.

That said, I've tried in this paper to be helpful to those who may wish to take the risks. I surveyed the very most basic stories that the data seem to tell, and pointed out the standard coincidences that may be telling us a lot about what is happening, but may turn out once again to have been just coincidences. I've described 5 distinct conceptions that in one flavor or another pretty well survey the collection of mental images that have so far been carried by those who've attempted models. One may be right, or something entirely different may be more appropriate. It's at least vital to realize that a conception comes first, followed by a simplified model of details. I've also included a long list of questions directed at observers. Some have partial answers, some one wouldn't know today quite how to approach. But it is a list that students of the soft x-ray background, interstellar absorption lines, possible instrumentation, and the heliosphere may wish to review from time to time, just to see whether they can figure out how to be more helpful. There is another list for modelers, things the models must address, however-so-flimsily if necessary, because there are strong observational constraints (and stronger ones coming) on what can and cannot be present in the local ISM. To that I've added a few remarks concerning x-ray emission coming from beyond the Local Bubble, and another few on how x-ray emission from within the solar system might be contaminating what we see. That last bit is new, exciting, and possibly wrong, but it is an example of the ongoing wariness I believe one has to take toward the facts in the case. By the way, Dieter, it really was a great meeting.

1 The Challenge

The fundamental challenge in modeling the Local Bubble is to find a source for approximately 200 solar luminosities of diffusely distributed soft x-ray radiation in a volume containing roughly a million stars.

One possibility is that there is million Kelvin gas at normal interstellar pressure filling an irregular volume of roughly 100 pc extent. The thermal radiation from the optically thin gas would then supply the emission. The cooling time of the hot gas has been estimated at about 2×10^7 years and one presumes that the age and lifetime of the bubble is some modest fraction of that. Assuming this configuration, modelers then concentrate on how the low density region might have come into being, what sort of boundary it has, what reheated the region most recently, when that occurred, and whether it is likely to occur again before the region's demise. (A volume of radius 100 pc probably has one or more random supernovae within it during a bubble lifetime.) Cox and Reynolds (1987) present a review.

But there may be other possibilities. One looks for continuous inputs that might keep the gas hot, perhaps leading to a steady-state phase of hot gas whenever the density gets low (H. J. Habing, 1969, private communication). Solar type stellar winds inject energy with sufficient vigor, but far too many would seem to be required. Consider however, the following coincidence: The Galactic energy flux in cosmic rays in the solar neighborhood is about 1 to 2×10^{-5} erg cm^{-2} s^{-1}, providing about 1000 to 2000 solar luminosities from an area of radius 100 pc. The escape time of the cosmic rays is about 2×10^7 years. Within the uncertainties, the (local) power requirements and timescales of the Local Bubble and cosmic rays are comparable, though there is no obvious connection.

2 Three More Coincidences

The intensity and broad band "colors" of the x-ray emission, interpreted with a plasma radiation code, yield measures of the temperature T and emission measure $n_e^2 R$ of the diffuse hot gas in the Local Bubble. Combining that with stellar absorption measures of the extent of the low density region around the Sun, to get R, yields an rms electron density measure and thus a pressure estimate. Knowledge of the volume and pressure allow us to estimate the thermal energy content of the bubble. Knowledge of the temperature and radial scale provide a rough estimate of the thermal conduction flux from the bubble to its boundary. From this exercise, three coincidences appear that illuminate and possibly confound modeling of the bubble: 1) The thermal pressure in the bubble is approximately the weight per unit area of the interstellar medium, the total interstellar average pressure at midplane, a factor of 3 to 10 higher than thermal pressures in clouds. 2) The thermal energy content of the bubble is about the energy of 1 supernova. 3) The possible thermal conduction flux to the edge, about $10^{-6}(T/10^6\mathrm{K})^{7/2}(100\mathrm{pc}/R)$ erg cm^{-2} s^{-1}, is approximately the same as the thermal radiation flux inferred from the x-ray emission.

Although the extreme sensitivity to temperature makes it difficult to insist that the third coincidence is meaningful, it is possible to argue that an evolving hot bubble would tend to this state. If one supernova worth of energy occupies a large bubble at normal interstellar pressure, its radius will be of order 100 pc. If it has a temperature well in excess of 10^6 K, thermal conduction will evaporate material from the surroundings, raising its density and lowering its temperature, until the conduction flux is just able to be radiated by the denser cooler material in the outer parts of the bubble. At that point its temperature will be of order 10^6 K and its lifetime about 10^7 years. In short, apart from the problem of creating the low density region in the first place, what we see is about what one might expect.

Except that what one might expect depends fundamentally on one's preconceptions of conditions in the ambient ISM, and the physical processes which are important. Is the galactic disk mostly at very low density and fairly hot, is it mostly warm intercloud gas, do clouds thermally evaporate when surrounded by hot gas, is thermal conduction active within the hot gas or quenched by a tangled magnetic field, do motions within the ambient medium so strongly distort a bubble in 10^7 years that it is meaningless to consider such lifetimes? The ambient ISM is far from understood, making it very difficult to know how to model the Local Bubble as an event within it (see Cox 1996a,b for thoughts on modeling the general ISM).

3 Four Conceptions, Maybe Five

At the meeting, I described what I thought were the four distinct views of the Local Bubble on which models have been based. For completeness, I now think it is important to include a fifth view.

3.1 The Hot Phase Model

In this model, much of interstellar space is assumed to be filled with million Kelvin gas at normal interstellar pressure, but our ability to see beyond 100 pc or so is limited by the local distribution of denser gas which readily absorbs the soft x-rays. The view was originally proposed by Cox and Smith (1974) and later explored in some detail by Jakobsen and Kahn (1986) and Kahn and Jakobsen (1988). In this visibility bounded view, there is no distinct Local Bubble, only the parts of the hot phase we are able to see from the solar location.

3.2 An Active SNR in an Already Low Density

If the ambient interstellar density in our vicinity were about 0.01 cm^{-3}, a single supernova remnant with an age of about 10^5 years, with the Sun inside, would resemble what we see. This differs substantially from the more quiescent bubble picture developed from the coincidences above (and modeled as an old Multi-SNR below) in that there is still a strong shock at the edge, heating gas to 10^6 K; the x-ray emitting gas is recently heated, underionized, and moving radially outward at about 300 km s^{-1}. This scenario was originally proposed by McKee and Ostriker (1977) within the context of their SNR dominated ISM model as a means of achieving x-ray emitting temperature gas locally, their hot phase being generally too cool for x-ray emission. It has been modeled in various versions since by Cox and Anderson (1982), Arnaud et al. (1984), and Edgar (1986).

3.3 An Old Multi-SNR

If the ambient interstellar density is closer to 0.1 cm^{-3}, a sequence of supernovae in the same region is required to build the low density cavity to the necessary size and then to reheat the interior. Once the most recent explosion's shock has reached the bubble boundary and reverberated a time or two, the evolution is slow and closely approximates a quiescent bubble of cooling hot gas. If thermal conduction is not severely quenched and the ambient medium not too active, the bubble tends toward a structure consistent with the observed coincidences. This scenario is similar to the development of a superbubble except that the recurring supernovae are infrequent (at perhaps one to a few million year intervals), only a few in total number, and the most recent is long enough in the past that it has lost its separate identity. The overall pressure in the bubble is comparable to that in the ambient medium. Any shell has long since rebounded to ambient density; the velocity of either expansion (if relatively young) or contraction (if older) is small, about 10 km s^{-1}. The idea for such a scenario derived from a single old quiescent SNR model by Innes and Hartquist (1984), and has been explored further by Cox and Snowden (1986), Cox and Reynolds (1987), Edgar and Cox (1993) and finally modeled in hydrodynamics and ionization with a sequence of explosions by Smith (1995) and presented in this volume by Smith and Cox (1997). Come to think of it, the single old quiescent SNR (in an already low density region) model of Innes and Hartquist is a sixth conception, in addition to having motivated the multi-SNR one.

3.4 An Exploded Cloud

In this scenario, roughly 10 supernovae occur very close together in time in a high density cloud, either close to the boundary or in a small cloud. The supernovae heat the cloud to very high temperature and pressure and it expands freely into a much lower density environment. The expansion lowers the temperature dramatically but the high ionization is frozen in. The expansion is subsequently slowed by the surroundings, but is dense enough that any reheating is transient. The present remnant is cool gas at moderately low density filling the local region. Its electron density is the same as the average in the ISM, explaining the dispersion measures of nearby pulsars, its thermal pressure is similar to that of the Local Cloud, its EUV emission is very low, and its x-rays arise from the recombination of highly ionized heavy elements (Breitschwerdt and Schmutzler, 1994).

3.5 Part of the Sco-Cen Association Superbubble

Noticing that Loop I is probably the edge of the Sco-Cen superbubble, that it occupies a very large angular size on the sky, and that the filaments of Local Fluff passing over the Solar System have velocities similar to that of the approaching hydrogen between us and the Sco-Cen Association leads to a different view. This scenario proposes that the Local Bubble is actually part of the Sco-Cen superbubble, perhaps preferentially expanded in a low density (interarm) direction. This view has been taken by Bochkarev (1987) and Frisch (e.g. this volume and references therein). It is not entirely clear how this scenario leads to hot gas surrounding the solar location while there is still a dense wall of neutral hydrogen about 70 pc away, between us and the Sco-Cen Association. No detailed model has been made, but the scenario is sufficiently complicated (read: realistic) that it would be difficult to imagine how to go about it. There seem to be two important aspects of this view, one that the Sco-Cen Association offers the only obvious nearby source of bubble power, there being no O or B stars in the Local Bubble itself. The other is the realization that the Local Bubble environment is far from simple; if nothing else it is at least partly wrapped around or tangent to the Loop I bubble, with possible free connections between their hot gas contents at high latitude (Cox and Reynolds, 1987).

This ends my survey of conceptions. The list is heavily tilted toward supernovae as the causative agent, but it may turn out that they were only the easiest thing to imagine or model.

4 Background and Observational Issues

There are many questions that need to be addressed observationally before we can have a confident interpretation of the Local Bubble. They include:

- Do the x-rays observed in the galactic plane definitely arise from outside the solar system?
- Do the x-rays have the spectral features of hot gas, or of recombining highly ionized cool gas, or is much of it an unexplained continuum?
- The gas phase elemental abundances in the hot gas, are they solar, depleted, or enriched?
- Is there evidence of edge brightening, suggesting emission concentrated at the outer surface rather than spread through the volume?
- Are there any regions of spectral anomaly that might suggest interaction between the Local Bubble and the hot gas of Loop I?
- Do spectral maps support the conclusion from broad band surveys that the Local Bubble is emission-bounded rather than visibility-bounded?
- Is there fast moving or hot O VI associated with the Local Bubble? Is the O VI associated with the bubble boundary consistent with quiescent hot bubble models?

- Is there any hint of boundary structure between the hot gas and the Local Cloud? Is the cloud evaporating and if so, how fast compared to the classical rate?
- In directions in which the hot gas appears to butt up against dense background material, what is the nature of the interaction? Is there any sign of evaporation or condensation? How do the high stage ions found in absorption compare with the x-ray spectral characteristics?
- Is there a region in which we can see the hot gas tangentially along a boundary, and if so does the spectral gradient confirm the conductive condensation picture?
- What more can we learn about regions of bizarre emptiness such as that toward epsilon CMa?

5 Issues Which Models Must Confront

A responsible attempt to model the Local Bubble should confront at least the following issues:

- What is the assumed environment prior to the generation of the Local Bubble and is that consistent with general ISM constraints?
- How likely is the Solar System to be found in a region of the sort envisioned? This is the Copernican question and is addressed by specification of the generating agent, estimation of the rate of creation of such regions and their lifetimes.
- How does the model produce the requisite x-ray surface brightness, the band ratios, and, when the long awaited observations are made, the spectrum?
- Is the model consistent with the degree of elemental depletion onto dust (or elemental enrichment) that will be measured from that x-ray spectrum?
- Is the model consistent with the strong limit on slow narrow O VI in the local vicinity and the O VI column densities and velocity profiles associated with the Local Bubble that should be available from FUSE?
- How does the model address the apparent displacement effects (the x-ray-NH anticorrelation): the sign of the effect, the color independence, the implication that the emission is in the volume rather than on the surface.
- How is the model consistent with the existence and persistence of the Local Fluff?

6 Three Remarks on X-Ray Emission from Beyond the Local Bubble

6.1 Breitschwerdt and Schmutzler's Wind Model

Breitschwerdt and Schmutzler (1994) also proposed a model for generation of high latitude x-ray emission from a galactic wind, the material initially having been injected by superbubbles into channels or chimneys leading out through the halo. A novel feature of their proposal was that some amount of the x-ray emission derived, as in their Local Bubble model, from recombination of high stage ions in expansion-cooled gas.

But, there is a simple upper limit to the amount of x-ray power available from delayed recombination. Each hydrogen atom escaping in the wind is accompanied by about 0.001 heavy ions capable of making x-rays in their recombination. Each of those ions can produce about 1 keV of x-rays in their recombination, for a total of about 2×10^{-12}ergs per hydrogen atom. In their model, 0.4 solar masses of material per year left in the wind, 4×10^{56} atoms per year, for a total recombination power of 10^{45} ergs per year, about 0.01% of the Galaxy's supernova power or 1% of the power density of the soft x-ray background. The bulk of the x-ray emission in their wind model is from thermal excitation of emission lines prior to the flow's cooling below roughly 10^6 K.

6.2 Halo Supernovae

There have been a variety of potential sources suggested for high latitude high stage ions and x-ray emission: a galactic wind, shocked diffuse galactic infall, wakes of infalling clouds, a galactic chromosphere, a quiescent galactic corona (e.g. Wang, this volume), microflare reconnection events, a general galactic fountain arising from a hot phase of the ISM, localized plumes arising through superbubble chimneys, etc. I would like to call attention to yet another potential contributor, halo supernovae, whose modeling has recently been undertaken by Robin Shelton (this volume).

6.3 The ROSAT Distant Component

It is my understanding that there are two ROSAT distant components, a hard one which seems to be concentrated around the galactic center and discussed in this volume by Wang and by Freyberg, and a softer one explored at moderately high latitude by shadowing experiments and described in this volume by Snowden. My comments are directed at the latter.

A reader familiar, for example, with the Cox and Reynolds (1987) review of the Local ISM will notice that my opinions concerning the characteristics, structure, and origin of the Local Bubble have evolved very little in the intervening years. The reason is found in Snowden's summary paper: ROSAT showed us a great deal about emission in the low lying halo, beyond the Local Bubble, but did little to alter our understanding of the bubble itself. A smaller fraction of the observed emission from high latitude is local, but the part that is local still has pretty much the original description.

But what is the "halo emission"? From Snowden et al. (1997), most of the excess emission is very nonuniform; it lies close to the galactic plane, and it tends to be found above active regions in the disk. My tentative summary is that it arises in discrete large scale regions in the thick disk of interstellar material, powered by events in the underlying thin disk. Referring to these very bright features as "halo emission" obscures the possibility that there may be a truly diffuse component at lower intensity that better deserves the name.

7 Solar System X-Ray Emission

In the early days of soft x-ray background studies, we tried very hard at Wisconsin to think of ways that a substantial x-ray contribution could be made within the solar system. We concerned ouselves with scattered solar x-rays, with colliding fast streams in the solar wind, with solar wind impacts on zodiacal dust, with the solar wind termination shock and the solar wind's magnetotail, and probably others I have since forgotten. We eventually concluded that because we could think of no major contributing processes there were none.

That, however, was before the discovery of "Long Term Enhancements" and the intense x-ray emission of comets!

Long Term Enhancements are the name given, I believe by Steve Snowden, to a time-variable component of the diffuse emission seen by ROSAT. It was a form of contamination that seemed uniform to within a factor of two over a great circle perpedicular to the Earth-Sun line, varying on a timescale of tens of hours. It looks just like the C-band x-rays from the sky, may come from closer than the moon (according to Snowden, private communication, it may have produced the dark side counts seen in the ROSAT picture of the moon), correlates with practically nothing and has an unknown origin. If the process producing this "emission" has a DC component, that component has not been subtracted from the soft x-ray background and would have been attributed to the Local Bubble. If such contamination were all of the emission seen in the galactic plane, only the higher latitude emission would be galactic and the Local Bubble would be two, one each above and below us in z. I do not offer these worries as a serious concern; the contribution would have had to remain stable for many years and been independent of observing instrument. I am concerned, however, that there is an undiscovered local process that can generate such contamination.

And along came Hyakutake, blazing away at 10^{-5} ergs cm^{-2} s^{-1}, compared to the soft x-ray background's 1 to 4 $\times 10^{-7}$ ergs cm^{-2} s^{-1}. And how has that been interpreted? Why, with the Breitschwerdt and Schmutzler mechanism! Or at least something similar. The solar wind has a particle flux of roughly $2 \times 10^8 (1\text{AU}/r)^2$ atoms cm^{-2} s^{-1}, with an average heavy element ionization energy of about 2×10^{-12} ergs per particle, making the potential x-ray recombination flux about $4 \times 10^{-4} (1\text{AU}/r)^2$ ergs cm^{-2} s^{-1}. By impacting the comet's halo, the solar wind encounters enough material to force a substantial fraction of the high stage ions to recombine, producing the observed flux (Cravens, 1997; Haberli, Gombosi, De Zeeuw, Combi, and Powell, 1997). It would be nice to have the confirming spectrum, but the mechanism is exceedingly plausible.

Could the same mechanism be responsible for the Long Term Enhancements, or to a DC level in the soft x-ray background? If all of the high stage ions in the solar wind are forced into recombination prior to reaching about 60 AU, their x-ray flux will be significant. (If a small fraction recombine closer, that too is significant.)

One would expect interplanetary transient events to be more localized in the sky than the LTEs, to have the wrong timescale, and not to be produced closer than the moon; but LTEs could potentially arise in interaction between the solar wind and the exospheric hydrogen of the geocorona. For the most part these two do not intermix, but the required efficiency is only 0.001. Above 10,000 km altitude h, the geocoronal density is roughly $2 \times 10^3 (10^4\text{km}/h)^2$ cm^{-3} (e.g. He et al., 1993). With the solar wind flux given above and a charge exchange cross section of 4×10^{-15} cm^2, the line of sight integrated x-ray emissivity from beyond 10 Earth radii would be of order 5×10^{-7} ergs cm^{-2} s^{-1}, easily comparable to the SXRB. Very likely the required 0.001 efficiency could occur via a variety of complications, but it is surprising that even this zeroth order estimate is of the required magnitude.

With this mechanism, one would expect the LTE intensity to depend on the current solar wind particle flux and ionization level, the exospheric hydrogen density profile, the relative orientation between the Sun-satellite-location-look-direction, and the degree to which the magnetospheric boundary allows solar wind access to the geocoronal hydrogen. The complexity of this dependency may have hidden it from correlations examined so far.

A DC level could potentially arise in interaction of the solar wind with the stream of interstellar material flowing through the Solar System. Assuming the interstellar stream density is roughly constant beyond 10 AU (within which it gets photoionized) at 0.1 cm^{-3}, and a charge exchange cross section of order 4×10^{-15} cm^2 yields a recombination probability of 5×10^{-3} per AU. Integrating the implied emissivity from 10 AU out yields an estimated x-ray flux of 2×10^{-7} ergs cm^{-2} s^{-1} for the upwind direction. As this is the observed level of emission in the soft x-ray background, it would appear that such a calculation needs to be made more carefully; it could well turn out that the spectrum of the soft x-ray background is locally contaminated with a significant recombination flux.

Should we be worried? Well, yes and no. The recombination spectrum calculated by Haberli et al. shows very little emission in the carbon band and could be ruled out as a major contaminant of the SXRB from even the crude pulse height distributions of the Wisconsin rocket data. In addition, there is a strong oxygen line predicted at 574 eV that would appear in the M band where there is no Long Term Enhancement and where shadowing experiments seem to show that the preponderance of the emission is quite distant. On the other hand, carbon band emission is expected to arise from silicon, sulfur, magnesium, and iron–elements that were simply not included in the analysis of cometary emission. So, the fact is, we could have a problem. Their abundances are lower, but the details need to be checked.

The anticipated distribution of interplanetary x-ray emission would probably resemble the distribution of scattered solar Lyman α off the same interstellar stream, a broad feature in the upwind direction which, if I remember correctly, shows parallax during the course of the year. It is a direction which has posed some problems in the past; roughly speaking it is difficult to see white dwarfs in the EUV in that direction (Barber et al., 1993) suggesting that absorbing interstellar material is not far away, leading us to wonder how there could be enough room to achieve the observed x-ray emission from the Local Bubble. Perhaps it has ultra-very local help. In the downwind direction there is a gravitationally focused column of interstellar helium which could similarly have an x-ray glow.

The material in the previous few paragraphs is new and tantalizing. Brief and exciting conversations with Walt Harris, Wilt Sanders, Ron Reynolds, Frank Scherb, and Dan McCammon contributed to its flavor, but much more thorough modeling will be needed to sort it out. It was fun, in any case, to consider yet another way in which a Breitschwerdt-and-Schmutzler-like notion may have some relevance to the SXRB.

Acknowledgements. This work was supported in part by NASA grants NAG5-3155, and NAG5-629. The manuscript benefitted greatly from suggestions made by Dan McCammon, Franz Kahn, Dieter Breitschwerdt, and Bob Benjamin.

References

Arnaud, M., Rothenflug, R. and Rocchia, R. (1984): in *The Local Interstellar Medium*, IAU Colloq. No. 81, eds. Y. Kondo, F.C. Bruhweiler, and B.D. Savage, p. 301, NASA CP-2345

Breitschwerdt, D. & Schmutzler, T. (1994): Nature, 371, 774

Barber, C.R., Warwick, R.S., Hodgkin, S.T., & Pye, J.P. (1993): MNRAS, 262, 289

Bochkarev, N.G. (1987): Astrophys. Space Sci., 138, 229

Cox, D.P. (1996a): in *Astrophysics in the Extreme Ultraviolet*, eds. S. Bowyer and R.F. Malina, Kluwer, p. 247

Cox, D.P. (1996b): in *The Interplay Between Massive Star Formation, the ISM and Galaxy Evolution*, eds. D. Kunth, B. Guideroni, M. Heydari-Malayeri, and T.X. Thuan, Editions Frontieres, p. 233

Cox, D.P. & Anderson, P.R. (1982): ApJ, 253, 268
Cox, D.P. & Reynolds, R.J. (1987): ARA&A, 25, 303
Cox, D.P. & Smith, B.W. (1974): ApJL, 189, L105
Cox, D.P. & Snowden, S.L. (1986): Adv. Space Res., 6, 97
Cravens, T.E. (1997): Geophys. Res. Lett., 24, 105
Edgar, R.J. (1986): ApJ, 308, 389
Edgar, R.J. &Cox, D.P. (1993): ApJ, 413, 190
Freyberg, M.J. (1997): *this volume*
Frisch, P. (1997) *this volume*
Haberli, R.M., Gombosi, T.I. De Zeeuw, D.L. Combi, & Powell, K.G. (1997): Science, 276,939
He, X., Kerr, R.B., Bishop, J, & Tepley, C.A. (1993): JGR, 98, p. 21, 611
Innes, D.E. & Hartquist, T.W. (1984): MNRAS, 209,7
Jakobsen, P. & Kahn, S.M. (1986): ApJ, 309, 682
Kahn, S.M. & Jakobsen, P. (1988): ApJ, 329, 406
McKee, C.F. & Ostriker J.P. (1977): ApJ, 218, 148
Shelton, R.L. (1997): *this volume*
Smith, R.K. (1995) PhD thesis, University of Wisconsin-Madison
Smith, R.K. and Cox, D.P. (1997): *this volume*
Snowden, S.L. (1997): *this volume*
Snowden, S.L., Egger, R., Finkbeiner, D., Freyberg, M.J. & Plucinsky, P.P. (1997): ApJ, *in press*
Wang, Q.D. (1997): *this volume*

Modeling the Local Bubble Using Multiple Supernova Remnants

R.K. Smith[1] and D.P. Cox[2]

[1] Laboratory for Astronomy and Solar Physics, NASA Goddard Space Flight Center, Greenbelt, Maryland 20771, USA

[2] Department of Physics, University of Wisconsin–Madison, Madison, Wisconsin 53706, USA

Abstract. We have modeled the Local Bubble (LB) using a one-dimensional hydrocode (ODIN) that can simulate multiple supernova remnants, with non-equilibrium ion evolution and dust. Our model assumes that the local interstellar medium was a cool (10^4 K) gas approximately 5-10 Myr ago; it was then disturbed by 2 or 3 supernovae exploding within 20-30 pc of each other over a period of 2-4 million years. The LB is the leftover hot gas from these explosions. The model predicts the x-ray emission from such a bubble, as well as ionic abundances for hot gas ions such as O VI. These are compared to the soft x-ray data from the Wisconsin all-sky survey and the ROSAT PSPC.

1 Introduction

The "displacement model" is a very successful explanation of the soft x-ray background (Snowden et al., 1990); it postulates the existence of a bubble of hot (10^6 K) gas in collisional ionization equilibrium, surrounding the Sun and creating the soft x-ray background. This hot gas "displaces" the neutral gas that might otherwise be present, and so this model dovetails nicely with the observations of a local "hole" in neutral hydrogen (Paresce 1984).

While successful, understanding the physical mechanism for creating this "Local Bubble" of hot gas has proven to be difficult. Innes & Hartquist (1984) fit the data with an analytic model of a superbubble, but they needed a very high explosion energy and external pressure. Numerical attempts have been made by Edgar (1986), who found that an active blast wave propagating in a cavity could generate the observed X-rays but over-produced OVI, and Edgar & Cox (1993), who attempted to model the Local Bubble as a old supernova remnant (SNR). This "old SNR" model failed because when a sufficiently large bubble was created it always had a temperature too low to create the observed X-rays, unless, again, the external ISM pressure was unrealistically high. To solve this problem, we assumed the Local Bubble is the long-lived remnant of a *series* of supernova explosions; such a remnant can easily be made to be both large and hot, and uses the inertia of the remnant walls to confine the bubble.

2 Method

We developed a one-dimensional Lagrangian hydrocode, ODIN, for this project; it is described more fully in Smith & Cox (1997). The code divides the ISM into mass parcels and follows each mass parcel throughout time, evolving the hydrodynamic parameters as well as the ionic abundances of the gas. The ionic abundances and the non-equilibrium cooling are calculated using the Raymond & Smith (1977) plasma spectral evolution code. Magnetic fields are included, but only as a pressure term and we assume that the field is frozen into the gas so that $|\mathbf{B}| \propto n$.

We included thermal conduction in ODIN, and found that the thermal conduction flux is one of the most important processes in fitting the observations. We used the Spitzer (1956) prescription for the thermal conduction flux:

$$F_{tc}(T) = c_{tc}\beta T^{5/2}\frac{\partial T}{\partial r} \qquad (1)$$

where T is the temperature of the gas and $\beta = 6 \times 10^{-7}$ cgs. c_{tc} is a unitless constant which allows for a simple "tuning" of the amount of thermal conduction in the code. We also included the effects of saturation as described by Cowie & McKee (1977).

3 Results

We ran the series of models shown in Table 1, designed to explore a reasonable range of parameter space. We considered models with 2 or 3 SN of standard ($\sim 10^{51}$ erg) energy; for the 2 SN models, the explosions were separated by one million years, and the same was true for the 3 SN models except the third explosion occurred 2 million years after the second. We assumed the interstellar medium was homogeneous, with temperature 2×10^4 K and densities of order $0.2 - 0.4\,\mathrm{cm}^{-3}$. Our goal was to find a range of parameters that would give a long-term fit to the observations.

After the final SN explosion, each "bubble" evolves similarly, rapidly reaching maximum size and collapsing. The ISM pressure ($\sim 2\times10^4\,\mathrm{cm}^{-3}\mathrm{K}$) exceeds the interior pressure throughout the collapse, and the rate of collapse ($\sim$ 10 km/s) is approximately the ISM sound speed. As Table 1 shows, the X-ray brightness but not hardness is affected by the ISM pressure. During the collapse the center of the bubble remains at a constant temperature while the edges cool, although the bubble collapse is not entirely due to cooling; the density inside the hot bubble does increase during the collapse. As a result, the x-ray emission increases in time, as shown in Figure 1, after the impulsive peaks of the explosions.

We found that thermal conduction was necessary to smooth out the temperature distribution of the gas; this agrees with observations which find that ratios of soft x-ray bands such as the ROSAT R1 and R2 bands are generally

Table 1. Model parameters and general results

Model	N	$n\,(\mathrm{cm}^{-3})$	E_{51} (10^{51} ergs)	c_{tc}	B (μG)	Result
A	2	0.2	1.0	1/6	5.0	Reasonable fit for t = 5-8 Myr
B	3	0.4	1.0	1/6	5.0	Reasonable fit for t = 5-8 Myr
C	2	0.2	0.5	1	5.0	Too dim, too soft
D	3	0.2	0.5	1/30	5.0	Far too dim, otherwise good
E	3	0.4	1.0	1	5.0	Too bright, too soft
F	3	0.2	1.0	1	7.0	Too bright, too soft
G	3	0.4	1.0	1	7.0	Too bright, too soft

smooth (Snowden et al., 1997). However, including the full amount of thermal conduction as predicted by Spitzer (1956) reduced the central temperature so much that the calculated x-ray flux was consistently too soft, the same problem that plagued Edgar & Cox (1993) when using a single SNR. When we reduced the amount of thermal conduction, however, we were able to find acceptable fits.

We found two models, labeled A and B in Table 1, which fit many observed quantities; we will describe only B (A is similar). Figure 1 shows the emission as a function of time in four bands (from McCammon et al., 1983): the Wisconsin all-sky survey boron (B) band ($E \sim 0.188\,\mathrm{keV}$) and carbon (C) band ($E \sim 0.284\,\mathrm{keV}$), the Wisconsin Beryllium (Be) band ($E \sim 0.118\,\mathrm{keV}$), and the ROSAT 3/4 keV band. The brightness increases over time, and around $\sim 8\,\mathrm{Myr}$, 5 Myr after the last SN explosion, the calculated fluxes are all within the measured values. One important characteristic of these models is the time-invariance of the band ratios. The calculated C/B ratio is 2.5, equal to the measured ratio, and it is nearly constant with time, just as the measured value is nearly constant on the sky. Switching to the ROSAT data, the flux in the ROSAT 3/4 keV band is greater than that in the equilibrium displacement model, but is under the 2σ upper limit for the Local Bubble emission as seen towards MBM 12 by Snowden et al.(1993). Finally, the predicted column density of O VI in the bubble and its edge is $\sim 1.0 \times 10^{13}\,\mathrm{cm}^{-2}$, very close to the $1.6 \times 10^{13}\,\mathrm{cm}^{-2}$ calculated by Shelton & Cox (1994) to be in the Local Bubble.

Multiple supernova remnant models seem to be able to reproduce a broad range of observations of the Local Bubble. These models explain the observations as well as or better than the displacement model, and have the advantage of evolving from known astrophysical processes. The next step will be to include dust cooling and sputtering in the model, to see the effect of variable abundances on the results.

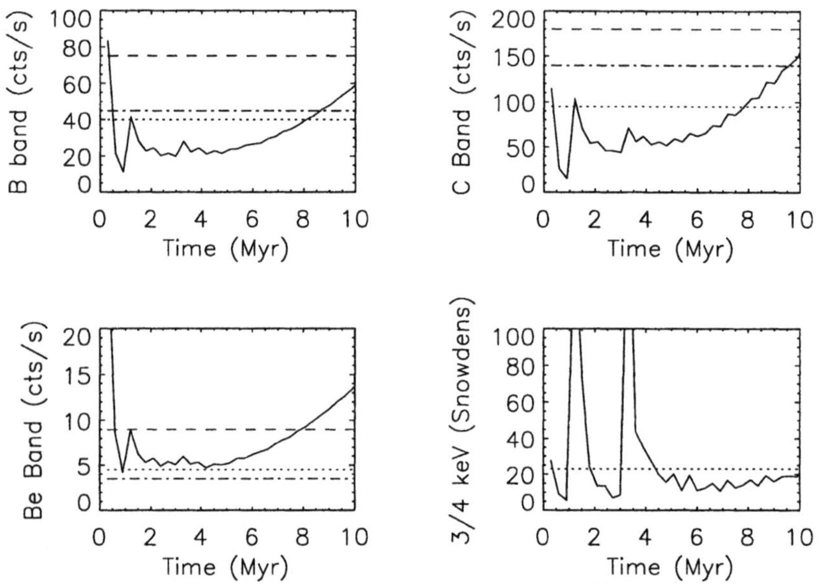

Fig. 1. The calculated emission from model B in the UW B, C, and Be bands, and the ROSAT 3/4 keV band emission. For the first 3 plots, the dotted line is the observed emission in the plane. The dashed line marks the emission towards the north Galactic pole, and the dot-dashed line is the south Galactic pole emission. In the ROSAT 3/4 keV plot, 1 Snowden = 10^{-6} counts s^{-1} $arcmin^{-2}$, and the dotted line marks the 2σ upper limit from the Local Bubble as measured towards MBM 12 (Snowden et al., 1993)

References

Cowie, L. L. & McKee, C. F. (1977): ApJ, 211, 135
Edgar, R. J. (1986): ApJ, 308, 389
Edgar, R. J. & Cox D. P. (1993): ApJ, 413, 190
Innes, D. E. & Hartquist, T. W. (1984): MNRAS, 209, 7
McCammon D., Burrows, D., Sanders, W. & Kraushaar, W. (1983): ApJ, 269, 107
Paresce, F. (1984): AJ, 89, 1022
Raymond, J. C. & Smith, B. W. (1977): ApJS, 35, 419
Shelton, R. L. & Cox, D. P. (1994): ApJ, 434, 599
Smith, R. K. & Cox, D. P. (1997): in preparation
Snowden S. L., Cox, D. P., McCammon, D., & Sanders W. T. (1990): ApJ, 354, 211
Snowden S. L., McCammon, D. & Verter, F. (1993): ApJ, 409, L21
Snowden, S. L. et al.(1997): ApJ, 485, 125
Spitzer, L. (1956): *Physics of Fully Ionized Gases*, (New York: Interscience)

Thin Layer Approximation in 3-D

G.S. Bisnovatyi-Kogan[1] and S.A. Silich[2]

[1] Space Research Institute, 84/32 Profsoyuznaya, Moscow, 117810, Russia
[2] Main Astronomical Observatory of the National Academy of Sciences of Ukraine 252650, Kiev, Golosiiv, Ukraine

Abstract. Equations are derived which describe the propagation of strong shocks in the interstellar matter, without any restrictions for symmetry, in a thin layer approximation (2.5 dimensions). Using these equations permits to calculate the propagation of shock waves from nonsymmetric supernovae explosions in a medium with arbitrary density distribution and the formation of superbubbles in galaxies.

1 Thin Layer Approximation

A thin shell approximation for discription of strong shocks is based on two simplifications. First, it is assumed that all swept-up intercloud gas accumulates into the thin shell just behind the shock front and moves with the velocity u. Second, the pressure distribution inside the cavity $P_{\rm in}(r,t)$ is taken to be uniform. The equations of mass and momentum conservation in spherically symmetric case may be expressed as follows (Chernyi 1957)

$$M = M_0 + 4\pi \int_0^R \rho(r) r^2 dr, \qquad \frac{d(Mu)}{dt} = 4\pi R^2 (P_{\rm in} - P) + Mg, \tag{1}$$

where M is the mass of the shell, M_0 is the ejected mass, R is the shock radius and u is the gas velocity behind the shock; $\rho(r)$ and P are the density and the pressure of the ambient gas, g is the external gravitational field. For the adiabatic blastwave without gravity

$$1)\, \frac{dR}{dt} = \frac{\gamma+1}{2} u \quad \text{or} \quad 2)\, \frac{dR}{dt} = u; \qquad E_0 = E_{\rm th} + \frac{1}{2} M u^2, \tag{2}$$

where $E_0 = \text{const}$ is the energy of the explosion, $E_{\rm th} = \frac{4\pi}{3(\gamma-1)} P_{\rm in} R^3$ is the thermal energy of the blastwave, and γ is the adiabatic index. In (2) the radius R is related to the shock front in the case 1) and to a sphere inside a thin layer in the case 2). Equations (1)–(2) have a simple solution for the homogeneous case if the swept-up mass is much greater than the ejected one: $R = \xi_0 \left[\frac{E_0}{\rho_0}\right]^{1/5} t^{2/5}$, where (Chernyi 1957, Bisnovatyi-Kogan and Blinnikov 1982)

$$1)\, \xi_0 = \left[\frac{75(\gamma-1)(\gamma+1)^2}{16\pi(3\gamma-1)}\right]^{1/5}, \quad 2)\, \xi_0 = \left[\frac{75(\gamma-1)}{8\pi}\right]^{1/5}. \tag{3}$$

Comparison with an exact self-similar (SS) solution (Sedov 1946) shows, that the case 2) gives much better precision, $\xi_0 = 1.033,\ 1.014\ 1.036$ for (SS), (1), (2) cases respectively at $\gamma = 1.4$ (molecular cloud); and similarily $\xi_0 = 1.15,\ 1.12,\ 1.15$ for $\gamma = 5/3$.

2 Three-Dimensional Shocks

Introduce, following Bisnovatyi-Kogan and Silich (1991), Silich (1992) (see also Palouš, 1990; Bisnovatyi-Kogan and Silich, 1995) a Cartesian coordinate system (x, y, z). Let m be the mass, $\mathbf{r}$ the radius-vector, $\mathbf{u}$ the velocity of a particular Lagrangian element of the shock, $\rho(x, y, z) = \rho_0 f(x, y, z)$ the ambient gas density, $\mathbf{n}$ the unity vector normal to the shock front, $\mathbf{g}$ the acceleration of the external gravitational field, $\mathbf{V}$ the velocity field of the undisturbed gas flow, Σ the surface area of the Lagrangian element, $m = \sigma\Sigma$, σ the surface density, and $\Delta P = P_{\rm in} - P$ the pressure difference between the hot interior and warm (cold) external gas. The pressure $P_{\rm in} = (\gamma - 1)E_{\rm th}/\Omega$ of the hot tenuous gas within the cavity is a function of the thermal energy, $E_{\rm th}$, of the bubble and the volume Ω. To describe the expansion of the shock we must introduce the surface area element and define the volume of any closed three-dimensional region. It is well known from differential geometry that any surface may be specified parametrically, with Cartesian coordinates at any point on the surface being a function of two parameters, λ_1 and λ_2: $x = x(\lambda_1, \lambda_2)$, $y = y(\lambda_1, \lambda_2)$, $z = z(\lambda_1, \lambda_2)$. Then the element of the surface area may be defined by the expression (Budak and Fomin, 1965):

$$d\Sigma = S(\lambda_1, \lambda_2) d\lambda_1 d\lambda_2\ , \tag{4}$$

$$S(\lambda_1, \lambda_2) = \left\{ \left[\frac{\partial(y, z)}{\partial(\lambda_1, \lambda_2)}\right]^2 + \left[\frac{\partial(z, x)}{\partial(\lambda_1, \lambda_2)}\right]^2 + \left[\frac{\partial(x, y)}{\partial(\lambda_1, \lambda_2)}\right]^2 \right\}^{1/2} . \tag{5}$$

Similar expressions can be obtained for Ω and $\mathbf{n}$. If parameters λ_1 and λ_2 are considered as the Lagrangian coordinates of the shock front, then the equations for 3-D shock propagation may be written for the mass $\mu = \sigma S(\lambda_1, \lambda_2)$ per unit of Lagrangian square on the surface of parameters (λ_1, λ_2), in a compact form, convenient for numerical integration

$$\frac{\mathrm{d}\mu}{\mathrm{d}t} = \rho\chi, \tag{6}$$

$$\frac{\mathrm{d}u_x}{\mathrm{d}t} = \frac{\Delta P}{\mu}\frac{\partial(y, z)}{\partial(\lambda_1\lambda_2)} - \frac{u_x - V_x}{\mu}\rho\chi + g_x, \tag{7}$$

$$\frac{\mathrm{d}u_y}{\mathrm{d}t} = \frac{\Delta P}{\mu}\frac{\partial(z, x)}{\partial(\lambda_1\lambda_2)} - \frac{u_y - V_y}{\mu}\rho\chi + g_y, \tag{8}$$

$$\frac{\mathrm{d}u_z}{\mathrm{d}t} = \frac{\Delta P}{\mu}\frac{\partial(x, y)}{\partial(\lambda_1\lambda_2)} - \frac{u_z - V_z}{\mu}\rho\chi + g_z, \tag{9}$$

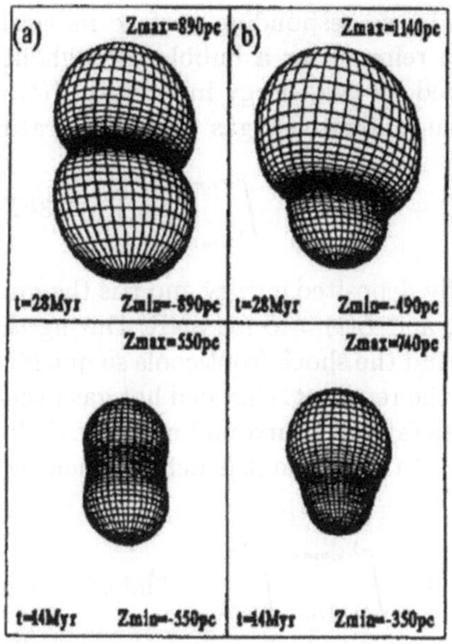

Fig. 1. Galactic superbubble morphology for different locations of the parent OB-association relative to the galactic plane. Left: the OB-association is at the midplane of the Galaxy. Right: the OB-association is 50pc above the Galactic plane.

Fig. 2. Scheme of the HI holes orientation in a spiral galaxy in the plane of view. Projection of the rotational axis to the plane of view is directed upward, determining a motion of the left side to the observer. The indicated direction of elongation of HI hole determines the orientation of the galaxy, shown in the figure.

$$\frac{\mathrm{d}x}{\mathrm{d}t} = u_x, \quad \frac{\mathrm{d}y}{\mathrm{d}t} = u_y, \quad \frac{\mathrm{d}z}{\mathrm{d}t} = u_z. \tag{10}$$

Here function χ is defined as follows:

$$\chi = (u_x - V_x)\frac{\partial(y,z)}{\partial(\lambda_1,\lambda_2)} + (u_y - V_y)\frac{\partial(z,x)}{\partial(\lambda_1,\lambda_2)} + (U_z - V_z)\frac{\partial(x,y)}{\partial(\lambda_1,\lambda_2)}. \tag{11}$$

Approximating the shock front by a number N of Lagrangian elements one gets a system of $7N$ differential equations for mass and momentum conservation. This set of equations is coupled with the equation for the gas pressure within the cavity, and the equation of total energy $E_{\mathrm{tot}} = E_{\mathrm{th}} + E_{\mathrm{k}} + E_{\mathrm{g}}$, consisting of the thermal energy of the hot bubble interior, kinetic and gravitational energies of the shell. The kinetic and gravitational energies of the shell are determined by corresponding surface integrals. Variations of the total energy $E_{\rm tot}$ of a remnant or a bubble throughout the adiabatic stage of evolution are defined by the energy input rate $L(t)$, kinetic and thermal energies of the swept-up interstellar gas with temperature $T(x, y, z)$:

$$E_{\rm tot} = E_0 + \int_0^t \left[L(t) + \frac{1}{2} \int_{\lambda_{1,min}}^{\lambda_{1,max}} \int_{\lambda_{2,min}}^{\lambda_{2,max}} \dot{\mu}(V^2 + 3kT/\eta) {\rm d}\lambda_1 {\rm d}\lambda_2 \right] {\rm d}t, \quad (12)$$

where E_0 is the initially deposited energy and η is the mean mass per particle. Here E_g is neglected, and $L(t) = 0$ for SNR. During the radiative phase of expansion the gas behind the shock front cools so quickly that it does not add to the total energy of the remnant. Rarefied hot gas inside the cavity expands adiabatically and accelerates the surrounding dense shell. The time-derivative of the thermal energy of the remnant is defined then by the equation, which is used instead of (12):

$$\frac{dE_{\rm th}}{dt} = L(t) - \int_{\lambda_{1,min}}^{\lambda_{1,max}} \int_{\lambda_{2,min}}^{\lambda_{2,max}} P_{\rm in} u_n S(\lambda_1, \lambda_2) {\rm d}\lambda_1 {\rm d}\lambda_2, \quad (13)$$

where u_n is the velocity component normal to the shock front. The set of $(7N + 1)$ equations is solved using Adams method of 12 order. Numerical calculations of supershell formation in the plane-stratified and differentially rotating Galactic disk have been performed by Silich *et al.* (1994) and are shown in Fig. 1. An hourglass remnant with a noticeable degree of deformation by the Galactic shear has developed. Formation of elongated superbubbles due to differential galactic rotation gives a possibility to determine unambiguously the direction of galactic rotation (Mashchenko and Silich, 1995). The means of this determination is shown in the Fig. 2.

References

Bisnovatyi-Kogan G. S. and Blinnikov S.I. 1982, Astron. Zh. **59**, 876 [Sov. Astron. **26**, 530 (1982)].

Bisnovatyi-Kogan, G. S. and Silich S.A., 1991, Astron. Zh. **68**, 749 [Sov. Astron. **35**, 370 (1991)].

Bisnovatyi-Kogan, G. S. and Silich S.A., 1995, Rev. Mod. Phys. **67**, 661.

Budak B. M. and Fomin S.V. 1965, Multiple integrals and series (Nauka, Moscow).

Chernyi G. G., 1957, Doklady Acad. Sci. USSR **112**, 213.

Mashchenko S. Ya. and Silich S.A. 1995, Astron. Zh. **72**, 660.

Palouš J., 1990, in The Interstellar Disk-Halo Connection in Galaxies, edited by H. Bloemen (Srerrewacht, Leiden) p. 101.

Sedov L. P., 1946, Dok. Akad. Nauk SSSR, **42**, 17.

Silich, S. A., 1992, Astrophys. Space. Sci, **195**, 317.

Silich S. A., Franco J. Palouš, J. and Tenorio-Tagle G. 1994, in Violent Star Formation from 30 Doradus to QSOs, edited by G. Tenorio-Tagle (Cambridge Univ. Press, Cambridge) p. 162.

Interaction of Interstellar Shocks with Dense Obstacles: Formation of "Bullets"

V.V. Gvaramadze

Abastumani Astrophysical Observatory, Tbilisi, Georgia

Abstract. The so-called cumulative effect take place in converging conical shock waves arising behind dense obstacles overtaken by incident interstellar shock. A significant part of energy of converging flow of matter swept-up by a radiative conical shock can be transferred to a dense jet-like ejection ("bullet") directed along the cone axis. Possible applications of this effect for star-forming regions (e.g., OMC-1) and supernova remnants (e.g., Vela SNR) are discussed.

1 Introduction

Recently in some galactic objects were discovered unusual linear structures radially diverging from the common origin. The most striking examples of this phenomenon were observed in the Orion molecular cloud (Allen & Burton 1993) and the Vela supernova remnant (Aschenbach et al. 1995). These structures look like wakes trailed behind dense "bullets" moving with supersonic velocities through the ambient medium. The fact that wakes have approximately equal projected extents suggests the simultaneous origin of "bullets", possible, in an explosive event. Allen & Burton (1993) and Aschenbach et al. (1995) suggested that the "bullets" were ejected directly from the central source of activity (young massive star and supernova star respectively). We propose (Gvaramadze 1997a,b,c) an alternative explanation based on the interaction of a spherical shock wave with dense spherical obstacles (e.g., cloudlets or stellar wind zones). The "bullets" result from cooling and condensation of a gas swept-up by converging conical shocks arising behind obstacles overtaken by the incident (spherical) shock (see Fig. 1). The role of obstacles consists only in a transient focusing of the incident shock.

2 Formation of "Bullets" in Radiative Converging Conical Shock Waves

A converging conical shock forms (see Fig. 1) when a strong shock overtakes a spherical obstacle. The height of the cone is approximately equal to the radius of the obstacle. The cone half-opening angle ν depends on the curvature of the incident shock; for plane shocks, $\nu \simeq 35°$ (e.g., Tenorio-Tagle & Rozyczka 1984).

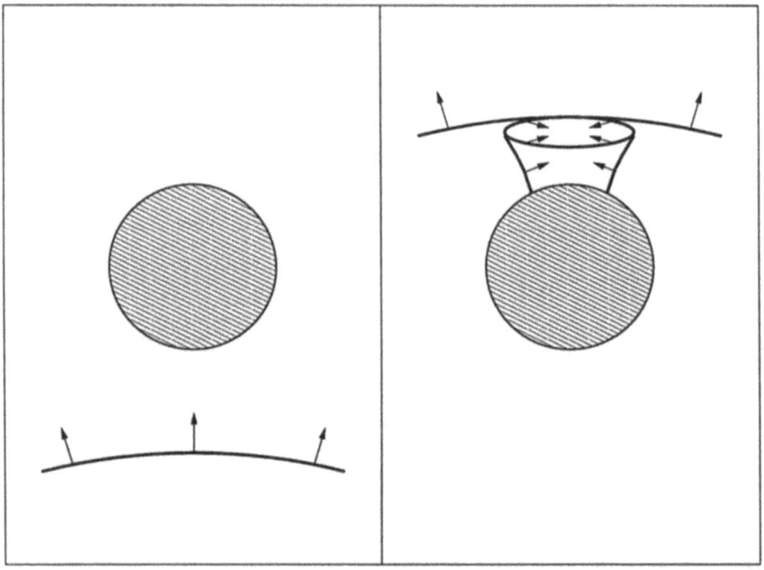

Fig. 1. Formation of a converging conical shock behind a dense obstacle.

An important property of a conical shock wave is that it creates, due to the Mach reflection, an axial pressure gradient which drives a high-velocity mass ejection along the cone axis (for details and references see Gvaramadze 1997a,b). Effects of such type are called cumulative (see Gvaramadze (1996) and references therein). The velocity of ejection (the so-called cumulative jet) is given by, $v_{\rm ej} \simeq U/\sin\nu$, where U is the velocity of the converging conical shock, ν is the cone half-opening angle; $v_{\rm ej} \simeq 1.7U$ for $\nu = 35°$. It is obvious that due to the finite dimensions of real conical flows, the formation of cumulative ejections is a transient process. Furthermore, the ejection in itself is unstable because the overpressure, created by the converging flow, tends to re-expand it. A significant step in understanding of formation of ejections in converging conical shocks was made by Tenorio-Tagle & Rozyczka (1984). They shown that the radiative cooling of shock waves favours the formation of long-living ejections ("bullets"). The sound speed inside a dense and cold "bullet" formed in a radiative conical shock can be much smaller than the velocity of directed motion. Such a "bullet" can propagate through the ambient medium without significant dispersion.

3 Formation of the Fan of Ejections in the OMC-1

The reader is referred to Gvaramadze (1997a,b), who describes the formation of the fan of ejections in the OMC-1 in detail.

4 Origin of X-ray Protrusions in the Vela SNR

The optical observations (e.g., Parker et al. 1979) and the X-ray observations (e.g., Aschenbach et al. 1995) of the Vela SNR reveal very complex and asymmetrical structure of this remnant (see Fig. 2), with some features extended up to $\simeq 5°$ from the center (associated with the Vela pulsar position). We suggest (Gvaramadze 1997c) that the shape of the Vela SNR might be explained as the result of interaction of the SN ejecta/shock with the preexisting wind-driven shell (WDS) created in the interstellar medium with a density gradient perpendicular to the Galactic plane. The mass of the WDS is a decisive factor, which determines the evolution of the SN shock. If the mass of the WDS is smaller than $\simeq 50M_{\odot}$, the SN shock overruns the shell and continues to evolve adiabatically (e.g., Franco et al. 1991). For more massive shells, the SN shock losses, due to the radiative cooling, a significant part of their energy and merges with the WDS. In this case, the SN remnant evolves as a radiative shock. These two cases can coexist in one and the same SN remnant if the WDS was blown-up in the density stratified medium. Apparently just this situation take place in the case of Vela SNR.

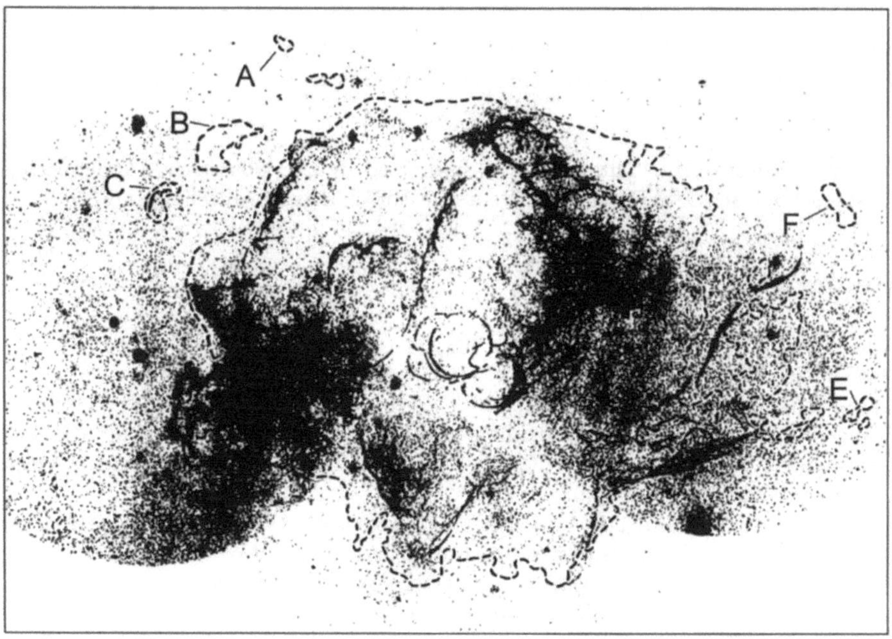

Fig. 2. Mosaic of [O III] λ 5010 filter photographs of the Vela SNR (Parker et al. 1979). The boundary of the ROSAT image of the remnant (Aschenbach et al. 1995) is outlined by dashed contours. Position of the Vela pulsar is indicated by a cross.

The number density of the interstellar medium, n_0, in the direction towards the Galactic plane can be estimated from the consideration of the dynamics of the "bullet" A (see Strom et al. 1995). It is connected with the number density inside the bullet, n_b, through the simple relation: $n_0 \simeq n_b r_b / L$, where r_b is the characteristic size of the "bullet" ($\simeq 1$ pc), L is the distance it has travelled. Provided the "bullet" was ejected from the SN explosion site, i.e. $L \simeq 50$ pc, and for $n_b \simeq 0.4 - 0.9\mathrm{cm}^{-3}$ (Strom et al. 1995, Aschenbach et al. 1995), it follows from the above relation that $n_0 \simeq 0.01 - 0.02\ \mathrm{cm}^{-3}$. However, n_0 is few times higher than it was estimated by Strom et al. (1995) if the "bullet" was formed at some distance (≥ 30 pc) from the center. We adopt the value $n_0 \simeq 0.03\ \mathrm{cm}^{-3}$. In this case, the mass of the WDS of radius $\simeq 30$ pc is $\geq 50 M_\odot$. This means that the SN shock cannot overrun the norteast half of the WDS faced towards the local density increase (see, e.g., Dame et al. 1987) and evolves as a radiative one. The interaction of this (radiative) half of the SN shock with dense obstacles (cloudlets, stellar wind zones) can produce cumulative ejections observed by Aschenbach et al. (1995) as X-ray "bullets" A, B and C. Note that the X-ray "bullet" C has an optical counterpart (see Fig. 2), consisting of two arc-like filaments resembling bow shocks.

The impact of the SN shock with the southwest (less massive) half of the WDS leads to its strong deformation and even disruption in places (see Fig. 2). A hot gas of SN interior escapes through the gaps on the SN shell and forms two outflows partially bounded by filamentary structures (see Fig. 2). These outflows appear as the X-ray protrusions E and F (Aschenbach et al. 1995).

Acknowledgements. The work was supported by the Deutscher Akademischer Austauschdienst (DAAD) and the Deutsche Forschungsgemeinschaft (DFG).

References

Allen D.A., Burton M.G. (1993): Nat 363, 54
Aschenbach B., Egger R., Trümper J. (1995): Nat 373, 587
Dame T.M. et al. (1987): ApJ 322, 706
Franco J., Tenorio-Tagle G., Bodenheimer P., Rozyczka M. (1991): PASP 103, 803
Gvaramadze V.V. (1996): In W.Kundt (ed.), Jets from Stars and Galactic Nuclei, Springer-Verlag, Berlin, p.92
Gvaramadze V.V. (1997a): In F.Malbet, A.Castets (eds.), Low Mass Star Formation – from Infall to Outflow, poster proceed. of IAU Symp. No. 182, p. 129
Gvaramadze V.V. (1997b): Astronomy Letters 23, No. 8, in press
Gvaramadze V.V. (1997c): in preparation
Parker A.R., Gull T.R., Kirschner R.P. (1979): An Emission-Line Survey of the Milky Way, NASA SP-434
Strom R., Johnston H.M., Verbunt F., Aschenbach B. (1995): Nat 373, 590
Tenorio-Tagle G., Rozyczka M. (1984): A&A 141, 351

Magnetorotational Mechanism: 2D Simulation

N.V. Ardeljan[1], G.S. Bisnovatyi-Kogan[2], and S.G. Moiseenko[2]

[1] Department of Computational Mathematics and Cybernetics, Moscow State University, Vorobjevy Gory, Moscow B-234 119899, Russia
ardel@redsun.cs.msu.su

[2] Space Research Institute, Profsoyuznaya 84/32, Moscow 117810, Russia
gkogan@mx.iki.rssi.ru, moiseenko@mx.iki.rssi.ru

Abstract. In the paper we present results of numerical simulations of magnetorotational model of explosion of magnetized cloud. For the simulation we used a specially developed 2D implicit numerical scheme on Lagrangian triangular grid with grid reconstruction. Our results show that due to the amplification of toroidal magnetic field component and transfer of angular momentum, a compression wave in the envelope of the cloud appears and moves through rapidly decreasing density background. This wave turns into a shock wave and pushes out part of the envelope of the star. Quantitative estimates of mass and energy carried away are given.

1 Introduction

One of the possible mechanisms of supernova explosions is magnetorotational mechanism, suggested by Bisnovatyi-Kogan (1970). It is based on the transformation of the gravitational energy to the energy of explosion by magnetic field. The poloidal magnetic field in differentially rotating star is twisted and toroidal component of the magnetic field appears which grows with time. When the force produced by the magnetic field becomes comparable to the gravitational force it pushes the matter of the star outwards.

Simulation of the collapse of magnetized rotating gas cloud in 2D has been done in number of papers, see for reference Ardeljan et al. (1996b).

Realistic magnetic fields in the stars are rather small (the ratio between magnetic and gravitational energies is about $10^{-6} - 10^{-8}$). Such weak magnetic fields are the main difficulty for the numerical simulations, because of the very small hydrodynamic time scale and large time scale of the magnetic field amplification.

Numerical simulation of such stiff problem requires application of the implicit numerical methods, which are free from Courant restriction on the time step.

2 Formulation of the Problem, Numerical Method

The problem of the collapse of rotating magnetized gas cloud is described by the set of magnetohydrodynamical equations with selfgravitation for gas with infinite conductivity.

For the initial conditions, we assume that the cloud is a rigidly rotating uniform gas sphere with the following parameters:

$$r = 3.81 \cdot 10^{16}\mathrm{cm},\ \rho = 1.492 \cdot 10^{-17}\mathrm{g/cm}^3,$$

$$M = 1.73 M_\odot = 3.457 \cdot 10^{33}\mathrm{g},\ \gamma = 5/3,\ u^r = u^z = 0,$$

$$E_{\mathrm{rot0}}/|E_{\mathrm{gr0}}| = 0.04, \quad E_{\mathrm{in0}}/|E_{\mathrm{gr0}}| = 0.01, \quad E_{\mathrm{mag1}}/|E_{\mathrm{in1}}| = 0.05.$$

For our simulations we used implicit conservative Lagrangian operator difference scheme on triangular grid with reconstruction (Ardeljan et al. (1996a)). Grid restructuring procedure allows us to overcome grid overlapping situation in gas flows with nonuniform contractions or expansions or flows with vortexes. Due to the implicitness of the scheme it was possible to use time steps, which are much larger then Courant time step.

3 Results

The initial number of grid knots was 5000. At the initial time moment ($t = 0$) magnetic field is "switched off", because it does not significantly influence the hydrodynamical collapse stage which is rather short in comparison with time of the amplification of the magnetic field.

After a few consequent contractions and expansions the cloud comes to the differentially rotating stationary state, at $t_1 = 5.153$ and consists of a rapidly rotating dense core and an extended light envelope. The angular velocity of the cloud at this time presents a function with maximum of gradient at the transition zone between core and envelope.

At the moment $t_1 = 5.153$ an initial magnetic field of the following configuration (Ardeljan et al. (1996b)) (Figure 1):

$$H_{r0} = F_r(0.5r, 0.5z - 2.5) - F_r(0.5r, 0.5z + 2.5)\,,\ H_{\varphi 0} = 0\,,$$

$$H_{z0} = F_z(0.5r, 0.5z - 2.5) - F_z(0.5r, 0.5z + 2.5)\,,$$

$$F_r(r,z) = k\left(\frac{2rz}{(z^2+1)^3} - \frac{2r^3 z}{(z^2+1)^5}\right)\,,\ F_z(r,z) = k\left(\frac{1}{(z^2+1)^2} - \frac{r^2}{(z^2+1)^4}\right)\,,$$

has been "switched on" with the coefficient $k = 0.43$. This magnetic field is symmetrical about the equatorial plane, it fulfills $\mathrm{div}\mathbf{H}_0 = 0$, but is not force-free. The choice of initial magnetic field configuration is connected with an attempt to avoid a singularity, which could create numerical problems. After the moment of "switching on" the magnetic field, the matter starts to move to the periphery of the envelope due to the magnetic force produced by

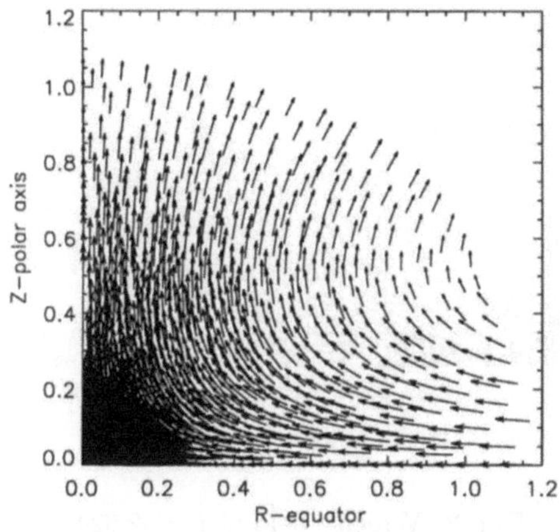

Fig. 1. Initial magnetic field at $t = 5.153$.

the initial configuration of the magnetic field, which leads to the appearance of a "finger" which grows at the outer boundary of the cloud. This artificial "finger" has no significant influence on the evolution of the inner parts of the cloud. Due to the quadrupole type of symmetry of the poloidal magnetic field the appearing toroidal component has 2 extrema. The first is situated at the equatorial plane in the zone between the core and the envelope and the second is situated above the equatorial plane close to the z-axis.

At $t = 10.234$ the toroidal part of magnetic energy reaches its maximum. While the total energy of the toroidal component of the magnetic field of the cloud is much smaller than its gravitational energy, the density of the toroidal magnetic energy in the regions near the extrema of H_φ becomes comparable to the density of the internal energy which pushes the matter out of the cloud. Starting from $t = 11.342$ the kinetic energy of the accelerated part of the envelope near equatorial plane becomes higher than its potential energy and can be ejected from the cloud. The ejection along the equatorial plane leads to a change of the shape of the cloud, which is stretched along r-axis (in r, z coordinates), (see Figure 2), related to the last computational moment $t = 32.634$.

At Figure 3 time evolution of the ejected mass of the envelope in percentage to the total mass of the cloud is presented. The ejected part of the matter contains about 0.09% of the gravitational energy of the stationary cloud after the collapse.

The model of a collapsing, rotating magnetized gas cloud, described above cannot be considered as a complete and final explanation of the mechanism of the supernova explosions. However the results of such simplified formulation give evidence in favour of the magnetorotational mechanism can as an explanation for the problem of the supernova explosions.

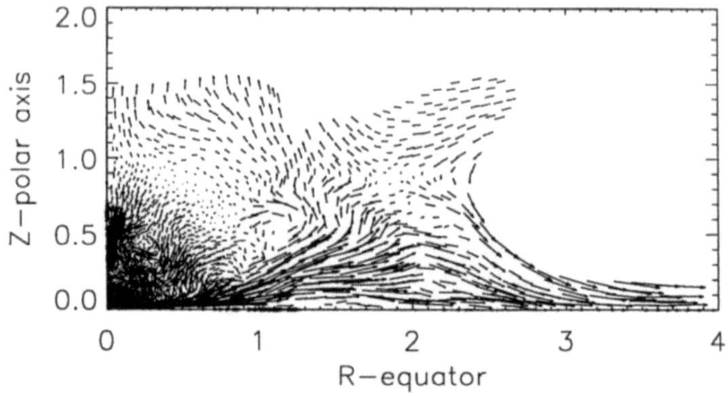

Fig. 2. Velocity field at $t = 32.624$.

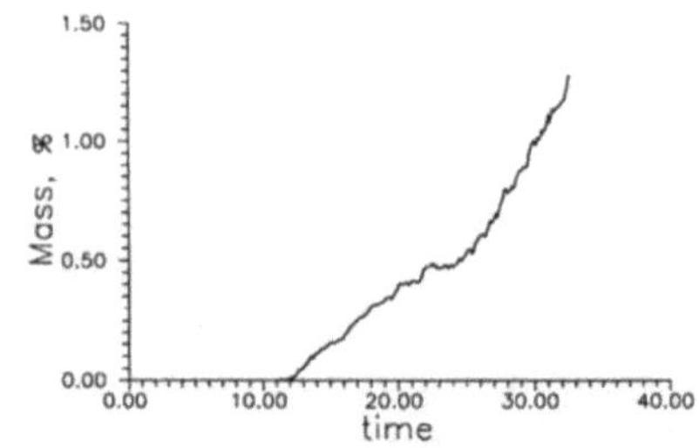

Fig. 3. Time evolution of the ejected mass in % to the total mass of the cloud.

Our recent simulations of the collapse of the star with more realistic equations of state and neutrino losses as in the paper by Ardeljan et al. (1987) show that the star contracts much stronger and rotates much more differentially, than for the simple equation of state used in this paper. We may expect a more efficient magnetorotational explosion.

Acknowledgements. S.G.M. and G.S.B.-K. are grateful to the partial support by the NSF grant AST-9320068, CRDF award # Rp1-173, and RFBR grants 96-02-16553, 97-02-26559. G.S.B.-K. thanks INTAS grant No.93-93. N.V.A. is thankful to the RFBR for the support by the grant 96-01-00838.

References

Ardeljan N.V., Bisnovatyi-Kogan G.S., Kosmachevskii K.V., Moiseenko S.G., 1996a, A&AS, 115, 573

Ardeljan N.V., Bisnovatyi-Kogan G.S, Moiseenko S.G., 1996b, Ap&SS, 115, 573

Ardeljan N.V., Bisnovatyi-Kogan G.S., Popov Ju.P., Chernigovskii S.V., 1987, AZh, 64, 761

Bisnovatyi-Kogan G.S., 1970, AZh, 14, 652

Part IV

Clouds, Ionized Gas, and Particles in the Local ISM

Coffee breaks with conversations ...

... and serious discussions.

High-Resolution Optical Observations of Diffuse Clouds

Daniel Welty

University of Chicago, Dept. of Astronomy & Astrophysics, 5640 S. Ellis Ave., Chicago, IL 60637, USA

Abstract. We describe results from high-resolution (Δv = 0.3-1.5 km s^{-1}) observations of absorption lines due to interstellar Na I, Ca II, K I, Ca I, and Ti II. At those resolutions, we can identify and characterize many (but not all) of the individual clouds along a given line of sight; complex spatial and/or velocity structure appears to be the norm. Both the spectra and the statistics of individual cloud properties derived from fits to the observed line profiles suggest that: 1) the five species are not entirely coextensive in the neutral ISM; 2) the typical separation between adjacent components (in Na I and Ca II) is ~1.2 km s^{-1}; 3) the median line widths (FWHM) for components in Na I, K I, and Ca I are $\lesssim$1.2 km s^{-1}; 4) larger line widths for Ca II, even for "corresponding" components at similar velocities, suggest that Ca II is more widely distributed; 5) line widths for clouds in the Galactic halo and in the local ISM (within 100 pc) are generally somewhat larger than for clouds in the general Galactic disk; and 6) components with very different properties (line width, relative abundances, overall column densities) are often separated by only several km s^{-1}. Observations of multiple systems and lines of sight in restricted regions reveal complex spatial structure and velocity coincidences even within 100 pc. These high-resolution optical spectra can be used to model lower resolution UV spectra of many other neutral and singly ionized species, enabling abundances and physical properties to be derived for individual interstellar clouds.

1 Introduction

One of the goals of studies of the interstellar medium (ISM) is the determination of the physical properties of the *individual* clouds which constitute the neutral component of the ISM — their temperatures, densities, relative abundances, electron fractions, internal turbulent motions, and structures — and how those properties differ for clouds in different environments. Accurate estimates of these quantities, for a significant set of individual interstellar (IS) clouds, are required to test any fundamental, global model of the ISM. Determination of individual cloud properties requires both high spectral resolution and access to the satellite UV, where the resonance lines of many important species are found. Systematic high-resolution observations of optical IS absorption lines were pioneered by Hobbs and collaborators starting nearly 30 years ago (e.g., Hobbs 1969; Marschall & Hobbs 1972; Hobbs 1984; see also Blades, Wynne-Jones, & Wayte 1980). Meyer (1994) has summarized some of the more recent moderate-to-high resolution optical and UV absorption-line studies of the ISM; Savage & Sembach (1996) have reviewed analyses of UV spectra obtained with the *HST* Goddard High-Resolution Spectrograph (GHRS; $\Delta v \sim 3.5$ km s^{-1} for the echelle modes). In this contribution, we describe results obtained from recent very high resolution ($\Delta v = 0.3$-1.5 km s^{-1}) observations of optical absorption lines due to IS Na I, Ca II, K I, Ca I, and Ti II toward stars (mostly) within 500 pc. As will be apparent, such high resolution is necessary to discern and characterize the individual IS clouds present along many lines of sight — e.g., to distinguish clouds associated with the Local Bubble from more distant clouds. High-resolution optical spectra are thus useful for interpreting lower resolution UV spectra obtained, e.g., with the GHRS — to obtain detailed abundances and physical properties for individual IS clouds, instead of line-of-sight or component blend averages.

2 Methodology and Goals

Figure 1 shows representative spectra of various IS absorption lines toward the bright B0 Ia star ϵ Ori, obtained with the coudé spectrograph on the McDonald Observatory 2.7m (Tull 1972), the Ultra-High Resolution Facility (UHRF) on the AAT (Diego et al. 1995), and the coudé feed and 2.1m echelle spectrograph at Kitt Peak. We use the method of profile fitting to determine column densities (N), line widths ($b \sim$ FWHM/1.665), and velocities (v) for the discernible individual components (which we assume correspond to distinct IS clouds) contributing to the observed profiles. Surveys of Na I (38 stars; Welty, Hobbs, & Kulkarni 1994) and Ca II (44 stars; Welty, Morton, & Hobbs 1996) have provided new information on the statistical properties of individual IS clouds:

- Distributions of individual cloud properties (N, b, v)
- Relationships among parameters (N vs. b, b_X vs. b_Y, N_X/N_Y vs. b)
- Limits on temperature (T) and turbulence (v_t) from line widths
- Information on cloud structure from comparisons of different species
- Suggestions of environmental and/or regional differences

By combining the high-resolution optical spectra with UV spectra from GHRS (Fitzpatrick & Spitzer 1997; Welty et al. 1997a), one can undertake detailed studies of individual lines of sight sampling diverse IS environments:

- Use the high-resolution optical spectra to model the UV spectra
- Determine detailed abundance/depletion patterns for individual clouds
- Determine local physical conditions (T, n_H, n_e) for individual clouds
- Compare depletions with local physical conditions

High-resolution optical spectra are also being used to study small-scale spatial structure in the ISM (Watson & Meyer 1996) and to identify, delineate, and characterize individual clouds in the local ISM (Crawford, Craig, &

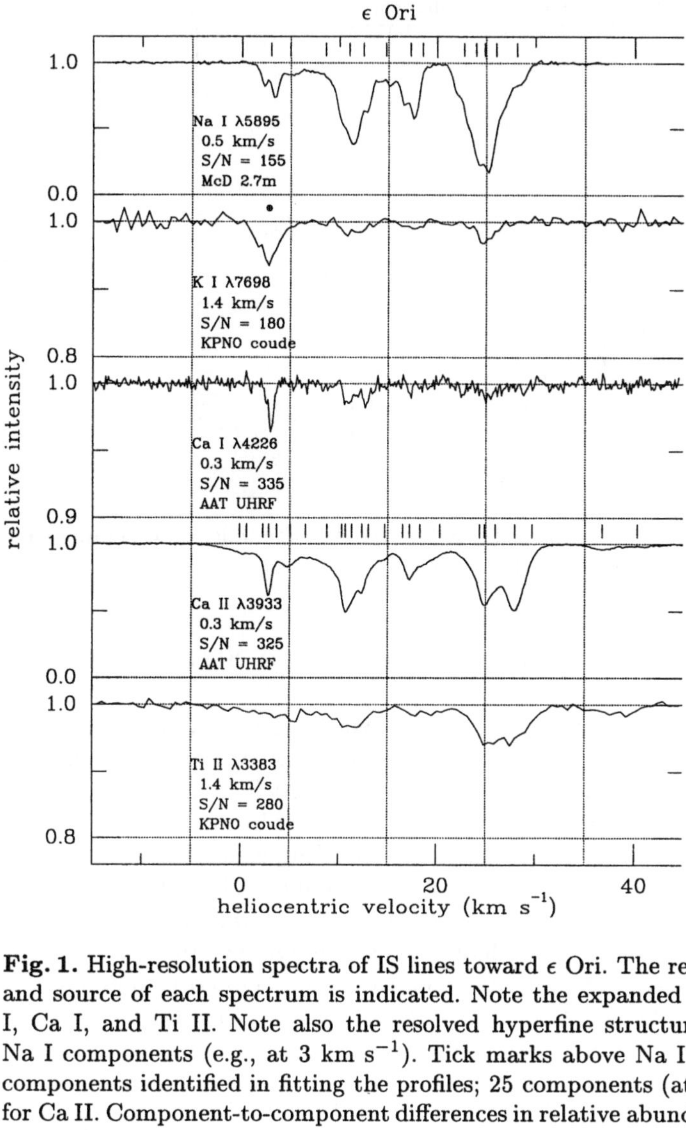

Fig. 1. High-resolution spectra of IS lines toward ϵ Ori. The resolution, S/N ratio, and source of each spectrum is indicated. Note the expanded vertical scale for K I, Ca I, and Ti II. Note also the resolved hyperfine structure in the narrowest Na I components (e.g., at 3 km s^{-1}). Tick marks above Na I and Ca II indicate components identified in fitting the profiles; 25 components (at least) are required for Ca II. Component-to-component differences in relative abundance and line width are evident.

Welsh 1997). While we focus in this contribution on high-resolution spectra of atomic species in diffuse clouds, we note that similar spectra of CH, CH^+, CN, and C_2 are allowing more detailed, component-by-component comparisons of molecular column densities, line widths, and velocities — thus providing more stringent constraints on the chemical processes operating in somewhat thicker IS clouds (e.g., Crane, Lambert, & Sheffer 1995; Crawford 1995; Sembach, Danks, & Lambert 1996).

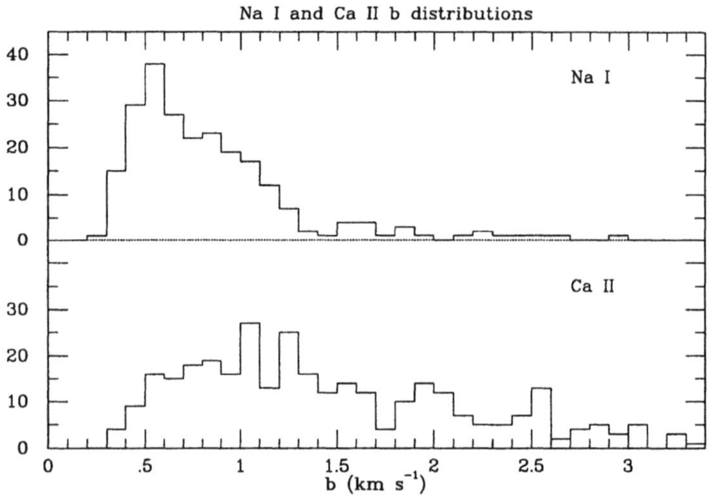

Fig. 2. Distribution of line width b ($\sim$ FWHM/1.665) for individual components in the Na I (upper) and Ca II (lower) surveys.

3 Temperature, Turbulence, and Cloud Structure

Measurements of individual component line widths, for a given species with atomic weight m, yield upper limits on the temperature and turbulence in the corresponding IS clouds, since $b = [(2kT/m) + 2{v_t}^2]^{1/2}$. If two species of significantly different m (e.g., Ca II and Na I) coexist in the same volume of gas, comparison of the widths of their corresponding components can allow estimates of the relative contributions of thermal and turbulent broadening. Comparison of the component widths for two species of similar atomic weight (e.g., Ca II and K I), on the other hand, can provide some indication of their relative spatial distribution. For example, if the widths of corresponding Ca II and K I components at the same velocity are different, then the two species cannot be identically distributed.

Figure 2 shows the distribution of b for the individual components seen in Na I (Welty et al. 1994) and Ca II (Welty et al. 1996). Even though the lines of sight sampled in the two surveys are very similar, the median b-values differ significantly: 0.73 km s^{-1} for Na I and 1.31 km s^{-1} for Ca II. Preliminary results for K I and Ca I suggest median b-values of about 0.65-0.70 km s^{-1} for both species. For all four species, the smallest b-values measured are in the range 0.3-0.4 km s^{-1}. Since the thermal contribution to the b-values, for T = 100 K, is 0.27 km s^{-1} for Na I and 0.21 km s^{-1} for Ca II, K I, and Ca I, the observed minimum and median b-values suggest either that the cloud temperatures are higher, that there is a significant turbulent contribution to the line widths, and/or that most narrow lines are obscured by blending. If 80 K is a representative temperature for the clouds detected in Na I, then at least 38% (and perhaps the majority) of those clouds have subsonic internal turbulent motions [assuming an isothermal sound speed $v_s \sim 0.7(T/80\ \mathrm{K})$]. If most of the gas traced by Ca II absorption is warmer than about 400 K, then the turbulence in that gas would also be subsonic.

Different subsets of the Na I and Ca II component samples can have different b-value distributions. Ca II components observed toward stars in the Galactic halo and components seen only in Ca II toward disk stars both have very broad distributions, with $b_{\mathrm{med}} \sim$ 1.6-1.7 km s^{-1}. Ca II components found within 100 pc are on average somewhat broader, with $b_{\mathrm{med}} \sim$ 2.1 km s^{-1} (e.g., Crawford et al. 1997). That median value is consistent with the $T \sim$ 6000-7000 K and $v_t \sim$ 1 km s^{-1} derived from UV spectra of very nearby clouds (Piskunov et al. 1997). There are, however, some cold components, with $b \lesssim 0.5$ km s^{-1}, both in the Galactic halo (e.g., Pettini 1988) and in the local ISM within 50 pc (Welty et al. 1994, 1996; Jenkins 1997).

Given the definition of the line-width parameter b, we would expect, if Ca II and Na I are similarly distributed, that $0.76 \leq b(\mathrm{Ca\ II})/b(\mathrm{Na\ I}) \leq 1.0$, where the lower limit corresponds to purely thermal broadening and the upper limit to purely turbulent broadening. From the Na I and Ca II surveys, we have identified 72 "corresponding" components for which the velocities agree to within 0.2 km s^{-1} and for which any significant near neighbor components also have similar velocities for both species. Contrary to the above expectation, b(Ca II) is often *greater* than b(Na I) for the individual corresponding components; the median b-values for the sub-sample of 53 best-determined corresponding components are 0.84 and 0.65 km s^{-1} for Ca II and Na I, respectively. The most straightforward explanation for this difference in b-values is that Ca II and Na I are not identically distributed, even though similar component velocities suggest they are associated (see also Blades et al. 1997). For a given velocity component, Ca II probably occupies a somewhat larger volume, characterized by a larger temperature and/or greater turbulence, than does Na I — likely related to enhanced grain erosion (and thus less severe Ca depletion) in the warmer, more diffuse gas outside the cloud cores (Barlow et al. 1995).

Detailed component-by-component comparisons of high-resolution spectra of Ca II, Na I, K I, Ca I, and Ti II provide additional insight into the structure of neutral IS clouds. The five species are characterized by different ionization and depletion behavior, and as a result appear to be distributed somewhat differently. Comparisons of high-resolution Ca II and Ti II spectra seem to show detailed correspondence in velocity and relative column density in many cases (Welty, Lauroesch, & Fowler 1997), consistent with a suggestion that Ca II and Ti II are often found in warm, diffuse gas, where both Ca and Ti are much less depleted than in cold, dense clouds (Crinklaw, Federman, & Joseph 1994; see also Stokes 1978). In such warm, diffuse H I gas, Ca II ($\chi_{\mathrm{ion}} \sim$ 11.9 eV) is presumably a trace ionization state, while Ti II ($\chi_{\mathrm{ion}} \sim$ 13.6 eV) is dominant. We note, however, that the observed line widths imply that at most 40% of the Ca II components could arise in gas with $T \gtrsim 6000$ K. In addition, there are relatively narrow Ca II components that correspond to narrow components seen in Na I and K I, for which Ti II is not similarly enhanced (see Fig. 1). The strong, narrow components in Na I and K I presumably arise primarily in the colder, denser cloud cores, since as trace ionization states their abundances should be proportional to $n_{\rm H}^2$. The Ca I absorption detected for some of those lines of sight also appears to arise in those narrow components (Welty, Hobbs, & Morton 1997), and the Ca I/Ca II ratio suggests that Ca II is the dominant form of Ca there. Apparently the transition of Ca II from a trace to a dominant species can substantially offset the strong depletion of Ca in the denser cloud cores. The interplay between two strongly variable effects — the Ca depletion and the Ca II–Ca III ionization balance — may explain the otherwise surprising similarity between the Ca II profiles and the profiles of other singly ionized species such as Zn II, Si II, and Fe II (which are typically dominant ionization stages in both warm and cold H I regions) seen in GHRS echelle spectra.

4 Complex Velocity and Spatial Structure

Examination of individual high-resolution, high S/N ratio spectra (e.g., Fig. 1) indicates that the component structures can be very complex. Furthermore, when the Ca II component structures are compared with those derived independently for Na I for the same lines of sight, we find that while a number of components do show good correspondence, there are also many which do not line up — which may indicate unresolved structure in one or both of the species. If the component velocities for the true complete sample of clouds (for either Ca II or Na I) are both uncorrelated and taken from a single Poisson distribution, then the distribution of velocity separations (δv) between adjacent components will be an exponential function of δv: $\ln[{\rm N}(\delta v)] = \alpha + \beta \times \delta v$. Figure 3 shows that the δv distributions are well-fitted by exponentials (with very similar slopes) at intermediate δv, as would be expected for uncorrelated component velocities. The slight excesses over the fits at higher δv are likely due to a population of higher velocity clouds. The fall-offs at small δv, however, suggest that unresolved component structure is present, even in these very high-resolution (Δv = 0.3-1.2 km s^{-1}) spectra. Extrapolations of the fits to $\delta v = 0$ suggest that we have discerned only about 60% and 40% of the true number of Na I and Ca II components, respectively, and that the true median δv for both species is about 1.2 km s^{-1}. We see no evidence for any non-random, preferential clumping of components, which would cause a steepening of the δv distributions, at least for $\delta v \gtrsim$ 1-2 km s^{-1}.

High-resolution spectra of Na I and Ca II have also revealed indications of significant small-scale spatial structure in the ISM. Meyer & Blades (1996) presented UHRF spectra of Na I and Ca II toward both members of the binary μ Cru, which has a projected separation of 6600 AU at a distance

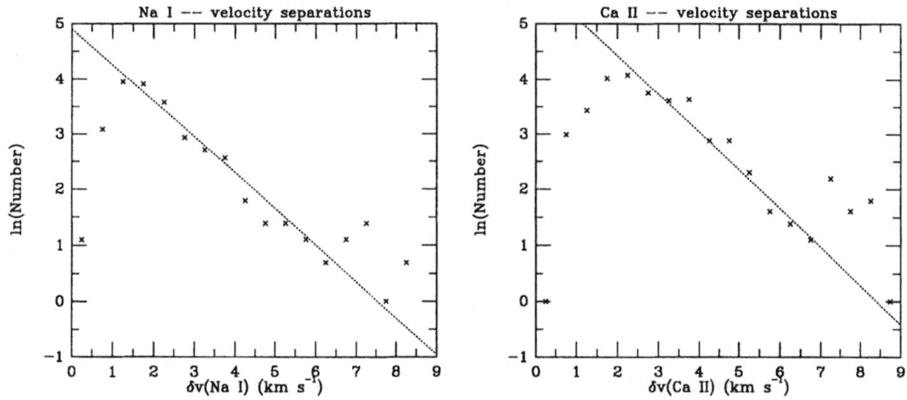

Fig. 3. Distribution of adjacent component separations [ln N(δv)] for the Na I (left) and Ca II (right) surveys. Dotted lines indicate linear fits for intermediate values of δv [corresponding to exponential fit to N(δv)].

of 170 pc. They found dramatic differences in N(Na I) (factors $\gtrsim$ 4) for three of the seven components in the Na I profiles, with corresponding (but smaller) differences for Ca II. In Na I spectra obtained at a resolution of 1.4 km s^{-1}, Watson & Meyer (1996) found differences in the line profiles for *all* of the 17 multiple systems observed, with projected separations of 500-30,000 AU. [By contrast, earlier lower resolution Ca II spectra of 12 binary systems had revealed only several cases with significant differences (Meyer 1994).] The differences seem to be most pronounced for Na I, and in particular for the narrowest components and/or those with higher Na I/Ca II ratios — consistent with the picture of cloud structure and species distribution discussed above. The differences in N(Na I) imply density contrasts of order 10^3 cm^{-3} — similar to those determined for the small (5-100 AU) neutral clouds inferred in several 21 cm studies (e.g., Frail et al. 1994). This apparently pervasive structure may be due to very cold ($T \sim 15$ K), dense filaments or sheets, embedded in warmer, less dense neutral gas and containing 10-30% of the total column density of cold, neutral gas (Heiles 1997).

The complex velocity structure found along many lines of sight and the small-scale spatial structure observed over small angular scales suggests that determining the detailed three-dimensional structure of the cold, neutral component of the ISM will be very difficult. Mapping the warmer, presumably more smoothly distributed material in the local ISM should be more feasible, but will still require sensitive, high resolution observations of many stars with well determined distances. High-resolution Ca II spectra of stars within 50 pc have revealed multiple components and velocity coincidences (i.e., components with similar velocities but different distances along a given line of sight) even within that restricted volume (Crawford & Dunkin 1995; Crawford et al. 1997; Welsh, Crawford & Lallement 1997; Frisch & Welty 1997).

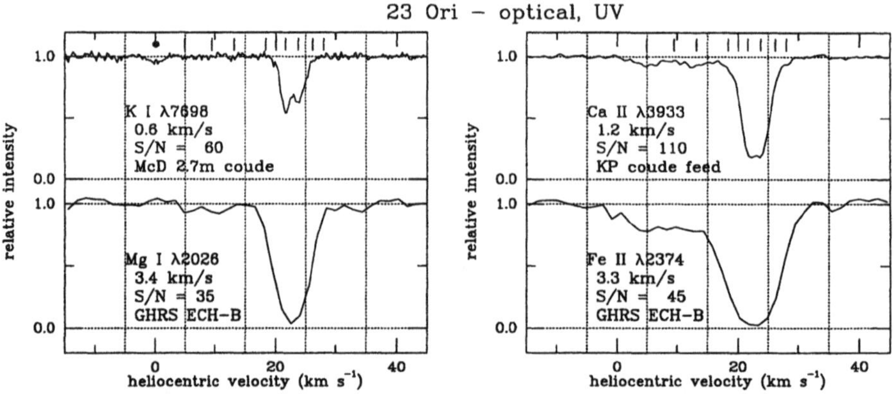

Fig. 4. Detailed component structure (tick marks) seen in high-resolution optical K I and Ca II spectra of 23 Ori cannot be discerned in lower resolution UV spectra of Mg I and Fe II obtained with GHRS. The optical data can be used to model the UV lines, however, to obtain detailed abundances and physical conditions for individual clouds.

5 Abundances, Depletions, and Physical Conditions

The statistics of component line widths and adjacent component separations imply that, for many lines of sight, much of the component structure will not be discernible, and many individual components will not be resolved, in UV spectra obtained with GHRS or STIS. Furthermore, the high-resolution optical spectra show that components with very different properties (e.g., overall column densities, line widths, Na I/Ca II ratios) are often separated by only several km s^{-1} along a given line of sight. Fortunately, however, the high-resolution optical spectra can be used to model the lower resolution UV spectra, so that abundances and physical conditions can be derived for individual IS clouds from the transitions observed in the UV spectra.

Figure 4 shows high-resolution optical spectra of K I and Ca II (upper) and GHRS echelle spectra of Mg I and Fe II (lower) observed toward the B1 V star 23 Ori (Welty et al. 1997a). Profile analysis of the high-resolution optical spectra of Na I, K I, Ca I, and Ca II indicates that the absorption near +22 km s^{-1} is due to two main components, separated by about 2.1 km s^{-1}, plus several weaker outlying components. The apparent similarity in b-values for the two strong components in Na I and K I suggests that the gas is cold, with the line widths dominated by turbulent broadening. For those two strongest components, the column densities for K I, Ca II, and Ca I can be used, together with the equation of ionization equilibrium, to estimate both the relative n_e and the relative column densities of many other neutral and singly ionized species observed (but not resolved) in the UV spectra. The absorption seen between 0 and +15 km s^{-1} is due to at least three weaker, somewhat broader components — probably warmer, lower density gas. Application of this component structure model to the GHRS echelle spectra yields individual component column densities for Zn II, Fe II, Si II, Cr II, Ni II, Mg I, S I, Si I, and Fe I, with formal uncertainties typically ±0.1 dex or better. For several neutral–first ion pairs in the two strong components, we find $n_e \sim 0.06 \pm 0.03$ cm^{-3}, assuming $T \sim 100$ K (from *Copernicus* H_2 data) and photoionization by the "typical" IS radiation field. Analysis of the C I fine structure excitation (Jenkins & Shaya 1979) yields a pressure $\log(n_H T) \sim 3.1$ cm^{-3}K, and thus $n_H \sim$ 10-15 cm^{-3} (for $T = 100$ K) and a total thickness of about 12-16 pc for the two main clouds. The fractional ionization $n_e/n_H \sim 5 \times 10^{-3}$ implied for those two clouds is a factor ~ 15 larger than expected from photoionization of heavy elements. If N(Zn II) is used to infer N(H) for the individual components, we can examine the depletions of different elements in the various components. The depletions in the components at 0 to +15 km s^{-1} are similar to those found for warm, low density gas. It is intriguing that the two strongest components toward 23 Ori have both lower n_H and less severe depletions than the main component(s) observed toward ζ Oph (Savage, Cardelli, & Sofia 1992). Analyses of additional lines of sight will be needed to test any possible general relationship between n_H and depletion — and thus to constrain the processes giving rise to the observed depletions.

Fitzpatrick & Spitzer (1997) have performed a detailed study of the line of sight to the halo star HD 215733, using high-resolution Ca II and H I data to help interpret GHRS echelle spectra of many other species. They find 23 neutral (H I) components between −95 and +15 km s^{-1}, with b(heavy elements) ranging from 0.5-6.6 km s^{-1}, $\log[N(\mathrm{H})]$ from 17.5-20.1 cm^{-2}, and T from $\lesssim 200$ K to $\gtrsim 1000$ K. The depletion of Fe is always at least a factor of 4, even at $|v_{LSR}| \gtrsim 50$ km s^{-1}. Estimates of n_e based on the ionization equilibria of C, Mg, S, and Ca appear to differ systematically, perhaps due to charge exchange or other processes; n_e/n_H in several cold clouds may be consistent with the electrons coming from photoionization of heavy elements.

We conclude with a few observations noted in comparing various quantities (individual component or integrated line-of-sight values):

- N and b show no obvious correlation, for either Na I or Ca II.
- Na I/Ca II shows a wide range for any $b \lesssim 1.5$ km s^{-1}, and Na I/Ca II $\lesssim$ 1.0 can occur for a wide range of b (e.g., even in gas with $T \lesssim 1000$ K).
- Na I and H I may be reasonably well correlated for $N(\mathrm{Na\ I}) \gtrsim 10^{11}$ cm^{-2}, but not for lower N(Na I).
- Ca I/Ca II shows little, if any variation with f(H_2) (fraction of H in molecular form, which should depend on n_H and radiation field). Does n_e not depend on n_H?
- Electron densities calculated from Ca I/Ca II, and depletions of Na, K, and Li inferred from those n_e, may be systematically in error.

Acknowledgements. The following have participated in the acquisition and/or analysis of the high-resolution spectra discussed above: L. Hobbs, D. Morton, P.

Frisch, V. Kulkarni, J. Lauroesch, J. Fowler, E. Fitzpatrick, M. Lemoine, J. Trapero. We appreciate the assistance of D. Doss (McDonald), J. Spyromilio and S. Ryan (AAO), and D. Willmarth (KPNO) in obtaining the spectra. We gratefully acknowledge support from NASA grant NAG5-3228.

References

Barlow, M. J., Crawford, I. A., Diego, F., Dryburgh, M., Fish, A. C., Howarth, I. A., Spyromilio, J., & Walker, D. D. (1995): MNRAS 272, 333
Blades, J. C., Sahu, M. S., He, L., Crawford, I. A., Barlow, M. J., & Diego, F. (1997): ApJ 478, 648
Blades, J. C., Wynne-Jones, I., & Wayte, R. C. (1980): MNRAS 193, 849
Crane, P., Lambert, D. L., & Sheffer, Y. (1995): ApJS 99, 107
Crawford, I. A. (1995): MNRAS 277, 458
Crawford, I. A., Craig, N., & Welsh, B. Y. (1997): A&A 317, 889
Crawford, I. A. & Dunkin, S. K. (1995): MNRAS 273, 219
Crinklaw, G., Federman, S. R., & Joseph, C. L. (1994): ApJ 424, 748
Diego, F. et al. (1995): MNRAS 272, 323
Fitzpatrick, E. L. & Spitzer, L. (1997): ApJ 475, 623
Frail, D. A., Weisberg, J. M., Cordes, J. M., & Mathers, C. (1994): ApJ 436, 144
Frisch, P. C. & Welty, D. E. (1997): in preparation
Heiles, C. (1997): ApJ 481, 193
Hobbs, L. M. (1969): ApJ 157, 135
Hobbs, L. M. (1984): ApJS 56, 315
Jenkins, E. B. (1997): these proceedings
Jenkins, E. B. & Shaya, E. J. (1979): ApJ 231, 55
Marschall, L. A. & Hobbs, L. M. (1972): ApJ 173, 43
Meyer, D. M. (1994): in The First Symposium on the Infrared Cirrus and Diffuse Interstellar Clouds, ed. R. M. Cutri & W. B. Latter (San Francisco: PASP), 3
Meyer, D. M. & Blades, J. C. (1996): ApJ 464, L179
Pettini, M. (1988): Proc. Astron. Soc. Aust. 7, 527
Piskunov, N., Wood, B. E., Linsky, J. L., Dempsey, R. C., & Ayres, T. R. (1997): ApJ 474, 315
Savage, B. D., Cardelli, J. A., & Sofia, U. J. (1992): ApJ 430, 650
Savage, B. D. & Sembach, K. R. (1996): ARA&A 34, 279
Sembach, K. R., Danks, A. C., & Lambert, D. L. (1996): ApJ 460, L61
Stokes, G. M. (1978): ApJS 36, 115
Tull, R. G. (1972): in Proc. ESO/CERN Conference on Auxiliary Instrumentation for Large Telescopes (Geneva: ESO), 259
Watson, J. K. & Meyer, D. M. (1996): ApJ 473, L127
Welsh, B. Y., Crawford, I. A., & Lallement, R. (1997): these proceedings
Welty, D. E., Hobbs, L. M., & Kulkarni, V. P. (1994): ApJ 436, 152
Welty, D. E., Hobbs, L. M., Lauroesch, J. T., Morton, D. C., Spitzer, L., & York, D. G. (1997a): in preparation
Welty, D. E., Hobbs, L. M., & Morton, D. C. (1997b): in preparation
Welty, D. E., Lauroesch, J. T., & Fowler, J. R. (1997c): in preparation
Welty, D. E., Morton, D. C., & Hobbs, L. M. (1996): ApJS 106, 533

The Line of Sight of ϵ Canis Majoris: Depletion and Extent of the Local Cloud

Cécile Gry[1,2] and Olivier Dupin[1]

[1] Laboratoire d'Astronomie Spatiale, CNRS, BP 8, F-13376 Marseille Cedex 12
[2] ISO Science Operation Center, PO Box 50727, E-28080 Madrid

Abstract. With new high resolution UV spectra of ϵ CMa we show that the gas column density in this sight-line is less than 4 10^{17} cm^{-2}, that the neutral gas density is less than 10^{-5} cm^{-3} after the first 3 parsecs, and that the Local Cloud seems to be almost undepleted and to extend to no more than 0.6 pc in this direction.

1 Introduction

The two lines of sight toward the early type stars ϵ Canis Majoris (l=240°,b=-11°) and β Canis Majoris (l=226°, b=-14°), located at respectively 132 pc and 153 pc (Hipparcos catalogue), are ideal to study diffuse clouds in the Local Bubble environment because of the very small amount of absorbing material they present relative to their length. In particular the early spectral type of ϵ CMa gives the unique opportunity to perform a complete study of the small cloud in which the Sun is embedded (LIC - Lallement and Bertin 1992), including all UV atomic lines, seldom observable toward the other (cooler) stars where the LIC is detected.
With the GHRS on board HST we have obtained high resolution Ech A ($\Delta\lambda/\lambda \sim 100\,000$) spectra of lines in the range 1180 Å- 1600 Å, that complete the pre-existing high resolution Ech B spectra at higher wavelengths (Gry et al. 1995). We will give here a rapid overview of the line of sight to ϵ CMa as studied from these new observations and we will focus on the description of the LIC.
The line of sight also includes an ionized component (C3) presenting interesting depletion and ionization patterns typical of previously shocked gas. It is discussed together with the β CMa sight-line in the accompanying paper by Dupin and Gry (this volume).

2 The Structure of the Line of Sight Toward ϵ CMa

Figures 1 and 2 show examples of the absorption profiles in the spectrum of ϵ CMa together with the best fit that allowed the determination of column densities, velocities and b-values for the considered species. In Figure 1, we show the strongest SiII line. It is not saturated, fact which demonstrates the extraordinarily low column densities. The two components marked LIC and

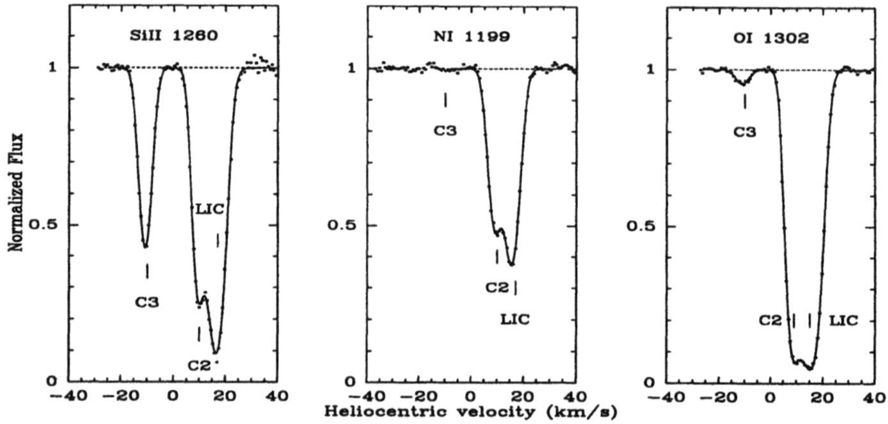

Fig. 1. Interstellar profiles in the GHRS-EchA spectra (dots) of ϵ CMa with their best fit. The strong SiII, NI and OI lines illustrate the weakness of the absorption in the sight-line and the low neutral gas content in C3.

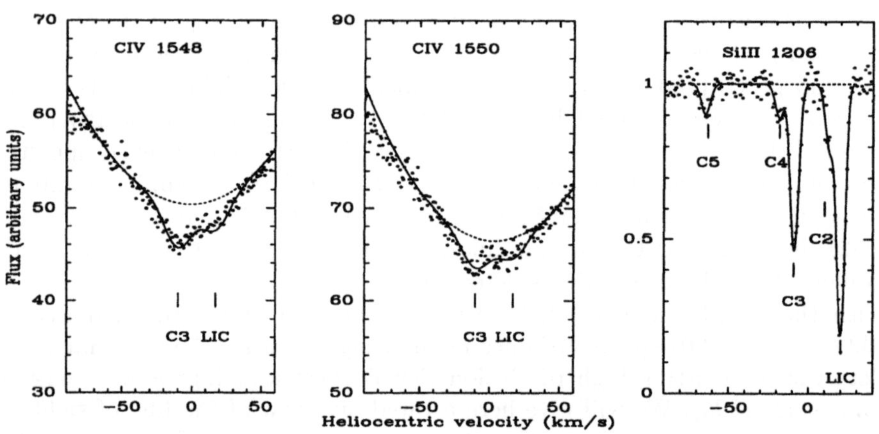

Fig. 2. High ionization species in the line of sight of ϵ CMa. The detection of CIV absorption at the velocity of the LIC and C3 suggests that we are observing an interface between the clouds and the hot bubble gas.

C2 are detected in MgII and FeII with very similar characteristics (Lallement et al. 1994 and Gry et al. 1995) in the spectrum of α CMa, located only 2.7 pc away from the Sun, and are therefore most probably very local. Apart from these local clouds another weak component -marked C3- is present in the ion absorption profiles. However, component 3 is not detected in any of the NI lines and shows a weak absorption feature in the strong OI line. This together with the most constraining depletion assumption implies that in component 3 the neutral gas column density is less than 5 $10^{15}\,\mathrm{cm}^{-2}$, and that in the

130 parsecs beyond the distance of Sirius, the mean neutral gas density is less than 1.2 $10^{-5}\,cm^{-3}$.

Another feature of this line of sight is illustrated in Figure 2 : the presence of highly ionized species. In SiIII the detection is clear, showing the same structure and b-value as the other less ionized species (C4 is only detected in the strongest lines and C5 only in SiIII and HI, where it is blended with the deuterium line). For CIV, the detection is also clear, with two absorption components detected in both doublet lines at the velocities of LIC and C3, with a temperature of about 150 000 K. However, they occur at the bottom of the broad and asymetric stellar line, and we cannot rule out at this stage that they could be due to features arising in the stellar wind of the star. In SiIV, there is only a 2-σ feature at the velocity of the LIC. The SiIV and CIV column densities and temperature derived from the fit of these features (see Table 1) are compatible with the prediction of Slavin (1989) model for the conduction layer between the Local Cloud and the hot gas supposed to fill the Bubble.

3 The Local Interstellar Cloud (LIC)

	log N at cm^{-2} (1)	log H derived with warm ISM depletion (2)	log H derived with minimum depletion (3)	ionization potential (eV) (4)
SII	$\leq$ 12.6	$\leq$17.3	$\leq$17.3	23.3
SIII	$\leq$12.3	$\leq$17.0	$\leq$17.0	34.8
NI	13.20	17.4	17.15	14.5
OI	14.15	17.5	17.2	13.6
MgII	12.48	17.55	17.23	15.0
FeII	12.13	17.97	17.18	16.2
SiII	12.64	17.73	17.28	16.3
SiIII	12.26	17.35	16.90	33.1
SiIV	11.0(2σ)	16.1	15.6	45.1
CIV	12.3	15.7	15.7	77.4

Table 1 column 1 lists the column densities derived for the Local Cloud from the fit of most available atomic lines in the spectrum of ϵ Canis Majoris. SII is a key element because its ionization potential makes it a tracer of both neutral and ionized gas and because, as it is usually observed with solar abundance in the ISM, it is traditionally considered to be undepleted. A very weak SII absorption feature is observed for the LIC, with a column density of 3 $10^{12}\,cm^{-2}$ given by the best fit and a strict upper limit of 4 $10^{12}\,cm^{-2}$. We can thus derive an upper limit for the total hydrogen column density in the LIC with this value and the sulfur solar abundance, -4.73 dex (Anders and Grevesse 1989). This gives $\log N(H) \leq 17.3$ cm^{-2}.

Similarly, we derive the hydrogen column density traced by the other species from their measured column densities and adopting the *mean depletion values in the warm ISM* (Jenkins, Savage and Spitzer (1986) for iron and magnesium, with the correction recommended by Fitzpatrick (1997) for magnesium, idem for silicon, Ferlet (1981) for nitrogen and Keenan (1985) for oxygen). OI, NI, SiII, MgII and FeII have ionization potentials lower than SII so they can not trace more gas than SII. The results are listed in column 2 of Table 1 and it is apparent that the above elements show no consistency when these depletion values are used, with a difference of almost a factor of 5 in the case of iron. This inconsistency could suggest that sulfur is somewhat depleted, but depetion in sulfur has never been observed to be larger than -.15 dex, so this depletion would have to be accompanied by a lower depletion of the other elements to have them all agree. Alternatively, if we consider the *minimum depletion* observed in the ISM for every species, i.e. no depletion for sulfur, nitrogen and oxygen and the minimum depletion values observed in the halo high velocity clouds and given by Fitzpatrick (1996) for magnesium, iron and silicon, we find (column 3) that the derived hydrogen column densities for all elements agree with each other to within 30 %. This suggests that the Local Cloud is only slightly depleted, implying that most species usually locked in the dust grains have been returned to the gas phase by some dust sputtering process.

With these depletion values we derive an HI column density of less than $2\,10^{17}$ cm^{-2} for the Local Cloud in the line of sight of ϵ Canis Majoris . With the approximate density $n_{HI} \sim 0.1$cm^{-3}, this implies that the extent of the Local Cloud in this direction is of the order of or less than 0.6 pc.

The second implication is that, adding the contribution of component 2, the total HI column density in the line of sight is less than $4\ 10^{17}$ cm^{-2}, considerably smaller than that derived from the study of the Extreme UV spectrum of the star (Cassinelli et al. 1995).

References

Anders E. and Grevesse N. (1989) : Geochim. Cosmochim. Acta, 53, 197.
Cassinelli J.P., Cohen D.P., MacFarlane J.J., et al. (1995) : ApJ. 438, 932.
Ferlet R. (1981) : A&A 98, L1
Fitzpatrick E.L. (1996) : ApJ 473, L55
Fitzpatrick E.L. (1997) : ApJ. in press
Gry C., Lemonon L., Vidal-Madjar A., et al. (1995) : A&A 302, 497.
Jenkins E.B., Savage B.D., and Spitzer L., (1986) : ApJ 301, 355.
Keenan F.P., Hibbert A., Dufton P.L. (1985) : A&A 147, 89.
Lallement R. and Bertin P. (1992) : A&A 266, 479.
Lallement R., Bertin P., Ferlet R., et al. (1994) : A&A 286, 898.
Slavin J. (1989) : ApJ 346, 718.

Neutral Clouds Within the Local X-ray Plasma

Jürgen Kerp and Jochen Pietz

Radioastronomisches Institut der Universität Bonn, D-53121 Bonn, Germany

Abstract. We present new results based on the analysis of *ROSAT* PSPC observations towards nearby neutral clouds. These clouds are detectable as deep soft X-ray absorption features within the $\frac{1}{4}$ keV energy band. Towards one of these clouds, optical absorption line measurements determine an upper distance limit of $D \leq 60$ pc. The presented data suggest that this filament is only a part of a much more extended local–cloud structure. Here we demonstrate that we most probably detected a local neutral atomic hydrogen "wall" in the direction of a low–velocity arc (Pietz 1994).

1 Basic information

Our view is focused towards the galactic sky region centered at $l \sim 85°$, $b \sim 37°$ (Fig. 1). This area partly covers an arc-like high H I column density structure (Pietz 1994). Here the low–velocity gas reaches H I column densities of about $N_{\rm HI} \simeq 5 \cdot 10^{20} {\rm cm}^{-2}$, and the H I 21-cm linewidth indicates kinetic cloud temperatures of $T_{\rm kin} \simeq 100$ K.
Towards the nearby low-velocity cloud LVC 88+36-2, part of the cloud structure of interest, Wennmacher et al. (1992) determined an upper distance limit of 60 pc. This suggests that this cloud may either be inside the Local Bubble, or form part of its boundary region. Kerp et al. (1993) analysed a pointed *ROSAT* PSPC observation towards LVC 88+36-2 and confirmed the suggestion that the cold, dense neutral hydrogen filament is embedded *within* the local X-ray plasma. They derived a three-dimensional structure of X-ray emitting and absorbing matter from the analysis of soft X-ray "colours". Because of the strong energy dependence of the photoelectric absorption cross section ($\sigma \propto E^{-3}$), the quantitative comparison of different energy bands provides a means of disentangling the various components within the interstellar medium. The analysis of Kerp et al. (1993) suggests a three-component model of the soft X-ray radiation transfer equation (MODEL I):

$$I_{\rm obs} = I_{\rm foregr} + I_{\rm backgr} \cdot e^{-\sigma \cdot N_{\rm HI}({\rm LHB})} + I_{\rm halo/extra} \cdot e^{-\sigma \cdot N_{\rm HI}({\rm total})} \quad (1)$$

Here, LVC 86+38–2 is embedded within the local X-ray plasma. A fraction of the local X-ray emission originates in front of the cloud ($I_{\rm foregr}$), while the residual local X-ray radiation is produced from beyond the low-velocity cloud ($I_{\rm backgr}$). Moreover, the H I 21-cm line data suggest that LVC 86+38–2 is only a part of a much more extended H I structure. We analysed four pointed PSPC observations towards this galactic sky area to search for evidence in soft X-ray colours that also these low-velocity clouds are probably located within the Local Bubble. We find that the above radiation transfer model provides a good quantitative fit to the observational data. This implies that also the very extended H I column density structure, which we called the "Basis" of the low-velocity arc, is embedded within the local X-ray plasma.

2 Alternative Explanation and Limits of the X-ray Data Analysis

We analysed the *ROSAT* PSPC data within the R1, R2, 1/4 keV and 3/4 keV energy bands. To improve the analysis we fitted the *ROSAT* PSPC R1:R2, 1/4 keV:3/4 keV band ratios as well as the 1/4 keV and 3/4 keV band count rate plotted versus the total $N_{\rm HI}$ column density (scatter diagrams) simultaneously. This method allows us to constrain the intensity of the individual soft X-ray source terms of Eq. (1) as well as the temperature of the local and distant X-ray plasma. The parameters of the extragalactic background are fixed for the representation by a power–law of $E^{-1.5}$ (Almaini et al. 1996) and a 1/4 keV intensity of $I_{\rm extra}(1/4\,{\rm keV}) = 2.3 \cdot 10^{-4}{\rm cts\,s^{-1}arcmin^{-2}}$ (Barber et al. 1996).
By fitting the X-ray data in all 4 diagrams simultaneously (Fig. 2) we determined, in the 1/4 keV band, for **MODEL I** (low–velocity clouds within the Local Bubble):

- $T_{\rm LHB} = 10^{5.8\pm0.1}$ K and $T_{\rm halo} = 10^{6.2\pm0.1}$ K,
- $I_{\rm foregr} = (3.5 \pm 0.5) \cdot 10^{-4}{\rm cts\,s^{-1}arcmin^{-2}}$ and $I_{\rm backgr} = (3.0 \pm 0.5) \cdot 10^{-4}{\rm cts\,s^{-1}arcmin^{-2}}$,
- $I_{\rm halo} = (10 \pm 1.5) \cdot 10^{-4}{\rm cts\,s^{-1}arcmin^{-2}}$.

An alternative approach assumes that the low–velocity clouds are embedded within the *warm neutral medium* (WNM) . In this case $I_{\rm backgr} = 0$ and the WNM forms the boundary of the local void of neutral matter (MODEL II). Here, we found that the data are fitted best by a mixture of the hot halo plasma $T_{\rm hotter\ halo} \simeq 10^{6.2}$ K (Pietz et al. and Wang, this volume) in addition to a cooler halo component with $T_{\rm cooler\ halo} \simeq 10^{5.9}$ K. For **MODEL II** we find:

- $T_{\rm LHB} = 10^{5.8\pm0.1}$ K, and a plasma mixture consisting of $T_{\rm cooler\ halo} = 10^{5.9\pm0.1}$ K and $T_{\rm hotter\ halo} = 10^{6.2\pm0.1}$ K,
- $I_{\rm local} = (3.5 \pm 0.5) \cdot 10^{-4}{\rm cts\,s^{-1}arcmin^{-2}}$,
- $I_{\rm cooler\ halo} = (30 \pm 5) \cdot 10^{-4}{\rm cts\,s^{-1}arcmin^{-2}}$ and $I_{\rm hotter\ halo} = (6 \pm 1.5) \cdot 10^{-4}{\rm cts\,s^{-1}arcmin^{-2}}$.

This model of the X-ray emitting and absorbing matter also fits the X-ray data well and, if substantiated, would have the following important consequences:

- The Local Bubble has an extent of less than 60 pc in this direction
- The WNM with $N_{\rm HI}({\rm WNM}) \simeq 1 - 3 \cdot 10^{20}{\rm cm}^{-2}$ is located close to the low–velocity clouds ($D_{\rm WNM} \geq 60$ pc).
- The emission measure of the halo component in MODEL II is **4** times higher than that of MODEL I.

3 Discussion

Because of the limited energy resolution of the *ROSAT* PSPC the X-ray data do not allow to distinguish between the two models considered above. Therefore, supplementary data are necessary to find the most likely interpretation of the observed situation. Optical absorption line measurements will be ideal to serve for this supplementary information.

Wennmacher (1994) estimated the distance of the WNM towards the low-velocity clouds distributed within the field of interest. He came to the conclusion that the bulk of the WNM has an average distance of about $D_{\rm WMN} \simeq$ 300 pc. This WNM structure has a neutral hydrogen column density ranging from $N_{\rm HI} = 1$ to $3 \cdot 10^{20}\,{\rm cm}^{-2}$. These values are close to the total neutral hydrogen column density observed in directions offset from the low–velocity clouds.

Quantitatively, this suggests that between the LVC 86+36–2 ($D_{\rm LVC} \simeq 60$ pc) and the WNM ($D_{\rm WNM} \simeq 300$ pc) there is only a weak increase of $N_{\rm HI}$. Hence, most likely the low-velocity clouds are really embedded within the local low–volume density region. In this sense MODEL I is the most plausible solution. Therefore, we can suggest that also the "Basis" is located within the Local Bubble. Furthermore, the thin H I 21-cm filaments, which are connecting the LVC 86+38–2 with the "Basis" in position and velocity, seem to be also local. Probably we found an H I 21-cm wall within the Local Bubble located at a distance between 50 – 60 pc. This model has to be confirmed by further optical absorption line measurements towards the low-velocity arc and LVC 86+38–2.

References

Almaini O. et al. (1996): MNRAS 282, 295
Barber C.R., Roberts T.P., Warwick R.S. (1996): MNRAS 282, 157
Kerp J., Herbstmeier U., Mebold U. (1993): A&A 268, L21
Pietz J. (1994): Diploma thesis, Universität Bonn
Wennmacher A., Lilienthal D., Herbstmeier U. (1992): A&A 250, L150
Wennmacher A. (1994): Ph. D. Thesis, Universität Bonn

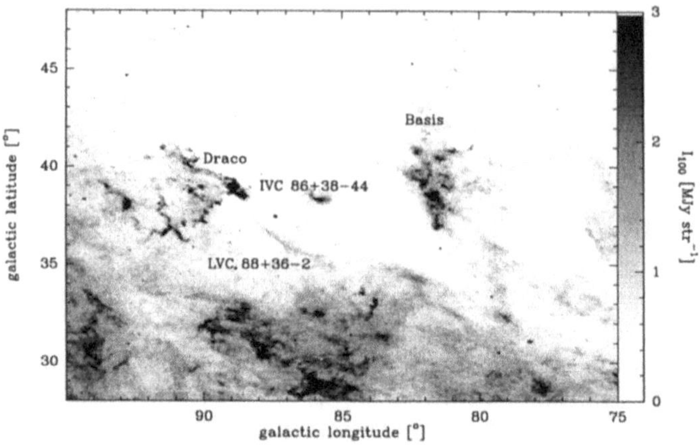

Fig. 1. *IRAS* 100μm map of the entire field of interest. The high-column-density clouds are located between $b = 35°$ and $42°$. The newly investigated field is called the "Basis" and is visible centered at $l \simeq 82°$ and $b \simeq 38°$. From this area the low–velocity arc emerges to high galactic latitudes. The low-velocity cloud LVC 86+38-2 is at a distance $D_{\mathrm{LVC}} \lesssim 60$ pc.

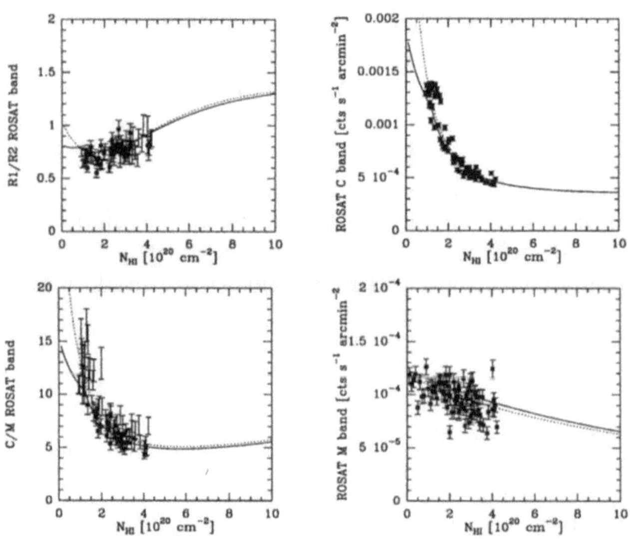

Fig. 2. Results of the analysis of four PSPC pointings towards the high-column-density filament centered at $l \sim 82°$, $b \sim 37°$ (Fig. 1). The X-ray data are simultaneously fitted for the R1:R2, 1/4 keV:3/4 keV band ratios and the 1/4 keV and 3/4 keV scatterdiagrams. The *solid* lines represent the best-fit of MODEL I (low–velocity clouds within the Local Bubble) and the *dotted* lines the best-fit of MODEL II (low–velocity clouds outside the Local Bubble).

The Evaporation of Nearby Clouds and the Soft X-ray Background

Jonathan D. Slavin

NASA/Ames Research Center, MS 245-3, Moffett Field, CA 94035-1000, USA

Abstract. In recent years the nature of the low density clouds within the Local Bubble has been increasingly well characterized. These clouds, including the one which surrounds the solar system, are embedded in the hot gas and therefore should be evaporating via thermal conduction. If several evaporation fronts exist within the Local Bubble, the emissivity and spectrum of the hot gas is significantly different from a single temperature, equilibrium ionization plasma. We explore models in which the the temperature, density and ionization in the hot gas are influenced by cloud evaporation and compare the results with the observed Soft X-ray Background.

1 Introduction

In the past decade, it has become increasingly clear that the simplest model for soft x-ray emission from the Local Bubble (LB) is not adequate. Observations at lower energies (Be Band, *EUVE* MW spectrometer) have indicated that the spectrum is harder in the region of 50 - 100 eV than the collisional ionization equilibrium, $T = 10^6$ K, Raymond & Smith model predicts. Higher spectral resolution observations by *DXS* (see Sanders, this volume) have confirmed that the simple models are not adequate to describe the emission.

At the same time, there has been mounting evidence that the Local Bubble, while of very low average density, does contain several low column density clouds (Frisch 1995 and references therein) including, of course, the Local Cloud in which the solar system is located. Warm ($T \approx 10^4$ K) clouds in a hot, soft x-ray emitting region such as the Local Bubble should evaporate via thermal conduction unless a special magnetic field geometry exists which prevents it. There are now observational indications that at least some nearby clouds do indeed have evaporative interfaces at their boundaries (Gry et al. 1995, Bertin et al. 1995).

Evaporating clouds radiate strongly in the EUV/soft x-ray spectral region. If the observed clouds in the LB are evaporating they could substantially alter the soft x-ray spectrum which is observed. This could in turn alter our interpretation of the soft x-ray background and potentially provide a solution to the puzzles of the SXRB spectrum.

2 Model

We explore some simple models of a hot bubble which contains evaporating clouds. Our models of evaporating clouds are detailed, containing most physical effects which should be operating. These include saturation of thermal conduction, radiative cooling, and non-equilibrium ionization (see Slavin 1989). Our model of the bubble as a whole on the other hand is very simple. We model the average emission observed in the bubble as the combination of "ambient" hot gas at some temperature, T_f, and emission from evaporating cloud boundaries. The average observed intensity is:

$$I_E = \langle \epsilon_{\rm cl} \rangle \ell_b f_{\rm cl} + \epsilon_{\rm hot} \ell_b (1 - f_{\rm cl}), \tag{1}$$

where $\langle \epsilon_{\rm cl} \rangle$ is the average emissivity in the cloud boundary regions, $\epsilon_{\rm hot}$ is the emissivity in the ambient hot gas, $f_{\rm cl}$ is the cloud boundary filling factor and ℓ_b is the total path length in the bubble. The average emissivity for cloud boundary emission is just

$$\langle \epsilon_{\rm cl} \rangle = \frac{\int_0^{R_f} \epsilon 4\pi r^2 \, dr}{\frac{4}{3}\pi R_f^3}, \tag{2}$$

where ϵ here is the emissivity at a given radial distance from a cloud center and R_f is the cutoff radius. R_f serves as an approximation to the mean distance between clouds (but due to cloud clustering, does not constrain $f_{\rm cl}$). We do not include the complex effects due to the interaction of the evaporative flows from adjacent clouds.

The cloud evaporation models have several parameters. These include the radius, $R_{\rm cl}$, temperature, $T_{\rm cl}$, density, $n_{\rm cl}$, and magnetic field, $B_{\rm cl}$, of the warm cloud and the reduction of conductivity due to magnetic field effects η. For this preliminary study we use the same values for these parameters in all the models: $R_{\rm cl} = 3\,{\rm pc}$, $T_{\rm cl} = 10^4\,{\rm K}$, $n_{\rm cl} = 0.2\,{\rm cm}^{-3}$, $B_{\rm cl} = 5\,\mu{\rm G}$, and $\eta = 1$.

Our models have two free parameters: R_f, the cutoff radius/mean cloud separation, and T_f, the temperature at R_f. We also use T_f as the temperature of the "ambient" hot gas. The model results are compared with observations of the soft x-ray background (SXRB) in the Wisconsin Be and B bands. Integrating equation (1) over the detector band response function for each of these bands, and choosing a given observed value for the count rate in the two bands, results in two equations with two unknowns, $f_{\rm cl}$ and ℓ_b which we then solve for.

For our current purposes we have used the values of 22.5 and 0.275 counts s^{-1} for the B and Be band count rates. The B band value is typical of those observed near the galactic plane (McCammon et al. 1983). The Be band value is below those actually observed but results in a Be/B band ratio which is the average of those observed (Bloch et al. 1986).

We calculate 12 models in all, one for $R_f = 10$, 20, and 30 pc for each of $\log(T_f) = 5.9$, 6.0, 6.1 and 6.2. Only for the models with $R_f = 10$ pc do we find values which fall in the ranges: $0 < f_{\rm cl} < 1$ and $0 < \ell_b < 200$ pc.

3 Results

The parameters for our "successful" models are listed in Table 1. Even for models for which reasonable values of ℓ_b and $f_{\rm cl}$ are found, other considerations may rule them out. One such is the C band count rate. The C/B band count rate ratio is confined to the range 1.9 – 3.3. This is enough to exclude the $\log T_f = 5.9$ model which has a ratio of 1.5. Another consideration is the path length. The $\log T_f = 6.2$ model path length of 120 pc may weigh against it. Although the dimensions of the local HI cavity are generally uncertain, there are at least some directions which indicate a path length smaller than this. In the low column density direction of the third quadrant, however, the cavity appears to extend for a path length of up to 200 pc.

Table 1. Cloudy Bubble Model Results

				x-ray band count rate[1]			
$\log T_f$	R_f (pc)	ℓ_b (pc)	$f_{\rm cl}$	$\varepsilon_{\rm cl}^{\rm Be}$	$\varepsilon_{\rm cl}^{\rm B}$	$\varepsilon_{\rm hot}^{\rm Be}$	$\varepsilon_{\rm hot}^{\rm B}$
5.9	10	32	0.36	0.111	0.816	0.0604	0.657
6.0	10	51	0.32	0.0786	0.570	0.0342	0.378
6.1	10	80	0.43	0.0506	0.370	0.0161	0.214
6.2	10	120	0.48	0.0354	0.250	0.00732	0.132

[1]Emissivity integrated over the band (Be or B) effective area ($\rm cm^2 sr$) in units of counts $\rm s^{-1}\ pc^{-1}$ for the cloud boundary or "ambient" hot gas. For the cloud boundaries the emissivity is the average over the volume out to the cutoff radius, R_f. The thermal pressure in the hot gas ranges from approximately 9000 to $10^4 \rm cm^{-3} K$ in the models.

An example model spectrum, calculated using the Raymond & Smith code (Raymond & Smith 1977, and updates) is presented in Figure 1. The spectral region shown is that for which there is the most evidence of discrepancy with the uniform temperature, $T = 10^6$ K, equilibrium ionization model. The spectrum clearly does differ substantially from the uniform bubble model with the same B band count rate. Particularly promising is the lower intensity of the iron line complex near 70 eV which appears, both from Be band pulse height analysis (Bloch 1988) and *EUVE* observations (Vallerga & Slavin, this volume) to be less than the uniform bubble model predicts.

4 Summary

We have created models in which Local Bubble contains clouds evaporating into the surrounding hot gas that match the soft x-ray data. While these models are greatly oversimplified, they are a first step towards incorporating

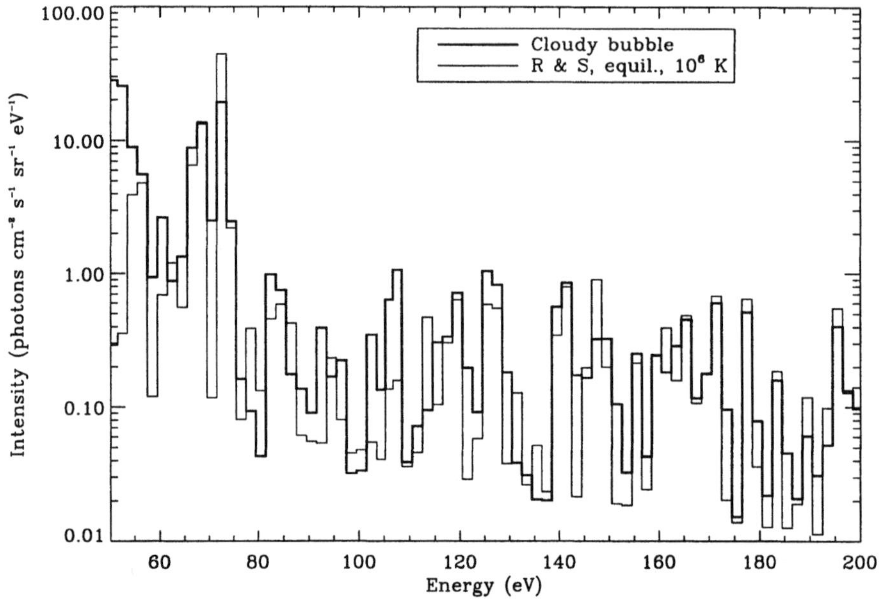

Fig. 1. An example of the low energy soft x-ray spectrum for a cloudy bubble model. For this model, $T_f = 10^{6.1}$. The model contains emission from the boundary regions of evaporating clouds as well as ambient hot gas. The thick line is the spectrum of a Raymond & Smith, equilibrium ionization, $T = 10^6$ K, model which produces the same B and Be band count rates as the cloudy bubble model.

the observed cloudy nature of the Local Bubble into models of the soft x-ray background emission.

References

Bertin, P., Vidal-Madjar, A., Lallement, R., Ferlet, R., & Lemoine, M., 1995, A&A 302, 889

Bloch, J. J., Jahoda, K., Juda, M., McCammon, D., Sanders, W. T., & Snowden, S. L. 1986, ApJ, 308, L59

Bloch, J. J. 1988, Ph.D. thesis, University of Wisconsin-Madison

Frisch, P. C. 1995, Space Sci. Rev., 72, 499

Gry, C., Lemonon, L., Vidal-Madjar, A., Lemoine, M., & Ferlet, R. 1995, A&A, 302, 497

McCammon, D., Burrows, D. N., Sanders, W. T., & Kraushaar, W. L. 1983, ApJ, 269, 107

Raymond, J. C. & Smith, B. W. 1977, ApJS, 35 419

Slavin, J. D., 1989, ApJ 346, 718

Diffuse Ionized Gas in the β CMa Tunnel

Olivier Dupin[1] and Cécile Gry[1,2]

[1] Laboratoire d'Astronomie Spatiale, CNRS, BP 8, F-13376 Marseille Cedex 12, France

[2] ISO Science Operation Center, PO Box 50727, E-28080 Madrid, Spain

Abstract. We present HST observations of the interstellar medium toward the star β CMa known to be located in a low density extension of the Local Bubble. Most of the matter in the sight-line is ionized and clumped in two main components. One of them, as well as one of the components detected toward ϵ CMa, is mostly ionized and only slightly depleted. Their ionization ratios are compatible with collisional ionization at T$\sim$25 000 K. These clouds could have been ionized by shocks related to the Local Bubble creation and also responsible of some dust grain sputtering.

1 Introduction

Copernicus (Gry et al. 1985), ground-based (Welsh 1991) and *HST* observations (Gry et al. 1995) have shown an extension of the Local Bubble in the direction of Canis Majoris. This tunnel free of neutral gas extends up to the stars β CMa ($l = 226°, b = -14°$, B1 II-III) and ϵ CMa ($l = 240°, b = -11°$, B2 I) located $\sim$150 pc away from the Sun, making these stars the brightest sources of the sky in the EUV (Cassinelli et al. 1995, 1996). These sight-lines with their long pathlength within the hot gas and their low column densities are the best opportunities to study the diffuse clouds embedded in the Bubble. Diffuse clouds elsewhere are often masked by larger absorption components.

First, we will give a short description of the β CMa sight-line and then we will focus on the physical characteristics of the two ionized clouds detected toward ϵ and β CMa.

2 Observations

The data for β CMa discussed here were obtained in December 1992 (cycle 2) by Vidal-Madjar and collaborators using the Goddard High Resolution Spectrograph onboard the Hubble Space Telescope. They consist of two sets of data at two different resolving powers. The Ech-B spectra (wavelengths from 1800 to 3000 Å) have a resolution of $\sim$ 3.5 km s^{-1} and present lines of Fe II, Mg II, Mn II, Mg I, Si II and Al III. The G160M spectra (from 1100 to 2000 Å) are less resolved ($\sim$ 15 km s^{-1}) and show lines of N I, O I, H I, D I, S II, Si II, Al II, C II, Si III, Si IV and C IV.

3 The Structure of the Sight-Line

3.1 Overview

The highly resolved Ech-B lines show four components, labelled A, B, C and D in order of increasing velocities (Figure 1). We focus on the components C and D for which column densities and b-values are well determined.
The medium resolution G160M data do not allow to separate the different components detected with the Ech-B, but they contain important lines for our study like SII and SiIII. The absorption feature in SiIII occurs at the velocity close to that of the component D while in NI and OI, it occurs at the velocity of component C. We thus conclude that D is mostly ionized while C is mostly neutral.
More details about this sight-line can be found in Dupin and Gry (1997).

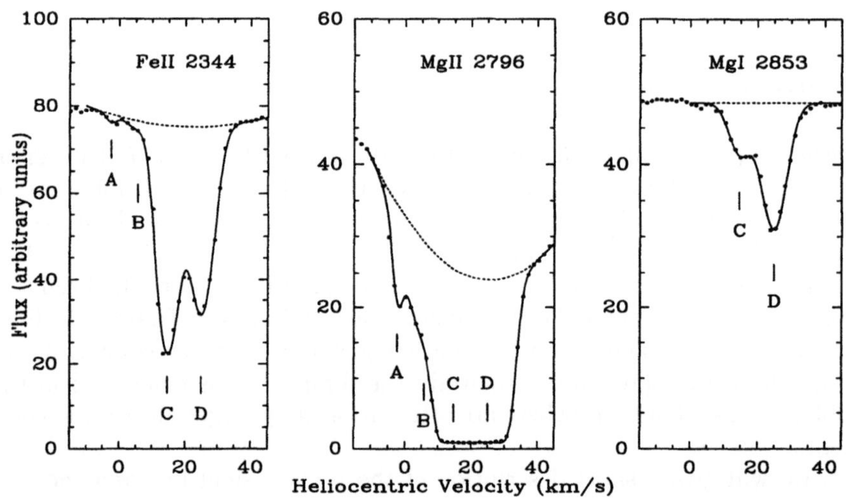

Fig. 1. Ech-B Fe II, Mg II and Mg I lines of β CMa. The dots are the observations, the dashed line the stellar continuum and the solid line the fit to the interstellar absorption profile

3.2 Temperature and Electron Density

Using the Fe II and Mg II b-values and thanks to the mass difference between these two elements we are able to separate the thermal and turbulent broadenings of the lines. Unfortunately, the uncertainties allow the Mg II b-value to be greater than that of Fe II and thus we only derive upper limits for the

Table 1. Column densities of important species (log N in at cm^{-2}) derived from the observations

	FeII	MgII	MgI	SiII	SiIII	SII
Cloud C	$13.07 \pm .04$	$13.48 \pm .08$	$10.90 \pm .03$	$13.75 \pm .03$	≤ 12.70	$14.01 \pm .17$
Cloud D	$13.00 \pm .05$	$13.77 \pm .06$	$11.40 \pm .05$	$13.87 \pm .02$	$14.59 \pm .51$	$14.35 \pm .10$

cloud temperatures. We find T≤9500 K for C and T≤13 500 K for D. For turbulent velocities, we find V_{turb}=3.1–3.7 km s^{-1} for C and V_{turb}=3.7–4.6 km s^{-1} for D, the upper values being achieved for purely turbulent broadening (i.e. for b(Mg)=b(Fe)). We also infer lower limits on temperature using the Mg I/Na I ratio (Pettini et al. 1977). Taking the Na I upper limit by Welsh (1991), we find T≳3000 K for C and T≳6000 K for D.
Assuming the ionization equilibrium between Mg I and Mg II, we derive the electron density in the clouds (see Gry et al. 1995). We find n_e= 0.03 – 0.61 cm^{-3} and n_e=0.02–1.34 cm^{-3} for the cloud C and D respectively. The large uncertainties come from the strong dependence of n_e on temperature.

4 Characteristics of the Ionized Components

We now focus our attention on the ionized cloud D and component 3 detected toward ϵ CMa (Gry and Dupin 1997), which present depletion and ionization similarities.

4.1 Depletion

As sulfur is usually undepleted in the interstellar medium, we derive the total hydrogen column density in cloud D from N(S II)+N(S III): we find $N(H_{tot})$=19.0–19.2 dex. On the other hand, we have a lower limit on the total silicon column density in cloud D: N(Si II)+N(Si III)≥14.34 dex. Thus, we derive a silicon gas-phase abundance in cloud D of $A(Si)=\log(N(Si)/N(H_{tot}))\geq -4.86$, and a silicon depletion of $D(Si)=A(Si)-A(Si)_{cosmic}\geq -0.41$. Similarly for ϵ CMa component 3 we have a lower limit on the silicon depletion of −0.5. The silicon depletion in these clouds is thus within a factor of 2 of the minimum depletion found in the halo high velocity clouds (Fitzpatrick 1996). Thus, component D and 3 are only slightly depleted.

4.2 Ionization

We derive an upper limit for the neutral gas column density in component 3 from OI and NI : N(H I)≤ 15.7 dex (Gry and Dupin 1997). For the total gas column density, we derive an upper limit from S II+S III and a lower limit from Si II+Si III: $17.06\leq\log(N(H_{tot}))\leq 17.38$. This cloud is thus more than

96% ionized. From the upper limit on N(H I) we can derive the maximum column densities expected for the species which have an ionization potential close to that of H I, and which are therefore commonly supposed to arise from H I region. The derived values for Si II, Mg II and Fe II derived with the solar abundances of the species, are about one order of magnitude greater than the mesured values. Note that if some depletion of these species is considered, the discrepancy would be even worse. These results show that H I is much more ionized than Si II, Mg II and Fe II: this can be explained by photoionization only with a very sharp radiation spectrum, very unlikely for any stellar photoionization source. We have similar results for cloud D although in this case the upper limit on the neutral gas can only be derived from the HI column density of the whole line of sight : N(HI)$\leq$ 18.34 dex (Gry et al. 1985 and Cassinelli et al. 1996).
We can however interprete this ionization in terms of collisional ionization. The ionization ratios of the species we observe are roughly compatible with collisional ionization at temperatures from 20 000 to 25 000 K, as given by Arnaud and Rothenflug (1985) and Sutherland and Dopita (1993). The temperature measured in the clouds by line fitting are however lower, of the order of 10 000 K for both components. We propose that the gas, previously collisionally ionized has already been cooling but is still in the process of recombining.
These clouds might thus give the evidence that shocks have been travelling in the area in the past; they could be the left-over from the blast-wave that created the Bubble.

References

Arnaud, M., Rothenflug, R. (1985): A&AS 60, 425

Cassinelli, J.P., Cohen, D.H., Mac Farlane, J.J., Hoare, M., Lynas Gray, T., Vallerga, J.V., Vedder, P.W., Welsh, B.Y., Hubeny, I., Lanz, T. (1995): ApJ 438, 932

Cassinelli, J.P., Cohen, D.H., Mac Farlane, J.J., Drew, J.E., Lynas Gray, T., Hubeny, I., Vallerga, J.V., Welsh, B.Y., Hoare, M. (1996): ApJ 460, 949

Dupin, O., Gry, C. (1997): A&A, submitted

Fitzpatrick, E.L. (1996): ApJ 473, L55

Gry, C., Dupin, O. (1997): these proceedings

Gry, C., Lemonon, L., Vidal-Madjar, A., Lemoine, M., Ferlet, R. (1995): A&A 302, 497

Gry, C., York, D.G., Vidal-Madjar, A. (1985): ApJ 296, 593

Pettini, M., Boksenberg, A., Bates, B., Rosemary, F., McCaughan, F., McKeith, C.D. (1977): A&A 61, 839

Sutherland, R., Dopita, M.A. (1993): ApJS 88, 253

Welsh, B.Y. (1991): ApJ 373, 556

Energetic Particles in the Local Bubble

G.E. Morfill and M.J. Freyberg

Max-Planck-Institut für extraterrestrische Physik, D-85740 Garching, Germany

Abstract. The current status of observations of energetic particles in the "local bubble" is reviewed. This includes primarily "direct" measurements of cosmic rays made in the Solar System, but also the "remote sensing" made possible by observing cosmic ray produced γ-rays in the nearby interstellar clouds. Since the energetic events responsible for the formation of our local bubble may also have produced copious amounts of cosmic rays, fossil records are examined to determine whether there is a corresponding signature. The observations show that: 1) the cosmic ray (proton) intensity is fairly homogeneous throughout the local bubble and its adjacent interstellar clouds, 2) there is some evidence for a "recent" local cosmic ray injection about 40,000 years ago, 3) on longer time scales (a few million years) the cosmic ray intensity was constant within a factor two, 4) there is apparently some "activity" in the Orion cloud, as evidenced by low energy γ-ray signatures, and 5) there are two unexplained observations – the variations in the energy spectra, in particular the significantly flatter spectrum of heavy cosmic rays (Fe) and the matter path length variation, which yields consistently larger path lengths for the lighter elements (H, He). It is suggested that these observations are compatible with two cosmic ray populations – an older one in equilibrium with losses from the galaxy and a younger one which is not yet strongly affected by losses. The latter could be a cosmic ray signature of the formation of the local bubble.

1 Introduction

Our "local bubble" is a somewhat complicated and inhomogeneous region in the galactic disk, which was probably formed a few million years ago through a series of supernova explosions, probably originating in the Scorpius-Centaurus Association (see Fig.1 of Frisch in this volume and the reviews by Breitschwerdt et al. 1996, Egger et al. 1996, and Frisch 1995). The total amount of energy liberated in such an event (or series of events) is of the order 10^{52} erg. Supernovae are believed to be the most important sources of cosmic rays (CRs) – at least for energies below about 10^5 GeV – with a few percent of the available energy "channelled" into energetic particles. The physical process believed to be responsible is "diffusive shock acceleration" (e.g. Axford et al. 1977, Krymsky 1977, Bell 1978a, b, Drury 1983, Völk et al. 1981, Morfill et al. 1984, Lagage & Cesarsky 1983, Ellison et al. 1981 for the early treatment of different aspects such as wave production, damping, energy losses, time dependence, the injection processes, maximum energy etc.).

The signature of these energetic events and the associated processes that marked the formation of the local bubble might still be present in CRs. Let us illustrate this with some numbers: If the total amount of energy in CRs is $E_{\rm CR} \sim 3 \times 10^{50}$ erg (3% of our assumed total energy), if the mean time of local bubble formation occurred $t_{\rm B} \sim 3 \times 10^6$ years ago and if the current size of the bubble is $R_{\rm B} \sim 200$ pc in diameter, then the CRs produced in the many supernova shock waves can fill this cavity completely, provided the diffusion coefficient is $\kappa > R_{\rm B}^2/t_{\rm B} = 10^{27}\,{\rm cm}^2\,{\rm s}^{-1}$. The CR energy density in the local bubble is then $\epsilon_{\rm CR} = 3E_{\rm CR}/4\pi R_{\rm B}^3 = 3 \times 10^{12}\,{\rm erg\,cm}^{-3}$ – or less, if the CRs have diffused further than $R_{\rm B}$. The obvious signature to look for is a local CR enhancement.

The energy spectrum of CRs is also characteristic. In diffusive shock acceleration one expects a power law with an exponent of about E^{-2} (strong shocks) or $E^{-2.3}$ (mixture of strong and weak shocks – see Bogdan and Völk, 1985). If the diffusion coefficient is energy dependent (κ increases with energy) and there is escape from our galaxy, then the high energy CRs are lost faster and the spectrum steepens some more. There is also a maximum energy (of about 10^6 GeV) that can be reached. This is due to a combination of the weakening of propagating shocks and the available time (see Forman & Morfill 1979, Morfill et al. 1984, Lagage & Cesarsky 1983). Thus the cut-off signature at 10^6 GeV should also be looked for. This then characterises the expected CR population in the vicinity of our solar system. If there were no other, more recent, CR sources in our local bubble, the CR intensity would be expected to be fairly uniform throughout. If there is also an older CR population, mixed and diffused over a correspondingly larger volume, then the "local bubble component" referred to above will simply be added on top of this older component.

The CR signature of the bubble forming "event" may be observable in other ways, too, e.g. in the traversed matter path length distribution (if there are young and old CR populations) and in time variations of fossil records. These may provide additional clues to the ones discussed above, i.e. the medium scale spatial CR intensity variation and the energy spectrum. These, therefore, are the CR signatures of interest, and the purpose of this review is to collect and analyse the available information.

2 Measurements in the Solar System

CRs have been studied using ground based and spacecraft detectors for many decades now. Below we summarize the most important findings:

In spite of being highly suprathermal with a mean energy in the GeV/ nucleon region (compared to the interstellar gas with a mean particle energy bracketing ~ 0.001 eV in the case of clouds, ~ 1 eV in the case of HI regions and ~ 1 keV in the case of supernova remnants), CRs nonetheless have roughly the same energy density as their "thermal" counterparts, i.e. $\sim 1.6 \times 10^{-12}\,{\rm erg\,cm}^{-3}$. This implies that energetically and dynamically they form an important component of the interstellar medium.

The highest energy CR particles measured so far (using extensive air-shower techniques) have several 10^{20} eV – this is equivalent to a BMW 735 moving at 1 km/hour (if you owned such a car you would never travel at this speed, of course, but it is theoretically possible).

The element abundance of CRs is roughly "cosmic", a strong indication that these particles originate from the interstellar medium and not some special source, which may have different abundances. Departures from the cosmic abundance occur, e.g.: Li, Be, B are enhanced by several orders of magnitude, as are the elements with atomic numbers 20 – 25, i.e. just below the Fe-group. This is readily explained by spallation reactions from C, N, O in the former and from Fe in the latter case.

The CR energy spectrum is a remarkable power law. From about 1 GeV per nucleon to about $10^6\,\mathrm{GeV\,n^{-1}}$ the spectral exponent is $E^{-2.75}$, above that until about $10^9 - 10^{10}\,\mathrm{GeV\,n^{-1}}$ it steepens to $E^{-3.0}$. The turn-up of the helium flux below $\sim 60\,\mathrm{MeV\,n^{-1}}$ is due to the additional flux of the anomalous ^{4}He component. The Fe-spectrum is harder than that for H and He (cf. Longair 1992). Further measurements in the experimentally difficult energy range from $10^3 - 10^7\,\mathrm{GeV\,n^{-1}}$ (Longair 1992) show that "heavy" CR particles, such as Fe, have similar fluxes as the "light" particles (H, He) above $\sim 10^4\,\mathrm{GeV\,n^{-1}}$ (or $\sim 10^5$ GeV/particle). Airshower simulations show in addition, that above $\sim 10^9\,\mathrm{GeV\,n^{-1}}$ (or $\sim 10^{10}$ GeV/particle) protons dominate again. The most likely explanation for this is photodisintegration of these ultra-high-energy particles by interactions with blue-shifted (in the particle's frame) photons from the cosmic microwave background.

The "age" of CRs can be determined in two ways. The most direct is to measure the abundances of radioactive isotopes, a second possibility is provided by the ratio of secondaries to primaries. Secondaries are, for instance, all those particles produced by spallation. Some of these are stable, so their abundances give a measure of the amount of material traversed by the CRs. Given an average density for the interstellar medium, it is then possible to determine the CR "age". However, uncertainties arise, because the volume occupied by CRs (i.e. the extent of the galactic halo) is not known too well. Basically, it is not clear how much time CRs spend on average in denser portions of the interstellar medium (e.g. the galactic disk) and how much time in the more tenuous regions (e.g. the halo). The amount of material traversed by CRs decreases with increasing energy. This indicates that energetic particles escape faster from the galaxy, i.e. the diffusion mean free path increases with energy. Combining secondary/primary ratios with radioactive isotope abundance measurements (in particular the ^{10}Be/^{9}Be ratio – see Simpson & Garcia-Munoz, 1988) gives a mean CR "age" of $\tau_{\mathrm{CR}} \sim 10^7$ years in the GeV-range. Compared to the age of our galaxy CRs are "young", compared to the age of the local bubble CRs are somewhat older or perhaps even contemporaries.

There are two possible interpretations for "young" CRs: 1) the production processes are a fairly recent phenomenon of galactic evolution, or 2) CRs have always been produced, but have a finite "life-time" – either getting "destroyed" (e.g. by energy degradation) or escaping from our galaxy. If CRs were destroyed by collisions with the interstellar gas, the secondary abundances would show this. In fact, the deduced path length distribution is exponential, with a mean value of only about $6\,\mathrm{g\,cm^{-2}}$, much less than the $\sim 70\,\mathrm{g\,cm^{-2}}$ required to significantly degrade relativistic CRs. Observed "anomalies" in the path length distribution of secondaries have so far not been interpreted as being connected to the local bubble (see e.g. Morfill et al. 1985 for a discussion).

Direct "fossil" evidence for CR existence can be obtained from tree rings (1000's years), from drill core measurements from the layerings of the polar ice caps (million years) and from meteorites (billion years). The result is that CRs have existed throughout time at roughly the same intensity. The enhancement expected in connection with the local bubble formation is not observed, at least not to the accuracy of the measurements ($\sim$ factor 2, see e.g. the reviews in Sonett et al. 1991), however, a more recent enhancement dated about 40,000 years ago (Sonett et al. 1987) suggests a "late" supernova, which may have exploded $\sim 10^5$ years ago. The enhancement marks the passage of the supernova shock wave across the solar system in this interpretation.

The conclusion is that CRs have been continuously produced in our galaxy for billions of years and escape into the intergalactic medium in a time scale of $\sim 10^7$ years. The implications for the local bubble will be discussed later.

3 Remote Sensing

We mentioned earlier that CRs interact with the interstellar gas. In these reactions π^0-mesons may be produced, which subsequently decay into γ-rays. Of course, there will be other γ-ray producing reactions (e.g. inverse Compton or Bremsstrahlung – see Fig.1), but it is nevertheless possible to isolate an energy window in γ-ray emission, from about $10^2 - 10^4$ MeV, where the π^0-decay dominates and where the γ-ray emission can therefore be used as a tracer of CR intensity.

A surprising conclusion from the analysis of the EGRET data (Strong et al. 1997) is the fact that the galactic γ-ray emission at high energies is significantly larger than that predicted using the local CR spectrum. In fact, a spectrum $\sim E^{-2.4}$ to $E^{-2.5}$ appears to be required for the galaxy as a whole. As we will discuss later, the CR spectrum can show local variations, especially if injection is impulsive, so that this observation may be a pointer to a local signature. For present purposes we are particularly interested in the emission from local clouds at the edge of our local bubble in addition to the global picture. Three pertinent investigations have been performed so far.

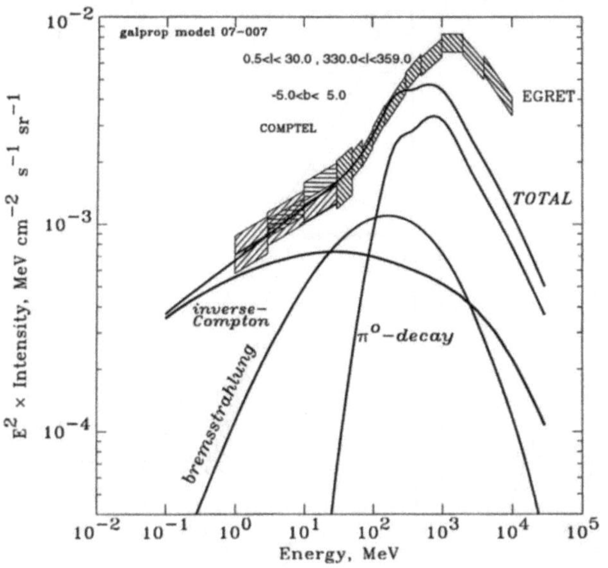

Fig. 1. The computed galactic γ-ray intensity for various processes and its comparison with experimental data (from Strong et al. 1997).

3.1 Ophiuchus

The Ophiuchus molecular cloud has long been regarded as a region of enhanced γ-ray – and therefore CR – activity. Excesses up to an order of magnitude above the local (solar, in situ) value were quoted. Careful analysis with the more sensitive EGRET instrument has shown that this enhancement is real, but that it is most likely caused by the quasar PKS 1622-253, which is located 20' away from the maximum likelihood position of the formerly identified Ophiuchus source (Hunter et al. 1994). The quasar signature can be identified fairly unambiguously because it is time variable. There is also, possibly, another (as yet unidentified) source, GRO J1631-27 in the vicinity. The resultant γ-ray production function in Ophiuchus is shown in Fig.2. Removal of the point sources increases the errors in the determination of the cloud's emission, of course.

3.2 Orion

The giant cloud complex in Orion was identified as a γ-ray source some years ago (Caraveo et al. 1980). Recent observations using EGRET were able to resolve both the Orion A and B clouds (Digel et al. 1995). The CO emission contours were compared with the γ-ray map of the same sky region. There was even a hint in the data that Mon R2 has been detected, although at a lower confidence level than the Orion clouds. The corresponding γ-ray emissivity

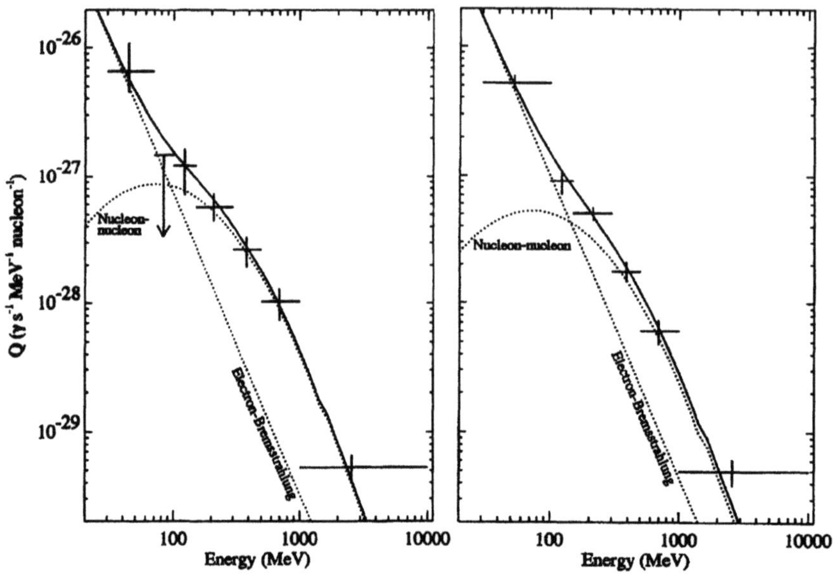

Fig. 2. Left: γ-ray production function in Ophiuchus derived from the fits to the separate energy ranges on the assumption of an energy spectral index of -2. The ranges are indicated by the horizontal error bars. The solid curve represents the best-fitting linear combination of the nucleon-nucleon and electron-bremsstrahlung production function used by Bertsch et al. (1993). The individual production functions are also shown as dotted curves (from Hunter et al. 1994).
Right: γ-ray production function in Orion (from Digel et al. 1995).

is shown in Fig.2. An excess in the $3-7\,\mathrm{MeV}$ γ-ray emission from Orion has been observed with COMPTEL (Bloemen et al. 1997, Bykov et al. 1996). One explanation is that this emission is due to Doppler-broadened C and O de-excitation lines, which occur at 4.44 and 6.13 MeV. A large flux of low energy (10's MeV) CRs is required. The interaction of such low energy CRs with the cloud would lead to ionisation, knock-on electrons and an X-ray bremsstrahlung flux that should be measurable. The observations performed by ROSAT show only marginally the required X-ray emission (cf. Dogiel et al. 1997, Schönfelder et al. 1997, and references therein).

3.3 Cepheus and Polaris Flare

Digel et al. (1996) examined the region $l^{II} = 100°$ to $130°$ and $b^{II} = -5°$ to $+32°$, which contains the Cepheus and Polaris Flares, cloud complexes clearly visible in CO emission. In addition they also examined the Perseus region. Utilising these new results as well as older studies, we have compiled two tables. Table 1 gives the γ-ray emissivity above $100\,\mathrm{MeV}$. This should be directly proportional to the CR intensity at the location (cloud) where the

Table 1. γ-ray emissivity (in $10^{-26}\,s^{-1}\,sr^{-1}$)

Local	(Strong et al. 1988)	1.8
Ophiuchus	(Hunter et al. 1994)	(2.4 ± 0.2)
Orion	(Digel et al. 1995)	(1.7 ± 0.1)
Cepheus & Polaris Flares	(Digel et al. 1996)	(1.8 ± 0.1)

measurement was made. These results therefore allow us to "remotely sense" the CR intensity at specific locations in a volume of radius $\sim$ 500 pc. An important quantity to know – in order to derive the γ-ray emissivities – is the conversion factor between CO emission ($K\,km\,s^{-1}$) and the hydrogen column density N_H (H-mols cm^{-2}). With the γ-ray emissivity, ϵ_γ (photons MeV^{-1} $s^{-1}\,n^{-1}$), the measured differential γ-ray flux of a cloud of area A at distance r is $F_\gamma = AN_H\epsilon_\gamma/4\pi r^2$ (photons $cm^{-2}\,sr^{-1}\,s^{-1}\,MeV^{-1}$) .

Table 2. Hydrogen column/CO column (units: 10^{20} H-mols $cm^{-2}/(K\,km\,s^{-1})$)

Ophiuchus	(Hunter et al. 1994)	(1.1 ± 0.2)
Orion	(Digel et al. 1995)	(1.06 ± 0.14)
Cepheus & Polaris Flares	(Digel et al. 1996)	(0.92 ± 0.14)
Perseus	(Digel et al. 1996)	(2.5 ± 0.9)
Galactic Average	(Bertsch et al. 1993)	1.6
High Latitude Cirrus	(Heithausen & Thaddeus 1990)	0.5

Since the CO emission of interstellar clouds is readily measurable, ϵ_γ can be determined (in principle) once the conversion factor is known. Various results are summarised in Table 2. Table 1 suggests a fairly uniform CR intensity, at least in a volume of 0.5 kpc or less, with possibly just the hint of an enhancement remaining at ρ-Ophiuchus. The quality of this statement is enhanced by the results compiled in Table 2.

4 Summary, Conclusions, and Speculations

In terms of the possible "cosmic ray signatures" of the processes that led to the formation of the local bubble, the available measurements to date give the following "answers":

1. Local Enhancement of Cosmic Rays: No evidence was found from γ-rays (to within 10%) of CR intensity variations in a region around the Sun of radius $\sim$ 0.5 kpc. Based on our earlier estimate of a CR energy injection of $\sim 3 \times 10^{50}$ erg some 3 million years ago, this requires a diffusion coefficient $\kappa > 2.5 \times 10^{28}\,cm^2\,s^{-1}$ – if the transport really was (is) diffusive. Note that

this constraint applies to protons only, because they provide most of the γ-rays.
2. Cosmic Ray Age: The inferred exponential path length distribution also implies an equivalent CR age distribution, with a mean of ~ 10 million years. The lack of a clear signature in the path length distribution (a smaller grammage) dating back ~ 3 million years could imply that our local bubble formed much earlier (this might not be compatible with the measurements of the local gas), or it implies that the "local" CRs from this event are sufficiently diluted in the galactic background population.
3. Cosmic Ray Spectrum: The observed spectral break occurs approximately at the highest energy computed for acceleration in multiple supernova remnant shocks (10^6 GeV/particle). This does not constitute a "signature" for our local bubble, however. It could and probably does, mean that shock acceleration is the dominant process for CR production in our galaxy generally, and that the local bubble is nothing special.
4. Time Variation: The identification of a supernova shock wave sweeping across the solar system about 40,000 years ago (Sonett et al. 1987) demonstrates that some kind of fossil records of such events should, and do, exist. The lack of identification of multiple events within a period of ~ 1 million years is explained as follows: polar ice cores records until now do not go far enough back in time, and meteoritic data do not yield a sufficiently fine time resolution since they only provide integral measurements. In other words, current fossil techniques cannot be expected to identify the event.
5. The Cosmic Ray Age, and Spectrum – Again: We mentioned earlier that the diffusive shock acceleration process yields a spectrum with a power law exponent of -2 to -2.3 (if weaker shocks are included). Further steepening to the observed value of -2.75 was attributed to energy dependent escape from our galaxy. However, this steepening is only obtained in the steady state ($t > \tau_{\rm CR}$). For impulsive injection, which occurred less than $\sim \tau_{\rm CR}$ ago, this is different. Obviously there will be some steepening because faster diffusive spreading of the more energetic particles dilutes them more. On the other hand, self-confinement (e.g. due to coupling onto the interstellar medium (e.g. via waves produced as a result of the streaming instability, when the CR energy density exceeds that of the gas – or plasma – by a sufficient amount) is energy independent, and would not affect the spectrum.

Well, there are two puzzling observations regarding CRs, which we have glossed over so far in order to "keep to the major issues". In doing so, we may actually have overlooked what could be a major issue:

The first observation, which we have simply stated as a fact, is the flatter Fe-spectrum, which is again a power law, but varies as $\sim E^{-2.5}$ from 10^3 – 10^7 GeV/particle. The second observation, which we have not discussed in detail so far, is the observation that the amount of material traversed by CRs decreases as the CR nucleus gets heavier, from about $10\,{\rm g\,cm^{-2}}$ for H to about $3\,{\rm g\,cm^{-2}}$ for the iron group.

So far, all attempts at explaining these observations (see e.g. Morfill et al. 1985 and references therein) have assumed local path length enhancements near the source region. These observations could also be explained using two CR components – one associated with the formation of the local bubble, the second averaging over many source regions in the galactic disk:

A: The lower matter path length derived for the heavy CRs (Fe) suggests that this is a younger component. Conversely, the greater matter path length derived for the light elements (H, He) suggests that this is an older component. The flatter spectrum observed for Fe suggests that the heavy CRs are not yet in equilibrium with losses from the galaxy, i.e. they are younger than 10^7 years. Conversely, the steeper spectra observed for the light elements (H, He) suggest that they are in equilibrium – which again would imply that they are older.

If we accept this "evidence", the question is, how do we account for a young local population of heavy CRs and an older non-local population of light CRs? And how do the GRO γ-ray data fit into this scheme, since they appear to require a flatter CR spectrum throughout the galaxy? Maybe the proton spectrum is the anomaly, not the heavies? If we follow this hypothesis, we would come to the following argument:

B: The equilibrium CR spectrum throughout the galaxy is a power law with an exponent $\sim E^{-2.5}$. We observe this locally only for the heavy elements, and must therefore conclude that these are "old" and equilibrated with losses. (Note that the He, CNO and Ne-Si spectra also appear harder than the proton spectrum, but less hard than the Fe spectrum). The protons have a softer spectrum. This could be a signature of time dependent shock acceleration, where different shocks of different time scales are superposed, each able to accelerate particles to a somewhat different upper energy cut-off, resulting – albeit fortuitously – in a steeper spectrum, or it could signify enhanced local losses.

Hypothesis A leads to the requirement that the local population was simply enhanced with heavy particles. Hypothesis B leads to the requirement that we have a "young" local proton component added on top of the uniform galactic CR component.

It is clear that investigation of local variations in CR production and losses as well as in transport appears necessary to understand the available information from γ-ray and "in situ" CR measurements. Being located in a local bubble of comparatively recent origin, and having the "remote sensing" capability of γ-ray measurements, gives us a reasonable chance to unravel the puzzles that observations have presented us with.

Acknowledgements. We wish to acknowledge the help in preparing this manuscript, both through discussions and otherwise, given by many of our colleagues at MPE. In particular, we wish to thank Andy Strong for some detailed information.

References

Axford W.I., Leer E., Skadron G., 1978, Proc. 15th Int. Cosmic Ray Conf. 11, 132
Bell A.R., 1978a, MNRAS 182, 147
Bell A.R., 1978b, MNRAS 182, 443
Bertsch D.L. et al., 1993, ApJ 416, 587
Bloemen H. et al., 1997, ApJ 475, L25
Bogdan T., Völk H.J., 1985, A&A 122, 129
Breitschwerdt D. et al, 1996, Space Sci. Rev. 78, 183
Bykov A.M., Bozhokin S.V., Bloemen H., 1996, A&A 307, L37
Caraveo P.A. et al., 1980, A&A 91, L3
Digel S.W., Hunter S.D., Mukherjee R., 1995, ApJ 441, 270
Digel S.W. et al., 1996, ApJ 463, 609
Dogiel V.A., Freyberg M.J., Morfill G.E., Schönfelder V., 1997, Proceedings of the 4th Compton Symposium, in press
Drury L.O'C., 1983, Rep. Progr. Phys. 46, 163
Egger R.J., Freyberg M.J., Morfill G.E., 1996, Space Sci. Rev. 75, 511
Ellison D.C., Jones F.C., Eichler D., 1981, Journal of Geophysics 50, 110
Ellison D.C., Drury L.O'C., Meyer J.-P., 1997, preprint (to appear in ApJ)
Forman M.A., Morfill G.E., 1979, Proc. 16th Int. Cosmic Ray Conf. 2, 70
Frisch P.C., 1995, Space Sci. Rev. 72, 499
Heithausen A., Thaddeus, P., 1990, ApJ 353, L49
Hunter S.D., Digel S.W., de Geus E.J., Kanbach G., 1994, ApJ 436, 216
Krymsky G.F., 1977, Dokl. Akad. Nauk SSR 234, 1306
Kulsrud R.M., Pearce W.D., 1969, ApJ 156, 445
Lagage P.O., Cesarsky C., 1983, A&A 125, 249
Longair M.S., 1992, High Energy Astrophysics Vol.1, 2nd ed., Cambridge UP
Morfill G.E., Forman M.A., Bignami G., 1984, ApJ 284, 856
Morfill G.E., Meyer P., Lüst R., 1985, ApJ 296, 670
Schönfelder V., Dogiel V.A., Freyberg M.J., Morfill G.E., 1997, these proceedings
Simpson J., 1983, in: Composition and origin of Cosmic Rays, ed. M. Shapiro (Reidel, Dordrecht), 1
Simpson J., Garcia-Munoz M., 1988, Space Sci. Rev. 46, 205
Sonett C.P., Giampapa M., Matthews M. (eds.), 1991, "The Sun in Time", Univ. of Arizona Press, Tucson
Sonett C.P., Morfill G.E., Jokipii J.R., 1987, Nature 330, 458
Strong A.W. et al., 1997, Proceedings of 4th Compton Symposium, in press
Völk H.J., Morfill G.E., Forman M.A., 1981, ApJ 249, 161

Low Energy Cosmic Rays Emitted by the Orion Complex into the Local Medium

V. Schönfelder[1], V.A. Dogiel[2], M.J. Freyberg[1], and G.E. Morfill[1]

[1] Max-Planck-Institut für extraterrestrische Physik, D-85740 Garching, Germany
[2] P.N.Lebedev Physical Institute, 117924 Moscow, Russia

Abstract. The excess of γ-ray emission in the energy range 3 – 7 MeV discovered by COMPTEL in the direction of the Orion complex may imply that there is a powerful source of low energy cosmic rays in the local galactic medium. Several interpretations of the excess have been suggested. One of them assumes that the emission is identified with nuclear de-excitation lines of excited ^{16}O and ^{14}C. To provide the excess the energy deposit in these nuclei should be as large as $\sim 10^{39}$ erg s^{-1}. If a comparable part of these nuclei escape into the local galactic medium they can play a significant role in the energy balance there.

If the excess is interpreted as continuous it can be produced by bremsstrahlung emission of fast 10 MeV electrons. Much less energy deposit in these electrons is necessary ($\sim 10^{36}$ erg s^{-1}) to generate the observed Orion flux. The intensity of these electrons even near Earth can be higher than that of the galactic cosmic ray electrons.

1 Introduction

One of the key problems of the local galactic medium is the question whether the conditions there are the same as in the other parts of the interstellar medium or whether this part of the Galaxy heated by several supernova explosions has rather peculiar parameters. This question is especially interesting in the context of the origin of cosmic rays since we do not know (for sure) whether the characteristics of cosmic rays observed near Earth differ (or not) from those in other parts of the Galaxy.

On the other hand one of the key assumptions of cosmic ray physics is that the parameters of galactic cosmic rays are not changed significantly in the local galactic medium. Only then it is possible to derive the cosmic ray life-time in the Galaxy from the chemical composition of cosmic rays measured near Earth.

Radio and γ-ray investigations seem to confirm this assumption since analyses of radio- and γ-ray background radiations do not show significant spatial variations of the spectrum and the density of galactic cosmic rays in comparison with that observed near Earth. On the other hand, γ-ray data may discover a near-by powerful source of cosmic rays with a luminosity as high as about $3 \cdot 10^{38}$ erg s^{-1} which may have a strong influence on the conditions in the local interstellar space. Below we discuss briefly this possibility.

2 Flux of Gamma-Rays from Orion

COMPTEL has detected strong 3 to 7 MeV γ-ray emission from the Orion complex with no significant evidence for emission outside this range. Its spectrum shows features that might be due to splitting of C and O nuclear interaction lines (Bloemen et al., 1997) The detected excess from Orion in the energy range 3 – 7 MeV has a value of

$$F_\gamma \sim (10.1 \pm 1.4) \times 10^{-5} \text{ph cm}^{-2}\text{s}^{-1} \tag{1}$$

To produce the observed excess in the energy range 3 – 7 MeV in the direction of Orion the photon flux should be about

$$L_\gamma \sim 3 \cdot 10^{39} \text{ ph s}^{-1}$$

or the luminosity in the 4.4 MeV line should be

$$F_\gamma \sim E_\gamma L_\gamma \sim 2 \cdot 10^{34} \text{erg s}^{-1}. \tag{2}$$

To satisfy the observational data obtained by COMPTEL and EGRET we should assume that the spectrum of accelerated particles is truncated at an energy E_m. Its value is less than 100 MeV/n and higher than 20 -30 MeV/n.

In the framework of the thick target model the total luminosity of cosmic ray nuclei L_n needed to produce the Orion γ-ray flux (in the energy range of nuclei $E_k < E_m$) is estimated as (see Bykov and Bloemen, 1994)

$$L_n \sim 10^{42} \text{ n s}^{-1}$$

or the energy luminosity of the Orion complex in these nuclei is

$$F_n \sim 8 \cdot 10^{38} \text{erg s}^{-1} . \tag{3}$$

3 Orion Cosmic Ray Nuclei in the Local Medium

The estimate (3) was derived under the assumption that the nuclei lose their energy by ionization only. On the other hand one cannot exclude that a part of the nuclei escape from Orion into the local galactic medium and reach the solar system. In this case their density near Earth is determined from the diffusion equation describing cosmic ray propagation in the Galaxy:

$$N_n = \frac{L_n}{4\pi r D} \tag{4}$$

where r is the distance between Orion and the Sun, D is the diffusion coefficient in the local galactic medium, and L_n is the production rate of nuclei in Orion.

From Eq.(4) we can easily estimate the flux of local nuclei if they are emitted by Orion, since we know almost all necessary parameters. The distance

$r \sim 500$ pc, the local diffusion coefficient D was determined by Dorman et al. (1984), from the anisotropy of very high energy cosmic rays ($E \sim 10^{12} - 10^{13}$ eV) near Earth: $D \simeq 10^{29}\mathrm{cm}^2\mathrm{s}^{-1}$. Energy variations of the diffusion coefficient were derived from the cosmic ray chemical composition. It was concluded that its value is approximately constant at energies $E < 1$ GeV and rises as $D \propto E^{0.6}$ at $E > 1$ GeV (see e.g. Berezinsky et al. 1990). Thus, extrapolating the diffusion coefficient value into the energy range of ≤ 1 GeV we obtain $D \sim 10^{27} - 10^{28}\,\mathrm{cm}^2\mathrm{s}^{-1}$. The average galactic diffusion coefficient was derived from analyses of the galactic non-thermal emission and the cosmic ray chemical composition (Berezinskii et al. 1990, Strong et al. 1997). Its value may be as large as $10^{29}\,\mathrm{cm}^2\,\mathrm{s}^{-1}$ at energies $E \leq 1$ GeV.

As we mentioned already the estimate (3) was derived under the assumption that the nuclei do not escape from Orion. If, however, a comparable flux of nuclei escapes into the local galactic medium, this results in an energy density of nuclei from Orion near Earth as high as $0.1 - 0.3\,\mathrm{eV\,cm}^{-3}$ even for a diffusion coefficient of $3 \cdot 10^{28}\,\mathrm{cm}^2\,\mathrm{s}^{-1}$. This energy density is comparable to that of galactic cosmic rays. Since the energy of cosmic rays in Orion is rather low they can play a significant role in the energy and ionization balances in the local medium.

4 10 MeV Electrons in the Local Medium from Orion

If the 3 to 7 MeV excess is interpreted as continuous emission, it can be produced by bremsstrahlung emission of fast electrons, which have an almost energy independent spectrum below 20 MeV and a cutoff at about 20 MeV. The energy needed to produce the Orion γ-ray excess is much less than in the case of the line origin. Only 10^{36} erg s^{-1} is necessary to produce the observed Orion γ-ray flux. If part of these electrons escape into the local interstellar medium, their density there may be significant and exceed that of the galactic cosmic ray electrons.

Direct observations of the local spectrum of cosmic ray electrons are restricted to energies above 1 GeV because of the modulation in interplanetary space, but contrary to the nuclear component the electron component can be derived at low energies from observations of the Galactic non-thermal radio and γ-ray emission.

Especially, the electron spectrum derived from the diffuse galactic γ-ray emission, measured by COMPTEL, was extended down to the energy of 1 MeV (Strong et al. 1997). This analysis has been performed in the framework of the diffusion model of cosmic ray propagation in the Galaxy. It was shown that the intensity of the Galactic electrons at energies of 10 MeV is about $2 \cdot 10^{-1}\,\mathrm{el\,cm}^{-2}\,\mathrm{s}^{-1}\,\mathrm{sr}^{-1}$.

A lower limit to the 10 MeV electron production by Orion can be derived in the framework of the thick target model from the Orion γ-ray excess, $L_e \sim 3 \cdot 10^{39}\,\mathrm{el\,s}^{-1}$. If a comparable flux of electrons escapes into the local galactic medium, this results in an intensity of electrons near Earth from Orion (using Eq.(4)) of about 1 to 10^{-2} el $cm^{-2}s^{-1}sr^{-1}$ depending on the value of the diffusion coefficient in the local Galactic space. In this respect the Orion source of 10 MeV electrons may play an essential role in the energy balance of the local galactic medium.

Acknowledgements. VAD is grateful to his colleagues from the Max-Planck-Institut für extraterrestrische Physik (Garching) for helpful and fruitful discussions. This work was partly supported by the Government of Russia through the Russian Foundation of Fundamental Researches under the grant No. 95-02-06422-a.

References

Berezinsky V.S., Bulanov S.V., Dogiel V.A., Ginzburg V.L., Ptuskin V.S., 1990, *Astrophysics of Cosmic Rays*, (ed. V.L.Ginzburg), North Holland

Bloemen H., Bykov A.M., Bozhokin S.V., Diehl R., Hermsen W., van der Meulen R.D., Oberlack U., Ryan J., Schönfelder V., Strong, A.W., Winkler C., 1997, ApJ, 475, L25

Bykov A.M., Bloemen H., 1994, A&A 283, L1

Dorman L.I., Ghosh A., Ptuskin V.S., 1984, Sov.Astron.Lett. 10, 345

Strong A.S., Bennett K., Diehl R., Hermsen W., Purcell W., Schönfelder V., Stacy J.G., Winkler C., Youssefi G., 1997, A&A Suppl. 120, 381

Structure Within the Local Bubble: Properties of Individual Clouds

J.E. Beckman[1], J. Trapero[1,2,3], J.R. Álamo[1], R. Génova[1], and I. Lundstrom[4]

[1] Instituto de Astrofísica de Canarias (IAC), La Laguna, Tenerife, Spain
[2] LAEFF, Villafranca del Castillo, Madrid, Spain
[3] Instituto de Astrofísica de Andalucía, Granada, Spain
[4] Lund Astronomical Observatory, Lund, Sweden

1 The Basic Methodology

Starting from the casual observation that neighbouring lines of sight to stars at similar distances in the LISM show abruptly varying column densities,we developed a technique for measuring properties of single clouds. Fig.1 shows an idealized version of this. The steps (Trapero et al. 1992, 1995) are:

– Correct for extinction the distances to local stars used, via B-V (where available), or our own measured NaI or KI absorptions.

– Measure diametral column density towards A from NaI or KI equivalent width and convert to hydrogen: N(H) [cm^{-2}].

– Measure cloud diameter, between E and E'.

– Estimate the number density, n(H) = N(H)/d [cm^{-3}].

– Use a canonical thermal pressure, nT = 3500 K cmThis at least gives the decision cool (<100K) or warm (>5000K).

– Estimate the cloud mass by integrating n(H) over the estimated volume.

Following up, cool cores within warm clouds, or warm envelopes around cool clouds can be identified using distance and velocity as tags. We have found empirical rules of thumb to guide us:

– The KI resonance at 7699Å is ideal for finding cool clouds; it is too weak to show up in warm clouds.

– NaI D doublet lines weaker than 5mÅ typify warm clouds. There are too many of these, by a factor 50, to come from cool cloud edges. This is confirmed by the "low velocity Routly-Spitzer effect" (see below).

2 Examples of Objects Measured

2.1 Cool Clouds: a Massive Local Cloud in Perseus

This section is a brief summary of results in 4 papers (Trapero et al. 1992, 1995, 1996; Serra et al. 1995). Strong HI column densities were inferred towards α 3h 40m, δ +48° via stellar KI absorptions. The above method yielded

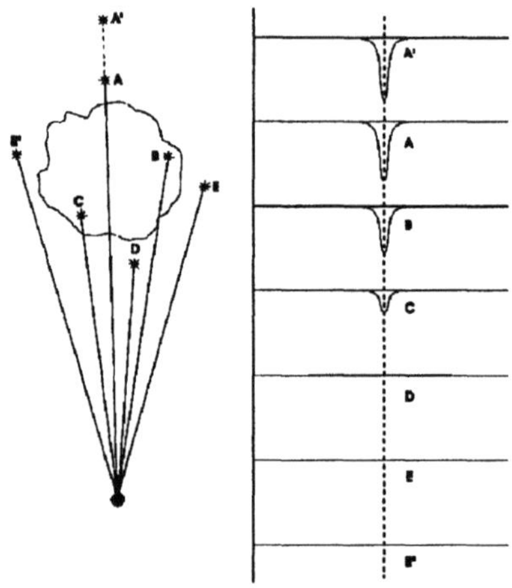

Fig. 1. Schematic showing how the parameters of a LISM cloud are inferred using lines of sight to stars around it (see text).

a cloud mass of $1300M_\odot$, temperature around 50K, and density $70cm^{-3}$. Given the velocity, we identified the cloud in absorption against the warmer background, using 21cm HI mapping, revising mass and temperature, and finding minima of 30K. Molecular cores were then found using CO mapping. Fig.2 is a map of HI temperature, from 21 cm data, and Fig.3 a map of a molecular core, whose mass is only a few $M_\odot$. The whole cloud lies within 120 pc of the sun. Numerous smaller but still massive ($M>200M_\odot$) clouds have been found via KI; they occupy less than 10% of the LISM volume, but form at least 50% of the mass.

2.2 Warm Clouds, with and without Cool Cores

Spectra obtained by the Lund team (see Lundstrom et al. 1991) using the ESO 1.4m CAT+CES with reticon detector have outstanding S:N characteristics, with values of up to 1000, allowing detections of sub-mÅ NaI D lines, ideal for exploring H column densities below 10^{18} cm^{-2}. A field in Sco-Oph, with several such clouds tagged by radial velocity, is shown in Fig.4. The warm cloud masses range from $1M_\odot$ to $150M_\odot$. and only one has a cool core, identified via KI, but this is not a statistically reliable sample. In an overlapping field 10° north of that in Fig.4 are six cool clouds, masses between $100M_\odot$ and $600M_\odot$, two thirds with warm envelopes. This warm cloud identification method, used so far over only a few % of the sky, has great potential. A well-known diagnostic, the ratio N(NaI)/N(CaII), was first postulated by Routly and Spitzer as showing post-shock gas,given its low value for high LSR velocities in the LISM. We show in Fig.5 this ratio v. N(NaI); all the

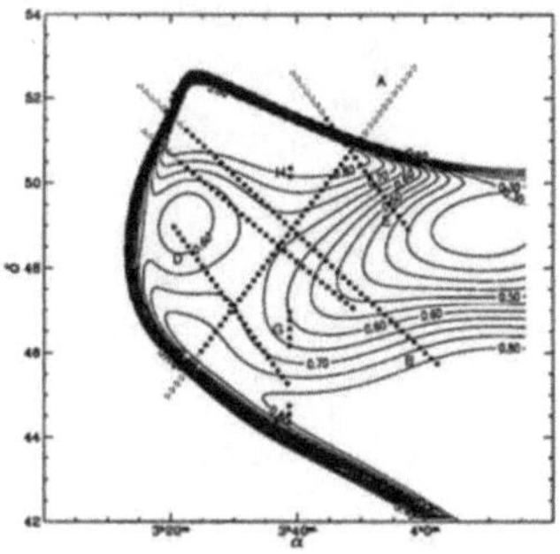

Fig. 2. Isotherm map (units of 50K) of the nearby cool cloud in Perseus, from 21 cm observations, using neural network methods to compensate a sparse sample (Serra Ricart et. al. 1995). Note the internal temperature minima.

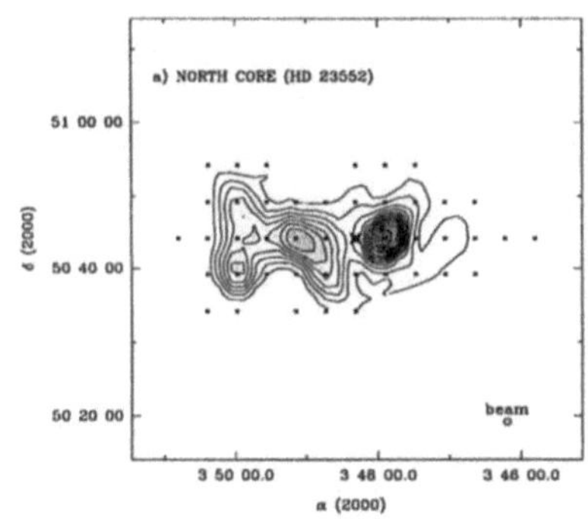

Fig. 3. Integrated CO emissivity over, in units of 0.2K kms^{-1} over a molecular core in the Perseus cloud. The star HD 23552, whose KI LISM absorption gave the original column density and velocity data is shown by a cross.

data is from low velocity clouds, with $V_{LSR} < 30$ km s^{-1}. Two regimes are seen, and low ratios correspond to low N(NaI) i.e. to warm, not cool clouds (whose signatures are all in the upper left of the Figure). At low velocity the effect is purely due to temperature difference, i.e. grain evaporation not shocks.

Acknowledgements. This work was carried out with the aid of grant PB94–11–7 of the Spanish DGYCIT.

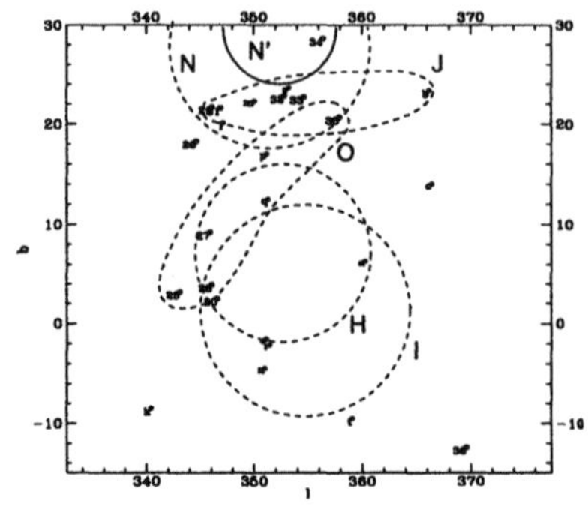

Fig. 4. Warm clouds identified using weak NaI absorption (dashed lines). In the present field only one of these (N N') has an identified cool core (solid line).

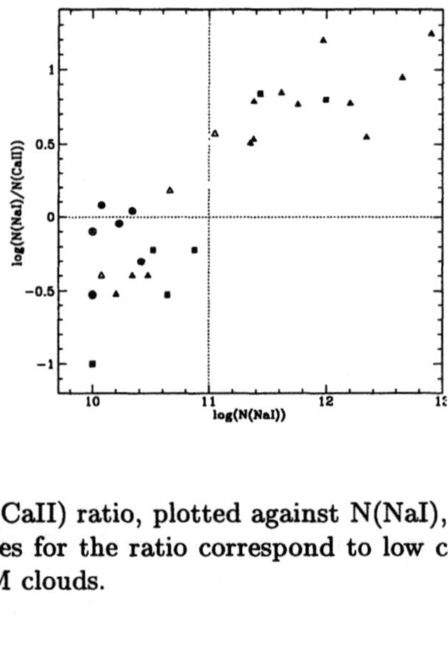

Fig. 5. The N(NaI)/N(CaII) ratio, plotted against N(NaI), for low velocity clouds in the LISM. Low values for the ratio correspond to low column densities, i.e. to warm (T>5000K) LISM clouds.

References

Lundstrom, I., Ardeberg, A., Maurice, E., Lindgren, H. (1991): A&A 91, 199
Routly, P.M., Spitzer, Jr.L. (1952): ApJ 115, 225
Serra-Ricart, M., Trapero, J., Beckman, J.E., Davies, R.D., Watson, R.A. (1995): AJ 109, 312
Trapero, J., Beckman, J.E., Génova, R., McKeith, C.D. (1992): ApJ 394, 552
Trapero, J., Beckman, J.E., Serra-Ricart, M., Davies, R.D., Watson, R.A., García López, R.J. (1995): ApJ 445, 231
Trapero, J. Sempere, M.J., Beckman, J.E., Hobbs, L.M. (1996): ApJ 457, 731

Kinematical Structure of the Local Interstellar Medium

R. Génova, J. E. Beckman, and J. Rodríguez Álamo

Instituto de Astrofísica de Canarias, Vía Láctea s/n, E-38200 La Laguna, Tenerife, Spain

Abstract. Observations of interstellar Na I in the spectra of 93 stars within 315 pc from the Sun show that it lies in a tunnel of gas moving away from Scorpio-Centaurus and is surrounded by gas moving *toward* the Galactic center.

Gas approaches the Sun from Scorpio-Centaurus expanding from (r, l, b)=(160 pc, 313°7, +28°2) with LSR velocity 15.3 km s^{-1}. The radius of this shell is 153 pc.

We identify these clouds:

D: velocity vector (v_d, l_d, b_d)=(+7.2 km s^{-1}, 305°1, −13°5), above and below the Galactic plane (GP) in the range of Galactic longitudes 357°–55°.

C: velocity vector (v_c, l_c, b_c)=(+11.5 km s^{-1}, 349°0, −35°2), above and below the GP in the range 30°$\leq l \leq$110°.

M: velocity vector (v_m, l_m, b_m)=(+21.9 km s^{-1}, 34°2, +1°5), above and below the GP in the range 100°$\leq l \leq$130°.

P: velocity vector (v_p, l_p, b_p)=(+13.8 km s^{-1}, 224°9, +5°4), above and below the GP from $l\sim$120° to the limit of our data at $l\sim$210°.

E: velocity vector (v_e, l_e, b_e)=(+16.8 km s^{-1}, 208°4, +6°2) in the range 160°$\leq l \leq$185° and −10°$\leq b \leq$−35°.

A: velocity vector (v_a, l_a, b_a)=(+12.9 km s^{-1}, 73°6, −5°6) towards the Galactic anti-center, below the GP.

I: velocity vector (v_i, l_i, b_i)=(+37.7 km s^{-1}, 132°8, −64°3) towards the Galactic anti-center, above the GP.

1 Historical Perspective

The X-ray emission ubiquitously observed in the Local Interstellar Medium (LISM) (Sanders et al. 1977) shows that the Sun and the nearest stars are surrounded by gas with the characteristics typical of the interiors of interstellar bubbles produced by supernova remnants and stellar winds, suggesting that the Sun is within a large interstellar bubble (Cox, & Reynolds 1987), consequently named "Local Bubble" (LB), whose detailed origin is still unknown. The interior of this LB is being modified by the effect of stellar winds and supernova explosions produced by stars that lie within the LB and, as new instrumental developments allow a deeper study of the shortest lines of sight, the complexity of the kinematics and spatial distribution of the local gas is becoming evident. This gas is moving away from a region of the sky occupied by the Scorpio-Centaurus OB association and its associated subgroups and it was naturally suggested that the supernova explosions and stellar winds produced by its most massive stars were the engines that set the gas in motion (Bertaux, & Blamont 1971; Adams, & Frisch 1977).

The first comprehensive attempt to explore the overall spatial distribution of the local gas closest to the Sun was made by Génova et al. (1990). These authors showed the presence of an asymmetry in the distribution of the local interstellar gas with high densities observed in the hemisphere that contains the Galactic center and low densities in the opposite Galactic hemisphere. The large-scale motion of the local gas toward the Galactic anti-center hemisphere previously observed by Crutcher (1982) was also detected in these low resolution data. When high-resolution optical data of the shortest lines of sight became available a new vision of the LISM kinematics emerged, showing that the local kinematical structure is not as homogeneous as previously thought. Several cloudlets were detected in the immediate vicinity of the Sun (Lallement, Vidal-Madjar, & Ferlet 1986; Lallement & Bertin 1992), nevertheless these cloudlets share a common motion toward the Galactic anti-centre hemisphere. The Sun happens to be submerged in one of the cloudlets of this ensemble, the *Local Cloud* (LC). The presence of this cloud was inferred from the early 70's through the detection of H and He atoms of interstellar origin flowing in the inner heliosphere (Bertaux, & Blamont 1971; Adams, & Frisch 1977) and it has now been detected in the spectra of the nearest stars (Lallement et al. 1994).

2 Recent Results

The presence of spheroidal shells approaching the Sun from the Scorpio-Centaurus association has been shown by Crawford (1991) and Génova et al. (1997a). The shell detected by Crawford (1991) is centered at $(l, b) \sim (320°, +10°)$ at 140 ± 20 pc from the Sun, its radius is 110 ± 10 pc, it is expanding with velocity 7 km s^{-1}. The shell detected by Génova et al. (1997a) is faster, its expansion velocity being 15.3 km s^{-1}, and is centered at $\sim$160 pc from the Sun at $(l, b) \sim (313°, +28°)$, its radius is $\sim$153 pc. These values place the centers of these shells well within the Scorpio-Centaurus association. Even if they may have a common direction of origin, the difference in their expansion velocities suggests that these shells may not be related, most probably they are shell-like filaments emitted from Scorpio-Centaurus at different epochs. In particular, the dimensions of the shell detected by Génova et al. (1997a) place its edge very close to the Sun in direction toward Scorpio-Centaurus, also its expansion velocity and the velocities of the local ensemble of clouds are of the same order; these facts suggest that the LC and its accompanying cloudlets are the leading edge of and expanding interstellar bubble that is engulfing the Sun.

A somewhat surprising result has been the recent detection of nearby interstellar clouds that show motions toward the Galactic center hemisphere.

Interstellar Na I absorptions detected in the spectra of 93 stars within 315 pc from the Sun observed from the 4.2m William Herschel Telescope on La Palma with a velocity uncertainty of order 1.0-1.5 km s^{-1} have shown the presence of the expanding shell from Scorpio-Centaurus described above as well as the 7 interstellar clouds whose velocity vectors and location in the sky are cited in the abstract (Génova et al. 1997a; Génova et al. 1997b). Four of these clouds — A, M, C and D — show motions toward the Galactic center hemisphere, which is in contradiction with the current ideas about the local gas kinematics. Three clouds — P, I, and E — show motions toward the Galactic anticenter hemisphere. Cloud P covers the largest area in the sky among these clouds, which seems to indicate that it lies close to the Sun. It also shows a velocity vector similar to the vector found by Crutcher (1982) in the nearby gas. Cloud E is kinematically similar to P, both clouds are seen simultaneously in several lines of sight toward the Pleiades, these are the clouds that give rise to the kinematical structure discussed by Gordon & Arny (1984) in the vicinity of this cluster.

3 The Emerging Picture of the Local Gas Kinematics

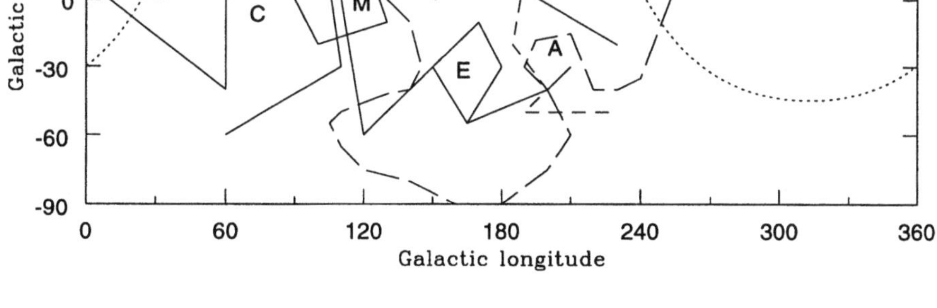

Fig. 1. The long-dashed line shows the limit between the low and high Mg II density regions identified by Génova et al. (1990) toward stars within the nearest 30 pc. The projection on the sky of the expanding shell identified by Génova et al. (1997a) is shown by the dotted line. Solid lines sketch the projections on the sky of the clouds detected close to the Sun by Génova et al. (1997a) and Génova et al. (1997b).

As shown in Figure 1 clouds P, I, and E are observed against the low Mg II density region. On the other hand, the clouds that show motions toward the Galactic center hemisphere are observed against the high Mg II density region. This morphology can be explained by means of the presence of two interstellar bubbles in the immediate vicinity of the Sun: one bubble is approaching the Sun from Scorpio-Centaurus, the nearby interstellar gas is being swept up and accelerated by the leading edge of this bubble. In the direction of the Galactic center we observe high densities produced by interstellar gas that has already accumulated in the shell of this bubble. In the opposite direction we observe a low density region toward which the former bubble is advancing. This region is in fact the interior of an older bubble whose shell is seen in the clouds that are moving toward the Galactic center hemisphere. Most absorptions in the direction of the Galactic anticenter are caused by gas very close to the Sun that has been accelerated by the Scorpio-Centaurus bubble and has already passed the Sun.

Although we give precise values of the velocity vectors of these clouds we feel that given the range of distances to the background stars used to determine those velocity vectors they represent average motions of the gas, nevertheless the existence of regions where the IS gas flows towards the Galactic center hemisphere is clearly shown in our data. Further observations, in particular precise determinations of the distances to the nearest edges of these clouds, are needed to confirm this view.

References

Adams T.F., Frisch P.C. (1977): ApJ 212, 300
Bertaux J.L., Blamont J.E. (1971): A&A 11, 200
Cox D.P., Reynolds R.J. (1987): ARA&A 25, 303
Crawford I.A. (1991): A&A 247, 183
Crutcher R.M. (1982): ApJ 254, 82
de Geus E.J. (1992): A&A 262, 258
Génova R., Molaro P., Vladilo G., Beckman J.E. (1990): ApJ 355, 150
Génova R., Beckman J.E., Bowyer S., Spicer T. (1997a): ApJ 484, (page not yet assigned)
Génova R., Beckman J.E., Rodríguez Álamo J. (1997b): ApJ submitted
Gordon K.J., Arny T.T. (1984): AJ 89, 672
Lallement R., Bertin P., Ferlet R., Vidal-Madjar A., Bertaux J.L. (1994): A&A 286, 898
Sanders W.T., Kraushaar W.L., Nousek J.A., Fried P.M. (1977): ApJL 217, L87

Radio, Millimeter and Infrared Observations of the Local Hot Bubble and Its Environment

Ulrich Mebold, Jürgen Kerp and Peter M.W. Kalberla

Radioastronomisches Institut der Universität Bonn, Auf dem Hügel 71, D-53121 Bonn, Germany

Abstract. We present a definition of the local void of neutral gas from observations in the radio frequency window. We question the concept of the Local Bubble in sense of a more or less spherical volume which is surrounded by a shell of denser gas. The concept of the Local Bubble is challenged by the discovery of numerous neutral, dense clouds inside the local void. The search for a "shell" around the suspected Bubble has resulted only in inconclusive findings so far. The sample of high latitude molecular clouds illustrates the situation particularly well. The statistical properties of their spatial distribution, e.g. the mean distance, seem to fit very nicely to the spatial extent of the Local Bubble. But a more detailed investigation shows that the concept of a bubble - in particular an expanding bubble - is not supported. We suggest that the local void is nothing more than a typical place in an interarm region of our Galaxy.

Finally, a discussion of the high latitude boundary of the local void does not give strong evidence for the concept of a bubble, that has once been in rapid expansion and is still showing signs of interaction with its environment. However, indications for interactions of IVCs or HVCs with their surroundings are found. These hint at the presence of a gaseous disk which is much more vertically extended than previously believed, or at a Galactic wind which may be blowing from the Galactic neighbourhood of the Sun.

1 Discovery of the Local Void, Some History

A void of neutral material in the interstellar medium (ISM) around the Sun was discovered by Frisch & York (1983) in studies of nearby early type stars. In 1968 Bowyer, Field and Mack discovered the soft X-ray background radiation. On the basis of a new extensive data-set McCammon et al. (1983) proposed that all of the observed soft X-ray radiation was emitted from within the local void of neutral matter. Their model of a Local Hot Bubble (LHB) was consistent with all the observations available until the launch of *ROSAT* (McCammon & Sanders 1990). The effect of the local void on the H I gas has been known for many years. The H I column densities projected vertically onto the Galactic plane show a depression of about 30% right at the location of the Sun (see Fig. 1). The depression derived at 90° and $z_0 = 400\,\mathrm{pc}$ is about 40% on the north side of the Galactic plane and about 30% on the southern galactic sky. This difference shows that local void is more prominent in the northern Galactic plane then in the southern. The plots for different sections show the assymetry of the local void with respect to the Sun, in particular its large extension towards the third galactic quadrant. In fact this assymetry hints at an interpretation of the local void as a feature which is related to the local interarm region, the region between the Sagittarius and the Perseus spiral arms of our Galaxy. If this interpretation is correct, the solar surroundings were nothing special but a typical place in an interarm region of our Galaxy. In contrast to this view Snowden et al. (1990) showed

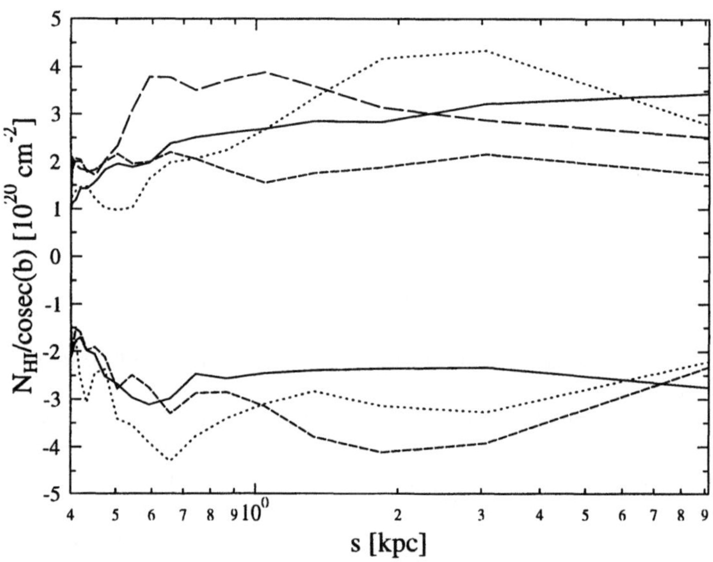

Fig. 1. Vertical projection of the total H I column density, $N_{HI} \times sin(b)$, versus the path length s through the Galactic layer, $\frac{z_0}{sin(b)}$, for $z_0 = 400$ pc. Positive latitudes for $N_{HI} \geq 0$, negative latitudes for $N_{HI} < 0$. Solid, dotted, dashed and long dashed lines indicate the first, second, third and fourth Galactic quadrant, respectively. The low values of $N_{HI} \times sin(b)$ at small path lengths, s, are indicative of the reduced column densities of the local void.

that this depression can be modelled by assuming that in a constant density environment H I gas is swept out of the region presently occupied by the local void. The sweeping mechanism may be a supernova blast wave or a strong stellar wind. This is the basic idea of the so-called "displacement" model. If all of the observed soft X-ray emitting plasma is placed in front of the first $N_{HI} \simeq 10^{19}\,cm^{-2}$ of H I column density along the line of sight and if a constant density and temperature is assumed for that plasma, then a length, R, of the line of sight through this plasma can be derived for each observed direction from the emission measure of the plasma. One finds an average length of $R \simeq 100$ pc for $n_e \simeq 10^{-3}\,cm^{-3}$, consistent with the radius of the local void.

2 Clouds Inside the Local Void

The very local clouds ($d < 10\,\mathrm{pc}$, say), sometimes called the local fluff, are not discernable in the H I 21-cm line or any other emission down to sub-mm wavelengths. Therefore the local fluff is not discussed here. In Table 1 we have collected some of the important astrophysical parameters of clouds which appear to be located in the local void. All of the four clouds are seen in H I and IR emission, one of them (LVC159-34+12, or MBM12) is detected in CO. All clouds, except LVC159-34+12, display a rather filamentary morphology with a ratio of the major to the minor axis of about 10 to 1. H I column densities are all around $N_{\mathrm{HI}} \simeq 1 \cdot 10^{20}\,\mathrm{cm}^{-2}$ with a rather small velocity dispersion implying kinetic temperatures of about 100 K or less. Distances of all clouds except IVC288-34+4 have been derived in two independent ways. First, interstellar absorption lines in stellar spectra constrains the upper-distance limit of the cloud; details of these distances can be found in the references given in the last line of Table 1. Second, soft X-ray radiation transport considerations can give an independent distance estimate if the X-ray volume emissivity of the plasma in the Local Bubble is assumed to be constant and if its full radius at the position of the cloud is known. The distance of 39 pc listed for LVC88+36-2 is derived from X-ray shadowing assuming a radius of 100 pc for the Local Bubble in this direction. Interstellar absorption line studies give an upper limit of 60 pc. Since the radius of the local void is about 100 pc in this direction, LVC88+36-2 is clearly located inside this void. As for LVC88+36-2, it has long been established that LVC159-34+12 (MBM12) is also located inside the local void. For LVC75+35+5 Kerp & Pietz (this volume) argue

Table 1. Properties of some local clouds. The star denotes quantities derived from CO line observations all others are based on H I 21-cm line measurements.

name	LVC 88+36-2	LVC 75+35+5	LVC 159-34+12	IVC 288-34+4
type of radiation	H I, *IRAS*	H I, *IRAS*	H I, *IRAS*, CO	H I, *IRAS*
morphology	filamentary	filamentary	cloud	filamentary
size: major × minor axis	$5^\circ \times 0^\circ.5$	$15^\circ \times 2^\circ$	$1^\circ \times 1^\circ$	$25^\circ \times 2^\circ$
position angle	45°	45°		0°
H I column density [cm^{-2}]	$< 1 \cdot 10^{20}$	$\leq 2 \cdot 10^{20}$	$\sim 1 \cdot 10^{21}$	$2 \cdot 10^{20}$
velocity dispersion [$\mathrm{km\,s}^{-1}$]	1.0	$\simeq 1.0$		~ 1.5
kinetic temperature, T_{kin} [K]	120	$\simeq 120$	~ 10	200
distance, d [pc]	39	50	65	130
total H I mass, M_{HI} [$\mathrm{M}_\odot$]	15	360	30	3000
volume density, n_{HI} [cm^{-3}]	80	200	1000	70
pressure $\frac{P}{k}$[$\mathrm{K\,cm}^{-3}$]	$1 \cdot 10^4$	$2 \cdot 10^4$	$1 \cdot 10^4$	$1 \cdot 10^4$
references	Wennmacher 1994	Kerp & Pietz this volume	Benjamin et al. 1996	Penprase et al. 1997

from both interstellar absorption lines and from soft X-ray absortion that it is probably located within the local void. IVC288-34+4 may be inside the Local Bubble inspite of its rather large distance of 130 pc because the radius of the bubble is also rather large in this direction.

Using distances, the angular sizes and the observed column densities, the clouds have the total H I masses given in line 10. Assuming circular symmetry, average volume densities are given in line 11. The temperature and density given for MBM12 were derived from molecular lines (Magnani et al ApJ 295,402). From the kinetic temperatures in line 8 and the volume densities in line 11 one can estimate the pressures given in line 12. The uncertainties of this estimate for the pressure are enormous. This is so because the line width may be dominated by turbulence or because the filament is a sheet seen edge-on rather than a worm-like structure. Then the density, the temperature and the pressure can be very much smaller. Nevertheless it is striking that the pressures derived here are all pretty similar and fall in the range of pressures which are models of the Galactic interstellar medium (e.g. Wolfire et al 1995).

3 The Quest for a Wall of the Bubble

A minimum H I column density of $N_{\rm HI} \leq 1 \cdot 10^{20}\,{\rm cm}^{-2}$ is found in various directions at the high latitude sky. Column densities as low as this represent gas which is in the form of the so-called intercloud medium. It has a velocity dispersion of $\sigma \sim 10\,{\rm km\,s}^{-1}$, is rather smoothly distributed on the sky and follows most closely the distribution of a plane-parallel layer (Mebold 1972, Falgarone & Lequeux 1973). H I in the form of clouds which are immersed in this medium dominates at lower latitudes (say $b < 40°$). In the presence of clouds the column density is typically more than doubled. These structural features imply that the local void is rather porous at high latitudes and may actually be open for dynamical interaction with the Galactic halo.

The most complete review of more than 100 molecular clouds at $|b| >$ 25 deg has recently been published by Magnani, Hartmann & Speck (1996). From the observed dispersion $\sigma \simeq 5.9\,{\rm km\,s}^{-1}$ of the centre velocities of these clouds the authors derive an average $< z >=$ 87 pc above the Galactic plane. This corresponds to an average distance of $< d >=$ 105 pc which coincides remarkably well with the average radius of the Local Bubble. Due to the lack of precise distance determinations is not possible to check whether the frequency distribution of clouds shows a peak at the radius of the Local Bubble or whether their space density is constant starting outwards at that radius. Therefore the coincidence of $< d >$ and the mean radius of the local void is nothing but a confirmation that we have locally a void of neutral interstellar material. From the global average of the centre velocities $< v >$ $= -0.3\,{\rm km\,s}^{-1}$ one can conclude that there is no significant expansion of these clouds. There is no indication that the concept of an expanding bubble applies to the void around the Sun.

The distribution of clouds on the sky shows a remarkable absence of molecular clouds at $|b| > 60°$. This could mean that the polar region is the place of a break-out of the LB into the halo of our Galaxy. If this were so, then we have to explain the fact that we do not see a coherent low column density structure in high Galactic latitudes. Rather we see various clouds or cloud complexes which we do not expect to find within the breakout region of a local Supernova event. This makes the concept of a bubble suspicious, once more.

Finally in this section we like to mention the structure of the ISM in the so-called Hercules Region (Lilienthal et al. 1992). This is a region bounded by $40° < l < 70°$ and $35° < b < 55°$. Distance estimates from interstellar absorption lines of the H I clouds in this region indicate excessive distance variations (some larger than a factor of two) on angular scales of a few degrees. This implies structures in 3-D space which resemble the finger of God, pointing to you and telling you that you are wrong! Here we can definitely state that a wall of the Local Bubble can not be found!

4 H I Features Beyond the Local Void

4.1 IVCs with Molecular Emission

There is a class of clouds in high latitudes, the so-called intermediate-velocity clouds (IVCs, $-15 > v_{\rm LSR} > -100\,{\rm km\,s^{-1}}$), which are most likely located outside the local void and frequently show signs of dynamical interaction with the plasma in the Galactic halo or the interface of the halo and the disk (Wesselius & Fejes 1973, Tenorio-Tagle & Bodenheimer 1988, Kuntz & Danly 1996, Moritz et al. 1997). There are three IVCs which show molecular line emission, IVC 92+38-22, IVC 135+53-45, and IVC 211+63-38.7. The first two of these are closely associated with HVCs, i.e. high latitude H I clouds at velocities $|v_{\rm LSR}| > 100\,{\rm km\,s^{-1}}$ and thus are candidates for an interaction of the kind specified before.

IVC 135+53-45 is part of the largest IVC complex in the north galactic hemisphere is the so-called IVC-arch (see Kuntz & Danly 1996). Interstellar absorption line studies indicate a distance of up to 1.7 kpc. IVC 92+38-22 which is frequently called the Draco Cloud (e.g. Mebold et al. 1985, Herbstmeier et al. 1996) is one of three high latitude molecular clouds which is found at intermediate negative velocities (Magnani, Hartmann & Speck 1996). Its distance is in dispute with estimates ranging between 300 pc and 800 pc (Lilienthal et al. 1991, Mebold et al. 1985). There are two particularly interesting aspects of IVC 92+38-22: first, it was the first interstellar cloud that revealed an X-ray shadow in the 1/4 keV soft X-ray background (Snowden et al. 1991). This can be used to derive the molecular hydrogen content in IVC 92+38-22. Second, it has a steep velocity gradient perpendicular to the boundary facing the Galactic centre. This is indicative of an interaction whose nature is not understood up to now.

Our group has studied the X-ray shadows of IVC 92+38-22 to derive a new and independent estimate of the so-called X-factor $X = \frac{N(H_2)}{WCO}$. This ratio is frequently used for the conversion of the $^{12}CO(1 \rightarrow 0)$ line integral into column densities of molecular hydrogen H_2. If this factor were strongly variable through-out the Galaxy, column densities of molecular hydrogen derived from this factor would be very unreliable. From the IRAS 100μm intensities, I_{100}, observed CO- and H I lines, and the assumption that the 100μm intensity is a linear measure of the total column density of hydrogen nuclei. Herbstmeier et al. (1993) find values in the range $(0.02 \cdot 10^{20} < X < 0.26) \cdot 10^{20} \frac{\mathrm{cm}^{-2}}{\mathrm{K\,km\,s^{-1}}}$. This is more than a factor 10 smaller than the standard X-factor of $X = 2.4 \cdot 10^{20} \frac{\mathrm{cm}^{-2}}{\mathrm{K\,km\,s^{-1}}}$. Moritz et al. (1997) analyze the anti-correlation of the soft X-ray count rate versus observed H I column density N_{HI} or I_{100}. They find absorbing column densities which are more than twice as large as the observed N_{HI}. At the edge towards low galactic latitudes and longitudes, up to 70% of the total neutral hydrogen is in molecular form. They also find an FIR emissivity per hydrogen nucleon of about $1.0 \cdot 10^{-20}\,\mathrm{MJy\,sr^{-1}\,cm^2}$, which is constant through out the Draco cloud. In contrast, the FIR emissivity per H I atom varies strongly across the Draco cloud. From these results the authors derive X-factors in the range $(0.17 < X < 0.52) \frac{\mathrm{cm}^{-2}}{\mathrm{K\,km\,s^{-1}}}$, values which are larger than those found by Herbstmeier, Heithausen & Mebold (1993) but which are still much smaller than those found for molecular clouds in the Galactic plane.

Moritz et al. (1997) find that the emission measure of the X-ray gas behind the Draco cloud is about 5 times larger than that of the Local Bubble if plasma temperatures of $T \simeq 10^{6.3}\,\mathrm{K}$ and $T \simeq 10^{5.85}\,\mathrm{K}$ are adopted for the background and the Local Bubble, respectively. Taking into account that the Draco clouds are located out of the Galactic disk, the large emission measure of the background is unequivocal evidence for a bright Galactic X-ray corona.

The second aspect of particular interest are sharp velocity gradients. In the Draco cloud is a steep velocity gradient of $\simeq 1\,\mathrm{km\,s^{-1}\,arcmin^{-1}}$ perpendicular to the steep edge facing the Galactic center direction (see Herbstmeier et al. 1996). This gradient, which could only be found with the high angular resolution provided by H I aperture synthesis observations with the WSRT and the VLA, shows up consistently for more than 2° along this edge. It may be caused by ram-pressure decelleration of the Draco cloud while it is penetrating the lower Galactic halo on its way towards the disk. The material exerting the ram pressure is either the intercloud H I gas located e.g. in the Lockman layer (Dickey & Lockman 1990) or else the ionized medium of the Reynolds layer (Reynolds 1991). Alternatively this gradient may be caused by a Galactic wind rising from the disk at high speed (e.g. Kahn 1989).

IVC 135+54-45 is the second of the high latitude intermediate-velocity clouds which shows molecular line emission, is associated with an HVC, and shows an X-ray shadow. Benjamin et al. (1996) derive the distance to this cloud complex as $d = (355 \pm 95)\,\mathrm{pc}$, which corresponds to $|z| = (285 \pm 75)\,\mathrm{pc}$. Estimating the effects of extinction, the most likely value could be as low as $d = 240\,\mathrm{pc}$. This distance puts the X-ray emission behind a total estimated atomic mass of $1600\,M_{\odot}$. The molecular mass of the cloud core, G 135.3+54.5, is probably much smaller.

Some evidence for an interaction of HVC 133+55-120 with this IVC has been derived by Weiss (1997) from H I aperture synthesis observations obtained with the WSRT. These data show a very close positional association of IVC and HVC and an orientation of the filamentary structure of the IVC H I emission which is suggestive of shocks located in the interface region between IVC and HVC. Also, the HVC shows a velocity gradient at one of its edges which appears to connect the emission of HVC to the velocity of the IVC. We call this phenomenon "velocity bridge" and interpret it as evidence for a decelleration of the HVC during the course of its approach towards the Galaxy (see below). The nature of the decellerating agent appears rather unclear at the moment.

Recent CO line observations of IVC 135+54-45 carried out with the Pico Veleta 30-m telescope by Weiss, Mebold & Heithausen (in prep.) indicate that the CO line intensities are rather large, i.e. that the so-called X-factor is very small, also for this IVC. If HVC 134+55-120 and IVC 135+54-45 do indeed interact, then both must be at essentially the same distance. Since the IVC is at a distance of about 350 pc this means that the HVC is exceptionally close to the Galactic plane. If the material in the HVC has the same heavy element abundance and the same total gas column density as the IVC then the HVC should show up in absorption as does the IVC. There is as yet no information about the distance of HVC 134+55-120 available. But it is of considerable interest that most IVCs do show IR emission, e.g. in the *IRAS* 100 and 60μm bands, while HVCs in general do not show any IR emission (Wakker & Boulanger 1986). This is also true in the cases of the IVCs and HVCs discussed in this section. The absense of IR emission could either be due to the absense of dust or the absence of a radiation field that does the heating of the dust. If HVC 134+55-120 indeed interacts with IVC 135+54-45 it must be the lack of dust because the radiation field is obviously strong enough to heat the dust in the IVC. Since interstellar absorption lines of HVCs observed towards bright nuclei of galaxies show only mariginal amounts of heavy elements (e.g. Wakker et al. 1996) the dust content of HVCs remains a critical point of the discussion of the HVC - IVC interaction scenario. If, in particular, HVC 134+55-120 does contain dust, then we would have to accept the conclusion that the dust is not heated by the Galactic radiation field because the HVC is not close to the Galactic plane. Then, the evidence for an interaction of the HVC and the IVC would be accidental in the present case.

4.2 High Latitude H I Gas with Large Velocity Dispersion

The indications for an interaction of the IVCs with their ambient medium discussed in the previous subsection obviously require an interaction partner of sufficient density. The densities $n = (10^{-2} - 1)\,\mathrm{cm}^{-3}$ are much higher than the densities expected from the standard z-distributions for atoms or electrons. Since these distributions predict densities of $n < 10^{-3}\,\mathrm{cm}^{-3}$ at the typical z-distances of the IVCs of, say 1 kpc, we asked ourselves whether or not there is evidence for gaseous constituents of the Galactic halo which have not been noticed up to date.

The Leiden/Dwingeloo H I Survey by Hartmann & Burton (1997) has been searched by Kalberla et al. (1997) and Westphalen et al. (1997) for H I gas that may be residing in the Galactic halo. These authors searched the high latitude regions of the Galaxy for the presence of H I emission with very large velocity dispersion. After smoothing the data over areas of 5° by 5°. they indeed found faint emission ($T_B \simeq 0.01\,\mathrm{K}$) with a velocity dispersion of $\sigma \simeq 60\,\mathrm{km\,s}^{-1}$ and a column density $N_{\mathrm{HI}} \simeq 2 \cdot 10^{19}\,\mathrm{cm}^{-2}$, the so-called "Large Velocity Dispersion" (LVD) component. Assuming hydrostatic equilibrium for the z-distribution of atoms and ions and assuming energy equipartition of gas, plasma, magnetic fields, and cosmic rays, Kalberla, Pietz & Kerp (this volume) derive a scale high perpendicular to the Galactic plane of $h_z \simeq 4.4\,\mathrm{kpc}$.

It is plausible that the LVD component is only the neutral fraction of the predominantly ionized plasma in the lower Galactic halo. The nature, in particular the degree of ionisation, of this plasma is not c lear at present. It is likely, however, that the LVD gas is intermixed (see Kalberla, Pietz & Kerp, this volume) with the plasma emitting the X-rays which have recently been detected in the Galactic halo (e.g. Pietz et al., this volume) and with the plasma that is responsible for the interstellar absorption lines of heavy ions observed in the UV and EUV range of spectrum (de Boer, this volume). If this is true, the neutral gas in the LVD emission component may represent density enhancements, something like density-"icebergs", in an otherwise ionized plasma of lower density. These enhancements may be similar to the H I features found within the local void (see section 4). They merge into a diffuse background when looked at with a 5° beam. Such a picture may lead us to speculate further that HVCs are nothing but anomalously large ones of these density-icebergs. Also, the enhanced X-ray emission associated with some HVCs (Kerp et al. 1996, Kerp et al., this volume) may be produced in the regions of higher density which pile up in front of an infalling HVC. In these regions the ram pressure of the HVC increases the emission measure of the compressed plasma and this may lead to an enhancement of the X-ray emission over what is emitted by the non-compressed plasma. Details of this emission mechanism are investigated and discussed in a recent paper by Lesch et al. (this volume).

Of course, this is a rather qualitative picture yet. It will require further investigations to substantiate and refine this picture or to replace it by a better one.

4.3 HVCs and HVC-Bridges

Finally in this Section we argue about the origin of the strong velocity gradients found recently close to the edges of HVCs by Pietz et al. (1996). These features have been termed "velocity bridges" by these authors. In spite of this term, it is completely open at the moment whether these bridges which show up in velocity space, actually refer to spatial connections between HVCs and any other of the prominent features in the H I spectra.

Although the first velocity bridges were discovered by Meyerdierks (1992) these bridges could only recently be established as fairly common features in HVCs. A significant measurement of their faint ($T_B \simeq 0.2\,\text{K}$) and very wide (half width $W \simeq 100\,\text{km}\,\text{s}^{-1}$) spectral signatures require observations of very high baseline accuracy and a very low level of confusion by instrumental stray radiation. At the moment this requirement and an adequate coverage of the sky is provided only by the Leiden/Dwingeloo H I Survey (Hartmann & Burton 1997). So far this survey has not yet been searched fully and systematically for velocity bridges. Pietz et al. (1996) have only searched the area of HVC complex C and list 11 such features in their Table 1. Herbstmeier et al. (1995) also find velocity bridges in the HVC complex M.

Typical parameters of these bridges are velocity ranges of $\Delta v \simeq 50$ km s^{-1}, the brightness temperatures and half widths are as given above and position shifts between the two ends of the bridges are between $0\overset{\circ}{.}5$ to 2°. Estimating the energy of the process responsible for producing the bridges from the parameters provided by Pietz et al. (1996) we find an energy content of about $5 \cdot 10^{49}$ ergs. This amounts to a few percent of the total kinetic energy content of the associated HVCs.

The parameters provided by Pietz et al. (1996) also allow us to estimate the time scale over which this energy content is built up. The velocity bridges may be regarded as the result of a mass loss of the corresponding HVCs in the past: The product of the gas mass, M_{VB}, at a certain velocity, v, in the velocity bridge times that velocity (in the rest frame of the HVC) is the momentum of that gas with respect to the HVC. If this momentum is divided by the path length, s, that the gas has travelled from the HVC we can calculate the relative mass loss rate of the HVC from $(\frac{\partial M_{\text{HVC}}}{\partial t})$: $M_{\text{HVC}} = (M_{\text{VB}} \cdot v) : (M_{\text{HVC}} \cdot s)$. Using $\frac{M_{\text{VB}}}{M_{\text{HVC}}} \simeq 0.05$ from the observations, $v = \frac{v}{2} \simeq 25\,\text{km}\,\text{s}^{-1}$, and estimating the path-length from the position shift between the two ends of the bridges of, say 1°, ($s = 60\,\text{pc}$) and a distance of 3 kpc, we get $(\frac{\partial M_{\text{HVC}}}{\partial t})$: $M_{\text{HVC}} \sim 7 \cdot 10^{-8}\,\text{a}^{-1}$, c orresponding to a time scale for the observed mass loss of about $\tau \sim 2 \cdot 10^{7}\,\text{a}$. This estimate is a lower limit to the true time scale because the value of s that we have used is smaller than its true value. We can see only the projection $s \times sin(\alpha)$ of the true path length, s, onto the sky. The true path length is probably much larger than the observed projection because selection effects favor the detection of velocity bridges with small α. The selection effect which is at work here is due to the fact that the observed brightness temperature of the bridges increases with decreasing α (Mebold et al., in prep.).

Adopting a value of $\tau > 3 \cdot 10^7$ a, we see that the velocity bridges are rather long-lived phenomena. This long time scale imposes constraints on to the mechanisms that may be producing velocity bridges. Pietz et al. (1996) propose three alternative mechanisms:
1. inelastic collisions of HVCs and IVCs with clouds at velocities close to zero,
2. continous sweeping of HVCs and IVCs by the Reynolds - or the Lockman layers, or
3. interaction with a Galactic wind. Selecting from these options we propose that slow and continous sweeping of the outer parts of a HVC in the gaseous Galactic halo is the most likely mechanism. A possible scenario which may lead to a decelleration of parts of an HVC in the Galactic environment has been modelled by Benjamin (1997). Considering a HVC as a conglomerate of clouds of different densities the drag force will lead to a spatial seperation of these clouds. Those with low density will stay behind, those higher densities will be in the lead. The free-fall time scale for a typical HVC starting at say 10 kpc under the drag force conditions of a gaseous Galactic halo is of the order of $3 \cdot 10^7$ a. This is in agreement with the time scale estimated for the mass loss associated with HVC velocity bridges. It is therefore conceivable that velocity bridges are indeed produced by the drag force of a gaseous Galactic halo on HVCs, which consist of a spectrum of clumps with different densities. Since this requirement is indeed met by HVCs (Wakker 1990) we consider it worth while to a more detailed investigation of the origin of HVC velocity bridges in the frame of the concept persented here.

All the HVCs that are associated with excess X-ray emission are also associated with velocity bridges. Given the long time scale for producing bridges in the scenario decribed above it is unlikely that the observed X-ray luminosity for HVCs with velocity bridges is being produced by the conversion of their kinetic energy into heat. As stated above it is more likely that the enhanced emission is due an increase of the emission measure caused by the compression of the ambient medium. This means that most of the energy comes from the physical state of the halo plasma and not from the kinetic energy of the HVC. – At any rate, it seems likely that velocity bridges are a test probe for the physical state of the plasma in the Galactic halo that is worth exploiting further.

5 Conclusion

It seems to us that the evidence for a local "bubble" provided by observations in other wavelength regimes, in particular in the soft X-ray band, can also consistently be explained by placing the local void in the interarm region between the Sagittarius and the Perseus spiral arms of our Galaxy. This void would then appear as one of many places in the interarm regions of our Galaxy, a proposition that has previously been put forward by other authors (see Hirth et al. 1991, Jacobsen & Kahn 1986).

The least well understood phenomenon presented in this paper is the extreme velocity gradients associated with a large fraction of the HVCs, called velocity bridges, and the moderate velocity gradients associated with some IVCs (see section 6.3). For IVCs, well defined gradients are found, e.g. for the Draco nebula (Herbstmeier et al. 1996) and for the IVC arc by Kuntz & Danly (1996). For HVCs these gradients could only be established (Pietz et al. 1996) with the very high sensitivity of the new Leiden/Dwingeloo H I survey (Hartmann & Burton 1997) although the first bridges were reported by Meyerdierks in 1992. The nature of these velocity gradients is essentially not understood. Obviously they represent rather energetic phenomena. Among the possible interpretations are inelastic collisions of IVCs and HVCs with Galactic gas at velocities close to zero, continuous sweeping of these clouds during their passage through a teneous medium in the Galactic halo, and interactions with a Galactic wind.

At any rate, HVCs with velocity gradients appear to be very interesting probes of their Galactic or extragalactic environment while the corresponding IVCs probe the interface between the Galactic disk and halo. Both probably hold interesting clues about their astrophysical origin and evolution. Given the importance of these clouds and their origin for problems like primordial gas replanishment and chemical evolution of the Galaxy or the phenomenon of a Galactic fountain, or given the ill understood physical conditions of the plasma and the gas in our halo, it seems very worth while to continue working on the nature of these phenomena.

Acknowledgements. We like to thank all scientists of the Radioastronomisches Institut der Universität Bonn and outside who have contributed to the results presented or discussed in the present paper. The work presented here was supported by DFG under project No. ME 745/17 - 1 (2) und DARA project No.50 OR 9203.

References

Benjamin R.A. (1997): "The Physics of Galactic Halos" eds. Lesch et al., Akademie-Verlag, p. 33

Benjamin R.A., Venn K.A., Hiltgen D.D., Sneden C. (1996): ApJ 464, 836

Bowyer C.S., Field G.B. & Mack J.E. (1968): Nat 217, 32

Dickey J. & Lockman F.J. (1990): ARAA 28, 215

Falgarone E. & Lequeux J, (1973): A&A 25, 253
Frisch P.C. &, York D.G. (1983): ApJ 271, L49
Hartmann Dap & Burton W.B. (1997): "Atlas of Galactic Neutral Hydrogen", Cambridge University Press
Herbstmeier U., Heithausen A. & Mebold U. (1993): A&A 272, 514
Herbstmeier U. et al. (1995): A&A 298, 606
Herbstmeier U. et al. (1996): A&AS 117, 497
Hirth W., Mebold U., Dahlem M. & Müller P. (1991), Ap&SS 186, 211
Jacobsen P. & Kahn S.M. (1986): ApJ 309, 682
Kahn F. (1989): "Structure and Dynamics of the Interstellar Medium", eds. Tenorio-Tagle et al. Proc. IAU Coll. No. 120, 474
Kalberla P.M.W. (1997): "The Physics of Galactic Halos" eds. Lesch et al., Akademie-Verlag, p. 3
Kerp J. et al. (1996): A&A 312, 67
Kuntz K.D. & Danly L. (1996): ApJ 457, 703
Lilienthal D., Hirth W., Mebold U. & de Boer K.S. (1992): A&A 255, 323
Lilienthal D., Wennmacher A., Herbstmeier U. & Mebold U. (1991): A&A 250, 150
Magnani L., Hartmann Dap & Speck B.G. (1996): ApJS 106, 447
McCammon D. & Sanders W.T. (1990): ARAA 28, 657
McCammon D., Burrows D.N., Sanders W.T. & Kraushaar W.L. (1983): ApJ 269, 170
Mebold U. (1972): A&A 19, 13
Mebold U. et al. (1985): A&A 151, 427
Meyerdierks H. (1992): A&A 253, 515
Moritz P. et al. (1997): A&A, submitted
Penprase B.E. (1997): in preparation
Pietz J. et al. (1996), A&A 308, L37
Reynolds R.J. (1991): IAU Symp. 144, "The disk-halo connection" ed. Bloemen H., p. 67
Snowden S.L., Cox D.P., McCammon D. & Sanders W.T. (1990): ApJ 354, 211
Snowden S.L., Mebold U., Hirth W., Herbstmeier U. & Schmitt J.H.M.M. (1991): Sci 252, 1529
Tenorio-Tagle G. & Bodenheimer P. (1988): ARAA 26, 145
Wakker B.P. (1990): Ph.D. thesis, University of Groningen
Wakker B.P. & Boulanger F. (1986): A&A 170, 84
Wakker B.P. et al. (1996): ApJ 473, 834
Weiss A. (1997): Diplomarbeit Universität Bonn
Wennmacher A. (1994): Ph.D. thesis University of Bonn
Wesselius P.R. & Fejes I. (1973): A&A 24, 15
Westphalen et al. (1997): "The Physics of Galactic Halos" eds. Lesch et al., Akademie-Verlag, p. 3
Wolfire M.G., Hollenbach D., McKee C.F., Tielens A.G.G.M. & Bakes E.L.O. (1995): ApJ 443, 152

Study of the LISM Using Pulsar Scintillation

N.D.R. Bhat, Y. Gupta, and A.P. Rao

National Centre for Radio Astrophysics (TIFR), Pune 411 007, India.

Abstract. We present here the results from an extensive scintillation study of twenty pulsars in the dispersion measure (DM) range $3 - 35$ pc cm^{-3} carried out using the Ooty Radio Telescope, to investigate the distribution of ionized material in the local interstellar medium (LISM). Our analysis reveals several anomalies in the scattering strength, which suggest that the distribution of scattering material in the solar neighborhood is not uniform. Our model suggests the presence of a low density bubble surrounded by a shell of much higher density fluctuations. We are able to put some constraints on geometrical and scattering properties of such a structure, and find it to be morphologically similar to the local bubble known from other studies.

1 Introduction

Propagation effects on pulsar signals, such as dispersion and scattering, probe the distribution of thermal plasma in the interstellar medium (ISM). Scattering studies of radio waves from pulsars, enable us to probe the electron density fluctuations in the ISM, which are presumably due to turbulence in the ISM (Rickett 1990). Not much is known about electron densities and their fluctuations in the LISM and in order to investigate this in detail, we have made extensive scintillation observations of twenty nearby pulsars. Reliable and accurate estimates of strength of scattering have been obtained and the results are used to study the distribution of electron density fluctuations in the LISM.

2 Observations and Data Analysis

The observations were made using the Ooty Radio Telescope (ORT) at 327 MHz. Within the sky coverage and sensitivity limits of the ORT, there are twenty nearby pulsars that were found suitable for studying the LISM. For these pulsars, the dynamic scintillation spectra – intensity variation in the frequency-time plane – were measured over 20–60 epochs, spanning $\sim$ 100 days during 1993–95. Such spectra display intensity scintillation patterns that fade over narrow frequency ranges and short time intervals (Fig. 1), arising from diffractive scintillation effects. We obtained dynamic spectra data with a frequency resolution $\approx$ 140 kHz and a time resolution $\sim$ 10 secs.

To quantify the average characteristics of scintillation patterns at any epoch, we have computed the two-dimensional auto co-variance function

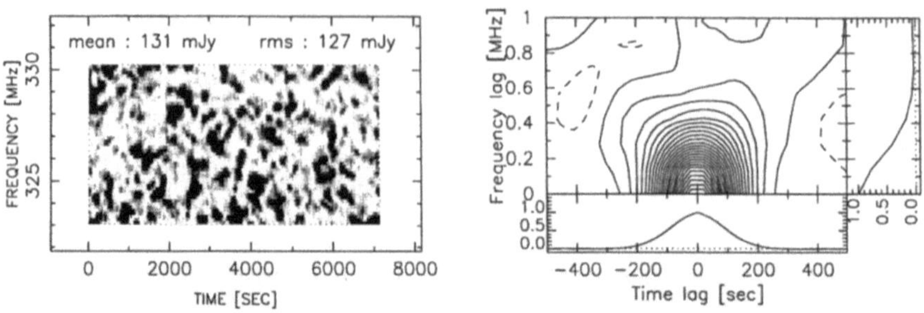

Fig. 1. The dynamic scintillation spectrum of PSR B0919+06 as observed on 24 May 1994, darker areas representing higher intensity values (left panel). The ACF is shown along with 1D cuts across zero lags of frequency and time (right panel).

(ACF) and fitted a two-dimensional gaussian to yield decorrelation bandwidth ν_d and scintillation time scale τ_d, which are the widths of the ACF along zero time and frequency lag axes respectively (Gupta *et al.* 1994). The large number of epochs of observations allow us to average out long-term fluctuations of these parameters, arising from refractive scintillation effects, giving an accuracy of 5−10%. From these estimates, we derive the line-of-sight averaged strength of scattering $\overline{C_n^2}$ (Cordes *et al.* 1985), given by

$$\overline{C_n^2} \propto \nu_{obs}^{\alpha} \, D^{-\alpha/2} \, \nu_d^{-(\frac{\alpha-2}{2})} \tag{1}$$

where ν_{obs} is the observing frequency, D is the distance estimate of the pulsar (based on the model given by Taylor and Cordes 1993), and α is the power-law index for electron density spectrum (Armstrong *et al.* 1995). We have assumed $\alpha = \frac{11}{3}$ in our calculations.

The derived values of $\overline{C_n^2}$ show roughly two orders of magnitude fluctuations (ranging from $10^{-4.8\pm0.02}$ to $10^{-3.1\pm0.02}$ $\mathrm{m}^{-20/3}$), which is about 10 times larger than earlier estimates and much larger than that predicted from current models for the $\overline{C_n^2}$ distribution in the galaxy (Cordes *et al.* 1985). Fig. 2.c shows these variations have a systematic trend, in that they are dominant for nearby (D $\lesssim$ 1.5 kpc) pulsars. In addition to this, we find several cases where pulsars at comparable DMs or distances show remarkably different scintillation characteristics. In order to study such effects, we consider two pulsars at similar DMs, having decorrelation bandwidths ν_{d_1} and ν_{d_2}, and define an anomaly parameter (A_{dm}) as

$$A_{dm} = \frac{(\nu_{d_1}/\nu_{d_2})_{observed}}{(\nu_{d_1}/\nu_{d_2})_{expected}} \qquad (\text{if } A_{dm} < 1, \text{ then } A_{dm} = \frac{1}{A_{dm}}) \tag{2}$$

A plot of the anomaly parameter against DM (Fig. 2.a) shows a systematic variation, implying density fluctuations are distributed in a structure that is local and is asymmetrically located relative to the sun. The anomaly parameter as a function of the distance (Fig. 2.b) shows a similar behaviour.

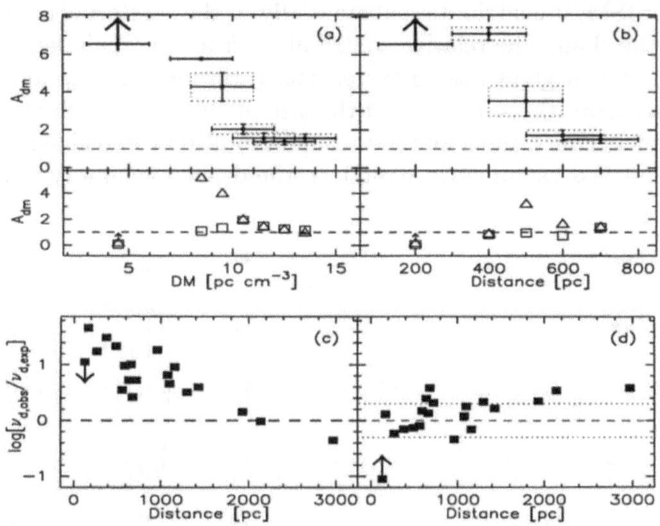

Fig. 2. The variations of the parameter $A_{\rm dm}$ are shown with DM (a) and with distance (b), where the expected values of ν_d are for a uniform scattering medium. The lower panels of (a) and (b) are for the 3-component medium (△ for uniform shell and □ for non-uniform shell). In (c), values of $\nu_{d,exp}$ are computed for a uniform medium and in (d), they are for the 3-component medium (non-uniform shell).

3 Interpretation of Results

We try to understand our observations by comparing the measured anomalies with anomalies predicted by specific density structures around the sun. Simple models consisting of an excess or deficiency of scattering material around the sun are unable to reproduce the observed trends of $A_{\rm dm}$ and $\nu_{d,obs}/\nu_{d,exp}$. To explain our observations, we need a 3-component scattering medium with a cavity surrounded by a shell of much higher density fluctuations embedded in the normal, large scale ISM. The sun has to be located away from the centre of the cavity. By adjusting the parameters of the model, we are able to generate a reasonable agreement with the observations. By further assuming that the strength of scattering from the shell decreases with z-height, we are able to improve the agreement (lower panels of Figs. 2.a and 2.b). The best fit values of the parameters of the local scattering structure are such that the observed trends of the parameters $A_{\rm dm}$ and ν_d (Fig. 2) are reproduced. The geometry of the structure is schematically shown in Fig. 3. For the size, constraints from our model are $a \sim 55$ pc, $b \sim 60$ pc and $250 < c < 300$ pc. The centre has an offset $r_c \sim 40$ pc from the sun, towards $210^o < l < 235^o$ and $-21^o < b < 21^o$. The density fluctuations are such that $10^{-5} < \overline{C_n^2} < 10^{-4.45}$ m$^{-20/3}$ for the inner cavity, $10^{-1} < \int_0^d C_n^2(l)\, dl < 10^{-0.57}$ pc m$^{-20/3}$ for the shell, (where d is the thickness) and $\overline{C_n^2} < 10^{-3.37}$ m$^{-20/3}$ for the outer ISM. Since the strength of scattering of the shell is much higher than those of the cavity and the outer ISM, it contributes substantially to the scattering of nearby pulsars outside the shell. The relative contribution from the shell decreases with the distance, which gives rise to systematic variations of A_{dm} and ν_d (Fig. 2). The offset from the sun (r_c) and the size in the galactic plane (a and b) are not uniquely constrained and our results can also be explained by a relatively bigger, but less asymmetric structure (the dashed geometry in Fig. 3).

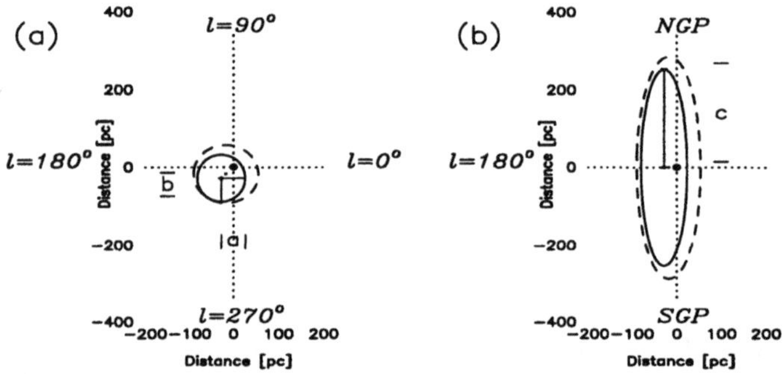

Fig. 3. The model for the local scattering structure shown as cuts through the galactic plane (a) and NGP–SGP plane (b). The solid geometry is for $r_c \sim 40$ pc, $a \sim 55$ pc and $b \sim 60$ pc, and the dashed one has $r_c \sim 25$ pc and $a \approx b \sim 75$ pc.

The exact relations between the scattering strengths and other properties of the medium are not well understood. If we assume $C_n^2 \propto n_e^2$, where n_e is the mean electron density, our observations would imply a density contrast $\sim$ 10–20 times between the shell and the ambient ISM in the case of a thin shell (~ 1 pc) and ~ 5–8 times for ~ 10 pc thick shell. The simple model considered here does not constrain the thickness of the shell. The pulsar PSR B0950+08 that has a parallax distance ≈ 130 pc is located, in our model, within the cavity implying $n_e \approx 0.02$ cm^{-3}, which is 4 times larger than its value from X-ray data (Snowden *et al.* 1990). Our constraints are, morphologically, in broad agreement with that of local bubble, obtained from X-ray, UV and HI data. Our results suggest that the bubble is surrounded by a shell of much higher density fluctuations and further investigations are required towards accurate estimates on more detailed characteristics of the shell boundary.

References

Armstrong, J.W., Rickett, B.J., Spangler, S.R. (1995): ApJ 443, 209
Cordes, J.M., Weisberg, J.M., Boriakoff, V. (1985): ApJ 288, 221
Gupta, Y., Rickett, B.J., Lyne, A.G. (1994): MNRAS 269, 1035
Rickett B.J. (1990): Ann. Rev. Astron. Ap. 28, 561
Snowden, S.L., Cox, D.P., McCammon, D., Sanders, W.T. (1990): ApJ 354, 211
Taylor, J.H., Cordes, J.M. (1993): ApJ 411, 674

Interstellar Scintillation of Radio Sources as a Probe for Investigations of the Local Interstellar Medium

N. Bochkarev[1] and M. Ryabov[2]

[1] Sternberg Astronomical Institute, Universitetskij prosp. 13, Moscow, 119899, Russia

[2] Odessa Observatory URAN-4 of Radioastronomical Institute of National Academy of Science, Pushkinskaya str. 37, Odessa, 270011, Ukraine

Abstract. A possibility of obtaining information on small scale inhomogeneities of the electron component of the local interstellar medium (LISM) is investigated using interstellar scintillations of extragalactic radio sources. We analyse Culgoora array observational data on variability of 190 extragalactic radio sources, covering most of the sky, at 80 and 160 MHz. The variability at time scales from 1 month to 15 years is interpreted as refractive interstellar scintillations in fast-moving nearby (less than 150 pc) hot gas near shock waves in the LISM. All-sky map of scintillation indices m averaged over 3–5 sources closest to one another shows several m maxima. Two of the 3 most pronounced maxima are probably connected with Loop I; the third one coincides with the soft X-ray (0.1–0.3 keV) background maximum near the South Galactic Pole. Other, less certain, m maxima probably correspond to the Orion star-formation region and to a soft X-ray maximum near the North Galactic Pole. The "free-of-gas" tunnel in the direction $l = 240°$ corresponds to low values of m . The estimated time scale of interstellar scintillations on the above-mentioned LISM structures is in agreement with that of the observed radio-source variations.

1 Introduction

Interstellar scintillations of radio sources resulting from small-scale inhomogeneities of interstellar matter's refractive coefficient (Rickett 1986) can yield information about plasma turbulence near shock fronts (Pikelner and Tsytovich 1969). Here, the thin scattering screen approximation may be applied.

2 Observational Data and Analysis

We used Culgoora (Australia) 1970–1984 array observations of extragalactic radio sources at 80 and 160 MHz (Slee and Siegman 1988). During 15 years, 412 sources covering homogeneously most of the sky (with the exception of a northern sky section) were observed repeatedly, in two or more 4-week-long series. 190 sources proved variable at least at one frequency and only 27, at both frequencies. Slee and Siegman (1988) give many arguments in favor of interstellar origin of the variations. They published measured scintillation variability indices m for the time scale of 1 month ($m1$) and from one to several years ($m12$).

The Slee & Siegman survey provides no information about the energy emitted by scintillating point components of individual sources. The m value for any source, however, depends not only on interstellar matter properties, but also on the source structure. Therefore, we have drawn an all-sky map for the interstellar scintillation indices m averaged over 3–5 neighboring sources (Fig. 1). Moreover, we used the averaged m values for the two time scales ($m1$ and $m12$) and the data for both frequencies (80 and 160 MHz). Typical angular distances between sources are about 15°. Thus, we hope the averaging will allow us to perceive interstellar formations of angular size not less than 20–30 degrees. We checked the sensitivity of our results to the input data used varying the data combinations ($m1$ or/and $m12$; 80 or/and 160 MHz) and the number of averaged sources, and also removing the data for the 3 sources with the extreme values of m. In all cases, the main maxima are present for the majority of the data combinations and conserve their positions within our angular resolution (about 20° – 30°), with slight variations of sizes and positions of the mapped structures .

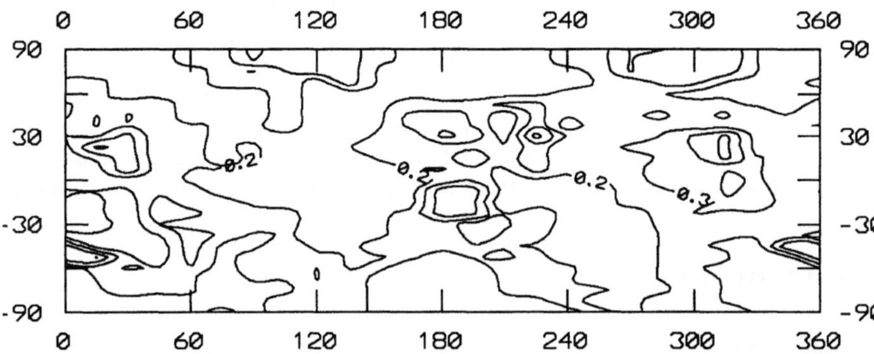

Fig. 1. The all-sky map of isolines of scintillation indices, m, for sources from Slee and Siegman (1988) survey, in galactic coordinates. The map is drawn using data averaged over 3 neighboring sources and over $m1$ and $m12$ for 80 MHz and 160MHz. Isolines are from $m = 0.2$, with 0.1 increment.

3 Results and Interpretation

We found 3 clear maxima, virtually independent of the data combinations and ways of averaging, in the all-sky distribution of interstellar scintillation

Table 1. Table 1. Galactic coordinates and possible identifications of the features in the map of interstellar scintillation index distribution

Features	Identification
Clear maxima	
1. $0° < l < 30°; 0° < b < 30°$	Loop I
2. $300° < l < 330°; 0° < b < 30°$	Loop I
3. $-20° < l < 20°; -60° < b < -40°$	Southern maximum of soft X-ray BG (B and C bands, McCammon et al. 1983)
Other possible maxima	
4. $170° < l < 220°; 30° < b < 45°$	Northern maximum of soft X-ray BG (B and C bands, McCammon et al. 1983)
5. $180° < l < 200°; -20° < b < 0°$	Orion star-formation region
6. $l \approx 120°$ and $300°$ $b > 75°$	Soft X-ray BG filament near NGP (B and C bands, McCammon et al. 1983)
Minimum of m	
7. $l \approx 240°$ low southern b	Tunnel free of interstellar extinction (Bochkarev 1987b, 1990,)

indices (Table 1). Three other maxima, though probably real, are not as definite as the three first ones. In all the cases, a minimum of m is present near the galactic longitude $l = 240°$ in the Southern hemisphere, but the data for this direction is too scarce to allow definite conclusions.

We assume that the revealed maxima of m can be identified with maxima of the soft X-ray background radiation distribution (B, C, and M bands in the survey by McCammon et al. 1983). The maxima 1 and 2 from Table 1 are overlapping a considerable part of the two brightest X-ray filaments in the M1 band (440–930 eV) map corresponding to Loop I, the old SNR expanding inside the coronal gas in a cavern formed around the Sco–Cen stellar association (Bochkarev 1987b, 1990,).

Another marked m maximum (number 3) obviously coincides with a soft X-ray background bright spot observed in the Southern hemisphere in B (130–188 eV) and C bands. According to Bochkarev (1987ab, 1990), the spot corresponds to the area of interaction of the outer part of the Local Cloud with the envelope surrounding the Sco–Cen association, namely to the position where the line of sight meets the relatively dense coronal gas near the shock front boundaries of the Sco–Cen superbubble. A similar northern maximum of m probably exists (number 4), but there is insufficient data for its direction in the Culgoora survey.

The maximum 5 is uncertain because its position and shape vary with data combinations. It might be identified with the densest northern part of the Ori–Eri superbubble, namely with the Orion star formation region.

The minimum 7 corresponds to a very old bubble, spread by differential rotation of the Galaxy within the Galactic quadrant III.

4 Discussion of Time Scale of Interstellar Scintillations

In accordance with the interpretation discussed in Section 3, the distance L to the maxima 1,2 (Loop I) is about 150 pc (Cox and Reynolds 1987; Bochkarev 1987b, 1990). The velocity of the shock wave is 530–580 km/s (Bochkarev 1990). Adopting typical angular sizes of the point components of extragalactic radio sources at 100 MHz $\theta = 0.2 - 0.5$ arcsec (see Janardhan and Alurkar 1993 and references in Slee and Siegman 1988), we find, in accordance with Rickett (1986), $\tau = L\theta/v = 0.1 - 0.7$ years, and the size of the scattering inhomogeneities $a = 20 - 70$ A.U.

For the maxima 3, 4, we assume the distance $L = 20 - 40$ pc (Bochkarev 1987ab, 1990) and v equal to the thermal velocity in coronal gas with the temperature $T = 10^6$ K: $v = 130$ km/s. In this case, $\tau = 0.1 - 0.3$ year.

In both cases (for all most definite m maxima), the estimated time scale of refractive interstellar scintillations is considerably shorter than that of the average scattering screen in the Galaxy (Rickett 1986; Slee and Siegman 1988) and is in good agreement with the observed radio-source variations.

5 Conclusions

A) Areas of the sky are found with intensified extragalactic radio-source interstellar scintillation at low radio frequencies (80 – 160 MHz), at time scales from months to years.

B) A correspondence between the areas mentioned in Conclusion A and soft X-ray background details of large angular size is found: two Loop I bright X-ray filaments; high Galactic latitude brightest spots in soft (B and C bands) X-ray background radiation. The direction of the "interstellar absorption-free tunnel" at the galactic longitude $l = 240°$ probably corresponds to a minimum of interstellar scintillation.

C) The estimated time scales $\tau = 0.1 - 0.7$ year at 100 MHz for interstellar scintillation in hot LISM structures are shorter than those for a typical scattering screen and correspond to Slee and Siegman (1988) observations.

References

Bochkarev, N.G. (1987a): Soviet Astron. 31, N 1
Bochkarev, N.G. (1987b): Astrophys. Space Sci. 138, 229
Bochkarev, N.G. (1990): Local Interstellar Medium, Moscow: Nauka Publ.
Cox, D.P., Reynolds, R.J. (1987): Ann. Rev. Astron. Astrophys. 25, 303
Janardhan, P., Alurkar, S.K. (1993): A&A 269, 119
McCammon, D. et al. (1983): ApJ, 269, 107
Pikelner, S.B., Tsytovich, V.N. (1969): Sov. Astron. 13, N. 1
Rickett, B.J. (1986): ApJ, 307, 564
Slee, O.B., Siegman, B.C. (1988): MNRAS, 235, 1313

Remarkable Details in the Local Bubble's HI Distribution

Natalija V. Bystrova

St.Petersburg Branch of the Special Astrophysical Observatory, Pulkovo, 196140 St.Petersburg, Russia

Abstract. We discuss details in the galactic HI distribution, which are relevant for studies of the Local Bubble. All results are based on the 21 cm Sky survey with the Large Pulkovo Radiotelescope. The antenna temperature scans were decomposed in two components and after the subtraction of the broad one two details in the residual are presented in this paper. The first is an estimate of the distance to the Sancisi-van Woerden filament in the Sco-Oph region. The second is the connection of a 21 cm feature in the southern galactic quadrant II with the negative-velocity elliptical brightness distribution around the South Pole of the Galaxy.

1 The HI Observations

Bystrova and Rachimov (1977) performed an HI 21 cm line survey with the Large Pulkovo Radiotelescope (LPR) - the "Pulkovo Sky Survey in the interstellar neutral Hydrogen Radio Line" (Bystrova and Rachimov (1977)). All sky visible for the LPR, i.e. between the declinations from $-29°$ to $+40°$ was covered by 29 drift scans 24 hours long each. The frequency resolution was 20 kHz and the spatial one in publications - one minute in RA. The antenna temperature scans were decomposed in two components to extract small scale structures and for the Pulkovo Sky Survey Atlas (Bystrova (1980)) maps were generated for the two components of the HI emission separately. The method consisted in a comparison of small details and the antenna temperature of a broad underlying component between adjacent drift scans in the measurement sequence. Fig.1 demonstrates examples of the RA drift scans from 12 to 24 hours for different declinations at fixed velocity with the approximate division on the two components. Because these components show different spatial characteristics including possibly different distance from the Sun both have to be compared with observations at other wavelengths separately.

This paper presents results on two details found in the residuals after subtraction of the broad component. First we analyse a region towards the Sco OB-2 association. There the broad structureless component can amount up to about 40% of the total signal. Then we present the surrounding of a filament near the south galactic pole.

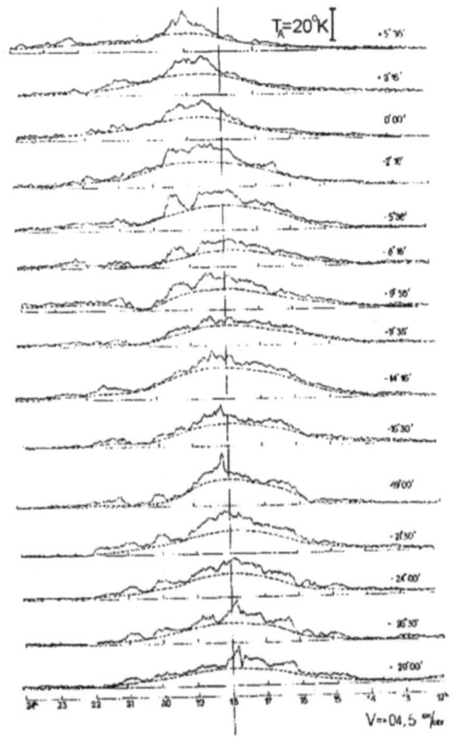

Fig. 1. Examples for HI drift curves of the St.Petersburg LPR Survey in RA-Dec coordinates

2 The Distance to the Sancisi – van Woerden Filament

In the direction of the Sco OB-2 association, next to the HII regions Sharpless 1 and 7, Sancisi and van Woerden detected an HI filament at the LSR-velocities around −11 km/s (Sancisi and Van Woerden (1970)). At these velocities the main HI profiles centered at about 0 km/s have weak components. This facilitated the determination of the borders on the sky of the S-vW detail. Studying the LISM (Crutcher (1982)) the S-vW detail was placed near the Sun at the distance of 10 – 20 pc. Figure 2 shows, however, morphological similarities between the S-vW detail and a feature at −6.1 km/s: the S-vW filament fills gap in the −6.1 km/s emission contours, see also Bystrova (1979). Sivan's object consists of the S1 and S7 nebulae and of two outer HII filaments(Sivan (1974)).The HI contours at −6.1 km/s follow these outer filaments and are most probably located at the same distance of about 170 pc as is then also the S-vW detail. I.e. the whole complex is located outside the Local Bubble. Cappa de Nicolau and Pöppel (1986) confirmed this result from their Gaussian decomposition of the HI profiles in this direction.

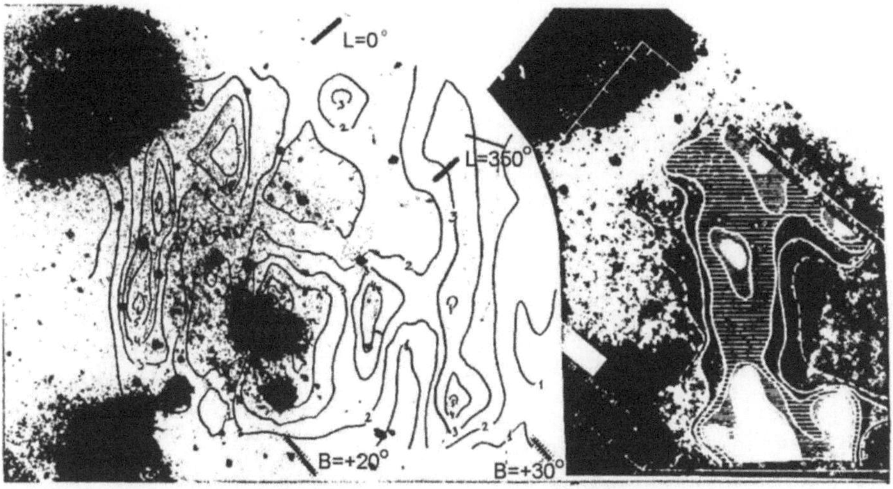

Fig. 2. HI contours of antenna temperatures at −6.1 (left) and about −11 km/s (right, the S-vW detail) overlayed on the optical image of the Sivan's object

3 The Neutral Hydrogen Feature Around the South Galactic Pole

According to the analysis by Frisch (1995) the 21 cm feature between $l = 60°$ to $170°$, $b = -30°$ to $-50°$ constitutes the Quadrant II boundary of the Local Bubble at about 40 – 80 pc from the Sun. This feature was observed in St.Petersburg and is presented in Fig.3. The LSR-velocities are −16.6 and −11.3 km/s where the detail is seen at best. But the measurements show that it is not a distinct feature. It is part of an ellipse shaped detail around the south galactic pole. The shape of the ellipse changes for other velocities. The full extent of the feature was observed in the southern hemisphere (Heiles and Cleary (1979)). The angular size of this ellipse is more than $80° \times 50°$. and the position angle of the major axis is about 130° – 310° in longitude. Till now the distance to this ellipse-shaped detail remained unknown. The results from Frisch (1995) can suggest that the entire ellipse is located at a distance around 40 – 80 pc. A future combined study of hydrogen and X-ray emission will possibly provide further information on the distance of the features in the second galactic quadrant and in the rest of the ellipse.This analysis will then test the proposed association between the detail and the rest of the ellipse around the south galactic pole.

Acknowledgements. The paper was done with the partial support by the State's scientific– technical program "The Astronomy", grant 1.3.1.3.

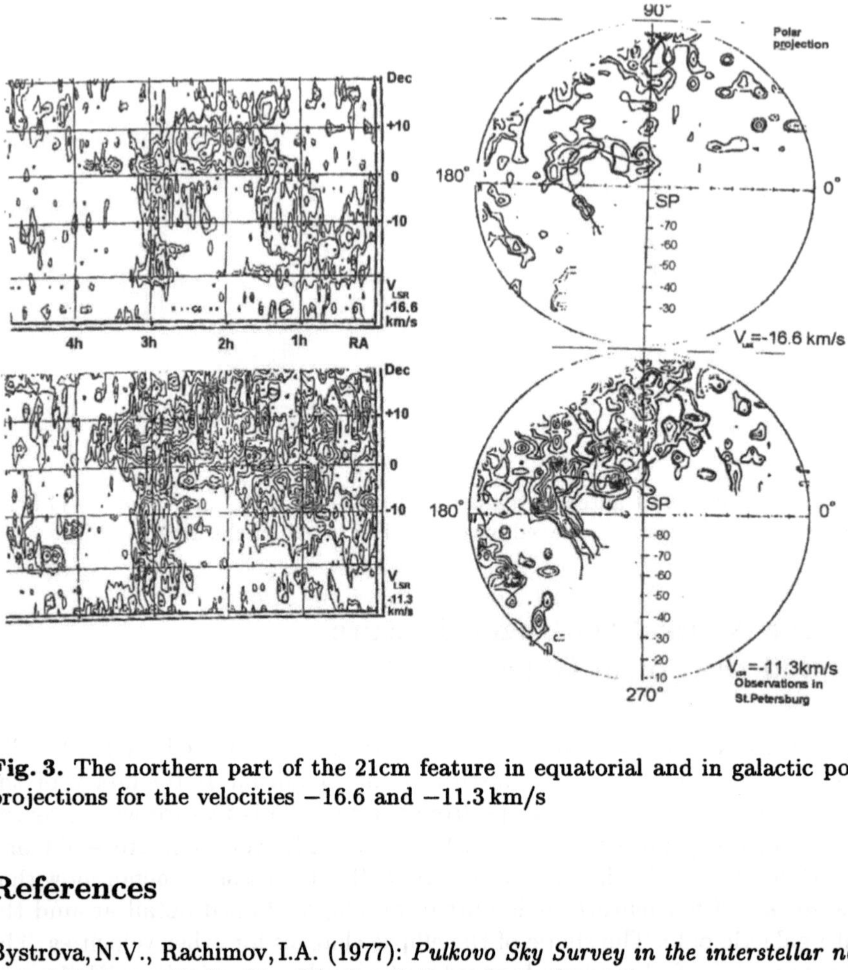

Fig. 3. The northern part of the 21cm feature in equatorial and in galactic polar projections for the velocities −16.6 and −11.3 km/s

References

Bystrova, N.V., Rachimov, I.A. (1977): *Pulkovo Sky Survey in the interstellar neutral Hydrogen Radio Line* (Nauka, St Petersburg, Russia)

Bystrova, N.V. (1980): *Contour Maps to the Pulkovo Sky Survey in the interstellar neutral Hydrogen Radio Line* (Nauka, St.Petersburg, Russia)

Bystrova, N.V. (1979): Astrophys. Issled. (Izv. SAO).**11**,236–243

Frisch, P.C. (1995): Space Science Reviews.**72**,499–592

Sancisi, R., Van Woerden, H. (1970): A&A.**5**,135–139

Crutcher, R.M. (1982): ApJ.**254**,82–87

Cappa de Nicolau, C.E., Pöppel, W.G.L. (1986): A&A.**164**,274–299

Heiles, C., Cleary, M.N. (1979): A&AS.**36**,95–127

Sivan, J.-P. (1974): A&AS.**16**,163–172

Pulsars and Bubble Dynamics

Ramen Kumar Parui

Aero Communication Station (ACS), Imphal Airport, P.O.-Tulihal, Imphal-795140, India

Abstract. Based on recent observation of the dominance of the flare generated solar wind streams over the co-rotating streams and a significant lowering in the cosmic ray intensity in the abnormal year of the solar activity period, I have predicted the occurrence of discernible changes in magnetic field strength and the thickness of the shell of the local bubble during high sunspots years of the solar cycle period. In order to observe the variation, both in the magnetic field strength and in the thickness of the compressed shell, a proposal of continuous monitor of these two parameters through the observation of Faraday Rotation Measure has been proposed with an emphasize to the observations in the period of next abnormal year of the solar cycle in near future.

1 Introduction

Radio observations of the interstellar medium (ISM) indicate a hierarchy of structures in the composition of the ISM masses from about $1\,M_\odot$ to $10^6\,M_\odot$ and with sizes from 10^{-4} to 100 pc. Dense shells around massive stars are formed when surrounding interstellar medium is swept up by the fast stellar winds of massive stars. Shells formed around single massive stars are called "bubbles" and the typical sizes of bubbles are $5-40$ pc. Recently Burrows and Guo (1996) identified some absorption features associated with the shell of the Local Bubble created by the fast stellar winds. Earlier, Vallee (1993) showed that the shell near the Sun is a magnetic shell and played an important role for enhancing the Faraday Rotation Measure of the more distant compact objects. Based on these above I have applied snowplow theory on the Local Bubble of the Sun in order to understand the stellar bubble dynamics. The snowplow theory of bubble formation indicates that the magnetic field strength in the shell has a link with the compressed shell thickness (Δr) by the relation (Vallee, 1993):

$$\mathbf{B}_{\text{shell}} = \mathbf{B}_{\text{outside}} \left[1 - \left(1 - \frac{\Delta r}{r_{\text{out}}} \right)^3 \right]^{-k} \tag{1}$$

where B_{outside} = magnetic field strength outside the shell in the general stellar medium, $\Delta r = r_{\text{out}} - r_{\text{in}}$, r_{in} and r_{out} being the inner and outer radii for this shell. $k = 1$ represents the matter in a shocked medium. Again Faraday Rotation (θ) is interrelated with the strength of the magnetic field B of the ISM through the relation:

$$\theta = \frac{C_V\, d}{\mu_0 \mu_r} B \tag{2}$$

where C_V = Verdet's constant, d = the length of the way through the magnetic field B, μ_r = relative permeability of the medium.

According to Weaver et al. (1977) the fast stellar winds are adiabatically shocked to $10^6 - 10^7$ K due to pressure in a homogeneous medium of ISM (Chu and Mac Low, 1996). These shocked stellar winds act like a piston which drives the expansion of the outer shell of swept-up ISM. This means that stellar winds play a critical role in the formation of thickness of the compressed shell. As the temperature rises up to $10^6 - 10^7$ K within the ISM due to the heating of the stellar winds, this may cause a variation in C_V and μ_r. Typically, the variations in C_V and μ_r in $10^5 - 10^6$ K are very small (Jackson, 1985, Cox and Snowden, 1986), we can neglect it.

A recent investigation (Pandeya et al, 1997) on high speed solar wind indicates its two varieties namely, co-rotating streams and flares generated streams depending upon its source of origin. This flare generated streams are short-lasting but its dominance has a significant contribution in lowering the solar wind speed on short-term basis during long-term solar cycle period. In this paper, I have presented the possible effect on the shell thickness and its magnetic field strength due to this low solar wind speed and its possible measure through Faraday Rotation Measure.

2 Recent Result on Solar Wind Streams

High speed solar winds are distinguished into two categories depending upon their locations (Mavromichalaki, 1988) namely:

- the co-rotating or coronal hole associated streams (CS) emitted by coronal holes. This is long-lasting high speed solar wind streams whose apparent tends to recur at intervals of $\sim$ 27 days;
- the flare generated streams (FGS) associated with strong active regions emitting solar flares. These are short lasting in character.

Studies on solar wind streams as well as cosmic ray transient variations are normally based on the fact that coronal hole associated streams are dominated during low solar activity period while dominance of flare generated streams during high solar activity period. In a recent study of solar activity during the period 1979 to 1990 (which covers the descending and ascending phases of solar activity cycle 22) Pandeya et al. (1997) found that solar wind velocity is one of the main responsible factors which produces changes in cosmic ray intensity and the flare generated streams, through sudden storm commencements in a short-term period, produce a significant lowering in solar wind speed, i.e. during high solar activity years, particularly in the abnormal year (i.e. 1989) of solar cycle 22 the flare generated high speed solar wind streams (FGS) were found larger than the co-rotating streams (CS).

3 Effects on Shell Thickness and Possible Detection

The dominance of flare generated streams over the co-rotating streams during the period of high solar activity may produce a change in shell thickness Δr as well as in the magnetic field strength of the shell. The main reason is that enhancements in the magnetic field of the inter-planetary medium are generally observed during the high solar activity period. As we know, magnetic fields can effectively confine bubbles, an enhanced magnetic field definitely produces a change in shell configuration i.e. in shell thickness of local bubble. As the duration between the occurrence of the Flare Generated Streams and its dominance over the co-rotating streams is ~days, these magnetic effects can be observed as an enhanced Faraday Rotation Measure of distant compact objects seen through the shell of the local bubble near the Sun.

Typical value of the change in Faraday Rotation is $\Delta\theta \approx 10^{-20}$ arcmin for a newly expanding shell of the bubble; and $\Delta\theta \approx 10^{-14}$ arcmin when the shell has expanded to sizes ~ parsecs with $C_V \approx 10^{-26}$ arcmin /Gauss/cm for the interstellar medium.

If the measurements will be done in the radio frequency band $\approx 10^8$ Hz then the above values could be increased by a factor $\approx 10^{14}$ which is still too less. Although it is an undetectable range at present but we have a hope that it can be possible with more advanced techniques of measurement in the near future.

As the change due to the enhanced magnetic field will be prominent particularly in the abnormal year, we therefore have to wait for the next abnormal year. In order to get an exact picture of the variation in shell thickness and its magnetic field strength, continuous observations throughout the solar activity cycle are very much essential, because only continuous observations can help us to correlate the effects of high speed solar wind variation on the bubble which ultimately offers the exact picture of bubble dynamics.

Acknowledgements. The author is grateful to Rodrigo Supper (MPE Garching) for his valuable comments and suggestions for improvement of this paper.

References

Burrows D.N., Guo Z., 1996, in "Proc. of the International Conference on *Röntgenstrahlung from the Universe* in Würzburg", eds. H.U. Zimmermann, J.E. Trümper & H. Yorke, MPE Report 263, p.221

Chu Y.-H., Mac Low M.-M., 1996, in "Proc. of the International Conference on *Röntgenstrahlung from the Universe* in Würzburg", eds. H.U. Zimmermann, J.E. Trümper & H. Yorke, MPE Report 263, p.241

Cox D.P., Snowden S.L., 1986, Adv. Space Res. 6 (2), 97

Jackson J.D., 1985, Classical Electrodynamics, Wiley Eastern, New Delhi, p.189

Mavromichalaki H., Vassilaki V., Marmatsouri E., 1988, Solar Phys. 115, 345

Pandeya A., Shrivastave P.K., Sabbah I., El-Borie M.A., 1997, Indian J. Phys. 71B, 455
Vallee J.P., 1993, ApJ 419, 670
Weaver R., McCray R., Castor J., Shapiro P., Moore R., 1977, ApJ 218, 377
Zirker J.D., 1977, Rev. Geophys. Space Phys. 15, 257

Part V

Magnetic Fields, Molecular Clouds, and Interstellar Bubbles

Carried away: The Local Bubble ...

... and some participants.

The Magnetic Field Near the Local Bubble

Carl Heiles

University of California, Berkeley, CA 94720, USA

Abstract. There are almost no direct observational indicators of the magnetic field inside the local bubble. Just outside the bubble, the best tracers are stellar polarization and HI Zeeman splitting. These show that the local field does not follow the large-scale Galactic field. Here we discuss whether the deformation of the large-scale field by the local HI shells is consistent with the observations. We concentrate on the Loop 1 region, and find that the field lines are well-explained by this idea; in addition, the bright radio filaments of Radio Loop 1 delineate particular field lines that are "lit up" by an excess of relativistic electrons.

1 Introduction: What Deforms the Local Magnetic Field?

The magnetic field inside the local bubble is very difficult to observe. There is not a single pulsar inside the local bubble for which Faraday rotation has been measured, nor are there any neutral clouds with which to measure Zeeman splitting. Although there are some observations of the polarization of nearby stars, they are all presented in tabular form except for Tinbergen's (1982) Figure 6 (see also Frisch 1991). This figure shows that stars within about 20° of $(l, b) \sim (0^\circ, -10^\circ)$ have polarizations that are consistent with both the direction of the local large-scale Galactic field (i.e., with field pointing towards $l \sim 90^\circ$) and, also, that field as it would be deformed by the North Polar Spur shell (see below). As we have heard at this meeting, there are models for the heliopause that predict a particular local field strength and orientation, but apart from the field orientation given by Tinbergen's map (which agrees with the prediction) there are no observational checks.

The magnetic field outside the local bubble is much easier to observe, most conveniently at high Galactic latitudes. Starlight polarization provides the plane-of-the-sky orientation (but not the direction) of $B_\perp$ weighted by extinction, synchrotron polarization provides the same for the radio-continuum-emitting regions, pulsar Faraday rotation and dispersion provide the mean line of sight field strength and direction $B_{||}$ for the Warm Ionized Medium (WIM), and Zeeman splitting provides $B_{||}$ in HI regions.

One of these four indicators, synchrotron polarization, is useful only if the Faraday rotation measure (RM) is known well enough, which is probably not the case in most regions. Consider the North Polar Spur as a well-studied example. Spoelstra (1971) mapped the RM's from polarization measurements of the diffuse radio continuum emission at several frequencies and found good agreement in the orientation of $B_\perp$ with starlight polarization for $b \gtrsim 40°$, but not at lower latitudes. However, his later map (Spoelstra 1984) of the radio polarization is based on a redetermination of the RM and doesn't allow this conclusion so clearly. In our opinion, the uncertainties in RM are too large to make a definitive statement.

We are left with starlight polarization, Zeeman splitting, and pulsar RM's. The local field, i.e. the field just outside the local bubble, differs from the large-scale Galactic field and has large-scale patterns that are correlated over fairly large angular scales, some tens of degrees.

Here we make a first attempt to quantitatively consider the following broad question: can this local magnetic field structure be explained as the large-scale Galactic field having been deformed by the local shells? This question was answered in semiquantitative terms by Weaver (1979), who eloquently described what we have in mind: "*... The numerous massive stars in the newly formed [Sco-Cen] association produced strong stellar winds. These inflated a bubble of gas and dust concentric with the Sco-Cen Association... stretched on the surface of the bubble into the filaments [and magnetic field lines] we see today...*"

This broad question, in turn, consists of four subsidiary ones. We provide the quick answers here, as we pose the questions, and elaborate in more detail below.

(1) *Question:* How do we best define the local shells? *Quick answer:* By HI maps at constant velocity, not by radio continuum loops.

(2) *Question:* What is the large-scale Galactic field? *Quick answer:* The mean magnetic field $\langle B_{||} \rangle \sim 2.2\ \mu$G and points towards $(l, b) \sim (80°, 0°)$. The total field strength is much larger, $B_t \sim 4.2\ \mu$G: the random field B_r dominates.

(3) *Question:* What are the observational data for the local field? *Quick answer:* Deformation of the field by shells produces easily recognizable fluctuations in the orientation of $B_\perp$ but not in the direction of $B_{||}$. This makes starlight polarization the best tracer of deformations, and they show clear disagreements with the orientation of the large-scale Galactic field $B_\perp$.

(4) *Question:* Do the local field data agree with the large-scale field as deformed by the local shells? *Quick answer:* Here we concentrate on the North Polar Spur: the agreement is very good in some regions, and the strongly emitting ridges in Radio Loop I follow the distorted field lines very well.

2 How Should We Define the Local Shells?

2.1 With Radio Continuum—No!

The four classical radio continuum loops are usually interpreted as the major tracers of large, local supernova shells (Berkhuijsen, Haslam, and Salter 1971). However, I want to emphasize that *with the exception of Radio Loop 1, none of these radio loops looks like a limb-brightened shell.* The arcs of Radio Loops 2, 3, and 4 are fat, with radial extents that are $\sim 1/3$ the loop radius. If they were limb-brightened shells, then the intensities inside the peripheries would be almost as large as those in the limb-brightened parts—but this is not the case. As we discuss below, Radio Loop 1 does have a brighter interior and is to some extent a limb-brightened shell, but it also exhibits a number of roughly concentric filaments at various radial distances inside its periphery, something that is difficult to achieve with nothing but limb brightening, even with a wrinkled boundary.

Furthermore, the radio loops are not reliable indicators of supernova shells. Radio Loop 1 is the most famous loop and it is certainly tracing a supernova shell which also exhibits a very well-defined expanding HI shell, soft X-ray emission from the interior, and an obvious structure in the famous map of starlight polarization by Mathewson and Ford (1970). However, the Eridanus loop also has the HI shell and X-ray emission and therefore deserves equal billing as a supernova shell—but it does *not* exhibit a well-defined radio loop structure. Finally radio loops 2, 3, and 4 have no reason, other than their radio continuum emission, to be billed as supernova shells. Moreover, these three loops are not very distinct, and there are a few other radio features that are just as clear but not classified as loops, one of the clearest having diameter $\sim 50°$ and centered near $(l, b) = (280°, -30°)$ (close to the Large Magellanic Cloud—almost certainly a pure accident!).

We conclude that the radio continuum loops are not very good tracers of interstellar shell structures, nor are they limb-brightened shells.

2.2 With HI—Yes!

Defining shells with the HI makes physical sense. The HI gas is affected strongly by stellar winds and supernova explosions. It expands, sweeping up the magnetic field with it into a shell; it slows down and stalls, with the HI remaining as a tracer until other shocks come along and jostle it again. HI shells become very clear and distinct in maps at different velocities. The grey-scale HI maps of Colomb, Pöppel, and Heiles (1980), as crude and poorly reproduced as they are, reveal a plethora of well-defined shells. The field strength $B_{||}$ in this neutral gas is derivable from HI Zeeman splitting and the orientation of $B_{\perp}$ from starlight polarization.

3 The Large-Scale Galactic Field

The recent least square fits to pulsar data by Rand and Lyne (1994) find $\langle B \rangle \equiv B_u \sim 1.4\ \mu$G, pointed towards $l \sim 88°$. However, while these results are most popular, they differ from those obtained from two indicators that sample larger volume fractions of space, synchrotron polarization and

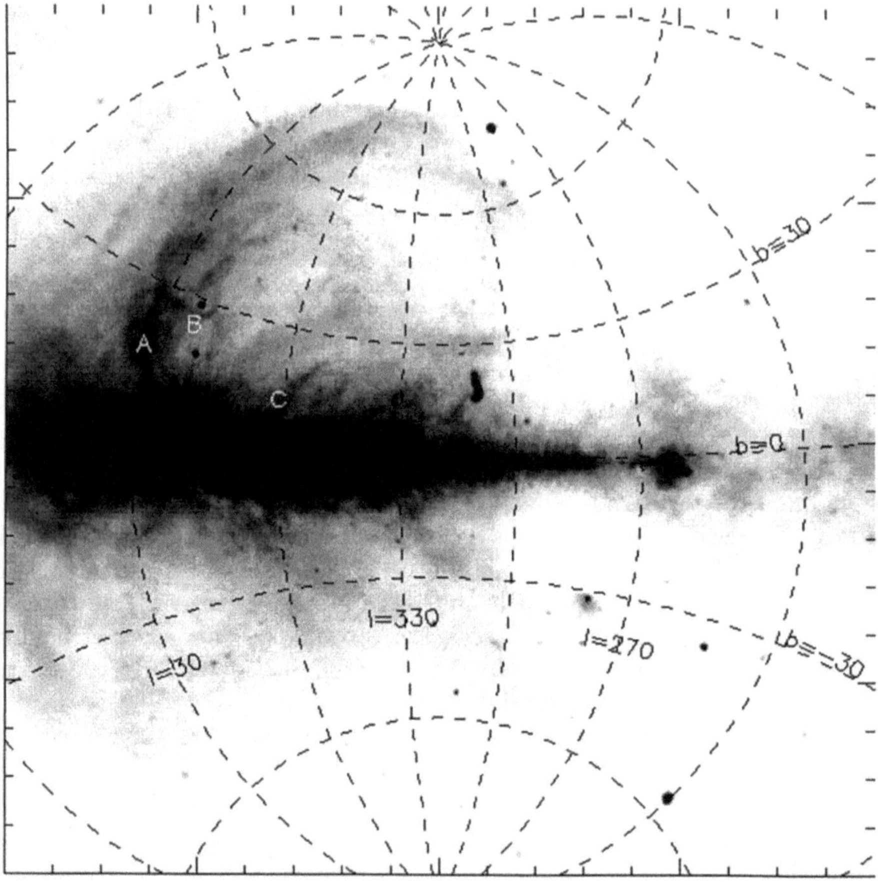

Fig. 1. The 408 MHz radio continuum from Haslam *et al.* (1983) (grey scale) and Galactic coordinate grid, in stereographic projection centered on $(l, b) = (320°, 5°)$. The dashed lines are lines of constant Galactic longitude and latitude in intervals of 30°. The letters A, B, and C label particular bright filaments discussed in the text.

starlight polarization. These more representative indicators provide $B_u \sim 2.2$ μG, pointed towards $l \sim 80°$ (Heiles 1995, 1996).

The random field strength B_r much exceeds the uniform component B_u, but by a factor that depends on whether the data are RM's (from the WIM) or other non-WIM indicators. Ohno and Shibata (1993) analyzed pulsar RM's and found $\frac{B_u}{B_r} \sim 0.28$. In contrast, the non-WIM $\frac{B_u}{B_r}$ as derived from starlight polarization (Jones, Klebe, and Dickey 1992) and synchrotron emissivity is at least twice as large. Moreover, the scale lengths for the random field variations differ, being much smaller in the WIM. Heiles (1995) considers all these data

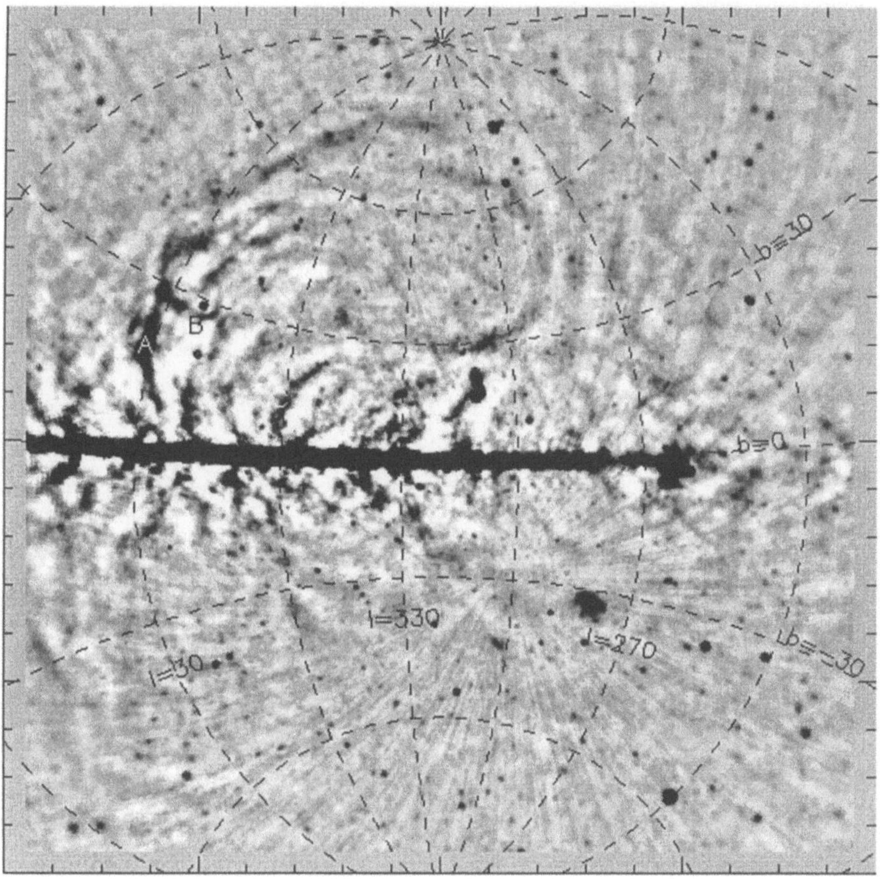

Fig. 2. High-pass filtered 408 MHz radio continuum; blacker is brighter. The stereographic projection and letters A, B, and C are identical to Figure 1.

together and believes that the pulsar data should receive less weight than they are usually given.

The RM data show disturbing large-scale residuals from the least square fit to the large-scale Galactic field. In Galactic quadrant 1 at $b > 0°$, the RM's are almost uniformly positive; in contrast, from the least square fit they should be negative. A similar situation holds for quadrant 3 at $b < 0°$. This disturbing discrepancy is the subject of a paper by Han et al (1997), who ascribe it to a large-scale Galactic dipole field caused by a dynamo of odd symmetry. However, the HI Zeeman splitting data—although they are too sparse to say definitively—do not seem to exhibit this discrepancy: the signs of the Zeeman-derived $B_{||}$ agree better with those of the large-scale

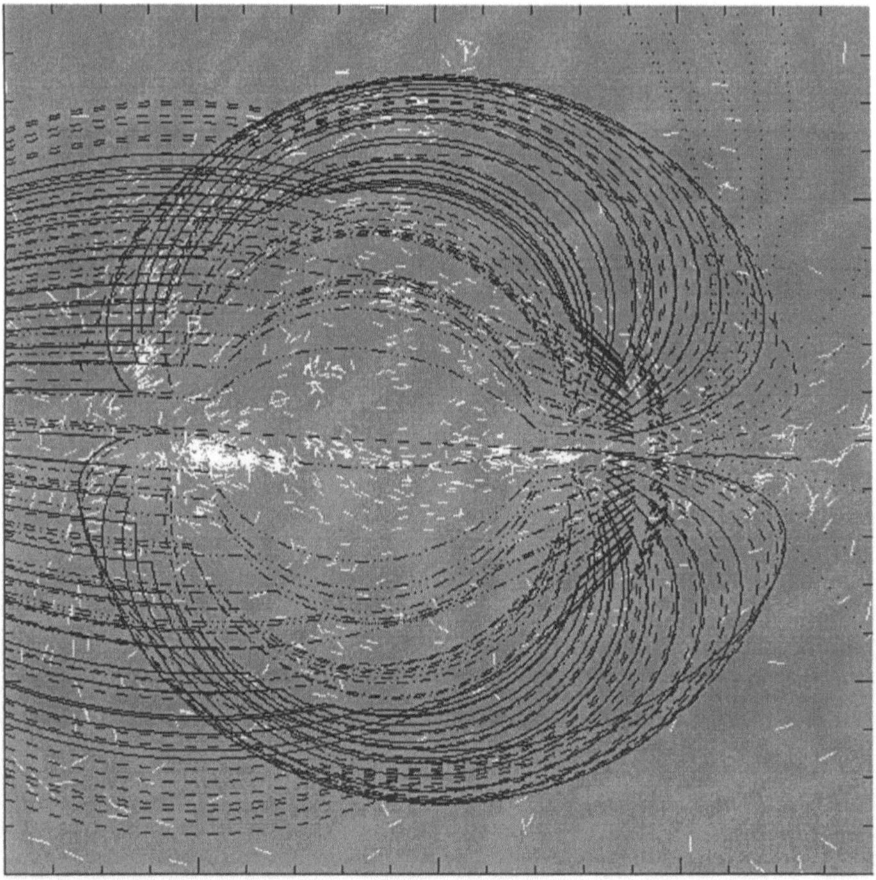

Fig. 3. The white "vectors" are starlight polarizations. The black lines represent our model of magnetic field lines deformed by the expanding HI shell, with the line type indicating the distance: the nearest field lines are dotted, then dashed, then solid, then dash-dot-dot, and the furthest dash-dot-dot-dot. The stereographic projection and letters A, B, and C are identical to Figure 1.

fit. Similarly, for the NPS deformed field model discussed below, the Zeeman splitting data agree well and the RM data do not.

Again, we believe that the WIM is not the best magnetic field tracer in most regions. In any case, for our problem of the deformed field in shells the starlight polarization results must be more relevant because they refer mainly to the HI shells, while the RM data refer to the WIM—which is not very well-correlated with the HI (Reynolds et al 1995). Thus we will ignore the RM data in the present discussion.

4 The Local Field as Distorted by the North Polar Spur

When an expanding spherical shell deforms field lines, the deflected lines follow lines of constant "longitude" on the sphere. Their appearance on the sky depends on the vantage point, like viewing the constant-longitude lines on a terrestrial globe from different vantage points.

When observed from the "magnetic pole" the spherically-deflected lines project radially onto the sky, and the deflections change neither the orientation of $B_\perp$ nor the sign of $B_{||}$. When observed from the "magnetic equator" the deformed lines look straight near the center and change gradually to being circumferential near the limbs; the undeformed $B_{||}$ is small, but the deformed $B_{||}$ is positive or negative depending on whether one observes the front, rear, left, or right hemispheres. From other vantage points the geometry becomes complicated, with surprising consequences. For example, field lines on the rear hemisphere can be almost perpendicular to those on the front. If the sphere subtends a large angle then the geometrical projections make things even more complicated. Some of these effects are visible in Heiles' (1997) study of the field structure in the Orion/Eridanus region and below in Figure 3.

Here we do not have enough space to discuss all the radio loops, so we concentrate on the physical structure associated with Radio Loop 1. Because this structure encompasses more than just the radio emission, we call the whole structure, including the radio loop, the "North Polar Spur" or NPS. Just outside of the radio loop lies the periphery of an expanding HI shell; in contrast to the radio loop, the HI shell extends below the Galactic plane to $b \sim -30°$. From an eyeball fit, the HI shell is centered near $(l, b) = (320°, 5°)$, significantly different from the radio loop center $(l, b) = (329° \pm 1.5°, 17.5° \pm 3°)$ (Berkhuijsen, Haslam, and Salter 1971). The fact that the HI center is lower than the radio center, and in particular as low as $b \sim 5°$, is important in our following discussion, where we match the pattern of deformed field lines to the bright filaments in Radio Loop 1.

Figure 1 shows a stereographic map of the 408 MHz radio continuum from the survey of Haslam *et al.* (1983), with a Galactic coordinate grid superposed. The stereographic projection is ideal for examining shells because it always maps circles into circles (but usually with displaced centers). The most intense radio emission arches up towards positive latitudes near $l \sim 30°$ and there are several roughly concentric filaments of different radius. The filaments lie roughly parallel to the stellar polarization (see Figure 3), suggesting that the filaments trace magnetic field lines. The letters A, B, and C trace particular bright radio filaments that we will discuss below.

We can highlight the radio filaments by using a high-pass filter ("unsharp masking") on the image, which rejects the broad background; this is shown in Figure 2, where blacker means more intense.

Figure 3 exhibits the large-scale Galactic field lines as they would be deformed by the expanding HI NPS shell. In this model, we assume that each unperturbed field line, with the gas, is swept up by the shock into the shell along a radial line emanating from the shell center at $(\ell, b) = (320°, 5°)$. The distances of the field lines from the Sun are indicated by their type, as explained in the figure caption.

We see that the pattern of radio filaments looks very similar to the pattern of the deformed lines, particularly near the upper left periphery of the shell. In particular, the sharp bend in the radio filament just above the letter B, where the deformed lines bend sharply from being nearly vertical to being horizontal as they approach the Galactic plane from above, is reproduced well.

This pattern matching, plus the many bright filaments that exist not only near the periphery, suggests that the bright radio filaments trace particular distorted magnetic field lines. This is a very strong indication that the brighter portions of Radio Loop 1 are not bright because of limb brightening. Rather, they are defined by distorted field lines that happen to be "lit up" by relativistic electrons. Whatever the physical cause, the effect is huge: the mean Galactic synchrotron emissivity near the Sun is ~ 7 K kpc^{-1} (Beuermann, Kanbach, and Berkhuijsen 1985) and the the bright filaments have at least several Kelvins pc^{-1}, $\gtrsim 500$ times higher!

It would be nice to find some reason why particular field lines are lit up by relativistic electrons. It may be that the bright lines are those that happen to run near dense regions, perhaps where the shock associated with the expanding HI shell traps and accelerates relativistic electrons more effectively. Dense gas clouds are highlighted by diffuse 100 μm IRAS emission, which is shown in Figure 4.

In Figure 1 we marked two of the brightest field lines with the letters A and C. Field line A is the brightest. It is morphologically similar to a fairly weak, but nevertheless distinct, feature on the 100 μm IRAS map. This IRAS feature lies outside the corresponding radio continuum feature, so we regard the association as "probable" but not definite. Comparison of the IRAS map with HI data shows that this feature has a velocity in the range $12 \rightarrow 22$ km s^{-1}, which places it somewhat beyond the tangent point of the HI shell; field lines at this distance (and also other distances) also have this bend. Similarly, field line C runs over a very bright feature on the IRAS map; this feature is the Ophiuchus molecular cloud region. This association of IRAS and radio is much clearer and we can regard it as "definite", although the pattern of distorted field lines in Figure 3 does not match very well. Ophiuchus has a velocity $\sim +8$ km s^{-1} and therefore lies on the far side of the HI shell.

The associations (one "probable", one "definite") of the bright radio filaments with these dense gas features suggests that the particular field lines that lie near the interaction region that exists between particular dense pock-

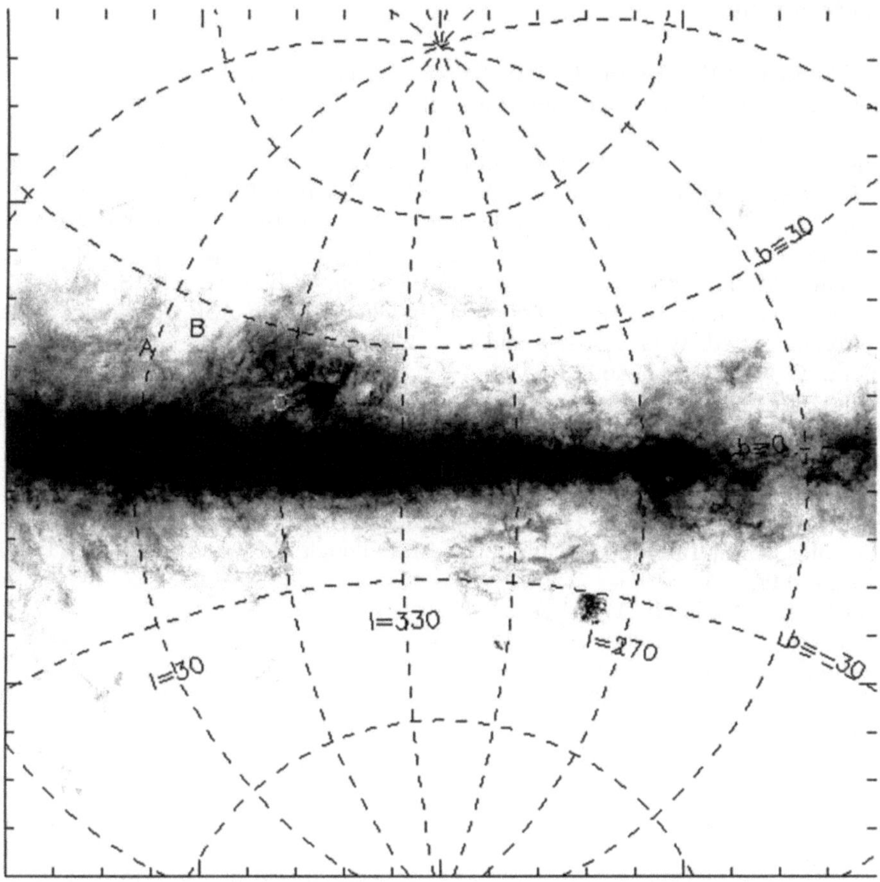

Fig. 4. Diffuse 100 μm IRAS emission. The stereographic projection and letters A, B, and C are identical to Figure 1.

ets of gas and the expanding shock are the ones where relativistic electron generation occurs particularly efficiently.

Finally, Figure 3 shows that the starlight polarization is well represented by the distorted field lines for $b \gtrsim 0°$ in the left-hand half of the NPS. This is encouraging as a positive answer to our fundamental question.

Acknowledgements. It is a pleasure to thank Doug Finkbeiner for his diffuse 100 μm IRAS map and his assistance in using it, and Dap Hartmann for assistance in using his Dwingeloo HI survey. This work was supported by a grant to the author from the National Science Foundation.

References

Berkhuijsen, E.M., Haslam, C.G.T., Salter, C.J. (1971): A&A, 14, 252

Beuermann, K., Kanbach, B., and Berkhuijsen, E.M. 1985, A&A, 153, 17

Colomb, F.R., Pöppel, W.G.L., and Heiles, C. (1980): A&AS 40, 47

Frisch, P.C. (1991): in *COSPAR Colloquia Series No. 1, Physics of the Outer Heliosphere*, S. Grzedzielski and E. Page (eds.), (London: Pergamon), p. 19

Han, J.L., Manchester, R.N., Berkhuijsen, E.M., and Beck, R. (1997): A&A, in press

Haslam, C.G.T., Salter, C.J., Stoffel, H., & Wilson, W.E. 1983, A&AS, 47, 1

Heiles, C. (1995): in *Polarimetry of the Interstellar Medium*, ASP Conference Series Volume 97, ed. W.G. Roberge and D.C.B. Whittet, p. 457

Heiles, C. (1996): ApJ, **462**, 316

Heiles, C. (1997): ApJS, in press

Jones, T.J., Klebe, D., and Dickey, J.M. (1992): ApJ, **389**, 602

Mathewson, D.S. and Ford. V.L. (1970): Mem. R. Astron. Soc., **74**, 139

Ohno, H. and Shibata, S. (1993): MNRAS, **262**, 953

Rand, R.J. and Lyne, A.G. (1994): MNRAS, **268**, 497

Reynolds, R.M., Tufte, S.L., Kung, D.T., McCullough, P.R., and Heiles, C. (1995): ApJ, **448**, 715

Spoelstra, T.A.Th. (1971): A&A, **21**, 85

Spoelstra, T.A.Th. (1984): A&A, **135**, 238

Tinbergen, J. (1982): A&A, **105**, 53

Weaver, H.F. (1979): in *The Large-Scale Characteristics of the Galaxy*, ed. W.B. Burton, p. 295

The Local Magnetic Field in the Milky Way

B. Uyanıker, E. Fürst, W. Reich, P. Reich, and R. Wielebinski

Max–Planck–Institut für Radioastronomie, Auf dem Hügel 69, D-53121 Bonn, Germany

Abstract. A radio continuum survey at medium Galactic latitudes with the 100-m telescope of the *Max-Planck-Institut für Radioastronomie* is being carried out at a center frequency of 1.4 GHz in total power and linear polarization. We present polarization and depolarization structures detected at medium latitititudes with no corresponding observable structure in total power emission. Existence of such kind of polarization structures imply that the emission, due to the Faraday modulation of the Galactic synchrotron foreground, occurs in thin sheetlike regions. This phenomenon is a sign for fluctuating local magnetic field.

1 Introduction

The Galactic area above +4° and below −4° has never been studied in a systematic way. Most large scale surveys, covering this area suffer from low angular resolution or they miss the diffuse emission as is the case for the VLA-survey (Condon et al. 1996). The knowledge of the magnetic field in the Galactic halo is still rather poor because of the lack of observations of the linear polarization at the proper angular resolution and sensitivity. To fill this gap new observations at 1.4 GHz are being carried out to cover the entire Galactic plane at medium Galactic latitudes ($|b| \leq 20°$) visible at Effelsberg. This survey is planned to reach a sensitivity at total power close to the confusion limit and a high sensitivity at linear polarization.

The observations are carried out with the two channel 1.3-1.7 GHz receiver installed in the primary focus of the Effelsberg 100-m telescope. The receiver is equipped with cooled HEMT amplifiers. The method of observation is to scan in Galactic latitude and longitude. The scanning speed is 4° per minute and the scan separation is 4′. The typical scan length is 10°. Each field was measured at least twice giving a total integration time per point of $\Delta t = 2$ s. A root mean square (r.m.s) value of 15 mK $\mathrm{T_B}$ (or 7.5 mJy/beam area) for the total power is reached. This value is close to the expected confusion limit. The sensitivity of polarization measurements is found to be about 8 mK $\mathrm{T_B}$ r.m.s. with $T_{\mathrm{B}}/S_\nu = 2.11$ mJy/mK. The half power beam width (HPBW) at this frequency is 9.′4. The polarization is corrected for instrumental polarization down to about 1%. The total power maps are calibrated to an absolute scale using the 1.4 GHz northern sky survey by Reich (1981) and Reich & Reich (1986) made with the Stockert 25-m telescope. The polarization data are calibrated to an absolute scale using the data by Brouw

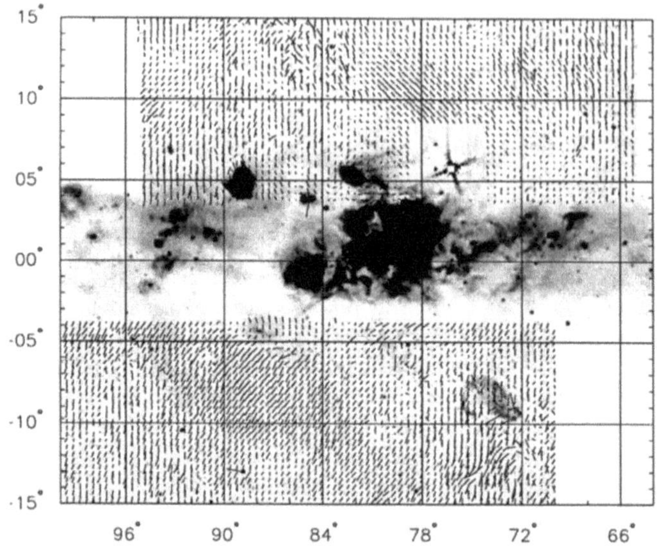

Fig. 1. A sample region from the 1.4 GHz continuum survey in gray scale. Electric field vectors are plotted as bars.

& Spoelstra (1976). A sample region from the survey is displayed in Fig. 1. Details of the survey are given by Uyanıker (1997).

In the following sections we present examples of the detected polarization structures and discuss their origin and importance for the understanding of the local magnetic field in the Milky Way.

2 Polarization Structures

Variations of the linear polarization are common at medium latitudes. These structures are close to highly polarized regions. In some examples the polarization intensity suddenly drops to a certain value and forms filamentary or loop like structures just in the middle of a highly polarized region. They are like lenghty slits in relatively high polarization regions. The electric field vectors of the linear polarized signal seem to follow these structures. A representative example of such a filament is given in Fig. 2.

Towards the Cygnus region, below the well-known supernova remnant Cygnus Loop, a polarization structure is detected (Fig. 3). The structure is about 6°.2 × 5°.6 in dimensions, in longitude and latitude respectively. Polarization vectors form two arc like structures touching each other. The magnetic field vectors form an elliptical shape giving the impression as if it is an ellipsoidal object. No corresponding variation in the total power emission has been seen.

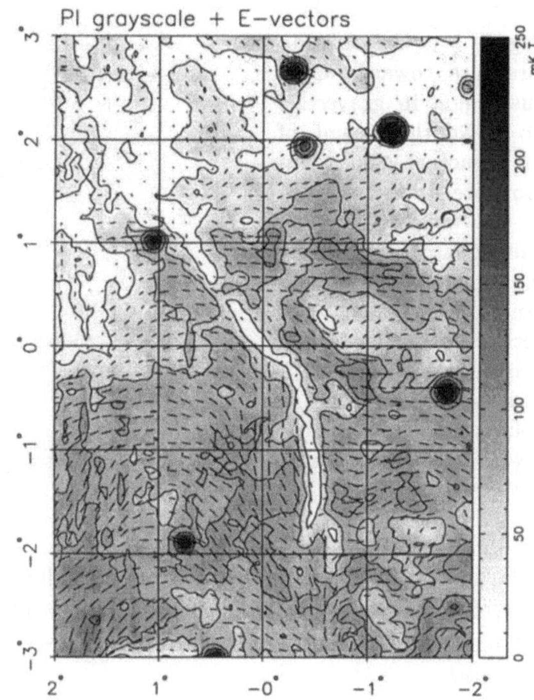

Fig. 2.
Depolarization filament towards Galactic anti-center. The filament is at least 4° in length. The upper part of the filament makes an angle of 40° with the north and the lower part is more or less perpendicular to the Galactic plane. Electric field vectors follow the structure. The width of the filament is about 10′ after correcting for the beam. This depolarization filament is most probably caused by the Faraday modulation of the local magnetic field. Relative coordinates are in units of degrees.

3 Discussion

It is known that polarization structures without corresponding variation in total power emission exist (see Wieringa et al. 1993). This is mainly due to the fact that at least half of the magnetic energy density in the interstellar medium is in the form of fluctuations (see Heiles 1996, Zweibel 1996). Since the polarized emission can not traverse great distances without beeing depolarized, the sources of polarized emission are generally rather nearby. It seems that the observed structures are closely connected with the fluctuating local magnetic field. In this way the filamentary structures of low polarization intensity may be interpreted as depolarization caused by Faraday modulation of synchrotron foreground. Therefore, at shorter wavelengths, due to the wavelength dependency of the Faraday rotation, these depolarization structures are expected to disappear or at least weaken. If however the structures are still observable at shorter wavelengths (than 21 cm), one should consider the interaction of magnetic fields, namely magnetic reconnection effects. On the other hand, depolarization can also result due to the finite beamwidth of the telescope (beam depolarization). If several regions of emission with different rotation measures simultaneously exist within the beam then the measured polarization is not a function of λ^2 (see Burn 1966). Alternatively, if the spectral index of the radiation varies within a small region then the

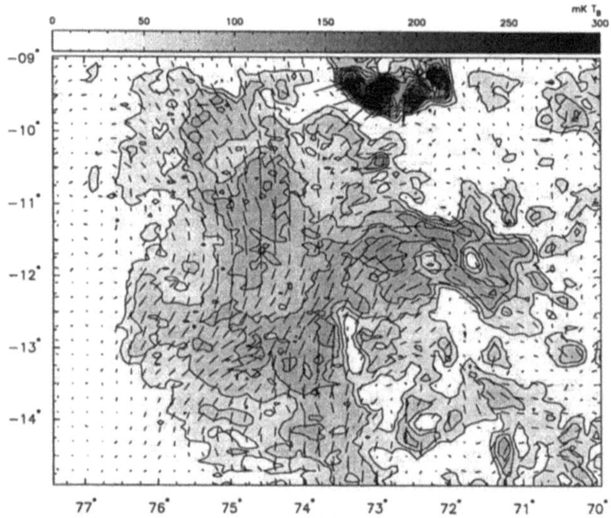

Fig. 3. Polarization structure towards Cygnus region in grayscale. To make the extent of the structure clear, contours are plotted as given in the wedge. The two uniform arches drawn by the electric field vectors can be seen easily. Magnetic field vectors, when plotted, form an elliptical shape. Percentage polarization varies between 5 to 16% across the structure.

polarization of the different components dominate at different wavelengths. Therefore, a detailed spectral information about these regions is necessary. We leave this question open until the observations at higher frequencies are available.

The interaction between the Galactic magnetic field and interstellar clouds leads to a variety of radio emitting structures, shells, filaments, and loops. Many of these features are linearly polarized. They extend up to large Galactic latitudes. They are very faint, an order of magnitude fainter than the large scale diffuse Galactic emission. They deform and compress the Galactic magnetic field. The detected polarization structures indicate that the local Galactic magnetic field is not constant. There are disturbances of the smooth background due to Faraday modulation of the magnetic field.

References

Brouw W.N. & Spoelstra T.A.Th. (1976): A&A Suppl. 26,129.
Burn B.J. (1966): MNRAS, 133, 67.
Condon et al.(1996): Priv. Comm.
Heiles C. (1996): ASP Conf. Ser. 97, eds. W.G. Roberge & D.C.B. Whittet, p.457.
Reich W. (1981): A&A Suppl. 48, 219.
Reich P. & Reich W. (1986): A&A 63, 205.
Uyanıker, B. (1997): Ph.D. Thesis, University of Bonn
Wieringa M.H. et al.(1993): A&A 268, 215.
Zweibel E.G. (1996): ASP Conf. Ser. 97, eds. W.G. Roberge & D.C.B. Whittet, p.486.

Imaging Polarimetry of Nearby Molecular Clouds

A.N. Ramaprakash, S.N. Tandon, and R. Gupta

Inter-University Centre for Astronomy and Astrophysics, Ganeshkhind, Pune - 411 007, India

Abstract. In order to study the role played by the magnetic fields in the dynamics of the interstellar medium (ISM) a two-channel imaging polarimeter (IMPOL) has been constructed. The instrument is capable of multi-wavelength observations in the visible and very near infrared (IR) wavelengths, has a *sensitive* CCD detector and a built-in acquisition and guidance unit. When used with a 1.2m, $f/13$ telescope IMPOL gives a field of view of 6 × 6 arcmin with an angular resolution of 2″ so that it is well-suited to observe both faint extended objects and moderately crowded stellar fields. The instrumental polarization is less than 0.05% and the accuracy of the measurements is primarily limited by photon noise in a typical observation. In this paper, we present a brief description of the instrument and observational techniques, as well as the results of a successful observation of a molecular cloud close to the open cluster IC 5146.

1 Introduction

The importance of polarimetry in understanding the structure of the interstellar medium (ISM) has been amply emphasized many times (e.g. Tinbergen 1996, review by Carl Heiles in this meeting). Even though optical polarimetry has been employed in astronomy since the 1950s (Hiltner 1949, Hall 1949), it became a regular part of the astronomers' toolkit only recently with the advent of more sensitive detectors and larger telescopes; polarimetry is always photon starved. Even after those advances the aperture photometric techniques were still very time-consuming, difficult to carry out, lacked angular resolution and produced only small samples (Sen & Tandon 1994).

With the introduction of the imaging methods in polarimetry the situation improved greatly. In the following section we briefly describe an imaging polarimeter (IMPOL) which uses a sensitive (liquid nitrogen cooled) CCD as the detector. In the third section the results of a successful observation of a molecular cloud close to the open cluster IC 5146 is presented to illustrate how the instrument can be employed to study the local ISM.

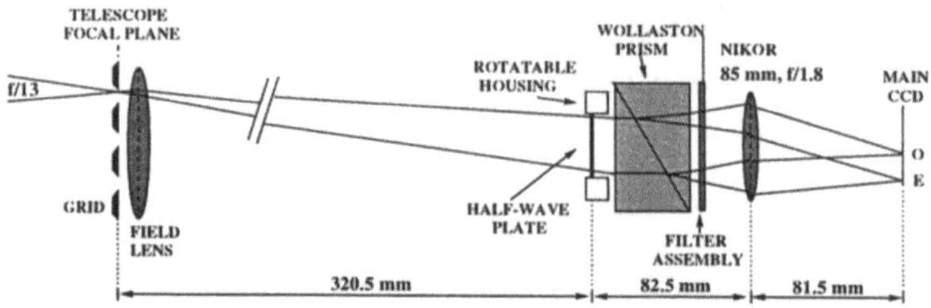

Fig. 1. Block schematic of the IMPOL optical system

2 The Polarimeter

Only the essential details of the instrument are presented here. For a more detailed description of the design, handling of errors and observational techniques, please refer to Ramaprakash et al. (1997).

As shown in Fig. 1, a single camera lens is used to produce focal reduction by a factor of about 3.8 and obtain the correct image scale at the CCD. A half-wave plate modulator and a beam-splitting Wollaston prism analyzer placed in the beam path enable measurement of linear polarization. The half-wave plate rotates the position angle of the polarization vector of the incoming light by twice the angle that itself has been rotated. This light is split by the Wollaston prism into two beams of linear polarization components in a coordinate system fixed with respect to the instrument. Since the Stokes parameters are defined with respect to this coordinate system, care has to be taken while mounting the instrument on the telescope to align it with the conventional reference axis on the sky, i.e., the celestial north-south axis.

The two beams (the so called ordinary and extraordinary beams) produced by the Wollaston prism propagate in slightly different directions separated by about 0.5°. Therefore the camera lens produces two images on the CCD for every point at the telescope focal plane. In order to avoid overlapping of the ordinary and extraordinary images of adjacent points, a set of properly spaced, parallel obscuring strips of the correct width are placed at the telescope focal plane.

The ratio R of the difference between the intensities of the ordinary (I_{o}) and the extraordinary (I_{e}) beams to their sum (Eq. 1) is by definition, the Stokes' parameter q, when the half-wave plate fast axis is aligned to the reference axis. Similarly when the half-wave plate is at 22.5°, the above ratio becomes the Stokes' parameter u.

$$R(\alpha) = \frac{I_{\mathrm{o}} - I_{\mathrm{e}}}{I_{\mathrm{o}} + I_{\mathrm{e}}} \ . \tag{1}$$

Now the fractional polarization (p) and the position angle (θ) can, in principle, be recovered from these two ratios as

$$p = \sqrt{q^2 + u^2} \qquad \text{and} \qquad \theta = \frac{1}{2}\tan^{-1}\left(\frac{u}{q}\right) \tag{2}$$

Since the above method allows determination of the Stokes parameter from a single exposure, it is free of the effects due to atmospheric scintillation, varying sky transparency, etc. To facilitate observations of faint objects that demand long exposures, the instrument is provided with an off-axis acquisition and guidance unit. Special precautions have been taken to minimize errors in the polarization measurement and it is seen that the polarimetric accuracy is limited by photon noise for typical observations. The instrumental polarization is found to be about 0.05%.

3 Observations

IC 5146 is an open cluster in Cygnus. Associated to it, there is a long, filamentary molecular cloud complex extending about 2° EW and about 0.5° NS in the Palomar prints. The extent and structure of this complex has been studied using ^{12}CO and ^{13}CO as tracers by Dobashi et al. (1992). Combining the classical star count methods and infrared colour excesses of reddened background stars, Lada et al. (1994) obtained detailed extinction maps of what they called the Northern Streamer part of the complex.

Here we present the results of imaging polarimetric observations on two fields of the Northern Streamer region, obtained using the IMPOL. The observations were carried out with the 1.2 m, Gurushikhar Infrared Telescope (GIRT) at Mt. Abu, Rajasthan, India[1]. In the polarization map shown in Fig. 2, we see that the magnetic field vectors form three distinct regions. The average position angle of the vectors in the northern periphery of the cloud is about 50°. In the southern periphery, for right ascention (RA) greater than $21^h46^m50^s$, the average position angle is about 55°, while below this RA it is about 69°. This matches quite well the contour maps of Lada et al. (1994), where it can be seen that the material distribution along the northern periphery is quite irregular and probably explains the twist in the magnetic field (see their Fig. 1a and b). On the other hand, for RA lower than $21^h46^m50^s$, the contours along the southern periphery are fairly aligned along EW, while those at the SE corner show kinks around a dense clump.

Future Prospects: The IMPOL can be very useful in the study of the local ISM. Gvaramadze (this meeting) and several others (e.g. Brand et al. 1983, Bertoldi & McKee 1990) have studied the effects of shock waves and ionization fronts on dense obstacles like cloudlets and clumps in the medium in which they propagate. The role of magnetic fields in providing pressure support to the obstacles against the invading front has been clearly realized, but not yet fully understood (Sridharan et al. 1996). Cometary globules and teardrops are supposed to evolve out from such interactions (Reipurth 1983).

[1] GIRT is operated by the Physical Research Laboratory, Ahmedabad, India.

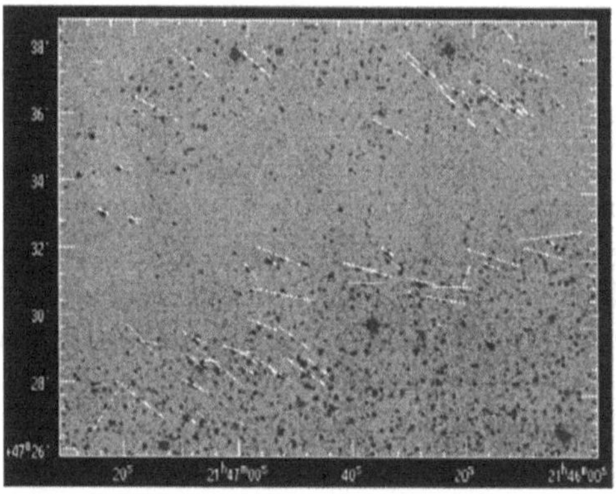

Fig. 2. Polarization map of two fields in the Northern Streamer region near IC 5146. The small horizontal vector on the NE corner is 1%

Observations of two cometary globules have been made using this instrument aiming to study the presence and nature of the magnetic fields along the tails. The analysis of this data is currently in progress.

A similar observation of the clouds found embedded in the hot plasma in the local ISM (e.g. Herbstmeier & Wennmacher, Beckman et al. in this meeting), can throw light on the magnetic field structure in these objects which might also explain what keeps these clouds from evaporating off, given the harsh environment they are in. Multi-wavelength polarimetry can be employed to determine the average grain sizes in these clouds.

Acknowledgements. The project has been funded by the Department of Science and Technology of the Government of India. The authors also wish to thank the Physical Research Laboratory, Ahmedabad for telescope time.

References

Bertoldi F., McKee C. F., 1990, ApJ 354, 529
Brand P. W. J. L., Hawarden T. G., Longmore A. J., Williams P. M., Caldwell J. A. R., 1983, MNRAS 203, 215
Dobashi K., Yonekura Y., Mizuna A., Fukui Y., 1992, AJ 104, 1525
Hall J. S., 1949, Sci 109, 166
Hiltner W. A., 1949, Sci 109, 165
Lada C. J., Lada E. A., Clemens D. P., Bally J., 1994, ApJ 429, 694
Ramaprakash A. N., Gupta R., Sen A. K., Tandon S. N., 1997, A&A (in press)
Reipurth B., 1983, A&A 117, 183
Tinbergen, J., 1996, in: Astronomical Polarimetry, (Cambridge Univ. Press)
Sen A. K., Tandon S. N., in: Instrumentation in Astronomy VIII, ed. D. L. Crawford, SPIE proceedings, Vol. 2198, part 1, p. 264
Sridharan T. K., Bhatt H. C., Rajagopal J., 1996, MNRAS 279, 1191

Distribution and Space Density of Soft X-ray Emitting Polars in the Solar Neighbourhood

H.-C. Thomas[1] and K. Beuermann[2,3]

[1] Max–Planck–Institut für Astrophysik, Karl-Schwarzschild-Str. 1, D-85740 Garching, Germany
[2] Universitäts-Sternwarte, Geismarlandstr. 11, D-37083 Göttingen, Germany
[3] Max–Planck–Institut für extraterrestrische Physik, Giessenbachstrasse, D-85740 Garching, Germany

The ROSAT All Sky Survey (RASS) was the first one performed with an imaging telescope in the soft X-ray regime and has led to the discovery of numerous new objects whose emission is dominated by soft X-rays. Among these are white dwarfs and a subclass of the cataclysmic variables (CVs), the Polars or AM Herculis binaries. From a pre-ROSAT census of only 17, the number of known sources of this class has increased to some 55 (Beuermann and Thomas 1993, Beuermann 1997). Distances or lower limits to the distance are available for some 35 of these, based on the detection or non-detection of the TiO-Features in their optical red spectra. The derived distances range from below 100 pc up to $\sim$ 600 pc, implying that many of these objects are located within the "Local Bubble" of low gas density in interstellar space. As the soft X-ray emission can be reasonably well represented by blackbody emission with a typical temperature of $kT_{\rm bb} \simeq 25\,$eV, spectral fits to the ROSAT PSPC spectra from either the All-Sky-Survey (RASS) or from subsequent pointed ROSAT observations allow to determine the foreground absorption column density in the direction of the polars.

Considering 9 polars with projected distances in the Galactic plane of $r < 200\,$pc, the derived mean atomic hydrogen density is 0.04 ± 0.01 H-atoms cm^{-3}. This result excludes 3 systems with substantially larger mean column densities, AM Her ($l, b, d = 79.9°, +25.9°, 90$ pc) with 0.32, RX J1313-33 ($l, b, d = 308.2°, +29.7°$, 170 pc) with 0.65, and RX J2022-39 ($l, b, d = 2.4°, -33.6°, >210$ pc) with 0.63 H-atoms cm^{-3}. Five of the 9 polars are located in the 3rd quadrant ($180° < l < 270°$), none in the 1st. In the southern part of the 4th quadrant 4 out of 5 polars indicate densities below 0.04 H-atoms cm^{-3} out to distances of 350 pc. The mean line-of-sight density for the four systems with $r > 400\,$pc is 0.25 H-atoms cm^{-3}. The crowding of very soft X-ray sources in certain parts of the sky noted, in particular, for hot white dwarfs (Warwick et al. 1993, Diamond et al. 1995) is similarily found for polars and confirms the conclusions on the structure of local interstellar space deduced from the other types of soft X-ray emitting stars as well as by other methods (e.g. Paresce 1984).

At high Galactic latitudes ($|b| \geq 49°$) the survey of polars is probably complete except for (i) a notable deficiency within the boundaries of the

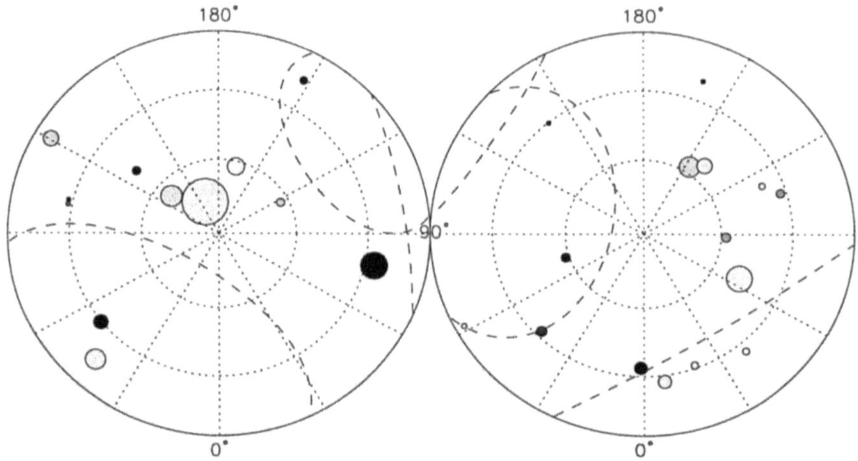

Fig. 1. Mean atomic hydrogen space densities along the lines of sight to the sources for which distances and PSPC spectra are available. The source positions refer to Galactic coordinates with positive latitudes for the left, negative latitudes for the right circle. The grayscale coding implies the following densities: white: 0, black: > 0.3 H-atoms cm^{-3}. The size is inversely proportional to the projected distance in the Galactic plane.

North-Polar Spur and possibly also the Galactic Loops II and III and (ii) a certain fraction of objects which were not detected during the ROSAT survey because they happened to be in an inactive state of low or discontinued accretion. Correcting for these losses in the counts of polars, yields estimates of the space density separately for long-period and short-period polars ($P_{\mathrm{orb}} > 3\,\mathrm{h}$ or $P_{\mathrm{orb}} < 3\,\mathrm{h}$, respectively). The two space densities differ by about an order of magnitude which reflects the more rapid evolution in time of the long-period over the short-period polars due to magnetic braking (momentum loss by a magnetized wind) and gravitational radiation, respectively. Assuming a Gaussian density distribution, the derived space density in the Galactic plane is $n_{\mathrm{o}} \simeq 6.1\,10^{-7}\mathrm{pc}^{-3}$. Corrected for losses and splitted according to orbital period the space density can be as high as $1.2\,10^{-6}\mathrm{pc}^{-3}$ for short-period polars and $1.7\,10^{-7}\mathrm{pc}^{-3}$ for long-period polars. The standard deviation of a Gaussian density distribution perpendicular to the Galactic plane was determined to be $\sigma \simeq 155\,\mathrm{pc}$, compatible with that derived for non-magnetic Cataclysmic Variables (Patterson 1984).

Our study of polars confirms the low mean particle density of the local ISM interspersed with some denser clouds. In some directions these low densities extend to large distances,

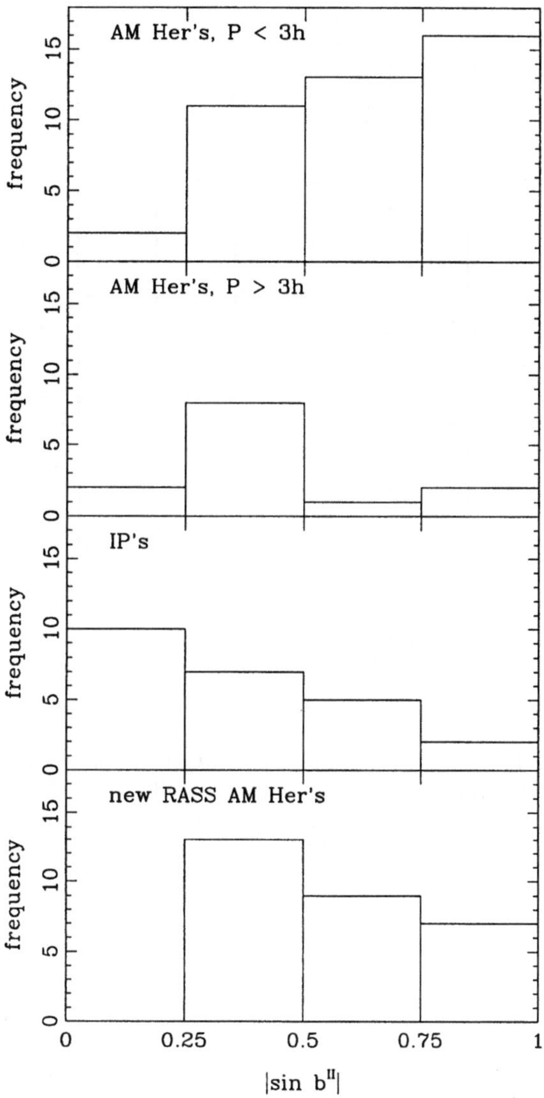

Fig. 2. Galactic latitude distributions of four subsets of magnetic cataclysmic variables: (i) short-period polars (AM Herculis binaries) with orbital period $P_{\rm orb} < 3\,\rm h$, (ii) long-period polars with orbital period $P_{\rm orb} > 3\,\rm h$, (iii) the subclass of intermediate polars (IPs), many of which possess harder X-ray spectra, and (iv) the polars newly discovered in the RASS. The lack of magnetic CVs at low Galactic latitudes is seen in all cases except that of the IPs. Derivation of the space density has to account for the severe Galactic absorption, therefore.

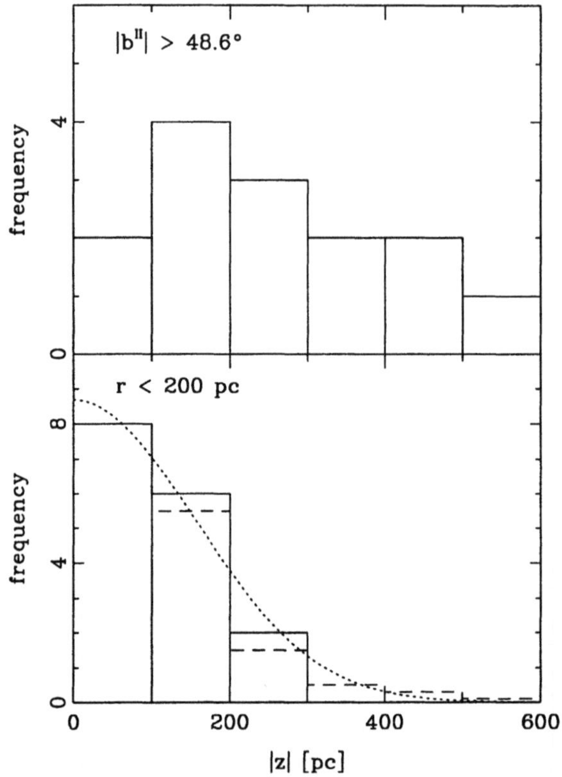

Fig. 3. *Upper panel:* The number of polars at high Galactic latitudes as a function of z. *Lower panel:* The number of polars within a cylinder of 200 pc radius perpendicular to the Galactic plane in intervals of 100 pc in $|z|$ (solid histogram). The same quantity for $|z| > 100$ pc as derived from the upper panel (dashed histogram). A Gaussion fit (dotted line) yields a standard deviation of $\sigma = 155$ pc.

References

Beuermann, K., and Thomas, H.-C. (1993): Adv. Space Res. 13(12), 115

Beuermann, K. (1997): "Magnetic cataclysmic variables: Observational and theoretical results", Proc. Int. Coll. Golden Jubilee of Tata Inst. Fund. Res., Mumbai, India

Diamond, C.J., Jewell, S.J., Ponman, T.J. (1995): MNRAS 274, 589

Paresce, F. (1984): AJ 89, 1022

Patterson, J. (1984): ApJ Suppl. 54, 443

Warwick, R.S., Barber, C.R., Hodkin, S.T., Pye, J.P. (1993): MNRAS 262, 289

Small Scale Structure in Nearby Molecular Gas

Edith Falgarone

Laboratoire de Radioastronomie, Ecole Normale Supérieure,
24 rue Lhomond, F-75231 Paris Cedex 05, France

Abstract. Recent observations at high angular resolution of molecular clouds of low column density have revealed the presence of a conspicuous net of small scale filamentary structures, visible in the ^{12}CO rotational lines only. In addition, the existence of unresolved structure at scales as small as $\sim$ 200 AU in space and/or velocity space is inferred from the spectral properties of the ^{12}CO and ^{13}CO emission. The resolved structures are part of the hierarchy of structures observed in molecular gas in the Solar Neighborhood and appear as non self-gravitating elements confined by an ambient pressure $P_0/k_B \sim 3 \times 10^4 \mathrm{cm}^{-3}$ K. We show why these structures might have their origin in the intermittent structures of turbulence in which viscous dissipation is concentrated in space and time.

1 A Short Overview of Past Evidence for Small Scale Structure in Molecular Clouds

As soon as, in the mid 70's, the first maps of interstellar clouds in the rotational lines of ^{12}CO appeared, it was inferred from the line excitation conditions and radiative transfer (Penzias et al., 1973) that lines were emitted by gas colder than $\sim$ 10K and as dense as 10^3 to 10^4 cm^{-3}. The ^{12}CO linewidths, $\Delta v \sim$ a few km s^{-1}, were strikingly non–thermal since the thermal velocity of CO at T=10K is 0.05 km s^{-1}.

Soon after, it was recognized that such large densities had to fill only less than one hundredth of the volume of molecular complexes because the gas column densities combined with the sizes indicated average gas densities not larger than $\sim$ 100 cm^{-3} within clouds of a few pc in size. Yet, lower average densities $\sim$ 10 cm^{-3} were derived for giant molecular complexes of size $\sim$ 50pc, (Pérault et al. 1985, Blitz, 1987). This was the starting point of the concept of clumpy molecular clouds. The clumps were then modelled as self–gravitating polytropes of about $\sim$ 1pc, bounded by the external pressure of a warm interclump medium of density a few tenth cm^{-3} and kinetic temperature of the order of 8000K (Falgarone & Puget, 1985). In this model, the density fall–off at the clump boundaries was generated by the thermal instability of the interstellar gas (Field, 1965). The interclump medium was heated by the drift of magnetic field lines attached to the clumps in their virialized motions in the gravitational potential well of the complex (Falgarone & Puget, 1986) .

At about the same epoch, the scale free properties of the hierarchy of structures of molecular clouds seen in the CO lines started to be unveiled (Larson, 1981), and scaling laws were built between the mass and size of structures, their internal velocity and size. Larson postulated that the scaling laws were reminiscent of the Kolmogorov law which characterizes the power spectrum of incompressible turbulence and suggested that they reflected the turbulent nature of the flows within molecular clouds. On the other hand, Chièze (1987) noticed that the hierarchy of observed structures coincided with the ensemble of the most massive isothermal polytropes (at different temperatures) which are gravitationally stable in an ambient pressure P_0.

Up to the mid 80's, a description was emerging, in which the structures of molecular clouds at all levels between $\sim$ 50pc down to $\sim$ 1pc were self–gravitating entities. The nature of the internal energy density was known to be dominantly non–thermal which demanded some refinement of Chièze's description. But the predictions were remarkably in agreement with the observations.

What have we learnt since then? One important step forward has been the discovery that the hierarchy extends down to scales two orders of magnitude smaller, $\sim$ 0.01pc, at which structures are no longer self-gravitating.

2 Recent Results from the first IRAM Key-Project

These recent progresses are examplified by the results obtained in the framework of the first IRAM key–project (Frisch 1997). The goal of this project was to map the environment of low mass dense cores at high angular resolution in the perspective of understanding some of the processes at work in the dissipation of the non–thermal support of molecular clouds. The selected targets are pieces of ordinary gas containing a low mass core with internal velocity dispersion close to thermal. Except for the dense cores themselves which have total H_2 column density $N_{H_2} \sim$ a few $10^{22}\,cm^{-2}$, the average column density of the fields mapped is quite low, $N_{H_2} \sim$ a few $10^{20}\,cm^{-2}$ at the parsec scale.

2.1 Resolved Structure

Figure 1 displays the maps of $^{12}CO(2\text{-}1)$, $^{13}CO(2\text{-}1)$ and $C^{18}O(1\text{-}0)$ line integrated emission in one of the fields studied, a high latitude cloud in the vicinity of the Polaris star. The pattern seen in the ^{12}CO line is quite different from that seen in the ^{13}CO and $C^{18}O$ lines. In ^{12}CO, there is almost no structure visible except for a few barely resolved hot spots while in the ^{13}CO and $C^{18}O$ lines an arc like structure which delineates high column density gas (the dense core) is clearly visible.

Figure 2 shows the large amount of small scale structure present in the $^{12}CO(2\text{-}1)$ maps as soon as the different velocities are separated from one

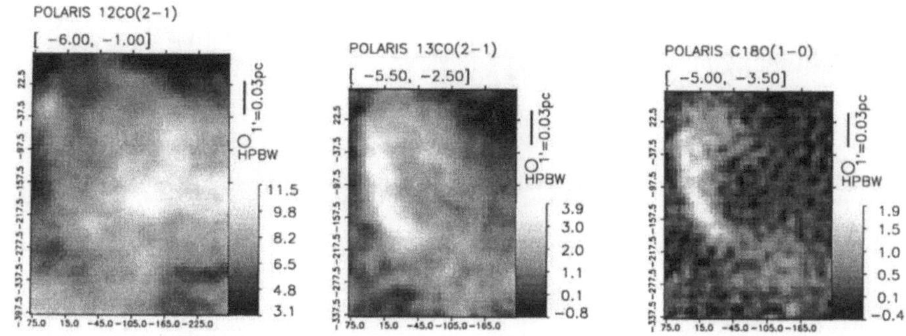

Fig. 1. Maps of line integrated area in the Polaris field. The different transitions are indicated above each panel. The linear scale is shown, together with the angular resolution (HPBW) and the intensity scale expressed in K km s^{-1}. The numbers between brackets give the velocity range of integration, in km s^{-1} . The coordinates offset are in arcsec.

another. The channel maps of Fig. 2 reveal a net of filamentary or more compact structures, not clearly connected to the arc visible in the rare isotopes maps. Note that some of the filaments have width at half-maximum of the order of the telescope resolution at that frequency, HPBW=14" or 0.007 pc or 1800 AU at the distance of the cloud (100 pc).

The discovery of such a small scale filamentary structure in the ^{12}CO lines was unexpected because the weak ^{12}CO emission observed at low angular resolution in cloud edges has long been thought to originate in a low density homogeneous medium that would subthermally excite the CO rotational lines. It is indeed a beam–diluted emission of bright small scale filaments, identified only in large maps observed at high angular resolution. A similar result has been found by Falgarone & Phillips (1996). Here, higher J ^{12}CO transitions are used to constrain the parameters of the gas emitting in a small filament detected in a cloud edge. Solutions range from dense $n_{H_2} = 10^3\,cm^{-3}$ gas at $T_k \sim 25$K to lower density gas $n_{H_2} \sim 200\,cm^{-3}$ at $T_k \sim 250$K, opening the possibility that, unlike cold gas in dense cores, these filaments seen in ^{12}CO are much warmer. These two sets of results also show that it is the gas which emits in the wings of the CO lines and has the broadest velocity coverage which has the highest level of small scale structure in ^{12}CO. This component is not visible in ^{13}CO, unless huge integration times are used.

2.2 Sub-beam Structure

In addition to the resolved small scale structure, the existence of structure at much smaller size scales is inferred from two very well defined properties of the CO line profiles disclosed in the above papers. These properties are illustrated in Figure 3 for the Polaris field. The first property, is the constancy of the

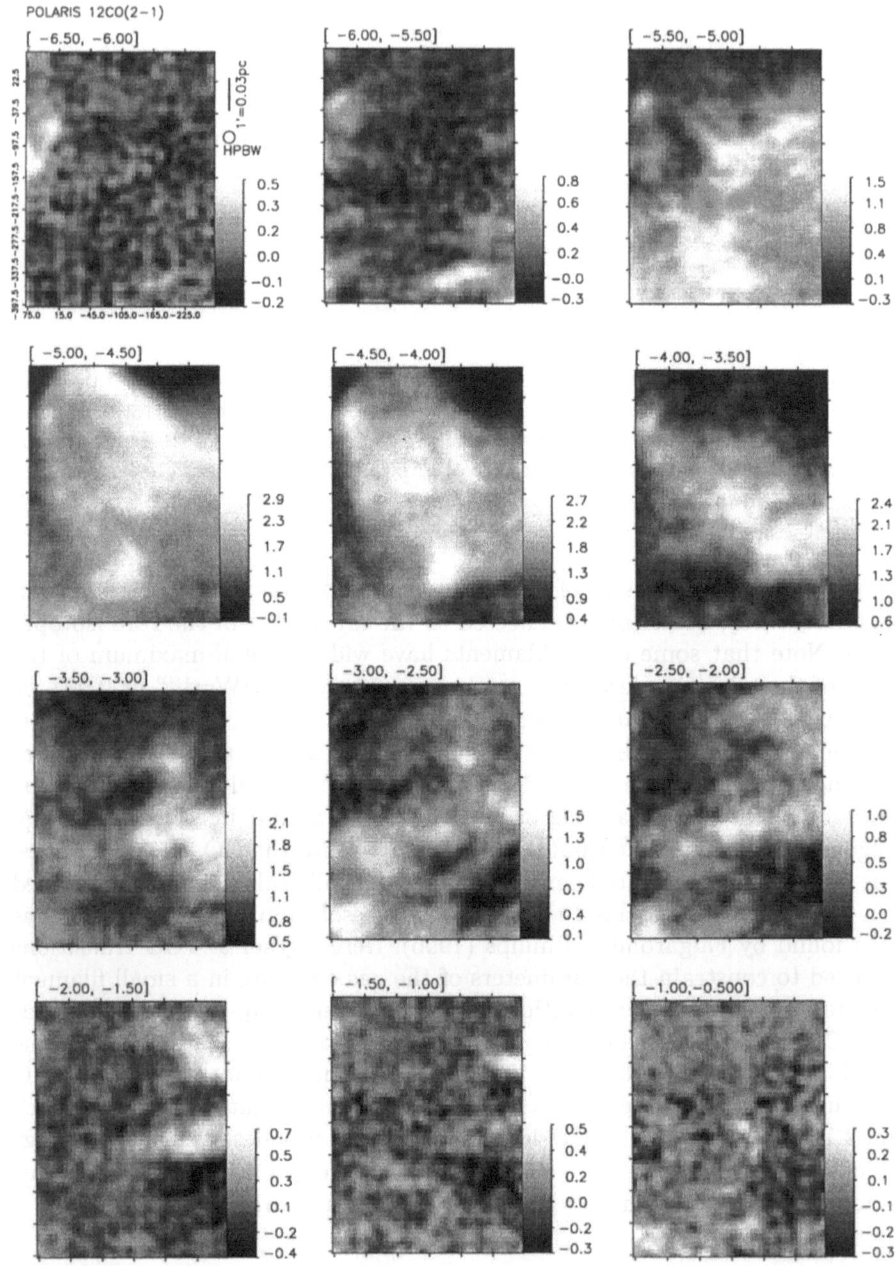

Fig. 2. Space–velocity maps of the $^{12}CO(2\text{-}1)$ emission in the Polaris field. The velocity intervals, in km s^{-1}, are given at the top of each panel and the grey scale calibration in K km s^{-1}, different for each velocity interval to emphasize the existence of the weak filamentary structure, is given at the right side of each map.

J=2-1 to J=1-0 line ratio R(2-1/1-0)∼ 0.65 across the line profiles and for the two ^{12}CO and ^{13}CO isotopic lines. The second property is the smoothness of the line profiles, although the channel width is as small as $0.05 \mathrm{km\,s^{-1}}$.

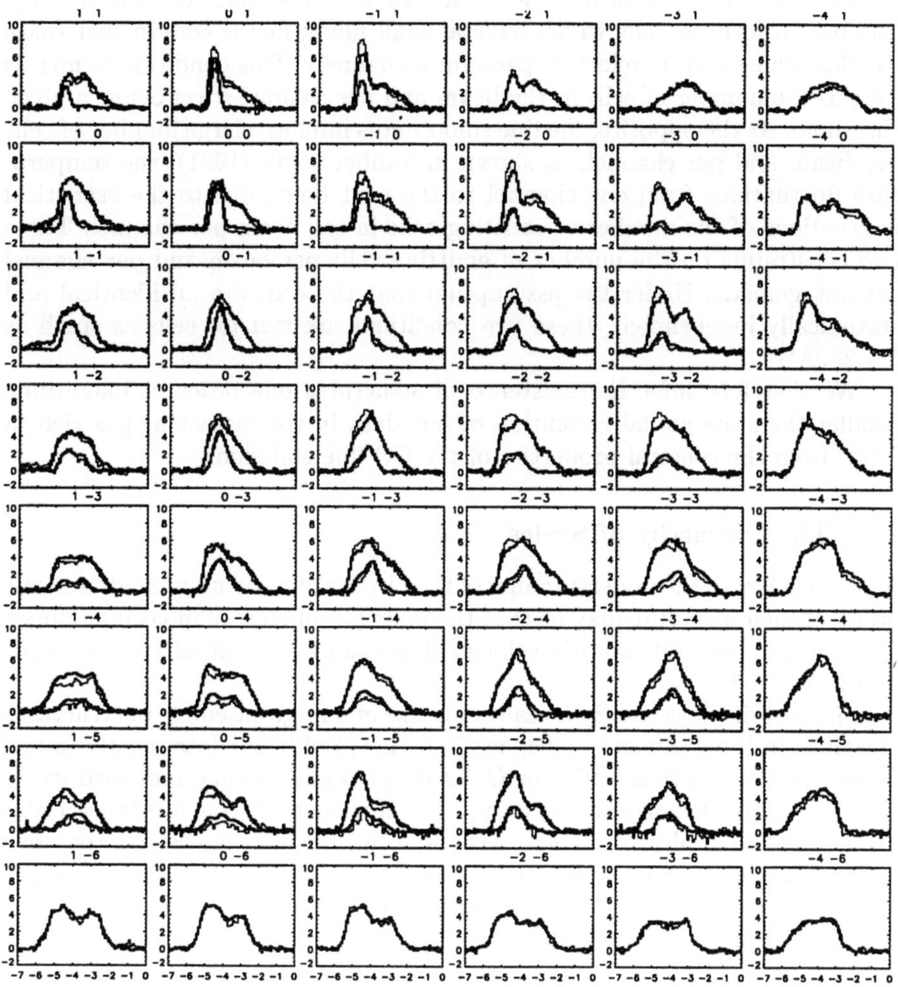

Fig. 3. Maps of ^{12}CO and ^{13}CO integrated spectra over 1' fields in Polaris. The temperature scale is antenna temperature corrected for error beam contributions. Thick lines are for the ^{12}CO and ^{13}CO J=1-0 transitions (same scale). The intensities of the $^{12}CO(2\text{-}1)$ and $^{13}CO(2\text{-}1)$ line profiles (thin lines) have all been divided by the *same* factor R(2-1/1-0)=0.65

From the constancy of the line ratio in space and velocity (*i.e.* across the line profiles) we derive that the excitation temperature of the rotational lines is independent of velocity *i.e.* there exists very little radiative coupling among regions on the same line of sight at neighboring velocities. These conditions for line formation in which there is little shadowing of the emitting regions in phase–space are called macroturbulent (see Martin, Sanders & Hills (1984) for more details). To ensure the small radiative coupling at each velocity, line has to form in cells which have a small filling factor *both* in real space (in the beam) and in velocity space (in a channel). This condition is met as long as the number of cells in the beam and per channel stays below a given threshold. At the opposite, the line smoothness implies a large number of cells per beam and per channel, as shown in Tauber et al. (1991), the temperature fluctuations from one channel to the next being due to the statistical fluctuations of the number of emitting cells in the beam per channel. These two constraints on the number of emitting cells per beam and per channel are antagonistic. Under the assumption that the cells are all identical and statistically independent, these two conditions are met for cells as small as $l \sim 200$AU.

We therefore infer the existence of structures one order of magnitude smaller than the actual resolution of our data in the molecular gas rich in ^{12}CO from the spectral properties of the CO line emission.

2.3 The Hierarchy of Scales

A possible key to the understanding of the physics of molecular clouds and the origin of their structure may be that the main parameters of molecular clouds (their mass, size and internal velocity dispersion) are related by power-law relations.

Figure 4 displays the H_2 mass versus size of a large ensemble of structures identified in the 4-dimensional space $T_L(x, y, v_z)$ of ^{12}CO data mostly (a few values are derived from ^{13}CO or C^{18}O lines). T_L is the line temperature at velocity v_z and beam position (x, y). The masses and sizes are those given by the authors, unless some scaling was necessary to take the different size definitions into account. The structures are identified in molecular surveys of the central parts of the Galaxy (stars: Dame et al. (1986), open triangles: Solomon et al. (1987)) and of the third quadrant (open hexagons: May et al. (1997)), in the Rosette (crosses) and Maddalena (open squares) molecular clouds (Williams et al., 1994), in non–star-forming clouds (solid triangles: Falgarone & Pérault (1987), solid squares: (Falgarone et al., 1992), tripods: Lemme et al. (1995), asterisks: Herbertz et al. (1991)), in ρ Ophiuchus (solid hexagons: Loren (1989)) and in the Polaris high latitude cloud (data from the IRAM key–project described above, starred triangles: Heithausen et al. (1997)). In addition, the average values obtained for a set of 10 representative dense cores correspond to the large solid squares, Wang et al. (1995). We have added masses and sizes derived from submillimeter continuum observations of the dust for another set of low mass dense cores (solid squares of intermediate size, Ward–Thomson et al. (1994)) and those derived from CS(5-4) observations of massive dense cores in Orion (large open squares, Lada et al. (1997)).

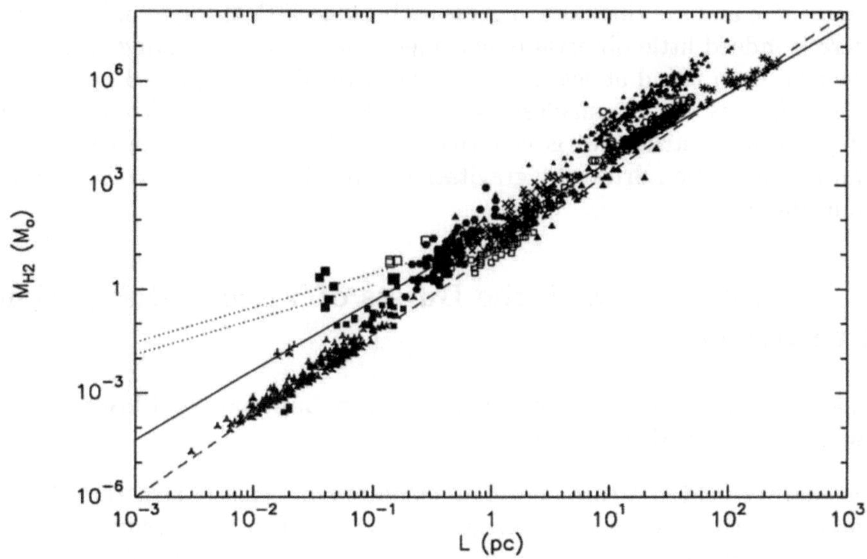

Fig. 4. H_2 mass versus size for structures in galactic molecular clouds. The symbols are explained in the text.

In Figure 4, the full line traces the scaling predicted by Chièze (1987):

$$\frac{M}{L^2} = 44.5\,(\frac{P_0/k_B}{3 \times 10^4\,\mathrm{cm^{-3}\ K}})^{0.5}\,\mathrm{M_\odot\,pc^{-2}}$$

for a pervading pressure $P_0/k_B = 3 \times 10^4\,\mathrm{cm^{-3}}$ K. This pressure is the same as that due to the weight of the HI gas layer at mid–plane in the Galaxy (Boulares & Cox, 1990), also equal to the observed average non–thermal kinetic pressure at small scale in non star forming regions of the Solar Neighnorhood (Falgarone et al., 1992). The two dotted lines in Figure 4 represent the locus of self–gravitating isothermal spheres at $T_k = 10$K and 50K. The dashed line has a slope 2.3, interpreted by Elmegreen & Falgarone (1996) as a signature of the fractal structure of molecular clouds. It is also the slope found by Heithausen et al. (1997) in the ^{12}CO emission of the high latitude cloud in Polaris, over two orders of magnitude in size.

Many structures lie above the full line, *(i)* the densest regions of cores in Orion or in nearby clouds, suggesting that these cores, as expected, are not bounded by the ambient pressure of the Solar Neighborhood but by a larger pressure, exerted by their self–gravitating environment, *(ii)* the structures in the molecular ring complexes (Solomon et al. (1987)) suggesting that the ambient pressure there is $\sim 10^2$ larger than that in the Solar Neigborhood.

Many structures, especially the smallest, lie below the full line. They have masses much smaller than the marginally stable self–gravitating structures of same size and are therefore in pressure balance with the external pressure. There is indeed little observational evidence that self–gravitating structures have ever been found at scales below that of the dense cores ~ 0.1 pc. If the mass estimates for the smallest scales are correct and the existence of non self–gravitating structures is confirmed, it will prove that a process other than fragmentation driven by gravitational instability is at work in shaping the medium.

3 Possible Link with the Bursts of Viscous Dissipation of turbulence

3.1 The Intermittency of Turbulence in Laboratory Flows and Numerical Simulations

The velocity field in a turbulent fluid exhibits more spatial structure than a random velocity field in all of the quantities related to increments or derivatives of the velocity field. Vorticity filaments in a turbulent field are more intense and more coherent spatially than in a random velocity field of same power spectrum, She et al. (1990). These vortices are observed in laboratory experiments, (Douady et al., 1991) and are thought to be responsible for the property of intermittency of turbulence, at the origin of the non–Gaussian probability distribution functions (PDFs) of the velocity increments, derivatives and vorticity (Vincent & Meneguzzi (1991), Porter et al. (1994), Anselmet et al. (1984), Tabeling et al. (1996)). Their diameter ranges between the dissipation scale and the Taylor microscale.

The regions of large vorticity in turbulent flows are associated with those of large dissipation rate of kinetic energy and the two subsets of space where these quantities are concentrated are highly structured and intermittent (Vincent & Meneguzzi (1991), Porter et al. (1994)). The dissipation of turbulent kinetic energy via the viscous process of elastic collisions between atoms and molecules eventually turns this energy into heat. The rate at which the gas is heated by this process is as large as the intensity of the intermittent event is large, *i.e.* the local dissipation rate may reach large values for events far in the non–Gaussian wings of the PDFs of velocity increments.

3.2 The Intermittency of Turbulence in Interstellar Gas

Experimental results obtained in flow experiments have been scaled to the interstellar medium, under the assumption that the statistical flow properties in turbulence only depend on the kinetic energy transfer rate. In atomic clouds, the size of the coherent vortices is estimated to range between ~ 20 AU and 0.03 pc. The local heating rate, due to bursts of turbulent viscous dissipation, has been found to exceed all the others heating rates in atomic clouds, by orders of magnitude. The lifetime of the intermittent structures is long enough to heat the gas to temperatures large enough to trigger chemical reactions with activation barriers and endothermic reactions (Falgarone & Puget,1995). Time–dependent calculations show that, as expected, the abundances of molecules like OH, H_2O, CH^+, HCO^+, and therefore CO, in these regions can be orders of magnitude larger than in the surrounding colder gas (Falgarone et al., 1995). The recent discovery of unexpected amounts of such molecules in interstellar gas of low extinction might be seen as a signature of intermittent dissipation of turbulence (Lucas & Liszt, 1996). Taking the effects of magnetic field into account yet reproduces better the observed correlation between OH and HCO^+ (Joulain et al., 1997). The intermittency of turbulence creates singularities in interstellar gas. The bursts of viscous dissipation locally heat the gas, triggering the formation of molecular species which in turn increase the radiative cooling of the gas, inducing condensation.

These structures may be too small to be observed individually, although laboratory experiments and simulations show that they are bunched together in space, Porter et al. (1994). Signatures of regions of enhanced vorticity have been searched for in molecular line profiles. Line profiles have been synthesized at different epochs in the evolution of a decaying run of compressible turbulence (Falgarone et al., 1994) and in a random velocity field with a Kolmogorov power spectrum (Dubinski et al., 1995). The results are very comparable except for the existence of very broad and weak line wings in the profiles present only in the turbulent case. These weak line wings are associated to regions of large vorticity within the beam (Lis et al., 1996). Further, the shape of the line profiles is statistically similar to observed line profiles only when the bulk of the turbulent kinetic energy is contained in solenoidal modes, not in compressible modes. This result suggests that the molecular line profiles of interstellar clouds trace vortices rather than shocks.

The above elements do not prove yet that any kind of small scale structure is created in the cold interstellar medium by the intermittent dissipation of turbulent energy. The kind of structures associated with the intermittent viscous dissipation of supersonic turbulence should be observed in all the turbulent components of the interstellar medium. One anticipated signature of this process is the existence of warm filamentary structures temporarily embedded within colder gas.

References

Anselmet, F., Gagne Y., Hopfinger E.J. & Antonia R.A. (1984): JFM 140, 63.
Blitz L. (1987): *Physical Processes in Interstellar Clouds*, eds. Morfill & Scholer, (Reidel, Dordrecht), 35

Boulares A. & Cox D., (1990): ApJ, 365, 544.
Chièze J.-P. (1987): A&A171 225
Dame T.M., Elmegreen B.G., Cohen R.S. & Thaddeus P., (1986): ApJ 305, 892.
Douady S., Couder Y. & Brachet M.E., (1991): Phys. Rev. Letters, 67, 983.
Dubinski J., Narayan R. & Phillips T.G., (1995): ApJ 448, 226.
Elmegreen B.G. & Falgarone E. (1996): ApJ 471, 816.
Falgarone E. & Puget J.-L. (1985): A&A 142, 157.
Falgarone E. & Puget J.-L. (1986): A&A 162, 235..
Falgarone E., & Pérault M., (1987): *Physical Processes in Interstellar Clouds*, eds. Morfill & Scholer, (Reidel, Dordrecht), 59
Falgarone E., Puget J.-L., & Pérault M. (1992): A&A, 257, 715
Falgarone E., Lis D.C., Phillips T.G., Pouquet A., Porter D.H. & Woodward P.R., (1994): ApJ, 436, 728.
Falgarone E. & Puget J.-L. (1995): A&A, 293, 840
Falgarone E., Pineau des Forêts G. & Roueff E., (1995): A&A, 300, 870.
Falgarone E. & Phillips T.G. (1996): ApJ 472, 191.
Falgarone E., Panis, J.-F., Heithausen A., Pérault M., Stutzki J., Puget J.-L., & Bensch F., (1997): A&A, in press.
Field G.B. (1965): ApJ 142, 531.
Heithausen A., Bensch F., Stutzki J., Falgarone E., Panis J.-F. (1997): A&A, submitted.
Herbertz R., Ungerechts H. & Winnewisser G. (1991): A&A 249, 483.
Lada E.A., Evans N.J. II & Falgarone E. (1997): ApJ in press.
Joulain K., Falgarone E., Pineau des Forêts G. & Flower D. (1997) A&A submitted
Larson R. (1981): MNRAS 194 809.
Lemme C., Walmsley C.M., Wilson T.L. & Muders D. (1995): A&A, 302, 509.
Lis D.C., Pety J., Phillips T.G. & Falgarone E. (1996): ApJ 463, 623.
Loren R.B. (1989): ApJ 338, 902.
Lucas R. & Liszt H. (1996): A&A, 307, 237
Martin H.M., Sanders D.B. & Hills R. (1984): MNRAS 208 35.
May J., Alvarez H. & Bronfman L. (1997): A&A in press.
Penzias A.A., Solomon P.M., Jefferts K.B. & Wilson R.W. (1973): ApJL 174, L43
Pérault M., Falgarone E., Puget J.-L. (1985):A&A 152, 371.
Porter D.H., Pouquet A. & Woodward P.R. (1994): Phys. Fluids, 6, 2133.
She Z.S., Jackson E., Orszag S.A. (1990): Nature, 344, 226.
Solomon P.M., Rivolo A.R., Barrett J. & Yahil A. (1987) ApJ, 319, 730.
Tabeling P., Zocchi G., Belin F., Maurer J. & Willaime H. (1996): Phys. Rev. E, 53, 1613
Tauber J., Goldsmith P.F. & Dickman R.L.(1991): ApJ 375, 635.
Vincent A. & Meneguzzi M., (1991): J. Fluid Mech., 225, 1.
Wang Y., Evans N.J. II, Zhou S. & Clemens D.P. (1995): ApJ 454, 217.
Ward-Thompson D., Scott P.F., Hills R.E. & André P. (1994): MNRAS, 268, 276.
Williams J.P., de Geus E.J. & Blitz L. (1994): ApJ 428, 693.

ORFEUS-I Observations of Molecular Hydrogen in the Galactic Disk

W. Van Dyke Dixon, Mark Hurwitz, and Stuart Bowyer

Space Sciences Laboratory and Center for EUV Astrophysics,
University of California, Berkeley, California 94720-5030, U.S.A.

Abstract. We present measurements of interstellar H_2 absorption lines in the continuum spectra of seven early-type stars in the Galactic disk at distances between 1 and 4 kpc. The spectra, obtained with the Berkeley EUV/FUV spectrometer on the ORFEUS telescope in 1993 September, have a resolution of 3000 and statistical signal-to-noise ratios between 20 and 80. We determine column densities for each observed rotational level and derive mean excitation temperatures and proton density limits for the H_2 clouds along each line of sight. The gross properties of the H_2-bearing clouds (e.g., column density, spatial density, cloud size) are consistent with those derived from *Copernicus* observations, though our lines of sight are much longer, with lower average reddenings and neutral gas densities. We find that the molecular fraction of the neutral hydrogen remains ~ 0.1 out to distances of 4 kpc in the Galactic disk.

1 Introduction

The hydrogen molecule (H_2) plays a central role in a variety of processes that significantly influence the chemical and physical state of the interstellar medium (ISM). From observations by *Copernicus* and other spacecraft-borne observatories, a picture has emerged in which the bulk of interstellar H_2 lies in clouds with densities between ~ 10 and a few 1000 cm^{-3}, diameters less than a few tens of parsecs, and column densities $\gtrsim 10^{20}$ cm^{-2}, which allow the rapid formation of H_2 on dust grains and provide self-shielding against dissociating photons (Wilson and Walmsley 1989). In order to determine whether this model, developed from observations of relatively nearby stars ($d \lesssim 1$ kpc), holds in other parts of the Galactic disk, we observed seven disk stars ($|z| < 300$ pc) at distances out to 3.9 kpc using the Berkeley spectrometer on the ORFEUS[1] telescope in 1993 September.

[1] Based on the development and utilization of ORFEUS (Orbiting and Retrievable Far and Extreme Ultraviolet Spectrometers), a collaboration of the Astronomical Institute of the University of Tübingen, the Space Astrophysics Group of the University of California at Berkeley, and the Landessternwarte Heidelberg.

Table 1. Target Summary

HD	l	b	Sp. Type	$E(B-V)$	d (pc)	$v \sin i$ (km s^{-1})	T (s)	S/N[a]	Ref.
41161	165.0	+12.9	O8 V	0.20	1253	300	814	45.2	1,2
54911	229.0	−3.1	B1 III	0.14	1893	100:	2462	78.0	1,3
93129a	287.4	−0.6	O3 If*	0.54	3470	120	3045	36.2	4,5
94493	289.0	−1.2	B0.5 Iab/Ib	0.20	3327	145	1528	34.9	1,3
99857	294.8	−4.9	B1 Ib	0.33	3058	180	1383	22.3	1,3
99890	291.8	+4.4	B0.5 V:	0.24	3070	180	2532	39.7	1,3
104705	297.5	−0.3	B0 III/IV	0.26	3898	215	1788	36.3	1,3

[a]Statistical signal-to-noise ratio in a 0.2 Å bin averaged over the 1045–1060 Å band.
References: (1) Fruscione et al. 1994; (2) Jenkins 1978; (3) Savage and Massa 1987; (4) Walborn 1973; (5) Gies 1987.

2 Observations and Data Reduction

With an effective area of about 4 cm^2 and a resolution $\lambda/\Delta\lambda = 3000$ between 390 and 1170 Å, the Berkeley spectrometer is ideally suited for absorption-line studies of bright, far-UV sources. Far-UV spectra of seven disk stars (Table 1) were obtained on the ORFEUS-I mission. We model their interstellar absorption features using an ISM line-fitting package written by M. Hurwitz and V. Saba. Given the column density, Doppler broadening parameter, and relative velocity of a given species, the program computes a Voigt profile for each line, convolves the lines with a Gaussian to the instrument resolution, and uses the result as a transmission function by which to scale the model continuum. Where available, we use the Doppler parameters, relative velocities, and relative column densities of the principal Na I absorption components (Sembach et al. 1993; Walborn 1982). Otherwise, we assume a single absorption component with a Doppler parameter $b = 5$ km s^{-1}. We adopt 0.31 Å as the FWHM of the instrument point-spread function. Reference spectra selected from the *Copernicus* Spectral Atlas (Snow and Jenkins 1977) provide an estimate of the stellar continuum. Column densities derived for the $J'' = 0$ to 5 rotational levels of H_2 are presented in Table 2. We estimate uncertainties of 0.2 dex for $J'' = 0$ to 3 and 0.5 dex for $J'' \geq 4$.

3 Analysis

The mean excitation temperature of the clouds along each line of sight can be derived from the column densities $N(0)$ and $N(1)$ using the relation

$$\frac{N(1)}{N(0)} = \frac{g_1}{g_0} \exp\left(\frac{-E_{01}}{kT_{01}}\right) = 9 \exp\left(\frac{-170 \text{ K}}{T_{01}}\right) \tag{1}$$

Table 2. Column Densities

HD	$N(\text{H I})^{a}$	$N(H_2)$	$N(0)$	$N(1)$	$N(2)$	$N(3)$	$N(4)$	$N(5)$	$T_{01}{}^{b}$ (K)	n^{c} (cm^{-3})
41161	21.01	20.0	19.7	19.7	17.6	17.6	15.0	14.2	77	16
54911	21.13	19.6	19.3	19.3	17.4	16.9	14.5	13.9	77	8
93129a	21.40	20.1	19.7	19.9	17.2	16.9	15.9	16.1	98	6
94493	21.11	20.1	19.7	19.8	17.7	16.2	15.0	15.1	86	13
99857	21.31	20.2	19.8	20.0	18.5	18.2	16.2	15.4	97	9
99890	20.93	19.6	19.2	19.3	17.7	17.6	16.0	14.7	86	11
104705	21.11	20.0	19.7	19.7	17.5	16.6	14.6	...	77	13

Note.—All column densities are given as logarithms. Units are cm^{-2}.
[a] $N(\text{H I})$ from the compilation of Fruscione et al. 1994, except for HD 93129a, which is from Taresch et al. 1997.
[b] Mean H_2 excitation temperature derived from $N(1)/N(0)$; see Sec. 3.
[c] Lower limit to cloud proton (H I + $2H_2$) density; see Sec. 3.

(Shull and Beckwith 1982). We can estimate the proton (H I + $2H_2$) volume density n by assuming

$$\left(\frac{n}{50\ \text{cm}^{-3}}\right)^2 \simeq \left[\frac{N(H_2)}{10^{19}\ \text{cm}^{-2}}\right]\left(\frac{T}{80\ \text{K}}\right)^{-1}\left[\frac{N(\text{H I})}{10^{20}\ \text{cm}^{-2}}\right]^{-2} \qquad (2)$$

(Reach et al. 1994). The resulting densities are lower limits, because an unknown fraction of $N(\text{H I})$ lies outside of the H_2-bearing clouds. The temperatures and density limits presented in Table 2 are consistent with the values derived from *Copernicus* observations (Savage et al. 1977; Jura 1975).

4 Conclusions

Our data reflect the trends among $N(H_2)$, $E(B-V)$, the molecular fraction $f = 2N(H_2)/[N(\text{H I})+2N(H_2)]$, and the total hydrogen column $N(\text{H I}+H_2) = N(\text{H I})+2N(H_2)$ established by Savage et al. (1977; see their Figs. 4–6), indicating a common mechanism for H_2 production. From our column densities $N(\text{H}) = 1-3\times10^{21}$ cm^{-2} and proton densities $n \gtrsim 10$ cm^{-3}, we estimate path lengths (cloud diameters) on the order of tens of parsecs, consistent with standard models. The molecular fraction f remains ~ 0.1 out to distances of 4 kpc in the Galactic disk (Fig. 1). These results are not entirely expected, as our observations probe sight lines of exceptionally low average density: $N(\text{H I} + H_2)\ /\ d$ is only about 0.2 cm^{-3}, and the mean reddening $\Sigma E(B-V)/\Sigma r = 0.11$ mag kpc^{-1}. Our results are consistent with a model in which a significant fraction of the neutral ISM is confined to the small, low-density molecular clouds, even along low-density lines of sight.

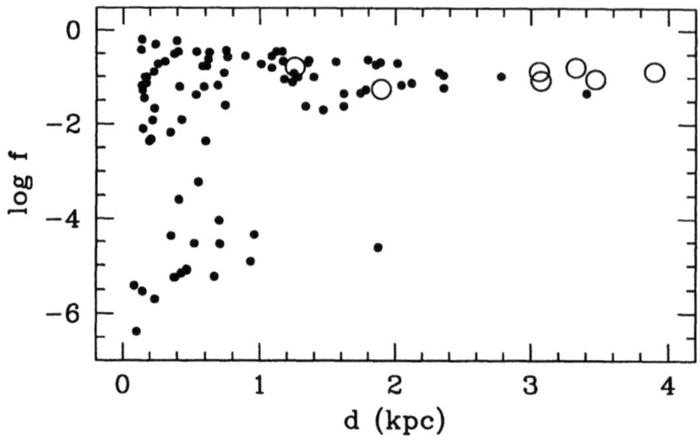

Fig. 1. The molecular fraction f plotted against stellar distance. Solid circles represent *Copernicus* observations (Savage et al. 1977), while the open circles are from this work. We find that $f \sim 0.1$ out to large distances in the disk.

Acknowledgements. This research has made use of the NASA ADS Abstract Service and the Catalogue Service of the CDS, Strasbourg, France. We thank J. Black for providing H_2 transition data in electronic format and acknowledge the many NASA and DARA personnel who helped make the ORFEUS-I mission successful. This work is supported by NASA grant NAG5-696.

References

Fruscione A., Hawkins I., Jelinsky P., Wiercigroch A. (1994): ApJS, 94, 127
Gies D. R. (1987): ApJS, 64, 545
Jenkins E. B. (1978): ApJ, 219, 845
Jura M. (1975): ApJ, 197, 581
Reach W. T., Koo B.-C., Heiles C. (1994): ApJ, 429, 672
Savage B. D., Bohlin R. C., Drake J. F., Budich W. (1977): ApJ, 216, 291
Savage B. D., Massa D. (1987): ApJ, 314, 380
Sembach K. R., Danks A. C., Savage B. D. (1993): A&AS, 100, 107
Shull J. M., Beckwith S. (1982): ARA&A, 20, 163
Snow, Jr. T. P., Jenkins E. B. (1977): ApJS, 33, 269
Taresch G. et al. (1997): A&A, submitted
Walborn N. R. (1973): ApJ, 179, 517
Walborn N. R. (1982): ApJS, 48, 145
Wilson T. L., Walmsley C. M. (1989): A&AR, 1, 141

Physical Association Between the Southern Coalsack and Chamaeleon-Musca Dark Clouds

J.B. Wagner Corradi[1,2], Gabriel A.P. Franco[1], and Jens Knude[2]

[1] Departamento de Física – ICEx – UFMG, Caixa Postal 702, 30 161-970, Belo Horizonte – MG, Brazil

[2] Niels Bohr Institute for Astronomy, Physics and Geophysics, Juliane Maries Vej 30, DK 2100, Copenhagen Ø, Denmark

Abstract. To investigate a possible physical association between the Southern Coalsack and the Chamaeleon-Musca (SCCM) dark clouds we have obtained *uvbyβ* photometry for 1017 stars covering the connecting area: $308° \geq l \geq 294°$ and $-20° \leq b \leq 5°$. Analysis of the various colour excess $E(b-y)$ vs. distance diagrams has indicated the presence of a local low absorption volume limited at 150 ± 30 pc from the Sun by an extended interstellar dust sheet-like feature, that is followed by a region where almost no additional reddening is measured for the next 350 pc. Combined with other data on the local ISM the existence of the dust sheet at an identical distance of the SCCM dark clouds have suggested that these clouds could be higher density regions associated to the diffuse lane of dust of the Local-Loop I bubbles' interface.

1 Introduction

A comparison of the colour excess $E(b-y)$ vs. distance diagrams for the Chamaeleon-Musca (Franco 1991) and the Southern Coalsack (Franco 1989) dark clouds shows great similarities. The jump of the colour excess to higher values occurs approximately at the same distance, and the observed minimum value of this rise is almost the same: $\Delta E(b-y) \approx 0\overset{\mathrm{m}}{.}100$. Although the clouds are apart by more than 15° these facts suggest that they might be dense condensations embedded in an extended interstellar structure, which may be related to the interface of the Local and Loop I Bubbles.

2 Reddening Distribution

In order to sustain the hypothesis of a physical association of the Southern Coalsack and the Chamaeleon-Musca dark clouds (SCCM), a new photometric programme was carried out. We used the Strömgren Automatic Telescope in La Silla (Chile) to obtain *uvbyβ* photometry for all SAO stars earlier than GO covering the connecting area: $308° \geq l \geq 294°$ and $-20° \leq b \leq 5°$. The programme resulted in accurately calibrated uvbyβ data for 1017 stars (Corradi & Franco 1995). The data were complemented by the *uvbyβ* measurements of 213 stars of a previous investigation towards the geometrical center of the Chamaeleon-Musca clouds (Franco 1992). Of the final sample, only 510 stars fulfilled the imposed selection criteria to calculate $E(b-y)$ and distance. Their (l,b) positions are given in Fig. 1.

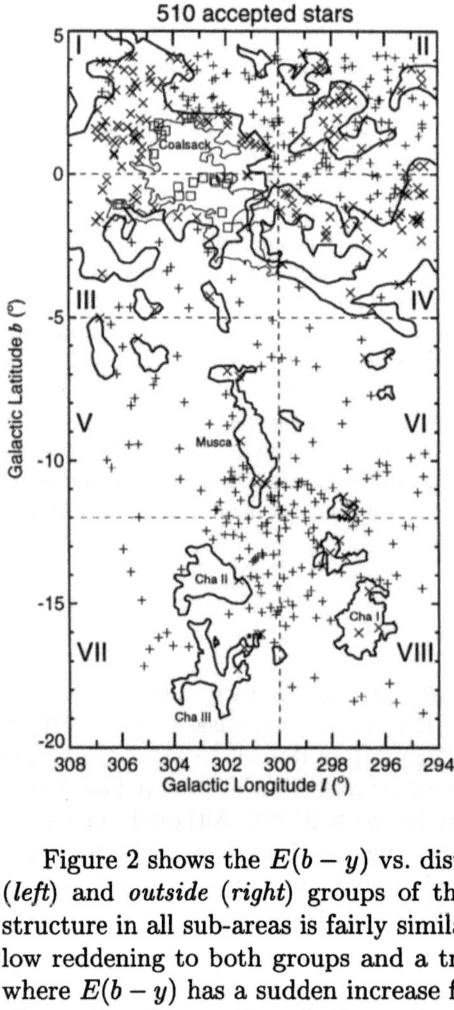

Fig. 1. Distribution of the 510 accepted stars. The thick contours are the lowest opacity level of the photographic Dark Clouds Catalogue (DCC) compiled by Feitzinger & Stüwe (1984). The thinner contour is the outer 2 K km s^{-1} velocity integrated CO emission for the Southern Coalsack (Nyman et al. 1989). Stars with line-of-sight *inside* and *outside* the DCC contours are represented by the (×) and (+) signs, respectively. Stars *inside* the CO contour are indicated by the (□) signs. The horizontal and vertical dashed lines delineate the eight sub-areas, indicated by the roman numbers, used to discuss the reddening distribution

Figure 2 shows the $E(b-y)$ vs. distance diagrams for stars in the *inside* (*left*) and *outside* (*right*) groups of the eight sub-areas ($d \leq 1$ kpc). The structure in all sub-areas is fairly similar, i.e., a foreground region with very low reddening to both groups and a transition region around 150 ± 30 pc, where $E(b-y)$ has a sudden increase forming a lower envelope of minimum absorption. These effects indicate that an extended interstellar structure is permeating the whole area.

As the complete range of $E(b-y)$ from $\approx 0\overset{m}{.}100$ to $0\overset{m}{.}300$, shown by the stars in a very narrow distance slot centered on 150 pc, does not change for another 350 pc the dust feature may be followed by another low reddening volume. Note that stars *inside* the CO contour follow the tendency of the others *inside* the DCC contours. The lower envelope of colour excesses does not show clear dependence with the galactic longitude, but may increase with the galactic latitude in the sense that $[E(b-y)_{\min}, b] = [0\overset{m}{.}050; 0°] \rightarrow$

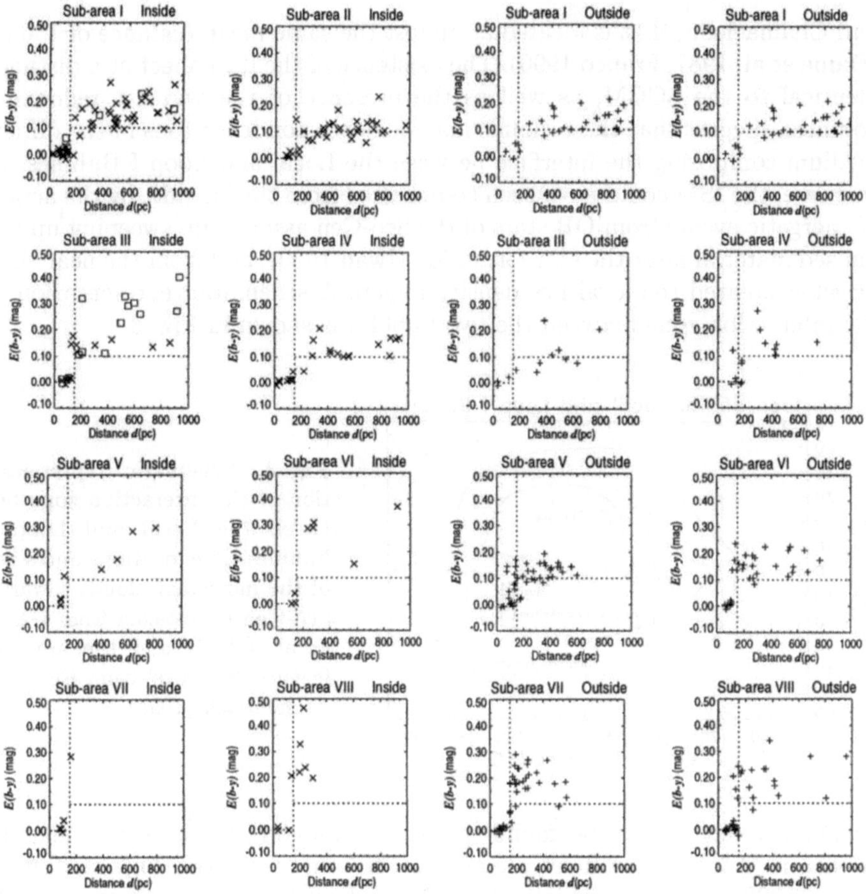

Fig. 2. $E(b-y)$ vs. d diagrams for the eight sub-areas. (*left*) to the *inside* group. (*right*) to the *outside* group. See the details in the text

$[0^{\rm m}.100; -8^\circ] \rightarrow [0^{\rm m}.150; -15^\circ]$. The quoted increase might suggest either an inhomogeneous sheet-like structure, that could be roughly perpendicular to the galactic plane, or a sheet-like structure of same column density, curved away from the Sun (Corradi et al. 1997).

3 Connection to the Local-Loop I Bubbles' Interface

Egger & Aschenbach have recently identified on the soft X-ray data a ring-like feature that may have formed at the region of significant compression between the Local and Loop I bubbles. In addition to this result, the molecular gas from $l \approx 290°$ to 360 deg, between $-25° \leq b \leq 25°$, seems to define a large complex of clouds, comprising ρ Oph, R Cra, G317-4, Lupus, Coalsack, Musca and Chamaeleon, that is located at almost the same mean distance of 150 pc (Dame et al. 1987, Franco 1990). The existence of the dust sheet at a distance identical to the SCCM, as well as the presence of the two low reddening volumes, suggest that these clouds may be dense condensations in the diffuse medium composing the interface between the Local and Loop I Bubbles. In this case, the ISM could have been compressed from the far side, by the action of energetic events from OB stars of the Sco-Cen association sweeping up the unused material after the star formation (Iwan 1980), and from the near side, by what created the local low density region. A schematic representation of the interaction zone between the two bubbles is shown in Fig. 3.

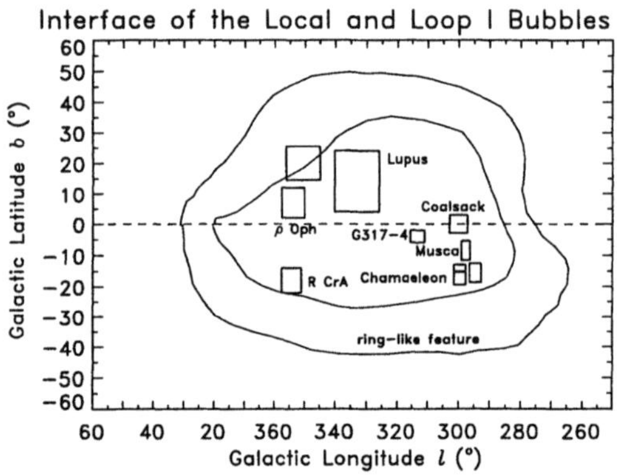

Fig. 3. Schematic representation of the interaction zone between the Local and Loop I bubbles. The locations and sizes of the molecular clouds towards Sco-Cen were taken from Dame et al. (1987) and the annular feature contours' from Egger & Aschenbach (1995)

Further details can be found in the full version of this work that will appear in A&A (Corradi et al. 1997). For our next work high-resolution ($R \approx 60\,000$) spectra of the interstellar NaI D lines for 72 B-type stars towards the SCCM clouds have been obtained. The data will allow us to distinguish the components of the proposed physical association, and combined with the known distance of the stars, to understand the gas kinematics of the interface.

References

Corradi W.J.B., Franco G.A.P. (1995): A&AS 112, 95
Corradi W.J.B., Franco G.A.P., Knude J. (1997): A&A in press
Dame T.M., Ungerechts H., Cohen R.S., et al. (1987): ApJ 322, 706
Egger R.J., Aschenbach B. (1995): A&A 294, L25
Feitzinger J.V., Stüwe J.A. (1984): A&AS 58, 365 (erratum: 63, 203)
Franco G.A.P. (1989): A&A 215, 119
Franco G.A.P. (1990): A&A 227, 499
Franco G.A.P. (1991): A&A 251, 581
Franco G.A.P. (1992): A&AS 93, 373
Iwan D. (1980): ApJ 169, 25
Nyman L.-Å., Bronfman L., Thaddeus P. (1989): A&A 216, 185

The Local Bubble, Local Fluff, and Heliosphere

Priscilla C. Frisch

University of Chicago, Dept. of Astronomy & Astrophysics, 5640 S. Ellis Ave., Chicago, IL 60637, USA

Abstract. The properties of the Local Bubble, Local Fluff complex of nearby interstellar clouds, and the heliosphere are mutually constrained by data and theory. Observations and models of the diffuse radiation field, interstellar ionization, pick-up ion and anomalous cosmic-ray populations, and interstellar dust link the physics of these regions. The differences between the one-asymmetric-superbubble and two-superbubble views of the Local Bubble are discussed.

1 Introduction

The Local Bubble, the Local Fluff complex of nearby interstellar clouds, and the heliosphere, are three astronomical phenomena with properties that can not be determined independently. The Local Bubble (LB) radiation field influences the ionization of both nearby interstellar gas and the heliosphere. The ionizations of nearby interstellar gas and pick-up ions in the heliosphere constrain this radiation field. The kinematics and properties of interstellar gas within and without the heliosphere can be compared to set the confinement pressure of the heliosphere and bow shock, and therefore the interstellar pressure. Interstellar grains observed within the solar system constrain the properties of dust in interstellar clouds. The morphology, properties and kinematics of nearby interstellar gas must be consistent with models of the origin of the soft X-ray background (SXRB).

The LB, Local Fluff and heliosphere make a science "pyramid", and the interrelation of the physical properties of these three regions is shown in Fig. 1. (The caption to Fig. 1 also summaries the abbreviations used in this paper.) In this context, the local interstellar matter (LISM) is the same as the "Local Fluff" complex of interstellar clouds (LFC) within about 35 pc of the Sun.

This conference, organized so beautifully by Dieter and his colleagues in Garching, is a timely sequel to two conferences in the last two decades on interstellar gas in the heliosphere, and on the local interstellar medium.[1] The initial linking of the Loop I superbubble, the LISM, and the neutral interstellar gas detected within the solar system was presented in a series of papers

[1] These two conferences were the meetings on "Interstellar Gas in Interplanetary Space, VI. MPAE Lindau Workshop", June 187–20, 1980, held in Lindau, Germany, and "Local Interstellar Medium, IAU Colloquium No. 81", held in Madison, Wisconsin June 4–6, 1984.

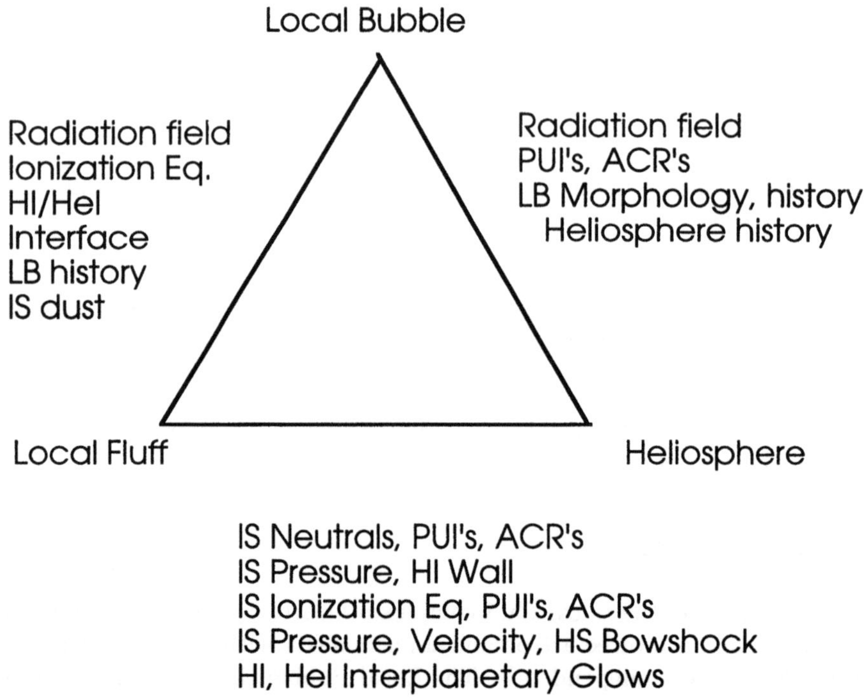

Fig. 1. This figure shows the links between the physics of the Local Bubble, Local Fluff, and Heliosphere. Abbreviations in the figure and text are: CSSS=cloud surrounding the solar system; LFC=Local Fluff complex; ISM=interstellar matter; LISM=local ISM; PUI=pick-up ions; IS=interstellar; IE=ionization equilibrium; ACR=anomalous cosmic rays; LB=Local Bubble; HS=heliosphere; SXRB=soft X-ray background; SCA=Scorpius-Centaurus Association.

culminating in Frisch (1981). We used the Copernicus satellite to conduct spectral observations of the interplanetary Lα glow caused by the resonance fluorescence of solar Lα emission off neutral interstellar gas which had penetrated the heliosphere[2] (Adams & Frisch 1977). I proposed these observations because I believed the cloud in front of α Oph had invaded the solar system, causing the interplanetary Lα glow. The star α Oph is located at a distance of 14 pc in the direction of the intense radio continuum source, the North Polar Spur[3]. My hunch evidently was right, and the velocities of the interplanetary glow emission feature projected to the direction of α Oph (–22.3 km s^{-1}), and Na^o and Ca$^+$ absorption features seen in α Oph (velocities –23.6 to –22.4 km s^{-1} for the reddest components, Welty et al. 1994, Crawford & Dunkin 1995) are nearly equal, although a possible velocity gradient

[2] This is also known as the "backscattered radiation".

[3] The North Polar Spur is the most intense section of the Loop I radiocontinuum superbubble feature.

is present in the LFC. At that time, the deceleration of interstellar gas at the heliopause was unknown (c.f. Lallement et al. 1993). The interstellar cloud fragment which surrounds the solar system is a member of the cloud complex seen in front of α Oph and other stars in the galactic center hemisphere of the sky. The enhanced abundances of interstellar refractory elements Ca^+, Fe^+ and Mn^+ seen in front of α Oph led me to also conclude that the Sun resided in the edge of the Loop I supernova remnant associated with the Scorpius-Centaurus stellar association. The sightline towards α Oph exhibits the strongest interstellar Ca^+ seen towards any nearby star. The need to reconcile the asymmetric distribution of interstellar gas within 30 pc of the Sun (e. g. Frisch 1996, Table 1) and the symmetry of the "bubble" inferred to explain the SXRB led to my suggestion that the Loop I supernova remnant had expanded asymmetrically into the low density interarm region surrounding the solar system. A different model forms the SXRB in an independent supernova explosion around the Sun (unrelated to Loop I) (e. g. Davelaar et al. 1980, Egger, this volume). These contrasting views are reviewed in Breitschwerdt et al. (1996). In this summary I will discuss:

- the heliosphere as a probe of the cloud surrounding the solar system (CSSS) and the Local Bubble (LB)
- the constraints LISM gas places on the LB by shadowing (or lack thereof)
- abundance and kinematical considerations, morphology and structure, and the historical heliosphere.
- My perspective of the LB.

One point needs to be emphasized strongly. There is an underlying question that must be answered, and that question is "What is the Local Bubble?" The region of space commonly referred to as the "Local Bubble" coincides with the interior of Gould's Belt and with the surrounding interarm region, therefore the answer to this question is not so obvious. Fig. 2 compares the distribution of interstellar clouds within 500 pc to the space motions of the Sun and CSSS. I adopt the view of defining the LB by its walls (Cox & Reynolds 1987)). Column densities $N(H^o)$ of $10^{19.8}$ cm^{-2} and 10^{20} cm^{-2} attenuate the B and C band radiation, respectively, by a factor of $\sim$3, giving a natural definition for the "walls" of the Local Bubble as the location where $N(H) \geq 10^{19.8}$ cm^{-2}, and the ISM opacity in the B-band exceeds unity.

2 The Heliosphere as a Probe of the LISM and LB

By way of background, I will summarize briefly the heliosphere (HS) structure and the properties of the CSSS. The overall structure of the heliosphere appears to be a two-shock structure, with an inner "termination shock" where the solar wind goes from supersonic to subsonic, the heliopause which is the stagnation surface between the solar wind and interstellar plasma components, and a bow shock surrounding the heliosphere. The CSSS has properties T$\sim$7,000 K, n(H^o)$\sim$0.2 cm^{-3} and n(p^+)=n(e^-)$\sim$0.1 cm^{-3}. A magnetic field is present, of unknown strength, but likely to be weak (B$\sim$1.5 μG, Frisch 1995, Gloeckler et al. 1997). The relative velocity between the Sun and CSSS is $\sim$26 km s^{-1}, approaching from the direction l$\sim$5^o, b$\sim$+16^o in the rest frame of the Sun. In the rest frame of the Local Standard of Rest, this corresponds to a cloud moving towards us at 19–20 km s^{-1} from the direction l$\sim$335^o, b$\sim$–2°. [4] Neutral interstellar atoms cross the heliosphere into the solar system, and turn into the pick-up ion population after ionization (by charge exchange with the solar wind and photoionization) and capture by the solar wind. The pick-up ions are accelerated (perhaps at the termination shock of the solar wind) and create the anomalous cosmic ray population, which propagates throughout the HS.

Heliosphere observations give direct data on the physical properties of the interstellar cloud which surrounds the solar system, the LFC, and radiation field within the LB. Listing the ways in which observations within the heliosphere serve as useful probes of the LISM and LB:

- The CSSS feeds interstellar neutrals into the HS. Thus, observations of pick-up ions and anomalous cosmic rays within the heliosphere provide direct information on the ratios of interstellar neutrals in the CSSS. The pick-up ion (PUI) and anomalous cosmic ray (ACR) data can thus be used to constrain interstellar ionization. Because the ionization levels are sensitive to the radiation field, in turn the PUI and ACR data prove to be a probe of the radiation field within the LB (see Slavin and Frisch, this volume, and Frisch and Slavin 1996). The elements He, Ne, H, O, N, C, and Ar have been observed in either the pick-up ion or anomalous cosmic ray populations. Since C^o is a subordinate ionization state of carbon in the LISM, the C/O ratio in the PUI and ACR populations yield an estimate of the ionization of the CSSS (Frisch 1994).
- Observations of Lα and 584 A backscattered radiation from interstellar H^o and He^o inside of the solar system, respectively, yield information on the temperature, velocity and density of the interstellar cloud which feeds neutrals into the solar system (e.g. Quemerais et al. 1996, Flynn et al. 1997, Adams & Frisch 1977, Scherer et al. 1997).
- Observations of Lα absorption from the pile-up of interstellar H^o outside of the heliopause stagnation surface, due to the charge-exchange coupling of interstellar H^o and protons, constrain the fractional ionization of the surrounding cloud and the Mach number of the bow shock of the solar system. The absorption from this pile-up must be included in Lα profile fitting for absorption lines in nearby stars, for good D/H ratios. In the α Cen direction, an outer heliosphere model with Mach number=0.9,

[4] Note that since the Sun is immersed in this flowing cloud, and the Sun itself moves through space, the solar motion must be removed from the observed upwind direction in order to get the true space velocity of the CSSS in the Local Standard of Rest.

$n(H^o)$=0.14 cm^{-3}, $n(e^-)$=0.1 cm^{-3}, T=7600 K, V=26 km s^{-1} yields the best fit, to within the limited parameter range considered, to observations of Lα absorption towards α Cen (Gayley et al. 1997).
- Interstellar dust observed by Ulysses and Galileo constrain the mass distribution of interstellar dust grains in the CSSS, with a mean mass of 3 x 10^{-13} gr corresponding to a grain radius of 0.3 cm^{-3} for silicate density 2.5 gr cm^{-3} (Baguhl et al. 1996).

3 LISM Constraints on the Local Bubble and Heliosphere

Models of the conductive interface between the LFC and LB plasmas, when compared to PUI, ACR, and interstellar absorption line data, give direct information on the LB radiation field, LFC conductivity, magnetic field, and other physical quantities (Slavin 1989, Slavin and Frisch, this volume). The morphology, abundance patterns, ionization, density and temperature of the cloud fragments which constitute the Local Fluff cloud complex give direct information on the physical history of the LISM. For instance, the enhanced abundance patterns seen in the LFC, in comparison to abundances seen in cold interstellar clouds, are interpreted as indicating that nearby interstellar material has been processed through a shock front (Frisch 1996). The velocity of the LISM gas indicates an outflow of gas from the Scorpius-Centaurus Association (SCA), suggesting, together with the anomalous abundances, that the LFC is part of the superbubble associated with star formation in this region. The flow of interstellar gas is seen in thirty-six interstellar Ca^+ absorption components seen in 17 nearby stars yield a relatively coherent flow velocity of –0.1 ± 2.2 km s^{-1} in a rest frame defined by the heliocentric velocity vector l=6°.2, b=+11°.7, V=–26.8 km s^{-1} (Frisch 1997a).

The asymmetry of the LISM gas, within about 35 pc, belies the symmetry inferred for the local component of the SXRB. Table 1 illustrates this well known property. This asymmetry, and the fact that clouds in the LFC flow outwards from the SCA, must be explained by any theory on the origin of the Local Bubble. Note that because of this asymmetry, most of the mass within 35 pc is in the galactic center hemisphere of the sky.

One important question: Can the LFC shadow the 0.25 keV SXRB and thereby give us information on the spatial distribution of the emitting plasma? Typical column densities to the "edge" of the LFC in several directions are given in Table 1, where the edge distance is defined as $N(H^o)/0.20$. The column densities are estimates in some cases, based on $N(Ca^+)/N(H^o)=10^8$, and an average cloud space density of 0.20 cm^{-3} is used. From this, it can be seen that most of the interstellar gas in the LFC cloud complex has column densities too low to provide significant shadowing of the SXRB – i. e. well below the $10^{19.8}$ cm^{-2} value needed to shadow the B-band. The exceptions are the α Oph (l=36°, b=+23°, d=14 pc) and HD149499B (l=330°,

Table 1. Column Densities through Local Fluff

l,b(deg)	log N(H)[a]	d(pc)[b]	l,b(deg)	log N(H)[a]	d(pc)[b]
28,+15	18.86	10	240,–11	17.9	1.1
101,65	17.85	1.0	289,–54	18.90	1.1
163,+5	18.24	2.7	295,+46	18.53	4.8
214,+13	17.88	1.1	330,–7	19.00	14
214,–60	18.70	0.7	350,–53	18.49	4.4

[a] Log N(H)=log N(H^o+H^+). See Frisch & Welty 1997, Frisch 1997b, Napiwotzki et al. 1995 for original references. [b]d(pc) is the distance to the LFC edge, which is calculated for n(H^o)=0.20 cm^{-3}.

b=–7°, d=37 pc) sightlines, with column densities logN(H^o)$\sim$19.57, $\sim$19.00 cm^{-2}, respectively. The α Oph column density is not well known. In these directions 10% – 50% percent attenuation of the B-band emission may be expected. These stars are in the direction of the dominant shadow (due to H^o filaments composing the Loop I shell, Cleary et al. 1979) in front of the X-ray emission from Loop I. The star HD149499B is located within $\sim$10° of the LFC flow direction in the Local Standard of Rest. Hardening of the SXRB is expected, in agreement with the observations of a dipole gradient pointed towards l=168°.7, b=11°.2 (Snowden et al. 1990). The plasma emitting in the 0.10–0.18 keV region in the upwind hemisphere must be in front of the nearest 6×10^{19} cm^{-2} hydrogen column density. This plasma could, therefore, be behind the upwind cloud with no significant shadowing. Alternatively, the prevalence of very small structure in the ISM allows the possibility that the plasma and cooler clouds forming the LFC are interspersed.

4 Journey of the Sun Through Space

Our improved understanding of the morphology and kinematics of nearby ISM in comparison to the space trajectory of the Sun permit a deeper understanding of the historical changes in the galactic environment of the Sun, and the effect those changes have on the heliosphere. From Fig. 2, we see that within the past $\sim$100,000–200,000 years the Sun emerged from the void of the surrounding interarm region and entered the LF complex of clouds. Within the past 10,000 years, and perhaps within the past 2,000 years, the Sun appears to have entered the cloud in which it is currently situated (Frisch 1997a). The physical properties of these clouds constrain the configuration and properties of the heliosphere. It is notable that the space velocities of

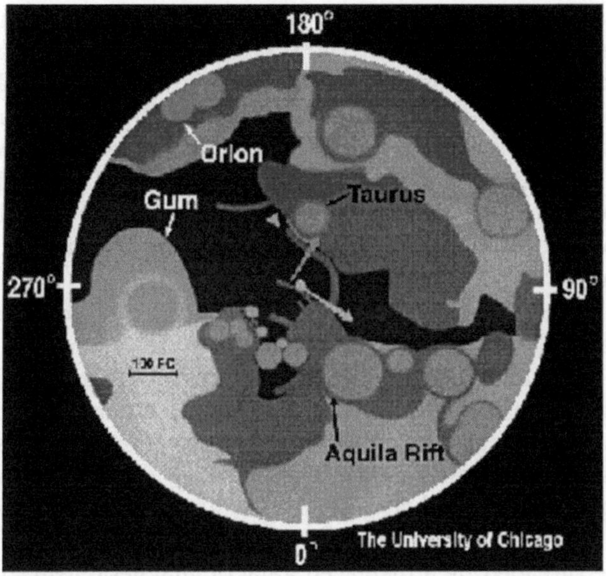

Fig. 2. The distribution of interstellar molecular clouds (traced by the CO 1->0 115 GHz rotational transition,Dame and Thaddeus 1985) and diffuse gas (traced by E(B-V) color excess due to the reddening of starlight by interstellar dust Lucke 1978) within 500 pc of the Sun are shown. The round circles are molecular clouds, and the shaded material is diffuse gas. Interstellar matter is shown projected onto the galactic plane, and the plot is labeled with galactic longitudes. The distribution of nearby interstellar matter is associated with the local galactic feature known as "Gould's Belt", which is tilted by about 15–20° with respect to the galactic plane. Thus, the ISM towards Orion is over 15° below the plane, while the Scorpius-Centaurus material (longitudes 300°–0°) is about 15–20° above the plane. Also illustrated are the space motions of the Sun and local interstellar gas, which are nearly perpendicular in the LSR velocity frame. The three asterisks are three subgroups of the Scorpius-Centaurus Association. The three-sided star is the Geminga Pulsar. The arc towards Orion represents the Orion's Cloak supernova remnant shell. The other arcs are illustrative of superbubble shells from star formation in the Scorpius-Centaurus Association subgroups. The smallest (i.e. greatest curvature) shell feature represents the Loop I supernova remnant.

the Sun and CSSS are nearly perpendicular, so that these ages are highly sensitive to uncertain assumptions about cloud morphology and kinematics.

5 Origin of the Local Bubble–One vs. Two Bubbles

What is the Local Bubble? There is no agreement on the answer to this question. Ask an X-ray astronomer and they will probably tell you that it is the physical location of the 10^6 K plasma from a recent local supernova explosion that emits radiation in the 0.1–0.18 KeV B-Band and 0.25 keV C-Band. Ask a radio astronomer, and they may be confused because the Sun is located in an interarm region between the Orion spiral arm and the Local Arm, which is a short (~1 kpc long) spur projecting from the Orion Arm. Spiral arms are traced out by strings of molecular clouds and star-forming regions, and Fig. 2 illustrates the nearby molecular clouds as defined by observation of CO. Interarm regions are regions with very low densities of interstellar matter.

Historically, the LB concept is a mixture of the view that a separate local supernova explosion formed the SXRB (e. g. Davelaar et al. 1980, Cox & Reynolds 1987) and the view that the superbubble formed by the successive epochs of star formation in the Scorpius-Centaurus Association have expanded asymmetrically into the low density interarm region surrounding the Sun (Frisch 1981, Frisch 1995). The Davelaar et al. view has been updated by Egger and Aschenbach (1995, EA, also Egger, this volume), who attribute the conventional H^o filaments which bound the Loop I supershell, and which are threaded by a displaced galactic magnetic field, to a collision between the supernova around the solar system and the Loop I superbubble.

The salient properties of the asymmetric superbubble model for the LB are (see Frisch, 1995, for more details):

- The three epochs of star formation (4–15 Myrs ago) in the SCA each created superbubble structures, with the later structures evolving within the cavities formed by the earlier events. In the asymmetric superbubble view these shells will have expanded asymmetrically into the low density interarm region surrounding the Sun. The ISM surrounding the superbubbles was initially asymmetric because of the local arm–interarm configuration, and the Aquila Rift molecular cloud between $l \sim 20^o$ and $l \sim 40^o$.
- The Loop I supernova remnant, ~250,000 years old (Borken & Iwan 1977), is confined because it expanded into, and ablated material from, the Aquila Rift molecular cloud. This can be seen in the galactic interval $l=17^o$–27°, $b=0^o$–10° when the configuration of the Aquila Rift CO cloud (Dame and Thaddeus 1985) is compared with the narrow neck region of the North Polar Spur (Sofue and Reich 1979). The Aquila Rift molecular cloud is the node region where all of the superbubble shells from the three epochs of star formation in the SCA, as well as the most recent supernova explosion creating the North Polar Spur, converged after plowing into the molecular gas and decelerating. I propose here that the soft X-ray emission associated with the North Polar Spur itself occurs in a position consistent with the formation of a galactic fountain with a footprint in the disrupted Aquila Rift molecular cloud.
- The characteristic filamentary structure seen in the H^o gas, which defines the annular ring attributed to the merged bubbles, appears to be due to confinement by the ~5 μG magnetic field embedded in the filaments

(Cleary et al. 1979) and does not require the explanation of merging bubbles. The superbubble shell boundaries portrayed in Fig. 2 represent the 21 cm filaments which are seen at negative galactic latitudes between l$\sim$40^o and l$\sim$180°, and which are threaded by the ambient magnetic field as is seen by Zeeman splitting and stellar polarization measurements.

- The LFC is part of the expanding superbubble shell from the formation of the Upper Scorpius subgroup 4–5 Myrs ago. Towards larger galactic latitudes, l=350° to l=40°, corresponding to the eastern boundary of Loop I, the expansion of the shells was impeded by collision with the Aquila Rift molecular gas. At lower longitudes, or the western boundary of the Loop I superbubble, expansion proceeded more freely and earlier shells expanded past the solar location, reheating the low density gas in the anti-center interarm region.
- The Loop I supernova remnant has expanded inside of the Upper Scorpius subgroup superbubble, and encountered denser ambient gas than did the previous superbubble shells because of the proximity to the denser Aquila Cloud. The upwind direction in the Local Standard of Rest, l$\sim$335^o, b$\sim$–2°, represents a direction offset from the center of the Loop I supernova remnant by about 20°.

The main source of disagreement between the asymmetric-superbubble versus symmetric-superbubble views is the requirement that in the latter scenario, a separate supernova explosion in the anti-center hemisphere explains the SXRB. In the former view, the source of the SXRB is the ambient plasma inside of the superbubble shells formed by the first star formation epochs in the SCA reheated by unspecified shocks and energetic radiation that would propagate freely through the very low density material. In the two-bubble scenario, a coherent dense wall of neutral hydrogen at 40–70 pc, with N(H^o)$\sim$10^{20} cm^{-2} is postulated to separate the two bubbles, but there is no observational evidence for such a wall within 50 pc covering the central regions of the Loop I bubble. Counter-examples are easily found. For example, comparing the stars β Cen (l=312^o, b=1^o, d=161 pc, log N(H^o)$\leq$19.5 cm^{-2}, Fruscione et al. 1994), and HD 149499 B (l=330°, b=–7°, d=37 pc, log N(H^o)=19.0 cm^{-2}) show that over 30% of the nearby gas is associated with the LFC. The absence of X-ray emission from the Loop I interior may be due to the fact this is an evacuated cavity. An adequate model for the SXRB emission mechanism is required to establish its origin (Sanders, this volume) and help resolve these differences.

The two scenarios agree on the distance of the H^o 21 cm filaments, which are established by reddening measurements. The LFC is in the interior of the ring, but it has distance <40 pc. I believe the LFC is gas which is part of an expanding superbubble shell from the SCA.

Acknowledgements. This research has been supported by NASA grant NAGW-5061.

References

Adams, T. F., & Frisch, P. C. 1977, ApJ 212, 300
Baguhl, M., Grun, E., and Landgraf, M. 1996, Space Science Reviews 78, 165
Borken, R. J., & Iwan, D. C. 1977, ApJ, 218, 511
Cleary, M. N., Heiles, C., & Haslam, C. G. T. 1979, AAS, 36, 95.
Breitschwerdt, D., Egger, R., Freyberg, M. J., Frisch, P. C., & Vallerga, J. V. 1996, Space Science Reviews 78, 183
Cox, D. P., & Reynolds, R. J. 1987, ARAA 25, 303
Crawford, I. A. & Dunkin, S. K. 1995, MNRAS 273, 219
Dame, T. M., Thaddeus, P. 1985, ApJ 297, 751
Davelaar, J., Bleeker, J. A. M., & Deerenberg, A. J. M. 1980, AA 92, 231
Egger, R. J. & Aschenbach, B. 1995, AA 294, L25.
Flynn, B., J., Vallerga, J., Dalaudier, F., and Gladstone, G. R. 1997, submitted to J. Geophys. Res.
Frisch, P. C. 1981, Nature 293, 377
Frisch, P. C. 1994, Science 265, 1423
Frisch, P. C. 1995, Space Science Reviews 72, 499
Frisch, P. C. 1996, Space Science Reviews 78, 213
Frisch, P. C. 1997a, submitted to Science
Frisch, P. C. 1997b, Proceedings of ACE Workshop, in press
Frisch, P. C., & Slavin, J. D. 1996, Space Science Reviews 78, 223
Frisch, P. C. & Welty, D. E. 1997, in preparation
Fruscione, A., Hawkins, I., Jelinsky, P., Wiercigroch, A. 1994, ApJS, 94, 127.
Gayley, K. G., Zank, G. P., Pauls, H. L., Frisch, P. C., & Welty, D. E. 1997, ApJ, 487, 259
Gloeckler, G., Fisk, L. A., and Geiss, J. 1997, Nature, 386, 374.
Lallement, R., Bertaux, J.-L., & Clarke, J. T. 1993, Science 260, 1095
Lucke, P. B. 1978, AA, 64, 367.
Napiwotzki, R., Hurwitz, M., Jordan, S., Bowyer, S., Koester, D., Weidemann, V., Lampton, M., & Edelstein, J. 1995, AA 300, 300
Scherer, H., Fahr, H. J., and Clarke, J. T. 1997, AA, in press
Snowden, S. L., Schmitt, J. H., Edwards, B. C. 1990, ApJ 364, 118
Quemerais, E., Bertaux, J.-L., Sandel, B. R., & Lallement, R. 1996, AA 308, 279
Slavin, J. D. 1989, ApJ 392, 718
Sofue, Y., Reich, W. 1979, AASS 38, 251
Welty, D. E., Hobbs, L. M., & Kulkarni, V. P. (1994): ApJ 436, 152

Measurements of the Size and Pressure of the Local Bubble Towards Eridanus

D.N. Burrows[1] and Z. Guo[2]

[1] Department of Astronomy & Astrophysics, Pennsylvania State University, 525 Davey Lab, University Park, PA 16802, USA
[2] Goddard Space Flight Center, Code 631, Greenbelt, MD 20771, USA

Abstract. We have measured the size and pressure of the Local Bubble in the direction of the constellation Eridanus using *ROSAT* PSPC observations of soft X-ray "shadows" cast by neutral clouds against the bright X-ray emission from the Orion-Eridanus superbubble. We discuss the technique and present our results for the size of the Local Bubble in the directions of our shadowing fields. Using these distances and the measured soft X-ray emission measures, we derive the Local Bubble's thermal pressure. For four lines of sight in Eridanus, we obtain pressures of about 1.2×10^4 K cm^{-3}, in good agreement with measurements in several other directions.

1 Introduction

Since the discovery of soft X-ray shadows in Draco (Snowden et al. 1991, Burrows & Mendenhall 1991), X-ray shadows observed by the *ROSAT* PSPC have been used to measure the brightness of the galactic halo, investigate cloud properties, study the Local Bubble, and set limits on the extragalactic diffuse background. Over the past several years, we have used *ROSAT* All-Sky Survey data and pointed observations to study the X-ray emission from the Orion-Eridanus superbubble (Snowden et al. 1995, Guo et al. 1995, Guo et al. 1997). Here, we summarize our results on the size and pressure of the Local Bubble in this direction.

2 Technique

This work is based on a multiwavelength investigation of the Orion-Eridanus superbubble, incorporating data in four wavelengths:

1. IR: *IRAS* 100μm maps of the Orion-Eridanus superbubble trace the cold cirrus clouds with an angular resolution of about 5′.
2. X-ray: *ROSAT* PSPC observations (both scanning and pointed) were compared with the IR maps to find soft X-ray shadows. The X-ray pointed observations also provided multicomponent spectral modeling which gave

the emission measure of the unabsorbed component produced in the Local Bubble.

3. 21 cm: N_H maps of ROSAT pointed fields were made using the 140 foot Green Bank telescope of NRAO. We scanned through the velocity channel maps to find velocities corresponding to the IR/X-ray shadow morphology.
4. Optical: Na D absorption line spectroscopy towards stars in these fields allowed us to place limits on the distance to the absorbing clouds, based on the presence or absence of optical absorption features at the velocities found from the N_H maps.

The outcome of this analysis is a series of distance estimates to the absorbing clouds. By studying these absorbing features over a large solid angle, and by comparison with Hα measurements (Reynolds & Ogden 1979), we have concluded that the pervasive velocity component observed in 21 cm emission, Hα emission, and Na D absorption in the LSR range +5 to +10 km/s is probably produced by material in the shell of the Local Bubble, while the components observed in the -20 to -30 km/s range originate in the near shell of the Orion-Eridanus superbubble (Guo et al. 1995, Guo et al. 1997).

We note evidence for interaction between the two shells in a field at $(\ell, b) = (200°, -47°)$, where both velocity components appear in the optical data over a narrow distance interval (Guo et al. 1995). Further evidence supporting an interaction between these bubbles comes from a 21 cm cloud in this field which has identical morphology in both the +4 and -25 km/s channels, suggesting that it is being crushed between the shells (Guo et al. 1995).

3 Results

In our previous paper (Guo et al. 1995), we showed that the field at $(\ell, b) = (200°, -47°)$ has a well-determined distance of about 159 pc for both the Local Bubble shell and the Orion-Eridanus shell. The local emission measure determined from the spectral fit for this field was 0.0057 cm^{-6} pc, using a solar abundance CIE plasma code (Raymond & Smith 1977) with a temperature of 10^6 K. The pressure inferred for this line of sight is 1.2×10^4 cm^{-3} K, in good agreement with the value found towards MBM12 by Snowden et al. 1993.

In Guo et al. 1997 we extend this result with the addition of three new fields for which we have obtained reasonably well-constrained distance estimates for the Local Bubble shell. These results are summarized in Table 1. They show a trend of decreasing shell distance with increasing latitude, but the derived pressure is quite uniform with a mean value of about 1.2×10^4 cm^{-3} K. For comparison, we include two measurements of the Local Bubble pressure by other authors using a similar technique.

Table 1. Local Bubble Density & Pressure

(ℓ, b)	X-ray EM (cm^{-6} pc)	d (pc)	N_e (cm^{-3})	P/k (cm^{-3} K)
(190°, −48°)	0.0050	177	0.0053	1.1×10^4
(200°, −47°)	0.0057	159	0.0060	1.2×10^4
(200°, −43°)	0.0045	130	0.0059	1.2×10^4
(206°, −40°)	0.0049	101	0.0070	1.4×10^4
(159°, −34°)†	0.0024	65	0.006	1.3×10^4
(166°, −32°)‡	0.0077	< 40	> 0.014	$> 1.9 \times 10^4$

† Snowden et al. 1993
‡ Bowyer et al. 1995

4 Discussion

The fact that independent measurements towards clouds in different parts of the sky agree on a pressure consistent with that needed to support the weight of the interstellar medium (Cox & Snowden 1986) is encouraging, but should be considered in light of two important caveats.

First, the pressures reported here are model-dependent. Specifically, the emission measures used to obtain the density (and hence, the pressure), assume collisional ionization equilibrium (CIE) conditions in a plasma with solar abundances, a condition known to be unlikely on the basis of DXS spectra of the soft X-ray diffuse background (Sanders et al. 1993). For example, if the gas is depleted but is still in CIE, a higher emission measure is required to produce the observed soft X-ray intensity. The pressure, which scales as the square root of the emission measure, will increase accordingly. Nonequilibrium (NEI) models will also produce different emission measures (and hence different pressures). Even different CIE models produce different results: Berghöfer reported at this meeting that the EUV pressure given in Table 1 for the (166°, −32°) field, which was based on spectral models by Landini & Monsignori-Fossi (1990), is in good agreement with the other fields listed in Table 1 if a Raymond & Smith plasma code is used instead.

The second important caveat concerns the distance estimates. We obtain the plasma density by dividing the emission measure by the path length, assuming uniform emissivity within the Local Bubble. If the emissivity is peaked at the edge of the cavity or in cloud boundaries, then we may be seriously underestimating the thermal pressure. Even if the volume emissivity is uniform, these pressures hinge on our distance estimates to the edge of the Local Bubble, and on the assumption that the boundary of the cavity surrounding the Sun is the same as the boundary of the hot plasma (an assumption we believe is violated in the direction of the "Welsh tunnel", where the cavity extends to great distances without a corresponding increase in the soft X-ray intensity). Finally, the distances we have used are based on stellar distances obtained by spectral parallax, a notoriously unreliable technique (but the best available to us at the time). We are in the process of revisiting these results using more reliable Hipparcos distances; the revised distances and pressures will be incorporated into Guo et al. 1997.

References

Bowyer, S., Lieu, R., Sidher, S. D., Lampton, M., & Knude, J. 1995, Nature, 375, 212

Burrows, D. N., & Mendenhall, J. A. 1991, Nature, 351, 629

Cox, D. P.,& Snowden, S. L. 1986, Adv. Space Res., 6, 97

Guo, Z., Burrows, D. N., Sanders, W. T., Snowden, S. L., & Penprase, B. E. 1995, ApJ, 453, 256

Guo, Z., Burrows, D. N., Penprase, B. E., & Sanders, W. T. 1997, ApJ, submitted

Landini, M., & Monsignori-Fossi, G. C. 1990, A&AS, 82, 229

Raymond, J. C., & Smith, B. W. 1977, ApJS, 35, 419 (and unpublished updates used in XSPEC code)

Reynolds, R. J., & Ogden, P. M. 1979, ApJ, 229, 942

Sanders, W. T., et al. 1993, Proc. SPIE: EUV, X-ray and Gamma-Ray Instrumentation in Astronomy IV, 2006, 221

Snowden, S. L., Burrows, D. N., Sanders, W. T., Aschenbach, B., & Pfeffermann, E. 1995, ApJ, 439, 399

Snowden, S. L., McCammon, D., & Verter, F. 1993, ApJ, 409, L21

Snowden, S. L., Mebold, U., Hirth, W., Herbstmeier, U., & Schmitt, J. H. M. M. 1991, Science, 252, 1529

Mapping Cool Gas in the ISM with *ROSAT*

I.B. Hutchinson, R.S. Warwick, and R. Willingale

Department of Physics and Astronomy, University of Leicester, University Road, Leicester, LE1 7RH, UK

Abstract. We have investigated the distribution of cool gas within the local interstellar medium (<150 pc) using data derived from both the ROSAT X-ray and EUV all-sky surveys. The EUV/X-ray spectra for a sample of 89 white dwarf stars and 200 active late-type stars have been analysed to give an estimate of the gas column density along each object's line of sight. In order to extend the sky coverage we have also included in the analysis 190 late-type stars detected only in the X-ray band. The derived N_{HI} measurements confirm the size of the local low density bubble as typically 50-100 pc and also show that in the general direction of Loop I there is a substantial excess of absorption at a range of distances from 10-150 pc.

1 Introduction

Over the last twenty years a detailed picture of the interstellar medium in the vicinity of the Sun has been built-up on the basis of the optical and UV absorption features observed in the spectra of nearby stars and from studies of the soft X-ray background (see Frisch 1997). In particular, there now exists a large database of hydrogen column density measurements, N_{HI}, which can be used to trace the distribution of cool gas at distances from a few out to many hundreds of parsecs (Fruscione et al. 1994). One feature of the local cool gas distribution which has been highlighted by the recent EUV all-sky surveys carried out by the *ROSAT* WFC and *EUVE* (Pounds et al. 1993; Pye et al. 1995; Bowyer et al. 1994 ; Barber et al. (1996)) is an apparent nearby "wall" of absorption observed in the general direction of the Galactic centre as revealed by the relative paucity of EUV source detections in this region (e.g. Warwick et al. 1993). Here we use a combination of EUV and X-ray data derived from the ROSAT all-sky survey to investigate the distribution of absorbing gas in the direction of this wall.

2 Data Analysis

The first part of our study is based on the EUV sources listed in the *ROSAT* WFC 2RE catalogue (Pye et al. 1995). The properties of a total of 89 white dwarf stars drawn from this survey have been discussed recently by Marsh et al. (1997). These authors list N_{HI} values for each source derived from fitting detailed atmospheric models to a combination of WFC S1 (60–140Å) and

S2 (112-200 Å) filter measurements and *ROSAT* PSPC (5.2-100 Å) data. We also cross-correlated the 2RE catalogue with the public version of the *ROSAT* PSPC bright source catalogue (RBSC) and selected a sample of 200 late-type stars common to both lists. The data from the RBSC were available in the form of three hardness ratios and a broad band count-rate, which we combined with the S1 and S2 measurements to give effectively five energy channels. Spectral models for the emission from late-type stars typically include a column density and two thermal plasma components, requiring in total five spectral parameters. For our purpose we set the two temperatures and also the ratio of the normalisations of the thermal components to values typical of the data set as a whole and then performed a spectral fitting analysis to derive an estimate for N_{HI} for each source. The sky distribution of the white dwarf stars and late-type stars which comprise the WFC sample is shown in Fig. 1.

In an attempt to obtain greater sky coverage and to increase the distance range sampled we have also extended our study to include all the sources in the RBSC catalogue which cross-correlated with known late-type stars contained in the Hipparcos Input Catalogue. For these sources we have only three effective energy channels, but again we were able to estimate the line-of-sight N_{HI} using the spectral model as described above. Since the PSPC-only measurements are insensitive to N_{HI} values less than $\sim 3 \times 10^{19}$ cm^{-2}, for the present work we restrict our attention to stars at distances greater than 50 parsecs. The sky distribution of the PSPC-only sample of 190 late-type stars is shown in Fig.2.

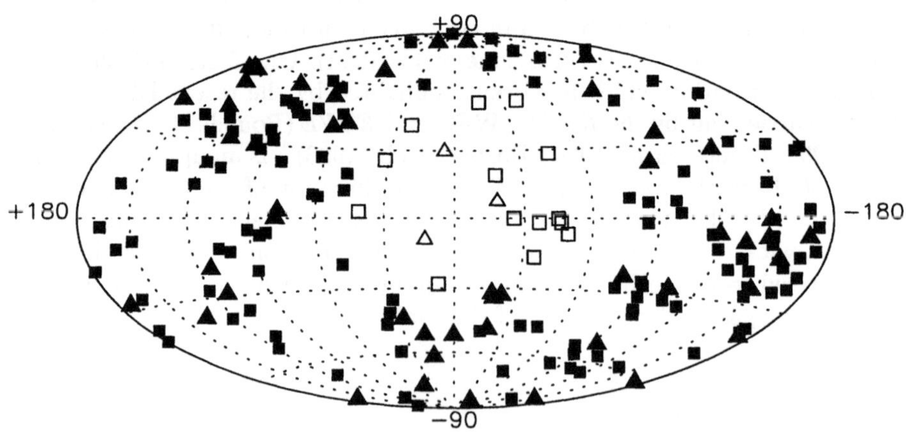

Fig. 1. The sky distribution in galactic coordinates of the white dwarf stars (triangles) and late-type stars (boxes) in the WFC sample. Stars inside the the nominal wall region are shown as open symbols.

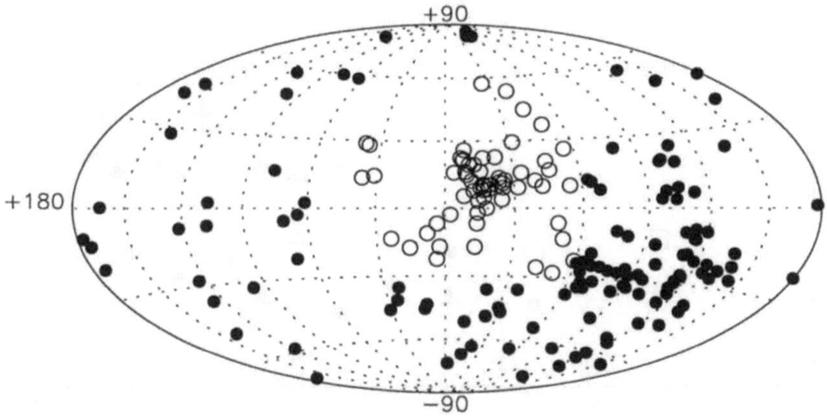

Fig. 2. The sky distribution in galactic coordinates of the late-type stars in the PSPC-only sample. Stars within the wall region are shown as open circles.

3 The 'Wall' of Absorption Towards the Galactic Centre

The sparse number of EUV source detections in the general direction of the Galactic centre has been noted in a number of studies. Based on a detailed analysis of the statistics pertaining to the *ROSAT* WFC survey, Warwick et al. (1993) concluded that the effective edge of the local bubble in this region is anomalously close to the Sun i.e. at ~ 10 pc as opposed to a more typical radius for the bubble of 80–100 pc). The presence of relatively nearby clouds in this part of the sky is evident from the optical/UV absorption studies of Frisch & York (1983) and Paresce (1984). The region also coincides with the Loop I superbubble (Egger 1997). Here we define the wall region as that bounded by $45° > l > 300°$ and $-30° < b < 60°$.

Figure 3 shows the median value of N_{HI} measured in various distance ranges for sources selected inside and outside of the wall region. (We use median values since each individual N_{HI} measurement is subject to fairly large errors, a more detailed analysis of which is outside the scope of the present paper.) The trend in the data in Fig. 3 is clear, namely that the column density measurements show a significant up-turn at distances of 50–100 pc in many directions. The wall region is characterised by excess column density compared to the all-sky norm but the data do not identify any specific cloud structure in this direction.

4 Discussion

The combination of EUV and X-ray data from the *ROSAT* all-sky survey for nearby stars provides us with a method of studying the distribution of

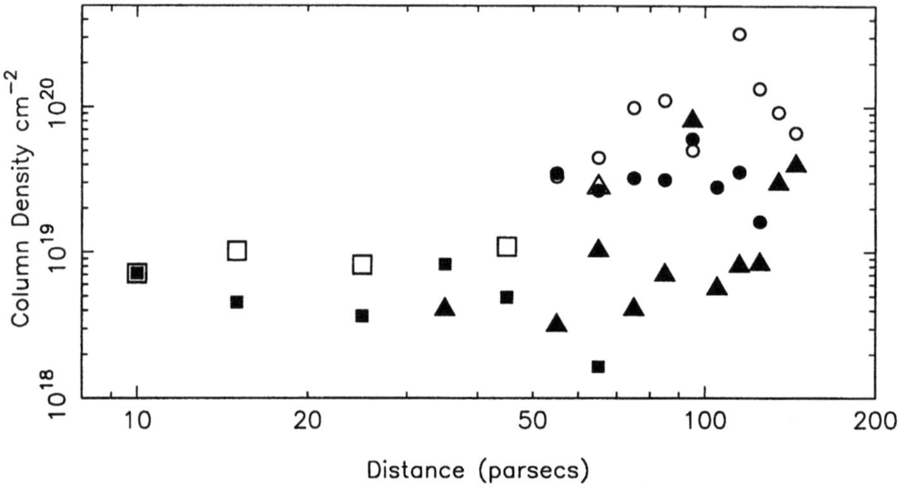

Fig. 3. The median value of the column density, N_{HI}, measured for the various source samples in different distance ranges. The late-type and white dwarfs stars from the WFC survey are shown as squares and triangles respectively, whereas the late-type stars drawn only from the RBSC are shown as circles. The open (filled) symbols correspond to those stars located within (outside) the region of the wall.

cool gas in the vicinity of the Sun. This approach can supplement the information available from the optical and UV absorption line studies. Here we have presented a very preliminary analysis emphasising a selected area of the sky (the general region of the Galactic centre) but the data are relevant to the global distribution of absorbing material over distance scales from 10 to several hundred parsecs. Future work will make use of the refined stellar distances and parameters available from the Hipparcos mission.

References

Bowyer, S., et al. (1994): ApJS, 93, 569

Bowyer, S., et al. (1996): ApJS, 102, 129

Egger, R., (1997): these proceedings

Frisch, P., York, D., (1983): ApJ, 271, L59

Frisch, P., (1997): these proceedings

Fruscione, A., et al. (1994): ApJS, 94, 127

Marsh, M., et al. (1997): MNRAS, 286, 369

Paresce, F., (1984): AJ, 89, 1022

Pounds, K., et al. (1993): MNRAS, 260, 77

Pye, J., et al. (1995): MNRAS, 274, 1165

Warwick R., Barber C., Hodgkin S., Pye J. (1993): MNRAS, 262, 289

Interaction of the Local Bubble with its environment

Roland Egger

Max–Planck–Institut für extraterrestrische Physik, Giessenbachstrasse, D-85740 Garching, Germany

Abstract. The Local Bubble (LB), a volume of roughly 100 pc radius filled with ionized gas of high temperature and low density, is not a unique object. It is obviously surrounded by several interstellar bubbles of similar properties. Some of them are as close as to undergo interaction with the LB. Our most prominent neighbour is the Loop I superbubble formed by the Sco-Cen OB association. It is shown that this huge shell is in collision with our Local Bubble. The product of this interaction is an annular cloud of dense neutral gas at the boundary between the two bubbles. Other examples for close neighbouring bubbles are the Gum nebula, the Eridanus bubble and possibly the two radio loops II and III.

1 Introduction

The structure of the diffuse soft X-ray background (SXRB, E < 2 keV) has been subject to controversial discussion ever since its discovery by Bowyer et al. (1968). On the one hand there is a very soft (C-band, < 0.3 keV) component which is more or less isotropic at low galactic latitudes ($b < 30^o$). This component is obviously not absorbed by any intervening neutral gas with column densities above $N_H = 10^{19}\mathrm{cm}^{-2}$ and therefore, attributable to a local diffuse X-ray source, most probably highly ionized gas filling the Local Bubble (LB). The properties and possible origin of this local interstellar bubble have been determined, modelled and discussed by various authors. A recent review is given by Breitschwerdt (1996) and Breitschwerdt et al. (1996). Two alternative models are either a remnant of a singular supernova event or a fossile superbubble formed by a stellar cluster dissociated long ago. Though the shape of this cavity is not entirely clear, there is general agreement on a mean radius of $\sim 100\,\mathrm{pc}$.

At higher galactic latitudes the C-band background is highly structured. This is in principle attributable to two possible origins: Firstly, galactic diffuse sources outside the LB and of varying brightness, shape and size cause strong variations in the measured intensity. Secondly, clouds of intervening neutral material cast shadows on distant diffuse sources. In the energy range above 0.5 keV this shadowing is less severe, apart from the very low latitudes, so that the source distribution is the dominant origin of structure. The two origins can be distinguished using tracers of the N_H (21 cm or infrared measurements). All Sky Surveys (e.g. Snowden et al., *this volume*) show several prominent large scale emission features. Some of them, through their large apparent size on the one hand and through their soft (unabsorbed) X-ray spectrum on the other hand, indicate that they reside in our immediate neighbourhood. Maybe the most impressive one of these is the North Polar Spur (NPS), a huge X-ray arc associated with radio Loop I. For a review of this object see Salter (1983). It has been shown that Loop I is most probably a superbubble formed by the nearby Sco-Cen OB association at ~170 pc distance, and that there is evidence for its interaction with the Local Bubble (Egger & Aschenbach 1995). Some details will be discussed below. Other prominent objects of the abovementioned kind are the Monogem Ring (Plucinsky et al. 1996), the Vela supernova remnant (Aschenbach et al. 1995) and the Eridanus X-ray enhancement. At least the latter one is suspected to show indications for an interaction with the LB (Guo et al. 1995). Other candidates for neighbouring interstellar bubbles are found in radio survey maps. The ones with the largest apparent diameters (~ 100°) are the radio loops II and III (Berkhuijsen 1971). These objects have to be very close, unless they have unrealistically large dimensions.

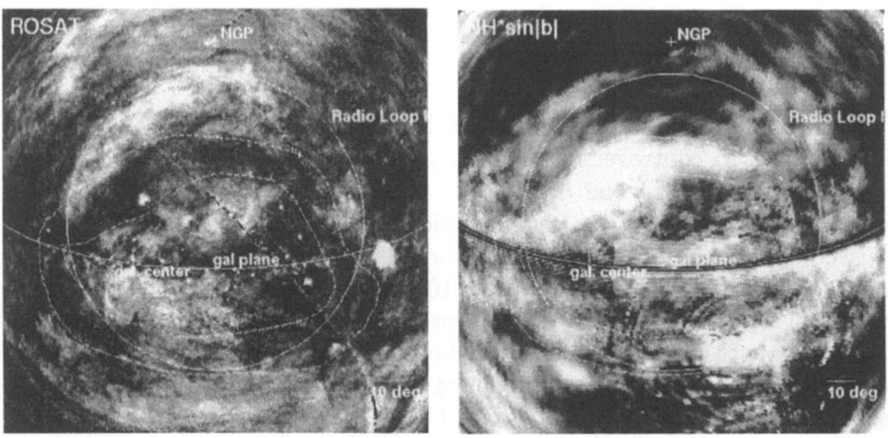

Fig. 1. *Left:* ROSAT Survey map centred on Loop I in the energy range 0.1–2.0 keV. The circle of 58° radius follows the radio continuum loop. The dashed lines outline the contours of the annular shadow. *Right:* HI map divided by cosec$|b_{II}|$; (see Plate 1).

2 Observations and Results

The ROSAT All Sky Survey in soft X-rays (Trümper 1983) is particularly well suited for studying faint extended sources. This is mainly due to the fast highly resolving X-ray optics (Aschenbach 1988) and the extremely sensitive low noise detector PSPC (Pfeffermann et al. 1986). The intrinsic angular resolution of the Survey is 1' and the energy resolution of $E/\Delta E \sim 2.5$ allows a limited spectral analysis. The data have been cleaned from non-cosmic contaminations, and point-like sources have been removed down to a flux threshold of 6×10^{-14} ergs cm^{-2}s^{-1} (Snowden et al. 1997). Figure 1 (*left*) shows a ROSAT Survey map of $160^o \times 160^o$ (E = 0.1 – 2.0 keV) centred on the apparent centre of radio Loop I ($l_{II} = 329^o$, $b_{II} = +17.5^o$) in equal area polar projection. The solid circle outlines a best circle fit to the radio loop of 58° radius. The X-ray morphology appears to be shell-like, well aligned to the radio loop, at least in the region associated with the NPS (upper left in figure 1). The rest appears rather fragmentary. Moreover, there is an elliptical annular shadow obscuring parts of the loop indicated by the dashed lines in figure 1). The ellipse is centred about $l_{II} = 335^o$, $b_{II} = 0^o$. Its long axis is roughly aligned with the galactic plane and has a length of about 127°, while the short axis is about 93°. The width of the ring is 15° on the average.

The right part of figure 1 shows an HI map of the same region (data from Dickey & Lockman 1990). The column densities have been divided by $cosec|b_{II}|$ in order to compensate for the contribution of the Galactic disc. This image clearly shows the HI shell with $N_H \sim 10^{20}$cm^{-2} enclosing Loop I at a radius of $\sim 68^o$. It also shows a fragmentary ring corresponding to the annular X-ray shadow with column densities up to $\sim 7 \times 10^{20}$cm^{-2}.

The central part of the annulus appears to be transparent for the broad X-ray band. Emission from the interior of the Loop I bubble is shining through the ring-shaped absorption feature (left part of figure 1). In the lowest ROSAT energy band R1 (0.1–0.2 keV), however, the interior of the annulus is almost opaque. This indicates the presence of a neutral gas wall of column density $N_H \sim 10^{20}$cm^{-2}, which corresponds to optical depth unity in this energy range.

2.1 ROSAT spectra of the North Polar Spur

Three cuts across the X-ray arc defining the North Polar Spur were studied as described in Egger (1993) and Egger (1995). The cuts run roughly along $b_{II} = 44^o$, 34° and 24°. Pulse height spectra were extracted from the PSPC data and analyzed using an absorbed three-component model. One component is the NPS thermal emission, the second component is the LB foreground thermal emission and the third component is an extragalactic power-law spectrum absorbed by the total galactic column density. The parameters of the latter two components were determined within a reference field just outside the emission region of Loop I. For the LB a temperature of $1.2(\pm 0.1) \times 10^6$ K and an emission measure of 3.0 cm^{-6}pc was found, resulting in an a particle density of 5.2×10^{-3}cm^{-3} (assuming collisional ionization equilibrium and a line of sight of 90 pc through the LB in this particular direction). The photon index of the extragalactic power-law component is $\Gamma = -2.5$ and the flux at 1 keV is 12.5 ± 0.3 cm^{-2}sr^{-1}s^{-1}keV^{-1} which is in agreement with previous results (cf. McCammon & Sanders 1990). The spectral studies of the three cuts across the NPS emission resulted in electron temperatures between 2.0 and 3.5×10^6 K and emission measures between 0.02 and 0.04 $cm^{-6}pc$. The absorption values are between 1 and $7 \times 10^{20}cm^{-2}$, about a factor of 2–5 below the total column densities in the corresponding directions. Assuming the X-ray NPS to be due to a spherical shock wave of 150 pc radius, shock densities between 0.01 and 0.02 cm^{-3} (increasing from high lo low b_{II}) and corresponding electron temperatures between 2.5 and 3.5×10^6 K at the shock front are derived.

3 Discussion

3.1 Loop I – an active superbubble

Stellar activity within the Sco-Cen OB association started $1 - 2 \times 10^7$ years ago (Bertiau 1958). Initially, the strong stellar winds of the young massive stars began to blow a big collective stellar wind bubble filled with hot gas of several 10^6 K. At the time when the massive stars reached their evolutionary time scale ($\sim 4 \times 10^6$ years) a cascade of supernova (SN) explosions began to occur in the interior of the pre-existing cavity taking over the blowing of the bubble from the decaying stellar wind activity. From then on the energy input (by SNe) can be regarded as fairly constant (see e.g. McCray 1987). The increasing lifetime of less massive stars is compensated by their greater number in the initial mass function (IMF) of the stellar cluster. Both follow approximately a power law with index about –1.5. This goes on until the age of about 5×10^7 years when the least massive stars that still die as SNe have reached the end of their lifetimes

If we now want to estimate the energy input by SNe per time for the present case we have to find out the number of SN explosions that have occurred within the Sco-Cen association up to now. In order to determine the parameters of the IMF we have to restrict ourselves to stars of spectral types later than B1, since these have lifetimes greater than 3×10^7 years and hence, are not affected by loss through SN explosions, yet. Integration of the IMF of the Sco-Cen association between the types B1 (M$\sim$ 14$M_\odot$) and B3 (M$\sim$ 9$M_\odot$), about the least massive stars that still may produce type II SNe, reveals

$$present\ N_{B1-3} = \left[\frac{C \cdot \tilde{M}^{-1.5}}{-1.5}\right]_9^{14} = 42 \quad \Rightarrow \quad C = 3510 \qquad (1)$$

where $\tilde{M}$ is the stellar mass in $M_\odot$. The total number of potential SNe is the integral up to the greatest possible masses, earlier than type B3:

$$initial\ N_* = N_{<B3} = \left[\frac{3510 \cdot \tilde{M}^{-1.5}}{-1.5}\right]_9^{60} = 82\,. \qquad (2)$$

Since the present number of stars earlier than type B3 is 44, already 38 SNe must have exploded. However, due to small number statistics, the true number may vary by plus or minus 10. Hence, we compute an average SN rate of

$$f_{SN} \approx \frac{82}{5 \times 10^7 \mathrm{yr}} = \frac{1}{600,000\,\mathrm{yr}} \tag{3}$$

The mechanical luminosity of the SNe is therefore

$$L_{SN} = f_{SN} \cdot E_0 = 5.3 \times 10^{37}\,\mathrm{erg/s}\,, \tag{4}$$

where $E_0 = 10^{51}$ erg is the canonical SN energy.

Now, we have the average "mechanical luminosity" of the stellar association and can handle the problem like a stellar wind blown bubble with a constant energy input per time. McCray & Kafatos (1987) derive the evolution of a superbubble from the stellar wind model of Weaver et al. (1977):

$$R_S = 269\ \mathrm{pc}\left(\frac{L_{38}}{n_0}\right)^{1/5} t_7^{3/5} \qquad V_S = 16\ \mathrm{km\ s^{-1}}\left(\frac{L_{38}}{n_0}\right)^{1/5} t_7^{-2/5} \tag{5}$$

and thus,

$$R_S = 17\ V_S\ t_7\ . \tag{6}$$

Here, R_S is the radius of the radiative shell in pc, V_S is its expansion velocity in km s^{-1} and t_7 the age of the bubble in 10^7 years. The model is compatible with the data of Loop I, if the age is about 10^7 years and the present expansion velocity is about 10 km s^{-1}. This is consistent with the literature quoted in the introduction where values between $V_S = 2$ km s^{-1} and $V_S = 40$ km s^{-1} can be found. Also the interior structure of the superbubble is similar to that of a stellar wind bubble in the radiative stage:

$$n_i \approx 1.5 \times 10^{-3}\,\mathrm{cm^{-3}}\,(N_* E_{51})^{0.2}\, n_0^{0.5}\, t_7^{-0.6} \left(1 - \frac{r}{R_S}\right)^{-0.4} \tag{7}$$

$$T_i \approx 1.1 \times 10^6\,\mathrm{K}\,(N_* E_{51})^{0.2}\, n_0^{0.1}\, t_7^{-0.2} \left(1 - \frac{r}{R_S}\right)^{0.4} \tag{8}$$

where n_i, T_i are interior density and temperature, n_0 is the density of the surrounding interstellar medium and E_{51} is the energy of a single supernova in units of 10^{51} erg (McCray & Kafatos, 1987). The measured column density of the HI shell around Loop I is $\sim 10^{20}\mathrm{cm^{-2}}$. With a shell radius of $R_S = 158$ pc this gives an initial ambient density of $n_0 \sim 0.63\,\mathrm{cm^{-3}}$. For comparison: Dickey & Lockman (1990) find an average density in the galactic plane of 0.57 cm^{-3}. Hence, the central density of the Loop I bubble would be $\sim 3 \times 10^{-3}\mathrm{cm^{-3}}$ which holds for a major fraction of the interior, while the central temperature would be $\sim 2.5 \times 10^6$ K.

The observed shell radius of 160 pc is by about 40 % smaller than the one predicted by the model. This can by no means be attributed to small variations of any of the contributing parameters, since their powers are so low. A possible explanation would be confinement by the ambient magnetic field. This is supported by the strong synchrotron radiation associated with Loop I that indicates field strengths of at least $B_0 = 5\ \mu$G. Slavin &Cox (1992) show, that bubbles in an ambient field of 5 μG have radii 30–50% smaller than those in an unmagnetized medium. Another possible reason for the smaller radius could be loss of interior thermal energy by mass load due to cloud evaporation (McKee & Ostriker 1977). Support is given to this view by the observation of clouds in the interior of Loop I (e.g. the R CrA complex, Wang, 1994) which, by their cometary shapes, actually show evidence for the interaction with blast waves from Sco-Cen supernovae.

The latest supernova that occurred within the Sco-Cen association must have produced a shock wave that crossed the hot tenuous interior rapidly until it was decelerated at the steep density gradient near the dense walls of the bubble. The result was heating the bubble walls to temperatures of 3–4 $\times 10^6$ K giving rise to the prominent X-ray feature of the North Polar Spur. It is reasonable to assume that the shock front has hit the walls no longer than 10^4 years ago, since, otherwise radiative cooling would have caused it to disappear out of the X-ray band. SNR evolution in a hot ambient medium is described by an expansion law like

$$\frac{V_S}{C_0} \propto \left(\frac{R_S}{R_0}\right)^{-3/2}, \tag{9}$$

where C_0 is the sound velocity and $R_0 \propto E_0/p_0$ is a characteristic radius (p_0 is the ambient pressure) (Gaffet 1978). It can be shown (see Egger 1994 and Egger 1995) that the temperatures and densities deduced from the X-ray observations are compatible with a supernova of $\sim 10^{51}$ erg that occurred somewhere within the Sco-Cen association about 2×10^5 years ago.

3.2 Interaction with the LB

If we consider the geometrical facts compiled in the introduction, it becomes obvious that the LB and the Loop I bubble have to have undergone some kind of interaction. A proof for the actual case of such an interaction is given by the annular X-ray shadow described above. An annular volume of dense neutral gas is predicted to form at the interaction zone between two colliding interstellar bubbles. Hydrodynamical computations (e.g. Yoshioka & Ikeuchi 1990) reveal that the density within the cool interaction ring is by a factor of 20 to 30 higher than that of the ambient medium. The observed neutral gas ring appears to be somewhat distorted and incomplete which may be due to deviations from spherical symmetry of either the LB or the Loop I superbubble or both. Apart from this, it resembles quite well the simulated scenario of the two colliding shells as seen from inside one of the two bubbles looking towards the other one (figure 2).

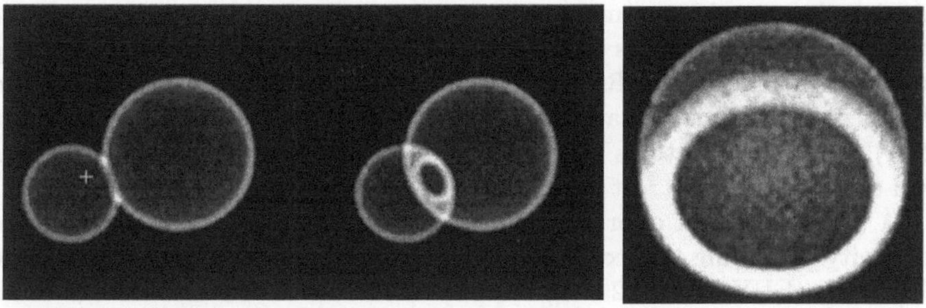

Fig. 2. *Left* : Projected column density of a simulation of two colliding bubbles seen from aside. *Middle* : The same, rotated by 30°. *Right* : The same as seen from inside. The viewing point is marked by the cross in the left image. The viewing direction is towards the larger bubble.

The distance of the interaction feature can be estimated by absorption line studies of nearby stars which are projected on the ring (figure 3, data compiled by Fruscione et al. 1994). It is found that the N_H jumps from less than $10^{20}\mathrm{cm}^{-2}$ to more than $7 \times 10^{20}\mathrm{cm}^{-2}$ near the distance of about 70 pc which places the dense gas between the LB and Loop I. The column density allows to estimate a particle density within the ring of $n \sim 15\,\mathrm{cm}^{-3}$. This exceeds the ambient density ($n_0 \sim 0.6\,\mathrm{cm}^{-3}$) by a factor of 25, in good agreement with the theoretical prediction.

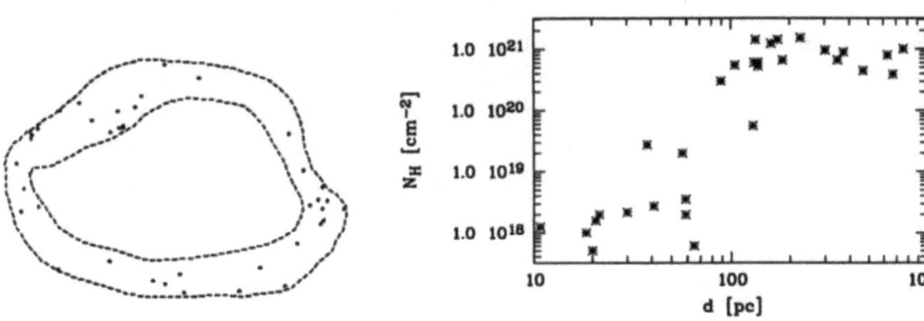

Fig. 3. *Left:* Positions of stars projected on the annular X-ray shadow. *Right:* HI column densities towards stars projected on the interaction ring. N_H rises to $\sim 7 \times 10^{20}\mathrm{cm}^{-2}$ near the distance of about 70 pc.

3.3 The wall

The simulations mentioned above also show that, if at least one of the bubbles has already reached the radiative stage of evolution prior to the collision, the two interiors would not merge but a dense wall would form between the bubbles. This corresponds to the observation that the interior of the HI ring is almost opaque in the R1 band. Further support is given by optical and UV spectral analysis of stars near the centre of Loop I (Centurion & Vladilo 1991). These observations reveal evidence for the presence of a neutral gas wall of $N_H \sim 10^{20}\mathrm{cm}^{-2}$ at a distance of 40 ± 25 pc.

A totally different approach to investigating the size and shape of the local cavity is probing the extinction of the interstellar medium in the extreme ultraviolet (EUV, 0.05–0.2 keV) by differential source counts. Warwick et al. (1993) performed such an analysis making use of the ROSAT Wide Field Camera source catalogue. They found that the apparent brightness and distribution of the EUV sources (mainly white dwarfs and late type stars) is compatible with a local cavity of 80 pc mean radius. There are two exceptional directions where the source counts indicate larger distances to the walls: the regions about $l_{II} = 200^o$, $b_{II} = -30^o$ ($\approx$ 100 pc) and $l_{II} = 120^o$, $b_{II} = +45^o$ ($\approx$ 120 pc). There is, however, a field on the sky that nearly exactly matches the region described above as the annular X-ray shadow, which is almost completely devoid of sources in the EUV band below 0.1 keV. This result again indicates the presence of a dense HI wall in the direction of Loop I that could be even as close as 10 pc.

It is still under discussion, if the interior of the LB is hot or warm or only lukewarm (see e.g. Breitschwerdt, *this volume*). In any case, the ongoing supernova activity within Loop I makes the pressure at least twice as high as that in the LB, even if we adopt the canonical $T_{LB} = 10^6$ K. This overpressure would cause the wall at the interaction zone bending and moving slowly towards us. The nearest part of it may have come already fairly close to the solar position (10–40 pc). Moreover, one would expect a gas "wall" of ~ 1 pc thickness and a density of 20–30 cm^{-3} to be unstable on a time scale of about 10^6 years (Kahn, F.D., *private communication*). A more thorough study of this aspect (Breitschwerdt & Egger, 1997) reveals the possibility of instabilities on the parsec scale. As long as the pressure driven state continues, the wall would not disintegrate but cloudlets of parsec size could be expelled from the instabilities towards the interior of the LB. Thus, it is likely that the Local Cloud or the Local Fluff of neutral gas surrounding the Sun within the next 10 pc (see e.g. Lallement, *this volume*) are a product of the interaction zone between Loop I and the LB.

3.4 Other bubbles in our neighbourhood

The Eridanus X-ray enhancement is also discussed to be a superbubble formed by the activity of young stars (Guo et al. 1995). The molecular gas associated with the near side of this bubble is only 160 pc away and its velocity structure suggests possible interaction with the LB. The Monogem Ring is a soft X-ray emitting shell of 25° diameter. Estimates of the distance to the centre vary between 300 and 1300 pc. Therefore, size and energy content are equally uncertain (Plucinsky et al. 1996). Though it is very close, the Monogem Ring seems not to be an immediate neighbour and any interaction with the LB is doubtful. The Vela supernova remnant has been shown to be embedded in a hot bubble confined by a shell associated with the Gum Nebula (Aschenbach et al. 1995). The X-ray spectrum of this "Gum-Bubble" reveals an equilibrium temperature of $\sim 10^6$ K. This is confirmed by a kinematic analysis of supernova explosion fragments propagating through the hot ambient medium. The distance to the Vela pulsar is estimated between 300 and 500 pc. The lower limit would place the near side of the "Gum-Bubble" close to the LB boundary. This is supported by the observation that the absorbing column density towards the Vela SNR is only about $10^{20}\mathrm{cm}^{-2}$.

The giant radio loops II, III and IV are other candidates for possible nearby superbubbles. Nowhere, however, is the situation as clear as in the case of Loop I. No obvious X-ray enhancement is associated with either of them. Loop IV is the only one that seems to be associated at least with an HI arc. This Loop is seen in projection through the interior of Loop I and therefore, it is most probably behind it. Loops II and III are nearly as large in apparent size as Loop I and therefore, probably close. However, they are crossing each other and thus, only one of them can be regarded as our immediate neighbour. It is however remarkable, that Loop II seems to be associated with a huge bulk of HI apparently filling the whole loop. This HI corresponds to a depression in the soft X-ray background. Possibly, it is due to neutral gas compressed between the LB and a fossil superbubble. This wall of neutral gas, on the other hand, would hinder any observation of X-ray emission if the interior of the neighbour bubbles is at equilibrium temperatures less or equal 10^6 K. This is subject of an ongoing study.

4 Conclusion

The solar neighbourhood is located within an interstellar bubble probably filled with hot gas of low density. The LB seems to be surrounded by numerous bubbles of similar properties. At least a few of them are filled with hot X-ray emitting gas due to ongoing stellar activity. The picture suggested by the observations is that of a "Local Interstellar Foam" made up of seven or more bubbles close to each other. It is well possible that this "foam" is associated with the local region of stellar activity defined as *Gould's Belt.* This indicates a high local filling factor of hot gas, maybe up to 50 %.

Acknowledgements

I would like to thank the referee for useful comments and hints. The ROSAT project has been supported by the German Bundesministerium für Zukunft (BMBF/DLR) and the Max-Planck-Gesellschaft (MPG).

References

Aschenbach, B., Applied Optics **27** (1988) 1404

Berkhuijsen, E.M., A.A. **14** (1971) 359

Bertiau, F.C., Bertiau, S.J., ApJ **128** (1958) 533

Centurion, M., Vladilo, G., ApJ **372** (1991) 494

Bowyer, C.S., Field, G.B., Nature **217** (1968) 32

Breitschwerdt, D., Space Science Reviews **78** (1996) 173

Breitschwerdt, D., Egger, R.J., Freyberg, M.J., Frisch, P.C, Vallerga, J.V., Space Science Reviews **78** (1996) 183

Breitschwerdt, D., Egger, R.J., A.A. (1997) (*in preparation*)

Dickey J.M., Lockman, F.J., ARA&A **28** (1990) 215

Egger, R.J., MPE Report **249** (1993) ISSN 0178-0719

Egger, R.J., 1995 in The Physics of the interstellar medium and the intergalactic medium, edts.: A. Ferrara, C. Heiles, C.F. McKee, P. Shapiro, Pub. Astron. Soc. of the Pacific Conf. Ser. **80** (1995) 45

Egger, R.J., Aschenbach, B., A.A. **294** (1995) L25

Fruscione, A., Hawkins, I. Jelinsky, P., Wiercigroch, A., ApJ suppl. **94** (1994) 127

Gaffet, B., ApJ **225** (1978) 442

Guo, Z., Burrows, D.N., Sanders, W.T., Snowden, S.L., Penprase, B.E., ApJ **453** (1995) 256

McCammon, D., Sanders, W.T., ARA&A **28** (1990) 657

McCray, R. in Spectroscopy of astrophysical plasmas, ets. A. Dalgarno, D. Layzer, Cambridge Astrophys. Series, Cambridge Univ. Press (1987) 255

McCray, R., Kafatos, M., ApJ **317** (1987) 190

McKee, C.F., Ostriker, J.P., ApJ **218** (1977) 148

Pfeffermann E., Briel, U.G., Hippmann, H. et al., Proc. SPIE, **733** (1986) 519

Plucinsky, P.P., Snowden, S.L.,Aschenbach, B., Egger, R.J., Edgar, R., McCammon, D., ApJ **463** (1996) 224

Salter, C.J., Bull. Astr. India **11** (1983) 1

Slavin, J.D., Cox, D.P., ApJ **392** (1992) 131

Snowden, S.L., Egger, R.J., Freyberg, M.J. et al., ApJ **485** (1997) 125

Trümper, J., Adv. Space Res. **4** (1983) 241

Wang, Q.D., in The Soft X-Ray Cosmos, AIP Conference Proceedings, ets. Schlegel, E.M., Petre, R., (1983) (New York, AIP) 301

Warwick, R.S., Barber, C.R., Hodkin, S.T., Pye, J.P., M.N.R.A.S. **262** (1993) 28

Weaver, R., McCray, R., Castor, J., Shapiro, P., Moore, R., ApJ **218** (1977) 377

Yoshioka, S., Ikeuchi, S., ApJ **360** (1990) 352

Bubbles in Collision: Hydrodynamical Models

A.D. Chernin

Sternberg Astronomical Institute, Moscow University, 119899 Moscow, Russia

Abstract. Computer simulations and laboratory experiments with shock waves are suggested for the study of collisions of bubbles or superbubbles. They describe how the colliding bubbles merge with the formation of an expanding common shell and a complex internal dynamical structure which includes two vortex tori and collimated outflows.

1 Introduction

Expanding bubbles in the interstellar medium can come into contact and interact with each other. The major physical mechanism of this process is the interaction of expanding shocks. We study it with the use of the Courant-Friedrichs analytical theory, new computer simulations and laboratory experiments with shock waves.

2 The Courant-Friedrichs Configuration and Its Evolution

The contact collision of shocks leads to the formation of a new regular dynamic structure in the interaction area studied in general terms by Courant and Friedrichs (1948). They analyzed a simple, but instructive example in which the fronts of the shocks are assumed to be spherical with equal radii. The structure includes the initial fronts, the annular Mach front, two tangential discontinuities and the reflected fronts. This configuration formed with these dynamical elements is not stationary and must evolve because of the interaction of these elements and their intrinsic transformations. As a result, the overall structure of the flow undergo considerable changes which, since a certain stage, can no longer be described by the Courant-Friedrichs theory.

With the same statement of the problem as in the Courant-Friedrichs picture, the supersonic collision of two expanding adiabatic shock fronts was studied in 3D (with axial symmetry) computer simulations (Voinovich and Chernin 1995). The method used in the computations was worked out at Ioffe Institute (Sankt Petersburg). The simulations give a picture of the advanced stages of the evolution of the Courant-Friedrichs structure.

The density of gas in the central area bounded by the Mach front and two reflected shocks increases ten times in the adiabatic regime. This dense and hot material forms a disc in the plane of symmetry of the flow. Later on the

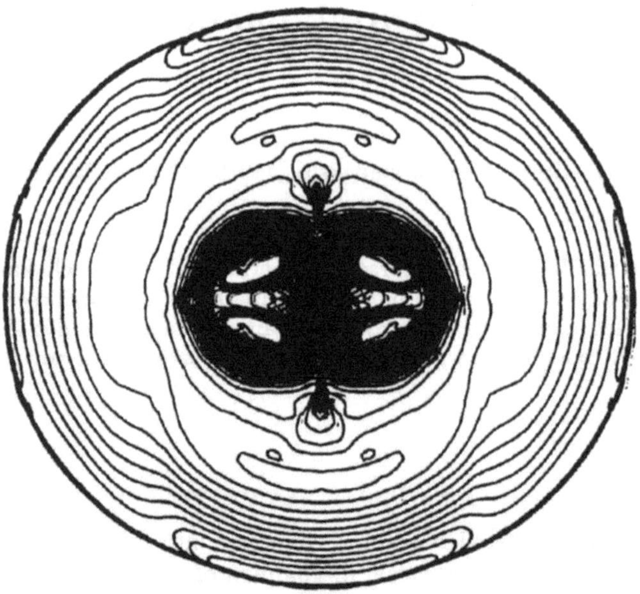

Fig. 1. Merging bubbles: the flow pattern indicated by density isolines.

disc transforms into a torus due to the gas outflows driven by the reflected shocks.

The simulations indicate also that the flows behind the reflected shocks move with increasing velocity when the shocks propagate into the area of the decreasing density in the cavities produced by the initial shocks. These flows reveal a tendency for a self-collimating behavior: while the density of the medium keeps decreasing along the direction of the shock propagation, the flows become more and more directed.

At the next stages, the secondary collimated outflows forms in the central area of the structure, and the central torus splits into two vortex tori, or toroidal vortexes. The vorticity here is mainly due to the transformation and decay of two tangential discontinuities of the Courant-Friedrichs structure.

In parallel with the evolution of their interaction area, two interacting bubbles as a whole merge, and the initial shocks together with the Mach front form a common envelope of more or less spherical shape which keeps expanding, as it is demonstrated by Fig.1. The high resolution structure of the central vortex tori and outflows from them at a late stage of the evolution is presented in Fig.2.

Fig. 2. The structure of the central tori and the outflows at a late stage of their evolution.

3 Laboratory Experiments

Can such a complex structure remain to be so regular and stable for long times? We have obtained a definite answer to this question from our laboratory experiments with shocks (Rybakov *et al.* 1997).

In the experiments, an adiabatic spherical shock front is produced by the explosion at the edge of a cooper needle under the action of a 80 ns long laser pulse of with an energy release of 12 J. The front collides with a glass plate, and the reflected shock forms. Its dynamic is followed on a high speed camera. The duration of the experiment is about 1 ms, and all the events are happening within a volume 5 cm across.

The structure and evolution of the reflected shock and the flow behind it are fairly similar, in general, to that in the Courant-Friedrichs theory and our computer simulations. It is most important that the experiments prove that the flow is stable, as a whole, against non-axisymmetric perturbations (which cannot be studied in the computer simulations because of the adopted symmetry). The picture of the flow keeps rather regular until its almost complete dissipation.

Unfortunately, we cannot control or even measure the strength of the 'initial' hydrodynamical perturbations that are present in the experiment. Perhaps it will be possible in a new set of experiments we plan now; if so, it would be interesting to compare these perturbations with what can be expected in real astrophysical situations.

4 Conclusions

The results described above provide, as we hope, new physical grounds for the studies of the collisions of expanding bubbles, their interaction and merging under the conditions when the process can be treated as adiabatic and radiative losses are not important. These conditions can be expected most commonly at early stages of the expansion of high-velocity shocks from supernova (see, for instance, Cox 1972, Lozinskaja 1986). Under these physical conditions, we can expect that the collisions and mergings of bubbles are accompanied by the formation of a complex and stable dynamical structure in the central area of the shock-shock interaction. It would be interesting to try to observe such structures – if not in the Local Interstellar Medium, then beyond it.

In the opposite extreme, when the cooling is very rapid, the collision of expanding shocks is almost completely inelastic, and the picture seems to be less dynamic: the merging bubbles produce a quasi-static central disc of increasing density with no prominent internal motions and an envelope formed by two initial expanding shock fronts without the Mach ring. The collisions of this type happen, perhaps, in the Large Magellanic Cloud where intersections of giant loops are seen (Meaburn 1980). This inelastic hydrodynamical process may be interesting for the physics of large-scale star formation observed in the regions like 30 Dor in the LMC (Chernin 1997).

Acknowledgements. I appreciate the kind support of the SOC.

References

Chernin A.D. (1997): in press

Courant R. & Friedrichs K.O. (1948): Supersonic Flow and Shock Waves, N.Y., Interscience

Cox D.P. (1972): ApJ 178, 159

Lozinskaja T.A. (1986): Supernova and Stellar Wind, Moscow, Nauka

Meaburn J. (1980): MNRAS 192, 365

Rybakov V.A., Artemiev V.I., Medvedyuk S.A., Voinovich P.A., & Chernin A.D. (1997): in press

Voinovich P.A. & Chernin A.D. (1995): Astr. Lett. 21, 926

The Volume Filling Factor of the WIM

Elly M. Berkhuijsen

Max–Planck–Institut für Radioastronomie, Auf dem Hügel 69, D-53121 Bonn, Germany

Abstract. A compilation of data on the volume filling factor f_v of the warm ionized medium (WIM) as a function of the local electron density n_e indicates that approximately $f_v \propto n_e^{-1}$ over 4 decades. This result is expected for a fractal ISM.

1 Introduction

The volume filling factors of various constituents of the interstellar medium (ISM) are of great interest for the understanding of the physics of this medium (see Kulkarni and Heiles (1988), McKee (1995) and Cox (1995) for recent reviews). In this paper I present observational data on the volume filling factor of the warm ionized medium (WIM) obtained from the literature and by myself.

The volume filling factor is usually defined as

$$f_v = \frac{\langle n_e \rangle}{n_e} = \frac{\langle n_e \rangle^2}{\langle n_e^2 \rangle} = \frac{\langle n_e^2 \rangle}{n_e^2} \ , \tag{1a}$$

where n_e is the local volume density of the WIM in cm^{-3} and brackets indicate averages. Pynzar (1993 and priv. comm.) has shown that for classical HII regions with $n_e \gtrsim 100\,cm^{-3}$ f_v can be approximated by

$$f_v \simeq \frac{\text{diameter of HII region}}{\text{distance to HII region}} \ . \tag{1b}$$

The variables in (1a) can be found from the emission measure $\mathrm{EM} = \int n_e^2 dl$, the dispersion measure $\mathrm{DM} = \int n_e dl$, and the rotation measure $\mathrm{RM} = 0.81 \int B_{\|} n_e dl$, where the line of sight l is in pc and the strength of the magnetic field along l, $B_{\|}$, in μG. Then

$$n_e = \mathrm{EM}/\mathrm{DM} \ , \tag{2}$$

$$\langle n_e \rangle = \mathrm{DM}/L = \mathrm{RM}/0.81 B_{\|} L \ , \tag{3}$$

$$\langle n_e^2 \rangle = \mathrm{EM}/L \ , \tag{4}$$

where L is the line of sight through the HII region considered. EM can be derived from observations of the Hα line, recombination lines and the thermal free-free radiation at radio wavelengths, DM from pulsar observations and RM from observations of pulsars or extragalactic radio sources.

2 Observations

Figure 1 shows a compilation of the presently available data on f_v as a function of n_e.

Cordes et al. (1991) constructed a model of the distribution of electrons in the Milky Way based on observations of EM, DM and the scattering measure SM. The model consists of 3 components of increasing density: I outer disk, II inner disk and III clumps, which have decreasing values of f_v. Values of n_e for the clumps of $\sim$ 1 pc size I took from Fig. 2 of Habing and Israel (1979).

Leahy (1987) analyzed the variations in RM between the lobes of radio galaxies. He argued that the observed variations must be due to irregularities in n_e of about 1 pc rather than to variations in $B_{\|}$ and gave the range of f_v and n_e shown by the box labelled L in Fig. 1. However, recently Minter and Spangler (1996) presented evidence that B and n_e may be correlated making Leahy's estimate less certain.

Pynzar (1993) was the first who plotted f_v as a function of n_e. He used EM and DM towards pulsars and tried to separate interarm (P1 in Fig. 1) and arm (P2) regions along the line of sight using the spiral arm model of Georgelin and Georgelin (1976). For the classical HII regions (P3) he used EM and f_v from (1b).

Heiles et al. (1996) derived f_v and n_e of the ELDWIM in a spiral arm region towards $l = 20°$ based on EM, DM and the model of the n_e distribution of Taylor and Cordes (1993).

Berkhuijsen (in prep.) calculated f_v by comparing the flux density of the thermal radio emission and that expected from the disk of thermal electrons as given by Reynolds (1991). A short description of this method is given in Berkhuijsen et al. (1997). The thermal radio emission at the solar radius ($R_\odot$ = 8.5 kpc) I calculated from the 408 MHz model of Beuermann et al. (1985) and the spectral index map of Reich and Reich (1988), and I used an exponential fit of one z-component to the two z-components of the electron density given by Reynolds (1991). For the solar neighbourhood I find $f_v = 0.038 \pm 0.010$ and $n_e = \langle n_e \rangle / f_v = 0.92 \mp 0.25\,\mathrm{cm}^{-3}$ at $z = 0$. This value refers to a mixture of arm and interarm regions and is consistent with the results of Pynzar (1993) and Heiles et al. (1996).

3 Results and Discussion

In spite of the large uncertainties the data in Fig. 1 are fairly consistent with each other. The line connecting the points of Berkhuijsen and Heiles et al. (1996) has a slope of -0.7 ± 0.4, in agreement with that of the line through the points in Pynzar's Fig. 2 of -0.71 ± 0.05 (eye estimate). However, the results of Cordes et al. (1991) suggest a slope close to -1. Thus f_v is approximately inversely proportional to n_e. This means that $\langle n_e \rangle = f_v n_e \simeq$ constant and about the same for regions of high and of low density.

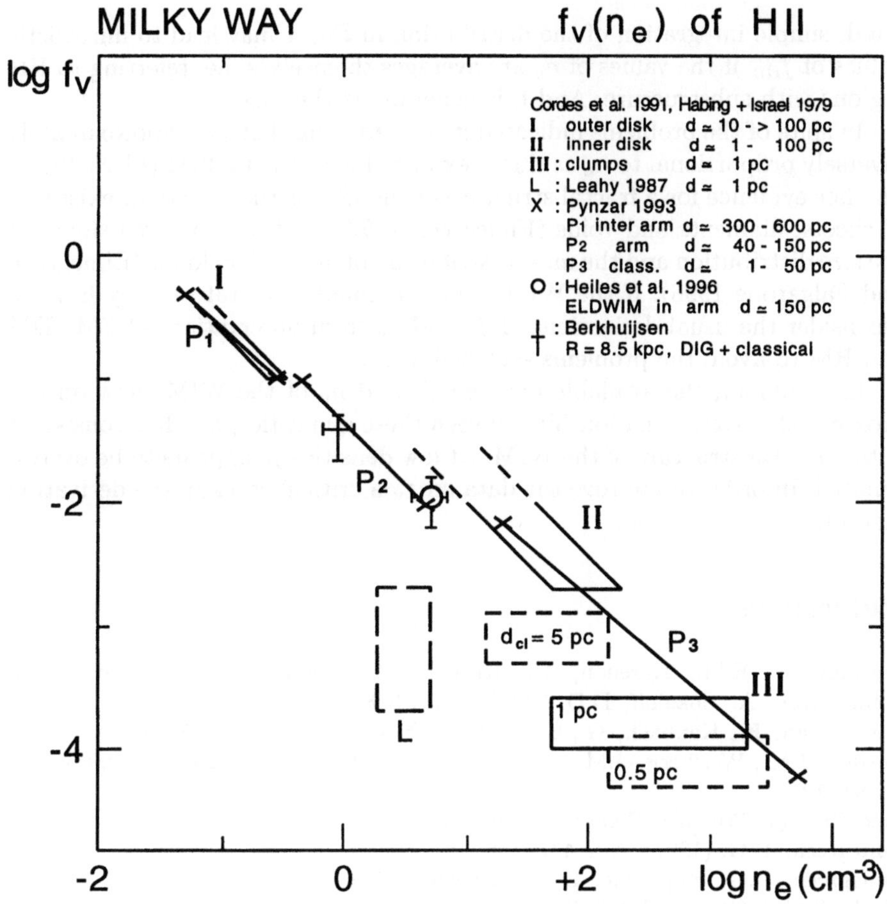

Fig. 1. The observed dependence of the volume filling factor f_v on the volume density of free electrons n_e.

On inspection of Fig. 1 two problems arise:

a) $f_v > 1$ for $n_e \lesssim 0.01\,\mathrm{cm}^{-3}$, thus no space is left for the hot interstellar medium (HIM). As the existence of the HIM seems well established f_v at low densities must be overestimated. This could happen if the electrons are distributed in sheets and filaments which occupy a smaller volume than spherical clouds. Sheets and filaments are indeed common in the ISM.

b) The filling factor of the total WIM, $f_{\rm HII}$, is the integral over $\log_e n_e$ of the function derived from Fig. 1. For the functions mentioned above integration yields $f_{\rm HII} = 0.005$ for the classical HII regions ($100 < n_e < 10^4\,\mathrm{cm}^{-3}$), but $f_{\rm HII} > 1$ for the diffuse gas ($0.01 < n_e < 100\,\mathrm{cm}^{-3}$) which is impossible. This again shows that at low densities f_v is overestimated. On the other hand, simple integration of the distribution in Fig. 1 may lead to unrealistic values of f_{HII} if the values of n_e are averages themselves, i.e. referring to HII regions with substructure. And this generally is the case.

In spite of the problems indicated it is interesting that f_v is approximately inversely proportional to n_e as this is expected for a fractal ISM (Fleck 1996). Further evidence for a fractal structure of the ISM is the abundant existence of sheets, filaments and voids (Elmegreen 1997), and the observed slopes of the size distribution and the mass distribution of molecular clouds (Elmegreen and Falgarone 1996). If the WIM is predominantly fractal we may have to reconsider the usual derivations of f_v and n_e from observations of EM, DM and RM to avoid the problems sketched above.

In summary, the available data on f_v and n_e of the WIM show an approximately inverse relationship between these quantities, which is consistent with a fractal structure of the WIM. At low densities f_v appears to be overestimated. In order to improve the data on f_v a critical check of the derivation methods seems required.

References

Berkhuijsen, E.M., Horellou, C., Krause, M., Neininger, N., Poezd, A.D., Shukurov, A., Sokoloff, D.D. (1997): A&A 318, 700

Beuermann, K., Kanbach, G., Berkhuijsen, E.M. (1985): A&A 153, 17

Cordes, J.M., Weisberg, J.M., Frail, D.A., Spangler, S.R., Ryan, M. (1991): Nat 354, 121

Cox, D.P. (1995): ASP Conference Ser. 80, 317

Elmegreen, B.G. (1997): ApJ 477, 196

Elmegreen, B.G., Falgarone, E. (1996): ApJ 471, 816

Fleck, R.C. (1996): ApJ 458, 739

Georgelin, Y.M., Georgelin, Y.P. (1976): A&A 49, 57

Habing, H.J., Israel, F. (1979): ARA&A 17, 345

Heiles, C., Reach, W.T., Koo, B-C. (1996): ApJ 466, 191

Kulkarni, S.R., Heiles, C. (1988): in Galactic and Extragalactic Radio Astronomy, eds. G.L. Verschuur, K.I. Kellermann, 2nd edition, p. 95

Leahy, J.P. (1987): MNRAS 226, 433

McKee, C.F. (1995): ASP Conference Ser. 80, 292

Minter, A.M., Spangler, S.R. (1996): ApJ 458, 194

Pynzar, A.V. (1993): Astron. Rep. 37, 245

Reich, P., Reich, W. (1988): A&A 196, 211

Reynolds, R.J. (1991): Proc. IAU Symp. 144, p. 67

Taylor, J.H., Cordes, J.M. (1993): ApJ 411, 674

The Influence of the Local Bubble on the Ionization of the Local Interstellar Cloud

Jonathan D. Slavin[1] and Priscilla C. Frisch[2]

[1] NASA/Ames Research Center, MS 245-3, Moffett Field, CA 94035-1000, USA
[2] University of Chicago, Department of Astronomy and Astrophysics, 5460 S. Ellis Avenue, Chicago, IL 60637, USA

Abstract. The ionization of the the Local Interstellar Cloud (LIC) is quite unusual compared with most warm ($T \sim 10^4$ K) interstellar gas. Hydrogen and helium are both partially ionized with helium surprisingly more ionized than hydrogen. Directly observed ionization sources including nearby stellar EUV sources and the diffuse emission of the Soft X-ray Background (SXRB), do not provide enough ionization and heating to account for both the ionization state and temperature of the LIC. We propose that an evaporative boundary between the LIC and the hot gas of the Local Bubble can provide the necessary ionizing radiation. Results of detailed models of the emission from the interface are presented and shown to be adequate to explain the observations.

1 Introduction

Recent measurements including both *in situ* measurements (see Frisch 1995 and references therein) and line of sight (column density) observations using *EUVE* (see Vallerga 1996) have shown that the Local Interstellar Cloud (LIC) has $N_{\rm HI}/N_{\rm HeI} \approx 10-14$, $X_{\rm H} \lesssim 0.4$, and $X_{\rm He} \lesssim 0.5$ (where $X_{\rm H} \equiv n_{\rm H^+}/n_{\rm H\,tot}$, and $X_{\rm He} \equiv n_{\rm He^+}/n_{\rm He\,tot}$). The velocity structure of the nearest interstellar gas is complex, but the lowest observed HI column densities observed are consistent with a column density for the local cloud alone of $\sim 5-10 \times 10^{17}\,{\rm cm}^{-2}$.

Both this absolute level of ionization of H and the relative ionization of H and He are unlike that observed in the warm ionized regions in the ISM. Reynolds (1989) finds that H must have a high degree of ionization, $X_{\rm H} > 0.67$, due to the lack of observable [O I] λ6300 Å emission associated with observed Hα emission. The low level of He ionization in the WIM, $X_{\rm He} \lesssim 0.27 X_{\rm H}$, is determined from the weakness of observed the He I λ5876 (Reynolds & Tufte 1995).

The directly observed sources of ionizing radiation fall into two categories: stellar EUV sources and diffuse soft x-ray emission. The former have all been observed by *EUVE* and the combined spectrum from the brightest sources has been presented by Vallerga (1996). The spectrum is unexpectedly dominated by the two B stars, ϵ CMa and β CMa. The most important part of the SXRB for ionization of the LIC is the low energy Be and B band radiation.

Vallerga (1996) has shown that the stellar EUV sources are not capable of providing the observed He ionization. We show below that emission from a 10^6 K, collisional ionization equilibrium plasma with emission measure sufficient to explain the SXRB, is also incapable of accounting for the observed ionization.

2 Radiation from the Boundary of the LIC

Most models of the Local Bubble assume it contains hot gas (but see Breitschwerdt & Schmutzler 1994) in order to explain the SXRB. If this is the case, then a warm cloud such as the LIC should be evaporating via thermal conduction. In the evaporative interface, gas is heated to temperatures intermediate between the cloud temperature, $T \approx 7000\,\mathrm{K}$, and the hot gas temperature, $T \sim 10^6\,\mathrm{K}$. At these temperatures the gas radiates strongly in the EUV. The mean ionizing photon energy of the interface radiation is $\bar{E} = 33\,\mathrm{eV}$ (376Å), making it efficient at ionizing He^0. Figure 1 shows a comparison of the stellar EUV field (de-absorbed to the edge of the LIC) and the spectrum from an evaporating cloud boundary model.

We have created models of the evaporative boundary which are similar to those of Slavin (1989). We assume steady flow evaporation and spherical symmetry and include the effects of radiative cooling, non-equilibrium ionization and saturation of heat flux. The spectra (as well as necessary ionization, recombination and cooling rates) are calculated using the Raymond & Smith plasma emission code (Raymond & Smith 1977 and updates).

The parameters for the model are as follow: $R_{cl} = 3\,\mathrm{pc}$, $n_{cl} = 0.22\,\mathrm{cm}^{-3}$, $T_f = 10^6\,\mathrm{K}$, $B_0 = 5.26\,\mu\mathrm{G}$, $\eta = 0.5$, where T_f is the temperature at an outer cutoff radius (30 pc), B_0 is the strength of the tangential magnetic field and η is the conductivity reduction factor. The mass loss rate for this model is $\dot{M} = 0.413\,M_\odot\,\mathrm{Myr}^{-1}$.

3 Ionization of the LIC

We take the cloud boundary spectrum we have generated and combine it with the stellar EUV spectrum and additional soft x-ray emission from hot gas to use as the flux incident on the LIC. To calculate the ionization in the cloud we employ the radiative transfer/thermal equilibrium code CLOUDY (Ferland 1996). CLOUDY calculates the detailed radiative transfer, including absorption and scattering, of the incident field and the diffuse continuum and emission lines generated within the cloud. The thermal and ionization balance is calculated at each point within the cloud.

Figure 2 shows the ionization resulting from the cloud boundary model as revealed by the column density ratio $N_{\mathrm{HI}}/N_{\mathrm{HeI}}$ vs. depth into the cloud. Also

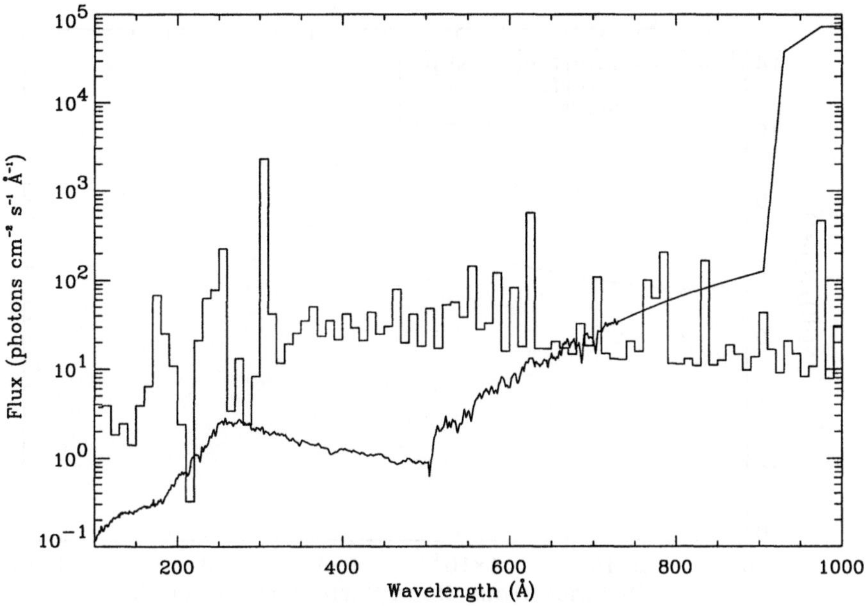

Fig. 1. Comparison of combined spectrum of stellar EUV sources with the model spectrum from an evaporative boundary on the LIC. The model spectrum is the histogram, binned at 10Å resolution. The stellar spectrum has been de-absorbed by $N_{\rm HI} = 9 \times 10^{17}\,{\rm cm}^{-2}$ to approximate the flux incident on the face of the LIC. The stellar spectrum between 730Å and 912Å is an extrapolation.

shown in the figure is the ionization which results from only including the radiation from the stellar EUV sources ("stars only") or from only including the boundary radiation ("interface only"). It is clear that without the interface radiation, the observed He ionization relative to H, $\langle N_{\rm HI}/N_{\rm HeI}\rangle \approx 10 - 14$, cannot be achieved. In addition, the heating rate without the interface radiation is too low, leading to a predicted cloud temperature of $T \lesssim 6000\,{\rm K}$.

An additional conclusion of our work on the ionization is that dust plays a critical role in the thermal balance of the cloud. Without dust we cannot achieve the observed temperatures, $T \approx 7000\,{\rm K}$, in the LIC. Both through grain photoelectric heating and reduced cooling due to depletion, the presence of dust raises the temperature.

4 Summary

An evaporative interface at the boundary of the LIC appears to be the only source of ionizing radiation capable producing the observed level of He ionization. Even with this source of radiation, however, there is insufficient heating

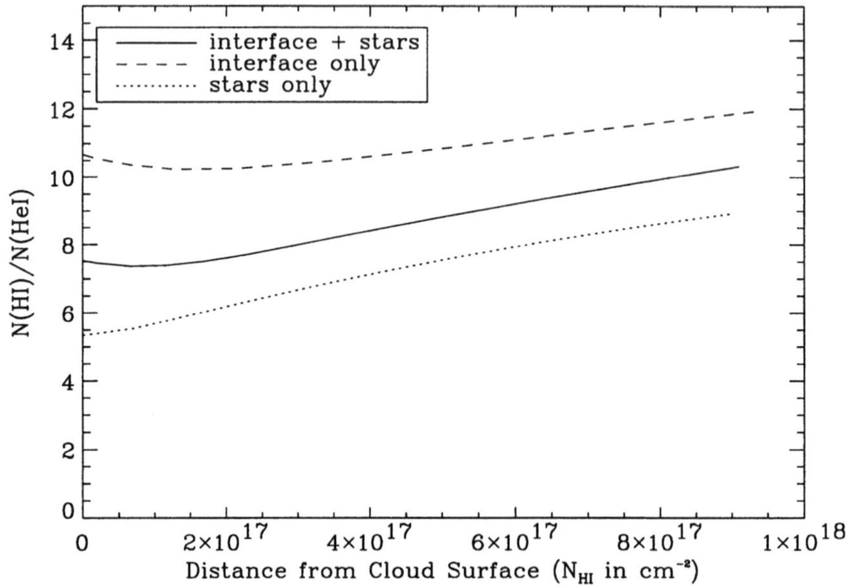

Fig. 2. Ratio of hydrogen to helium column density vs. depth into the cloud (expressed in terms of neutral hydrogen column density) for different incident fluxes. Here "stars" refers to the stellar EUV source flux. "Interface" refers to the radiation from the evaporative boundary between the LIC and the hot gas of the Local Bubble.

unless dust is present. More directly observed sources of radiation, i.e. stars and the SXRB, provide neither the ionization nor the heating necessary to produce the ionization and temperature observed in the LIC. The model of the ionizing radiation field will face tighter constraints as the ionization of the cloud is better determined by *in situ* measurements and line of sight observations towards nearby stars.

References

Breitschwerdt, D. & Schmutzler, T., 1994, Nature, 371, 774

Ferland, G. J. 1996, *Hazy, a Brief Introduction to Cloudy*, University of Kentucky Department of Physics and Astronomy Internal Report

Frisch, P. C. 1995, Space Sci. Rev., 72, 499

Raymond, J. C. & Smith, B. W. 1977, ApJS, 35 419

Reynolds, R. J. & Tufte, S. L. 1995, ApJ, 439, L17

Reynolds, R. J. 1989, ApJ, 345, 811

Slavin, J. D., 1989, ApJ 346, 718

Vallerga, J. V. 1996, Space Sci. Rev., 78, 277

Part VI

Distant Hot Gas, SXRB Fluctuations, Dust, γ-rays

Different points of view during the coffee break ...

... and the Wednesday afternoon excursion to Nymphenburg Castle.

Modeling the Galactic $\frac{3}{4}$ keV X-ray Background

M.J. Freyberg and J.H.M.M. Schmitt

Max–Planck–Institut für extraterrestrische Physik, D-85740 Garching, Germany

Abstract. We have analyzed the ROSAT PSPC all-sky survey maps of the soft X-ray background (SXRB) in the 3/4 keV band. One approach was to study the large-scale distribution of the X-ray emission with a multipole analysis. Here a significant dipole toward the galactic center region was found. This is interpreted in terms of variation of distant X-ray emission, e.g. galactic halo. Also a small-scale structure analysis of the 3/4 keV X-ray sky has been performed and a new analytic fluctuation probability distribution has been derived. No significant excess over the expected extragalactic point source contribution has been found. Finally, X-ray colours have been used to investigate spectral variations of the SXRB.

1 Introduction

At energies above approximately 2 keV most of the observed X-ray background is certainly of extragalactic origin except for a rather narrow galactic ridge seen out to energies of a few keV (cf. Worrall et al. 1982), while in the 1/4 keV band a major fraction of the X-rays originates within the next hundred parsec (e.g., the reviews by McCammon & Sanders 1990, Fabian & Barcons 1992, Breitschwerdt et al. 1996). At energies around and above 1 keV the decomposition of the diffuse background into galactic and extragalactic components is much less clear. Extrapolating from the 1/4 keV band and from higher energies into the 3/4 keV band an excess is found. However, at high galactic latitudes where "deep surveys" with imaging X-ray telescopes have traditionally been performed (cf. Giacconi et al. 1979, Hasinger et al. 1993), the dominant background component is still of extragalactic origin (cf. Schmitt & Snowden 1990).

In the following a report is given on the analysis of the ROSAT 3/4 keV band. The distance to the origin of the detected soft X-rays is related to their energy as the interstellar absorption cross section varies as $E^{-8/3}$ (away from absorption edges). The higher the energy the more distant emission is detected with ROSAT. Table1 summarizes the relation of the optical depth $\tau = 1$ for various ROSAT PSPC energy bands for Raymond-Smith-type plasma with $\log T\,[\mathrm{K}] = 6.0 - 6.4$. Abundances of cold and hot matter used in our modeling were according to Morrison & McCammon (1983) and Raymond & Smith (1977), respectively.

Band	Channels	Column density for $\tau = 1$
R1	8 – 19	8 – 10.5 $\times 10^{19}\,\mathrm{cm}^{-2}$
R2	20 – 41	1 – 1.5 $\times 10^{20}\,\mathrm{cm}^{-2}$
R4	52 – 69	6 – 12 $\times 10^{20}\,\mathrm{cm}^{-2}$
R5	70 – 90	1 – 1.5 $\times 10^{21}\,\mathrm{cm}^{-2}$
R6	91 – 131	1.8 – 2.7 $\times 10^{21}\,\mathrm{cm}^{-2}$
R7	132 – 201	3 – 6 $\times 10^{21}\,\mathrm{cm}^{-2}$

Table 1. Optical depth $\tau = 1$ for various energy bands for Raymond-Smith-type plasma with $\log T\,[\mathrm{K}] = 6.0 - 6.4$ for the ROSAT PSPC. The PSPC channels multiplied with 10 eV give the approximate energies.

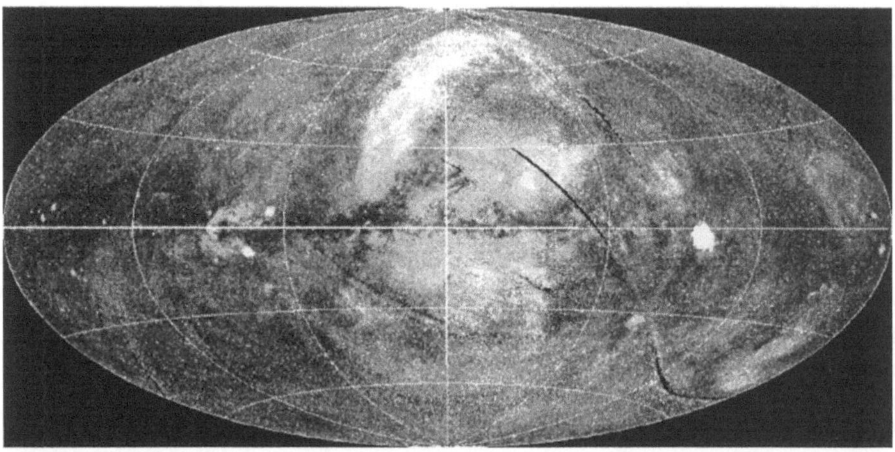

Fig. 1. RGB image of the SXRB using the new survey maps (Snowden et al. 1997). The three energy bands, 1/4, 3/4, and 1.5 keV, are colour-coded with red, green, and blue, respectively. As in all our maps we use galactic coordinates with increasing longitudes to the left and $l = 0°$ in the center; (see Plate 3).

2 X-ray Colours

One way to characterize spectral parameters of X-ray emission is in terms of X-ray colours. This is especially useful for soft X-rays as the absorption cross section strongly depends on energy (Tab.1). Figure 1 shows a 3-colour map of the ROSAT PSPC all-sky survey. Three energy bands, 1/4, 3/4, and 1.5 keV, are colour-coded with red, green, and blue, respectively. At higher galactic latitudes soft emission dominates while in the galactic plane this band is highly absorbed and only hard X-rays are present.

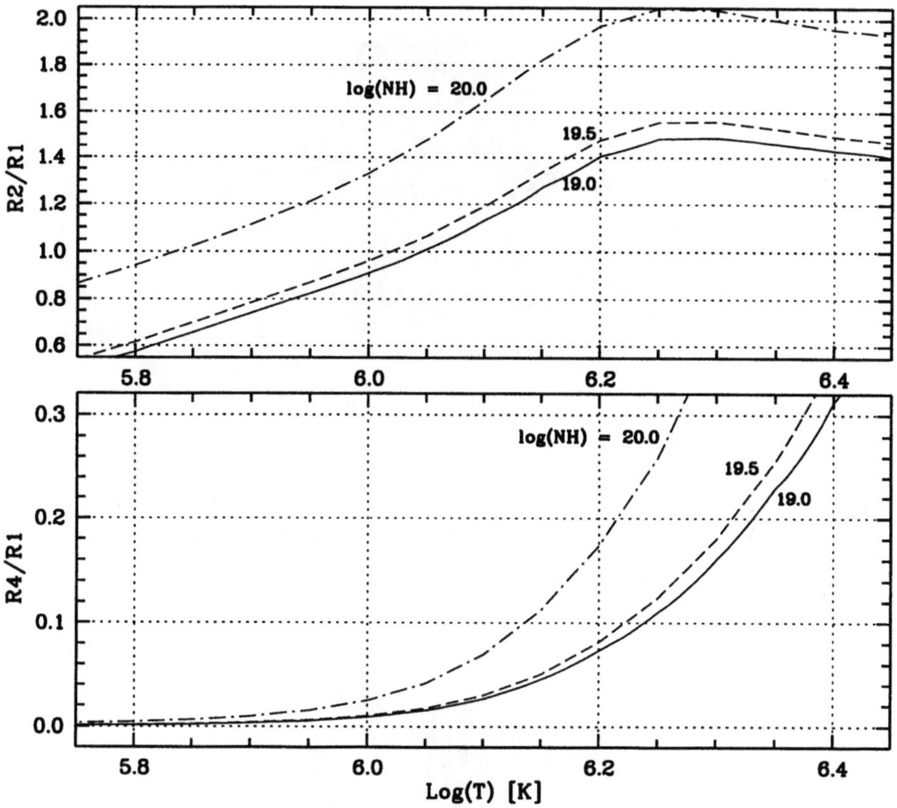

Fig. 2. a) R2/R1 ratio as a function of plasma temperature for three representative column densities N_H. b) R4/R1 ratio similar to above.

The ratio of the Wisconsin B and C band survey maps showed only small variations of C/B with large variations of absorbing neutral hydrogen column densities N_H (McCammon et al. 1983). This led to the conclusion that most (if not all) X-rays in the 1/4 keV band originate in front of the bulk of absorbing matter. For the ROSAT PSPC corresponding bands are the R1 and R2 band, respectively. In Fig.2a we show the dependence of R2/R1 as a function of plasma temperature for 3 column densities, $\log N_H\,[\mathrm{cm}^{-2}] = 19.0, 19.5, 20.0$. Similarly, Fig.2b gives the ratio R4/R1 for these column densities. Snowden et al. (1998) discuss the R2/R1 ratio in detail and try to disentangle local and distant contributions to the 1/4 keV band.

The R4/R1 ratio is compared to the R2/R1 ratio more sensitive to distant emission. While a typical value for R2/R1 is 1.1 (consistent with $\log T\,[\mathrm{K}] \sim 6.05$ and $\log N_H\,[\mathrm{cm}^{-2}] \sim 19.5$), an average R4/R1 at low galactic latitudes in the galactic anti-center direction is 0.13. This is not consistent with a low local temperature, but would require $\log T\,[\mathrm{K}] > 6.25$. At the moment

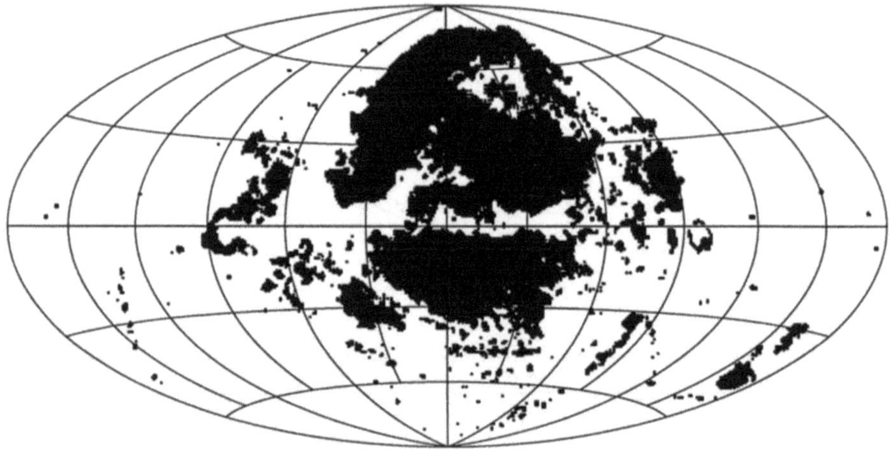

Fig. 3. Distribution of regions with R1 $< 100 \times 10^{-6}\,\mathrm{cts\,s^{-1}\,arcmin^{-2}}$ and R4 $> 80 \times 10^{-6}\,\mathrm{cts\,s^{-1}\,arcmin^{-2}}$. A median filter had been applied to the maps.

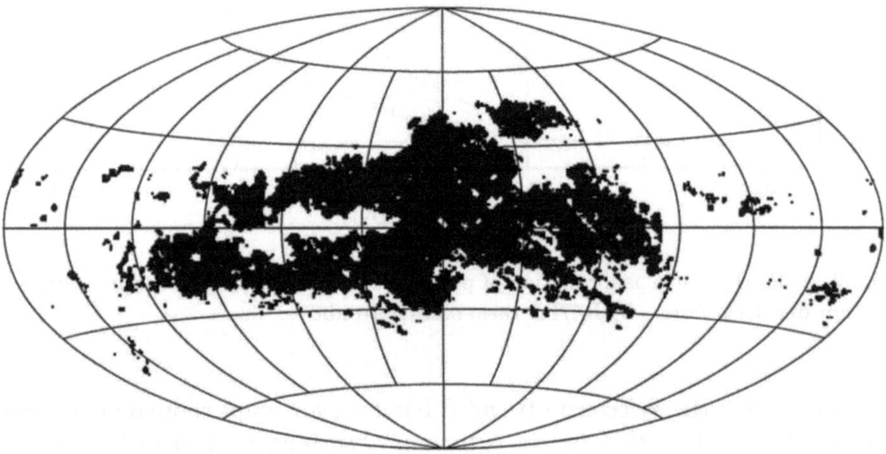

Fig. 4. Distribution of regions with R1 $< 200 \times 10^{-6}\,\mathrm{cts\,s^{-1}\,arcmin^{-2}}$ and corrected (local bubble subtracted, residual de-absorbed) R4 $> 50 \times 10^{-6}\,\mathrm{cts\,s^{-1}\,arcmin^{-2}}$.

X-ray shadows in the 3/4 keV band – e.g. MBM 12 (Kuntz et al. 1997) – do not rule out such a high local temperature, however, they are also not inconsistent with $\log T\,[\mathrm{K}] \sim 6.25$. Or phrased differently, observational data do not rule out either possibility of local or distant origin of the galactic R4 (and R5) band emission. If the 3/4 keV background were merely due to point-like sources (e.g., variations in the extragalactic spectrum or galactic source population), then these could be traced by a fluctuation analysis (cf. Sect.4). The overall spatial distribution of the 3/4 keV emission is discussed

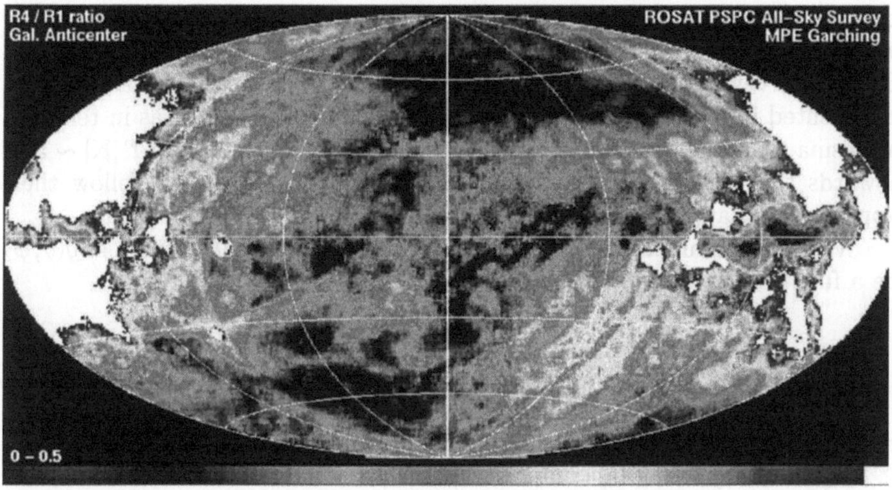

Fig. 5. R4/R1 ratio centered at $l = 180°$, the colour bar ranges from 0.0 to 0.5; (see Plate 4).

in Sect.3, where variations are most likely due to a hot galactic halo. Using the R4/R1 ratio spectral changes in the soft X-ray background (SXRB) are visible, which may reflect spatial emission variations but may also be explained by a non-isothermal distant component (like in the model by Wang 1997). In Fig.3 we show the part of the sky that shows low R1 with $R1 < 100$ but high R4 with $R4 > 80$. Snowden et al. (1997) interpret this component as galactic bulge emission. If we assume a standard local (hot bubble) model with $\log T\,[\mathrm{K}] \sim 6.05$, and subtract local contributions and de-absorb the residual emission for galactic absorption we obtain a map like in Fig.4. Here the points denote regions with $R1 < 200$ and $R4_{\mathrm{mod}} > 50$ (where $R4_{\mathrm{mod}}$ denotes the local bubble subtracted and de-absorbed $R4$ intensity). Obviously, this is still concentrated toward the galactic center, but extends also into the second quadrant. The region at $l \sim 240°$ seems to lack R4 band emission. The Sco-Cen superbubble contributes to the northern central region ($l^{\mathrm{II}} > 20°$). Assigning the structure of X-ray colours to a distant hot plasma this can be converted into temperatures of $\log T\,[\mathrm{K}] \sim 6.35$. Finally, Fig.5 shows a sky map of the R4/R1 ratio centered on $l = 180°$, where the spectral change from center to anti-center can clearly be seen.

3 Large-Scale Structure: Harmonic Analysis

To characterize the large-scale structure of the SXRB the intensity distribution can be expanded in spherical harmonics. The strengths of the individual orders then gives information about isotropic (monopole, e.g. extragalactic background) and non-isotropic components (dipole, quadrupole, etc.). Such a dipole was found in a harmonic analysis of the ratio of the Wisconsin C and B band survey maps (Snowden et al. 1990): this colour gradient was found to be oriented in the galactic plane, one possible interpretation was in terms of a plasma temperature gradient from the galactic anti-center ($\log T\,[\mathrm{K}] \sim 5.9$) towards the galactic center region ($\log T\,[\mathrm{K}] \sim 6.2$). Here we follow their approach and apply the same method on the ROSAT 3/4 keV band.

We expand a function g into a set of orthogonal basis functions. Let $g(\theta, \phi)$ be a function on the unit sphere, then

$$g(\theta, \phi) = \sum_{l=0}^{\infty} \sum_{m=-l}^{m=+l} A_{lm} Y_{lm}(\theta, \phi) . \tag{1}$$

Here $Y_{lm}(\theta, \phi)$ denote the normalized spherical harmonics (e.g., Abramowitz & Stegun 1965) and A_{lm} complex coefficients. $l = 0$ describes a monopole, $l = 1$ a dipole and so forth. We want to fit a model (certain orders l) to a measured quantity $g(\theta, \phi)$ such as the ROSAT 3/4 keV corrected count rate. This fit is performed by minimizing χ^2,

$$\chi^2 = \int \frac{\mathrm{d}\Omega}{\sigma^2(\theta, \phi)} \left[g(\theta, \phi) - \sum_{l=0}^{L_{\max}} \sum_{m=-l}^{m=+l} A_{lm} Y_{lm}(\theta, \phi) \right]^2 . \tag{2}$$

The integration over the solid angle turns into a sum in case of a discrete quantity $g(\theta, \phi)$. the weight functions $\sigma(\theta, \phi)$ contain statistical and systematic uncertainties. $L_{\max}$ is the maximum multipole order to be considered. Using real $g(\theta, \phi)$ only A_{lm} with $m \geq 0$ are independent quantities. We use galactic coordinates $\phi = l^{\mathrm{II}}$ and $90° - \theta = b^{\mathrm{II}}$. Equation (2) can be solved by inversion of a matrix of order $(L_{\max} + 1)^2$ (see also Snowden et al. 1990; Freyberg 1994, 1996).

We excluded all those regions of the sky that appeared to be non-typical for the SXRB (e.g., North Polar Spur, Cygnus, Vela) as well as the galactic plane, the galactic center and anti-center regions. Also a minor foreground contribution has been subtracted. To suppress small-scale effects larger spatial bins have been chosen ($\sim 7 - 28$ deg^2). In the 3/4 keV band a significant dipole (strength $\leq 10\%$) has been found in the general direction of the galactic center, $l^{\mathrm{II}} \sim 345°$, $b^{\mathrm{II}} \sim +6°$ (note, that this region had been excluded). The dipole was very stable against further omissions of regions in the fitting procedure. This may be interpreted in terms of a large-scale temperature gradient in the *local* component like in Snowden et al. (1990). However, one can also assume *distant* origin and assign this dipole to a galactic X-ray halo. Higher multipole moments could in principle constrain the shape of such a halo. Unfortunately, quadrupole and even higher moments seem to have reached the noise level which means that these contributions may be affected by small-scale fluctuations like unresolved sources, dim supernova remnants etc. Moreover, no formally acceptable fits could be achieved ($\chi^2_{\mathrm{red}} > 5$) which could be caused by an underestimation of systematic effects (and thus too small uncertainties σ). In a similar analysis the 1.5 keV band data did not show any significant dipole, all moments but the monopole were of approximately equal strength. This may be caused by the intrinsically high fraction of small-scale structure due to the presence of extragalactic sources.

4 Small-Scale Structure: Fluctuation Analysis

4.1 Introduction

The usual approach in studies to determine the nature of the diffuse X-ray background is to image a fraction of the sky to very low flux levels, to detect and identify all X-ray sources and to construct the so-called $\log N - \log S$ function, i.e. the cumulative distribution function of the number of sources per steradian with fluxes in excess of S, i.e. $N(>S)$. Integration of $\frac{dN(>S)}{dS}$ then yields the contribution of the detected sources to the total background.

Having exploited the information contained in the detected sources the question remains whether the not detected can provide any additional information. This problem was first studied by radio astronomers; within rather large beams of the radio telescopes used at that time there would typically be more than one source. Scheuer (1957) developed a formalism how to retrieve information on the number density of sources below the detection limit. The basic idea here is to look at the background fluctuation spectrum. If the background is made up of many individually faint sources, the background fluctuation spectrum will be quite different from the case when the background is made up of a relatively smaller number of stronger sources. Later, the same formalism was applied to X-ray astronomy by Scheuer (1974). Following Freyberg (1994) a fully analytic computation method is presented to study fluctuations in the diffuse galactic SXRB assuming a power-law source luminosity distribution.

4.2 The Fluctuation Probability Distribution

Let a source population with a cumulative flux distribution of the form

$$N(>S) = K\,(S/S_0)^{-\beta} \tag{3}$$

be given, where $N(>S)$ denotes the number of sources per steradian with flux exceeding the level S, S_0 is an arbitrary flux constant, β is the power law index, and K a normalization constant. The integrated flux $\Phi(>S)$ of sources with flux exceeding S and distributed as in Eq.(3) is given by

$$\Phi(>S) = \int_S^\infty d\tilde{S}\,\tilde{S}\,\nu(\tilde{S}) \tag{4}$$

with the differential relation $\nu(S) \sim \frac{dN(>S)}{dS}$. Inserting Eq.(3) we obtain

$$\Phi(> S) = S_0 \, K \, \frac{\beta}{\beta - 1} \, (S/S_0)^{1-\beta}. \tag{5}$$

Clearly, $\Phi(> S)$ diverges at faint flux levels (Olbers paradox), and therefore Eq.(3) must cut off at some flux level $S_{\rm cut}$. If the diffuse cosmic X-ray background flux $\Phi_{\rm cosm}$ per steradian were known and assumed to be entirely composed of discrete sources, the required cutoff flux could be calculated from $\Phi(> S_{\rm cut}) = \Phi_{\rm cosm}$,

$$S_{\rm cut} = S_0 \left(\frac{\beta - 1}{\beta} \frac{\Phi_{\rm cosm}}{K \, S_0} \right)^{1/(1-\beta)} . \tag{6}$$

In applications of X-ray astronomy the fundamental observable is not the X-ray flux, but the (integer) number of counts n measured in an effective exposure time $t_{\rm eff}$ and an effective solid angle $\Omega_{\rm eff}$. We assume n to be Poisson distributed with some (real) mean $C_{\rm exp}$. Depending on the prevailing instrument characteristics and source spectrum a (single) source with flux S gives rise to $C_{\rm exp} = S \, t_{\rm eff} \, f$ with a suitable flux-to-count conversion factor f. Hence Eq.(3) can be immediately expressed as

$$N(> C) = K \, \Omega_{\rm eff} \, (C/C_0)^{-\beta}, \tag{7}$$

with $N(> C)$ denoting the number of sources in the solid angle $\Omega_{\rm eff}$ with an expected number of counts in excess of C and $C_0 = S_0 \, t_{\rm eff} \, f$. The expected number of counts in a given beam is given by the superposition of sources distributed according to Eq.(7). The probability distribution of such a sum of random variables, i.e., the fluctuation spectrum, can be most easily calculated with characteristic functions (cf., Scheuer 1957, Scheuer 1974). Denoting by $p_1(C)$ the differential distribution of expected counts (i.e., $p_1(C) \sim C^{-(\beta+1)}$ for $C > C_0$), the characteristic function $p_1(\omega)$ of C is given by

$$p_1(\omega) = \int_{C_0}^{\infty} p_1(C) \, \exp(i\omega C) \, dC \, . \tag{8}$$

As long as $C_0 > 0$ and $\beta > 1$ this integral converges and yields $p_1(\omega) = \beta \, (-i\omega C_0)^{\beta} \, \Gamma(-\beta, -i\omega C_0)$,where $\Gamma(a, z)$ denotes the complementary incomplete Γ-function as usual. For convenience let us define a modified Γ-function in which a factor z^{-a} is absorbed, i.e. $\hat{\Gamma}(a, z) \equiv z^{-a} \Gamma(a, z)$. Furthermore, $p_N(\omega) \equiv \exp\{N[p_1(\omega) - 1]\}$, where in this definition N is an arbitrary constant, which in our case becomes $N = K\Omega_{\rm eff}$. The probability $\hat{p}_N(C)$ of obtaining the number of counts C is given by the Fourier transform of $p_N(\omega)$: $\hat{p}_N(C)$ is the probability distribution of the expected number of counts which is not observed directly in X-ray astronomy, but rather convolved with a Poisson distribution, and so statistical fluctuations due to counting statistics may enlarge or diminish the "intrinsic" background fluctuations. Let therefore the expected total number of (unfluctuated) background counts per beam be given by B. This background can be thought of being composed of the mean cosmic (truly) diffuse background as well as backgrounds produced locally (i.e., particle events and scattered solar X-rays). The total background $B_{\rm total}$ consists of a fluctuating and non-fluctuating background component, $B_{\rm total} = B + C$. Interpreting B as the mean value of the unfluctuated background and C as the fluctuated (Poisson) mean background, both measured *per pixel*, the probability $p(n)$ to observe n counts in a beam is then given by

$$p(n) = \int_0^\infty dC \left[\frac{(B+C)^n}{n!} \, \mathrm{e}^{-(B+C)} \right] \hat{p}_N(C) \,. \tag{9}$$

$p(n)$ finally represents the probability to observe n counts in the solid angle $\Omega_{\rm eff}$, given the parameters B, β, K and C_0. The integrals in Eq.(9) can be solved using the residue theorem (cf. Freyberg & Schmitt 1998). Also a recurrence relation can be obtained such that the $\hat{\Gamma}$-function has to be computed just once. Since the probability distributions can be computed analytically this method is suitable for a wide range of applications. Freyberg & Schmitt (1998) discuss the fluctuations in the new SXRB all-sky maps with $12'$ resolution and bright point sources masked (Snowden et al. 1997).

For practical use we rearranged the parameters of the distribution, the mean $\langle B_{\rm total}\rangle$ of the total background in counts, and the fraction $\xi = \frac{\langle C\rangle}{\langle B_{\rm total}\rangle}$ of the fluctuated background $\langle C\rangle$ of the total background $\langle B_{\rm total}\rangle$. Tests of this method have been performed with non-cosmic background components such as the PSPC particle background ($\beta \sim 4.2$ and $\xi < 0.01$) and scattered solar X-rays. Also pointed observations (much longer exposure) can be used for this analysis. In the 3/4 keV band best maximum likelihood fit parameters at $b^{\rm II} \sim 20°$ were $\beta \sim 1.6$ and $\xi \sim 0.12$. At higher galactic latitudes ξ tends to increase while at lower latitudes ξ decreases. At 1/4 keV we obtained $\beta \sim 2.0$ and $\xi \sim 0.05$, at 1.5 keV $\beta \sim 1.5$ and $\xi \sim 0.35$. In the 3/4 keV band similar values are expected from the extragalactic background. Therefore it appears unlikely that much of the excess 3/4 keV X-rays is due to point sources. Besides point-like ***sources*** also small-scale ***absorption*** variations (clouds, holes) enter the fluctuation distribution. While extra clouds cause lower counts than expected holes in the interstellar medium can be related to "sources". The angular scales of $1'$ used in our analysis of pointed observations can be converted to spatial scales of 1 pc if we assume a distance of 60 pc. Recent observations and models of the local neutral interstellar medium have shown inhomogeneities even on scales < 0.01 pc (e.g., Diamond et al. 1989, Marscher et al. 1993, Meyer & Blades 1996, Elmegreen 1997, Vergely et al. 1997). This has not been discussed here in detail and will be presented in a later work.

5 Summary

In our short report we have given an overview of the analysis of the 3/4 keV SXRB. In particular, we have developed a new computational scheme for fluctuation analysis. Point sources do not appear to dominate the 3/4 keV band excess. Multipole analysis has revealed a dipole similar to the C/B ratio found in the Wisconsin surveys. We favour spatial variations of a distant hot plasma component (galactic halo). However, X-ray colours do not exclude additional temperature changes of the distant component or a multi-component distant emission.

References

Abramowitz M., Stegun I. (eds.), 1965, *Handbook of Mathematical Functions*, Dover, New York

Breitschwerdt D., Egger R., Freyberg M.J., Frisch P.C., Vallerga J.V., 1996, Space Sci.Rev. 78, 183

Diamond P.J., Goss W.M., Romney J.D., Booth R.S., Kalberla P.M.W., Mebold U., 1989, ApJ 347, 302

Elmegreen B.G., 1997, ApJ 477, 196

Fabian A.C., Barcons X., 1992, ARA&A 30, 429

Freyberg M.J., 1994, PhD Thesis, LMU München

Freyberg M.J., 1996, in "The Physics of Galactic Halos", H. Lesch, R.-J. Dettmar, U. Mebold, R. Schlickeiser (eds.), Akademie-Verlag Berlin, p.117

Freyberg M.J., Schmitt J.H.M.M., 1998, *in preparation*

Giacconi, R., Bechtold, J., Branduardi, G., Forman, W., Henry, J.P., Jones, C., Kellogg, E., van der Laan, H., Marshall, H., Murray, S.S., Pye, J., Schreier, E., Sargent, W.L.W., Seward, F., Tananbaum, H.: 1979, ApJ 234, L1

Kuntz K.D., Snowden S.L., Verter F., 1997, ApJ 484, 245

Lukacs E., 1970, *Characteristic Functions*, 2nd ed., Griffin, London

Marscher A.P., Moore E.M., Bania T.M., 1993, ApJ 419, L101

McCammon D., Burrows D.N., Sanders W.T., Kraushaar W.L., 1983, ApJ 269, 107

McCammon D., Sanders W.T., 1990, ARA&A 28, 657

Meyer D.M., Blades J.C., 1996, ApJ 464, L179

Morrison R., McCammon D., 1983, ApJ 270, 119

Raymond J.C., Smith B.W., 1977, ApJSS 35, 419

Scheuer P.A.G., 1957, Proc.Cambridge Philos.Soc. 53, 764

Scheuer P.A.G., 1974, MNRAS 166, 329

Schmitt J.H.M.M., Snowden S.L., 1990, ApJ 361, 207

Snowden S.L., Schmitt J.H.M.M., Edwards B.C., 1990, ApJ 364, 118

Snowden S.L., 1996, in "Proc. of the International Conference on *Röntgenstrahlung from the Universe* in Würzburg", eds. H.U. Zimmermann, J.E. Trümper & H. Yorke, MPE Report 263, 299

Snowden S.L., Egger R., Freyberg M.J., McCammon D., Plucinsky P.P., Sanders W.T., Schmitt J.H.M.M., Trümper J., Voges W., 1997, ApJ 485, 125

Snowden S.L., Egger R., Finkbeiner D.P., Freyberg M.J., Plucinsky P.P., 1997, ApJ (*accepted*)

Wang Q.D., 1997, these proceedings

Worrall D.M., Marshall F.E., Boldt E.A., Swank J.H., 1982, ApJ 255, 111

Vergely J.-L., Egret D., Freire Ferrero R., Valette B., Köppen J., 1997, in "Hipparcos – Venice '97", ed. B. Battrick, ESA SP-402, p.603

Tracing Small-Scale Fluctuations in the Soft X-ray Background

R. Warwick[1], I. Hutchinson[1], R. Willingale[1], K. Kuntz[2], and S. Snowden[3]

[1] Department of Physics and Astronomy, University of Leicester, Leicester, UK
[2] University of Maryland, College Park, MD 20742, USA
[3] NASA/Goddard Space Flight Center, Greenbelt, MD 20771, USA

Abstract. An overlapping set of *ROSAT* PSPC observations made in a region of very low Galactic foreground column density, has been used to investigate variations in the soft X-ray background on angular scales of $15' - 5°$. In the 1/4 keV band there is a clear inverse correlation of the count-rate with the line-of-sight hydrogen column density. However, after correcting for this absorption effect, strong residual fluctuations remain in the data, with an amplitude which is significantly larger than that due to the counting statistics or the confusion of unresolved discrete sources. In contrast a similar analysis for the 3/4 and 1.5 keV *ROSAT* bands shows no evidence for an excess signal. The most likely origin of the 1/4 keV fluctuations would seem to be in a patchy distribution of $\sim 10^6$ K gas in the Galactic halo.

1 Introduction

A major achievement of the *ROSAT* mission has been the detailed mapping of the spatial structure in the soft X-ray background (SXRB) both on an all-sky basis and in pointed observations. This observational progress has, in turn, led to considerable new insight into the nature and origin of the background radiation, particularly in the softest accessible band at 1/4 keV. The observed diffuse flux in the 1/4 keV band is now known to be comprised of at least three components, emission from 10^6 K plasma residing in the Local Bubble, thermal emission at a similar temperature associated with the Galactic Halo and extragalactic flux of uncertain spectral form, representing the summed emission from discrete sources including broad-line QSOs. Numerous attempts (e.g. Burrows & Mendenhall 1991; Snowden et al. 1991; Wang & Yu 1995; Snowden 1997) have been made to separate the foreground emission (produced in the Local Bubble typically within ~ 100 pc of the Sun) from the halo/extragalactic flux via the shadows cast on the latter by cold interstellar clouds (a gas column density of $N_H \sim 10^{20}$ cm^{-2} gives an optical depth of unity at 1/4 keV). One such study (Snowden et al. 1994) considered the diffuse 1/4 keV emission observed in the primary Lockman Hole region (an area of ~ 300 deg^2 centred on $l \approx 147°, b \approx 54°$) where the Galactic line-of-sight column density, N_H, falls to a global minimum value. It was found that, although the observed 1/4 keV intensity shows a clear anticorrelation with N_H, there is very considerable scatter in the correlation plot. Here we focus on the origin of this scatter. A preliminary version of this work has been reported earlier by Barber, Warwick & Snowden (1996).

2 Data Reduction and Analysis

The present investigation is based on a set of nine overlapping *ROSAT* PSPC observations targeted at a region near $l \approx 161°, b \approx 51°$, in the south-west corner of the primary Lockman Hole. The $N_{\rm H}$ in the surveyed region is relatively uniform ($6 - 10 \times 10^{19}$ cm^{-2}) except in one or two localized areas where it increases to a maximum of 14.0×10^{19} cm^{-2}.

The data from each observation were reduced using standard background subtraction, exposure correction and flat-fielding procedures (Snowden 1995). After correcting for field-to-field baseline variations, the 1/4 keV PSPC images were mosaiced together, smoothed with a $4'$ radius circular top-hat function and bright sources removed. The observed 1/4 keV background intensity was then correlated with N_H (we use the same N_H data as Snowden et al. 1994). As in the earlier study of Snowden et al. (which encompasses a much larger sky area but with significantly lower exposure) these parameters show a clear inverse correlation consistent with the progressive absorption of halo/extragalactic flux with increasing N_H. In order to investigate the residual scatter in the correlation plot we corrected the original PSPC images for the effects of absorption (by adding back signal in accord with the best-fitting absorption model) and then heavily smoothed the source-excluded image with a $15'$ radius top-hat function. The resulting map is shown in Fig. 1.

The peak-to-peak signal in the residual fluctuations apparent in Fig.1 on a scale of $\sim 15' - 30'$ amounts to $\sim 15\%$ of the average surface brightness. In principle such fluctuations could arise from counting statistics, from the confusion effects of discrete sources just below the source exclusion threshold or from variations attributable to the diffuse SXRB. The magnitude of the first effect can be determined from the typical exposure time of ~ 10 ksec whereas the confusion noise can be estimated from the source counts measured in the 0.5–2.0 keV *ROSAT* band (e.g. Hasinger et al. (1993)), assuming a standard spectral conversion to the 1/4 keV band (in fact we use a power-law spectrum with an energy index $\alpha = 1.0$). In practice we have used Monte Carlo simulations of the mosaiced image in Fig.1 to determine the expected level of scatter and find that the observed fluctuations are very significant (i.e. $> 4\sigma$) with an amplitude roughly a factor two greater than the estimates based solely on counting statistics and confusion noise.

As a check the analysis was repeated for the 3/4 keV and 1.5 keV ROSAT bands (where the effects of Galactic absorption are negligible). In this case the Monte Carlo simulations showed that the level of the observed fluctuations were consistent with the predicted scatter. This would seem to rule out the clustering of discrete sources as a source of the 1/4 keV excess fluctuations (see also Soltan et al. 1996; Carrera, Fabian & Barcons 1997).

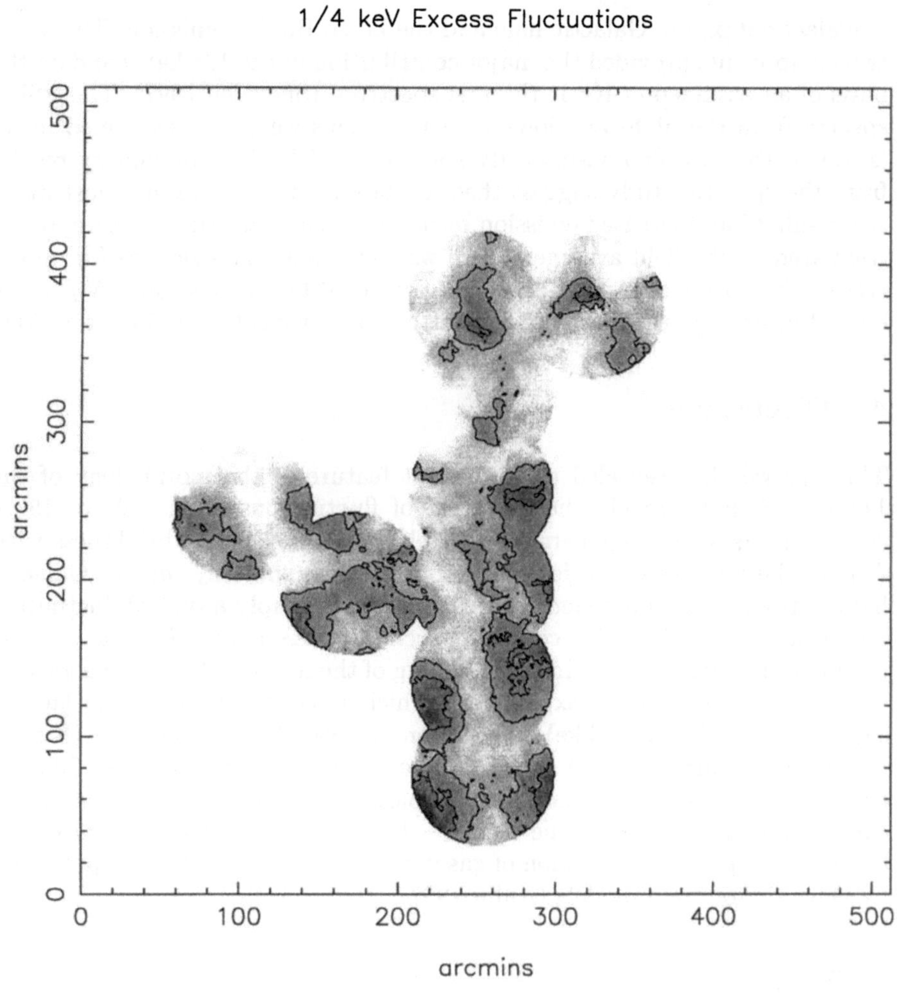

Fig. 1. The *ROSAT* PSPC image of the residual 1/4 keV fluctuations after correcting for the best-fitting absorption model. The grey scale encompasses surface brightnesses from $1050 - 1350 \times 10^{-6}$ count s^{-1} $arcmin^{-2}$ (white to black). The contours correspond to surface brightnesses of 1150 and 1200×10^{-6} count s^{-1} $arcmin^{-2}$ and accentuate the bright fluctuations.

3 Spectral Signature of the 1/4 keV Fluctuations

The spectral characteristics of the fluctuations may provide a further clue as to their origin. Based on the 1/4 keV background image shown in Fig. 1 we have extracted PSPC spectra for the high-, mid- and low-intensity regions. The spectrum corresponding to the mid-intensity regions was then fitted with a model including spectral components representative of the extragalactic flux, the Galactic halo and the Local Bubble emission. The latter two components provided the major contribution in the 1/4 keV band on the basis of an assumed $\sim 10^6$ K thermal spectrum (Snowden 1997). The PSPC spectra from the high- and low-intensity regions were then examined in relation to the best-fit mid-intensity spectral model. The preliminary results from the spectral study suggest that the bright fluctuations are most likely the result of an increased emission measure in the Galactic halo component compared to the field average. There was also marginal evidence for an excess of absorption over and above that predicted by the measured N_H in the low-intensity regions. However, this latter point needs further investigation.

4 Discussion

This analysis has revealed an important feature of the morphology of the 1/4 keV SXRB, namely the existence of fluctuations on a scale of $15' - 30'$ superimposed on a general anticorrelation of the surface brightness with N_H (of the form expected for the absorption of a spatially smooth Galactic halo/extragalactic component). The observations imply a typical fluctuation amplitude of $\sim 50 \times 10^{-6}$ count s^{-1} $arcmin^{-2}$. As noted above a possible origin of the fluctuations is in the clumping of the hot $\sim 10^6$ K plasma located in the Galactic halo, an explanation which is consistent with the lack of significant small-scale 1/4 keV fluctuations at lower Galactic latitude (Kuntz, Snowden & Warwick 1997). If we assume a typical clump extent of $20'$, the clump lifetime can be estimated on the basis of the sound crossing time; this calculation suggests a lifetime of $10^5 - 10^6$ years. It is plausible that these clumps relate to the circulation of gas in the lower Galactic halo as predicted by Galactic fountain models (Kahn 1997).

References

Barber C.R., Warwick R.S., Snowden, S.L. (1996): MPE Report 263, 319
Burrows D.N., Mendenhall, J.A. (1991): Nature, 351, 629
Carrera F.J., Fabian A.C., Barcons, X. (1997): MNRAS, 285, 820
Hasinger, G., et al. (1993): A&A, 275, 1
Kahn, F.D. (1997): these proceedings
Kuntz K.D., Snowden S.L., Warwick R.S. (1997): these proceedings
Snowden S.L. et al. (1991): Science, 252, 1529
Snowden, S.L. et al. (1994): ApJ, 430, 601
Snowden S.L. (1995): Cookbook for Analysis Procdures for *ROSAT* XRT/PSPC Observations of Extended Objects and the Diffuse Background
Snowden, S.L. (1997): these proceedings
Soltan A., Hasinger G., Egger R., Snowden S., & Trümper J. (1996): A&A, 305, 17
Wang Q.D., Yu K.C. (1995): AJ, 109, 698

Intensity Variations of the Soft X-ray Background: the Boundary Structure of the Local Hot Bubble at Low Galactic Latitudes

S. Park[1], J.P. Finley[1], and S.L. Snowden[2,3]

[1] Department of Physics, Purdue University, 1396 Physics Building, West Lafayette, IN. 47907, USA
[2] NASA/Goddard Space Flight Center, Code 662, Greenbelt, MD. 20771, USA
[3] Universities Space Research Association, USA

Abstract. 42 *ROSAT* PSPC pointed observations in the Galactic plane ($l \sim 4° - 26°$) are mosaicked in order to study the spatial structure of the X-ray emitting gas in the Local Hot Bubble (LHB). Degree scale X-ray intensity variations are detected at the ±10% level in the $\frac{1}{4}$ keV band, which imply a likely influence from a clumpy boundary shell of the LHB in the observed $\frac{1}{4}$ keV band X-ray background. The possible origins of such a clumpy boundary structure of the LHB are discussed.

1 Introduction

In the Galactic plane, the $\frac{1}{4}$ keV band soft X-ray diffuse background (SXRB) is expected to originate within the LHB due to the substantial absorption cross-section of the ISM (e.g., $\tau \sim 1$ at ~30 pc, assuming n(H)~1 cm^{-3}). This "isolation" from the contribution of any flux of a more distant Galactic origin allows for the study of the detailed structure of the LHB by searching for the $\frac{1}{4}$ keV band X-ray intensity variations at various angular scales. Here we report detection of degree scale variations of the $\frac{1}{4}$ keV band SXRB in the Galactic plane ($l \sim 4° - 26°$), which implies an influence by a shell-like boundary structure of the LHB on the observed $\frac{1}{4}$ keV band SXRB.

2 Data

An R1L and R2 band (Snowden et al. 1994) mosaic of 42 *ROSAT* PSPC pointed observations are used in this study. All identified non-cosmic backgrounds (~22% of the total counts) are modeled and subtracted from the data as described in Snowden et al. (1994). The detected point sources and possible enhancements by SNRs and X-ray binaries are removed and the relative offsets between overlapping pointings are corrected (~9%). The final mosaic (Figure 1) covers ~60 degree^2 (l~4° – 26°, b~-3° – +2°) with an average exposure of ~8 ks. With a 10′ binning, an average of 6% statistics per pixel is achieved in the R1L+R2 band (Park, Finley, & Snowden 1997 for a detailed description of the data).

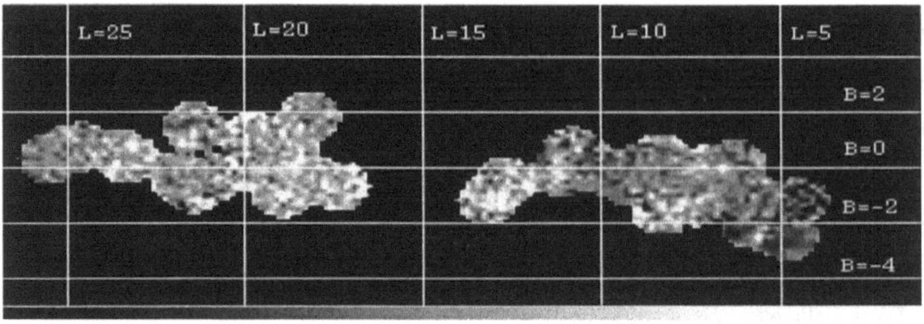

Fig. 1. The $\frac{1}{4}$ keV band (R1L+R2) mosaic of the 42 *ROSAT* PSPC pointings in the Galactic plane. The pixel size is 10′ and the data have been smoothed. The gray-scale ranges 0 – 800 × 10^{-6} counts s^{-1} $arcmin^{-2}$.

3 Analysis and Results

In order to search for X-ray intensity variations along the Galacticplane, the data are first integrated across the plane in Galactic latitude, b, to create 10′ columns, typically within ±2° from the plane. With this integrated 1-D binning, the R1L+R2 band intensity variation along the plane is displayed in Figure 2a. The average flux is ~400 × 10^{-6} counts s^{-1} $arcmin^{-2}$. The X-ray intensity is spatially variable (reduced χ^2 >5 about the mean) at the ±10% level. A spatial Fourier transform reveals degree scale variations with an ~5.5° scale being the most prominent in all bands (Figure 2b). Sub-degree scale variations are investigated with a 2-D autocorrelation function (ACF). The ACF at angular scales ≤3° is displayed in Figure 3. The ACF in both the R1L and R2 bands indicates little correlation (formal errors include zero) at angular scales of ≤3°. The difference of the ACF between the three bands (R1L, R2, and R1L+R2) is not significant and lies within the statistical uncertainties. The average hardness ratio (1.35) of the mosaic implies a plasma temperature of ~$10^{6.1}$ K with no absorption which is consistent with that of Snowden et al. (1997) for emission from the LHB.

4 Discussion

Possible origins of the detected variations are discussed below.

4.1 Magnetic Rayleigh-Taylor (R-T) Instability

A SNR in the adiabatic phase is R-T unstable. Assuming an ~10^5 year old blast wave in a pre-existing local cavity (n~0.004 cm^{-3}) "reheated" by a SN explosion, the critical wavelength of the magnetic R-T instability, λ, can be estimated (Table 1) with a typical Galactic midplane magnetic field B =

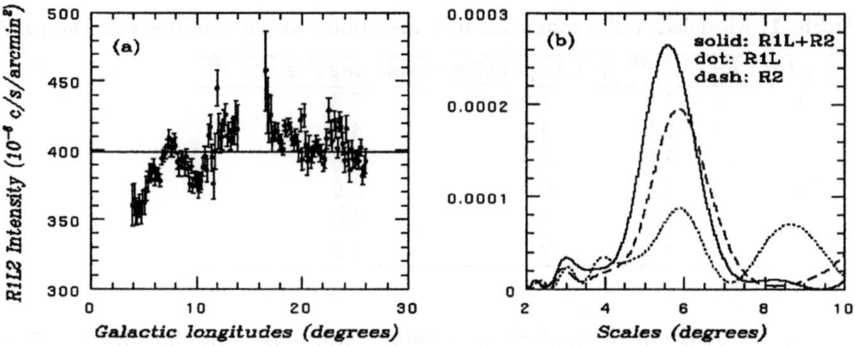

Fig. 2. (a) The $\frac{1}{4}$ keV band intensity along the Galactic plane. Each point presents a 10′ column integrated across the plane. The horizontal line is the mean. (b) Spatial Fourier transform of the R1L, R2, and R1L+R2 band X-ray intensities .

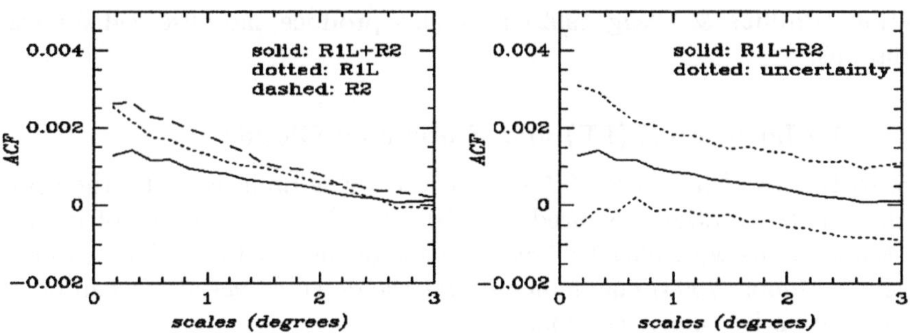

Fig. 3. The ACF of the $\frac{1}{4}$ keV band X-ray intensity for angular scales $\lesssim 3°$. The left panel shows the ACF for the R1L, R2, and R1L2 bands. The right panel shows the R1L2 band ACF and its formal uncertainty.

5 μG, a plasma temperature of the LHB T$\sim 10^{6.1}$ K, and a stored thermal energy $E_{TH} = 1 - 3 \times 10^{50}$ ergs (see Park, Finley, & Snowden [1997] for the detailed calculation). The range of λ is remarkably consistent with the detected $\sim$5.5° scale variation given the inaccuracies embedded in the simple spherical blast wave model. With this model the detected $\pm$10% variation at $\sim$5.5° scale can be attributed to the path-length variation of the wave-like boundary structure due to the magnetic R-T instability.

4.2 Inhomogeneous ISM

If the blast wave of the LHB is in the radiative phase ($\tau \gtrsim 10^6$ yr), the boundary shell can become clumpy due to inhomogeneities in the ISM. For

Table 1. Modeled λ of a magnetic R-T instability at the boundary of the LHB

R_{LHB} (pc)	E_{TH} (10^{50} erg)	λ (pc)	Observable angular size (°)
50	1.0	1.9	2.2
	2.0	4.1	4.7
	3.0	5.8	6.6
75	1.0	0.8	0.9
	2.0	1.7	2.0
	3.0	2.5	2.9

example, absorption by typical interstellar cloudlets (angular size 2 – 5 pc, internal density 1 – 3 cm^{-3}, Heiles 1967) are reasonable to produce detected $\lesssim$5.5° scale variation at ±10% level.

4.3 Emission Variations at the Boundary Layer

The emission variation at the LHB boundary such as electron density fluctuation (Phillips & Clegg 1992) may also produce the observed intensity variations.

4.4 The Local Fluff (LF) and Embedded Clouds

Due to the low N_H ($\lesssim 3 \times 10^{18}$ cm^{-2}), the LF is unlikely to be the source of the ±10% variation (required $N_H \sim 10^{19}$ cm^{-2}). Observations of local interstellar clouds with $N_H \gtrsim 10^{19}$ cm^{-2} inside of the LHB (e.g., Kerp, Herbstmeier, & Mebold 1993) may indicate that the detected variations can be due to absorption by embedded clouds.

5 Conclusions

Degree scale intensity variations (at ±10% level) in the $\frac{1}{4}$ keV SXRB are detected in the Galactic plane. The origin of these small-scale variations is most likely the presence of R-T instability and/or clumpy cooler ISM at the boundary of the LHB. Variations due to absorption by embedded clouds cannot, however, be ruled out.

References

Heiles, C. 1967, ApJS, 15, 97
Kerp, J., Herbstmeier, U., & Mebold, U. 1993, A&A, 268L, L21
Park, S., Finley, J. P., & Snowden, S. L. 1997, ApJ, 491, in press
Phillips, J. A. & Clegg, A. W. 1992, Nature, 360, 137
Snowden, S. L. et al. 1994, ApJ, 424,714
Snowden, S. L. et al. 1997, ApJ, submitted

Small-Scale $\frac{1}{4}$ keV Fluctuations Due to the Local Hot Bubble

K.D. Kuntz[1], S.L. Snowden[2], and R.S. Warwick[3]

[1] University of Maryland, College Park, Maryland, 20742, USA
[2] USRA/GSFC, Greenbelt, Maryland, 20771, USA
[3] University of Leicester, University Road, Leicester LE1 7RH, United Kingdom

Abstract. To determine fluctuations in the $\frac{1}{4}$ keV emission due to the LHB, without contamination due to sources within the galactic disk, the Galactic halo or the extragalactic background, we have formed a mosaic of PSPC observations towards the Hyades, a direction where distant emission is effectively blocked by a Galactic column of many optical depths. After removing the point sources from this image, we find the $\frac{1}{4}$ keV emission to be smooth on scales from 6′ to a few degrees.

1 Introduction

Barber, Warwick, & Snowden (1995) discovered small-scale fluctuations in the $\frac{1}{4}$ keV emission in the UK medium deep survey region, high galactic latitude field. These fluctuations have amplitudes of 10% to 20% of the total $\frac{1}{4}$ keV background, and have scale-sizes of about 20′ (Warwick et al, 1997). To characterize these fluctuations, and to isolate the component of the $\frac{1}{4}$ keV background from which they arise, it is necessary to measure the spectrum of fluctuations from all of the "foreground" background components. We present here a measure of the small-scale fluctuations due to the Local Hot Bubble (LHB).

To determine the contribution due to the LHB, without contributions from galactic halo or extragalactic components, it is necessary to observe the LHB in the direction of background masses of neutral hydrogen optically thick to $\frac{1}{4}$ keV X-rays. The Hyades cluster, which contains many overlapping PSPC fields, lies in the direction of the Taurus-Auriga molecular clouds. The $\frac{1}{4}$ keV opacity ($N_H \times \sigma_{eff}$) of the Taurus-Auriga clouds in the direction of the PSPC observations is greater than ~ 5, effectively blocking extragalactic and halo $\frac{1}{4}$ keV emission.

2 Data

We have mosaicked all of the overlapping PSPC pointed observations towards the Hyades cluster. Individual exposures were reduced using the ESAS software (Snowden 1995) in the manner described by Kuntz, Snowden, & Verter

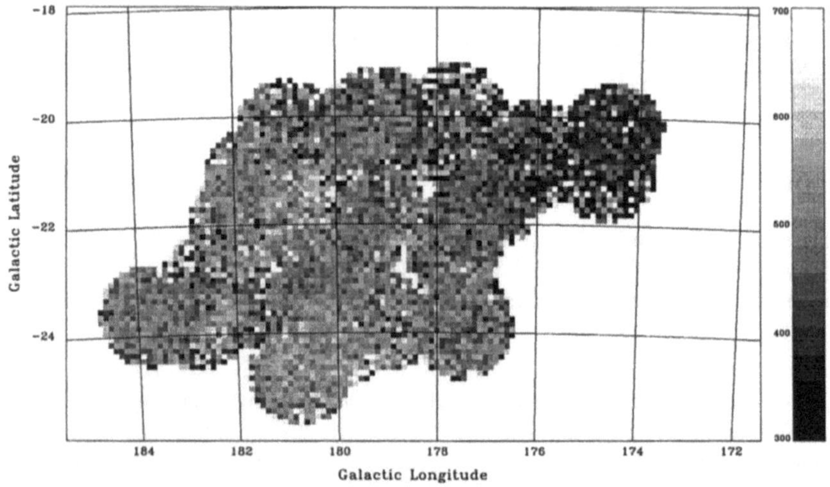

Fig. 1. $\frac{1}{4}$ keV (band R1L+R2) image of the Hyades cluster after point source removal. The pixel size is $6' \times 6'$, the units are 10^{-6} counts s^{-1} $arcmin^{-2}$. The pixel-to-pixel variation is indicative of the statistics; regions with high pixel-to-pixel variation have low exposure. The "bright" spots at (ℓ, b)~(178°,-21°) and (ℓ, b)~(178.5°,-23°) are "holes" in the mosaic.

(1997) and references therein. Point sources were removed from the individual exposures which were then mosaicked together in the manner described by Snowden (1995). The mosaic covers nearly 35 deg^2, contains exposures from < 1 ks to ~ 32 ks, and is shown in Figure 1.

3 Analysis

Although the $\frac{1}{4}$ keV image shows some large scale structure, there is little structure on scales smaller than a degree. We have employed two methods for characterizing the smoothness on these scales; the autocorrelation function (ACF) and the σ-normalized distribution.

We have calculated the autocorrelation function for the point-source-removed image with $6' \times 6'$ pixels. As can be seen in Figure 2, the overall trend of the ACF is consistent with the large scale structure, i.e., a peak at about (ℓ, b) = (180.5°,-21°), and lower emission regions at (175°,-21°) and (184°,-23°). There is power on scales smaller than a few degrees, but the slope between 0 and 2° implies that it is due to large-scale structure rather than to small-scale fluctuations.

The ACF is not an effective tool for scales smaller than a few times the pixel size. To more carefully test the existence of fluctuations at smaller scales, we model the pixel-to-pixel variations. The comparison at different pixel sizes places limits upon the non-statistical fluctuations at various scales.

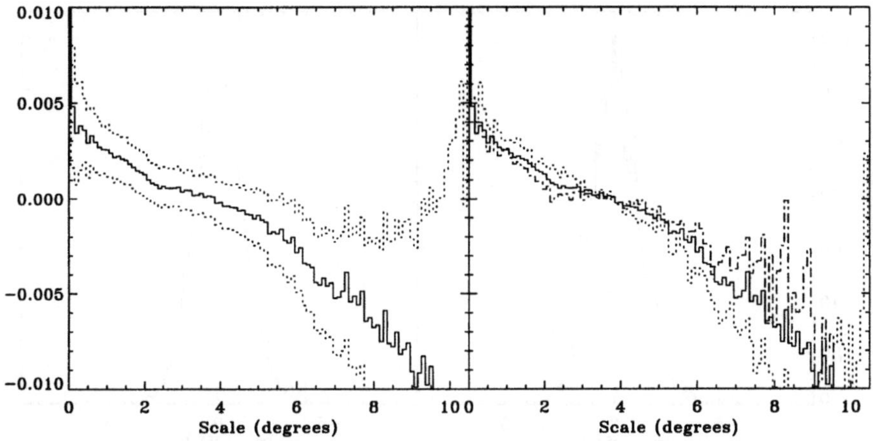

Fig. 2. *Left:* The ACF for the mosaic in the band R1L+R2. The dotted lines show the formally propagated uncertainty. *Right:* The ACF for the band R1L+R2 (solid), compared to the ACF for bands R1L (dashed) and R2 (dot-dashed).

We define a σ-normalized distribution for an image with a count rate $\rho(x,y)$ and uncertainty $\sigma(x,y)$ to be the distribution of the quantity

$$\frac{\rho(x,y) - \overline{\rho(x,y)}}{\sigma(x,y)}, \tag{1}$$

where $\overline{\rho(x,y)}$ is the mean of the eight pixels adjacent to position (x,y). The σ-normalized distribution is shown in Figure 3.

To model the expected distribution we have taken the exposure map and multiplied it by the mean count rate. To this image is added the same modeled non-cosmic background (scattered solar X-rays, long-term X-ray enhancement, etc.) used in the reduction. (This is a model "raw" image.) We then create an image where an individual pixel value is randomly drawn from a Poisson distribution whose mean is the model "raw" count at that pixel. From this image is subtracted the non-cosmic background; the result is divided by the exposure time. Point sources thought to exist below the detection threshold can be then added to produce a model image, however we find that the addition of only a very small flux in point sources can be tolerated before the model diverges greatly from the data.

The data and the model do not agree in a statistical sense, but, we can test the effect of small-scale fluctuations. The right-hand panel of Figure 3 shows that the addition of 2.5% fluctuations (pixel-to-pixel variations) on $6'$ scales are clearly incompatible with the data. The difference becomes greater with pixel size. More careful modeling of the $20'$ fluctuations such as those found by Barber, Warwick, & Snowden (1995), must be done before their existence in the LHB can be confidently excluded.

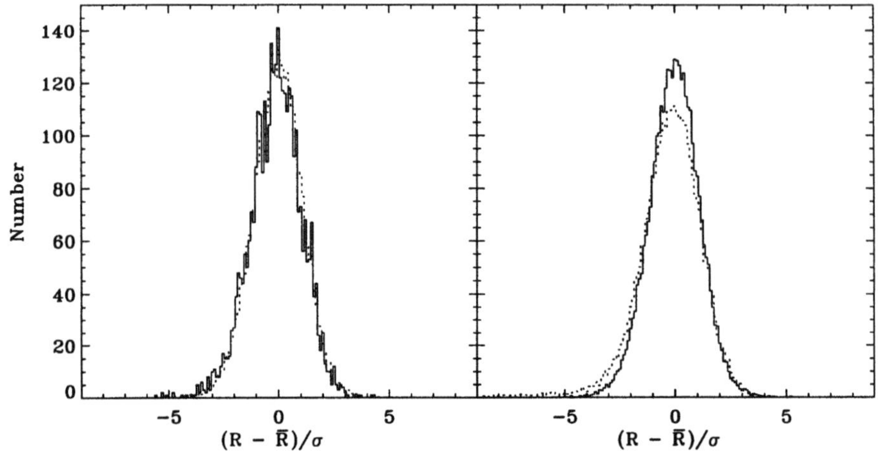

Fig. 3. *Left:* The σ-normalized distribution for the mosaic in the band R1L+R2 (solid line). The pixel size was $6' \times 6'$. The model distribution (dotted line). *Right:* The model distribution (solid line). The model distribution with an additional 2.5% fluctuation (dotted line). Model distributions are the mean of 10 trials.

4 Discussion

We have found the $\frac{1}{4}$ keV emission from the LHB to be smooth on scales from $6'$ to $\sim 1°$, 0.13 pc to ~1.3 pc at the edge of the LHB (d ~ 70 pc). If fluctuations were caused by a change in the path-length to the LHB wall, a 0.13 pc wide fluctuation could have a depth of no more than 1.8 pc. Wider fluctuations will be more tightly constrained. Thus, there are no low-column-density "tunnels" on this scale, as might be expected were the LHB optical depth bounded. This result may also place constraints on the degree to which edge brightening may contribute to the LHB emission, as well as constraints upon the small scale structure of H I embedded in the LHB. We continue this work to more tightly constrain small-scale fluctuations.

References

Barber, C., Warwick, R. S., Snowden, S. L. (1995): *Röntgenstrahlung from the Universe*, eds., Zimmermann, H.U., Trümper, J.E., & Yorke, H., 319.

Kuntz, K.D., Snowden, S.L., Verter, F. (1997): ApJ, in press

Snowden, S. L. (1995): Cookbook for Analysis Procedures for *ROSAT* XRT/PSPC Observations of Extended Objects and the Diffuse Background.

Warwick, R., Hutchinson, I, Willingale, R., Kuntz, K., & Snowden, S. (1997): this volume.

Fluctuations in the Diffuse Soft X-ray Background; Probably of Galactic Halo Origin

T.J. Sumner, S.D. Sidher, J.J. Quenby, M. Hernandez, A. Mian, R. Jesseit, and M. Gambhir

Astrophysics Group, Blackett Laboratory, Imperial College, Prince Consort Road, London, SW7 2BZ, UK

Abstract. Soft X-ray background data from the *ROSAT* PSPC have been fitted by a model including emission from the local bubble, embedded hot gas within the disk, hot gas in the Galactic halo and extra-galactic emission, together with local and disk absorption. In *all* directions a halo component at $10^{6.2}$ K (0.2 keV) is required for an acceptable fit. The halo emission measure shows spatial variability and the global trend suggests a disk-like planar rather than more extended spherical distribution. New values for the emission measure within the local bubble are derived.

1 Introduction

Detailed spectral modelling is being used to establish the mean and fluctuating components of the local bubble, embedded disk and halo contributions to the soft X-ray background (Sidher et al. 1996). *ROSAT* shadowing data (Burrows and Mendenhall, 1991, Snowden et al. 1991) have cast doubts on early models of the $<$ 2 keV background, based either on absorption of a 3×10^6 K halo or on purely local bubble emission with cavity walls at varying distances. Sidher et al. 1996 showed that the minimal model giving an acceptable spectral fit must include an extra-galactic power law component, a hot gas Galactic halo component at $\log T = 6.2$ and a local bubble at $\log T = 5.9$. In some directions, such as the Lockman hole region (Sidher et al. 1997), an additional component is needed which may indicate embedded disk emission from wind or supernova driven bubbles. Radio data and cosmic ray propagation studies suggest an extended galactic halo and a sufficiently hot and extensive gas component is needed for stability (Bloemen 1987). Kahn, 1994 reviews the fountain model in which regions of upward flow, driven by SNRs, are balanced by others returning to the disk. Cosmic rays may drive the flow far out (eg Breitschwerdt et al., 1993). The 0.5-2 keV soft X-ray spatial auto-correlation function has been used to study the contribution of various types of source to the extra-galactic background (eg Soltan et al. 1996, Barcons and Fabian, 1989). The possibility of substantial Galaxy related spatial fluctuation has not been considered. Suto et al. (1996) show that a Local Group halo origin for the 1 keV background could account for the microwave background quadrupole anisotropy through the Sunyaev-Zel'dovich effect, although Pildis and McGaugh, 1996 disagree using data from other poor galaxy groups.

2 Spectral Modelling

Modelling is done using any number of concentric shells, each of which may contain a mixture of hot emitting gas and cold absorber. Details may be found in Sidher et al. 1996. The radiative power loss function of Landini and Monsignori Fossi, 1990 is used for all emitting components except the extragalactic background where a power law is used. Predicted spectra are convolved with the instrument performance. 'Clean' observational data are extracted from archived *ROSAT* PSPC data using a procedure designed to minimize contamination from particle backgrounds. From these, images are displayed and binned background spectra are recovered from within regions devoid of obvious sources. Observed and predicted spectra, identically binned, are then using a χ^2 test. Best fit models and their associated free parameter ranges are assessed using the criteria of Lampton et al., 1976.

3 Results and Discussion

3.1 Local Bubble and Galactic Halo Emission Measure

Fits have been performed on 78 *ROSAT* observations including the Draco complex, the Lockman hole and M87 (Sumner et al. 1997). For the local bubble a fixed temperature ($\log T = 5.9$, Lieu et al., 1992) has been used for all directions and the best fit emission measures (EMs) are shown on Fig. 1. The radius of each circle is proportional to the EM which ranges from 0.0028 to 0.011 cm^{-6} pc. The mean value of all pointings is 0.006 cm^{-6} pc. The temperature and EM of the halo component are free parameters in the fits. Apart from the M87 region, the best fit halo temperature is constant (to within the fitting step size of $\Delta\log T$) across the sky and is given by $\log T = 6.2$. The best fit EMs are shown in Fig. 2. The radius of each circle is proportional to the EM which ranges from 0.0096 to 0.083 cm^{-6} pc. The mean value of all pointings is 0.024 cm^{-6} pc. There is a tendency for the EM to increase towards the galactic equator, which we have previously shown (Sidher et al. 1996) is consistent with a disk-like distribution (following Bloemen 1987).

3.2 Local Group Halo Model

In Fig. 3 solid contours show the expected variation in halo EM assuming it has a spherical distribution about the Local Group centre with a density profile following a speculative dark matter distribution (see model of Sidher et al. 1996 with $p = 2$). Contours labels are in units of 0.0032 cm^{-6} pc. The model predicts an decrease in EM in all directions away from the centre of the Local Group. This trend is not consistent with the results.

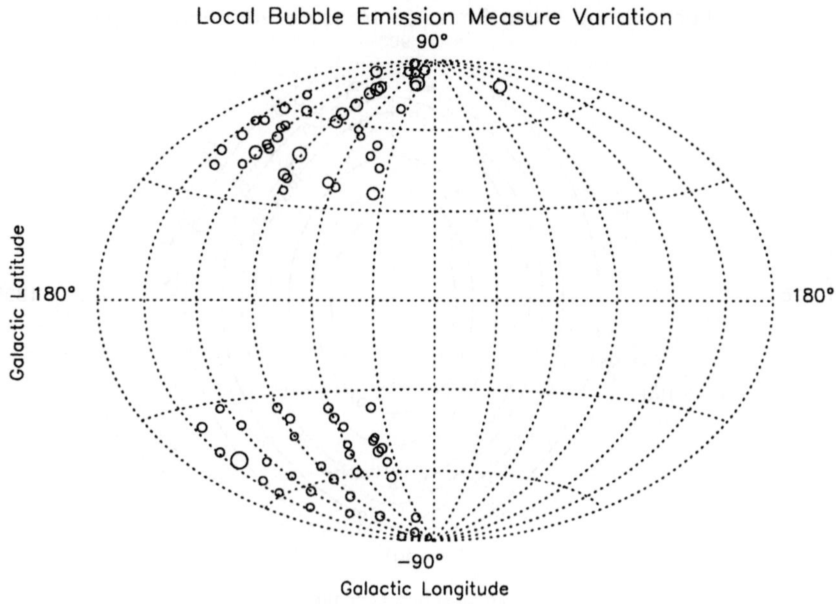

Fig. 1. Local Bubble emission measures

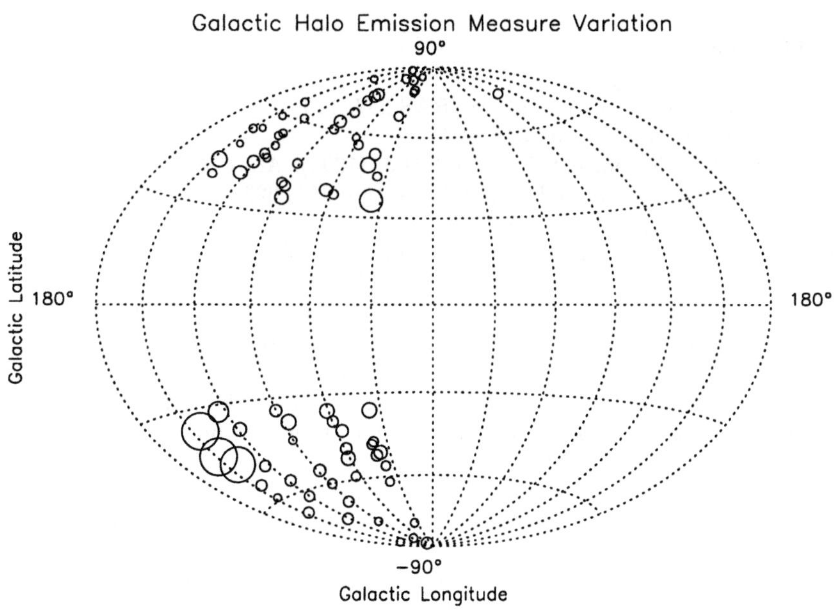

Fig. 2. Galactic Halo emission measures

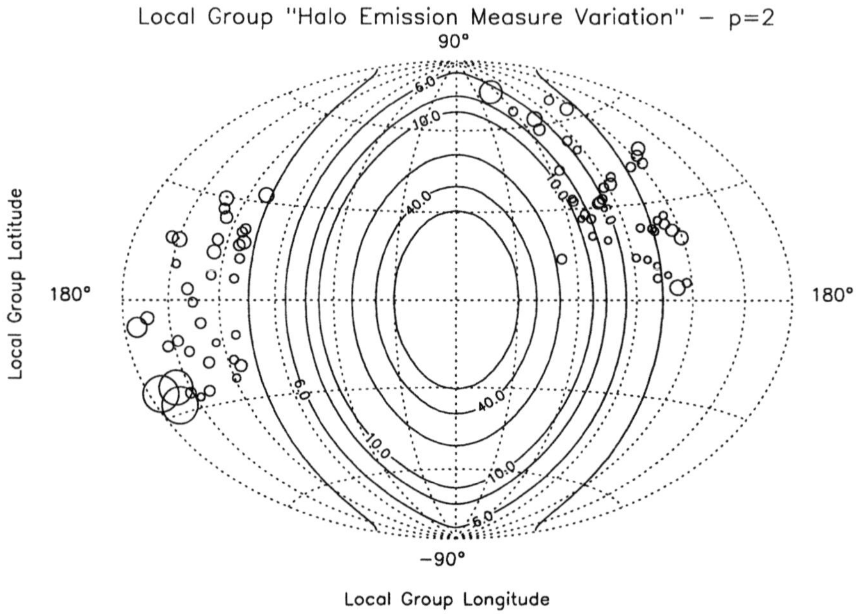

Fig. 3. Local Group Halo emission measure

References

Barcons X., & Fabian A.C. (1989): MNRAS 237, 119
Bloemen J.B.G.M. (1987): ApJ 322, 694
Breitschwerdt D., McKenzie J.F., & Völk H.J. (1993): A&A 269, 54
Burrows D.N., & Mendenhall J.A. (1991): Nature 351, 629
Kahn F.D. (1994): Ap&SS 216, 325
Lampton M., Margon B., & Bowyer S. (1976): ApJ 208, 177
Landini M., & Monsignori Fossi B.C. (1990): A&A 82, 229
Lieu R., Quenby J.J., Sidher S.D., Sumner T.J., Willingale R., West R.G., Harris A.W., Snowden S.L., & Bickert K. (1992): ApJ 397, 158
Pildis R.A., & McGaugh S.S. (1996): ApJL 470, 77
Sidher S.D., Sumner T.J., Quenby J.J., & Gambhir M. (1996): A&A 305, 308
Sidher S.D., Sumner T.J., & Quenby J.J. (1997): In preparation
Snowden S.L., Mebold U., Hirth W., Herbstmeier U., & Schmitt J.H.M.M. (1991): Science 252, 1529
Soltan A.M., Hasinger G., Egger R., Snowden S., & Trümper J. (1996): A&A 320, 685
Sumner T.J., Sidher S.D., & Quenby J.J. (1997): In preparation
Suto Y., Makishima K., Ishisaki Y., & Ogasaka Y. (1996): ApJL 461, 33

The Radiation Transfer of Soft X-rays

Jürgen Kerp and Jochen Pietz

Radioastronomisches Institut der Universität Bonn
Auf dem Hügel 71, D-53121 Bonn, Germany

Abstract. We discuss the link between the halo plasma temperature and the power-law spectral index of the extragalactic background radiation. This link is of strong influence for the derivation of the Galactic halo intensity distribution. In principal, we can distinguish between two combinations of Galactic halo plasma temperature and power-law slope. The first combination consists of a halo plasma of $T_{\rm halo} = 10^6$ K and an E^{-2} approximation of the extragalactic background radiation. The second combination is $T_{\rm halo} = 10^{6.2}$ K and $E^{-1.5}$. Both combinations are in agreement with recent observational results, thus it is not feasible to discriminate between both models on the basis of X-ray data available. But, the soft X-ray background intensity distribution in the 1/4 keV and 3/4 keV *ROSAT* energy bands differs significantly. The $T_{\rm halo} = 10^6$ K, E^{-2} allows a patchy 1/4 keV intensity distribution while the $T_{\rm halo} = 10^{6.2}$ K, $E^{-1.5}$ predicts a much smoother intensity variation since the hotter halo plasma accounts for a significant fraction of the 3/4 keV background radiation.

1 The Soft X-ray Background Models

ROSAT data established three soft X-ray background (SXRB)source terms. The first one is the local X-ray gas. The second source term is the extragalactic X-ray background radiation. Hasinger et al. (1993) determined within the 0.5 – 2.0 keV *ROSAT* energy band a power-law index ($E^{-\Gamma}$) of $\Gamma \simeq 2.0$ for the bright discrete sources. Almaini et al. (1996) found for fainter sources a power-law index of $\Gamma \simeq 1.5 \pm 0.1$ within the *ROSAT* energy window ranging from 0.1 keV to 2.0 keV. The exact value of this power-law slope is of high importance, because it determines the derived temperature, and in consequence the halo intensity distribution. The Galactic halo plasma is the third component of the SXRB. The plasma is located beyond the galactic neutral hydrogen layer (Kerp & Pietz 1996).

2 The Method

The *ROSAT* data allow only to differentiate between two main combinations of plasma and the extragalactic background radiation. A $T_{\rm halo} = 10^6$ K plasma can only fit the observational data in addition to an extragalactic power-law spectrum of $\Gamma \simeq 2$, while a $T_{\rm halo} \simeq 10^{6.2}$ K plasma fits the data well in combination with a $\Gamma \simeq 1.5$ spectral slope.

To show this we fitted the scatter diagrams (1/4 keV and 3/4 keV versus N_{HI}) as well as the energy band ratios (R1:R2 and 1/4:3/4 versus N_{HI}) of deep pointed *ROSAT* PSPC data simultaneously. In Figure 1 we present the fits to the data. The dashed line marks the $T_{halo} = 10^6$ K, $\Gamma = 2$ combination while the solid line represents the $T_{halo} = 10^{6.2}$ K, $\Gamma = 1.5$ combination. The R1:R2 band ratio is better represented by the hotter Galactic X-ray halo. But within the statistical uncertainties both models fit the data very well. The main difference between both combinations is hidden in the *unabsorbed* X-ray intensities. In Figure 2 we plotted the amount of photons emitted from each SXRB source normalized to the total number of photons emitted from all three soft X-ray background sources. Obviously in the 1/4 keV energy range there is no significant difference between a ($T = 10^6$ K, $\Gamma = 2$) and ($T = 10^{6.2}$ K, $\Gamma = 1.5$) combination. But in the 3/4 keV energy band (Fig. 2 *upper panel*) 56% of the observed 3/4 keV emission is produced by the $T_{halo} = 10^{6.2}$ K plasma while in case of the cooler $T_{halo} = 10^6$ K plasma only 23% of the observed 3/4 keV emission originates from the Galactic X-ray halo. This significant difference has several consequences, since the higher the contribution of the Galactic halo plasma to the 3/4 keV radiation the smoother the intensity distribution of the Galactic X-ray halo. This is the reason why the 3/4 keV X-ray background intensity distribution is much smoother than the 1/4 keV one. Therefore, the hotter the Galactic X-ray halo plasma the smoother its intensity distribution.

3 Results

Based on fits to the scatter diagrams it is *not* feasible to distinguish between a cooler Galactic halo in addition to a steeper extragalactic power-law index ($T_{halo} = 10^6$ K, $\Gamma = 2$) and a hotter halo plasma in combination with a flatter extragalactic power-law slope ($T_{halo} = 10^{6.2}$ K, $\Gamma = 1.5$).
In our model the $T_{halo} = 10^6$ K, $\Gamma = 2$ combination predicts a 1/4 kev background intensity of $I_{extra}(1/4\,\mathrm{keV}) = 5 \cdot 10^{-6}$ cts s^{-1} arcmin^{-2} which is consistent with the upper limit intensity value derived by Cui et al. (1996). The $T_{halo} = 10^{6.2}$ K, $\Gamma = 1.5$ combination however is consistent with $I_{extra}(1/4\,\mathrm{keV}) = 2.3 \cdot 10^{-6}$ cts s^{-1} arcmin^{-2} which was derived by Barber et al. (1996). Nevertheless, all recent investigations of the extragalactic intensity contribution to the soft X-ray background emission do not allow to differentiate between both combinations.
One key to discriminate between both combinations is given by the analysis of the extragalactic background radiation (e.g. Gendreau et al. 1995). According to Almaini et al. (1996) the contribution of faint emission line galaxies increases towards fainter flux limits. This new population of faint X-ray galaxies reveals a flatter averaged spectral slope than the bright discrete X-ray sources studied by Hasinger et al. (1993). This indicates that the

$T_{\rm halo} = 10^{6.2}$ K, $\Gamma = 1.5$ is the most likely combination for the representation of the diffuse soft X-ray background radiation.

References

Almaini O. et al. (1996): MNRAS 282, 295
Barber C.R. et al. (1996): MNRAS 282, 157
Cui W. et al. (1996): ApJ 468, 117
Gendreau K.C. et al. (1995): PASJ 47, 5
Hasinger G. et al. (1993): A&A 275, 1
Kerp J., Pietz J., in "Röntgenstrahlung from the Universe", eds: Zimmermann H.U., Trümper J.E., Yorke H., MPE–Report 263, p. 329

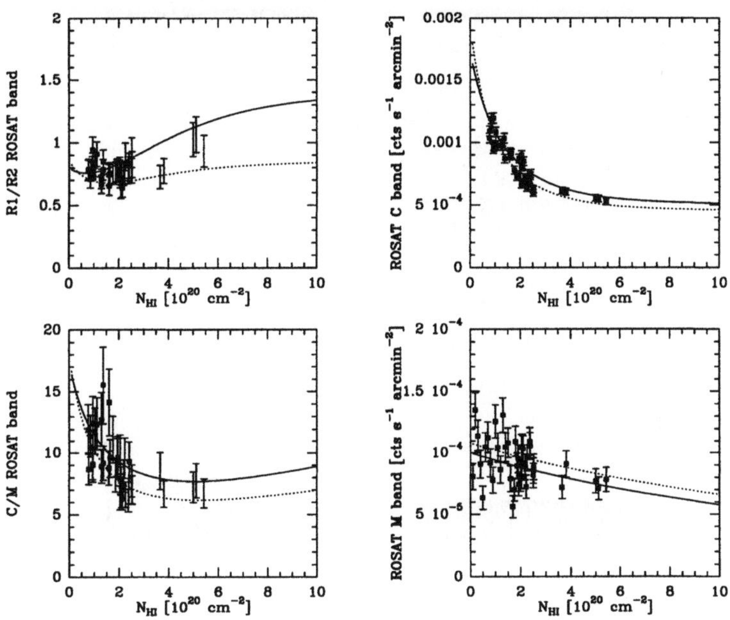

Fig. 1. Several PSPC pointings were analysed towards a high galactic latitude cloud $l \sim 90°$, $b \sim 60°$. Plotted is the 1/4 keV (C-band) and 3/4 keV (M-band) emission as well as the R1:R2 and C:M band intensity ratio versus the neutral atomic hydrogen column density ($N_{\rm HI}$). The dashed line denotes the superposed emission of a Galactic halo plasma with $T_{\rm halo} = 10^6$ K combined with the extragalactic background radiation described by a power-low E^{-2}. The solid line marks an alternative approach of $T_{\rm halo} = 10^{6.2}$ K in combination with $E^{-1.5}$.

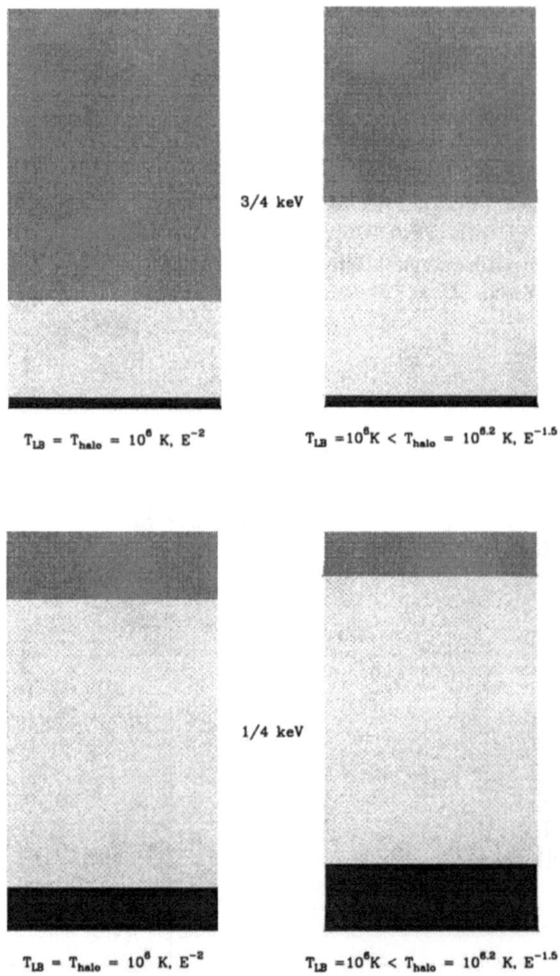

Fig. 2. The number of photons produced by the local X-ray plasma (bottom), Galactic X-ray halo (middle), and extragalactic X-ray background (top) normalized to the total number of emitted soft X-ray photons. The lower panel displays the 1/4 keV energy range while the upper panel shows the situation in the 3/4 keV energy band. On the left-hand side the combination $T_{\mathrm{halo}} = 10^6$ K, E^{-2} is displayed in each panel, while on the right-hand side $T_{\mathrm{halo}} = 10^{6.2}$ K, $E^{-1.5}$ is shown. Obviously the combinations are very similar in the 1/4 keV band while the combinations differ significantly in their Galactic halo contribution to the 3/4 keV energy band.

Equilibrium Ionization and Non-Equilibrium Ionization Plasma Models

Hans Böhringer

Max–Planck–Institut für extraterrestrische Physik, Giessenbachstrasse, D-85740 Garching, Germany

Abstract. This paper provides a brief review on ionization and radiation models of an optically thin, hot plasma and their application to the interpretation of observations of the Local Bubble. The basic structure and atomic physics ingredients of the models are outlined. Some general characteristic results of ionization equilibrium and non-equilibrium calculations are shown. Recent progress in plasma model codes is reviewed. The interpretation of the observations of absorption lines and emission spectra from the Local Bubble is discussed in the framework of ionization equilibrium and non-equilibrium ionization models with particular reference to the fast adiabatic cooling model of Breitschwerdt and Schmutzler.

1 Introduction

The lack of observed absorption towards nearby stars (see York & Frisch 1983 and P. Frisch in this volume) and the structure of the neutral hydrogen distribution as observed in 21 cm (see C. Heiles this volume) gives a first impression that the solar system is located in a partly evacuated cavity that might result from an expanding hot superbubble in which we are located. Convincing evidence for a Local Hot Bubble and a physical understanding of this environment come from the observation of soft X-ray emission and UV absorption lines. It is the interpretation of these EUV and X-ray data that leads to a knowledge of physical parameters like temperature, plasma density, and pressure in our local galactic environment. A physical model of the Local Bubble can therefore be tested in the first place by its ability to reproduce the spectral features observed in the UV and soft X-rays.

In this paper we review the principles of plasma emission models and their application to the Local Bubble observations. It turns out that the observed features can very roughly be reproduced by two different model sets: ionization equilibrium models with a fairly hot temperature of the order of 10^6 K and fast cooling non-equilibrium ionization models with temperatures as low as a few 10^4 K. An illustration of the differences between the two model families is the main goal of this paper.

2 The Basic Structure of Plasma Emission Models

The atomic physical processes involved in the plasma models comprise the processes controlling the ionization of the constituents of the interstellar plasma and events leading to the excitation and radiative deexcitation of atomic levels of the ion species. For the hot interstellar medium (ISM) the case is greatly simplified in two ways by the fact that the plasma is very tenuous. Firstly, the plasma is "optically thin" such that a photon once emitted has a negligible chance to be reabsorbed (except for hydrogen Ly-α at low temperatures). This further implies that the radiation spectrum has no feedback to the ionization structure. Secondly, the collision frequency is so low that excited states are deexcited by radiation rather than by collisions even for "strongly forbidden" transitions. The line emission rate is thus given by the rate of collisional excitation. The latter simplification is called "coronal approximation" and certainly holds for electron densities $\leq 10^4$ cm^{-3}.

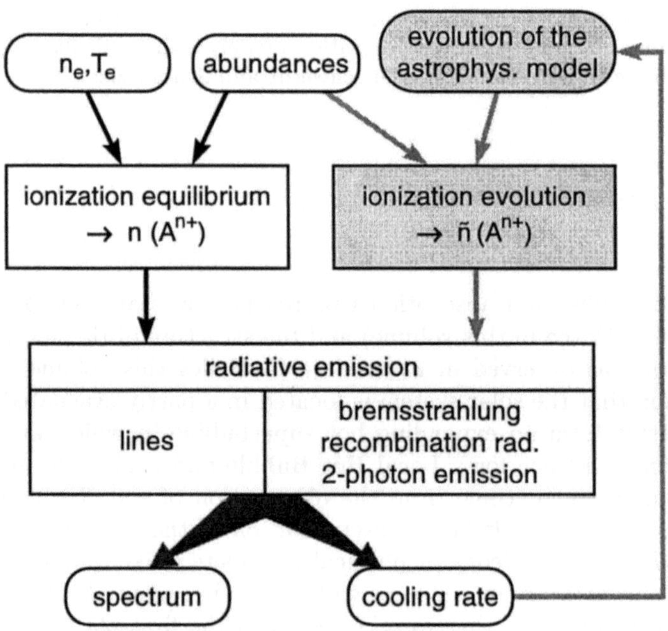

Fig. 1. Schematics of the basic structure of plasma emission models: the white boxes and black arrows outline the structure of equilibrium models; the grey boxes indicate the modifications necessary for time dependent non-equilibrium calculations.

The basic structure of plasma emission models is sketched in Fig. 1. For ionization equilibrium models the only input parameters are the plasma temperature and the abundances of the elements. The first part of the model consists of the calculation of the ionization structure. In equilibrium, for each consecutive ion pair the ionization rate has to equal the recombination rate as given by equ. (1). This implies the solution of the set of linear equations to determine the relative abundance of each ion of an element:

$$\begin{aligned}[I^{n+}]\, n_e\, (C_{ci} + C_{ai}) + [I^{n+}]\,([H^+]\, X_{ce1} + [He^+, He^{++}]\, X_{ce2}) = \\ [I^{(n+1)+}]\, n_e\, (C_{rr} + C_{dr}) + [I^{(n+1)+}]\,([H]\, X_{ce3} + [He]\, X_{ce4}) \qquad (1)\end{aligned}$$

Several processes contribute to the ionization and recombination rate. The ionization comprises direct electron impact ionization (C_{ci}), electron impact excitation to an autoionizing state with subsequent autoionization (C_{ai}), and ionizing charge exchange with H^+, He^+, and He^{++} (with the rates X_{ce1} and X_{ce2}, respectively). Direct ionization is the dominant process. Recombination includes radiative recombination (C_{rr}) and dielectronic recombination (C_{dr}). In the second processes the recombining collision complex is energetically stabilized by the excitation of a second bound electron which will subsequently radiate the excess energy. This is a resonant process which naturally has a strong energy dependence. It is most important for the more complex ions with a large number of electrons (e.g. Fe). It can dominate the recombination process in certain energy (temperature) ranges. Also charge exchange with H and He (rates X_{ce3} and X_{ce4}) contribute to the effective recombination.

Once the density of each relevant ion species is determined by the elemental abundances and the set of equations (1), the spectral contribution of each ionic species can be calculated. There are four processes to be considered, line emission from collisionally excited states and three continuum emission processes: free-free radiation, recombinative radiation, and two-photon emission. The latter process originates from collisional excitation of energy levels with forbidden transitions which can only radiatively decay by the simultaneous emission of two photons (in particular transitions from nS to $(n-1)S$ orbitals).

All these processes governing the ionization structure and spectral emission involve the collision of electrons with ions with rate equations of the basic form:

$$R = n_e\, n_I\, C_{Ix} = n_e^2 \left[\frac{n_I}{n_E}\right] \left[\frac{n_E}{n_e}\right] C_{Ix}(T)\,, \qquad (2)$$

where n_E and n_I are the densities of the element and ion, respectively, and C_{Ix} are the rate coefficients of the relevant processes. Thus all rates are proportional to the plasma density parameter, n_e^2, and otherwise depend only on the relative abundances and the temperature dependent rate coefficients. Therefore the ionization structure and the form of the emission spectrum can be evaluated independently from the plasma density. The normalization of the emitted spectrum is then obtained from the emission measure, E, calculated for the astrophysical model as

$$E = \int n_e^2 \; dl \tag{3}$$

The strategy for the calculation of non-equilibrium ionization models is also sketched in Fig. 1. Briefly, the ionization structure is determined by the ionization history: ionization and recombination rates for all relevant ion pairs and the resulting change in the ionization structure have to be calculated along with the astrophysical model in a time dependent way. The radiation can still be calculated independently once the ionization structure is determined. But if radiative cooling is important in the astrophysical model, the total cooling power of the emitted spectrum has also to be calculated simultaneously with the evolving dynamical model.

The calculation of the ionization-radiation model is a very intense bookkeeping task of the rates of all the important atomic processes. Several widely used plasma emission codes have been developed over the last two or three decades which by today keep track of thousands of atomic collision rates. Pioneering work was in particular done by Cox, Tucker, Raymond and Smith. There are currently four openly distributed, frequently used plasma codes which experience a continuous development:

- **RS, RSB code:** by Raymond & Smith (1977), Raymond 1988, Brickhouse, Raymond & Smith (1995a), Raymond & Brickhouse (1996)
- **Mewe, MEKA, MEKAL, SPEX codes:** by Mewe et al. (1981, 1985), Gronenschild & Mewe (1978), Mewe (1990, 1991), Kaastra & Mewe (1994), Mewe, Kaastra, & Liedahl (1995)
- **Masai code:** by Masai (1984), Masai & Kato (1987), Masai (1994a,b)
- **LF code:** by Landini and Monsignori Fossi (1970, 1990, 1991), Monsignori Fossi & Landini (1994).

A recent discussion on a comparison of these codes is given in the Legacy Journal (Brickhouse et al. 1995b). More codes have been constructed, that were used mostly by the authors themselves as for example Cox & Tucker (1969), Tucker & Koren (1971), Cox & Daltabuit (1971), Cox (1972), Shapiro & Moore (1976, 1977), Kato (1976), Gaetz & Salpeter (1983), Böhringer & Hensler (1989), Schmutzler & Tscharnuter (1993), and Sutherland & Dopita (1993).

3 Ionization Structure of Equilibrium Models

Some generally interesting results concern for example the calculated ionization structure for oxygen as a function of temperature. Results taken from Böhringer & Hensler (1989) are shown in Fig. 2. The lithium-like oxygen ion OVI, which is marked in the plot, has a single outer electron in the L-shell and is characterized by a very large photon absorption cross section (in the UV). It also has a very peaked abundance profile as a function of temperature.

Both properties are very important: the discovery of this ion in absorption in the ISM with the Copernicus satellite (Jenkins & Meloy 1974) has led to the implication of the existence of a hot ISM phase with coronal temperatures. The narrow range over which this ion appears abundant (see Fig. 2) implies a plasma temperature of about $2 - 5 \cdot 10^5$ K for the local ISM.

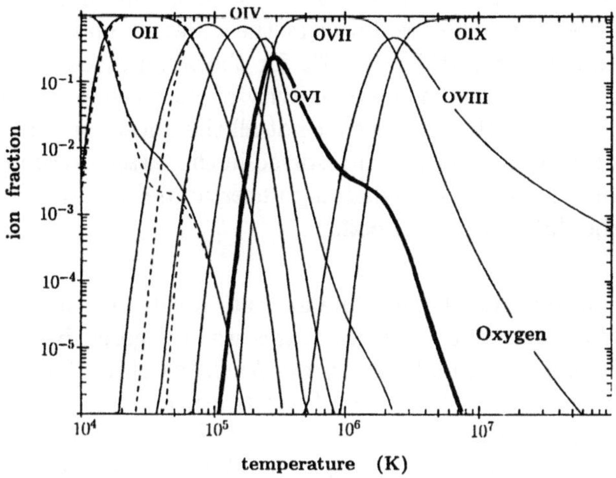

Fig. 2. Equilibrium ionization structure of oxygen as a function of temperature calculated from the data of Arnaud & Rothenflug (1985). The thick line marks the abundance of O^{5+}, (OVI). The dashed lines show the calculations without charge exchange reactions.

Similarly the ions NV and CIV - also lithium-like - have been observed in the line of sight through the galactic halo (e.g. Savage & DeBoer 1981, Savage & Massa 1987). They indicate temperatures in the interval $1.5 - 2.5 \cdot 10^5$K and $0.8 - 1.3 \cdot 10^5$K, respectively.

4 Recent Progress in Plasma Codes

How reliable are the results for theoretically calculated spectra? One may get an idea of the accuracy of the models by inspecting the effects that two recent major revisions had on the results from plasma codes. The remaining uncertainties in the present models may be of similar order in the worst case.

The first improvement concerns the calculation of the ionization structure of the heavy elements, in particular iron. Arnaud & Raymond (1992) have compiled new experimental data from crossed beam techniques, photoionization cross sections, and theoretical calculations for dielectronic recombination and recalculated a scheme for the ionization and recombination structure of iron. The results are major changes in the ionization structure of iron as a function of temperature. The differences are mostly due to the uncertainties in the very complex dielectronic recombination process of iron. With these changes some of the puzzles in the ion abundances deduced from solar X-ray spectra can be explained, but also some puzzles remain (see Arnaud & Raymond 1992).

Masai (1997) has studied the effects of the use of different ionization-recombination data sets from Arnaud & Rothenflug (1985), Masai (1994), Kato, Masai, Arnaud (1991), Arnaud & Raymond (1992) for spectra of a plasma with a temperature around 1 keV and about solar abundances. He finds that the differences between the results from the different codes are: $\Delta(T_e) \sim 15\%$ for the temperature, $\Delta(Z) \geq 30\%$ for the abundances, and there are up to factors of 2 variations in the emissivity of the plasma in the spectrum around 1 keV. Masai (1997) argues that not all the uncertainties in the ionization structure of iron may have been resolved in the latter revisions and therefore one may conclude that the remaining uncertainties could be of almost similar order as the differences outlined.

The second revision concerns the L-shell line excitation rates of the element iron. The reconsideration of the excitation rates was triggered by the study of early spectra of galaxy clusters taken by the ASCA satellite. Fabian et al. (1994) found that the Fe (3-2) and Fe (4-2) transition line blends could not be fitted simultaneously by single- or multi-temperature models in the high quality spectrum taken of the Centaurus galaxy cluster. Reconsideration of the transition probabilities of the iron L-shell lines by Liedahl et al. (1995) showed that the major problems lay in the inaccuracy of the previously calculated atomic data. Recalculation of the line strength could solve the major discrepancies and led to the development of the MEKAL code (Mewe, Kaastra, & Liedahl 1995). The results are now included in most modern codes. The effect of the improvements are changes in the intensities of iron line blends around 1 keV of up to a factor of two.

These improvements have been driven by the wish to understand the observed astrophysical spectra and have essentially focussed in the element iron. The progress goes on line with the technical improvement of the observations. For the earlier X-ray spectroscopy at higher energies with the EXOSAT or GINGA observatories, the K-shell lines of iron were most important and easy to model. With ASCA the other heavy element K-shell lines are well observed and the iron L-shell lines are prominent in the observational window. Hence the need to understand the more complicated iron L-shell physics. With further progress in the detectors down to lower energies the L-shell lines of elements like S, Si, Mg, Ne, O and eventually the M-shell lines of Fe are becoming important. DXS observations of such lines were reported by Sanders at this conference and new calculations of the line strengths of the relevant ions are in progress by Liedahl et al.. There is great hope to gain a better understanding of the physical state of the Local Bubble from the detailed analysis of such EUV and very soft X-ray spectra.

5 Non-Equilibrium Ionization Models

There are several types of equilibria that have to be established for ionization equilibrium models to apply: (i) electrons and ions have to be thermalized to unique electron and ion temperatures, T_e and T_i; (ii) T_e and T_i have to come to equilibrium (which may not be the case e.g. in young supernova remnants); (iii) the ionization and recombination rates have to come to a balance. This requires that the atomic collision processes are faster than any other change of the considered environment. The first two processes are relatively fast and controlled by Coulomb collisions or even faster plasma effects. It is the third type of deviation from equilibrium that is of concern here in the context of the Local Bubble.

In particular the cooling of the ISM plasma can be faster that the recombination of some ions with important cooling lines. In this case the ionization structure is not equal to the ion abundances calculated for ionization equilibrium. As an example we can show this effect for the recombination of CV: $CV + e \rightarrow CIV + h\nu$ for which $t_{recomb}(3 \cdot 10^5\text{K}) = \alpha_{recomb}^{-1} \cdot n_e^{-1} \sim 20000\,\text{yrs}\left(\frac{1\ \text{cm}^{-3}}{n_e}\right)$ which is longer than the plasma cooling time $t_{cool} = \frac{1.5nkT}{n_e n_p \Lambda(T)} \sim 6600\ \text{yrs}\left(\frac{1\ \text{cm}^{-3}}{n_e}\right)$. In a cooling plasma the recombination of CV will lag behind and the plasma will contain less of the strongly emitting lithium like CIV. The paper by Sutherland & Dopita (1993) shows a very nice diagnostic plot in Fig. 15 which gives a comparison of the recombination times versus the cooling time for a large set of relevant ions.

In early calculations by Shapiro and Moore (1976), for example, the authors have shown that in the cooling of hot plasma (starting with equilibrium conditions well above 10^6 K) non-equilibrium effects set in at temperatures below about $3{-}5{\cdot}10^5$ K. These effects are strong enough to change the cooling rate of the plasma as shown in Fig. 3. The non-equilibrium model contains more highly ionized ion species with less electrons in the outer shells which can provide the strongest emission lines. Therefore the non-equilibrium ionization model provides a less fast cooling than calculated for the equilibrium case. The retarded recombination also leaves its trace in the plasma spectrum. There is a substantial energy release connected with the delayed recombining ions which shows up in a saw-teeth continuum structure in the upper energy part of the spectrum.

A more recent and more detailed calculation of the ionization structure of cooling plasma is given by Schmutzler and Tscharnuter (1993). One interesting aspect seen in the results of their calculations is the change in the ionization structure of C, N, and O. Fig. 4 shows for example the ionization structure of oxygen (which should be compared to the equilibrium ionization structure shown in Fig. 2). Above, with only the equilibrium ionization models at hand, we have argued that the abundant observation of the ion OVI in the ISM implies that there is an important, widespread hot ISM phase with

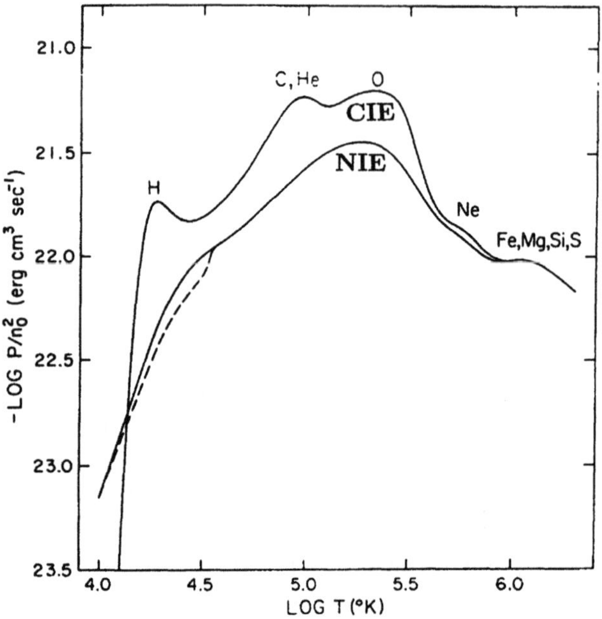

Fig. 3. Cooling coefficient as a function of T_e for the non-equilibrium ionization calculation of cooling plasma from Shapiro & Moore (1976).

a temperature in the narrow range $2 - 5 \cdot 10^5$ K. Now in Fig. 4 we note that the "diagnostic ion" OVI is abundant over quite large range of temperatures from about $4 \cdot 10^5$ K to temperatures as low as 10^4 K. Also the ion $OVII$ (the He-like species) which is also observed in the local ISM (see contribution by Bowyer et al.) can be found at low temperatures in non-equilibrium models. In a similar way the ions NV and CIV, observed in the galactic halo, lose their diagnostic power to indicate a very hot galactic halo.

The opposite effect of deviation from ionization equilibrium is observed in models with extra heat input to the plasma, where the heating rate can be faster than the ionization rates. Shapiro and Moore (1977) show such models for example for coronal loops with sudden heating. For the ISM this may be relevant for conductively heated interfaces between the hot ISM and embedded clouds. In a model for such interfaces Böhringer & Hartquist (1989) showed that non-ionization equilibrium structures can occur at the intermediate temperature ranges (few 10^5 K) which are most important for their effective radiative emission. The principle results of these type of models are characterized by low-ionization stage ions in a hotter environment with an increased contribution of emission lines to the total radiation as compared to continuum processes. Some strong and dominant emission lines can be boosted by up to about two orders of magnitude (e.g. Fig. 6 in Böhringer & Hartquist 1989).

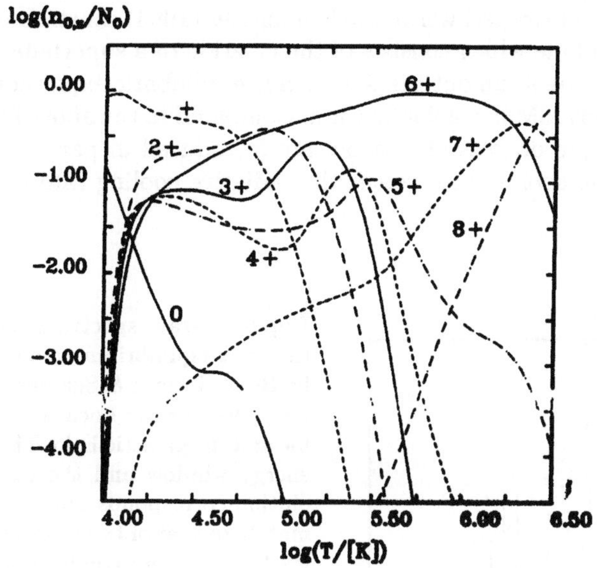

Fig. 4. Non-equilibrium ionization calculation for the ionization structure of Oxygen for cooling plasma from Schmutzler & Tscharnuter (1993).

6 Models for the Local Bubble

To approximately account for the soft X-ray and UV observations of the ISM two families of models are invoked: models near ionization equilibrium at high temperatures (around 10^6 K) and models that start at high initial temperatures (above 10^6 K) but experience a rapid cooling disturbing the ionization balance.

For the first case, early models, as for example Cox & Anderson (1982), have considered supernova explosions to explain the soft X-ray observation with typical parameters of a gas temperature of $\sim 10^6$ K, a pressure of $P/k \sim 10^4$ cm^{-3} K, and an extent of one to several 100 pc. More recent models also consider the case of a superbubble with an energy input more than an order of magnitude higher than that of a single supernova (e.g. Innes & Hartquist 1984, Edgar & Cox 1993). While in the first model the bubble is fairly young with an age of the order of 10^5 yr, the superbubbles can be inflated by subsequent energy releases and have a much larger age. For the younger bubbles even non-equilibrium ionization effects in the sense of too low ionization stages for the ambient temperature have to be considered (see Cox & Anderson 1982).

In summary these "$\sim 10^6$ K-models" can reproduce the observed Be, B, C band data in soft X-rays. Two problems have been outlined for these models: they produce little emission in the M-band (see Cox & Anderson 1982, Edgar & Cox 1993) unless the temperature is raised considerably above 10^6 K; They imply a large pressure of the LISM which exceeds the pressure deduced for the local cloud by an order of magnitude (e.g. Breitschwerdt 1996).

As an alternative model Breitschwerdt and Schmutzler (1994) constructed a history of the Local Bubble which consists of the creation of a superbubble in a dense cloudy environment, an out-break with a fast adiabatic expansion, and the escape of some hot ISM in a galactic wind. Compared to the above hot Bubble models, some more degrees of freedom are introduced in particular the ratio of the adiabatic cooling rate versus the radiative cooling rate as a function of time.

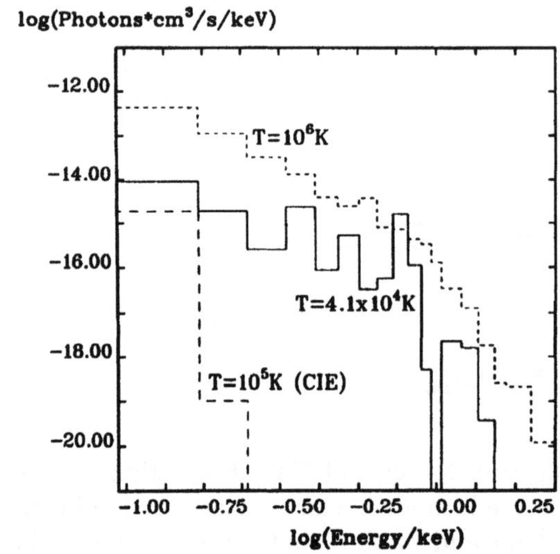

Fig. 5. X-ray spectrum of the non-equilibrium model by Breitschwerdt & Schmutzler (1994) for the Local Bubble and the galactic halo. The energy window and the resolution corresponds approximately to that of the ROSAT PSPC. The non-equilibrium model for a temperature of $4 \cdot 10^4$ K is compared to a 10^6 K model and an equilibrium model at 10^5 K. The non-ionization equilibrium model mimics the coronal temperature model in spite of its low temperature.

With appropriately chosen model parameters (initial temperature $= 2.5 \cdot 10^6$ K, initial ambient density up to 10^4 cm^{-3}) one finds results that can fit the observations for a present ambient temperature of the Local Bubble as low as a few 10^4 K. These solutions are characterized by large deviations from ionization equilibrium with most highly ionized species from the initial temperature plasma still present. A large fraction of internal energy of the plasma is stored in the highly ionized species, which can subsequently be released by recombinative radiation. The emission spectra are characterized by an unusually large contribution from recombination radiation, such that the high energy part of the spectrum is completely dominated by recombination photons. Due to this high energy recombination tail the emission spectrum of $4 \cdot 10^4$ K non-equilibrium plasma can mimic a spectrum of 10^6 K plasma in its rough overall shape and hardness of the radiation as shown in Fig. 5. Such a model can roughly reproduce the observed spectra in the soft X-ray and EUV regime as shown in Fig. 6 (taken from Breitschwerdt & Schmutzler 1994).

As shown in this paper, the introduction of fast cooling non-equilibrium ionization models opens a larger spectrum of scenarios for the explanation of the Local Bubble observations. Firm model predictions can, however, be made. One can in the future differentiate between the various models when sufficiently high resolution spectroscopic data become available resolving the lines and recombination radiation features.

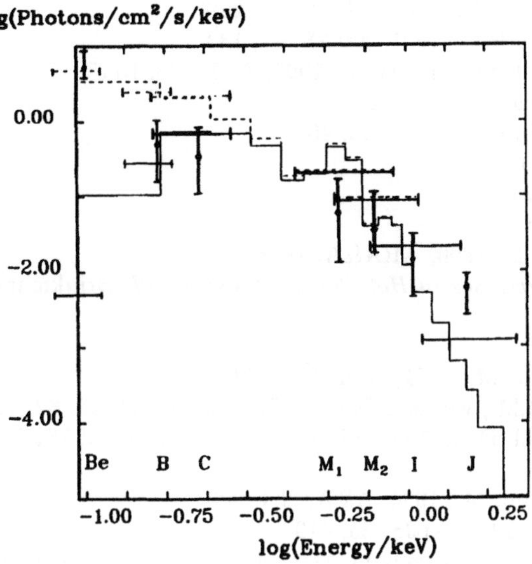

Fig. 6. Comparison of the results for the emission spectrum of the Local Bubble and a galactic wind from the model of Breitschwerdt & Schmutzler (1994) with the observations in the various EUV and soft X-ray bands. Good agreement can be found for appropriately chosen model parameters (for details see Breitschwerdt & Schmutzler (1994).

References

Arnaud, M. & Rothenflug, R., 1985, A&AS, 60, 425
Arnaud, M. & Raymond, J., 1992, ApJ, 398, 394
Böhringer H. & Hartquist, T.W., 1987, MNRAS, 228, 915
Böhringer H. & Hensler, G., 1989, A&A, 215, 147
Breitschwerdt, D., 1996, Space Sci. Rev., 78, 173
Breitschwerdt, D. & Schmutzler, T., 1994, Nature, 371, 774
Brickhouse, N.S., Raymond, J.C., Smith, B.W., 1995, ApJS,
Brickhouse, N. et al., 1995, LEGACY, Journal of HEASARC at NASA GSFC, 6, 4
Cox, D.P., 1972, ApJ, 178, 143
Cox, D.P. & Tucker, W.H., 1969, ApJ, 157, 1157
Cox, D.P. & Daltabuit, E., 1971, ApJ, 167, 113
Cox, D.P. & Anderson, P.R., 1982, ApJ, 253, 268
Edgar R.J. & Cox, D.P., 1993, ApJ, 413, 190
Fabian, A.C., Arnaud, K.A., Bautz, M.W., Tawara, Y., 1994, ApJ, 436, L63
Frisch, P.C. & York, D.G., 1983, ApJ, 271, L59
Gaetz, T.J. & Salpeter, E.E., 1983, ApJS, 52, 155
Gronenschild, E.H.B.M. & Mewe, R., 1978, A&AS, 32, 283
Innes, D.E. & Hartquist, T.W., 1994, MNRAS, 209, 7

Kaastra, J.S. & Mewe, R., 1994, LEGACY , Journal of HEASARC at NASA GSFC, 3, 16
Kato, T., ApJS, 30, 397
Kato, T., Masai, K, Arnaud, M., 1991, NIFS-DATA-14, Nat. Inst. Fusion Res., Nagoya
Landini, M. & Monsignori Fossi, B.C., 1970, A&AS, 6, 468
Landini, M. & Monsignori Fossi, B.C., 1990, A&AS, 82, 229
Landini, M. & Monsignori Fossi, B.C., 1991, A&AS, 91, 183
Liedahl, D.A., Osterhold, A.L., Goldstein, W.H., 1995, ApJ, 438, L115
Masai, K., 1984, Astrophys. Space Sci., 98, 367
Masai, K. & Kato, T., 1987, Phys. Lett. A, 123, 405
Masai, K., 1994a, ApJ, 437, 770
Masai, K., 1994b, JQSRT, 51, 211
Masai, K., 1997, A&A, preprint
McCammon, D. & Sanders, W.T., 1990, ARA&A, 28, 657
Mewe, R., 1990, in *Physical Processes in Hot Cosmic Plasmas*, W. Brinkmann et al. (eds.), Kluwer, p. 39
Mewe, R., 1991, A&AR, 3, 127
Mewe, R. & Gronenschild, E.H.B.M., 1981, A&AS, 45, 11
Mewe, R., Gronenschild, E.H.B.M., van den Oord, G.H.J., 1985, A&AS, 62, 197
Mewe, R., Kaastra, J.S., Liedahl, D.A., 1995, LEGACY, Journal of HEASARC at NASA GSFC, 6, 16
Monsignori Fossi, B.C. & Landini, M., 1994, A&A, 284, 900
Raymond, J.C. & Smith, B.W., 1977, ApJS, 35, 419
Raymond, J.C., 1988, in Hot Thin Plasmas in Astrophysics, R. Pallavicini (ed.), Kluwer, Dordrecht, p. 3
Raymond, J.C. & Brickhouse, N.S., 1996, Astrophys. Space Sci., 237, 321
Savage, B.D. & DeBoer K.S., 1981, ApJ, 243, 460
Schmutzler, T. & Tscharnuter, W.M., 1993, A&A, 273, 318
Shapiro, P.R. & Moore, R.T., 1976, ApJ, 207, 460
Shapiro, P.R. & Moore, R.T., 1977, ApJ, 217, 621
Shull, J.M., 1981, ApJS, 46, 27
Sutherland, R.S. & Dopita, M.A., 1993, ApJS, 88, 253
Tucker, W.H. & Koren, M., 1971, ApJ, 168, 283

Infrared Emission from Interstellar Dust in the Local Interstellar Medium

W.T. Reach and F. Boulanger

Institut d'Astrophysique Spatiale, Bâtiment 121,
Université Paris Sud, F-91405 Orsay cedex, France

Abstract. In this contribution, we discuss some topics in the study of dust in the local interstellar medium. The spectrum of local clouds has recently been measured using *ISO* and *COBE*, confirming in general the models of excitation and cooling of dust illuminated by the the interstellar radiation field. We discuss in some detail the spatial correlation of interstellar dust and gas and the idea that infrared emission traces the total column density of the interstellar medium, with the particular application to the formation of H_2. We also show that dust is present in the nearby Loop I shell, with properties similar to average, suggesting that the walls of the Local Bubble would also have relatively 'normal' dust content.

1 Introduction

Dust grains play crucial roles in both the thermodynamics and chemistry of the interstellar medium, but their role in the local bubble is not well known. In general, dust is observed to be well-mixed with interstellar gas for a wide range of physical conditions; however, grains are suspected to be efficiently destroyed in fast interstellar shocks, such as are thought to have formed the 'local bubble'—by which we mean the boundaries and interior of the region of low H I column density and of significant X-ray emissivity surrounding the Solar System to ~ 100 pc. Here we address the question of whether dust grains exist in the local bubble and what their properties might be. The Ulysses spaceprobe has already detected interstellar dust flowing through the Solar System (Grün et al. 1990, Landgraf et al. 1997), so understanding the nature of interstellar dust is no longer a 'remote' astrophysical issue but rather it makes a direct connection to the Solar System. The diffuse interior of the local bubble itself (including any 'local interstellar cloud') has too low column density for its infrared emission or starlight extinction to be directly observed, but some relatively dense clouds have been identified with distances that place them in the bubble (Kerp, Herbstmeier, & Mebold 1993). Furthermore, the extent of the 'local bubble' is comparable to the scale-height of neutral gas in the disk. Therefore, emission from neutral phases of the interstellar medium at relatively high galactic latitudes (say $|b| > 30°$) has a significant probability of being part of the local bubble.

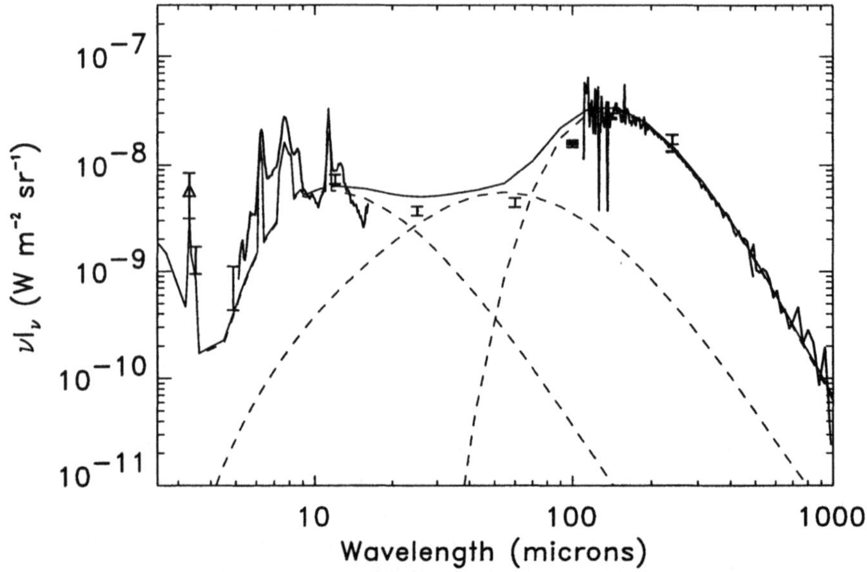

Fig. 1. The infrared emission spectrum of interstellar dust heated by the interstellar radiation field. The brightness here corresponds to an interstellar gas column density of 10^{20} cm^{-2}. The *continuous thin curve* is a model consisting of three dust components, whose individual contributions—discussed in the text—are shown as *thin dashed lines*. The observed spectrum of interstellar dust is shown as *solid curves*; the far-infrared spectrum is from *COBE*/FIRAS, and the mid-infrared spectrum is from ISOCAM. Points with error bars are from *COBE*/DIRBE broad-band observations and the *Arome* balloon experiment.

2 Origin of Infrared Emission

Interstellar dust absorbs energy from starlight, and it emits this energy in the infrared. No other heating or cooling mechanisms have been found to be of comparable importance for dust grains, under typical interstellar conditions, and it is doubtful that any other mechanism (such as collisional heating by hot plasma) is important in the local bubble. The distinctive pattern of starlight extinction by interstellar dust has allowed the determination of the size distribution, and to some extent the composition, of dust particles (Mathis, Rumpl, & Nordsieck 1977). Knowing the average value of the interstellar radiation field (Mathis, Mezger, & Panagia 1983), the infrared emission can be calculated by a simple thermal balance for each grain size. However, it came as somewhat as a surprise, when the *Infrared Astronomical Satellite* (Low et al. 1984) systematically observed the infrared emission from the diffuse interstellar medium at 12, 25, 60, and 100 μm wavelength, that the straightforward estimate under-predicted the 12 and 25 μm brightness by orders of magnitude. The *IRAS* observations led to two significant advances. First, it was realized that small grains are not in equilibrium with the relatively slowly arriving ultraviolet photons of the interstellar radiation field. A non-equilibrium model, where the dust temperature rises significantly after the absorption of an ultraviolet photon, can explain some of the bright mid-infrared emission—in particular at 25 μm (Draine & Anderson 1985). But the small size of the particles that are required to reproduce the 12 μm emission makes them essentially large molecules, of which a likely candidate, polycyclic aromatic hydrocarbons (PAH), was suggested as the carrier (Leger & Puget 1984).

The infrared emission spectrum calculated by a 'modern' model for the interaction of interstellar dust with the interstellar radiation field is shown in Figure 1. The model shown here, taken from Désert, Boulanger, & Puget (1990), includes three components of the grain population:

- *big grains* in thermal equilibrium with the radiation field dominate the far-infrared emission;
- *very small grains* stochastically heated by ultraviolet photons dominate the emission between 30 and 80 μm; and
- *polycyclic aromatic hydrocarbons (PAHs)* produce the lines and shortest-wavelength infrared continuum.

The optical properties and size distributions in this model are based on laboratory data (optical properties of silicate and graphite), astronomical data (selective extinction and broad-band infrared colors), and a string of astrophysical inferences. Some of these inferences, for example the assignment of the far-ultraviolet extinction to PAHs and the assignment of the 2200 Åbump to very small grains, remain to be tested. But the separation of the dust into three components based on characteristic size of the particle, where the interaction between the particles and the radiation field changes from classical equilibrium (big grains) to stochastic heating (very small grains) to quantum mechanical (carrier of the infrared emission bands), is likely to be a characteristic of all successful models (cf. Dwek et al. 1997).

3 Spectroscopy of Interstellar Dust

3.1 Far-Infrared Spectrum

We may now compare the models that can successfully match the broad-band infrared emission with detailed spectroscopic observations. In Figure 1, we show recent spectral observations from *COBE* and from *ISO*. The far-infrared spectrum shown here was obtained by a comparison of the *COBE* FIRAS observations with the H I column density over a large fraction of the sky (Boulanger et al. 1996a). The FIRAS observed the entire sky in a 7° beam, with very stable and calibration and good sidelobe rejection—important for measuring the faintest parts of the sky. The broad-band points in Figure 1 were obtained by correlating the *COBE* DIRBE (Boggess et al. 1992) skymaps with the H I map. It is immediately evident that the far-infrared spectrum is very smooth (apart from the bright spectral line of gaseous C^+, which was excised from the spectrum shown here), and a single blackbody at 17.5 K is a very good fit to the high-latitude spectrum (Boulanger et al. 1996a, Reach et al. 1995). The observed spectrum is in quite good agreement with the models.

3.2 Mid-Infrared Spectrum

Another significant advance in the observational study of the local interstellar dust is due to the mid-infrared camera aboard *ISO* (Kessler et al. 1996). The field of view of the camera is much smaller than that of the *COBE* or H I observations (compare the 7° FIRAS beam to the 3′ ISOCAM field of view), so all-sky or even wide-field observations are out of the question. But the sensitivity of the ISOCAM is enough to detect the emission from nearby reflection nebulae and molecular clouds, even with the circular variable filter, which allows spectrophotometric maps to be made. In Figure 1 is the 4.5–17 μm spectrum of a cloud near ρ Oph, scaled appropriately to account for the high radiation field and column density (Boulanger et al. 1996b). We see that the mid-infrared emission from the diffuse interstellar medium is dominated by spectral lines at 6.2, 7.7, 8.6, 11.3, and 12.7 μm, and the spectrum is almost exactly the same in radiation fields ranging from the average interstellar radiation field (Chamaeleon clouds) to 50 times this (ρ Oph) to 10^3 times this (NGC 7023). In very strong and hard radiation fields, the mid-infrared spectrum does change (Roelfsema et al. 1996), but there is a nearly 'universal' spectrum away from hot stars. The fact that the spectrum is dominated by the emission bands clearly demonstrates that the interaction of the smallest particles in the interstellar medium is quantum mechanical. The energy absorbed in these quantum mechanical modes is not trivial in comparison with that absorbed by larger grains; to produce the observed infrared emission, the PAH must absorb about 1/3 as much energy from the interstellar radiation field as the grains. The abundance requirements of the infrared band carriers is also non-trivial, with current estimates indicating about 20% of the total interstellar C being locked in PAH (Dwek et al. 1997).

4 Relationship Between Interstellar Dust and Gas

The fact that the interstellar dust and gas are well mixed is demonstrated by the relatively tight correlation of infrared emission with H I 21-cm column density for lines of sight away from star forming regions and other molecular clouds. This correlation is not just a reflex of the fact that both the dust and gas are components of the disk—which would cause both to scale (to first order) as the cosecant of galactic latitude—but rather the dust and gas are correlated in detail. This was shown on large scales by dividing both the infrared emission and H I column density maps (at half-degree resolution) by the cosecant of latitude; these normalized maps exhibited the same correlation as the overall emission (Boulanger and Pérault 1988). Examples of the infrared-H I correlation are shown in Figure 2. A similar comparison of high-latitude microwave emission (from the *COBE* DMR) with infrared maps suggests that free-free emission from diffusely-distributed H II is also correlated with dust (Kogut et al. 1996)—although the separation of dust between H I and H II phases remains to be explored. On smaller angular scales, where high-resolution H I observations have been performed, the correlation with infrared emission remains strong (Joncas, Boulanger, & Dewdney 1992) and with similar slope, suggesting that the dust-to-gas ratio is relatively constant (at the factor of two level).

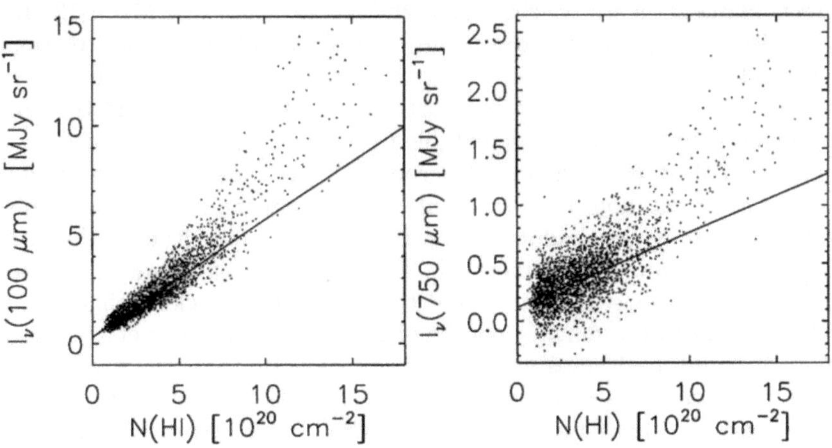

Fig. 2. Infrared surface brightness at 100 μm (*COBE* DIRBE broad filter) and 750 μm (*COBE* FIRAS spectra averaged 600–900 μm) as a function of H I column density for the high-latitude sky (from Boulanger et al. 1996a). The straight lines are linear fits to the points with $N(\mathrm{H\ I}) < 4.5 \times 10^{20}$ cm^{-2}. At higher column densities the infrared emission is systematically higher than the extrapolation of the linear fits, probably due to emission by dust in molecular clouds.

We explored the constancy of the dust-to-gas ratio by comparing the *COBE* DIRBE 100 μm map to the Leiden-Dwingeloo H I total column density in a network of patches distributed over the entire sky. It is clear that there are patches of 'excess' emission, which we will discuss below, in places where the H I column density is high. To evaluate the dust-to-gas ratio, we excluded pixels (which were half-degree in size) where the column density was greater than 3×10^{20} cm^{-2}. There is a range of infrared emission per unit H I column density, due to dust temperature variations from region to region (Désert, Bazell, & Boulanger 1988, Reach et al. 1997). This effect is especially important at 100 μm (and shorter) wavelength, where the interstellar spectrum is relatively sensitive to temperature.

4.1 Importance of Cloud Shadows

One perennial problem in studies of the interstellar medium is the lack of distance information. For high galactic latitudes, one generally assumes that the clouds are within a scale height, $H/\sin|b|$, of the Sun. The scale height is determined from observations of more distant gas at low latitudes and near the terminal velocity of galactic rotation (Lockman 1984). In some cases, the extinction (or Na or Ca line absorption) of light from stars at a range of distances allow one to bracket the distance to the clouds (cf. Welty, this conference). Thus some clouds are known to be relatively local, such as the 'finger' filament (Kerp, Herbstmeier, & Mebold 1993) and MBM 12 (Hobbs et al. 1988). A significant advance in understanding the spatial relationship of the interstellar clouds and the local X-ray emitting medium was made possible by *ROSAT*. The sky is relatively bright in X-rays, and the absorption by even small clouds is detectable in the soft X-rays to which *ROSAT* is sensitive. If there is pervasive 'background' X-ray emission, as expected from a galactic corona (and, for very local clouds, from the local bubble emission itself), then an X-ray shadow is expected to be present either in the *ROSAT* all-sky survey (for places where the X-ray emission is bright) or in deeper pointed observations. It is clear that the comparison of cloud shadows, both in X-rays and in the number counts of galaxies (Finkbeiner et al., this conference), to infrared maps will shed a new light on local interstellar clouds. The optical shadows (due to dust) trace essentially all the dust, and the X-ray shadows (due largely to He and metals) trace essentially all the gas. Thus the X-ray optical depth and the optical extinction should differ in detail from the infrared and H I maps, depending on the dust-to-gas ratio and the H ionization state for a particular line of sight. At this meeting it was shown in several presentations (e.g. Snowden; Mebold; Warwick; Wang) that X-ray shadows are evident essentially everywhere and from a wide range of cloud sizes. Because of the small-scale structure of the interstellar medium, and the relatively high resolution of *ROSAT*, the X-ray maps were often compared to infrared maps such as the *IRAS* 100 μm maps. Our experience with the correlation between infrared emission and H I supports the idea that infrared emission is a tracer of the total column density of gas. However, molecular clouds have a significantly lower infrared color temperature than atomic clouds. At half-degree angular resolution, high-latitude molecular clouds have a dust temperature of 14.7 K compared to atomic clouds at 17.0 K (Reach et al. 1997); the corresponding 100 μm emission per unit total gas column density is 3.8 times higher in atomic gas compared to molecular gas. At the higher angular resolution of *IRAS* and *ROSAT*, one will resolve even colder cloud cores; therefore, one must be cautious in the comparison of X-ray shadows (or other measures of extinction) to the 100 μm emission.

4.2 Formation of Molecular Hydrogen

At column densities exceeding 5×10^{20} cm^{-2}, the infrared emission per unit H I column density is noticeably higher than at lower column densities. This is seen both in the correlation of large-scale emission (Fig. 2), within individual clouds, and for random lines of sight through clouds (Savage et al. 1977). A particularly favorable example for an individual cloud is G236+39, which appears on the *IRAS* maps as a large, centrally concentrated cloud. This cloud has been mapped in the 21-cm line at the highest resolution possible using single-dish telescopes, the Arecibo 305-m. The infrared emission correlates perfectly with the 21-cm column density in the peripheries of the cloud, but toward the center there is an infrared peak significantly stronger than the H I peak. As there is no reason to suspect that the center of this cloud is heated (in fact the 60 μm emission is relatively depressed in the center of the cloud), then either there is a source of dust in the center of the cloud (which is highly unlikely because there are no 12 μm sources there), or the infrared emission is due to dust associated gas in molecular form (Reach et al. 1994). Using the *COBE* data at 100, 140, and 240 μm, we confirmed that the core of this cloud is *not* heated, and it is only slightly colder than other dust in the region. Balancing the formation and destruction rates of H_2 (van Dishoeck and Black 1986), we obtain the molecular column density of a uniform-density slab as follows:

$$N(\mathrm{H}_2) = 4 \times 10^{19} \chi^{-2} \left(\frac{n}{100 \ \mathrm{cm}^{-3}}\right)^2 \left(\frac{T}{80 \ \mathrm{K}}\right) \left[\frac{N(\mathrm{H\ I})}{10^{20} \ \mathrm{cm}^{-2}}\right]^2 \ \mathrm{cm}^{-2}, \quad (1)$$

where N denotes column density, n volume density, T the gas temperature, and χ the strength of the radiation field relative to the average ISRF at the Solar circle (Mathis, Mezger, & Panagia 1983). Then the total column density of gas should have a quadratic dependence on the H I column density. (This applies only on the square-root portion of the H_2 ultraviolet absorption curve of growth; at very low column densities the relation is more complicated.) The quadratic dependence was observed for G236+39, which is an isolated cloud with relatively gradual internal structure. But in general the quadratic dependence is only expected for a uniform-density region. There are two phases of atomic gas (Kulkarni & Heiles 1988), and H_2 will only form in the colder, higher-density phase; thus the total column density will be

$$N(\mathrm{H})_{tot} = N(\mathrm{H\ I}) + 2N(\mathrm{H}_2) = N(\mathrm{H\ I}) + \alpha N(\mathrm{H\ I})^2_{cold}. \quad (2)$$

For a large patch of sky, both cold and warm H I contribute comparable column densities, so the scatter diagram of infrared surface brightness *versus*

H I column density has a linear lower envelope and roughly quadratic upper envelope, with individual lines of sight falling in between.

The fact that infrared emission traces the *total* interstellar column density, while the 21-cm line traces the atomic gas, means that regions with 'excess' infrared emission are likely locations for molecular and ionized gas. This idea has been used to create an all-sky survey of the molecular gas (Désert, Bazell, & Boulanger 1988), which we are currently updating using improved infrared and H I surveys. At high latitudes, the molecular gas is not diffusely distributed, because the radiation field is too strong outside of relatively dense clouds. Instead, a finite number of complexes of molecular clouds exist. Many of these have already been discovered as diffuse patches of extinction on the Palomar plates and confirmed as molecular using millimeter observations of CO lines (Magnani, Blitz, & Mundy 1985). The infrared excess technique adds some more clouds, which may have undetectable extinction or CO lines. We have confirmed that the most significant infrared excess clouds are really regions of enhanced column density, rather than regions of warmer dust: the far-infrared colors of these regions (form the DIRBE at 100, 140, and 240 μm maps) indicate that the infrared excess clouds are significantly *colder* than average. The combination of enhanced infrared emission, relative to H I, and relatively colder dust both indicate regions relatively shielded from the interstellar radiation field, where molecules can form. As mentioned above, the significant temperature difference between atomic and molecular gas, even at half-degree angular resolution, means that the infrared emission—especially at 100 μm and shorter wavelengths—is not a *linear* tracer of the total column density, but this effect can be calibrated with multi-wavelength far-infrared observations (as exist on *COBE* and are planned for *Planck surveyor*).

4.3 Shock Processing of Local Dust

The large-scale distribution of interstellar matter consists of filaments and loops that are clearly evident in both H I 21-cm and infrared maps. If the Solar System is within a 'local bubble,' then we can hope to understand processes that affect dust grains by studying the other, similar structures. Fast shocks can destroy large grains and shatter almost any grain, significantly modifying the size distribution (Jones, Tielens, & Hollenbach 1996). When large grains are destroyed, the mass of very small grains must increase; comparing to the infrared emission model shown in Figure 1, we see that the 25–60 μm emission should be enhanced relative to that at 100–240 μm. Hints of this effect have been seen in the infrared colors of clouds with a range of velocities (Heiles et al. 1988). Should the dust in the 'local bubble' have the strongly-shocked or the average size distribution?

The nearby Loop I or North Polar Spur sheds some light on this issue. The Spur is clearly evident in the nonthermal radio continuum, H I 21-cm, and infrared 60–240 μm maps. We extracted profiles through the Spur at each wavelength and the H I map—they are very similar—and measured the following infrared colors:

$$I_\nu(12)/I_\nu(100) \simeq 0.031 \pm 0.005,$$
$$I_\nu(25)/I_\nu(100) \simeq 0.04 \pm 0.01,$$
$$I_\nu(60)/I_\nu(100) = 0.19 \pm 0.02,$$
$$I_\nu(240)/I_\nu(100) = 1.3 \pm 0.1,$$
$$I_\nu(100)/N(\mathrm{H\ I}) = 0.54 \pm 0.04 \ \mathrm{MJy/sr}/10^{20}\ \mathrm{cm}^{-2},$$

where I_ν is the surface brightness of Loop I minus a baseline due to more distant dust. These colors are indistinguishable from those of the average interstellar interstellar medium as shown in Figure 1. The 12 and 25 μm brightness profiles are rather uncertain because the brightness of Loop I is less than 0.1% of the zodiacal light brightness at these wavelengths; however, even if the values listed here are considered as upper limits, it is still clear that the mid-infrared brightness of Loop I is *not* enhanced. Furthermore, the 100 μm emission per unit gas column density is indistinguishable from the average value, so the dust-to-gas mass ratio is the same. We conclude that either (1) the dust in the walls of Loop I has *not* been significantly processed by fast shocks, or (2) interstellar grains are significantly more resistant to sputtering and shattering than has been estimated. The same conclusions will likely apply to the walls of the 'local bubble:' we expect the grain properties to be not much different from those of the average interstellar medium.

References

Boggess, N. W., Mather, J. C., Weiss, R., Bennett, C. L., Cheng, E. S., Dwek, E., Gulkis, S., Hauser, M. G., Janssen, M. A., Kelsall, T., Meyer, S. S., Moseley, S. H., Murdock, T. L., Shafer, R. A., Silverberg, R. F., Smoot, G. F., Wilkinson, D. T., & Wright, E. L. (1992): ApJ 397, 420

Boulanger, F., Abergel, A., Bernard, J. -P., Burton, W. B., Désert, F. -X., Hartmann, D., Lagache, G., Puget, J. -L. (1996a): A& A 312, 256

Boulanger, F., Reach, W.T., Abergel, A., Bernard, J.P., Cesarsky, C.J., Cesarsky, D., Désert, F.X., Falgarone, E., Lequeux, J., Metcalfe, L., Pérault, M., Puget, J.L., Rouan, D., Sauvage, M., Tran, D., Vigroux, L. (1996b): A& A 315, 352

Boulanger, F., & Pérault, M. (1988): ApJ 330, 964

Cesarsky, D., Lequeux, J., Abergel, A., Pérault, M., Palazzi, E., Madden, S., Tran, D. (1996): A& A 315, L305

Draine, B. T., & Anderson, N. (1985): ApJ 292, 494

Désert, F.-X., Bazell, D., & Boulanger (1988): ApJ 334, 815

Désert, F.-X., Boulanger, F., & Puget, J.-L. (1990): A& A 237, 1

Dwek, E., Arendt, R. G., Fixsen, D. J., Sodroski, T. J., Odegard, N., Weiland, J. L., Reach, W. T., Hauser, M. G., Kelsall, T., Moseley, S. H., Silverberg, R. F., Shafer R. A., Ballester, J., Bazell, D., Isaacman, R. (1997): ApJ 475, 565

Grün, E., Gustafron, B., Mann, I., Baguhl, M., Morfill, G. E., Staubach, P., Taylor, A., & Zook, H. A. (1990): A& A 286, 915

Heiles, C., Reach, W. T., & Koo, B.-C. (1988): ApJ 332, 313
Hobbs, L. M., Penprase, B. E., Welty, D. E., & Magnani, L. (1988): ApJ 327, 356
Joncas, G., Boulanger, F., & Dewdney, P. E. (1992): ApJ 397, 165
Jones, A. P., Tielens, A. G. G. M., & Hollenbach, D. J. (1996): ApJ 469, 740
Kerp, J., Herbstmeier, U., Mebold, U. (1993): A& A 268, L21
Kessler, M. F., Steinz, J. A., Anderegg, M. E., Clavel, J., Drechsel, G., Estaria, P., Faelker, J., Riedinger, J. R., Robson, A., Taylor, B. G., Ximenez de Ferran, S. (1996): A& A 315, L27
Kogut, A., Banday, A. J., Reach, W. T., Hinshaw, G., & Bennett, C. L. (1996): ApJ 460, 1
Kulkarni, S., & Heiles, C. (1988): in *Galactic and Extragalactic Radio Astronomy*, (Springer), p. 95
Landgraf, M. et al. (1997): this volume
Leger, A., & Puget, J. L. (1984): A& A 137, L5
Lockman, F. J. (1984): ApJ 283, 90
Low, F. J., Young, E., Beintema, D. A., Gautier, T. N., Beichman, C. A., Aumann, H. H., Gillett, F. C., Neugebauer, G., Boggess, N., Emerson, J. P. (1984): ApJL 279, L19
Magnani, L., Blitz, L., & Mundy, L. (1985): ApJ 295, 402
Mathis, J. S., Rumpl, W., & Nordsieck, K. H. (1977): ApJ 217, 425
Mathis, J. S., Rumpl, W., & Nordsieck, K. H. (1977): A& A 128, 212
Reach, W. T., Dwek, E., Fixsen, D. J., Hewagama, T., Mather, J. C., Shafer, R. A., Banday, A. J., Bennett, C. L., Cheng, E. S., Eplee, R. E., Jr., Leisawitz, D., Lubin, P. M., Read, S. M., Rosen, L. P., Shuman, F. G. D., Smoot, G. F., Sodroski, T. J., Wright, E. L. (1995): ApJ 451, 188
Reach, W. T., Koo, B.-C., & Heiles, C. (1994): ApJ 429, 672
Reach, W. T., Wall, W. F., Odegard, N., Weiland, J., Arendt, R. G., & Hauser, M. G. (1997): in preparation
Roelfsema, P. R., Cox, P., Tielens, A. G. G. M., Allamandola, L. J., Baluteau, J. -P., Barlow, M. J., Beintema, D., Boxhoorn, D. R., Casinelli, J. P., Caux, E., Churchwell, E., Clegg, P. E., de Graauw, T., Heras, A. M., Huygen, R., van der Hucht, K. A., Hudgins, D. M., Kessler, M. F., Lim, T., Sandford, S. A. (1996): A& A 315, L289
Savage, B. D., Bohlin, R. C., Drake, J. F., & Budich, W. (1977): ApJ 216, 291
van Dishoeck, E. F., & Black, J. H. (1986): ApJS 62, 109

Cirrus as a Probe for the Galactic Bubble Bath

M.-A. Miville-Deschênes[1,2], F. Boulanger[1], and W.T. Reach[1]

[1] Institut d'Astrophysique Spatiale, Orsay, France
[2] Département de Physique, Université Laval, Québec, Canada

Abstract. Galactic filaments called *cirrus* are used here to study the froth structure of the interstellar medium. Using the Leiden/Dwingeloo HI survey we observe that a significant fraction of intermediate latitude cirrus are distant clouds ($R_{Gal} < 6$ kpc) and that some cirrus extend far above (or below) the Galactic plane (1 kpc). Among them, a supershell centered at $l = 15°$, $b = -12°$ is studied.

1 Introduction

In the 80s, the IRAS all-sky survey revealed that the infrared (IR) sky is highly structured. Filaments called *IR-cirrus* interwine at all Galactic latitudes. This filamentary picture of the whole sky is corroborated by HI observations. There is indeed a general good spatial correlation between the diffuse far IR emission of dust and the diffuse H I.

Observing this filamentary structure over the whole sky may suggest that it reflects the local environment. Nevertheless filamentary clouds seen near the Galactic plane may be huge structures seen at great distance. Studying these large and distant clouds may help us to understand how the Galactic interstellar matter is organized on large scale, what is its morphology, distribution and dynamics, how the stellar cycle acts on the ISM and what is the nature of the disk-halo interconnection. Having a distant view of large structures may also give strong insights about the structure of the local bubble and the local environment. In that context we propose here to study distant cirrus, or more generally distant filamentary ISM structures.

2 Distant H I

The Leiden/Dwingeloo survey (Burton & Hartmann (1994)) is used as kinematic distance indicator to find distant structures at intermediate latitudes ($|b| \leq 20°$). The H I column density at a given radius is obtained with the velocity information given by the Doppler shift of the 21 cm line assuming that the gas motions are dominated by the global Galactic rotation. Then the H I datacube (l, b, v) can be transposed in the (l, b, R_{Gal}) space using a $v \Rightarrow R_{Gal}$ relation based on the Galactic rotation curve (Burton et al.(1992)).

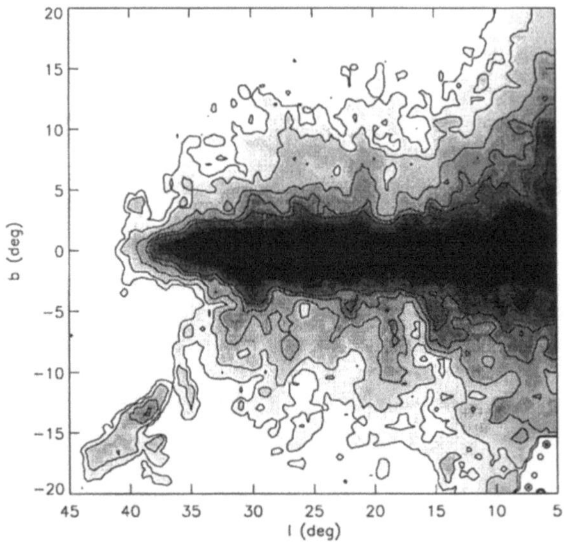

Fig. 1. Distant HI ($2 < R_{Gal} < 6$ kpc); contours are 1.9, 3.8, 9.4, 18.8 and $28.3 \times 10^{19} cm^{-2}$

We show in figure 1 the H I emission located in the galactocentric radius range $2 \leq R_{Gal} \leq 6$ kpc. There is no significant emission at $l > 40°$; on first approximation, the kinematic of the H I gas located in the inner portion of the Galaxy is well described by the Galactic rotation curve used here. The H I emission is concentrated around the galactic plane ($b = 0°$), but significant emission is observed at all latitudes up to $|b| = 15° - 20°$. There is artificially strong H I emission at intermediate latitudes for $l < 10°$ due to velocity crowding (Burton et al.(1992)).

Elongated structures are apparent in many places , for examples (l,b) = (19°,-8°), (30°,-5°), (14°,-5°), (22°,5°), ... Most of these structures seem to be perpendicular to the galactic plane and look connected to it. Some of them were described as worms by Koo et al.(1992); in particular the (30°,-5°) feature (GW30.5-2.5 in Koo et al.(1992)) could be a Galactic blowout. Besides these filaments, the distant, intermediate-latitude H I emission is also characterized by a faint and diffuse component that extends far above (or below) the Galactic plane. A cloud found at (l,b) = $(20°, -15°)$ and $R_{Gal} =$ 5 kpc is at $z \approx 1.1$ kpc under the plane. This should be compared to the 200 pc scale height of the Galactic H I (Lockman (1984)). Thus this gas is very far from the main H I disk. We should ask ourselves: what put this gas there and what maintains it at this location ? A Galactic fountain (Kahn (1997)) or chimneys (Ikeuchi (1997)) may be invoked to explain the presence of high-z

H I . A study of the physical parameters of individual distant clouds may help us to understand the z distribution and the morphology of the gas, and the importance of dust in a disk-halo circulation scenario.

3 Sagittarius Supershell

We will now focus on a pair of elongated filaments: $(19°, -8°)$ and $(14°, -5°)$. We show in figure 2 a) a sum of selected velocity channels ($20 < v_{LSR} < 69$ km s^{-1}) of that region. Each row of this image has been divided by sin(b) to highlight fainter structures at higher $|b|$. Two bright north-south filaments are evident. Both seem to have a common origin in the Galactic plane at ($l = 17°$) where Myers et al. (1986) found 4 active H II regions and 2 molecular clouds ($M > 10^6 M_\odot$). There are also fainter H I clouds at higher $|b|$ that seem to be kinematically associated with the brighter filaments, forming a cone-like structure as a whole. This large structure has been partly identified by Heiles (1979) as GS 016-06+43 (it goes up to $b \approx -10°$ - see figure 2 a)). From an analysis of the velocity field, no net expansion of the supershell is detected. All the individual substructures have a velocity peak near 45 km s^{-1} and a kinematical distance of $d_\odot = 3.9$ kpc ($R_{Gal} = 4.8$ kpc) is adopted. The total H I mass of the walls of the supershell (for $b < -4°$) is $M_{HI} \approx 10^6 M_\odot$ and the maximum extent under the Galactic plane is $z_{max} \approx 1.1$ kpc. In spite its larger size, this supershell can be qualitatively compared to other supershells recently discovered (Normandeau & Dewdney (1996) and Maciejewski et al. (1996)) and to the theoretical description of Ikeuchi (1997).

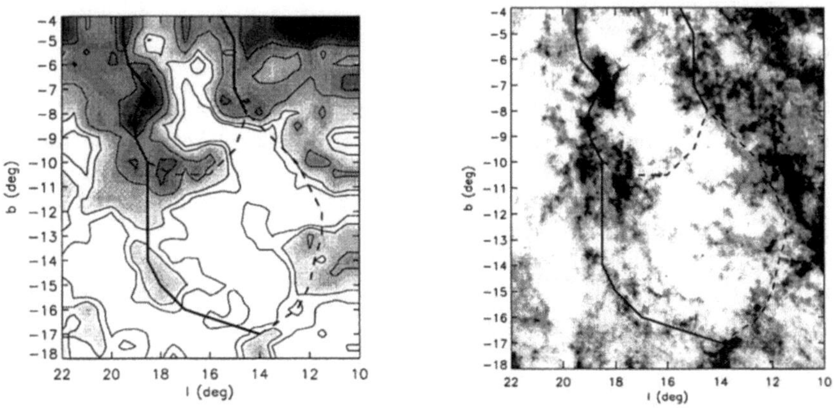

Fig. 2. The Sagittarius Supershell: **a)** Integrated HI emission divided by sin(b) ($20 < v_{LSR} < 69$ km s^{-1}). The contours are 1.5, 1.9, 2.8, and 3.8$\times 10^{19}$. **b)** IRAS 100 μm emission (each row is divided by the median of the row)

The IRAS 100 μm emission (each row has been divided by the median of the row) of the same region is shown in figure 2 b). The walls of the supershell are apparent, especially the one at $(19°, -9°)$. The higher angular resolution of IRAS allows us to observe the fragmentation of the walls; the filaments are composed of smaller filaments and nodes.

The emission at 60 and 100 μm is highly correlated all over the walls of the supershell with $I_{60\mu m}/I_{100\mu m} = 0.33$. This value reflects a small excess of very small grains that may be attributed to energetic processes associated with the creation of the supershell that grind down large grains into smaller ones (Jones et al. (1996)). The average 100 μm emissivity of the H I walls is in accordance with the mean value at $R_{Gal} = 5$ kpc (Bloemen et al. (1990)).

We also looked at the ROSAT all sky survey and found an X-ray emission pattern corresponding to a background source shadowed by the supershell. The X-ray shadowing and IR observations confirm that the walls at $b < -5°$ are mostly composed of neutral gas. A more detailed physical analysis of the sagittarius supershell will be the subject of a forthcoming paper.

4 Conclusion

The filamentary structure of the ISM may reveal two realities: 1) large gas structures blown away from the plane by successive supernova, stellar winds and HII regions and 2) a perturbed and complex medium that could, in certain cases, be the small scale structure of the first one. Thus, local cirrus should be considered as small scale structures of larger complex like the local bubble or Loop I. It is also possible that some of the local cirrus may be high-z clouds. Therefore cirrus may reveal the ISM all-scale froth built up by more-less violent dynamical processes that forge the Galactic bubble bath.

References

Bloemen, J. B. G. M, Deul, E. R, & Thaddeus, P. (1990): A&A 233, 437.

Burton, W. B, Elmegreen, B. G, & Genzel, R. (1992): *The Galactic interstellar medium.* Springer Verlag.

Burton, W. B & Hartmann, D. (1994): ApSS 217, 189.

Heiles, C. (1979): ApJ 229, 533.

Ikeuchi S. (1997): these proceedings

Jones, A. P., Tielens, A. G. G. M., & Hollenbach, D. J. (1996): ApJ 469, 740.

Kahn, F. D. (1997): these proceedings

Koo, B.-C., Heiles, C. & Reach, W. T. (1992): ApJ 390, 108.

Lockman, F. J. (1984): ApJ 283, 90.

Maciejewski, W., Murphy, E. M., Lockman, F. J. & Savage, B. D. (1996): ApJ 469, 238.

Myers, P. C, Dame, T. M, Thaddeus, P, & Cohen, R. S. (1986): ApJ 301, 398.

Normandeau, M. & Dewdney, P. E. (1996): Nature 380, 687.

A Limit on Galactic Extinction Not Correlated with Far IR Emission

Douglas P. Finkbeiner[1], David J. Schlegel[2], and Marc Davis[1]

[1] University of California at Berkeley, Departments of Physics and Astronomy, 601 Campbell Hall, Berkeley, CA 94720
[2] University of Durham, Department of Physics, South Road, Durham DH1 3LE, United Kingdom

Abstract. We have reprocessed the IRAS and DIRBE full-sky 100 μm maps, using H I as a zero-point, to produce a map of Galactic extinction. This map is cross-correlated with APM galaxy counts over the full APM region, including dusty regions of the sky not used to compute the APM correlation function in Maddox et al.(1990a). The angular correlation function, $w(\theta)$, of galaxy counts is computed with and without the extinction correction, and is used to suggest that our map accounts for $\sim 90\%$ of the dust-induced correlation in the APM region on scales from $2 - 10°$. This is the first time that a limit has been placed on the correlation function of the extinction *not* represented in a given map.

1 Introduction

The Burstein & Heiles (1978, 1982) reddening map has been the standard method of estimating Galactic extinction for nearly 20 years. At the meeting, I presented an alternative method based on $100 - 240$ μm emission from dust, using H I emission for a zero-point. The calibration from infrared emission to $A(B)$ extinction is based on the Mg_2 - color relation for elliptical galaxies (Faber et al.1989), and will be reported in Schlegel et al.(1998).

Because dust affects galaxy counts in a non-trivial way (Heiles) we use a likelihood analysis to determine the right coefficient for correcting APM galaxy counts. We will use this, along with information about the underlying correlation of the APM counts, to put a limit on how much extinction is not detected by our map. In addition to placing limits on the interstellar medium within our Galaxy, such a limit is critical for studying the large-scale structure in the universe. In addition, we claim that the limit applies to any dust associated with a layer of H II in our Galaxy.

2 Data

The SFD Map: In Schlegel et al.(1998) we will present a reprocessing of the IRAS and DIRBE all-sky $100 - 240$ μm maps, which trace the interstellar medium. We have improved the ISSA maps (Wheelock et al.1994) with a

Fourier destriping method, and have removed 20,000 confirmed point sources. Spatial scales larger than 1 degree have been replaced with data from DIRBE (0.7° beam) using an appropriate scaling. The DIRBE 25 μm map is used as a spatial template for zodiacal light, but is modulated in a 3 parameter model that minimizes dependence of the dust-gas residual upon ecliptic latitude. The 100 – 240 μm channels of the DIRBE maps are used to recover the temperature of the dust, and convert the emission to column density. The composite map is smoothed to a $\sim 6'$ roughly-Gaussian beam, and is known as the Schlegel-Finkbeiner-Davis (SFD) map.

APM Galaxy Survey: The APM galaxy survey (Maddox et al.1990b,c) covers a large region of the southern sky at high Galactic latitudes, and provides a background distribution of galaxy counts that are extinguished by dust in our Galaxy. The dust extinction, measured by the SFD maps, can be calibrated by studying the statistical covariance between the APM and SFD maps: dusty regions have increased dust emission and diminished galaxy counts. The large number of galaxies in the APM survey (2.8×10^6) provides excellent statistics.

3 Likelihood Analysis

The calibration of extinction in the APM galaxy maps is very amenable to a maximum-likelihood solution. In this analysis, we treat the underlying mean galaxy counts, $\bar{N}$, the extinction normalization, p, and the star leakage parameter, q, as model parameters.

Each APM pixel, or cell, has an expected number of galaxies drawn from the underlying galaxy distribution, modified by the extinction in that cell. In addition, each cell contains a number of stars which have been confused with galaxies – this usually happens because two stars are very near each other. We compute the likelihood of measuring the observed number of galaxies, N_i in cell i for a grid of star and dust models. The likelihood of a model is the product of the likelihoods of each independent cell. We choose the model with the highest likelihood.

The underlying galaxy distribution is well sampled in the high Galactic latitudes of the APM maps. The area at $|b| > 50°$ is a "clean" region largely free of Galactic dust contamination. The observed distribution of galaxies in the clean APM region is well fit by a "counts in cells" function (Saslaw & Hamilton 1984) which gives $f(N_i|\bar{N})$ where N_i is the observed number of counts in cell i, and $\bar{N}$ is the mean number of counts per cell of the entire APM region. This function is similar to a Poisson distribution, but takes the spatial correlation of galaxies into account.

Because dust will reduce the number of counts in a cell, we now define the expected mean number of counts in a cell with dust to be

$$\bar{N}' = \bar{N}10^{-\gamma g p I_{100}}$$

where g is a geometrical/completeness factor ($g = 0.6$ for Euclidean space), p is a ratio of 100 μm flux to $A(B)$, I_{100} is 100 μm flux, and γ is a parameter of order unity that depends on the specifics of the APM survey. In these terms, $A(B) = pI_{100}$.

For point sources, $\gamma = 1$, but there are several effects that can cause γ to deviate from unity for galaxies. Extinction changes the size of a galaxy's isophotes, and may cause galaxies to be confused with stars more easily at the faint end of the survey. Also, the non-linearities of photographic film can cause the behavior of galaxies to differ from that of stars. See Heiles (1976) and Santiago et al.(1996) for a discussion of this effect.

Because of these ambiguities, the APM analysis alone cannot break the degeneracy between p, g, and γ. The product γgp that we fit in the likelihood analysis is appropriate for the APM survey, but lacking a separate measurement of g and γ, this is not a measurement of p or $A(B)$.

4 Correlation Function

A main motivation for the APM survey was to measure the angular correlation function, $w(\theta)$. We display $w(\theta)$ in Figure 1a, calculated for the following cases: uncorrected counts (w_{raw}), corrected counts (w_{corr}), clean counts (w_{clean}), and dust (w_{dust}). The "clean" counts are counts in a region where our dust correction has little effect. Also displayed in Figure 1a is $w_{dust} + w_{clean}$. Correlation functions obey a simple addition rule, $w_{A\times B} = w_A + w_B$ for two maps, A and B. Therefore, the extinction not accounted for by the SFD map is $w_? = w_{corr} - w_{clean}$. We call this the "excess correlation" in the corrected map, and compare it to the "excess correlation," $w_{raw} - w_{clean}$, in the raw map in Figure 1b.

5 Conclusions

We have placed a constraint on the spatial correlation of extinction *not* present in the SFD map. Such extinction must be either small or uniform. If it is uniform, it may be constrained by stellar reddening measurements, as in Schlegel et al.(1998). Although we have attempted to correct for dust temperature, our assumptions may break down near shock fronts, GMCs, and other regions of unusual temperature. However, at high latitude, the SFD map appears to account for $\sim 90\%$ of the observed extinction (See Fig. 1b). This limit includes extinction from dust uncorrelated with HI such as dust in the Reynolds layer.

Acknowledgements. The COBE data sets were developed by the NASA Goddard Space Flight Center under the guidance of the COBE Science Working Group

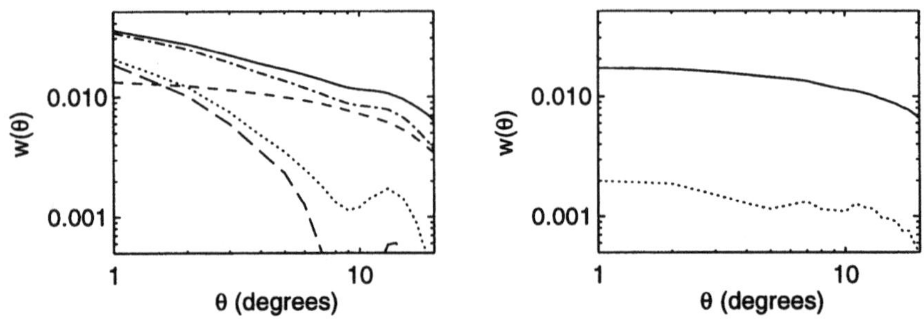

Fig. 1. The solid line in Fig. 1a is the correlation function of the entire APM region. The upper dashed line is dust, and the lower dashed line is the clean APM region (corrected), taken to be the underlying galaxy correlation. The dotted line is the correlation of the full APM region after correcting for dust and stars. The dash-dot line is the sum of the dust and clean lines. In Fig. 1b, the solid and dotted lines are the excess correlation of uncorrected and corrected counts, respectively.

and were provided by the NSSDC. We wish to thank Sherry Wheelock of IPAC for providing ISSA coverage maps, the joint IRAS Science Working Group for all IRAS data, Steve Maddox for APM maps, and Carl Heiles for invaluable discussions and guidance. DF acknowledges support from an NSF Graduate Fellowship. This work was supported by NASA grant NAG5-1360.

References

Burstein, D. & Heiles, C. 1978, ApJ, 225, 40
Burstein, D. & Heiles, C. 1982, AJ, 87, 1165
Maddox, S.J., et al.1990a, MNRAS, 242, 43p
Maddox, S.J., et al.1990b, MNRAS, 243, 692
Maddox, S.J., Efstathiou, G., & Sutherland, W.J. 1990c, MNRAS, 246, 433
Hartmann, D. & Burton, W.B. 1996, Atlas of Galactic Neutral Hydrogen, Cambridge University Press.
Faber S.M., et al.1989, ApJS, 69, 763
Heiles, C. 1976, ApJ, 204, 379
Santiago, B.X. 1996, et al.ApJ, 461, 38
Saslaw, W.C. & Hamilton, A.J.S. 1984, ApJ, 276, 13
Schlegel, D. J., Finkbeiner, D. P. and Davis, Marc 1998 *in prep*
Wheelock, S.L. et al.1994, IRAS Sky Survey Atlas: Explanatory Supplement, JPL Publication 94-11 (Pasadena: JPL)

Dust in Hot Environments: Giant Dusty Galactic Halos

Andrea Ferrara

Osservatorio Astrofisico di Arcetri, Largo Fermi 5, 50125 Firenze, Italy

Abstract. I review some of the evidences for dust in the Local Bubble and in galactic halos and show that a general mechanism based on radiation pressure is capable of evacuating dust grains from regions dominated by massive star energy input and thus originate huge dusty halos. A Monte Carlo/particle model has been developed to study the dust dynamics above HII chimneys and the results, among other findings, show that dust can travel several kpc away from the plane of the parent galaxy. The cosmological implications of extragalactic dust are briefly outlined.

1 Introduction

The evolution of dust in hot environments is a very important topic in astrophysics with a variety of implications. There are several excellent reviews on the subject (Draine & Salpeter 1979, Seab 1987, McKee 1989, Dwek & Arendt 1992) - and in particular on the issue of grain survival - to which I defer the interested reader for a complete treatment. Instead, in the spirit of a highlight talk, I would like to focus on and isolate one particular effect, namely the dynamics of grains in a hot environment, and show how this might affect the structure and evolution of a galaxy and perhaps even our understanding of the distant universe, as other speakers at this Colloquium have emphasized (see Finkbeiner's contribution).

In this sense, three "hot" environments - the Local Bubble (LB), Galactic Halos and the Intergalactic Medium (IGM) - central to the subject of this Colloquium are probably different aspects of the same phenomenon: the feedback of star formation and supernovae on the interstellar medium of a galaxy, whose manifestation is the disk/halo interaction. The central point of this paper is that the disk/halo interaction - and in particular its dusty form - is indeed crucial to understand several galactic properties and also the distant universe. The relatively well studied local environment might then represent the Rosetta stone to interpret the interstellar processes occurring during the epoch of galaxy formation.

2 Dust in the Local Bubble

Contrary to what one might expect, not much is known yet about dust in the LB. The most firm results come from a pioneering paper by Tinbergen (1982). This author observed that stars closer than $\simeq$ 35 pc from the sun in the direction of the Galactic Center show appreciable polarization (polarization degree of the order of 0.02 %). From the same data Frisch (1995) found no correlation between the polarization degree and the the distance to the sample stars. This could indicate that the dust is closer than the nearest star, and therefore located in the neutral cloud embedding the sun, commonly named the Local Fluff. One might also be led to speculate that the amount of dust outside this very small region of a few parsecs size is very limited. Another interesting feature is the absence of polarization in the anticenter hemisphere, which could be explained by a lack of grain alignment and/or a different configuration of the magnetic field. It has to be pointed out, though, that Leroy (1993) did a very similar study to the one of Tinbergen, failing to confirm the above discussed polarization; however, this could be accounted for by the lower sensitivity of Leroy's data. New promising perspectives for the investigation of dust in the local environment are opened by ongoing infrared emission studies (Reach, this conference) and in-situ experiments (Landgraf, this conference).

The absence of polarization outside the very local region might have different interpretations. If one is inclined to attribute this effect to a real deficiency of grains rather than to other reasons, such as a poor alignment, then the issue of grain destruction or evacuation should be addressed. Grains in a hot gas are essentially destroyed via thermal sputtering, i.e. collisions with ions or electrons with Maxwellian velocity distribution. The hot gas in the LB is now thought to have temperature $\sim 10^6$ K and number density $n \sim 5\times10^{-3}$ cm^{-3}. For these figures, the sputtering time calculated using the recent sputtering yields by Tielens et al. (1994), for a 0.01 μm grain is $\sim 10^8$ yr, much longer than the estimated age of the LB. Dust destruction can have been caused by the supernova shock that has generated the hot bubble, whose mechanical luminosity has been inferred to be equal to $\sim 2.8 \times 10^{36}$ ergs s^{-1} by Frisch (1995). However, in spite of the still poorly understood underlying physics, it seems unlikely that the efficiency of grain destruction in a shock can be higher that 10% (McKee 1989). Grains are more likely to be destroyed behind radiative shocks, by the combined effects of a greatly enhanced gas density and betatron acceleration that increases the grain Larmor frequency. There is clear evidence of supersonic motions in cloudlets inside the Local Fluff, and collisions among them necessarily produce shocks. Ricotti, Ferrara & Miniati (1997) have shown that supersonic cloud-cloud collisions are inelastic for a wide range of parameters, which, stated in an alternate manner, implies that the corresponding shocks are radiative; this result has now been confirmed by extensive hydro and MHD numerical simulations performed by Miniati et al. (1997). Thus, the presence of dust in a 5 pc sphere around the sun, confirms that the dust destruction has not been very efficient. This forms a basis for our hypothesis that dust can have been *evacuated* from the LB rather than destroyed locally. The LB is an example, and probably far cry from being the most spectacular one (see Normandeau et al. 1996), of the effects of (multi)–supernova explosions resulting in the vertical tunnelling of hot gas inside HI walls – in brief a galactic "chimney". It is reasonable to expect that the global effects of a collection of such objects might profoundly influence the structure and properties of the ISM of galaxies and of their halos. A very brief summary of some observational evidence that might hint at the relevance of supernova energy injection in shaping the dust distribution in galaxies is given in the next Section.

3 Observational Facts on Dust Large-Scale Distribution

The tremendous observational improvement occurred in the last few years has substantially challenged some of our long-standing prejudices about the distribution of dust in galaxies. Dust has long been believed to be confined in a very thin disk, with approximately the same horizontal extension as disk stars. However, this does not appear to be the case from the first dedicated observations which tackle the problem exploiting the range of wavelengths appropriate for the detection of cold dust, i.e. FIR and sub-mm bands. Davies et al. (1997), from an analysis of 140 μm and 240 μm DIRBE observations of the Galaxy demonstrated very convincingly the existence of an extended (scale height ~ 0.5 kpc), cool ($T \sim 18 - 22$ K) dust component. They favour the interpretation that "...this dust may be supported high above the plane by radiation pressure". Dust therefore seems to be present in the lower halo of the Milky Way at least at the same distance as the HI gas in the Lockman layer. Sofue et al. 1994, pointed out the presence of a variety of dust structures (arcs, loops, bubbles, streamers) in the halo of NGC 253 extending into the first 3 kpc of the halo of that galaxies. This limit is merely being set by instrumental sensitivity. This adds to the already well established evidences of similar phenomena in several edge-on galaxies (where the identification is definitely easier); for a review see Dettmar (1992). Not only is the vertical distribution in spirals more extented than previously thought, but similar conclusions are emerging also for the one in the disk plane. Observations of NGC 6949 in the ISO 200 μm band have shown that a cold dust component exists which is considerably more extended compared to that measured by IRAS in the same galaxy and sampling a warmer component. Zaritsky (1994), studying the B and I colors of distant galaxies seen through the halo of two nearby spirals, concluded that background galaxies at smaller projected separations are statistically redder than those in the outer regions. This fact suggests the existence of an extended dust halo with a derived scale length of 31 ± 8 kpc. This figure is in remarkable agreement with the prediction of

Heisler & Ostriker (1988), who required a scale length of 33 kpc to account for the observed quasar number counts.

4 Grain Dynamics Above Bubbles

There are two main mechanisms able to evacuate dust from the interior of a supernova driven interstellar bubble and inject it into the halo: (i) a convective flow (i.e. a "galactic fountain", Shapiro & Field [1976] or a chimney, Norman & Ikeuchi [1989]) in which the gas heated by a supernova blast wave becomes buoyant and raises into the halo carrying along the dust grains that are coupled with the gas; (ii) a wind driven by the radiation pressure above the clusters of young OB stars. In order to work efficiently, the first mechanism requires the occurrence of the so-called "blow-out" in which the gas of the shell created by the explosion is effectively reaccelerated and injected into the halo. However, recent studies have demonstrated that this phenomenon, at least in the Galaxy, should be much rarer than previously thought, either because the growth of the shell is inhibited by the presence of a magnetic field (Tomisaka 1990; MacLow & Norman 1992) or by effects related to non-coeval star formation (Shull & Saken 1995), which essentially result in a time dependent mechanical luminosity, in turn producing lower expansion velocities. Here we explore the second possibility, i.e. dust is injected into the intergalactic space by radiation pressure. It is interesting to note that through this mechanism most of the dust is evacuated from the region surrounding the star cluster *prior* to supernova explosions. As the latter take place, only some dust will be still close enough to the center of the star cluster to be reached and processed by the expanding shock. Thus, at least part of the grain population will have a higher chance of survival. This effect will occur no matter how big the association is (and also for an isolated supernova) since grains will be always pre-exposed to the radiative flux of the stars eventually turning into supernovae.

To substantiate the above points and to understand the radiation-driven dynamics of grains above these so-called "HII chimneys" - vertical, density bounded ionization structures (Dove & Shull 1994) allowing Lyc photons to escape the disk and penetrate into the halo - we have performed a set of numerical simulations of the above "dusty chimneys" based on a mixed Monte Carlo/particle approach. A complete description of the calculation and of the detailed results will be presented in Ferrara & Shull (1997). Here we present the main features of the model and discuss some results relevant to the present topic.

We assume that dust grains are immersed in the time-dependent radiation field of a stellar association containing N OB stars (typically $N = 100$). The spectrum of the radiation field has been adapted from the results of Sutherland & Shull (private communication) who calculate the evolution of the composite radiation spectrum from the most updated stellar models.

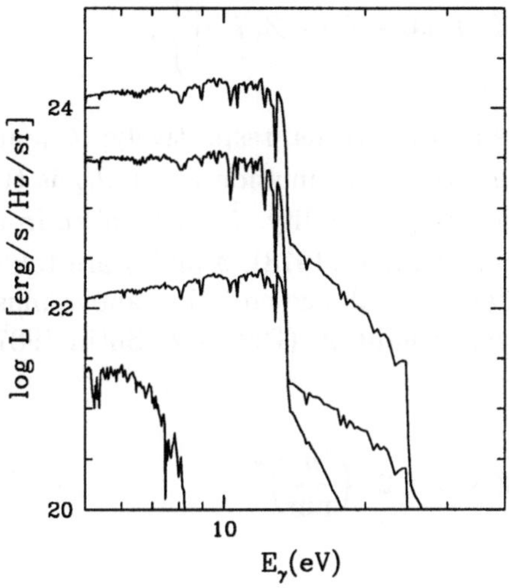

Fig. 1. Time evolution of the radiation spectrum above an OB association containing $N = 40$ OB stars as a function of photon energy. From top to bottom the curves refer to the evolutionary times $17, 29, 1, 42$ Myr, respectively.

The evolution of the radiation spectrum for a poor association with $N = 40$ is shown in Fig. 1. The luminosity increases up to a maximum and then decreases and fades away as the stars evolve into supernovae. At $t = 42$ Myr time, the luminosity, L_ν, of the radiation field at 5 eV is roughly 3 orders of magnitude lower than the maximum intensity $L_\nu \sim 10^{24}$ erg s^{-1} Hz^{-1} sr^{-1}, reached at $t \sim 17$ Myr. The spectrum is characterized by a Lyc break of about 1.5 decades, and by a flat distribution below the Lyman limit. Our simulations end at $t \sim 50$ Myr, when there is essentially no power in the radiations to drive the grains.

We take the vertical distribution of the gas, $n_g(z)$, from Dickey & Lockman 1990; the dust distribution, $n_d(z)$, is assumed to be initially exponential with a scale height $z_d = 120$ pc. The initial position and radius of each (spherical) grain are extracted via a Monte Carlo procedure from the parent distributions, $n_d(z)$ and MRN, respectively. We then follow the dynamical evolution of the grain ensemble as driven by radiation, gravity and drag (viscous + coulomb) forces; the grain charge is also calculated consistently solving the detailed balance equation in which both collisional and photoelectric charging rates are included. The grain charge Z is given by the solution of

$$n_p \left\{ \left(\frac{8kT_e}{\pi m_p} \right)^{1/2} \left[\tilde{J}_{\nu_p,\tau_p}(a, Z, T_e) s_p + \delta_p(a, Z, T_e) \right] + \tilde{J}_{pe}(a, \mathcal{F}_\nu) \right\} =$$

$$n_e \left\{ \left(\frac{8kT_e}{\pi m_e} \right)^{1/2} \left[\tilde{J}_{\nu_e, \tau_e}(a, Z, T_e) s_e - \delta_e(a, Z, T_e) \right] \right\},$$

where the index p,e stands for protons and electrons, respectively; $\tilde{J}_{\nu,\tau}$ are the collisional charging rates; δ are the secondary emission rates; $\tilde{J}_{pe}$ is the photoelectric charging rate; s is the sticking probability. $\tilde{J}_{\nu,\tau}$ are taken from Draine & Sutin (1987); $\tilde{J}_{pe}$ is taken from Draine (1978); δ and s are taken from Draine & Salpeter (1979). The detailed balance equation above must be solved together with the field emission condition (Draine & Sutin 1987) that limits the value of the charge:

$$-1 - 0.7 \left(\frac{a}{\mathrm{nm}} \right)^2 < Z < 1 + 21 \left(\frac{a}{\mathrm{nm}} \right)^2 . \tag{1}$$

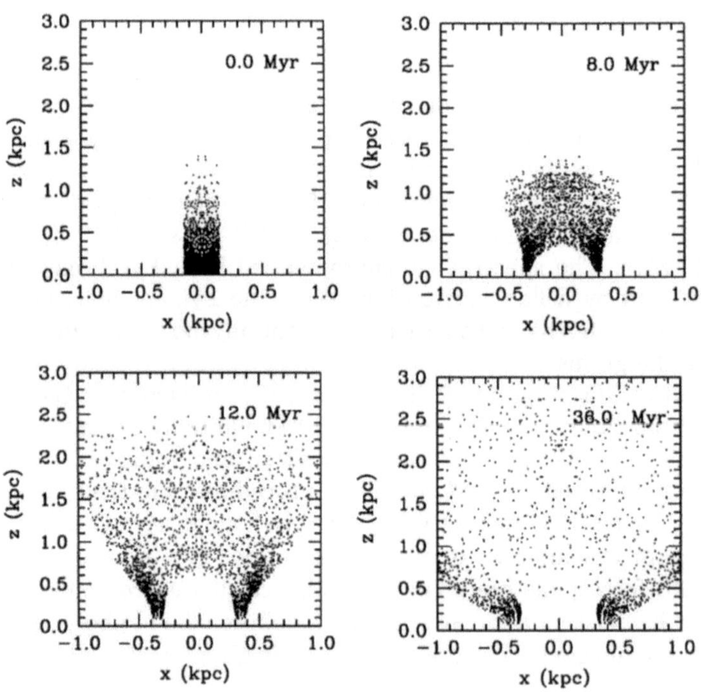

Fig. 2. Spatial distribution of dust grains for an association (located at the origin of the axes) containing $N = 500$ OB stars at the evolutionary times shown in the upper corner of each panel

Fig. 2 illustrates the spatial distribution of the dust grains (all sizes $a = 10 - 250$ nm are shown) for a $N = 500$ OB stars association at various evolutionary times. An inspection of Fig. 2 shows the presence of a central cavity completely devoid of dust swept by the very strong radiation pressure. In addition, the configuration of the region affected by the radiation field resembles now an inflated bubble in which a relevant fraction of grains has been pushed well out of the main disk of the Galaxy, at $|z| > 3$ kpc; it is also possible to identify the filamentary structures that form the lateral walls of the dusty chimney. A vertical cut through the origin (i.e. above the center of the association) shows that the z-distribution of the dust becomes much flatter with time than the original exponential one. Thus, the dust is rather homogeneously distributed by the radiation pressure in the portions of the halo above the associations; moreover, this implies that the dust-to-gas ratio has to be larger in the upper halo than in the disk. Finally, since the drag force becomes vanishingly small because the gas density is rapidly decreasing with height, some grains can continue their journey for several kpc before they feel the gravitational pull of the Galaxy. The velocities achieved by the grains are rather high, and they can exceed (for a $N = 500$ OB stars association) 100 km s^{-1}. Smaller grains tend to move faster since, neglecting drag forces, the ratio of the radiation-to-gravitational force is $\propto a^{-1}$.

5 A Possible Scenario

If dust is present in galactic halos it could be responsible for several important detectable effects. In addition to the standard direct observations discussed in the Introduction involving either the obscuration of distant object by the grains or their FIR emission, a number of indirect but nonetheless intriguing consequences of dust can be individuated and used to put constraints on any theory of galactic halos.

The simple sketch presented in Fig. 3 helps building a possible scenario which might be quite common among star-forming spiral galaxies at any epoch and includes both the effects discussed above and additional ones. The gaseous disk of the Milky Way is now understood to consist of at least two components: a neutral one, which extends roughly up to $z \sim 0.5$ kpc (including the so-called Lockman layer), and an ionized component (Reynolds layer) whose vertical extent is less clear but of the order of 1 kpc. An ordered magnetic field component with strength $|\mathbf{B}| \sim 3\mu$G in the plane of the disk is known to be present (Heiles 1995). As we have seen above, OB associations will produce (before SN explosions) a strong ionizing flux and an HII chimney, pushing the grains into the halo by radiation pressure. Due to the presence of the magnetic field, and due to the higher density of the gas and hence enhanced drag, grains located initially at low-z will be prevented from being evacuated, *if they have a non-zero charge*. This last point is very important, since in spite of the fact that we have not included the Lorentz

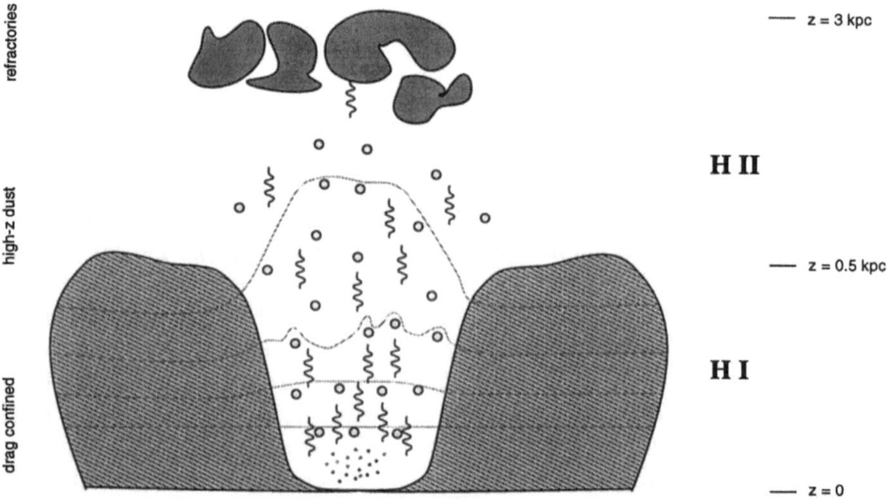

Fig. 3. Schematical representation of the various processes occurring above a HII chimney discussed in the text

force in our simulations, nevertheless our accurate determination of the grain charge shows that considerable time intervals are spent by grains in a neutral charging state: this is due to the balance between photoelectric and collisional charging. Thus, at least intermittently, grains will be free to move even in presence of a magnetic field above HII chimneys. During the periods in which they are charged, grains are trapped and spiral around field lines. However, grains located at higher altitudes ($z \sim 0.3 - 0.4$ kpc) will experience a much reduced drag force and therefore the pressure of the dust "fluid" will be felt by the magnetic field. This pressure adds up with the standard thermal and cosmic rays one and will enhance the Parker instability, thus originating a vertical component of **B**. The criterion for instability when a dust pressure term $\delta = p_r/p_{th}$ is added to the magnetic field, $\alpha = p_B/p_{th}$, and cosmic ray, $\beta = p_{cr}/p_{th}$ pressure (where p_{th} is the thermal pressure of the gas) becomes in the long wavelength limit:

$$\gamma < \frac{(1+\alpha+\beta+\delta)^2}{1+\beta+\delta+(3/2)\alpha}, \tag{2}$$

where $\gamma = 5/3$ is the ratio of the specific heats. Since $\delta \gg 1$ above HII chimneys, radiation pressure can profoundly modify the topology of the magnetic field. Grains spiralling along the vertical field lines can now flow into the halo and eventually be sputtered (nonthermal sputtering is likely due to the high grain velocities), releasing the refractory elements detected by absorption lines experiments through the halo and known to have a scale height of the order of a few kpc, depending on the species (Edgar & Savage 1989; Lipman & Pettini 1995; see Ferrara et al. 1991 for a detailed calculation). This implies that the size distribution of grains can deviate substantially from a MRN distribution in the halo, with an excess of small grains. Finally, collisions with the gas atoms will put grains into rotation and, being charged, they will emit radiation with a spectrum sharply peaked at a cutoff frequency typically in the range 10-100 GHz (Ferrara & Dettmar 1994). The estimate of the global dust mass loss due to this effect from galactic disks depends on the detailed modelling of the star formation rate in the galaxy and other galactic properties and will be given in Ferrara & Shull (1997).

6 Cosmological Implications of Dusty Halos

There are several consequences of the presence of dust in the halos of galaxies and perhaps in the intergalactic medium. The most obvious one concerns the determination of the deceleration parameter, q_0; indeed, this parameter is extremely sensitive to the assumptions concerning the extinction. In an important paper, Margolis & Schramm (1977) have estimated that the correction due to EGD to q_0 is

$$(\Delta q_0) = 2.8 \times 10^3 \left(\frac{A_v}{1 \text{ mag Mpc}^{-1}} \right) h^{-1}, \tag{1}$$

where A_v is the visual extinction and $h = H_0/100 \text{ km s}^{-1}\text{Mpc}^{-1}$. Thus, as little as 2×10^{-4} mag Mpc^{-1} absorption would cause a measurement error of $(\Delta q_0) \sim 1$.

Extragalactic dust (i.e. dust outside the main body of galaxies) might also affect QSO number counts (Heisler & Ostriker 1988). The ratio of the total cross section of dusty absorbers to the total sky area in a $\Omega = 1$ universe at redshift z is

$$\mathcal{F}(z) = \frac{2\pi R(0)^2 n(0) c}{3H_0}(1+z)^{3/2} \tag{3}$$

where $n(0)$ and $R(0)$ are the local number density and size of the absorbers. Substituting typical numbers for ($n(0) \sim r_0^{-3} \sim 0.008h^3 \text{ Mpc}^{-3}$, where $r_0 \sim 5h^{-1}$ Mpc is a correlation scale for galaxies and $R(0) \sim 100$ kpc) halos associated with galaxies, we conclude that $\mathcal{F}(z) = 1$ at redshift $(1+z) = (2/h^2)^{2/3} = 1.58h^{-1/3}$. Thus high redshift objects are likely to suffer substantial extinction by intervening systems. As a final remark, the release of heavy elements following grain sputtering in the halo might be relevant to the interpretation of QSO absorption lines in metal line systems.

Acknowledgements. I deeply acknowledge the constant inspiration provided by the friend Prof. Lyman Spitzer, Jr. "from whom I have learned a lot". I also thank Mike Shull, my collaborator in this project.

References

Davies, J. I., Trewhella, M., Jones, H., Lisk, C, Madden, A. & Moss, J. 1997, preprint
Dettmar, R.-J. 1992, Fund. Cosm. Phys., 15, 14
Dickey & Lockman 1990, ARA&A, 28, 215
Dove, J. B. & Shull, J. M. 1994, ApJ, 430, 222
Draine, B. T. 1978, ApJSS, 36, 595
Draine, B. T. & Salpeter, E. E. 1979, ApJ, 231, 77
Draine, B. T. & Sutin, 1984, ApJ, 320, 803
Dwek, E. & Arendt, R. G. 1992, ARA&A, 30, 11
Edgar, R. J. & Savage, B. D. 1989, ApJ, 340, 762
Ferrara, A, Ferrini, F. Franco, J. & Barsella, B. 1991, ApJ, 381, 137
Ferrara, A. & Dettmar, R.-J. 1994, ApJ, 427, 155
Ferrara, A., & Shull, J. M. (1997), in preparation
Frisch, P. C. 1995, Sp. Sc. Rev., 72, 499
Heiles, C. 1995, in The Physics of the Interstellar Medium and Intergalactic Medium, eds. A. Ferrara et al., ASP Conf. Series, 80, 507
Heisler, J. & Ostriker, J. P. 1988, ApJ, 332, 543
Leroy, J. L. 1993, A&A, 274, 203
Lipman, K. & Pettini, M. 1995, ApJ, 442, 628
Mac Low, M. M. & Norman, M. L. 1992, BAAS, 181, 6706
Margolis, S. H. & Schramm, D. N. 1977, ApJ, 214, 339
McKee, C. F. 1989, in Interstellar Dust, IAU Symp. 135, eds. L. J. Allamandola & A. G. G. M. Tielens, (Kluwer: Dordrecht), 431
Miniati, F., Jones, T., Ferrara, A. & Ryu, D. (1997), ApJ, submitted (astro-ph/9706208)
Norman, C. A. & Ikeuchi S. 1989, ApJ, 345, 372
Normandeau, M, Taylor, A. R., & Dewdney, P. E. 1996, Nature, 380, 687
Ricotti, M., Ferrara, A., & Miniati, F. 1997, ApJ, in press, astro-ph/9702143
Seab, C. G. 1987, in Interstellar Processes, eds. D.J. Hollenbach & H. A. Thronson
Shapiro, P. R. & Field, G. B. 1976, ApJ, 205, 762
Shull, J. M. & Saken, J. M. 1995, ApJ, 444, 663
Sofue, Y., Wakamatsu, K. & Malin, D. 1994, AJ, 108, 231
Tielens, A. G. G. M., McKee, C. F., Seab, C. G. & Hollenbach, D. J. 1994, ApJ, 431, 321
Tinbergen, J. 1982, A&A, 105, 53
Tomisaka, K. 1990, ApJL, 361, 5
Zaritsky, D. 1994, AJ, 108, 1619

In Situ Measurements of Interstellar Dust

M. Landgraf and E. Grün

Max-Planck-Institut für Kernphysik
D-69029 Heidelberg, Germany

Abstract. We present the mass distribution of interstellar grains measured *in situ* by the Galileo and Ulysses space probes as cumulative flux. The derived *in situ* mass distribution per logarithmic size interval is compared to the distribution determined by fitting extinction measurements. Large grains measured *in situ* contribute significantly to the overall mass of dust in the local interstellar cloud. The problem of a dust-to-gas mass ratio that contradicts cosmic abundances is discussed.

1 Introduction

Dust alters the conditions of the diffuse medium, e.g. it provides a dominant heating mechanism via photo-electric heating (Slavin & Frisch, 1997). So understanding the size distribution is important for modeling the diffuse interstellar medium of which the local bubble is an example. Extinction measurements give information about the size distribution of grains in the diffuse medium and hints at their composition (Massa & Savage, 1989). The extinction curve can be fitted by a mixture of graphite and silicate grains with a power-law size distribution (MRN distribution) $n(a) \propto a^{-q}$ in the interval $0.005\ \mu\text{m} \leq a \leq 0.25\ \mu\text{m}$ with an exponent of $3.3 \leq q \leq 3.6$ (Mathis et al., 1977). This optically active population can explain all depletions (Spitzer, 1978). No larger grains, that are optically gray, are needed. Interstellar dust was first detected in situ by the dust detector on board the space probe Ulysses in 1992 after Jupiter flyby (Grün et al. , 1993). The detector, that is identical to the one on board the Galileo space probe, is an impact charge detector and measures the mass, impact velocity, and impact direction of the impacting grain as described by Grün et al. (1992). The finding was confirmed by the data collected by the Galileo dust detector. Galileo reached Jupiter in December 1995 where planetary dust is dominant, so the Galileo measurements range from mid 1993, when Galileo left the inner solar system, to the end of 1995. Ulysses dust data is available up to March 1996. The criteria to identify interstellar impacts in both Ulysses and Galileo data were given by Baguhl et al. (1996). The impact rate measured by Ulysses is 0.45 day^{-1} which translates into a flux of $1 \cdot 10^{-4}\ \text{m}^{-2}\ \text{s}^{-1}$. Interstellar grains provide the dominant dust flux in the outer solar system.

Analysis of the directional information (Baguhl et al., 1995) indicates a stream of grains entering the solar system from a direction that is compatible with the upstream direction of interstellar helium (Witte et al., 1993).

2 Fit to Cumulative Mass Distribution

The grains detected *in situ* are larger than the ones needed to explain the extinction curve. We check if the mass distribution is just an extrapolation of the MRN population, ignoring the fact that this would contradict cosmic abundance considerations (Spitzer, 1978). To avoid binning effects, we investigate the cumulative mass distribution of flux $F(m)$. This is the flux of grains with masses larger than a given mass.

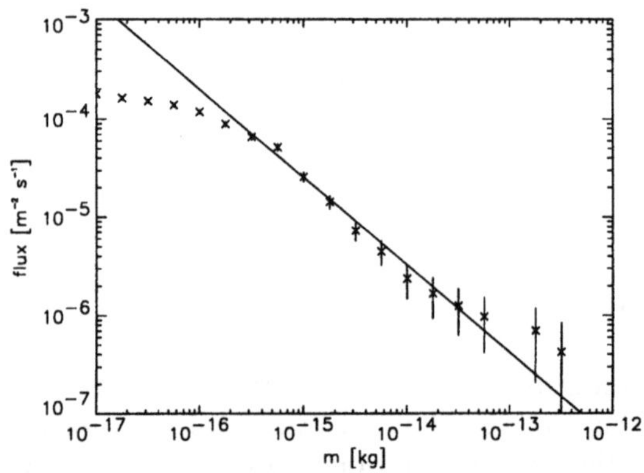

Fig. 1. Cumulative mass distribution of grains detected by Ulysses and Galileo including power-law fit. Error-bars indicate statistical errors.

For the MRN distribution we expect $F_{\mathrm{MRN}}(m) \propto m^{-q'}$ with $q' = 0.83$. Fig.1 shows the cumulative mass distribution of all particles detected by Ulysses and Galileo together with the power-law fit to masses larger than 10^{-16} kg, where the distribution was not altered by filtration as described by Levy and Jokipii (1976). The fit is weighted to match the points with good statistics. An exponent of $q' = 0.90$ fits the data best with $\chi^2 = 9.4$. χ^2 is well below the 1σ-limit of $\chi^2_{1\sigma,12} = 14$ for 12 degrees of freedom. Due to bad statistics for grains with large masses, the fit is not too sensitive, but indicates a steeper slope than the MRN extrapolation. The χ^2 for the MRN value is close to the 1σ-limit $\chi^2_{\mathrm{MRN}} = 14$. Slopes of $q' > 1$ are ruled out on the 1σ-level with $\chi^2_{q'>1} = 20$. Such a steep drop-off would not lead to a problem with cosmic abundances when extrapolating to large masses since the small grains would then contribute more to the overall mass than the large ones.

3 Comparison to Extinction Data

We compare the *in situ* distribution to the distribution fitted to extinction measurements by Kim et al. (1994). Since extinction is measured along long

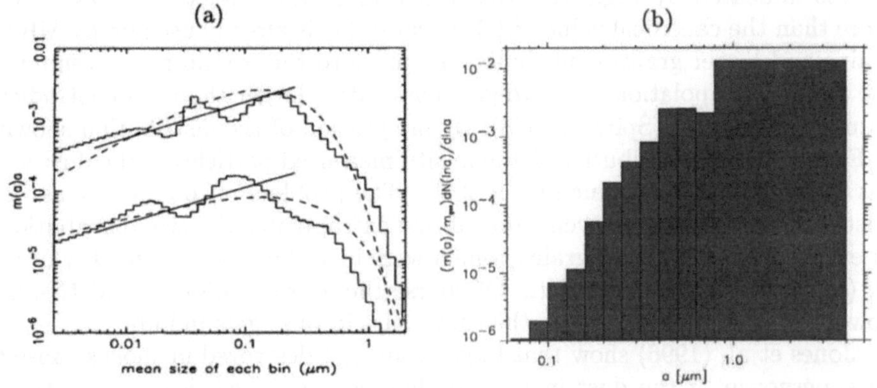

Fig. 2. (a) Differential distribution of mass in silicate (upper histogram) and graphite grains (lower histogram, scaled down by a factor 10) per logarithmic size interval normalized to the mass of hydrogen in the same volume (taken from (Kim et al., 1994), the abscissa is the mean grain size in the bin and the ordinate is identical to the ordinate in (b).). The solid line is the MRN distribution and the dashed line is the PED- (Power-law with Exponential Decay) fit. (b) Mass distribution per logarithmic size interval of *in situ* data.

lines of sight through the diffuse medium and *in situ* measurements are very local compared to this, one has to be careful interpreting this comparison. The fits to extinction measurements rely on a dust model with a size distribution and different grain compositions. Kim et al. (1994) give the mass per logarithmic size interval normalized to the mass of hydrogen in the same volume as shown in Fig.2 (a) assuming the canonical value of 0.1 H-atoms cm^{-3} for the gas density in the diffuse medium. We calculate the *size* distribution $N(\ln a)$ from the *mass* distribution given in Sect.2 by assuming homogeneous, spherical grains with a bulk density of $\rho_{ISP} = 2.5$ g cm^{-3} (Burns et al., 1979). From the size distribution the differential mass distribution per logarithmic size interval $(m(a)/m_{gas})dN(\ln a)/dlna$, which is identical to the function given by Kim et al. is calculated. The comparison of Fig.2 (a) with Fig.2 (b) shows, that the *in situ* distribution does not reproduce the steep drop-off fitted to the extinction measurements. The *in situ* distributions show, that there is much mass in grains larger than 0.2 μm.

4 Discussion

We have shown that the existence of big ($a > 0.2$ μm) interstellar grains is evident. If the size distribution of small grains is extrapolated to larger sizes as was indicated by Fig.2 (b), this will lead to dust-to-gas mass ratios of more than the canonical value of 1%, because 1% is already used up by MRN grains and larger grains contribute even more to the overall mass assuming the MRN extrapolation. Dust-to-gas mass ratios larger than 1% contradict cosmic abundances (Spitzer, 1978). By integration of the distribution shown in Fig.2 (b), the contribution of the in situ measured particles to the dust-to-gas mass ratio is determined to be 2.2%. The problem of too much mass in dust in the VLISM gets even worse if one extrapolates the size distribution to even larger interstellar grains, which were identified recently by Taylor et al. (1996) in radar meteor data. Of course the extrapolation of a MRN-like power-law is limited otherwise the total mass in dust gets infinite.

Jones et al. (1996) show that large grains get destroyed in shocks caused by supernovae. If the dust in the LIC has been shocked in the past, than there should be much more smaller grains, and if not, the dust has to be younger than the typical time-scale for returning the dust mass into the gas phase by supernova shocks which is $\tau_{\rm SNR} \approx 10^8$ a (Jones et al., 1996).

References

Baguhl M., Hamilton D.P., Dermott S.F., et al. 1995 *Science*, 268, 1016

Baguhl M., Grün E., & Landgraf M. 1996 *Page 165 of:* von Steiger, R., Lallement, R., & Lee, M.A. (eds), *Space Sci. Rev.*, 78, Kluwer Academic Publishers

Burns J.A., Lamy P.L., & Soter S. 1979 *Icarus*, 40, 1

Grün E., Fechtig H., Hanner M.S., Kissel J., Lindblad B.-A., Linkert D., Maas D., Morfill G.E., & Zook H.A. 1992 *Space Sci. rev.*, 60, 317

Grün E., Zook H.A., Baguhl M., et al. 1993 *Nature*, 362, 428

Grün E., Baguhl M., Fechtig H., Kissel J., Linkert D., Linkert G., & Riemann R. 1995a *Planet. Space Sci.*, 43, 941

Grün E., Baguhl M., Divine N., et al. 1995b *Planet. Space Sci.*, 43(8), 953

Grün E., Baguhl M., Divine N., et al. 1995c *Planet. Space Sci.*, 43(8), 971

Jones A.P., Tielens A.G.G.M., & Hollenbach D.J. 1996 *Astrophys. J.*, 469, 740

Kim S.-H., Martin P.G., & Hendry P.D. 1994 *Astrophys. J.*, 422, 164

Levy E.H., & Jokipii J.R. 1976 *Nature*, 264, 423

Massa D., & Savage B.D. 1989 *Page 3 of:* Allamandola, L.J., & Tielens, A.G.G.M. (eds), *Interstellar dust* IAU Symp., vol. 135, Kluwer Academic Publishers

Mathis J.S., Rumpl W., & Nordsieck K.H. 1977 *Astrophys. J.*, 280, 425

Slavin J., & Frisch P. 1997 *this volume*

Spitzer L. 1978 *Page 153 of:* Physical Processes in the Interstellar Medium, New York: Wiley Interscience

Taylor D., Baggaley W.J., & Steel D.I. 1996 *Nature*, 380, 323

Witte M., Rosenbauer H., Banaszkiewicz H., & Fahr H. 1993 *Adv. Space Res.*, 13, (6)121

Dust Composition in the Low Density Medium Around Spica

F. Zagury, A. Jones, and F. Boulanger

Institut d'Astrophysique Spatiale Université Paris-Sud, Bâtiment 121,
F-91405 Orsay Cedex, France

Abstract. Dust grains in low density gas are subjected to sputtering and shattering in fast supernovae shocks. Although there has been extensive modelling of the dust destruction, still little is known about the grains which survive. Jones et al. (1996) have modeled the effect of the destruction processes on the size distribution of grains. They predict that fast J shocks are efficient in grinding large dust grains into smaller particles. We present observations of the dust emission around the nearby star Spica. Comparison of the observations to the infrared emission of 'standard' cirrus suggest that the abundance of very small grains is significantly enhanced.

1 Introduction

Depletion studies of the grain forming elements (C, Mg, Fe, Si, O, etc...) show that interstellar grains are partly destroyed in the low density phases of the interstellar medium. We present a first study of the infrared emission around Spica. Spica, a B1 III-IV star, is surrounded by a vast 18° in extent HII region of low density, $n \sim 0.5\,\mathrm{cm}^{-3}$ (Reynolds 1985, York and Kinahan 1978). The presence of the star enhances the dust infrared emission, which becomes comparable to that of denser cirrus and can be separated from the background cirrus emission.

2 Spica Situation Relative to the Local Bubble

Spica (Galactic coordinates l=316.11, b=50.84) is at 80pc from the Sun, in the direction of the Scorpius Centaurus association. It is probably located close on the outer side of the cavity blown by the association, defined by Loop I: the column density of gas ($N_H \sim 10^{19}\mathrm{cm}^{-2}$) towards the star is lower by an order of magnitude than that measured for stars inside the bubble (Centurion and Vladilo 1991). On the other hand its distance from the Sun puts Spica near the edge and outside the Local Bubble. In figure 1 the HI emission to the East of Spica might be the HI shell around Loop I which would clearly prove that Spica is outside the Scorpius Centaurus Bubble. The low NaI/CaII ratio on the line of sight toward Spica suggests that the region has been crossed by shock waves (Centurion and Vladilo 1991).

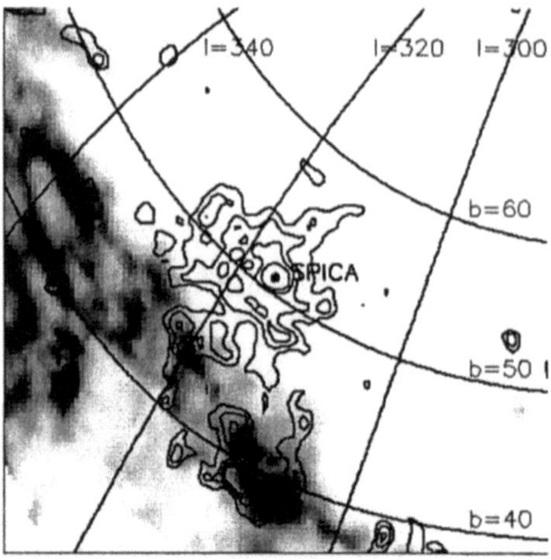

Fig. 1. HI EMISSION AROUND SPICA
VLSR ranges from -25 km/s to 25 km/s. Bright HI emission appears dark on the image. Contours are 100μm DIRBE emission from which the correlated part of the HI gas has been subtracted. The contour values are : 0.5, 0.8, 1.1, 1.4 MJy/sr. Some HI and 100μm emission are still correlated. This might be due to the presence of molecular gas in the brightest HI features.

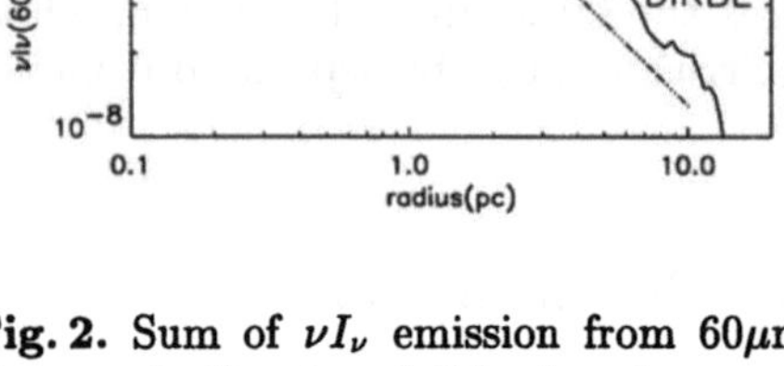

Fig. 2. Sum of νI_ν emission from 60μm to 100μm for IRAS et DIRBE versus distance to the star. A $1/r$ slope is compared to infrared emission.

3 Infrared Emission of the HII Region Around Spica

In figure 1 superposed on the HI map are the 100μm emission contours, after subtraction of that part correlated with the HI emission. The infrared emission associated with the HI gas was estimated by scaling the spatial distribution of the HI emission taken from the Leiden-Dwingeloo survey (Hartmann and Burton 1995) with the average IR/HI ratio measured at high Galactic latitude (Boulanger et al. 1996). The residual infrared emission is centered on Spica and comes from the HII region around the star. In figure 2 the integrated emission from 60 to 100μm, νI_ν, is plotted versus the distance (in pc) to the star for both IRASand DIRBE data. It follows a 1/r law, expected for the emission from an homogeneous medium heated by a central star. This attests that the infrared emission we observe does come from Spica's HII region.

4 Modelling the Infrared Emission

Desert et al. (1990) have modelled the emission of infrared cirrus. This model includes 3 components : big grains, very small grains (VSG) and Polycyclic Aromatic Hydrocarbons (PAH). Big grains have a black body like emission and are responsible for part of the 60μm emission and all the emission above 100μm. VSGs are transiently heated particles which radiate between 25μm and 60μm. PAHs make the infrared emission at shorter wavelengths. We apply this model to Spica's HII region. The star's luminosity determines the radiation field and is that of a black body at a temperature of 27000K (from Spica's spectral type) whose intensity can be found by the visual magnitude of Spica. We assume the medium to be homogeneous with n=0.5cm^{-3} derived from Hα observations.

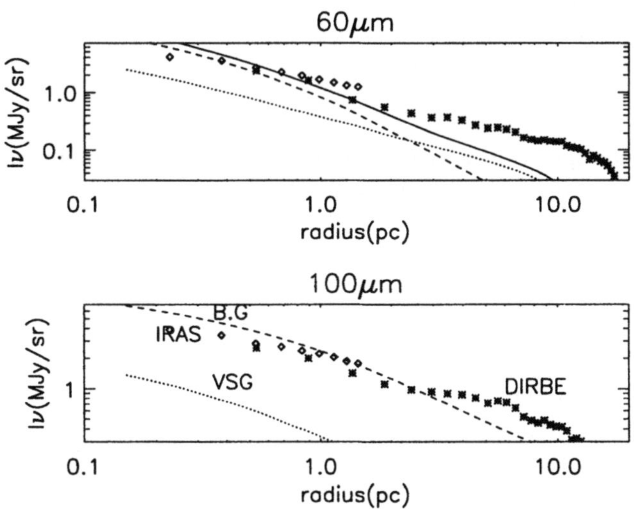

Fig. 3. MODEL OF THE INFRARED EMISSION AROUND SPICA
Dust emission at 60 and 100μm for a medium of density 0.5 cm^{-3} in Spica's radiation field is plotted versus the distance from the star. For comparison DIRBE and IRAS emission are also plotted.

5 Discussion

In figure 3 the model and the data are compared. While big grains can account for the 100μm emission, the 60μm predicted emission is too low at $r \geq 4$ pc. In figure 4 we present DIRBE spectrum 5 pc from the star and the relative contributions of PAH, VSG and big grains as set by the model. These figures convincingly demonstrate that it is necessary to enhance the VSG emission by a factor of 2 to 3 to reproduce the observed spectrum. Since Spica is located in a low density region of the interstellar medium at the edge of the Loop I Bubble it is tempting to associate the enhanced abundance with a shock as predicted by Jones et al (1996) .

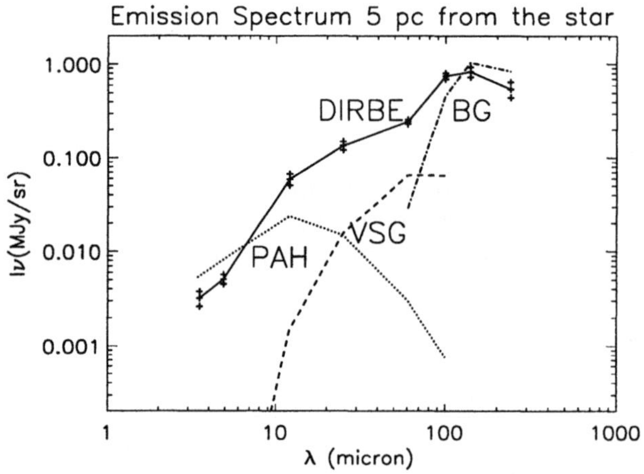

Fig. 4. DIRBE emission spectrum, 5 pc from the star, is compared to that expected from an average cirrus of the Galaxy (our model).

References

Boulanger, F., et al. (1996): A&A 312, 256
Centurion, M., Vladilo, G. (1991): ApJ 372, 494
Desert, F. X., Boulanger, F., Puget, J.L.(1990): A&A 237, 215
Hartmann, D., Burton, W.B (1995): Atlas of Galactic HI emission. Cambridge University Press
Jones, A.P., Tielens, A.G.G.M., Hollenbach, D.J. (1996): ApJ 469, 740
Reynolds, R.J. (1985): AJ 90, 92
York, D.G., Kinahan, B.F. (1979): ApJ 228, 127

^{26}Al in the Local Interstellar Medium

J. Knödlseder[1], K. Bennett[5], H. Bloemen[3], R. Diehl[2], W. Hermsen[3], U. Oberlack[2], J. Ryan[4], V. Schönfelder[2], and P. von Ballmoos[1]

[1] Centre d'Etude Spatiale des Rayonnements (CNRS/UPS), BP 4346, 31028 Toulouse Cedex, France
[2] Max-Planck-Institut für extraterrestrische Physik, 85740 Garching, Germany
[3] SRON-Utrecht, Sorbonnelaan 2, 3584 CA Utrecht, The Netherlands
[4] Space Science Center, University of New Hampshire, Durham NH 03824, U.S.A.
[5] Astrophysics Division, ESTEC, ESA, 2200 AG Noordwijk, The Netherlands

Abstract. We estimate the 1.8 MeV luminosity of the Sco-Cen association due to radioactive decay of ^{26}Al to $(4 - 15)\,10^{-5}$ph cm^{-2} s^{-1}. We propose a low surface brightness, limb brightened bubble for the 1.8 MeV intensity distribution. The detectibility of this distribution with existing γ-ray telescopes is discussed.

1 Introduction

Gamma-ray line astronomy is a young, promising discipline which is on the way to become a powerful diagnostic tool of nuclear astrophysics. It allows the unambiguous identification of isotopic species in the interstellar medium (ISM) by their spectral fingerprints: the characteristic nuclear de-excitation lines. The history of recent Galactic nucleosynthesis activity can be studied by measurements of the 1.809 MeV line arising from the decay of radioactive ^{26}Al. Possible sources of ^{26}Al are core collapse supernovae (SNe), metal rich novae, and massive stars with strong stellar winds (see review by Prantzos & Diehl 1996). The distribution of ^{26}Al was mapped by the γ-ray telescope COMPTEL aboard the Compton Gamma-Ray Observatory (CGRO). The 1.8 MeV all-sky map (Oberlack et al. 1996) clearly shows that almost all emission is concentrated in the Galactic plane. The irregular structure of the COMPTEL image with some intermediate-latitude features, and the appearance of distinct emission in the nearby Vela and Cygnus regions, had led to speculations about a source distribution, where a global Galactic nucleosynthesis glow underlies emission from relatively few localized source regions with particular recent nucleosynthesis activity (Diehl et al. 1996; Oberlack et al. 1996). Local ^{26}Al sources had been proposed before: Morfill & Hartquist (1985) suggested a SN event in the solar vicinity, Blake & Dearborn (1989) proposed SNe in the Sco-Cen association – the OB association nearest to the Sun – as possible origin of the observed ^{26}Al. The purpose of this paper is to revisit the ^{26}Al contribution from Sco-Cen based on recent nucleosynthesis calculations and new observational constraints on the Sco-Cen history, and in view of the latest COMPTEL 1.809 MeV measurements.

2 Loop I

Berkhuijsen et al. (1971) summarize observational evidence that Loop I, a giant radio continuum loop centered on Sco-Cen, was created by supernova explosions in the association. Fejes & Wesselius (1973) observe a HI shell surrounding the radio continuum loop some 5°-15° outside the best-fitting small circle. Using ROSAT X-ray data, Egger (1993) developed a more detailed scenario. He claims that Loop I (the radio structure and surrounding HI shell) is a superbubble formed by stellar winds and SN explosions of the stars in Sco-Cen; a recent supernova ($2\,10^5$yr ago) within the superbubble may have re-heated the gas, leading to the observed X-ray emission.

Based on Egger's model we estimate the ^{26}Al production of Sco-Cen from two components: ^{26}Al from the recent *re-heating* supernova and ^{26}Al from the *older* supernovae (and Wolf-Rayet stars) which formed the Loop I superbubble. The ^{26}Al yield of the recent SN can be estimated from the earliest spectral type B0V in the association which corresponds to an initial progenitor mass of 15-20 $M_{\odot}$. According to nucleosynthesis calculations for type II SN about $(3-9)\,10^{-5}$ $M_{\odot}$ of ^{26}Al is expected for such a star (Timmes et al. 1995). The ^{26}Al production of the events which formed the superbubble is estimated by means of an analytic OB association evolution model which predicts the ^{26}Al output of an association as function of the association age. Stars in the mass interval 10-40 $M_{\odot}$ explode as type II SN and release ^{26}Al into the ISM at the end of their life. Stars more massive than 40 $M_{\odot}$ are assumed to exhibit a Wolf-Rayet (W-R) phase during which they eject ^{26}Al into the ISM by stellar winds. Stellar and W-R lifetimes were taken from Schaller et al. (1992). ^{26}Al yields for W-R stars as function of initial stellar mass were taken from Meynet et al. (1997), type II SN yields from Timmes et al. (1995).

The resulting ^{26}Al light curve for Sco-Cen is shown in Fig. 1 for different initial mass function (IMF) slopes Γ and upper mass limits M_{up}. The IMF was normalized to 42 stars with spectral type between B3 (7 $M_{\odot}$) and B1 (13 $M_{\odot}$) (Bertiau & Bertiau 1958). The general feature of the light curve is a short luminosity peak between 5 and 7 Myr after the formation of the assocation due to the explosion of massive stars as supernovae. The peak is preceded by a small bump due to ^{26}Al ejection by W-R stars and followed by a tail up to 21 Myr due to less massive supernova events. After 21 Myr all stars more massive than 10 $M_{\odot}$ exploded as supernovae, hence the supply of potential ^{26}Al sources is exhausted – ^{26}Al decays exponentially ($\tau_{26} = 1.04\,10^6$yr). It is clear from Fig. 1 that the actual age of the association is the most crucial parameter in the ^{26}Al yield estimate – the slope of the IMF being of minor importance. Taking an IMF slope of $\Gamma = -1.5$ and the age of Sco-Cen between 10-20 Myr (as estimated from the most massive member, Antares) results in an ^{26}Al yield of $(4-20)\,10^{-5}$ $M_{\odot}$. We also applied the OB evolution model to the data of de Geus (1992) who determined membership and age for each of the three subgroups of Sco-Cen separately. Combining

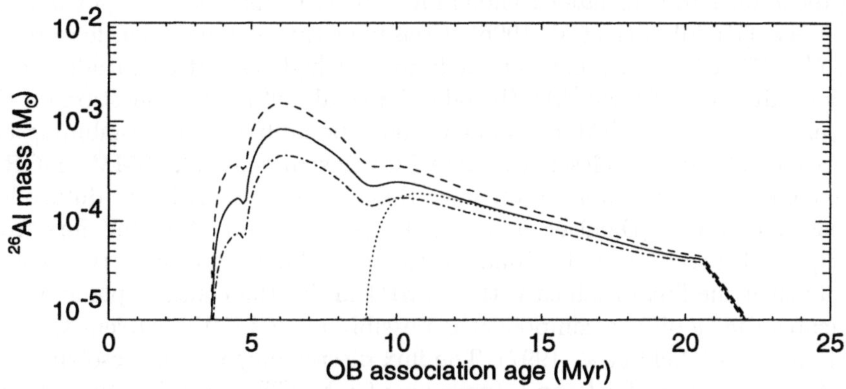

Fig. 1. ^{26}Al light curves for the Sco-Cen OB association. Solid: $\Gamma = -1.5$, $M_{up} = 60$ $M_{\odot}$; dashed: $\Gamma = -1.0$, $M_{up} = 60$ $M_{\odot}$; dashed-dotted: $\Gamma = -2.0$, $M_{up} = 60$ $M_{\odot}$; dotted: $\Gamma = -1.5$, $M_{up} = 20$ $M_{\odot}$.

the ^{26}Al light curves of the different subgroups gives a today ^{26}Al mass of $(5-10)\,10^{-5}$ $M_{\odot}$. The processed material from the early SN explosions which created the superbubble was probably swept up by subsequent explosions and could now be in a shell on the wall of the superbubble. The ejecta of the most recent event, however, are expected to fill the bubble more homogeneously due to turbulent mixing in the remnant interior (Tenorio-Tagle et al. 1991). This scenario can be translated into a characteristic signature in the angular distribution of 1.8 MeV γ-rays: a low surface brightness bubble would be surrounded by a circular emission limb. The 1.8 MeV limb would dominate the intensity distribution since ^{26}Al is concentrated in a much smaller region on the sky. Taking the distance to Sco-Cen of 170 pc as the center of the bubble, a bubble radius of 160 pc and a shell thickness of 10 pc yields expected 1.8 MeV fluxes of $(3-11)\,10^{-5}$ph cm^{-2} s^{-1} and $(1-4)\,10^{-5}$ph cm^{-2} s^{-1} for the shell and the bubble component, respectively.

The above scenario, however, is based on a spherically symmetric SNR. The ROSAT X-ray image of the Vela SNR shows deviations from a spherical shell, which have recently been interpreted as high-velocity supernova ejecta (Aschenbach et al. 1995). If Loop I obeys a similar morphology, presence of supernova-generated ^{26}Al far outside the classical Loop I boundary would be possible.

3 Observations

The observation of diffuse low-intensity 1.8 MeV emission as expected from Loop I is difficult with COMPTEL. Simulations show a tendency of our imaging techniques to translate extended low-intensity emission into spot-like image noise (Knödlseder et al. 1996). Consequently weak emission features in the COMPTEL all-sky map at medium and high Galactic latitudes could be an indication of local ^{26}Al (Knödlseder et al. 1997). The question on the presence of a diffuse ^{26}Al component could be addressed by combining observations of various telescopes: large FOV instruments like SMM or GRIS give complementary information since they are more sensitive to diffuse low intensity emission (Diehl et al. 1997). Indeed, the total 1.8 MeV flux from the general direction of the Galactic Center is higher for these two instruments then the flux obtained with COMPTEL for the Galactic plane, which indicates that a diffuse component is possibly missed in the current COMPTEL analysis (Diehl et al. 1997). The flux discrepancy may be resolved with our ^{26}Al predictions for Loop I. Improving COMPTEL's sensitivity to diffuse emission, and simultaneously analyzing data of COMPTEL, SMM and GRIS are in progress.

Acknowledgements. J. Knödlseder is supported by the European Community through grant number ERBFMBICT 950387. The COMPTEL project is supported by the German government through DARA grant 50 QV 90968, by NASA under contract NAS5-26645, and by the Netherlands Organisation for Scientific Research NWO.

References

Aschenbach, B., et al. 1995, Nature, 373, 587
Berkhuijsen, E.M., et al. 1971, A&A, 14, 252
Bertiau, F.C. & Bertiau, S.J. 1958, ApJ, 128, 533
Blake, J.B., & Dearborn, D.S.P. 1989, ApJ, 338, L17
de Geus, E.J. 1992, A&A, 262, 258
Diehl, R., et al. 1997, AIP Conf. Proc., in press
Diehl, R., et al. 1996, A&AS, 120C, 321
Egger, R. 1993, PhD thesis, MPE Report 249
Fejes, I. & Wesselius, P.R. 1973, A&A, 24, 1
Knödlseder, J., et al. 1996, SPIE, 2806, 386
Knödlseder, J., et al. 1997, Proc. 2nd INTEGRAL Workshop, p. 55
Meynet, G., et al. 1997, A&A, 320, 460
Morfill, G.E. & Hartquist, T.W. 1985, ApJ, 297, 194
Oberlack, U., et al. 1996, A&AS, 120C, 311
Prantzos, N. & Diehl, R. 1996, Phys. Rep., 267, 1
Schaller, G., et al. 1992, A&AS, 96, 269
Tenorio-Tagle, G., et al. 1991, MNRAS, 251, 318
Timmes, F.X., et al. 1995, ApJ, 449, 204

^{26}Al Radioactivity in the Galaxy

R. Diehl[1], U. Oberlack[1], J. Knödlseder[5], K. Bennett[4], H. Bloemen[2], W. Hermsen[2], J. Ryan[3], V. Schönfelder[1], and P. von Ballmoos[5]

[1] Max-Planck-Institut für extraterrestrische Physik, 85740 Garching, Germany
[2] SRON-Utrecht, Sorbonnelaan 2, 3584 CA Utrecht, The Netherlands
[3] Space Science Center, University of New Hampshire, Durham NH 03824, USA
[4] Astrophysics Division, ESTEC, ESA, 2200 AG Noordwijk, The Netherlands
[5] Centre d'Etude Spatiale des Rayonnements, 31028 Toulouse Cedex, France

Abstract. ^{26}Al radioactivity is believed to originate predominantly from massive stars, ejected into interstellar medium in wind phases and/or supernova events. With its million-year decay time, penetrating γ-rays from ^{26}Al decay measure the massive-star history averaged over a time scale of $\simeq$million years, thus extending times cales accessible otherwise. The COMPTEL 1.809 MeV all-sky data from 5 years of observations show irregularities and features at intermediate latitudes, which may have a more local origin ($\simeq$ 1 kpc). We find that the large scale emission can be characterized by a Galactic scale height of $\simeq$ 130 pc, and a Galactocentric scale radius of $\simeq$ 5 kpc, with features from spiral structure. Catalogues from massive-star related objects do not significantly improve the description of COMPTEL data above this. Emission associated with nearby structures such as the Gould Belt, Loop I, or stellar aggregates, is indicated, yet cannot be clearly detected. Combined with our imaging results, this suggests that ^{26}Al yields from massive star ensembles depend on specifics of those stars and their history. Further ^{26}Al γ-ray studies are underway to help mapping of the massive star history in the solar vicinity.

1 Introduction

Massive stars in the nearby region of the Galaxy have been detected through a variety of observables, such as their optical and UV brightness and unique spectral lines from the stellar photosphere, coronal X rays, and emission over a wide range of frequencies from their wind-blown bubbles and their interaction with the ambient gas. Even if stars cannot be detected directly due to obscuration, regions of high space density of massive stars have been identified indirectly, through, e.g., HII regions, Hα emission, infrared emission from massive protostars and circumstellar dust, and supernova remnants visible in radio through X rays. More directly, however, observation of nucleosynthesis products became now feasible, through penetrating gamma-rays emitted during decay of radioactive trace elements which are freshly produced (Prantzos & Diehl 1996). ^{26}Al radioactivity is believed to originate predominantly from massive stars, ejected into interstellar medium through the wind phases and/or supernova events. With its million-year decay time, gamma-rays from

^{26}Al decay measure the massive-star formation history averaged over a time scale of several million years, thus extending time scales accessible otherwise.

2 Gamma-Ray Results

The COMPTEL all-sky data from 5 years of observations (Figure 1) have been translated into an image in the 1.809 MeV gamma-ray line, using the Maximum Entropy deconvolution method and a background model based on simultaneous measurements in adjacent energy bands (Oberlack et al. 1997). The image shows dominant emission from the plane of the Galaxy, yet also contains irregularities and features at intermediate latitudes, which may have a more local origin ($\lesssim$1 kpc). Therefore these have been investigated for correlation with formation sites of massive stars in the nearby region of the Galaxy, after accounting for the Galaxy-wide emission to first order. We find

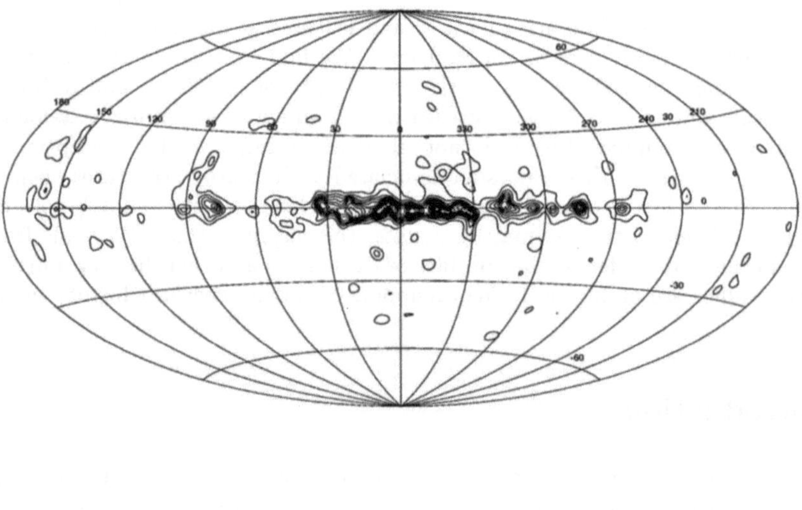

Fig. 1. COMPTEL all-sky image of 1.8 MeV emission (Oberlack et al. 1997)

that the large scale emission can be characterized by a Galactic scale height of 130 ($^{+100}_{-75}$) pc, and a Galactocentric scale radius of 5 ($^{+1.5}_{-0.8}$) kpc, with features from spiral structure(Diehl 1997). We fit a set of plausible large scale models to COMPTEL data: geometrical disk models with exponential scale height and Galactocentric scale radius dependancies (Diehl et al. 1995), spiral arm models derived from free electron measurements with exponential scale height as parameter (Taylor & Cordes 1993), molecular gas data as measured in CO (Dame et al. 1987), and stellar light- through dust-dominated measurements of the whole Galaxy in the infrared (Sodroski et al. 1994). Globally, we find that exponential disks are among the best generic models of our measurement, insignificantly inferior to the best-fitting spiral model based on free electrons; therefore we quote the scale radius determined from optimizing this model. Note that this may be biased by up to 1.5 kpc to high values due to structures such as the Cygnus/Vela/Carina regions. For determination of the scale height, we use the slightly better fitting spiral arm model, and optimize the fit through scale height variation. Note also here that this assumes a generic scale height value for the Galactic plane emission, averaging specific features along the plane.

There are emission features far outside the inner Galaxy regime, in the Cygnus, Vela, and Carina regions, each region by itself detected at significances $\geq 5\sigma$. This suggests significant deviations from a large scale symmetric and smooth ^{26}Al emission pattern. Those regions have been studied for candidate sources, each of them found to have special characteristics, which are compatible with a localized enhancement of star formation/^{26}Al: the Cygnus region with the Cygnus superbubble as remnant of possibly 60 recent supernovae as well as several Wolf Rayet stars along the line of sight (del Rio et al. 1996), the Vela region with the Vela supernova remnant and the closest Wolf Rayet star in the system 'γ Vel' (Oberlack et al. 1994), and the Carina region with η Car but also the largest space density of young open clusters in the Galaxy (Knödlseder et al. 1996). We take this as evidence that massive stars are among the most promising candidate sources of ^{26}Al, rather than novae or AGB stars.

Therefore we invert the argument, and attempt to constrain the massive-star population in the Galaxy and specific regions by the measurement of radioactive ^{26}Al ejected through their winds and supernovae. Globally the estimated total mass of ^{26}Al produced in the Galaxy is consistent with the star formation rate (or supernova rate) of the Galaxy (Timmes et al. 1997). A more specific search for correlation with catalogues from massive star related objects was not successful, however: neither Wolf Rayet star (van der Hucht et al. 1988) or OB star catalogues (Garmany & Stencel 1992), nor catalogued radio supernova remnants (Green 1996) do significantly improve the description of COMPTEL data above our first-order model. We note that object catalogues are strongly biased in all cases, however: the completeness limit for massive star catalogues does not extend beyond $\simeq$3kpc, and radio supernova remnants can only be detected in less confused directions. Therefore, since our main ^{26}Al emission arises from the inner Galaxy and integrates over all source regions throughout the Galaxy, we can expect that a dominating correlation for our measurement cannot be found with these catalogues; ^{26}Al measurements rather may provide a more realistic map of massive stars. Yet, for nearby regions ($\leq$ 500pc), specifically those outside the general direction of the inner Galactic plane, additional emission at intermediate or high latitudes may be identified by modelling plausible candidate sources in more detail:

The Gould Belt has been found to describe a local structure of O stars deviating from the plane of the Galaxy, modelled by a disk with $\simeq 20°$ inclination to the plane (Comeron et al. 1994). The origin and precise inclination and center of the Gould Belt are subject to some discussion, depending on modelling it from O stars or HI gas structures. We adopt a geometrical model encompassing the O star population of the belt, described as a Gaussian belt (θ=22.3°, ϕ=284.5°). This model indicates a slight improvement of our fit ($\simeq 2\sigma$), yet insufficient to claim detection. We note however that testing an "inverted Gould Belt" (i.e., a model belt reflected on the Galactic plane) yields an *anti*correlation of similar significance.

The closest OB association to the Sun, the Sco-Cen association, is related to one of the most prominent large scale radio structures on the sky, the North Polar Spur/Loop I, through supernova activity originating from the most massive stars of this association. The last supernova event may have occurred as late as $\simeq 2\ 10^5$y ago, enriching the Loop I bubble with nucleosynthesis products including ^{26}Al. Being so nearby, this could result in low surface-brightness 1.809 MeV emission barely detectable by COMPTEL (see Knödlseder et al., this volume). Our search with a geometrical model of Loop I (depositing the ^{26}Al in a thin outer shell, due to the age of the supernova) reports a minor hint ($\simeq 1.5\sigma$) only.

The ^{26}Al production from the Sco-Cen association as a whole is under study: ages of stellar subgroups lend themselves to separatable features on the sky, detailed HI maps help to clarify how those may relate to low-significance structures of the 1.8 MeV image (Blaauw 1964, deGeus 1992).

References

Blaauw A. 1964, Ann.Rev.Astr.Astroph. 2, 213
Comeron F., Torra J., Gomez A. 1994, A&A, 286, 789
Dame T.M., et al., ApJ, 322, 706
deGeus E.J. 1992, A&A, 262, 258
del Rio E., et al. 1996, A&A, 315, 237
Diehl R., et al. 1995, A&A, 298, 445
Diehl R., et al. 1997, AIP Conf. Proc., in press
Garmany C.D., Stencel R.E. 1992, A&AS94, 214
Green D.A., http://mrao/cam.ac.uk/surveys/snrs/
Knödlseder J., et al., A&AS, 120, 4, 327
Oberlack, U., et al. 1994, ApJ Suppl., 92, 443
Oberlack, U., et al. 1997, AIP Conf. Proc., in press
Prantzos N., Diehl R. 1996, Phys.Rep. 267, 1, 1
Sodroski T.J. et al. 1994, ApJ, 428, 638
Taylor J.H., Cordes J. M. 1993, ApJ, 411, 674
Timmes F., Diehl R., Hartmann D. 1997, ApJ, 479, 760
van der Hucht K., et al., 1988, A&A, 199, 217.

Part VII

Gas in Superbubbles and in the Galactic Halo

Where's the Local Bubble? Here ...

... or there? Everywhere!

Evolution of Evolution of Superbubbles

Satoru Ikeuchi

Department of Earth and Space Science, Faculty of Science, Osaka University, Toyonaka, Osaka 560, Japan

Abstract. In the first place, I will summarize the evolution of our picture of the ISM over the past several decades in order to see the importance of superbubbles for ISM structure. In accordance with the progress of observations, the picture of the ISM has dramatically changed from the 1960s. Then, I will review the evolution of theoretical models of superbubbles including the idea of chimney structures of the ISM.

1 Evolution of ISM Pictures

Until the 1960s, the picture of the interstellar medium (ISM) was very simple and static. The ISM is composed of Cold Clouds (CNM) confined by ambient Warm Ionized Medium (WIM), originally proposed by L.Spitzer (1956) and formulated by Field, Goldsmith & Habing (1969). These two components of the ISM were thought to be in pressure equilibrium, so this model is called the Static Two-Phase Model. At that time, since only optical lines and the HI 21 cm radio line were available to observe the ISM, the observable ranges of temperature and density of the ISM were limited to T= $10^2 \sim 10^4$ K and n= $0.1 \sim 10^2 \mathrm{cm}^{-3}$ as is shown in Figure 1.

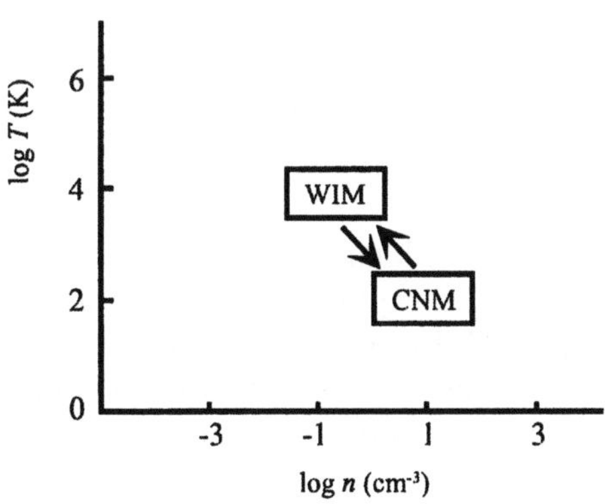

Fig. 1. Static Two - Phase Model

In the 1970s, many new phases of the ISM were discovered, such as highly ionized gas seen in OVI by the COPERNICUS satellite and the hot gas seen in soft X-ray emission by sounding rockets. At the same epoch, giant molecular clouds (GMCs) were discovered by CO observations. These discoveries made a strong impact on studies of ISM structure and star formation. The dynamically exchanging picture of the ISM composed of Hot Ionized Medium (HIM), WIM and CNM were proposed by Cox & Smith (1974), who have firstly indicated the importance of hot phase formed by multiple supernova remnants, and Shapiro & Field (1976) who have discussed the possibility of disk-halo connection through an individual supernova remnant, galactic fountain. Finally, McKee & Ostriker (1977) have formulated the Three-Phase Model. I remember that Professor Hayakawa (1979) proposed the Local Hot Bubble picture by indicating the anticorrelation of the intensity of soft X-ray background and that of HI 21 cm. As is seen in Figure 2, the hot gas phase is thermally unstable and its cooling time is shorter than 1 Myr. Therefore, the three-components of the gas must be mutually interchanging. The driving force for this interchange seems to be energy input by supernova explosions. The analogy of worms and fountains was suggested for the cold gas component and the hot ambient gas.

In the 1980s, this dynamical picture of the ISM was established by discoveries

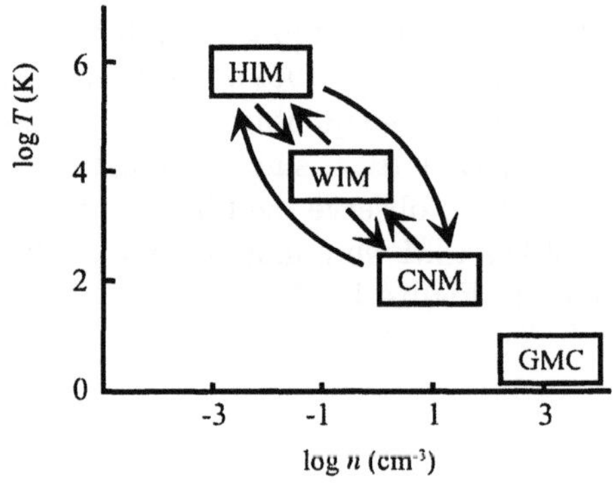

Fig. 2. Three-Phase Model

of superbubbles in Cygnus, Orion-Eridanus and Gum Nebulae. Superbubbles are formed by cumulative supernova explosions in an OB association, where massive star formation is active. McKee et al (1984) have explored the evolution of superbubbles within two-phase medium, taking account of photoevaporation of cold clouds. Moreover, McCray & Kafatos (1987) have proposed the self-propagating star formation due to the shock compression by supershells. Since the gas in the galactic disk is plane-stratified, the superbubble predominantly expands perpendicular to the disk and finally breaks through to the galactic halo. This gives rise to the picture of disk-halo connection, and a hot gaseous halo is naturally expected. In 1985, F.Lockman and J.Bregman organized a small workshop entitled "The Galactic Gaseous Halo" at Greenbank. After the session, we discussed the possibility of disk-halo connection while drinking wine in front of the fireplace. Just then, I got the idea of the Chimney Model (Ikeuchi 1988, Norman & Ikeuchi 1989) for superbubbles, which break through to the halo from the disk. The analogy is as follows: The fireplace is an OB association, and the firewood is OB stars and supernovae. The chimney wall is composed of cold gas expelled by supernova explosions. The smoke is hot gas heated by stellar winds and supernovae,and the falling acid rain is cooled gas in the halo, which is polluted by metals. The importance of superbubbles for the evolution of large scale structures in the ISM is clearly reviewed by Tenorio-Tagle & Bodenheimer (1988).

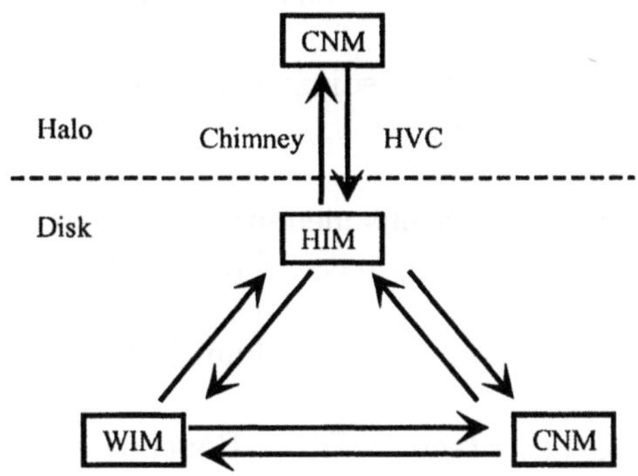

Fig. 3. Disk-Halo Connection

In the 1990s, the ISM picture extends from the Local Hot Bubble, superbubbles and worms to extragalactic gaseous halos probed by QSO absorption lines, and the observations extend from the radio, infrared, optical to the EUV and X-ray portions of the spectrum. Moreover, we can discuss the evolution of the ISM in relation to galaxy formation and evolution in the near future. That means that the structure of the local ISM is connected with cosmology. We may say that the study of the ISM has a bright future.

2 Evolution of Models of Superbubbles

There are two possibilities for the formation of superbubbles. One is cumulative supernova explosions in OB associations. In this case, massive stars form sequentially from giant molecular clouds, make giant HII regions and stellar wind bubbles, and then explode as supernovae within previous supernova remnants (SNRs). This type of superbubble is naturally expected in spiral arms. They look like chimneys because the supperbubbles are standing perpendicular to the disk and ample cold gas is associated with them.

The other possibility for the formation of superbubbles is the collision and merging of neighboring SNRs. In other words, a new supernova explosion occurs outside of another SNR. This type of superbubble can be expected in the inter-arm regions. The local superbubble corresponds to this type. Though I worked on these two cases in the past, I will concentrate to the former type in the following.

The historical evolution of Models of Superbubbles can be summarized as follows:

(a) Spherical Superbubble in an homogeneous medium (Tomisaka, Ikeuchi & Habe 1981).
(b) Superbubble in a plane-stratified medium (Tomisaka & Ikeuchi 1986, McLow, McCray & Norman 1989).
(c) Superbubble with magnetic field of
  (i) 2D calculation in an homogeneous medium (Tomisaka 1992),
  (ii) 2D calculation including the gravity of disk (Kamaya, Shibata & Mineshige 1996),
  (iii) 3D calculation in a plane-stratified medium (Tomisaka 1997).
(d) Superbubble in
  (i) differential rotating disk (Tenorio-Tagle, Bodenheimer & Rozyczka 1987, Bisnovatyi-Kogan & Silich 1995),
  (ii) non-coeval star formation history (Shull & Saken 1995),
  (iii) two-phase medium (Silich et al. 1996).

I shall give a brief overview of the important findings of these models. (I apologize that I could not mention many other works.)

(a) Spherical Superbubble in a homogeneous medium.

For any phenomenon, we should start from the simplest model in order to see the underlying physical processes and to confirm if the basic idea works. For wind bubbles in a homogeneous medium, the similarity solution had been found (Weaver et al 1977) and applied. In an OB association, supernova explosions occur within a wind bubble (Bruhweiler et al 1980). In that case, the supernova energy is not efficiently used for expansion of the remnant because the ambient medium is so rarefied that the kinetic momentum is not much transferred to the expanding shell. The similarity solution is not applicable to such a case. Therefore, it is necessary to calculate numerically the expansion of a superbubble and find the conditions for formation of a superbubble with the radius of several hundreds parsecs, such as is observed in Cygnus and Orion-Eridanus. Tomisaka, Ikeuchi & Habe (1981) first explored the expected supernova rate in a typical OB association, and showed that such a huge superbubble can be naturally formed in an OB association within several Myr. At the same time, we presented a simple analytic formula for the expansion law of a superbubble that is useful for estimating the size and age of a superbubble as functions of ejected energy and ambient gas density.

(b) Superbubble in a plane-stratified medium.

In the next step, we must study a more realistic model for comparison with observations. The gas density in the disk is not homogeneous but plane-stratified, although the exact distribution law is not well known. For this calculation, we must develope a 2D hydrodynamic code. Tomisaka & Ikeuchi (1986) have presented the first results of elongated superbubbles that stand perpendicular to the disk. In this 2D calculation, unknown parameters such as the height of the center of OB association and the scale height of the ambient medium must be varied. When the expanding front breaks through to the halo, the superbubble structure resembles a Chimney, as seen in Figure 4.

This break-through is very important for the disk-halo connection and the formation of hot gaseous halo. The definition of the Chimney is the break-through superbubble beyond the Reynolds layer, and if it does not blow out we may call it a supershell or worm (Heiles 1990). As a simple criterion, we presented the condition for break-through that the superbubble accelerates into the halo because of the rapid decrease of ambient gas density, that is, $d\ln Z_{up}/d\ln t = 1$, where $Z_{up} = Z_{OB} + R_s$ and Z_{OB} and R_s are the height of OB association and the radius of shock front, respectively. This criterion is usually satisfied when $Z_{up} \sim 2$ - 3 times of the scale height of the gas distribution. This is valid even now, and we consider this as the Chimney formation criterion. Recently, a chimney structure has been observed in the W4 region (Normandeau, Taylor & Dewdney 1996). We may expect more chimneys in nearby OB associations.

(c) Superbubble with magnetic field.

(i) 2D calculation in an homogeneous medium.

Since the strength of the general magnetic field in the disk is estimated to be several microgauss, the magnetic pressure is comparable to the gas pressure of the ISM, i.e., $P_{ISM} \simeq P_B = 10^{-12}(B/5\mu G)^2$ dyn cm^{-2}. Therefore, we must include the magnetic pressure for the evolution of a superbubble. Naturally, the magnetic pressure works to suppress the expansion of a superbubble and its break-through to the halo is highly hampered. By using a 2D magnetohydrodynamic code, Tomisaka (1992) has attempted to calculate the evolution of a superbubble in a homogeneous medium with a uniform magnetic field. As expected, the magnetic field strongly suppresses the expansion

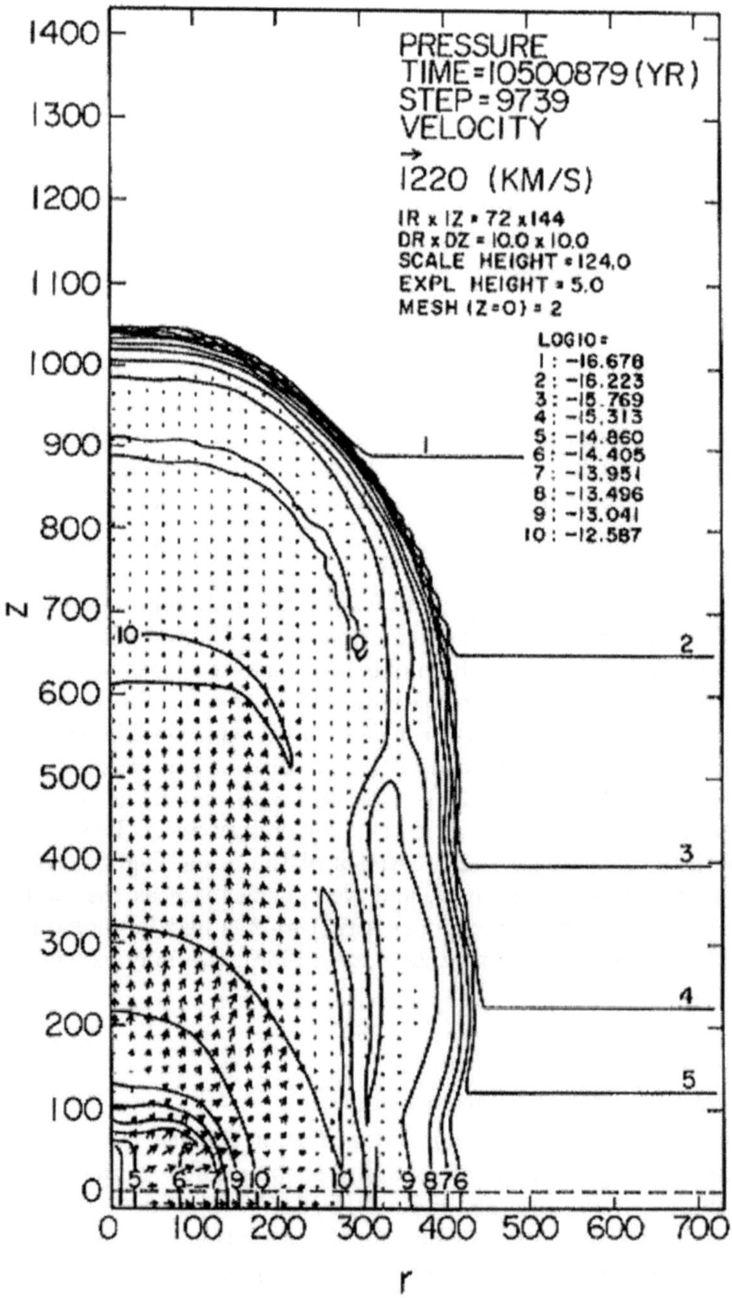

Fig. 4. Chimney formation in a plane-stratified medium

perpendicular to itself. Thus, break-through is rarely expected unless sequential supernova explosions occur well away from the midplane of the disk. Since OB associations are rare in such a region, break-through of superbubbles to the halo is not expected when the ambient medium is homogeneous and the gravity of disk is not included.

(ii) 2D calculation including the gravity of the disk.

Kamaya, Shibata and Mineshige (1996) have presented results of 2D calculations with magnetic field, including the constant gravity perpendicular to the magnetic field. Though their calculation is limited to the case of adiabatic expansion, they have shown that the Parker instability is triggered by the inflated magnetic field and that it gives rise to the break-through of a superbubble into the halo. Although the radiative cooling must be considered, this provides another possibility for the formation of Chimneys.

(iii) 3D calculation in the plane-stratified medium.

Tomisaka (1997) has extended his calculation to the 3D case in order to include the ambient gas density distribution. He assumed a magnetic field parallel to the disk, which has a density decreasing with height. His results are as follows: When the magnetic field is constant, the density stratification does not accelerate the expansion of a superbubble in the Z-direction because the magnetic pressure dominates at high Z and suppreses the break-through. However, if the magnetic field is proportional to the square root of the ambient gas density, which means the Alfvén speed is constant, the expansion is accelerated due to the rapid decrease of ambient pressure with height in the halo. In this case, break-through can be expected even if the supernova explosions occur close to the midplain of the disk.

(d) Superbubbles under realistic conditions.

In recent years, the evolution of superbubbles has been explored under much more realistic conditions. I shall briefly summarize several results. The first is the effect of differential rotation of the galactic disk (Tenorio-Tagle, Bodenheimer & Rozyczka 1987; Bisnovatyi-Kogan & Silich 1995). Since the radius of a superbubble becomes larger than 100 pc at the final stage, the difference of rotation speeds between the innermost and outermost edges of a superbubble as measured from the galactic center becomes comparable to the expansion speed. This causes an elongation of the initially circular cross section of the superbubble. This phenomenon is observed in M31, where some of the HII regions, which frequently contain OB associations and may be windbubbles, are observed to have elliptical cross sections. The second is the effect of the two-phase nature of the ambient medium (Silich et al 1996). If diffuse clouds are distributed around a superbubble, they are compressed by the shock wave and evaporated by thermal conduction. As a result, the ambient gas density increases and radiative cooling is enhanced. This effect works to shorten the lifetime of a superbubble. The third is the effect of noncoeval star formation (Shull & Saken 1995). Since the star formation in an OB association is not coeval, the energy input rate by stellar winds and supernova explosions is time-varying. Generally, the peak of star formation results in faster shell growth and affects the dynamics of a superbubble.

3 Remarks on the Future

As seen above, superbubbles play an important role in the structure and evolution of the ISM. Here, I shall briefly summarize the future direction of research in relation to the evolution of galaxies, paying special attention to the role of superbubbles. The first point is that the disk-halo connection through chimneys will be very important for the dynamo mechanism of the galactic magnetic field, cosmic ray propagation in the galaxy, and energy storage in the halo. These processes should be reexamined in consideration of superbubbles. As an example, Ferrière (1992a,1992b,1993) has calculated the effect of superbubbles to the galactic dynamo. The second point is that the mass and momentum circulation between the disk and halo might affect the evolution of a galaxy. The mass mixing would give rise to a decrease in the metallicity gradient, and the energy and angular momentum supply from the disk to the halo would drive the evolution of the gaseous halo. This may be probed by QSO absorption lines, such as CIV and MgII. The third point is that the picture of the ISM in our galaxy can be applied to the evolution of young galaxies and starburst galaxies. At present and in the near future, we can get many more samples of high redshift galaxies that show various stages of evolution. In order to analyze such new phases of galaxy evolution, the theoretical and observational implications of the ISM in our galaxy will be very instructive. The Local Bubble is thus strongly connected with cosmology.

References

Bisnovatyi-Kogan, G.S., Silich, S.A. 1995, Rev. Mod. Phys. 67, 661.
Bruhweiler, F.G., Gull, T., Katatos, M., Sofia. S. 1980, ApJ 238, L27.
Cox, D.P., Smith, B.W. 1974, ApJ 189, L105.
Field, G.B., Goldsmith, D.W., Habing, H.J. 1969, ApJ 155, L49.
Ferrière, K. 1992a, ApJ 389, 286.
Ferrière, K. 1992b, ApJ 391, 188.
Ferrière, K. 1993, ApJ 404, 162.
Heiles, C. 1990, ApJ 354, 483.
Hayakawa, S. 1979, PASJ 31, 621.
Ikeuchi, S. 1988, Fund. Cosm. Phys. 12, 255.
Kamaya, H., Shibata, K., Mineshige, S. 1996, ApJ. 458, L25.

McCray, R., Kafatos, M., 1987, ApJ 317, 190.
McKee, C.F., Ostriker, J.P. 1977, ApJ 218, 148.
McKee, C.F., VanBuren, D., Lazareff, B., 1984, ApJ 278, L115.
Mac Low, M.-M., McCray, R., Norman, M.L. 1989, ApJ 337, 141.
Norman, C.A., Ikeuchi, S. 1989, ApJ 345, 372.
Shapiro, P.R., Field, G.B. 1976, ApJ 205, 762.
Shull, J.M., Saken,J.M., 1995, ApJ 444, 663.
Spitzer, L. 1956, ApJ 124, 20.
Silich, S.A., Franco,J., Palous,J., Tenorio-Tagle, G. 1996, ApJ 468, 722.
Tenorio-Tagle, G., Bodenheimer, P., Rozyczka, M. 1987 A&A 182, 120.
Tenorio-Tagle, G., Bodenheimer, P. 1988, ARA&A 26, 145.
Tomisaka, K., Ikeuchi, S. 1986, PASJ 38, 697.
Tomisaka, K. 1997, in preparation.
Tomisaka, K., Ikeuchi, S. 1986, PASJ 38, 697.
Tomisaka, K.,Ikeuchi, S., Habe, A. 1981, Ap.Space Sci. 78, 273.
Weaver, R., McCray, R., Castor, J., Shapiro, P., Moore, R. 1977, ApJ 218, 377.

Gravitational Fragmentation in Expanding Shells

Ch. Theis[1], S. Ehlerová[2], J. Palouš[2], and G. Hensler[1]

[1] Institut für Astronomie und Astrophysik, Universität Kiel, 24098 Kiel, Germany
[2] Astronomical Institute, Acad. of Sci. of the CR, 14131 Prague 4, Czech Republic

Abstract. We investigate the gravitational fragmentation in expanding shells by applying an instability 'thermometer' similar to the Toomre parameter for instabilities in self-gravitating disks. For Sedov-like evolving systems the onset of instability is mainly depending on the density of the ambient medium and the sound speed of the shell matter, whereas the energy injection rate is less important. Shells evolve towards gravitational instability, if the density gradient of the ambient medium is shallower than an isothermal profile, otherwise they become more stable. For density gradients flatter than $\propto r^{-1}$, the fragmentation becomes non-linear on the same time scale as the gravitational instability needs to start. In a homogeneous ambient medium the typical size of gravitationally unstable shells is 1 kpc for a gas density of $n = 1\,\mathrm{cm}^{-3}$ and decreases to 10 pc for $n = 10^4\,\mathrm{cm}^{-3}$.

1 Introduction

At the end of their life time massive stars strongly influence the interstellar medium (ISM) by the metal and energy input of supernova (SN) explosions. This produces supersonically expanding thin shells which sweep up the ambient gas. These shells might become gravitationally unstable, and therefore form fragments and finally stars, i.e. they could act as the agents of induced or propagating star formation. Many analytical and numerical calculations of SN induced shocks have revealed that they are unstable to the Vishniac or the Rayleigh–Taylor instability at some stages of their evolution (Vishniac, 1983; Mac Low & McCray, 1988; Mac Low et al., 1989; Vishniac & Ryu, 1989). On the other hand, in most studies the gravitational instability was considered to be practically negligible because of the low masses of the swept-up material for typical shell radii and ISM densities. However, this assumption fails when dealing with large superbubbles of the size of several 100 pc or with SN events in dense giant molecular clouds (GMC). In both cases, the mass in the shell reaches 10^4 – $10^6\,\mathrm{M}_\odot$ and gravitational fragmentation can become important. This paper investigates the gravitational instability and the fragmentation of shell matter in spherical systems applying a stability parameter – similar to Toomre's Q – to self-similar solutions for expanding shells. In an accompanying paper by Ehlerová et al. in these proceedings the implications for large-scale superbubbles in a realistic galactic environment including differential rotation and an inhomogeneous ISM are discussed.

2 Gravitational Fragmentation

Onset of gravitational instability. According to Elmegreen's (1994) linear perturbation analysis the maximum growth rate ω of a transverse perturbation in a shell is given as

$$\omega = -\frac{v}{R}\left(-3 + \sqrt{1 + \frac{8}{\xi^2}}\right) . \tag{1}$$

An instability occurs, if $\omega > 0$ or the dimensionless stability parameter fulfills

$$\xi \equiv \frac{\sqrt{8} v c}{\pi G R \Sigma} < 1 . \tag{2}$$

R is the radius of the shell with a mass column density Σ. v denotes its expansion velocity relative to the ambient medium, c is the sound speed within the shell, and G is the constant of gravity. A second necessary condition is related to the finite size of the shell: The wavelength of the growing perturbation must not exceed the size of the shell. However, for the systems investigated here this criterion is always fulfilled, if $\xi < 1$ and, therefore, we can restrict the stability analysis to Eq. (2).
Fragmentation Time. The fragmentation starts at an 'instability time' t_b when Eq. (2) is fulfilled for the first time. In the linear stage of the growing instability we follow the evolution, until the fragmentation integral $I_f(t) \equiv \int_{t_b}^{t} \omega(t')dt'$ reaches unity. This defines the 'fragmentation time' t_f, when the fragments become strongly nonlinear, and the linear approximation is no longer valid.

3 Homogeneous Ambient Medium

Onset of instability. Using a dimensional analysis, Sedov (1959) derived a solution for supersonic, spherical expansion of a shell into the homogeneous medium in case of a single instantaneous energy input and vanishing external pressure. For the case of interstellar wind bubbles Castor et al. (1975) derived the solution which describes the structure and evolution of the shocked stellar wind region. This case is characterized by steady energy and mass input rates. The solution gives the relation between the radius of the shell R, the supernova rate $\dot{N}_{\mathrm{SN}}$, the mass density of the ambient medium ρ_0, and the expansion time t as follows (E_{SN} is the energy release per supernova):

$$R(t) = \left(\frac{125}{154\pi}\right)^{1/5} \left(\frac{\dot{N}_{\mathrm{SN}} \cdot E_{\mathrm{SN}}}{\rho_0}\right)^{1/5} t^{3/5} \tag{3}$$

where n_0 is the number density of particles in the ambient medium. The expansion velocity v of the shell is given by the time derivative of Eq. (3)

and the surface density Σ can be calculated assuming that the total mass $m(R) = 4\pi/3 \cdot \rho_0 R^3(t)$ inside the shell is swept up. Thus, we can calculate the instability parameter ξ at each time:

$$\xi(t) = \frac{24\sqrt{2}}{5} \left(\frac{18711}{64000\,\pi^4}\right)^{1/5} \cdot \frac{c}{G} \left(\frac{1}{\rho_0^4\ \dot{N}_{\rm SN} \cdot E_{\rm SN}}\right)^{1/5} t^{-8/5} \tag{4}$$

Eq. (4) shows that the system is always gravitationally stable prior to t_b and becomes unstable at later times.

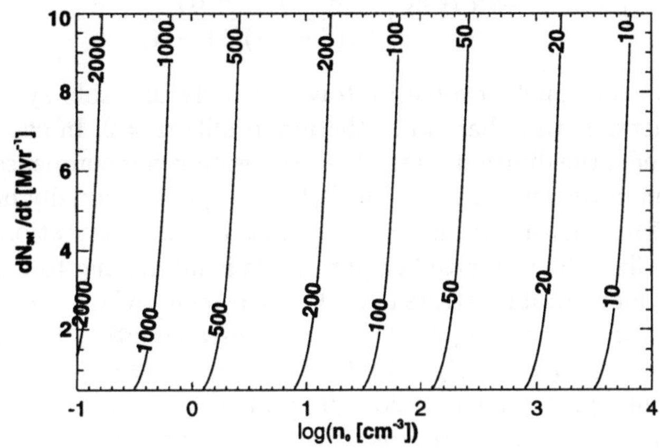

Fig. 1. The contour lines show the radius of the shell (in pc) at t_f as a function of the supernova rate $dN_{\rm SN}/dt$ and the number density n_0 for a homogeneous ambient ISM. The sound speed inside the shell was assumed to be $c = 1\,{\rm km\,s^{-1}}$.

Non-linear Stage of Fragmentation. In order to determine the fragmentation time t_f, we apply the relation $\xi(t) = (t/t_b)^{-8/5}$ together with the Sedov solution (3) to Eq. (1). From the condition $\int_{t_b}^{t_f} \omega(t)\,dt = 1$, we get

$$\frac{5}{3} = \int_1^{x_f} \frac{-3 + \sqrt{1 + 8x^{16/5}}}{x}\,dx \tag{5}$$

where $x_f \equiv t_f/t_b$. It should be noted, that the ratio of the fragmentation time t_f and the instability time t_b does neither depend on the density of the ambient medium nor on the energy injection rate and the sound speed in the shell. For a homogeneous medium a numerical solution of Eq. (5) gives $x_f \approx 2.03$. The time scales themselves are also almost independent on the energy injection rate, but depend on the density of the ambient medium. For a typical density in the solar neighbourhood of $n = 1\,{\rm cm^{-3}}$, the fragmentation time is of the order of (5-6)$\cdot 10^7$yr. The corresponding radii of the shells are shown in Fig. 1: For $n = 1\,{\rm cm^{-3}}$ we obtain radii of 600 pc to 1.1 kpc for sound speeds c within the shell between $1\,{\rm km\,s^{-1}}$ and $5\,{\rm km\,s^{-1}}$, respectively.

4 Inhomogeneous Ambient Medium

In order to extend our analysis to inhomogeneous systems, we investigated spherical systems with a radial density gradient: $\rho(r) = \rho_0 \cdot r^{-\alpha}$. Again a self–similar solution for the expansion of the shell can be found for $\alpha < 3.5$, and we get the instability parameter

$$\xi(t) = \sqrt{\frac{8}{\pi^2} \cdot \frac{c}{G\rho_0} \cdot \frac{3(3-\alpha)}{5-\alpha} \cdot K^{\alpha-1} \cdot t^{\frac{4(\alpha-2)}{5-\alpha}}} \tag{6}$$

$$K^{5-\alpha} \equiv \frac{E_{\mathrm{SN}}\,\dot{N}_{\mathrm{SN}}}{\rho_0} \cdot \frac{(3-\alpha)(5-\alpha)^3}{6\pi(7-2\alpha)(11-\alpha)}. \tag{7}$$

Eq. (6) shows that shells evolve to gravitationally unstable systems for all density gradients flatter than an isothermal profile $\alpha = 2$. In case of steeper density gradients the dilution due to the expansion is not compensated by the agglomeration in the shell and the shells become gravitationally more stable. For an isothermal density profile ξ is constant, i.e. the gravitational stability of the shell is fixed by the initial conditions. It is interesting to note that the Rayleigh–Taylor instability starts to evolve for systems which are steeper than $\alpha = 2$. Therefore, expanding shells become unstable either to gravitational fragmentation or to the Rayleigh–Taylor instability (except for the marginally stable isothermal profile). The ratio t_f/t_b is almost constant for $\alpha < 1.2$, and goes to infinity for $\alpha = 2$, where t_b is either zero or infinity. Generalizing these results to non–spherical systems like exponential galactic disks, one expects that the whole shell is only gravitationally unstable, if it evolves into an ambient medium with a density gradient shallower than an isothermal profile. If the radius of the shell is larger than two scale–heights in z–direction before gravitational fragmentation starts, the over–pressure inside the shell will be transformed mainly into motion in z–direction, which decelerates the motion of the shell in the galactic plane and, therefore, stops further agglomeration of material there. Thus, the shell should be stable against gravitational fragmentation, if it has been grown to two scale–heights without gravitational fragmentation. However, then the Rayleigh–Taylor instability will develop.

References

Castor, J., McCray, R., & Weaver, R., 1975, ApJ, 200, L107
Elmegreen B. G., 1994, ApJ, 427, 384
Mac Low, M.-M. & McCray, R., 1988, ApJ, 324, 776
Mac Low, M.-M., McCray, R. & Norman, M. L., 1989, ApJ, 337, 141
Sedov, L., 1959, Similarity and Dimensional Methods in Mechanics, Academic Press, New York
Vishniac, E. T., 1983, ApJ, 274, 152
Vishniac, E. T. & Ryu, D., 1989, ApJ, 337, 917

Star Formation in Expanding Shells: When and Where

S. Ehlerová[1], J. Palouš[1], Ch. Theis[2], and G. Hensler[2]

[1] Astronomical Institute, Academy of Sciences of the Czech Republic, Boční II 1401, 141 39 Prague, Czech Rep.
[2] Institut für Astronomie und Astrophysik, Univ. Kiel, Germany

Abstract. The fragmentation of expanding shells and subsequent star formation are analyzed using an analytical model and computer simulations. We discuss the role of the sound speed in the ambient interstellar medium and the influence of the finite thickness of the gaseous disk.

1 Introduction

Fragmentation of dense gaseous shells expanding around OB associations and subsequent formation of molecular clouds is one of the ways how the star formation propagates in galaxies. The amount of the energy input from stellar winds and supernovae in an OB association, density distribution of the interstellar medium, random motions in the ambient medium, galactic rotation and K_z force are main factors influencing the evolution of shells. The relation of these properties to the propagating star formation is discussed in the present contribution.

2 Fragmentation in the Thin Shell Approximation

Initially, a multi-supernova remnant follows a quasi-adiabatic stage. Once radiative losses become important a thin and cold shell forms. It expands sweeping the ambient medium until it slows down to a velocity comparable to the random motions in the interstellar medium. The latter phase of expansion can be described by an infinitesimally thin shell approximation (Ostriker & McKee, 1988; Bisnovatyi-Kogan & Silich, 1995).

An analytical self-similar solution is in this paper complemented by a numerical computer code (Ehlerov et al, 1997). It gives us an opportunity to study effects of the stratified distribution of the interstellar medium, galactic rotation, K_z force and cooling from inside the hot bubble.

Using the linear perturbation theory, Elmegreen (1994) described fragmentation in an expanding thin shell. A density fluctuation in the shell can grow if two conditions are fulfilled:

1) growth rate condition given in Equation (2) by Theis et al. in these proceedings, and
2) dimension condition $\eta = \pi G \Sigma R / c > 2\pi$,

where R is the diameter of the shell, Σ is the shell column density, c is the sound speed inside the shell, and G is the constant of gravity.

The time, when the two conditions are fulfilled for the first time, is called the instability time t_b. For times $t > t_b$ we define the fragmentation integral I_f:

$$I_f(t) = \int_{t_b}^{t} \omega(t')dt', \tag{1}$$

where ω is the maximum growth rate of a transverse perturbation in the shell. At the time t_f, when the fragmentation integral $I_f(t_f) = 1.0$, fragments are in a non-linear phase of the evolution and the linear approximation of growth is no longer valid. At this time fragments are well developed, subsequently forming new clouds.

To form molecules fragments must be self-shielded against the ionizing radiation. It happens when

3) The self-shielding condition $N \geq N_{crit} = 5 \times 10^{20} Z_{\odot}/Z\,\mathrm{cm}^{-2}$ is fulfilled.

(Franco & Cox, 1986), where N is the particle column density of the shell, Z is the metallicity.

In a static, homogeneous medium with a negligible external pressure fragmentation conditions 1) and 2) may be rewritten to an analytical self-similar solution as functions depending on the energy input rate from the OB association, density and sound speed of the ambient medium (Theis et al., these proceedings).

However, the analytical solution is restricted to spherical or cylindrical symmetries. The spherical case is probably a good approximation for shells inside GMCs or shells in galaxies with thick gaseous disks. In spiral galaxies with thin gaseous disks the cylindrical symmetry may apply but there the galactic differential rotation plays a role. Also, some time after the formation of an OB association, which may be shorter than time t_b, all massive stars have exploded as supernovae and the energy input is stopped. Another problem, which is not taken into account in the analytical solution, concerns the shell expansion velocity v. When v decelerates to the sound speed in the ambient medium c_{amb}, the shell begins to propagate as a sound wave. The blast wave approximation is no longer valid. The way how to take mentioned effects into account is to use numerical computer simulations (Ehlerováet al, 1997).

3 Sound Speed and Stratified Medium

We have already indicated the importance of the sound speed in the ambient medium. If the shell reaches the subsonic expansion velocity before the time

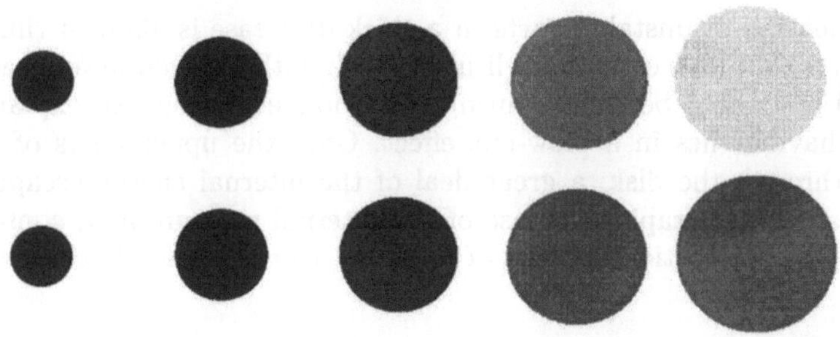

Fig. 1. Numerical simulations: shells in a homogeneous medium with a density of 1 cm^{-3} and a sound speed of 5 $km\,s^{-1}$ (upper panel) and 10 $km\,s^{-1}$ (lower panel) at times 10, 20, 30, 40 and 50 Myr. Black color indicates stability, grey instability.

t_b, it never fragments. Even if the fragmentation starts, the fragments may re-expand and the shell remains stable. Illustrative case is given in Fig. 1, which shows two shells with the same energy input into the ambient medium with the density of 1 cm^{-3}. The two cases differ in the sound speed in the ambient medium, which is 5 $km\,s^{-1}$ or 10 $km\,s^{-1}$. In the first case the shell completely fragments while in the second case it reexpands and remains stable.

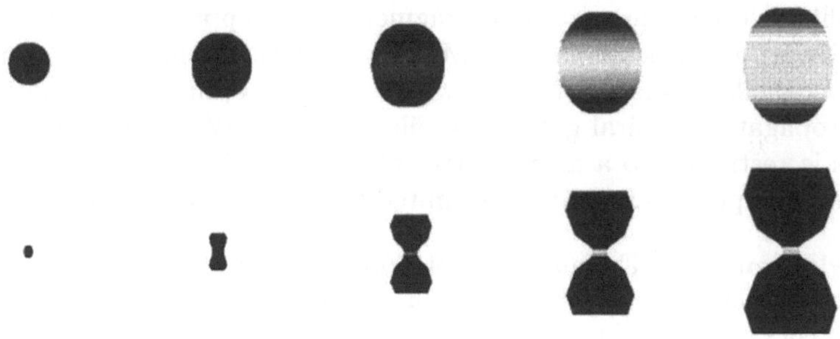

Fig. 2. Numerical simulations: shells expanding in a thick disk (upper panel) and thin disk (lower panel) at times 10, 20, 30, 40 and 50 Myr. Black color indicates stable parts, grey color unstable ones.

Shells in galaxies evolve in a gaseous disk of a certain thickness. Different thicknesses lead to different fragmentation properties, which is demonstrated in Fig. 2 showing the evolution of two shells: one in a thick gaussian disk (the scale height of 500 pc), the other in a thin gaussian disk (the scale height of 200 pc). At the time when I_f in the most unstable parts reaches the value 1.0, nearly the whole surface of the shell in the thick disk is unstable, while in the thin disk the shell develops only a tiny instability ring. In absolute values, a z-dimension of unstable parts in a thick disk case is about 4 times larger than in a thin disk case. A shell in the disk with an even lower scale height ($\sim$ 100 pc) would be stable during the whole evolution. An explanation of this behaviour lies in a blow-out effect. Once the upper parts of the shell break through the disk, a great deal of the internal energy escapes to the halo leading to a rapid decrease of the internal pressure and, consequently, to a rapid deceleration of parts of the shell near the galactic plane.

4 Conclusions and Summary

The sound speed in the ambient medium is one important quantity influencing the fragmentation. A critical value dividing the fragmentating and non-fragmentating shells lies between $(5 - 10)\ \mathrm{km\,s^{-1}}$, which is in the range of random velocities measured in galaxies. Another important quantity is the thickness of the HI disk: in thin disks, shells are gravitationally stable or develop only a narrow instability belt near the galactic equator; in thick disks, a great part of the shell is unstable and new clouds may form also off the galactic plane. This corresponds to conclusions presented by Theis et al. in these proceedings based on analytical self-similar solution: in a thick gaussian disk the shell fragments in shallow parts of the density profile, where is the density gradient flatter than an isothermal profile, whereas in a thin gaussian disks the density gradient increases to values larger than an isothermal profile before the gravitational fragmentation happens. We can conclude that in dwarf galaxies similar to LMC, SMC or HoII, the fragmentation in expanding shells occurs nearly everywhere. Therefore, the star formation can easily propagate. In spiral galaxies similar to Milky Way, the fragmentation of shells is restricted to a narrow strip at the galactic plane. Thus the star formation can propagate only to a limited portion of the galaxy.

Acknowledgements. The authors gratefully acknowledge financial support by the Grant Agency of the Czech Republic under grant no. 205/97/0699 and by the Deutsche Forschungsgemeinschaft under grant no. 436 TSE 17/5/96.

References

Bisnovatyi-Kogan G.S. & Silich S.A., 1995, Rev. Mod. Phys. 67, 661
Ehlerová S., Palouš J. Theis Ch. & Hensler G., 1997, submitted to A&A
Elmegreen B.G., 1994, ApJ 427, 384
Franco J. & Cox D.P., 1986, ApJ 273, 243
Ostriker J.P. & McKee C.F., 1988, Rev. Mod. Phys. 60, 1
Theis Ch., Ehlerová S. & Palouš J., 1997, this volume

X-ray Emission from G359.1-0.5

Roland Egger[1] and Xuejun Sun[2]

[1] Max–Planck–Institut für extraterrestrische Physik, Giessenbachstrasse, D-85740 Garching, Germany
[2] Beijing Astronomical Observatory, Chinese Academy of Sciences, 100080 Beijing, P.R.China

Abstract. We report on the detection of X-ray emission from the supernova remnant (SNR) G359.1-0.5 in a deep ROSAT PSPC pointed observation. The diffuse emission is well confined within the radio shell of the SNR. Its spectrum can be represented by a thermal plasma model at T$\sim$ 1 keV and $F_x \sim 2\times10^{-11}$ erg cm^{-2} s^{-1} (0.1–2.0 keV) modified by the column absorption of $N_H \sim 3 \times 10^{22}$ cm^{-2}. This result supports the view that the source is at a large distance consistent with that from radio observation, and helps to resolve a puzzle about the SNR in previous observations.

1 Introduction

G359.1-0.5 has been of particular interest due to, among others, its possible association with *the Snake* or *the Mouse*, two members of a class of non-thermal radio filaments near the galactic centre (see Uchida et al. 1992). Estimates of the distance of G359.1-0.5 are important to support or eliminate the associations. However, there are contradictions between radio and X-ray observations. The radio result indicates a distance within the 3 kpc ring around the Galactic Center while a ROSAT All Sky Survey (RASS) study of the source would suggest its nearby location (Reich & Schlickeiser 1992, Uchida et al. 1992). We intend to study the SNR with a deep ROSAT PSPC pointed observation and present our results and discussion in the following sections.

2 Observations and Results

A search for serendipitous observation of SNR G359.1-0.5 reveiled that ROSAT sequence No. 400275 is the longest PSPC observation toward any direction where the object is within the Field of View (FOV). The SNR is about 20' off-axis where the sensitivity and spatial resolution of the telescope are modest which is enough for our proposed study although a support structure of the PSPC window just crosses the SNR. The total exposure time is 27,770 seconds from which about 27,255 seconds are left after rejecting the intervals

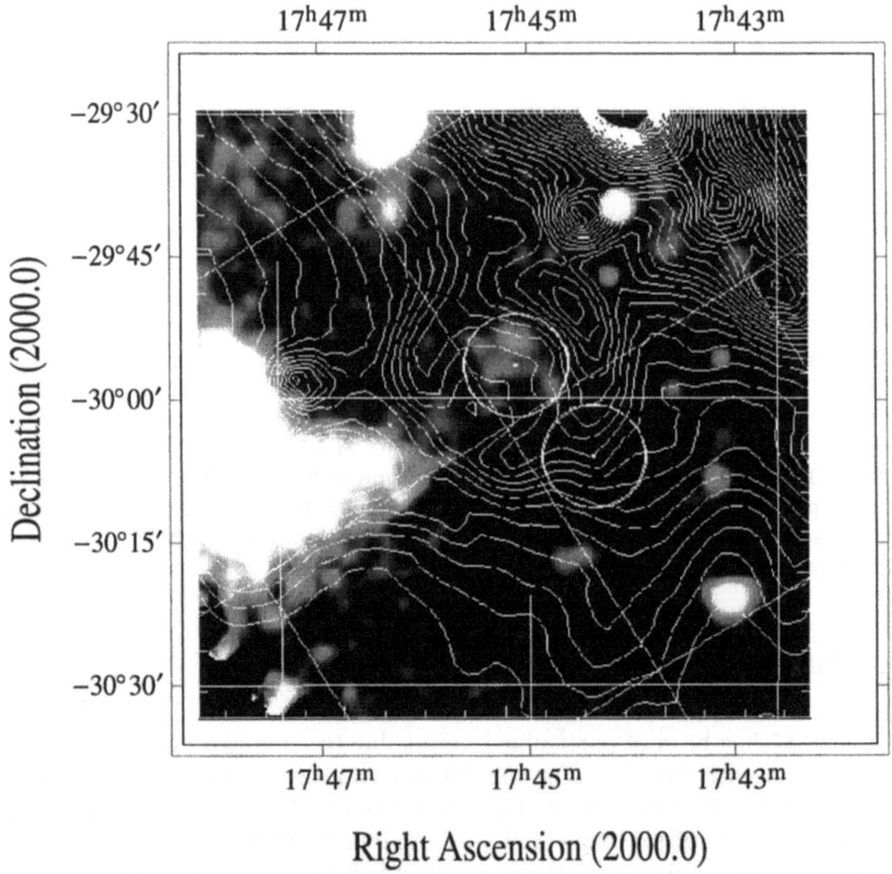

Fig. 1. X-ray intensity superimposed by 11 cm radio contours. The diffuse X-ray emission is enclosed by the radio shell of G359.1–0.5 at the figure center. Circles indicates the source and background region in spectral analysis. The diagonal coordinate lines crossing near the center are l=359° (lower left to upper right) and b=−0.5° (upper left to lower right).

with high background. The effective exposure at the object's detector position is between 11 and 19 kiloseconds taking into account obscuration by the wobbled window support structure.

The data analysis was carried out using the Extended Scientific Analysis System (EXSAS, Zimmermann et al. 1994). In the image of the soft energy band of 0.1–0.4 keV, there is neither evidence of point-like nor diffuse emission within the radio shell of G359.1–0.5. We therefore concentrate on the hard energy band. Fig. 1 shows the X-ray intensity map superimposed by the 11 cm radio contours. A highly significant diffuse emission feature is well confined within the shell structure of the SNR at the center of the figure. The circle enclosing the source region contains a total of 737 counts in the PSPC channel range 52–201, while the background region (lower right circle) contains 441 counts. Thus, the net source flux is 2.0×10^{-4}cts arcmin^{-2}s^{-1} which corresponds to a signal-to-noise-ratio of 14 sigma.

Fitting the X-ray spectrum with a thermal plasma model modified by neutral hydrogen absorption resulted in the reduced χ^2 of 1.2 (d.o.f. = 11, Fig. 2 (left)). This is not unacceptable considering the complexity of the background around the direction of the source and the influence of the detector support structure. The resultant parameters are T$\sim$ 1.0 keV for its temperature, $F_x \sim 2.0 \times 10^{-11}$ erg cm^{-2} s^{-1} for its flux in 0.1–2.0 keV and $N_H \sim 3.0 \times 10^{22}$ cm^{-2} for the absorption, respectively. It shall be noted that the temperature is not well determined due to the limited energy resolution and energy range of the instrument. Figure 2 (right) shows a χ^2 contour plot with 1, 2 and 3 σ confidence levels. It can be seen that the uncertainty is considerable. In particular, the temperature is not very well limited towards high values.

3 Discussion

The diffuse X-ray emission from G359.1–0.5 is slightly smaller in extent, than and seemingly well confined within, the radio shell. It was found that around the radio shell, there exists a dense molecular cloud and an HI ring. The interior of the radio shell, on the other hand, is relatively free of neutral material (Uchida et al. 1992). This is in accordance with the feature of the X-ray emission which is produced by hot plasma and is relatively easy to be detected within the SNR. The X-ray feature to the south-east which crosses the SNR shell near its south-eastern rim is most probably not associated with G359.1–0.5 at all. We suppose that it is an extended portion of the SNR G359.0–0.9.

Thus, the puzzle about the distance arisen from contradictory observations in the past can be resolved. The column density of G359.1–0.5 is large, 3×10^{22} cm^{-2}, which is not in conflict with the source distance near the Galactic center suggested by the radio observations. Its X-ray luminosity is then about $L_x \sim 2 \times 10^{35}$ erg s^{-1} in the 0.1–2.0 keV band assuming 8.5 kpc distance.

References

Reich, W., Schlickeiser, R., (1992): A&A 256, 408
Uchida, K., Morris, M., Yusef-Zadeh, F., (1992): AJ 104, 1533
Zimmermann, H. et al., (1994): "EXSAS User's Guide", MPE Report 257

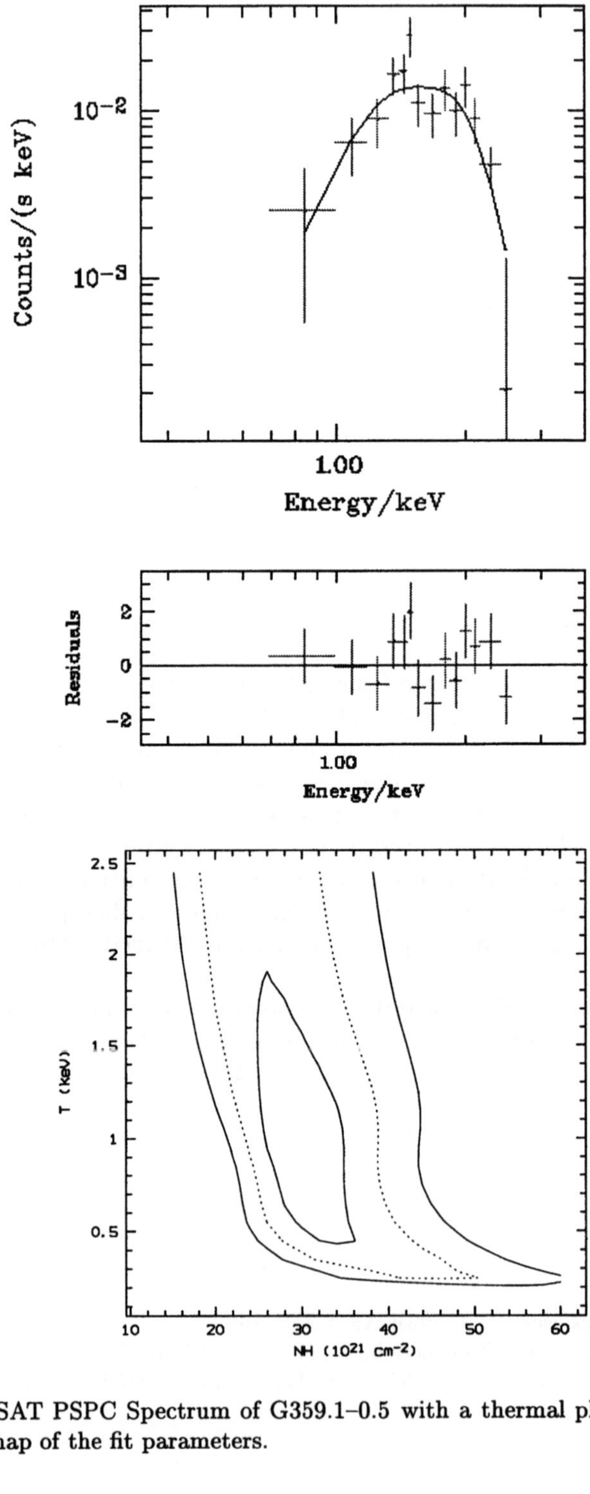

Fig. 2. ROSAT PSPC Spectrum of G359.1–0.5 with a thermal plasma fit and χ^2 contour a map of the fit parameters.

Origin and Evolution of the Cepheus Bubble

N.A. Patel[1], P.F. Goldsmith[2], M.H. Heyer[3], R. Snell[3], and P. Pratap[3]

[1] SAO, Cambridge, MA, USA
[2] NAIC, Cornell University, Ithaca, NY, USA
[3] FCRAO, University of Massachusetts, Amherst, MA, USA

Abstract. The Cepheus bubble is a giant shell with a diameter of about 120pc, discovered by Kun (1987) from IRAS sky flux maps. We have imaged this $10^o \times 10^o$ region in the CO J=1-0 line, using the FCRAO Quarry system with sample spacing equal to the 45" FWHM beam width. We estimate the total molecular gas mass to be $\sim 10^5$ solar masses and a total kinetic energy $\sim 10^{51}$ ergs. We find 49 O and B0-B2 type stars in the Cepheus bubble region, most belong to the CepOB2 association. 120 IRAS point sources, selected according to color criteria representative of young stellar objects, fall within our map. Many of these sources are associated with the globules and filamentary structures seen in the CO emission, similar to the situation found in IC1396 (Patel et al. 1995). These stars, which are still associated with dense molecular gas plausibly represent the third and youngest generation of stars in this region.

1 Introduction

The Cepheus bubble is similar in size to the Local Bubble. Its relative proximity (distance $\sim$ 800 pc, Ábráham et al. 1993) makes it a good object for a molecular line study in order to understand the structure and kinematics of the gas associated with the bubble and the phenomenon of sequentially triggered star formation. Here we briefly present some of the results from our CO mapping of this region, with the hope of providing an example of a bubble that can be compared in physical characteristics and phenomena with those of the Local Bubble.

2 Observations & Results

The observations consist of CO 1-0 emission line mapping carried out using the Five College Radio Astronomy Observatory 14m telescope at New Salem, Massachusetts, with the QUARRY focal plane array receiver. The 1 MHz per channel filterbank provided a velocity coverage of about 100 km/s centered upon -10 km/s V_{LSR}. Observations were made in position switching mode. Integration time was varying between 20 to 30 seconds per pointing, resulting in an rms noise of about 0.1 to 0.2 K. The total area covered by the map is about 72 square degrees, sampled every 50" with a FWHM resolution of 45", resulting in about 387,000 spectra using a total telescope time of about 260 hours. The mapping boundaries were based upon the IRAS emission maps. Figure 1. shows the peak intensity map with antenna temperature in the velocity range -28 – +9 km/s V_{LSR}. The features in Figure 1 which appear like filaments or globules, show values of antenna temperature typically between 2 and 8 K, whereas the diffuse clouds appear significantly fainter. In the latter, we are likely to be under-resolving the CO lines with our 3 km/s velocity resolution.

We have selected O and up to B2 type stars from the Position and Proper Motions Catalog (Roeser & Bastian 1988) and found 49 stars, including 6 O type stars belonging to the CepOB2 association. The O stars themselves appear to form a smaller ring within the larger CO and infrared ring. We have also selected IRAS points sources according to their IRAS color criteria as suggested by Beichman, Boulanger and Moshir (1992) to represent young stellar objects, as in our previous work on IC1396 (Patel et al. 1995). We found 120 IRAS point sources in the field of the Cepheus bubble satisfying these color criteria. 50 of them are directly coincident upon the globules and filamentary clouds. The diffuse clouds are noticeably lacking in IRAS sources.

From the 21 cm line observations of neutral atomic hydrogen, made at the Dominion Radio Observatory's 26 m telescope, we estimate a total atomic gas mass to be of the same order as the molecular gas: about 3×10^5 $M_\odot$. Within the uncertainties involved in the determination of both CO and HI masses, we can conclude that they are of comparable masses. Thus, the total gas mass associated with the Cepheus bubble is about 4×10^5 $M_\odot$. The total kinetic energy in the system of molecular clouds associated with the Cepheus bubble is about 10^{51}ergs (Patel et al. 1997).

3 Interpretation

Since the center of the bubble is well above the galactic plane, the expansion of the shell due to supernova explosion must proceed in an asymmetrical way. Numerical calculations taking into account of this effect, have been presented by Silich (1996). (Scc also Bisnovatyi-Kogan, these proceedings).

The Cepheus bubble region has been influenced by energy input from a previous generation of massive stars as suggested by the following observations. (1) As noted by Kun et al. (1987), the existence of the run-away star, λ Cep suggests that its binary companion could have exploded as a supernova somewhere in the central region of the Cepheus bubble. Another run-away star, 68 Cyg is also likely to be arising from Cep OB 2 region (Blaauw 1961). (2) The O stars (and many B stars) appear to form a ring on their own, in Figure 1, which is smaller than the Cepheus bubble. (3) Around Cep OB2, there appears a significant population of evolved massive stars such as μ Ceph (M2), ν Ceph (A2), VV Ceph (M2), belong to an older association with an age of about 10^7 years (Simonson and van Someren Green 1976) (4) The properties of dust in and around IC1396 suggest that it has been pre-heated due to shocks presumably due to supernovae (Clayton & Fitzpatrick 1987). (5) As noted in our work on IC1396 (Patel et al. 1995), the region outside of the ring of globules also appears to be very effectively evacuated of neutral gas; which is consistent with an event preceding the onset of the formation of the exciting star of IC1396.

The massive stars of this previous generation, would also have led to an expanding shell of gas in the parent molecular cloud in which they were born. This shell, initially expanding adiabatically, would gradually become gravitationally unstable, and lead to the condensations resulting in formation of the second generation of stars (Comerón & Torra 1994; McCray & Kafatos 1987).

We estimate the time scale for this instability as well as the size, velocity of expansion and the collected mass of the shell at which the instability occurs, by solving numerically the equation of motion of the shell following Mazurek (1980) and Maddalena & Morris (1987). We obtain $10\,\mathrm{cm}^{-3}$ for the molecular gas density and $8.0 \times 10^{47}\,\mathrm{s}^{-1}$ for the luminosity of the driving source, for such a shell to become unstable on reaching a radius of about 30 pc which is roughly the radius of the ring defined by the O type stars in Figure 1. The instability occurs at an age of about 7 Myrs and the mass accumulated by such a shell is about $6.0 \times 10^4\,\mathrm{M}_\odot$.

4 Conclusions

The observational characteristics of the Cepheus bubble are consistent with stellar wind and supernovae energy input due to an earlier generation of OB stars, over a period of about 10^7 yrs, which roughly agrees with the age of NGC 7160 (Leisawitz et al. 1989). CepOB2 may have formed as a consequence of gravitational instability in the expanding shell around this 1st generation of OB stars. The molecular cloud kinematics show evidence for expansion away from the galactic plane. A third, and youngest, generation of low to intermediate mass stars appear to be forming in the globules and filamentary clouds of the Cepheus bubble.

References

Ábráham, P., Kun, M., Balázs, Holl, A. & Fronto, A., 1993, A&A 268, 230
Beichman, C. A., Boulanger, F. & Moshir, M., (1992), ApJ 386, 248
Blaauw, A., 1961, Bull. Astron. Inst. Neth., 15, 265
Clayton, G.C. & Fitzpatrick, E.L., 1987, AJ 92, 1
Comerón, F. & Torra, J., 1994, ApJ 423, 652
Kun, M., Balazs, L. G. & Toth, I., 1987, Astrophys. Sp. Sci. 134, 211
Leisawitz, D., Bash, F.N. & Thaddeus, P., 1989, ApJSS 70, 731
Maddelena, R.J. & Morris, M., 1987, ApJ 323, 179
Mazurek T.J., 1980, A&A 90, 65

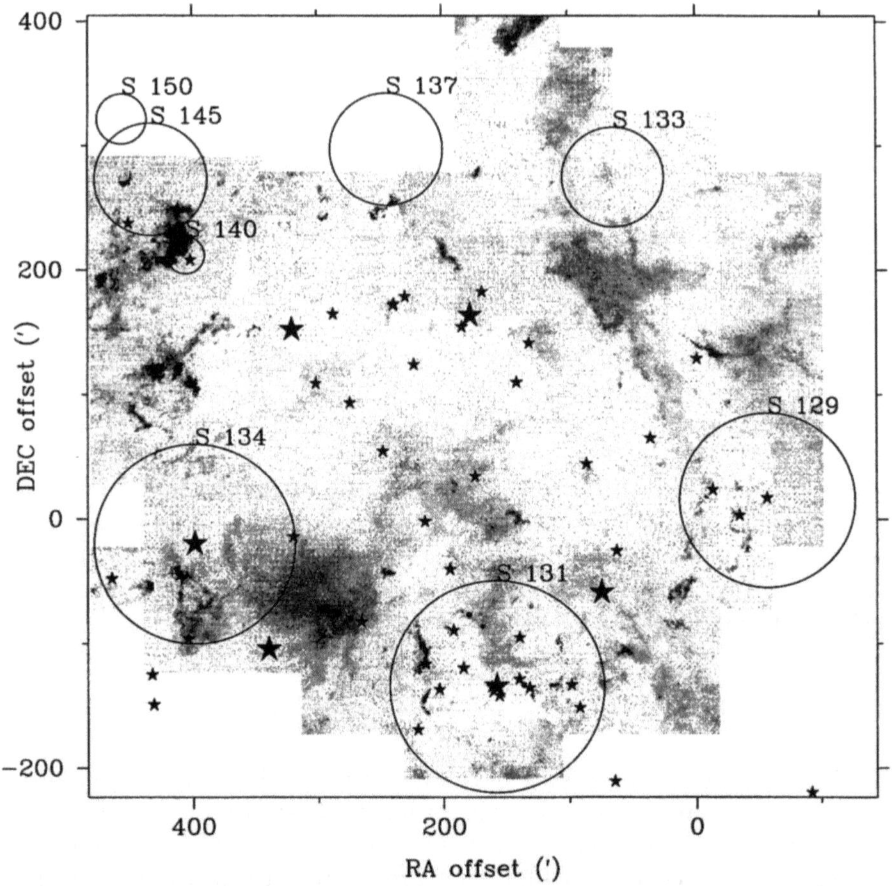

Fig. 1. Peak intensity of CO 1-0 emission. The grey scale represents antenna temperature values scaled linearly between 0 and 3.5 K. The strongest emission occurs at the S140 region and globule A of IC1396, where the line peak antenna temperature is about 10 K. A map of the integrated emission appears similar to this figure and is therefore excluded. The position offsets are measured from $\alpha(1950) = 21^h18^m00^s, \delta(1950) = 59^o30'00"$, near S129.

McCray, R., & Kafatos, M., 1987, ApJ 317, 190
Patel N.A., Goldsmith, P.F., Snell, R.L., Hezel T., Xie, T., 1995, ApJ 447, 721
Patel N.A., Goldsmith, P.F., Heyer, M.H., Snell, R.L., Pratap, P., 1997, in preparation.
Roeser, S. & Bastian, U., 1988, A&ASS 74, 449
Silich, S. .A., 1996, ApJ 468, 722
Simonson, S.C.III, & van Someren Greve, H.W., 1976, A&A 49, 343

ASCA SIS X-ray Observations of the Wind Blown Bubble NGC 6888

Matthias Wrigge[1], You-Hua Chu[2], Eugene A. Magnier[3], and Yuichi Kamata[4]

[1] Hamburger Sternwarte, Gojenbergsweg 112, D-21029 Hamburg, Germany
[2] Astronomy Dept., Univ. of Illinois, 1002 W. Green St., Urbana, IL 61801, USA
[3] Astronomy Dept., Univ. of Washington, FM-20, Seattle, WA 98195, USA
[4] Nagoya University, Furo-cho, Chikusa-ku, Nagoya 464-01 Japan

Abstract. We present ASCA SIS observations of the wind-blown bubble NGC 6888. Because the ASCA SIS is sensitive to higher energy photons and has a higher spectral resolution compared to the ROSAT PSPC, we are able to detect a $T \approx 8 \times 10^6$ K plasma component besides the $T \approx 1.5 \times 10^6$ K component known from previous PSPC observations. The existence of a high-temperature component, the observed limb-brightened X-ray surface brightness profile, and the observed level of X-ray surface brightness cannot be satisfactorily explained by currently available models. Reducing heat conduction at the contact discontinuity may raise the central temperature and produce a limb-brightening; however, the expected X-ray surface brightness is still considerably higher than the observed surface brightness.

1 Introduction

The fast stellar wind of an early-type star sweeps up the surrounding medium into a dense shell, of which the central cavity is filled with a hot, tenuous, X-ray emitting plasma. This structure is called a wind-blown bubble. Of all known bubbles, NGC 6888 is the most well-studied. The physical conditions of the hot interior of NGC 6888 have been analyzed by Wrigge at al. (1994) using ROSAT PSPC observations; a plasma temperature of 1.5×10^6 K is derived. This analysis might have missed a very hot plasma component because of the ROSAT PSPC's soft energy response and low spectral resolution. Therefore, we have obtained ASCA SIS observations of NGC 6888.

2 Results and Discussion

The ASCA-SIS and ROSAT-PSPC count rate spectra of the southern region of NGC 6888 are shown in Fig. 1a and a smoothed ASCA-SIS image of NGC 6888 is shown in Fig. 1b. We have fitted Raymond & Smith's (1977; hereafter RS) optically-thin thermal-equilibrium plasma emission models to the combined SIS and PSPC spectra. We find that it is not possible to fit the combined spectra with a single-temperature plasma component. An additional, high-temperature component contributing fluxes to the higher energy

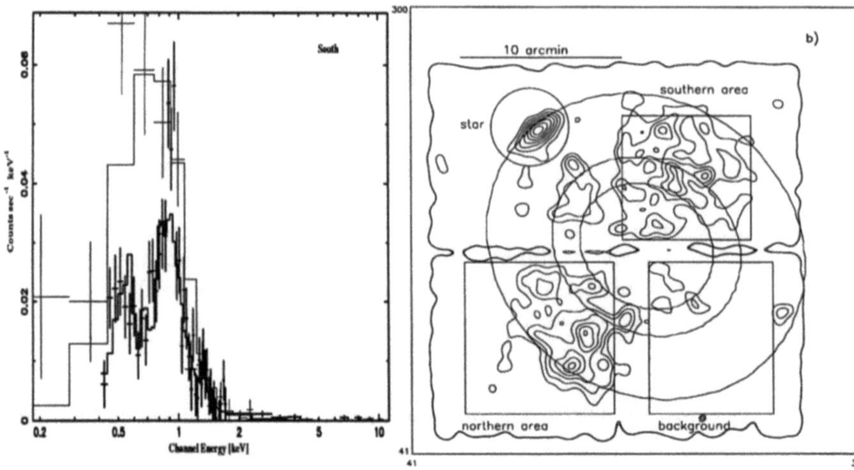

Fig. 1. **(a)** Count rate spectrum of the southern lobe of NGC 6888. The crosses are observation, and the fitted spectra are in lines. The ASCA-SIS spectrum is plotted in thick lines, and the ROSAT-PSPC spectrum in thin lines. **(b)** An ASCA-SIS image of NGC 6888 smoothed to 80″ resolution. North is at 103° counterclockwise from the top. The first contour level corresponds to 6.8×10^{-4} counts s^{-1} $arcmin^{-2}$, 1.5σ above the background of $\approx 4.8 \times 10^{-4}$ counts s^{-1} $arcmin^{-2}$. The step size between adjacent contours is 1σ. The contours running horizontally near the center of the image mark the gap between the SIS 0 and SIS 1 detectors. The bright point source to the west of NGC 6888 (at the upper left corner of the graph) is the nearby G8V star HD192020 unrelated to the nebula.

channels is needed to yield acceptable χ^2 values. The two temperature components are $T \approx 1.5 \times 10^6$ K and $\approx 8 \times 10^6$ K . The bubble structure of NCG 6888 has been analytically and numerically modeled by García-Segura & Mac Low (1995a, 1995b; hereafter GM1 and GM2). Adopting the density and temperature structures provided by GM1 and adopting the plasma emission given by RS, we have modeled the X-ray spectra of NGC 6888 using the empirically determined physical parameters of the stellar wind and the expanding shell of NGC 6888 (see Wrigge et al. 1994). The modeled X-ray spectrum is then convolved with the instrumental response of ASCA-SIS for a direct comparison with the observed count rate spectrum. The modeled ASCA-SIS spectrum exhibits one peak, which can be reasonably well fitted by a $\approx 1.5 \times 10^6$ K plasma emission component. This is clearly uncompatible with the observed spectrum, which requires an additional high-temperature component. We have also modeled the X-ray surface brightness profile of NGC 6888, using the above method. Despite the stronger emission expected in the dense outer region, the derived X-ray surface brightness profile shows a "centrally filled" appearance. The reason for the "centrally filled" appearance is that the softer X-ray emission from the cooler outer parts of the bubble is more heavily absorbed, as the absorption cross-section increases rapidly with

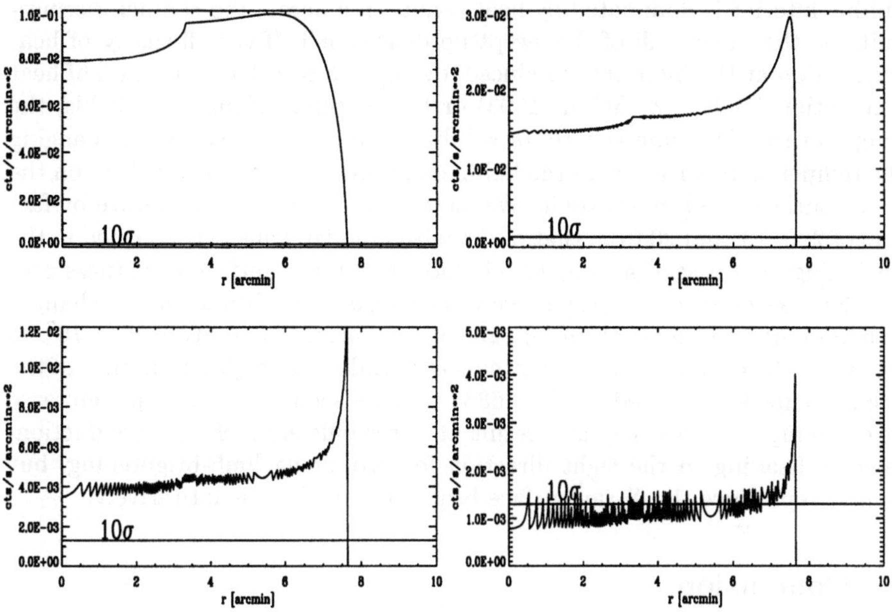

Fig. 2. Expected surface brightness for models with decreased thermal conduction efficiency, i.e. increased temperature scale. The temperature inside the bubble is increased by a factor of 2 (upper left), 5 (upper right), 10 (lower left), and 20 (lower right) from the initial model.

decreasing plasma temperature, especially for temperatures below 1.5×10^6 K. The "centrally filled" profile contrasts sharply the observed limb-brightened X-ray surface brightness profile of NGC 6888, shown in Fig. 1b. Lowering the foreground absorption column density N_H might help reconcile the difference between the expected and observed X-ray surface brightness profile. The X-ray spectral fits give an N_H in the range of $(3 \ldots 3.8) \times 10^{21}\text{cm}^{-2}$, which is consistent with that derived from the optical extinction of NGC 6888 and a typical Galactic gas-to-dust ratio. Lowering this absorption column density by a factor of three increases the expected count rates by roughly a factor of three; however, the expected surface brightness profile still does not show significant limb-brightening as in the observed surface brightness profile. To produce an X-ray surface brightness profile that matches the observed profile, we have used the analytical model of GM1 with a range of shell mass, M_{shell} = 4 and 40 $M_\odot$, and shell expansion velocity, v_{exp} = 75, 100, and 200 km s^{-1}. However, none of these models can produce the observed limb-brightening. As the numerical calculations of GM2 reproduce the bubble dynamics of NGC 6888 remarkably well, we will consider GM2 models and assume that the bubble of NGC 6888 is pressure driven as in the "energy conserving" phase of expansion. We further assume that the temperature structure of the bubble interior is dominated by heat conduction across the contact discontinuity at the inner wall of the swept-up dense shell. If the efficiency of heat conduction at the interface is reduced, owing to a possible saturation of heat conduction (Dalton & Balbus 1993) or the presence of magnetic fields, the temperature of the bubble interior will be higher. Thus, we consider changing the temperature structure of the bubble interior and examine its effect on the X-ray surface brightness profile. We increase the central temperature by factors of 2, 5, 10, and 20 from that of the initial model, which is described in the beginning of this section. Fig. (2a-d) show the X-ray surface brightness profiles for these four cases, respectively. The surface brightness profile changes to a limb brightened profile in Fig. 2b with a central temperature of 4×10^7 K. However, the overall surface brightness is considerably higher than that of the brightest peaks observed in NGC 6888. It thus seems that a simple increase of the temperature scale, as a result of lower efficiency of heat conduction, may be heading in the right direction for producing limb-brightening, but cannot reproduce the X-ray surface brightness profiles quantitatively.

3 Conclusion

Using ASCA-SIS observations, we have detected a high-temperature component of $T \approx 8 \times 10^6$ K besides the $T \approx 1.5 \times 10^6$ K plasma in NGC 6888. In contrast to what is expected, the high temperature gas resides in the periphery of the nebula; no significant emission is detected from the central regions of the bubble. Both the northern part and the southern part show similar properties. Current models (GM1 and GM2) cannot reproduce the limb brightened appearance of NGC 6888. Varying input parameters within plausible ranges does not alleviate the problem. Ad hoc variations of the central plasma temperature of the bubble and a corresponding decrease in density may produce limb-brightened profiles; however, quantitative agreement is still not achieved. Only two Galactic bubbles have been detected in X-rays, NGC 6888 and S308. As we discussed above, the X-ray spectra and morphology of NGC 6888 cannot be satisfactorily modeled, while S308 is too faint for detailed analysis of its spectral and morphological properties. Future deep X-ray observations of more wind-blown bubbles are needed to study the physical conditions of their hot interior.

References

Dalton W. W., Balbus S. A., 1993, ApJ 404, 625
García - Segura G., Mac Low M. - M., 1995a, ApJ 455, 145 (GM1)
García - Segura G., Mac Low M. - M., 1995b, ApJ 455, 160 (GM2)
Raymond J.C., Smith B.W., 1977, ApJS 35, 419 (RS)
Wrigge M., Wendker H.J., Wisotzki L., 1994, A&A 286, 219

New Measurements of the ^{4}He Abundance in Galactic HII Regions

M.R. Cioni[1,2], S. Cortiglioni[3], M. Orsini[1], E. Palazzi[3], G.G.C. Palumbo[1,3], R.L. Sorochenko[4], and A.P. Tsivilev[4]

[1] Univ. of Bologna-Italy
[2] Univ. of Leiden-The Netherlands
[3] I.Te.S.R.E./CNR-Bologna-Italy
[4] P.N. Lebedev Institute-Moscow-Russia

Abstract. Preliminary results of the ^{4}He relative abundance determination from Radio Recombination Lines observations at different frequencies in the Orion, Rosette and W3 HII regions are presented. The Orion HII region has been mapped at both 22 GHz and 36 GHz with the same beamwidth (2 arcmin) using the Medicina and Puschino radio telescopes respectively. The RRLs parameters, together with their variation with frequency and with distance were determined by centering the map on the star θ^1OriC. Three positions were observed in the Rosette HII region at 8.3 GHz leading to the first detection in this region of the transition He92α. The derived ^{4}He relative abundance is considerably greater than the ones obtained from previous measurements. The W3 HII region was observed at 36 GHz and the ^{4}He/H value derived was compared with previous measurements performed with higher spatial resolution.

1 Introduction

More than 90% of the observed helium (Hoyle & Tayler, 1964) must have been produced at the pregalactic epoch and it is commonly supposed to be produced by Big Bang nucleosynthesis. The observations of H and He Radio Recombination Lines (RRLs) in galactic and extragalactic HII regions gives the possibility to measure the primordial helium abundance with a 10% accuracy and provide a better understanding of the astrophysical processes acting inside the observed regions. With these aims a mid-term program has been started involving Italian and Russian groups which are using the two radio telescopes of Medicina (32 m), Bologna (Italy), and Puschino (22 m), Russia. In the following sections the first preliminary results of such a program are reported.

2 Observations

22.4 and 36.5 GHz maps of Orion were made with the Medicina and Puschino radio telescopes respectively, the two different $n\alpha$ transitions of RRLs, 66α

(22GHz) and 56α (36.5 GHz), were observed with the same beam width (2 arcmin). The mapped grid has been centered on θ^1OriC ($\alpha = 5^h32^m49^s.0$, $\delta = -5^o25'16''.0$). At 22 GHz three rings separated by 2 arcmin were observed for a total of 13 different positions in addition to the central one. An autocorrelator spectrometer with bandwidth of 20-25 MHz was used at the Medicina radio telescope to simultaneously observe the H and the He lines; it was possible to gain a factor of two in integration time by using the two available polarizations. The instrumental configuration provided a 0.65 km/s velocity resolution allowing enough accuracy to measure the expected line width (20-30 km/s), typical for our target RRLs in HII regions. The adopted observational technique was the position switching mode, the total integration time is greater than 40 minutes on each position. Both the Orion mapping at 36.5 GHz and W3a observations at 36.5 and 34.6 GHz were made with the RT22 at Puschino. A filter bank spectra analyzer with a 4.1 km/s (36.5 GHz) and a 4.3 km/s (34.6 GHz) velocity resolution was used.

The Medicina radio telescope was also used to observe three positions in the Rosette HII region at 8.3 GHz which corresponds to the 92α RRL transitions. The bandwidth to detect both H and He lines was set at 12.5 MHz with a velocity resolution of 0.88 km/s. Due to the weakness of the lines a long on source integration time was necessary (greater than 4 hours on each observed position) to obtain good S/N. A standard data reduction procedure was followed to get the final spectra: the spectra obtained at each position, or ring in the case of the Orion observations, were averaged together after a first quality check to look for interferences; then an high order (6-9) polynomial was substracted as baseline.

3 Results

3.1 Orion

By mapping the center of the Orion HII region a radial distribution of the ^{4}He/H relative abundance (y^+) has been obtained at two frequencies. The resulting two values for y^+ at each map position agree very well within the errors. The high spectral resolution, combined with a good signal to noise ratio, allows to separate the C66α which is blended with the He66α and generally pollutes the ^{4}He abundance determination (Figure 1a). Figure 1c shows the derived y^+ abundances as a function of the radial distance from the map center (θ^1OriC). Under the 'standard' assumption that helium is almost completely ionized in HII regions and that the contribution due to y^{++} is negligible (Churchwell et. al, 1974) the primordial helium abundance can be determined using the relation: $Y = Y_p + Z * dY/dZ$ (Y: total helium abundance; Y_p: primordial helium abundance; Z: heavy elements abundance).

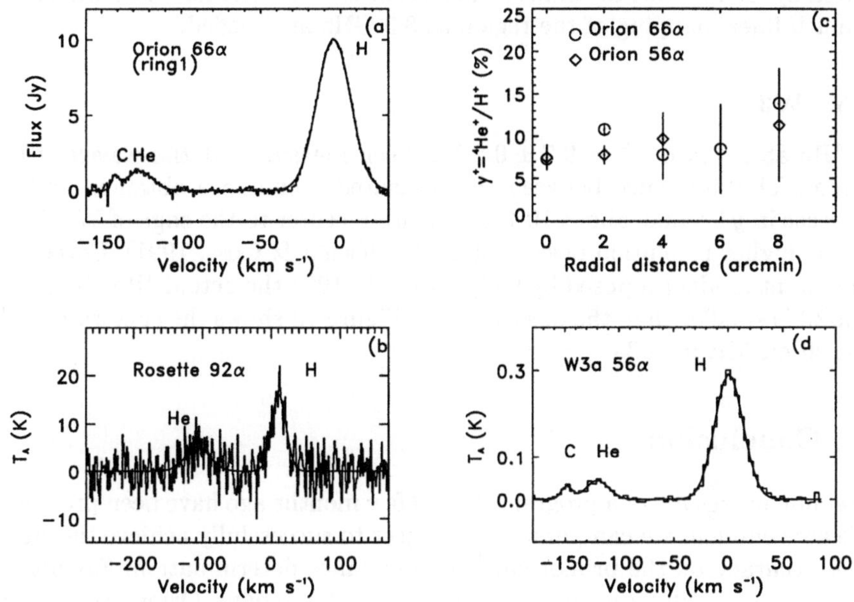

Fig. 1. 1a - spectrum of the transition 66α of ring1 in the Orion HII region; 1b - spectrum of the transition 92α of one position ($\alpha = 6^h28^m15^s.7$, $\delta = +4^o44'05''$) in the Rosette HII region; 1c - distribution of the relative abundance y$^+$ for both transitions (56α and 66α) in the Orion HII region; 1d - spectrum of the transition 56α in the W3a HII region.

3.2 Rosette

The 8.3 GHz frequency was chosen for the Rosette observations for two main reasons. First of all no RRLs observations had been previously performed at this high frequency towards this target. The second reason was to better understand the instrumental behaviour and sensitivity limits at this frequency in view of our future goal to observe the $^3He^+$ hyperfine transition at 8.665 GHz. Three positions have been observed in Rosette, the same ones done by Celnik (1985) at lower frequency (5 GHz). A V_{LSR} dependence versus the quantum number has been detected and a new physical model, which proposes a T_e gradient, was built from the observed line parameters (Cortiglioni et al., 1997). Such a model shows that the expansion of the nebula modifies the helium profile more than the hydrogen profile especially towards the center. The importance of measuring accurately the helium line parameters lies in the role they have in determining the nebula expansion. Figure 1b shows the spectrum of the transition 92α taken at ($\alpha = 6^h28^m15^s.7$, $\delta = +4^o44'05''$) in the Rosette nebula. The derived value for y^+ is much higher than the one found by Celnik, but to confirm the measured value further observations and a much finer mapping of the region at 8.3 GHz are needed.

3.3 W3

A ^{4}He abundance $y^+ = 9.7 \pm 0.8$ has been measured at the wavelength of 8 mm. The difference between the measured abundance obtained and the increase in y^+ value, about 10% to 40% from center to the edge of W3, is due to the higher resolution observations (Roelfsema & Goss, 1991). Corcerning the recent model proposed by Gulyaev et al., 1997 the actual ^{4}He abundance (8.9%) is smaller than the observed one. Figure 1d shows the spectrum of the transition 56α in W3a.

4 Conclusion

Preliminary results of a program started few months ago have been presented. We are confident we can pursue the project to successfully achieve the stated 10% accuracy on the primordial ^{4}He abundance determination. Further observations are still necessary on all our target HII regions, Orion, Rosette and W3 using both radio telescopes of Medicina and Puschino.

In particular towards Rosette we will also increase the number of mapped positions to better understand the chemical and physical conditions in the region and its dynamics.

In addition, the excellent sensitivity reached at 8.3 GHz in the Rosette spectra (rms less than 3 mK with an on source integration time of about 14 hours) lead to optimistic expectations for the ^{3}He$^+$ abundance determination by measuring the hyperfine ^{3}He$^+$ line at 8.665 GHz. Observations have being continued at Medicina towards W3 aimed at confirming the detection of Balser et al. (1994).

References

Balser , D. S., Bania, T. M., Brockway, C. J., Rood, R. T., Wilson, T. L. (1994): ApJ 430, 667

Celnik, M.E. (1985): A&A 144,171

Churchwell, E., Mezger, P.G., Huchtmeier, W. (1974): A&A 32, 283

Cortiglioni, S., Cioni, M.R., Palazzi, E., Palumbo, G.G.C., Tsivilev, A.P. (1997): MNRAS, submitted

Gulyaev, S. A., Sorochenko, R. L., Tsivilev, A. P.(1997): Pis'ma Astron. Zh., 23, 191-198

Hoyle, F.R.S., Tayler, R.J. (1964): Nature 203, 1108

Roelfsema, P.R., Goss, W.M. (1991): A&AS 87, 177

Schmidt-Burk (1981): in Nuclear Astrophysics, Proc. Intern. School of Nuclear Physics, Ed. Wilkinson, Oxford: Pergamon Press 6, 295

Ultraviolet Absorption Line Studies of Halo Gas

Klaas S. de Boer

Sternwarte, University of Bonn, Auf dem Hügel 71, D-53121 Bonn

Abstract. The status of halo studies using UV absorption lines is assessed. Si II absorption is essentially only seen from gas with $z \leq 1$ kpc, and shows a modest flow toward the galactic disk. C IV shows diffuse absorption in general only from gas with $z \geq 1$ kpc with predominantly negative velocities. Three exceptional cases of gas showing C IV absorption without absorption by neutral components are discussed. Outflowing gas has not been detected thusfar.

1 History

The first mentioning of UV absorption by O VI, N V, and C IV in the literature occurs in 1956. Basing himself on knowledge of the discovery of 'high-velocity clouds' (HVCs) by Guido Münch at MtWilson, considerations of cloud stability led Spitzer (1956) to predict the existence of a galactic corona. He did not cite Münch's first publication on the subject (Münch 1952) but referred to 'Münch, in preparation', a paper which ultimately appeared as Münch & Zirin (1961). For more details see de Boer (1989).

Spitzer then started to press for the realization of a space observatory. With the COPERNICUS (launched in 1973) and the Hubble Space Telescope (launched in 1990) he realised his early dream. In addition, the International Ultraviolet Explorer (IUE, launched 1978 and functioning until 1996), and more recently the Hopkins Ultraviolet Telescope (HUT), the Extreme Ultraviolet Explorer (EUVE), and the Orbiting Retrievable Far and Extreme Ultraviolet Spectrometer (ORFEUS) enabled to detect many features of the ultraviolet properties of interstellar gas in general and that of the local environment and the halo in particular.

Today we use not so much the word corona as Spitzer did, basing himself on the high temperatures of the solar corona, but instead the word halo. Halo derives from the Greek word for threshing ground. Threshing grounds are round. One can also imagine that light chaf blown into the air leads to roundish clouds lighting up in the sunlight (see de Boer 1989).

2 Detectability, Metal Abundances

Absorption line studies of halo gas have over the years been limited in their success mostly because of the lack of UV-bright background sources *in* the

Milky Way halo. Stars available are the odd B-type stars suspected to be massive main-sequence like stars (but there is nagging doubt about the correctness of this typing; see for normal stars e.g. Keenan 1997) and the subdwarf stars of sdB, sdO, and pAGB type. Here the sdB stars are optimal being sufficiently hot to have clean spectra yet of still acceptable absolute magnitude (see e.g. de Boer et al. 1994).

For useful studies the stars have to be close to each other in the sky and at the same time be spread well in distance in order to sample the halo gas with distance. Furthermore, an extragalactic light source helps to verify the presence of absorbing agents in principle (van Woerden 1997). The most sensitive absorption lines for neutral clouds are those of C II at 1335 Å and of MgII near 2800 Å (for details see, e.g., de Boer 1997).

The gas in the halo of the Milky Way contains all the usual elements in abundance. This was demonstrated originally by Savage & de Boer (1979) by the presence of absorbing HVCs (at +60 and +120 $\mathrm{km\,s^{-1}}$) on the line of sight to the LMC containing C II, Fe II, MgII, SiII, and the like. Since then measurements in spectra of more distant extragalactic objects have corroborated the significant metal content of HVC gas (see review by Savage & Sembach 1996). In fact, both for Mg and Si the abundance in halo gas lies, compared to galactic gas (see review by de Boer, Jura & Shull 1987), in the galactic range of values. The highest element abundances quoted for diverse HVCs are [Mg/H] > −0.59 (Bowen et al. 1994) and [Si/H] > −0.80 (Danly, Albert & Kuntz 1993). For the gas in the Magellanic Stream, which may be a mix of halo gas and Magellanic Cloud gas, these abundances are [Mg/H] > −0.49 (Savage et al. 1993) and [Si/H] > −0.70 (Lu, Savage & Sembach 1994).

3 Neutral Diffuse Gas in the Halo

The neutral diffuse halo gas manifests itself in the common absorption lines. The best studied lines are those of Si and Mg, being numerous and among the strongest and thus allowing the determination of reliable column densities.

SiII lines have been investigated extensively, because the available ones allow to cover a large range in optical depth. The width of the SiII absorption toward halo stars is documented by Danly et al. (1992), reporting on their own and on archival IUE data. I have plotted that width in velocity units in Fig. 1, thereby selecting only data in the direction more or less toward the galactic pole. In this way one is not significantly hampered by effects of galactic rotation. The absorption range lies between approximately −60 and +40 $\mathrm{km\,s^{-1}}$, the centroid of the absorption being near −10 $\mathrm{km\,s^{-1}}$. The maximum and minimum velocity of absorption reported corresponds to $\tau \simeq 1$ in the SiII 1260 Å line, which equals $N(\mathrm{SiII}) \simeq 2 \cdot 10^{12}\ \mathrm{cm^{-2}}$, (or $N(\mathrm{H}) \simeq 6 \cdot 10^{16}\ \mathrm{cm^{-2}}$ if solar abundance is assumed). Does the observed width of 100 $\mathrm{km\,s^{-1}}$ represent temperature? Or is it turbulent motion? Since

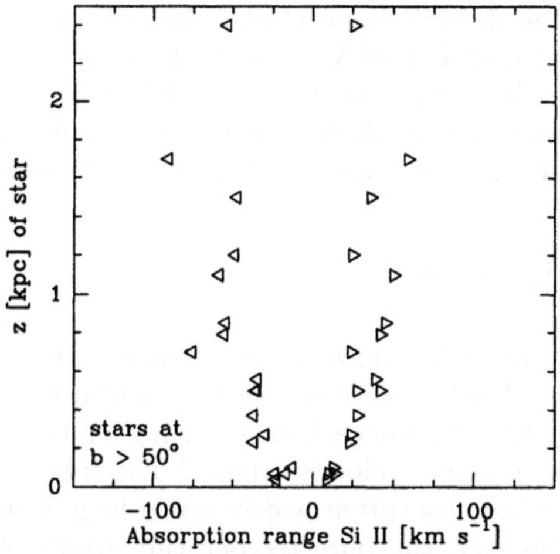

Fig. 1. The width of the interstellar absorption in SiII (triangles) is shown against the z-distance of the star used as background light source (see Sect. 3). The full width essentially stays constant for stars at $z > 0.8$ kpc indicating no further absorption beyond that z-distance. The average velocity of absorption lies rather near -10 km s^{-1} , indicating a gentle downward motion. Absorption by CIV sets in not before $z = 1$ kpc (see Sect. 5)

the full width of absorption does not significantly increase beyond $z \sim 800$ pc, one may conclude that SiII exists predominantly within a small layer. The mean velocity together with the distance show that neutral gas of the lower halo *on average* descends slowly toward the Milky Way disk.

4 Halo High Velocity Clouds

The quest to find the distance to HVCs has been at last successful. Before I summarize the results I like to emphasize (again) that the formal distinction between HVCs and IVCs at -100 km s^{-1} is artificial, since it does not take effects of galactic latitude and of galactic rotation into account. Yet, there are surely gas structures at 'intermediate' velocities at not too large z distance, as well as true 'high-velocity' clouds. But equally sure there exist gas clouds at large z having only small radial velocities, hiding in the absorption by local low velocity galactic gas.

Apart from the many lower limits (see for reviews with tabular material de Boer 1997 and Wakker & van Woerden 1997) there are now the upper limits to the distance of $1.2 \leq z \leq 4.4$ kpc for gas between Complex M II and M III (Danly, Albert & Kuntz 1993), of $z \leq 4.1$ kpc for gas at the edge of Complex C based on studies of stars in the direction to M 13 (de Boer &

Savage 1983; Bates et al. 1995; Shaw et al. 1996), of $z \leq 1.4$ kpc for Complex L (Albert et al. 1989), and of 25 kpc for the HVC in the foreground of the LMC (Savage & de Boer 1981; Molaro et al. 1993). Significant lower limits are $z > 2.4$ kpc toward Cloud A in Complex C I (Schwarz, Wakker & van Woerden 1995), and $z > 3$ kpc toward Cloud A III (Wakker et al. 1996).

5 Highly Ionized Gas

Absorption by CIV normally shows up in spectra of stars only when $z > 1$ kpc (see Danly et al. 1992). There the velocity width is larger than that of SiII at low z. Does this difference indicate a higher temperature? Or is it a width dominated by a real velocity spread? The latter is the more likely explanation because the absorption widths come from sight lines extending over a few kpc. Moreover, the pure temperature width of CIV absorption lines is 20 km s^{-1} for $2 \cdot 10^5$ K, this temperature being the one of maximum abundance of C^{3+} in thermal equilibrium gas. If the gas were to be at 10^6 K, the thermal width of CIV absorption would be about 50 km s^{-1} .

Furthermore, the CIV absorption line normally is smooth and does not show so much the component structure as the SiII lines do. Apparently the SiII absorption is more defined in velocity (in clouds?) whereas the CIV is rather due to diffuse material. However, temperature as well as the element's weight may contribute to the smoothness of the absorption.

NV is more difficult to detect than CIV, because the product of the NV absorption line f-value and N abundance is a factor of $\simeq$10 smaller than that of CIV. First successes of NV measurements are by Savage & Massa (1987), but one must note that their lines of sight run over substantial portions through the ever thicker disk in general galactocentric directions.

Collecting all the CIV and NV data available to them, Savage & Massa (1987) derived a scale height for these ions of 2-3 and 1-4 kpc, respectively. This can be related to the scale height of free electrons of 1-2.5 kpc (Reynolds 1989).

OVI has been detected with ORFEUS in the spectra of several galactic stars at limited distances from the disk (Hurwitz & Bowyer 1996), and toward an SMC star (Hurwitz et al. 1995). New high dispersion data have been taken Nov 1996 with ORFEUS but have not yet been analyzed in detail and thus nothing can be said about presence of OVI in relation to NV or CIV.

Dixon, Davidsen & Ferguson (1996) give evidence from HUT spectra for OVI *emission* by halo gas. The most significant detection is in the direction $l = 218.2°, b = 56.4°$, a region devoid of high velocity gas.

Information on the distribution of highly ionized species in z can be obtained from the structure of the absorption profile. When one adopts a model for the rotation of halo gas, and then translates the limits of each velocity bin into a distance bin, the column density derived from the absorption in that bin can so be transformed into a gas density for that distance bin. Doing so, Savage & de Boer (1981; their Fig. 6) demonstrated that C IV may extend out to $z > 10$ kpc, based on the assumption that the halo gas corotates with the disk. We know now that this is not the case (see below). We now also know that C IV may exist at other temperatures than that for the peak appearance in thermodynamic equilibrium (see e.g. Böhringer, 1997), but even then C IV may still be present with a constant fraction of all C over a large temperature range. Thus the $n(z)$-profile of C IV as published by Savage & de Boer (1981) most likely has to be modified only to a smaller z-extent due to slower rotation.

6 Motion of Halo Gas

From the survey of HVC detections at 21 cm by Giovanelli (1980) it is evident, that a substantial part of the observed velocity is the reflection of the galactic rotation of the Sun (see also Wakker 1991). However, more can be said.

In plots of v_{HVC} versus galactic longitude one finds that the sine-wave of the galactic rotation has at $l = 180°$ a velocity of $\simeq -100$ km s^{-1} . This means that the HVCs as a sample move at about that speed toward the rotation axis of the galaxy. Absorption lines toward the North Galactic Pole (star vZ 1128 in M 3) by de Boer & Savage (1984) show a velocity of $\simeq -70$ km s^{-1} , a motion toward the disk (see Fig. 1 for the slow flow of gas *near* the disk). Furthermore, on the sight line toward the globular cluster M 13 ($l = +60°, b = +45°$) the absorption velocity is $\simeq -80$ km s^{-1} , although velocities of $0 \leq v \leq +20$ km s^{-1} are expected with corotation (de Boer & Savage 1983). This means that in this direction the gas stays *behind* the disk rotation. Collectively these findings were used in a consistency test by Kaelble, de Boer & Grewing (1985) and they concluded that, on average, halo HVC gas is at $z = 3$ kpc and has velocities of up to 100 km s^{-1} toward the galactic rotation axis, of up to 100 km s^{-1} toward the galactic disk, and of only 100 km s^{-1} along with the disk rotation (see also de Boer, 1985).

7 Three Cases of Absorption by C IV Without Si II or H I

In most of the absorption spectra C IV is present at the same velocity as Si II, albeit with a more smooth profile (see e.g. toward SN 1987A, Savage et al. 1989). There are now three cases known in which C IV is present *without* detection of Si II or of H I.

Mrk 509. Sembach et al. (1995) found C IV at -270 km s^{-1} in the spectrum of Mrk 509 ($l = 36°, b = -30°$), no Si II at that velocity, and H I only at a position 2° away. The velocity is exceptional; from galactic rotation one would expect velocities ranging from 0 to +80 km s^{-1} for gas with $d < 16$

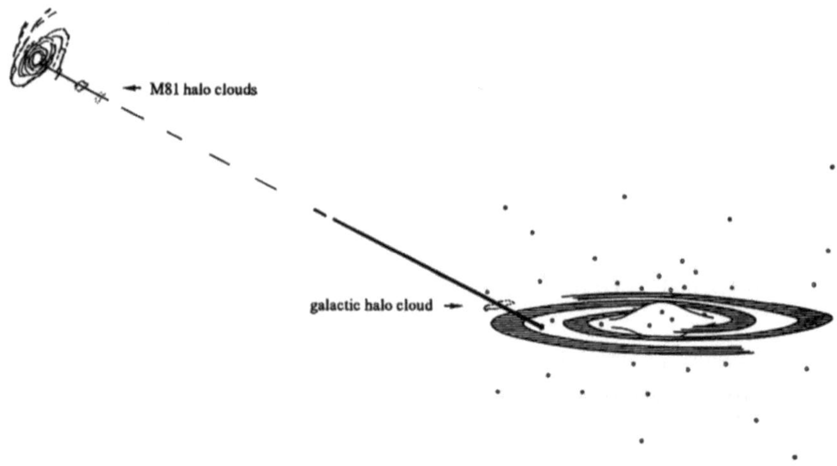

Fig. 2. The line of sight from the Milky Way to SN 1993J in M81 contains numerous clouds. Absorption near 0 $\mathrm{km\,s^{-1}}$ is due to gas in the solar vicinity, absorption near -50 $\mathrm{km\,s^{-1}}$ due to a local halo cloud, absorption near -120 $\mathrm{km\,s^{-1}}$ due to gas in M81. The absorption complex at velocities between $+120$ and $+240$ $\mathrm{km\,s^{-1}}$ is due to clouds with very small spatial extent but with high column densities of Ca II, MgII, and CIV, somewhere along the 3 Mpc line of sight, but most likely near M 81. For the exceptional absorption profiles see see Vladilo et al. (1993) and de Boer et al. (1993)

kpc, or at most down to $-110\,\mathrm{km\,s^{-1}}$ at larger distances. However, corotation can only be an hypothesis so far away from the disk.

HD 119608. Toward stars inside galactic radio Loop I and IV Sembach, Savage & Tripp (1997) detected CIV at -45 $\mathrm{km\,s^{-1}}$ $(l = 320°, b = +43°)$ without Mg II at 1239 Å. This gas most likely is part of thc galaxy since its velocity is similar to that of other absorptions found in this general direction of the halo. However, a final check should be made by obtaining a spectrum of the much more sensitive line of Mg II at 2795 Å.

SN 1993J. On the sight line of 3.25 Mpc toward SN 1993J in M 81 $(l = 142°, b = +41°)$ absorption was found in CIV at the usual galactic velocities and at $\simeq -120\,\mathrm{km\,s^{-1}}$ (the velocity of M 81). In addition, strong absorption was found at $+130\,\mathrm{km\,s^{-1}}$ in Ca II, in Mg II, and in CIV (Vladilo et al. 1993, de Boer et al. 1993, Vladilo et al. 1994, Bowen, Blades & Pettini 1994). Most extraordinary, the CIV absorption extended up to 200 $\mathrm{km\,s^{-1}}$, also in SiII, with the highest velocity cloud seen at $+240$ $\mathrm{km\,s^{-1}}$. These velocities are not expected with any reasonable model for the dynamics of gas of the Milky Way. In addition, there is no trace of H I in Effelsberg 21 cm spectra at these

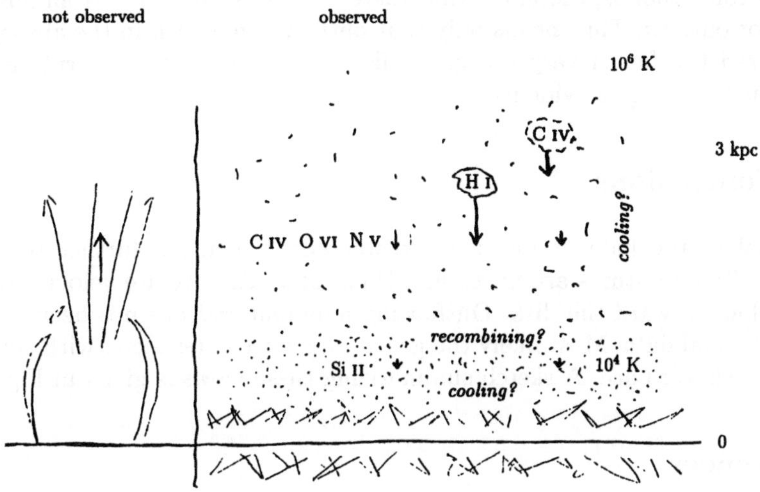

Fig. 3. The presence, the characteristics, and the flow of the material in the halo can be summarized as shown. Note that the sketch is a grand average but consistent with what is observed

velocities, implying that this gas can occupy only a very small fraction of the radio telescope beam, viz. that it is far away from us. The complex most likely represents HVCs falling toward M 81, or moving somewhere through intergalactic space (see Fig. 2).

8 Outflow of Gas?

The SiII absorption lines detected in spectra of halo stars at high galactic latitudes show a certain width in velocity. Taking these velocities at face value one might conclude that there is gas flowing away from as well as toward the disk. On the other hand, normal turbulence of clouds may account for a substantial fraction of the width. In the CIV absorption lines the width is mostly toward negative velocities. There is, to my knowledge, no detection of positive velocity CIV, except of course those structures which readily can be explained as due to normal galactic rotation (such as the CIV toward LMC stars).

Positive velocity clouds were detected toward M 81. These can, however, be anywhere on the line of sight and the more likely explanation is that they are high velocity clouds descending in the M 81 halo.

My conclusion is, therefore, that there is as yet no evidence from observations for outflow. This means only that outflow is not seen in the absorption lines used thusfar. If very hot gas is flowing out, the O VI absorption lines may the first to give evidence.

9 Conclusions

Material in the halo at low z is predominantly neutral, further out it is ionized. Temperatures are uncertain. High- and intermediate velocity clouds come down toward the disk. Outflowing halo material has not been seen in observational data. Most likely the gas at larger z rotates along with the disk but at a slower pace. A sketch summarizing these views is given in Fig. 3.

References

Albert C.E., Blades J.C., Morton D.C., Proulx M., Lockman F.J., 1989, in IAU Coll. 120, 'Structure and Dynamics of the Interstellar Medium', eds G. Tenorio-Tagle et al., Springer Lect. Notes in Physics No 350, p. 442

Bates B., Shaw C.R., Kemp S.N., Keenan F.P., Davies R.D., 1995, ApJ 444, 672

Böhringer H., 1997, these proceedings

Bowen D.V., Roth K.C., Blades J.C., Meyer D.M., 1994, ApJ 420, L 71

Bowen D., Blades J.C., Pettini M., 1995, ApJ 448, 662

Danly L., Albert C.E., Kuntz K.D., 1993, ApJ 416, L 29

Danly L., Lockman F.J., Meade M.R., Savage B.D., 1992, ApJS 81, 125

de Boer K.S., 1985, Mitteilungen Astron. Ges. 63, 25

de Boer K.S., 1989, in IAU. Coll. 120, 'Structure and Dynamics of the Interstellar Medium', eds. G. Tenorio-Tagle et al.; Springer Lect. Notes in Physics No 350, p. 432

de Boer K.S., 1997, in 'The Physics of Galactic Halos', eds H. Lesch et al., Akademie Verlag, Berlin, p. 89

de Boer K.S., Savage B.D., 1983, ApJ 265, 210

de Boer K.S., Savage B.D., 1984, A&A 136, L 7

de Boer K.S., Jura M.A., Shull J.M., 1987, in 'Exploring the Universe at Ultraviolet Wavelengths', eds Y. Kondo et al.; Dordrecht, Reidel; p. 485

de Boer K.S., Rodriguez Pascal P., Wamsteker W., Sonneborn G., Fransson C., Bomans D.J., Kirshner R.P., 1993, A&A 280, L 15

de Boer K.S., Altan A.Z., Bomans D.J., Lilienthal D., Moehler S., van Woerden H., Wakker B.P., Bregman J.N., 1994, A&A 286, 925

Dixon W.V.D., Davidsen A.F., Ferguson H.C., 1996, ApJ 465, 288

Giovanelli R., 1980, AJ 85, 1155

Hurwitz M., Bowyer S., Kudritzki R.-P., Lennon D.J., 1995, ApJ 450, 149

Hurwitz M., Bowyer S., 1996, ApJ 465, 296

Kaelble A., de Boer K.S., Grewing M., 1985, A&A 143, 408

Keenan F., 1997, in 'Third Conference on Faint Blue Stars', eds A.G.D. Philip et al., Davis Press, Schenectady, in press

Lu L., Savage B.D., Sembach K.R., 1994, ApJ 437, L 119

Molaro P., Vladilo G., Monai S., D'Odorico S., Ferlet R., Vidal-Madjar A., Dennefeld M., 1993, A&A 274, 505
Münch G., 1952, PASP 64, 312
Münch G., Zirin H., 1961, ApJ 133, 11
Reynolds R.J., 1989, ApJ 339, L 29
Savage B.D., de Boer K.S., 1979, ApJ 230, L 77
Savage B.D., de Boer K.S., 1981, ApJ 243, 460
Savage B.D., Massa D., 1987, ApJ 314, 380
Savage B.D., Sembach K.R., 1996, ARAA 34, 279
Savage B.D., Jenkins E.B., Joseph C.L., de Boer K.S., 1989, ApJ 345, 393
Savage B.D., Lu L., Bahcall J.N., Bergeron J., Boksenberg A., et al., 1993, ApJ 413, 116
Schwarz U.J., Wakker B.P., van Woerden H., 1995, A&A 302, 364
Sembach K.R., Savage B.D., Trip T.M., 1997, ApJ 480, 216
Sembach K.R., Savage B.D., Lu L., Murphy E.M., 1995, ApJ 451, 616
Shaw C.R., Bates B., Kemp S.N., Keenan F.P., Davies R.D., Roger R.S., 1996, ApJ 473, 849
Spitzer L., 1956, ApJ 124, 20
van Woerden H., 1997, these proceedings
Vladilo G., Centurión M., de Boer K.S., King D.L., Lipman K., Stegert J., Unger S.W., Walton N., 1993, A&A 280, L 11
Vladilo G., Centurión M., de Boer K.S., King D.L., Lipman K., Stegert J., Unger S.W., Walton N., 1994, A&A 291, 425
Wakker B.P., 1991, A&A 250, 499
Wakker B.P., van Woerden H., 1997, ARAA 35, 217
Wakker B., Howk C., Schwarz U., van Woerden H., Beers T., Wilhelm R., Kalberla P., Danly L., 1996, ApJ 473, 834

Simulations of Halo Supernova Remnants

R.L. Shelton

Laboratory for High Energy Astrophysics, NASA/Goddard Space Flight Center, Greenbelt, MD 20706, USA.

Abstract. High latitude observations find C^{3+}, N^{4+}, and O^{5+}, while shadowing by intervening hydrogen clouds indicates the presence of soft X-ray emitting gas in the halo of our galaxy. This paper reports on detailed hydrocode simulations of one of the possible source types: isolated supernova remnants evolving in the tenuous ambient medium of the halo. Their O^{5+} column densities are similar to those observed from the halo. The N^{4+}, C^{3+}, and X-ray properties are also simulated and discussed. The calculations may also be of interest for the Local Bubble, external galaxies, and non-thermal pressure estimations.

1 Introduction

The high latitude observations of C^{3+}, N^{4+}, O^{5+}, and soft X-rays are not completely explained, though several scenarios have been proposed and modeled. Multiple mechanisms may be at play and all should be considered. One of the possible contributors is the collection of isolated supernova remnants (SNRs) evolving in the halo. The simulations presented in this paper show that they harbor hot gas and UV ions (C^{3+}, N^{4+}, and O^{5+}) for nearly 2×10^7 yrs[1]. The supernova rates from Ferrière (1995) give about one supernova per 600,000 years per kpc^2 above $z \geq 300$ pc at our galactocentric[2] radius. Combining the rate with the lifetime yields about 30 halo SNRs per kpc^2, a number which is high enough to make them relevant to our understanding of the halo. Previously, Ferrière (1995) and Cioffi (1991) modeled halo SNRs. The simulations presented in this paper differ from those in method and breadth of predictions.

This paper is focused on characterizing the O^{5+}, N^{4+}, and C^{3+} and X-ray properties of isolated SNRs evolving in the halo. The main results are that the halo supernova remnants harbor much O^{5+}, and may explain roughly 3/4ths of that expected to reside in the halo. The average $N^{4+}$and C^{3+} column densities are lower. The young SNRs are X-ray bright, but the fluxes averaged over their entire lifetimes are much smaller. Isolated SNRs fill a very small fraction of the halo.

[1] Merging with the ambient gas does not shorten their lifetimes.

[2] A cone having a 50° angle from the zenith and truncated below $z = 300$ pc also has about one supernova every 600,000 years.

2 Modeling

Halo SNRs were simulated with a detailed, one-dimensional hydrodynamic computer code (Shelton, 1996) employing non-equilibrium ionization and recombination rates, thermal conduction, and magnetic pressure. The (unmodeled) density gradient should have minimal effect because the hot bubble diameters are $\sim 1/3$ of the density scale height and Maciejewski et al. (1996) found that SNRs evolving in moderate density gradients are nearly spherical.

The primary computer run used an explosion energy of 5×10^{50} ergs, ambient density (n) of 0.01 atoms cm^{-3}, ambient temperature of 10^4 K, and ambient non-thermal pressure (P_{B+CR}) equivalent to that of a 2.5 μG effective magnetic field. Based on the H and H^+ vertical distributions, the choice of n corresponds to $z = 1000$ pc. A variety of ionization conditions appear in the halo gas, but whether the warm phase or the hot phase is more plentiful is not well determined. The choices of ambient thermal and non-thermal pressures used here apply if the halo is cool and dominated by non-thermal pressures (see Boulares and Cox, 1990).

Although the abundances of gas-phase metals in the halo are not well determined, the halo gas-phase abundances may be closer to solar than are those in the disk (Savage and Sembach, 1996). Solar abundances are often used in SNR simulations of disk remnants and were used in these calculations.

3 Evolution and Predictions

Results for a Single SNR

A cool shell forms between 2.5 and 5×10^5 yrs. Afterward, the shock front expands faster than the hot bubble. Between the shock front and the periphery of the hot bubble lies a "thick" (large Δr), expanding, cool, weakly compressed shell. The hot bubble continues to expand for some time after shell formation, reaching its maximum radius of about 140 pc at about 2×10^6 years. The ensuing diminution is very slow. For example, the hot cavity extends to ~ 80 pc at 10^7 years.

At 10,000 yrs, the remnant is filled with hot ($> 10^7$ K) gas, but its ions are far from equilibrium. The atoms in the interior are ionizing up through the C^{3+}, N^{4+}, and O^{5+} states. As a result, very large column densities of UV ions exist in the young remnant. Their Doppler shifted velocities and thermal broadening can be large. Around the time of shell formation, the shock front becomes too weak to ionize the swept up ambient gas to the C^{3+}, N^{4+}, and O^{5+} states. Henceforth, these ions only derive from previously hot, recombining gas. Oxygen can recombine to O^{+4} before nitrogen recombines to N^{+3} or carbon recombines to C^{+2} and the gas cools before all of the carbon has recombined to the C^{+2} level. As a result, the C^{3+} extends out to greater radii in the cooling gas than does the N^{4+} or O^{5+}, and at very late times resides in cooled gas. For example, some C^{3+} remains at 18 million years, though the SNR had already cooled to the temperature of the ambient medium a million years earlier. The column densities are only weakly dependent on the age of the SNR, particularly after the first 5×10^4 yrs. The time integrals of the number of UV ions contained by the SNR are 8×10^{69} O^{5+}seconds, 7×10^{68} N^{4+}seconds, and 2×10^{69} C^{3+}seconds. Dividing by the time integral of the area covered by UV ions gives the time and impact parameter averaged column densities: 5×10^{13} $O^{5+}cm^{-2}$, 5×10^{12} $N^{4+}cm^{-2}$, and 1×10^{13} $C^{3+}cm^{-2}$. The soft X-ray flux peaks within the first 10^5 yrs. At the peak, a halo SNR produces ~ 4000 ROSAT 1/4 keV counts s^{-1} $arcmin^{-2}$, making it brighter than most areas within the northern and southern halos (Snowden et al., 1997). Early on, the ions are drastically underionized, with the ionization state varying within the remnant. But, the SNR cools faster than the X-ray ions can recombine. As a result, the gas becomes overionized and the spectra exhibit strong recombination edges (for example, the oxygen 739 eV edge is prominent in the spectra by 2.5×10^5 yrs). Combining the time integral of the volume filled by hot ($T \geq 10^6$ K) SNR gas with the explosion rate from Ferrière (1995) gives a negligibly small filling fraction at $z = 1000$ pc.

Estimates for the Population of Halo SNRs

Most of the supernova explosions occur at lower heights than $z = 1000$ pc and so are better confined by the greater ambient density and pressure. By running additional, computationally cheaper simulations for other z's, it was possible to estimate the fraction of volume filled by hot, halo SNR gas at several heights above the plane between about 300 and 2200 pc. The fraction is negligibly small. It is important to note that the volume filled by isolated SNR bubbles is much, much less than that outlined by SNR shock fronts because the radius to the shock front grows significantly beyond that to the bubble edge within the first 1/30 of the remnant's life.

Because of the computational expense of using non-equilibrium ionization and recombination rates, the additional runs do not predict the time dependent UV ion and X-ray properties. Very approximate estimates of the time and space averaged column densities and emission fluxes of the ensemble of SNRs above $z \sim 300$ pc can be gotten by combining the supernova explosion rate with the predictions for a single SNR at $z \sim 1000$ pc. The spatial distributions should be patchy. Using the UV ion-time integrals presented earlier, the predicted time and space average O^{5+}column density is 4.5×10^{13} cm^{-2}, which represents 3/4ths of the average observed column density (Hurwitz and Bowyer, 1996) after the Local Bubble contribution (Shelton and Cox, 1994) has been subtracted. The average N^{+4} and C^{+3} column densities from the simulations (4×10^{12} cm^{-2} and 1×10^{13} cm^{-2}, respectively) provide smaller fractions of the observed totals. The gap may be due to photoionization and differences in ion production between this run and those at greater heights[3]. A halo SNR is rich in UV ions for its entire lifetime but is bright in soft

[3] The scale heights of C^{3+} and N^{4+} are greater than that of O^{5+}.

X-rays for only the first 1×10^5 yrs. Thus, they cover much more of the sky with their UV ions than with their bright, soft X-ray emitting gas. The older halo SNRs contribute a patchy field of extremely dim X-rays having recombination edges and a softer spectrum. A rough estimate of the total X-ray contribution (found from the X-ray rates and the SN rate) is smaller than that found by Snowden et al. (1997), suggesting that either the model is in need of tweaking or other physical scenarios are also at work.

4 Pressure Estimations and the Local Bubble

The thermal pressure, P_{th}, calculated from the *observations* of O^{5+} is much larger than that calculated from the C^{3+}. There may be a natural, although scenario dependent, explanation. The simulations show that the O^{5+} lies in the hotter, thermal pressure dominated gas, while some of the C^{3+} lies in the cooler, non-thermal pressure dominated shell. As a result, P_{th} calculated from the O^{5+} emission and absorption *predictions* is much larger than that calculated from the C^{3+} results. This effect provides a method of estimating the lower limit to the non-thermal pressure ($P_{B+CR} \geq P_{th}(O^{5+})$ - $P_{th}(C^{3+})$).

Might the Local Bubble consist of the hot bubble of a SNR evolving in a low density, high non-thermal pressure medium? These simulations and those in Shelton (1996) show that until the cool shell forms, the SNRs are sufficiently X-ray bright and have B to C band ratios of ~ 0.4, in agreement with McCammon and Sanders (1990). Using O^{5+} as a criterion is more complicated because other sources (the Local Cloud and other clouds) may contribute. The O^{5+} column densities in the simulated SNRs are large enough to agree with Shelton and Cox (1994), but the O^{5+} velocities are too large to be observed by COPERNICUS (Jenkins, 1978) until after the shell forms.

Acknowledgements. Parts of this work were performed while the author held a National Research Council-NASA/GSFC Research Associateship.

References

Boulares, A., and Cox, D. P., (1990): ApJ 365, 544
Cioffi, D. F., (1991): In "The Interstellar Disk-Halo Connection in Galaxies", ed. H. Bloemen, p. 355
Ferrière, K. M., (1995): ApJ 441, 281
Hurwitz, M. and Bowyer, S., (1996): ApJ 465, 296
Jenkins, E., (1978): ApJ 219, 845
Maciejewski, W., Shelton, R. L., and Cox, D. P., (1996): AAS 188.4410M
McCammon, D., and Sanders, W. T., (1990): ARA&A 28, 657
Savage, B. D., and Sembach, K. R., (1996): ARA&A 34, 279
Shelton, R. L., (1996): Ph.D. Thesis, Dept. of Physics, Univ. of Wisconsin – Madison
Shelton, R. L., and Cox, D. P., (1994): ApJ 434, 599
Snowden, S. L., Egger, R., Finkbeiner, D., Freyberg, M. J., and Plucinsky, P. P., (1997): submitted to ApJ

Optical Studies of a Local Filament Toward the Magellanic Clouds and a Local IVC Region

B.E. Penprase[1], Q.D. Wang[2], and U. Herbstmeier[3]

[1] Pomona College, 610 N. College Ave, Claremont, CA 91711, USA
[2] Department of Astronomy, NWU, 2145 Sheridan Road, Evanston, IL 60208, USA
[3] MPI für Astronomie, Königstuhl 17, D-69117 Heidelberg, Germany

Abstract. We present new spectroscopic results which constrain the distance of a large IR emitting filament near the LMC, as well as some new results for a local region with widely distributed detections of Intermediate Velocity Clouds (IVCs). We present spectroscopic observations of stars toward both regions which constrain the distances and kinematics of the interstellar features. New optical high resolution ($\lambda/\delta\lambda \simeq 10^6$) spectra have been obtained for the star HD 22252, and are used to estimate approximate distances, temperatures, and pressures within the cloud. An unusual local region of the interstellar medium with several IVCs is also reported, with multiple components of absorption from -80 < $v_{\rm LSR}$ < 70 km s^{-1} seen toward several stars in the region.

1 Optical Detection of Absorption from a Local Filament Toward the Magellanic Stream

The filament in our study is a very bright object in the infrared, and extends to high galactic latitudes in the little-studied part of the local bubble toward the Magellanic Clouds (see Figure 1). Radio observations of the object have been obtained previously by McGee et al. (1986), but these observations were not able to provide an estimate of the distance to the cloud. Previous observations (Penprase et al. (1997a)) have enabled us to constrain the distance to 110 pc, and the mean value of extinction to A_v=0.53 magnitudes. The filament has mean galactic coordinates of l=280° and b=-45°. Examination of the galactic plane maps from Frisch (1995) and the high latitude maps of Snowden (1993) places the object within the local bubble, and as such the filament is a valuable probe of the temperatures, and pressures within the local bubble. If the local cavity is completely filled with X-ray emitting hot gas, and if pressure equilibrium prevails within the local bubble, then measurements of the pressure for this filament will provide an independent method of measuring the pressure of the hot component of the ISM. We discuss below some constraints on the pressure of the filament, and possible implications for the local bubble.

We have estimated the linear sizes of clumps within the filament, which range in size from 0.15 to 0.75 pc, based on the resolved features seen in *IRAS* maps of the region. The pointings in McGee et al. (1986) include a wide range

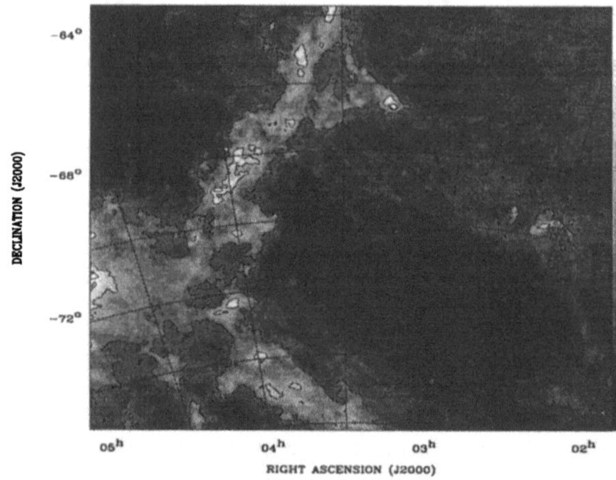

Fig. 1. *IRAS* 100 μm image of the filament, at the distance of 110 pc (Penprase et al. (1997a)), showing an X-ray shadowing region at center right.

of observed column densities, with 0.5×10^{19} cm^{-2} < N(HI) < 2.0×10^{20} cm^{-2}. For simplicity, we have adopted an average H I column density of $< N(HI) >$ = 10^{20} cm^{-2} from these H I emission measurements, and initial examination of the H I 21 cm absorption observations appear consistent with this estimate.

By combining our value of total H I within the filament with the computed linear dimensions, we may derive estimates of the density within the filament, which range from $50 < n_H < 200$ cm^{-3}, using a conservative estimate of the size of the clumps within the filament which corresponds to cloudlet dimensions of 0.15 to 0.60 pc. Recent observations (some presented at this conference) suggest that the cloudlets could be much smaller, and therefore the above density estimates are probably lower limits to the actual density within the cloud.

New spectroscopic observations have been obtained for both NaI D and Ca II K absorption using the UHRF at the AAT (Penprase et al. (1997b)), and with the resolving power of $\lambda/\delta\lambda = 10^6$, it is possible to constrain the temperature within the filament using the observed resolved linewidths of the optical absorption lines. One of the spectra from the AAT is shown in Figure 2, for the target star HD 22252 (d=120pc), which is at a distance of 120 pc. Analysis of linewidths suggests temperatures within this filament ranging from 200 < T < 750 K. Examination of the linewidths from the 21 cm absorption profiles (which were of lower resolution than the optical lines) results in a weaker upper limit of temperature consistent with this result, at T < 700 K.

The spectra suggest that the pressure nT of the cloud ranges from 10^4 < nT < 10^5 cm^{-3}K, which is comparable, and possibly higher than, the estimated pressures within the local bubble, which are typically reported to

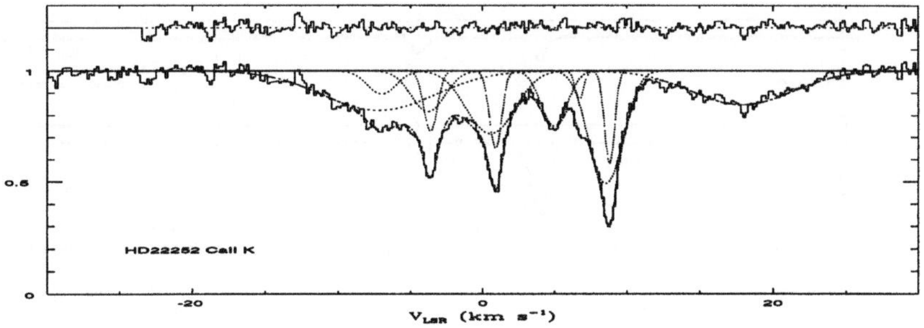

Fig. 2. AAT UHRF Ca II spectrum for HD 22252, showing over 10 separate components of absorption for a local bubble sightline.

be nT = 10^4 cm^{-3}K. Soft X-ray cloud shadowing for a portion of the filament has been detected in ROSAT PSPC data (Wang and Yu (1994)), and so the filament would be expected to be in contact with the hot gas within with the local bubble under the McKee-Ostriker model. Since it is likely that the filament has much smaller and denser structure than is visible within the *IRAS* data, the pressures of the filament could be significantly greater than the hot gas pressure within the local bubble.

If the pressure of the filament is higher than the hot gas pressure, the filament could be in transient dense state, and expanding into the local bubble. Alternatively, stability of the high pressure filament could imply a magnetically confined region of the interstellar medium. Further observations are being planned to resolve the small scale structure of the cloud, and to search for evidence of magnetic fields within the filament. For the upper range of our pressure estimates, the magnetic field required for confinement would be within the range of $B \simeq 70 - 140\,\mu$G. This strong magnetic field should be detectable through radio continuum observations.

2 Studies of a Local IVC Region

The second region we consider briefly is in the same quadrant of the galaxy, at galactic coordinates l=280°, b=11°. Adjacent to the region are two high velocity clouds, HVC279.5+10.7 and HVC279+13, which were observed to have HI velocities of +185 km/s. Since most HVCs appear to be moving towards the galaxy, these HVCs are somewhat anomalous. In an earlier paper (Penprase and Blades (1992)) it was found that the nearby star HD 93721 shows extremely complex optical absorption, with over 10 components of absorption between -65 < $v_{\rm LSR}$ < 75 km/s in the Ca II spectrum. While large deviation velocities from galactic rotation are expected in this direction, the detection of these features toward a star at d = 410 pc appears to imply highly unusual kinematics for this part of the interstellar medium.

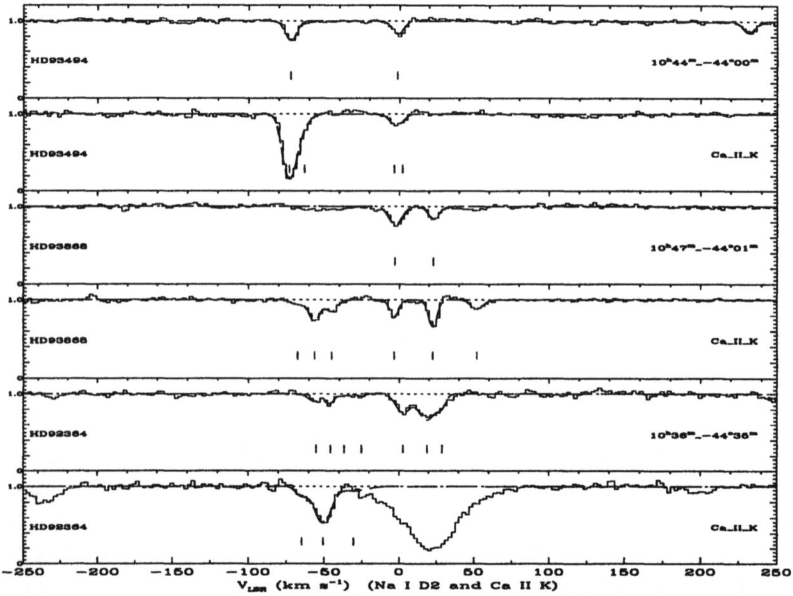

Fig. 3. Three pairs of Na I and Ca II absorption spectra for the stars HD 93494 (top), HD 93868 (middle), and HD 92364 (bottom), showing clear evidence of IVC components.

Figure 3 shows new high dispersion Na I and Ca II absorption spectra for three of five newly studied additional stars in the region. These spectra show high S/N detections of several additional IVCs, with $-85 < v_{\mathrm{LSR}} < 30$ km/s. The distances of the three stars in Figure 4 are (from top) 250, 260 and 350 pc, confirming that the IVCs originate from the local ISM. These spectra help preclude a circumstellar origin for these features, and give further evidence that these components result from a widespread disruption of the interstellar medium in the region.

References

Frisch, P. (1995): Space Science Rev. 72, 499.

McGee, R.X., Haynes, R.F., Grognard, J.M., and Malin, D. (1986): MNRAS 221, 543.

Penprase, B.E., and Blades, J.C. (1992): ApJ 391, 276.

Penprase, B.E., Lauer, J., Aufrecht, J., and Welsh, B. (1997): ApJ, *in press.*

Penprase, B.E., Blades, J.C., and Sahu, M. (1997): ApJ, *in preparation.*

Snowden, S.L. (1993): Advances in Space Research 13, 103.

Wang, Q.D., and Yu, (1994): AJ 109,698.

The Role of Magnetic Reconnection in the Interaction of High-Velocity Clouds and the Galactic Disk

Harald Lesch and Guido T. Birk

Institut für Astronomie und Astrophysik, LMU München, Scheinerstr. 1, 81679 München, Germany

Abstract. The boundaries of high-velocity-clouds that hit the Galactic plane are sources of strong X-ray emission indicating gas temperatures of a few million Kelvin. We show that numerous individual magnetic reconnection processes in the interaction region between an HVC and the ionized Reynolds layer may cause the gas heating via dynamical Ohmic dissipation. We present results of numerical plasma-neutral gas fluid simulations performed to study the dynamical reconnection process and to calculate the conversion of magnetic energy into heat.

1 Introduction

The partially ionized halo plasma is agitated by the activity in the disk (supernova remnants, stellar winds, bubbles and superbubbles, etc.) which results in Galactic fountains (cf. Kahn 1981) formed by hot plasma rising from the Galactic disk. At some height the plasma cools and may partially return to the disk in form of magnetized high-velocity clouds (HVCs) (Kahn 1991). Kerp et al. (1994, 1996), e.g., found clear evidence for the association of HVCs with strong X-ray emission indicating a plasma temperature of a few million degrees. In recent papers we argued that magnetic reconnection can play an important role in the heating of the edges of HVCs. Our analysis was based on the assumption of a fully ionized plasma (Zimmer et al. 1996, 1997). In this contribution we show that the reconnection process also works effectively in a more appropriate weakly ionized regime.

2 Numerical Model

We idealize the interaction of an HVC with the halo at a distance of about 1.5 kpc from the disk (cf. Kuntz and Dendy 1996) by considering two magnetized plasmas encountering with a velocity corresponding to the observed HVC-speeds of $100\,\mathrm{km\,s^{-1}}$ (for details see Zimmer et al. 1996). The first stage of the interaction is dominated by compression of the partially ionized plasma and the magnetic field perpendicular to the direction of motion of the cloud. Compression leads to a reversible increase of the plasma pressure and temperature in the boundary layer between the cloud and the disk. Due to internal motions inside the layer, the compressed field lines are stretched, twisted, strained and curled by plasma motions. Numerous current sheets will form in the whole interaction volume. The free energy stored in these current filaments can be converted mainly into heat by magnetic reconnection provided that the current densities grow supercritical thereby giving rise to anomalous dissipation (e.g. van Hoven 1985).

We model the impact of a magnetized partially ionized cloud onto the Galactic disk by numerically integrating the continuity, momentum, and energy fluid balance equations (where Ohmic dissipation is included; radiative losses prove to be unimportant) for the plasma and the neutral gas fluid as well as the induction equation (including convective and dissipative terms) that governs the dynamical evolution of the magnetic field (for details of the set of equations and the numerical procedure - an explicit difference scheme based on the Leapfrog algorithm - cf. Birk and Otto 1996). We concentrate on a single current filament to study one elementary reconnection process. The initial magnetic field is chosen as:

$$\mathbf{B} = B_0 \tanh\left(z - \frac{z_{\max} + z_{\min}}{2}\right) \mathbf{e}_y \tag{1}$$

A localized resistivity $\eta(j)$ is switched on when the current density, due to the compression caused by the impact, exceeds a critical value.

To model the inflowing plasma of the cloud the magnetic field and the velocities of the ionized ($\mathbf{v}$) and neutral ($\mathbf{v_n}$) components at $z = z_{\max}$ are chosen as:

$$\mathbf{B} = 0.9\, B_0\, \mathbf{e}_z \quad ; \quad \mathbf{v} = \mathbf{v}_n = \frac{-v_0}{\cosh(y/10)} \mathbf{e}_z \tag{2}$$

with $v_0 = 0.2$ and $B_0 = 1$ in normalized units. We choose for the plasma particle density, the magnetic field strength and the thickness of the current filament $n_{plasma} = 0.015\,\mathrm{cm}^{-3}$, $B = 17\mu\mathrm{G}$ and $L = 10^9\,\mathrm{cm}$.

3 Simulation Results

Magnetic flux as well as plasma and neutral gas momentum are transported into the 2D numerical domain with the dimensions $y \in [-15, 15]$ and $z \in [0, 40]$ in units of L. Fig. 1 shows the magnetic field configuration after 60 Alfvénic transit times τ_A. Later the initial current sheet becomes supercritical and the resistivity is switched on.The resulting magnetic field configuration after 90 Alfvénic transit times is illustrated in Fig. 2. The reconnection process, easily identified by the occurrence of an X-type configuration located at $y = 0$ and $z = 32$, works efficiently even in the considered case of a weakly ionized plasma ($n_{\mathrm{neutral}}/n_{\mathrm{plasma}} = 1000$). During the reconnection process magnetic energy is converted into heat via Ohmic dissipation (cf. Fig. 3). For the parameters chosen magnetic flux penetrates too quickly into the interaction region for the reconnection process to dissipate the newly incoming magnetic flux in order to reach a quasistationary state. Consequently, the

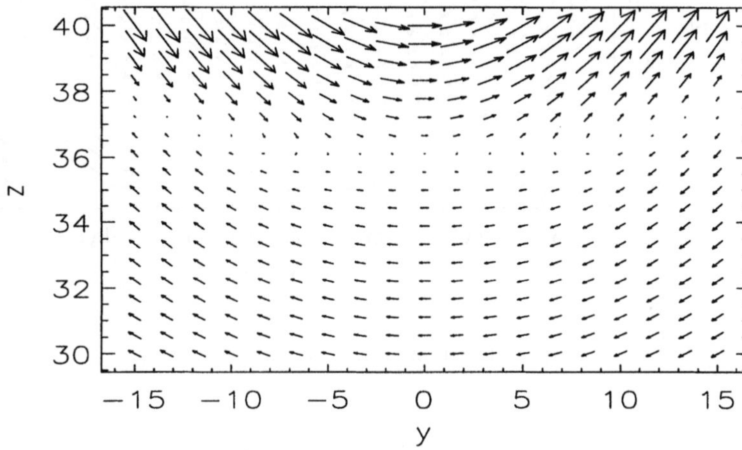

Fig. 1. The magnetic field after $t = 60\tau_A$.

overall magnetic energy also grows with time. In the proposed scenario as in the case of a fully ionized plasma regime (Zimmer et al. 1997) plasma should eventually be heated up to some 10^6 K during a huge number of individual reconnection processes operating successively at different locations which is in good agreement with observations as in HVC90.5+42.5-130 (Kerp et al. 1994).

Acknowledgements. This work was supported by the DFG through the grant ME 745/18-1.

References

Birk, G.T., Otto, A. (1996): J. Comp. Phys. 125, 513–525

Kahn, F.D. (ed.) (1981): *Dynamics of the Galactic Fountain* (Reidel, Dordrecht).

Kahn, F. (1991): *The Interstellar Disk-Halo Connection in Galaxies*, ed. H. Bloemen, IAU Symp. 144, 1–14

Kerp, J., Lesch, H., Mack, K.H. (1994): A&A 286, L13–16

Kerp, J., Mack, K.H., Egger, R. et al. (1996): A&A 312, 67–73

Kuntz, K.D., Dendy, L. (1996): ApJ 457, 703–717

van Hoven, G. (1985): *Unstable Current Systems and Plasma Instabilities in Astrophysics*, eds. M.R. Kundu and G.D. Holman, IAU Symp. 132, 263–271

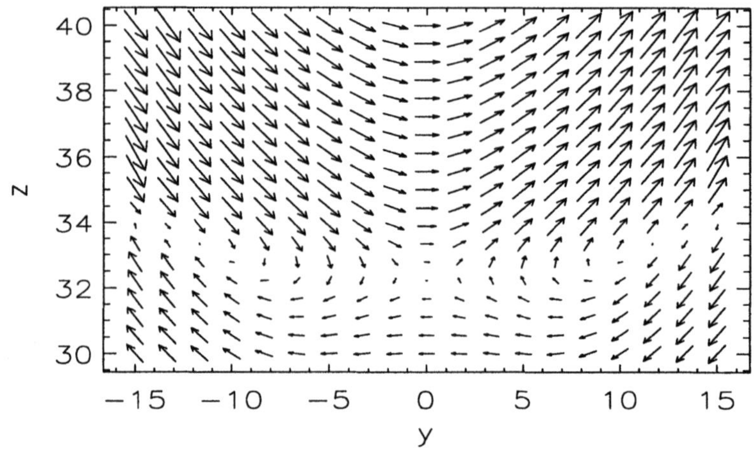

Fig. 2. The magnetic field after $t = 90\tau_A$.

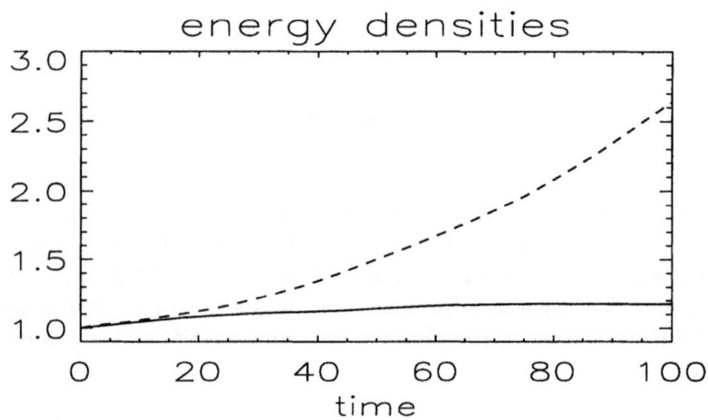

Fig. 3. The temporal evolution (normalized to the respective initial values) of the magnetic field energy density (solid line) and the thermal energy density (dashed line).

Zimmer, F., Birk, G.T., Kerp, J., Lesch, H. (1996): Astrophys. Let. Com. 34, 193–198

Zimmer, F., Lesch, H., Birk, G.T. (1997): A&A 320, 746–756

Part VIII

High-Velocity Clouds, Galactic Halo Models, Observations of the LMC

The Bavarian Evening ...

... a mixture of local culture and Spanish flair.

High-Velocity Clouds and their Soft X-ray Emission

J. Kerp[1], J. Pietz[1], P.M.W. Kalberla[1], W.B. Burton[2], R. Egger[3], M.J. Freyberg[3], Dap Hartmann[4], and U. Mebold[1]

[1] Radioastronomisches Institut der Universität Bonn, Auf dem Hügel 71, D-53121 Bonn, Germany
[2] Sterrewacht Leiden, P.O. Box 9513, NL-2300 RA Leiden, The Netherlands
[3] Max-Planck-Institut für extraterrestrische Physik, Postfach 1603, D-85740 Garching, Germany
[4] Harvard-Smithsonian Center for Astrophysics, 60 Garden Street, Cambridge, MA 02138, U.S.A.

Abstract. Diffuse excess 1/4 keV soft X-ray emission was found to be positionally correlated with the column density distribution of the high velocity cloud (HVC) complex C (Kerp et al. 1996). Here we point out that the detected diffuse X-ray emission is indeed associated with the HVC phenomenon. For this purpose we study the 1/4 keV radiation transfer as well as the H I column density distribution of HVCs and intermediate velocity clouds (IVCs) towards HVC complex C in detail. We present evidence that on arcmin scales the 3/4 keV soft X-ray emission is positionally anticorrelated with the HVC column density distribution of an individual HVC filament of complex C.

1 Introduction

Up to now HVCs have been detected, in emission, only in the line radiation of the neutral atomic hydrogen (for a comprehensive review see Wakker & van Woerden 1997). Recent analyses of *ROSAT* data suggest that HVCs also emit soft X-rays (Herbstmeier et al. 1995 and Kerp et al. 1994, 1995, 1996). Here we focus on HVC complex C towards high galactic latitudes. We present a comparative analysis of X-ray and H I 21 cm line data available in this region ($73° \leq l \leq 118°$, $38° \leq b \leq 63°$).
In the 1/4 keV energy regime one has to disentangle the X-ray emission of the individual soft X-ray sources which contribute to the observed X-ray background in order to reveal the excess emission associated with the HVCs. Especially the intermediate- and low-velocity gas determines the intensity distribution of the soft X-ray sky, by photoelectric absorption. Because of the much smaller absorption cross section in the 3/4 keV band such a decomposition is not necessary. We will demonstrate that it is feasible to separate the emission of the "normal" soft X-ray background (SXRB) from that of the excess emission associated with HVCs. In consequence we have to solve the radiation transport of soft X-ray photons through the galactic interstellar

medium (ISM) to identify sky areas which reveal deviations from this "normal" SXRB intensity distribution.
Here we focus on two aspects which are important to understanding the recent results on X-ray emission of HVCs. First, we will show that the *ROSAT* 1/4 keV all-sky survey data reveal soft X-ray emission which is positionally correlated with the HVCs of complex C. We show new H I 21 cm line maps of the total and IVC column density distribution and discuss the relation of the individual cloud populations to the appearance of the soft X-ray intensity distribution on the sky. Second, we will discuss in Sect. 3 the small scale positional association of an individual HVC filament with excess soft X-ray emission in the *ROSAT* 3/4 keV energy band. This is of importance because the 3/4 keV radiation can penetrate the galactic ISM without significant attenuation towards high galactic latitudes.

2 The Soft X-ray Background

It is a matter of discussion whether the SXRB has a patchy source intensity distribution or not (see also Pietz et al., Snowden, and Wang; this volume). For our purpose knowledge of the soft X-ray intensity distribution on large angular scales is vital to disclose "deviating" X-ray emission close to individual clouds.
The *ROSAT* mission established the existence of three main SXRB source components. The first one is the well known local X-ray radiation originating within the local void of neutral matter. Its radiation dominates the emission within the *ROSAT* 1/4 keV band, while it is nearly undetectable in the *ROSAT* 3/4 keV and 1.5 keV energy bands (Snowden et al. 1993).
Also well known is the radiation of the extragalactic X-ray background. The diffusely distributed extragalactic X-ray emission is the superposition of the radiation of individual, unresolved X-ray sources (Hasinger et al. 1993, Gendreau et al. 1995). The extragalactic X-ray background dominates the emission within the *ROSAT* 1.5 keV band and accounts for a large fraction of the observed 3/4 keV radiation but contributes only a few percent to the observed total 1/4 keV radiation.
Of greatest importance for the present paper is the intensity distribution of the third diffuse X-ray source, the Galactic X-ray halo. During the *ROSAT* mission it became obvious that a significant fraction of soft X-rays originate within the Galactic halo. Here we show that, consistent with the approach of Pietz et al. and Wang (both, this volume), the observational 1/4 keV and 3/4 keV X-ray data are in quantitative agreement with a *smooth* Galactic X-ray halo intensity distribution across the entire Galactic sky. This smooth Galactic X-ray halo intensity distribution together with the isotropic extragalactic background radiation gives us the opportunity to study the 1/4 keV X-ray properties of an entire HVC complex towards the high-latitude sky.

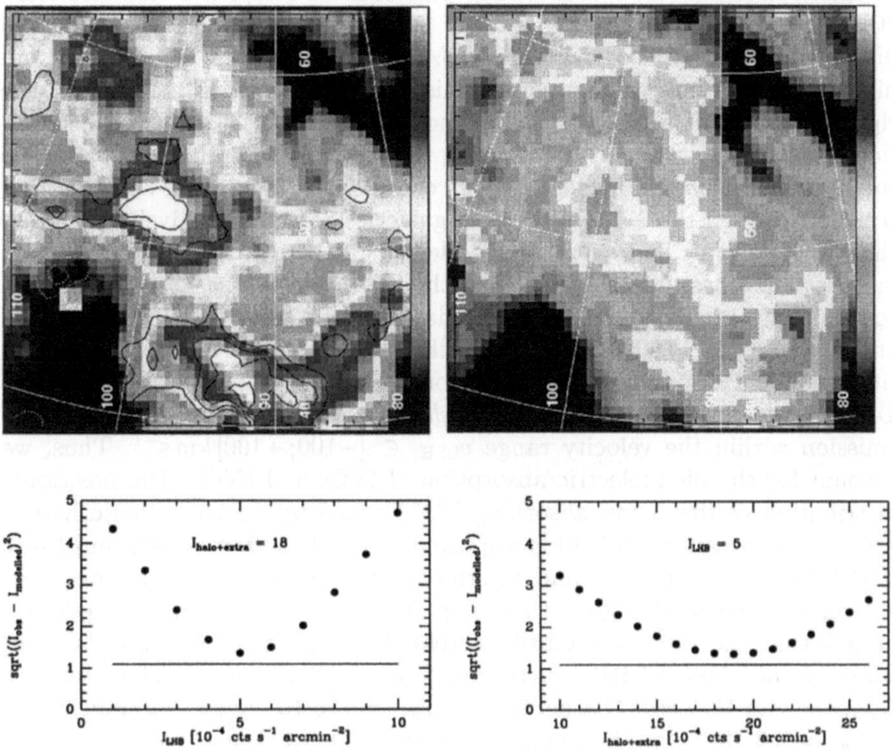

Fig. 1. *Top left:* Map extracted from the *ROSAT* 1/4 keV all-sky survey (Snowden et al. 1997) with an angular resolution of 48′. White denotes high X-ray intensities while black indicates weak X-ray emitting areas. The 1/4 keV minimum count rate displayed is $I_{1/4\,\mathrm{keV}}(\mathrm{min}) = 7 \cdot 10^{-6}\,\mathrm{cts\,s^{-1}\,arcmin^{-2}}$, the maximum displayed count rate is $I_{1/4\,\mathrm{keV}}(\mathrm{max}) = 18 \cdot 10^{-6}\,\mathrm{cts\,s^{-1}\,arcmin^{-2}}$. Superimposed as contour lines are the areas of excess emission (solid contours) and too weak X-ray emission (dashed contours). The contour levels start at 4σ and increase in steps of 2σ. *Top right:* Modelled 1/4 keV SXRB intensity distribution towards the same part of the galactic sky as shown *left*. The intensity scale is identical with that of the *left* map. To model this map we assumed $I_{\mathrm{local}} = 4.4 \cdot 10^{-6}\,\mathrm{cts\,s^{-1}\,arcmin^{-2}}$, $I_{\mathrm{halo}} = 16 \cdot 10^{-6}\,\mathrm{cts\,s^{-1}\,arcmin^{-2}}$, and $I_{\mathrm{extra}} = 2.3 \cdot 10^{-6}\,\mathrm{cts\,s^{-1}\,arcmin^{-2}}$. These SXRB source intensity values are assumed to be constant across the *entire* field of interest.
Bottom: To evaluate the statistical significance of our modelling of the 1/4 keV X-ray intensity distribution we calculated the field averaged standard deviation between the observed and modelled intensity distribution as a function of I_{local} and $I_{\mathrm{halo+extra}}$. The line marks the field averaged standard deviation of the *ROSAT* all-sky survey data.

Figure 1 presents a comparison between the observed and modelled SXRB intensity distribution towards a large part of HVC complex C (see also Kerp et al. 1996). On the left-hand side of Fig. 1 the *ROSAT* 1/4 keV image is displayed. Superimposed on the *ROSAT* grey scale map are the 4-σ and higher contour lines of the excess emission areas. For comparison, on the right-hand side of Fig. 1 we display the modelled SXRB intensity distribution towards the same sky area. Using the Leiden/Dwingeloo H I 21 cm line survey (Hartmann & Burton 1997) we can evaluate the soft X-ray radiation transfer through the X-ray absorbing galactic ISM quantitatively traced by the neutral atomic hydrogen. The modelled soft X-ray intensity distribution matches the observed one well with the exception of the areas where the contour lines are located. The X-ray dark regions (i.e. $l \sim 105°$, $b \sim 43°$, and $l \sim 88°$, $b \sim 60°$) are positionally associated with high column density areas of IVCs and low velocity clouds (LVCs, compare also Fig. 3 *left*). To calculate an image like Fig. 1 (*right*) we integrated the H I 21 cm line emission within the velocity range $v_{\rm LSR} \in [-100; +100]\,{\rm km\,s^{-1}}$. Thus, we account for the photoelectric absorption of IVCs and LVCs. The positional distribution of the X-ray absorbing ISM, traced by neutral atomic hydrogen, across the analyzed field is displayed in Fig. 1 (*bottom panel*) and Fig. 2 (*left-hand side*). The low and intermediate velocity gas column densities are represented in this grey scale map. We used the extremely simple approach that the 1/4 keV intensity across the entire field of interest is constant for *all* three SXRB source components. The derived intensities are $I_{\rm local} = 4.4{\cdot}10^{-6}\,{\rm cts\,s^{-1}\,arcmin^{-2}}$, $I_{\rm halo} = 16{\cdot}10^{-6}\,{\rm cts\,s^{-1}\,arcmin^{-2}}$ and $I_{\rm extra} = 2.3{\cdot}10^{-6}\,{\rm cts\,s^{-1}\,arcmin^{-2}}$ (Barber et al. 1996) and constant across the field of interest. The Galactic X-ray halo is the brightest X-ray source of all. That this simple assumption fits the observation well is demonstrated in Fig. 2 (*right*); here we plotted the 1/4 keV SXRB intensity distribution profiles averaged in galactic longitude and/or latitude. The dots with the corresponding 1 σ errorbars represent the observational X-ray data while the solid line, in each panel, marks the modelled SXRB intensity profile. These averaged intensity profiles indicate that our approach of constant intensities of the three source components fits the observed situation very well. We subtracted the modelled map (Fig. 1 *right*) from the observed SXRB intensity distribution (Fig. 1 *left*) to search for excess X-ray emission areas and evaluated the residuals in units of the statistical significance of the X-ray data. The contour lines start at a significance level of 4 σ and increase in steps of 2 σ. In absolute values 4 σ is equivalent to a count rate of $I_{1/4\,\rm keV}(4\sigma) \simeq 4 \cdot 10^{-4}\,{\rm cts\,s^{-1}\,arcmin^{-2}}$. With respect to an averaged observed X-ray intensity of $I_{1/4\,\rm keV}({\rm mean}) \simeq 12 \cdot 10^{-4}\,{\rm cts\,s^{-1}\,arcmin^{-2}}$ the brightest excess emission area at $l \sim 90°$, $b \sim 42°$ is two times brighter than the "normal" SXRB intensity.

We note that the H I 21 cm line emission includes the IVCs and LVCs as absorbers of soft X-ray photons but *not* the HVCs. We *exclude* the HVC as soft

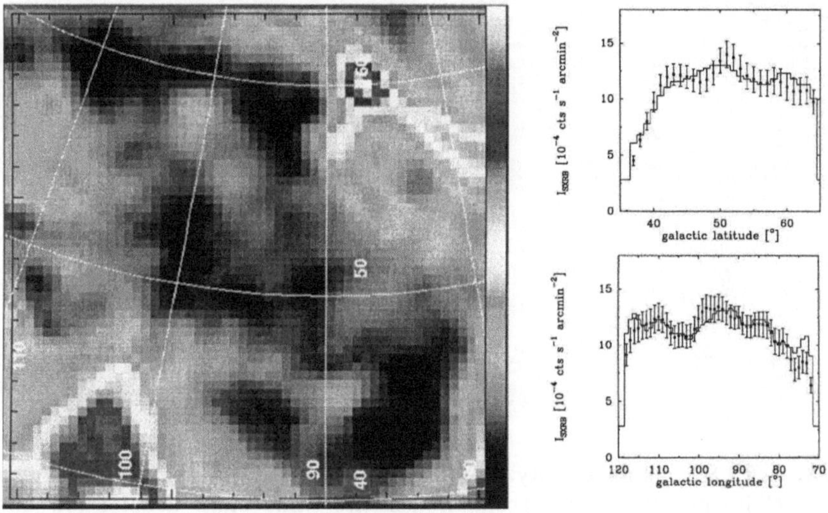

Fig. 2. *Left:* Neutral atomic hydrogen column density distribution derived from the Leiden/Dwingeloo H I 21 cm line survey (Hartmann & Burton 1997). The intensity scale is the same as in Fig. 1 with $N_{\rm HI}(\rm min) = 0.5{\cdot}10^{20}\,\rm cm^{-2}$ and $N_{\rm HI}(\rm max) = 2.0{\cdot}10^{20}\,\rm cm^{-2}$ The brightness temperature is integrated across the velocity interval of $v_{\rm LSR} \in [-100; +100]\,\rm km\,s^{-1}$. This map traces the distribution of the X-ray absorbing ISM. *Right:* Comparison of SXRB intensity profiles versus galactic coordinates averaged in galactic longitude and/or latitude. The dots with the corresponding 1 σ errorbars mark the observational data while the solid line in each panel marks the modelled SXRB intensity distribution (compare also Fig. 1 *right*). Obviously both intensity profiles match the observational data well and consequently the complex intensity pattern of the SXRB is entirely caused by photoelectric absorption because all X-ray source terms have a *constant* intensity across the field of interest.

X-ray absorbers; thus, at the position of the HVCs, our modelled SXRB map predicts a higher modelled X-ray intensity than we would obtain if the HVCs attenuate the Galactic X-ray halo emission also. Thus, our bias is towards the *non-detection* of HVC excess emission.

In Fig. 3 we compare the positional correlation of the column density distribution of the IVCs ($v_{\rm LSR} \in [-90; -25]\,\rm km\,s^{-1}$, Fig. 3 *left*) and the HVCs ($v_{\rm LSR} \in [-450; -90]\,\rm km\,s^{-1}$, Fig. 3 *right*) with the X-ray excess regions superimposed as contours on the corresponding grey scale maps. Obviously, the IVCs are not close positionally correlated to the excess X-ray emission regions, while the HVCs are nearby these areas of enhanced X-ray radiation. Especially, the area of $l \sim 90°$ and $b \sim 42°$ reveals positional correlation of the HVCs with excess X-ray emission. In Sect. 3 we will study this region using a pointed *ROSAT* observation and show that this area is also bright in 3/4 keV X-ray radiation. Because of the very different absorption cross sec-

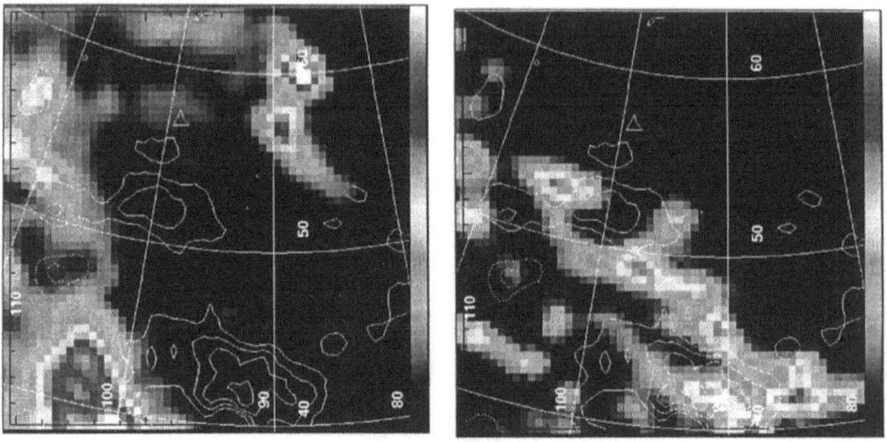

Fig. 3. *Left:* H I column density distribution of the IVCs across the field of interest. White denotes high H I column densities while black denotes low ones, with $N_{\rm HI}(\min) = 0.2 \cdot 10^{20}\,{\rm cm}^{-2}$ and $N_{\rm HI}(\max) = 2.0 \cdot 10^{20}\,{\rm cm}^{-2}$. The H I 21 cm line emission is integrated across the velocity interval $v_{\rm LSR} \in [-90; -25]\,{\rm km\,s}^{-1}$. The contour lines mark the 1/4 keV excess emission (solid lines) starting at the 4-σ level with steps of 2σ. *Right:* The HVC column density distribution with $N_{\rm HI}(\min) = 0.1 \cdot 10^{20}\,{\rm cm}^{-2}$ and $N_{\rm HI}(\max) = 1.0 \cdot 10^{20}\,{\rm cm}^{-2}$. The column density is derived within the velocity interval $v_{\rm LSR} \in [-450; -90]\,{\rm km\,s}^{-1}$; (see Plate 2).

tion of both energy bands this positional correlation demonstrates the quality of the performed radiation transfer calculation.
In conclusion, the close positional correlation between the HVCs of complex C and the excess X-ray emission region of the 1/4 keV *ROSAT* all-sky survey suggests that the HVCs are physically related to this soft X-ray enhancements.

3 The X-ray Emission of an Individual HVC Filament

The previous Section pointed out that we find a large scale positional correlation between 1/4 keV X-ray emission and the H I column density distribution of HVC complex C. Now we will demonstrate that HVCs reveal a detailed positional anticorrelation of H I column density distribution and 3/4 keV emission. This is of importance because the photoelectric absorption cross section of the galactic ISM is a factor of 10 lower in the 3/4 keV band than in the 1/4 keV range. For instance, the effective photoelectric 3/4 keV absorption cross section is $\sigma(N_{\rm HI} = 1.5 \cdot 10^{20}\,{\rm cm}^{-2}) = 9 \cdot 10^{-22}\,{\rm cm}^2$. This H I column density which is typically for the *warm neutral medium* (WNM) towards the northern galactic sky. It attenuates the distant 3/4 keV X-ray emission by only 13%. Here we study an HVC filament belonging to the HVC complex C

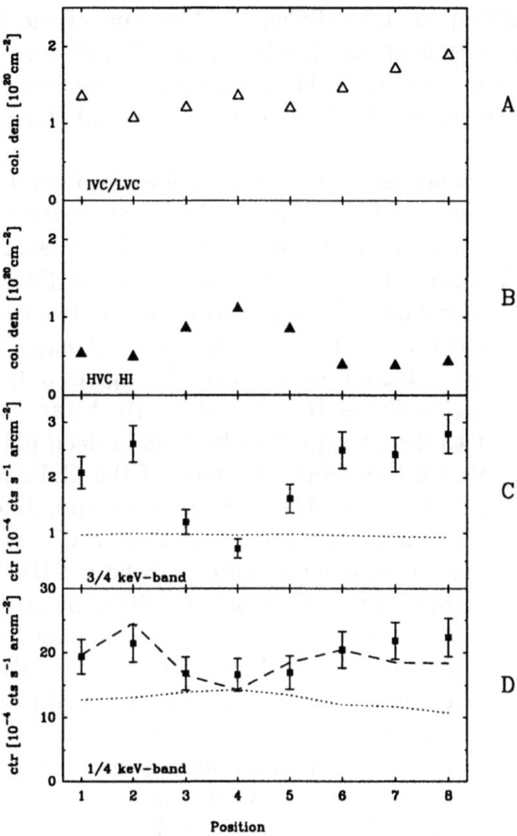

Fig. 4. Diagram showing the functional dependence of IVC/LVC and HVC column density as well as the *ROSAT* 3/4 keV and 1/4 keV count rate on positions along a linear slice across HVC 90.5+42.5–130 (see Fig. 1 of Kerp et al. 1994). The "normal" 3/4 keV and 1/4 keV SXRB intensity level is marked by the dotted line in the corresponding panel. In panel D (1/4 keV emission) the dashed line represents the modelled count rate applying the working hypothesis that the 3/4 keV excess plasma radiation produces also 1/4 keV emission. This constrains also the plasma temperature responsible for the HVC excess X-ray radiation to about $T \simeq 10^{6.2-6.3}$ K. A detailed discussion of this figure can be found in the text below.

and denoted as HVC 90.5+42.5–130 (Kerp et al. 1994, 1995). This filament has also an H I column density of $N_{\rm HI} = 1.2 \cdot 10^{20}\,{\rm cm}^{-2}$. Figure 4 shows a linear slice of 2° extent across the HVC 90.5+42.5–130 (compare also Fig. 1 of Kerp et al. 1994). In panel A the IVC/LVC column density is plotted as function of the position along the linear slice. This gas traces quantitatively the photoelectric absorption as discussed in Sect. 2. Panel B displays the HVC column density distribution along the linear slice. The peak column density is about $N_{HI}(HVC) \simeq 1.2 \cdot 10^{20} cm^{-2}$. The sum of the IVC/LVC and the HVC column density is about $N_{HI}(total) \simeq 2.7 \cdot 10^{20} cm^{-2}$ at the maximum. Panel C shows the *ROSAT* 3/4 keV intensity along this slice. A strong anticorrelation between the HVC column density and the 3/4 keV emission is detected.

Next we evaluate whether or not this obvious anticorrelation is caused by photoelectric absorption. If we sum the X-ray absorption of the WNM and the HVC filament we expect to detect a ~ 26% intensity variation in the *ROSAT* 3/4 keV data. The HVC may be so distant (Wakker & van Woerden 1997) that its column density distribution only attenuates the extragalactic background radiation with an intensity of $I_{extra}(3/4\,keV) \simeq 0.53 \cdot 10^{-4}\,cts\,s^{-1}\,arcmin^{-2}$ (Barber et al. 1996). This gives a 3/4 keV intensity attenuation of $\Delta I_{extra}(WNM + HVC) = 0.10 \cdot 10^{-4}\,cts\,s^{-1}\,arcmin^{-2}$ which is extremely difficult to detect significantly, even in deep pointed PSPC observations. If the HVCs are in front of a part of the Galactic X-ray halo, the depth of the 3/4 keV X-ray shadow increases correspondingly to the increasing amount of X-ray emitting plasma beyond the HVC. In the extreme case, all Galactic X-ray halo emission originates beyond the HVCs, the 3/4 keV radiation will be absorbed by the WNM and the HVC by $\Delta I_{extra+halo}(WNM + HVC) = 0.44 \cdot 10^{-4}\,cts\,s^{-1}\,arcmin^{-2}$ while the expected 3/4 keV intensity variation between ON and OFF the HVC is only about $\Delta I_{OFF-ON} = 0.24 \cdot 10^{-4}\,cts\,s^{-1}\,arcmin^{-2}$. But in any case the WNM absorbs the distant SXRB sources. In consequence, we do not expect to detect any significant 3/4 keV intensity variation towards the high galactic latitude sky.

But we observe three times more 3/4 keV emission close to the HVCs as radiated from the "normal" 3/4 keV X-ray background. The expected 3/4 keV intensity variation, caused by photoelectric absorption of the Galactic X-ray halo and the extragalactic background radiation by the total H I column density along the line of sight, is marked by the dotted line in panel C. The deviation between the "normal" 3/4 keV SXRB emission and the observational data yields the conclusion that we detected *excess* 3/4 keV emission. The excess 3/4 keV intensity is about a factor of two brighter than the "normal" 3/4 keV SXRB intensity variation, and a factor of about 10 larger than the expected intensity contrast even in the extreme case that the HVC is located in front of the entire Galactic X-ray halo emission.

Panel D shows the 1/4 keV intensity distribution along the linear slice. We see a similar intensity variation with position as detected in the 3/4 keV band. Here we have to account for the significant photoelectric absorption quantitatively traced by the IVC/LVC column density distribution, plotted in panel A. Again, the dotted line represents the "normal" 1/4 keV SXRB intensity distribution derived with the identical intensity values as used in Sect. 2. The low H I column density contrast along the analyzed slice yields also a low 1/4 keV intensity contrast. If we assume that the HVC rims are X-ray bright and transform quantitatively the 3/4 keV excess X-ray radiation to

the 1/4 keV energy regime, we predict a 1/4 keV intensity variation which is marked by the dashed line. This dashed line represents the observed 1/4 keV intensity variation well, suggesting that also the excess 1/4 keV emission is caused by the *same* plasma as the 3/4 keV excess emission close to the HVC. To perform this transformation $I_{3/4\,\mathrm{keV}} \rightarrow I_{1/4\,\mathrm{keV}}$ we subtracted the "normal" 3/4 keV intensity level of $I_{3/4\,\mathrm{keV}} \simeq 1.0 \cdot 10^{-4}\,\mathrm{cts\,s^{-1}\,arcmin^{-2}}$ from the individual data points along the slice and transformed the excess intensity to the *ROSAT* 1/4 keV band response. This excess 1/4 keV emission is attenuated by the IVC/LVC column density distribution shown in the lowest panel. Because the unabsorbed $I_{3/4\,\mathrm{keV}} : I_{1/4\,\mathrm{keV}}$ ratio is a very sensitive function of the plasma temperature, it varies between $I_{3/4\,\mathrm{keV}} : I_{1/4\,\mathrm{keV}} \Leftrightarrow 1 : 90$ for $T = 10^{6.0}\,\mathrm{K}$ and $I_{3/4\,\mathrm{keV}} : I_{1/4\,\mathrm{keV}} \Leftrightarrow 1 : 7$ for $T = 10^{6.3}\,\mathrm{K}$, we can constrain the plasma temperature range. We find a plasma temperature of about $T \simeq 10^{6.2-6.3}\,\mathrm{K}$ for the excess X-ray emission.
In summary, we find a detailed positional anticorrelation of 3/4 keV X-ray radiation with the H I column density distribution of HVC 90.5+42.5-130. This detailed negative correlation *cannot* be caused by photoelectric absorption, because the involved column densities are more than an order of magnitude too low for a significant attenuation of the 3/4 keV radiation. In absolute numbers, the observed 3/4 keV radiation is a factor of two brighter than the "normal" 3/4 keV SXRB radiation, and the observed 3/4 keV intensity contrast is a factor of 10 higher than expected by photoelectric absorption. These are clear indications that the detailed positional anticorrelation between the HVC column density distribution and the 3/4 keV radiation is caused by *excess* X-ray emission close to the HVC. Also in the *ROSAT* 1/4 keV energy band we find excess emission. This excess 1/4 keV emission can be quantitatively evaluated from the excess 3/4 keV emission using a plasma temperature of $T \simeq 10^{6.2-6.3}\,\mathrm{K}$ close to the HVCs.

4 Summary and Conclusion

The aim of the paper was to demonstrate that HVCs are emitters of soft X-rays. In Sect. 2 we analyzed the 1/4 keV SXRB intensity distribution towards the major part of HVC complex C. We showed that the 1/4 keV SXRB intensity distribution can be modelled well by a simple soft X-ray radiation transport approach, which assumes that across the entire field of interest the local, the Galactic X-ray halo, and the extragalactic component of the 1/4 keV radiation are constant. The success of modelling the observed SXRB intensity pattern indicates that the complex SXRB brightness distribution is due to photoelectric absorption of the galactic ISM, which is quantitatively traced by the neutral atomic hydrogen. The areas of excess emission are located close to the peaks of the HVC column density distribution.
In Sect. 3 we presented evidence that on arcmin angular scales the H I column

density distribution of a HVC is positionally anticorrelated to the 3/4 keV emission. We showed that this emission can only be interpreted in terms of excess 3/4 keV emission close to the rims of the HVC. Assuming that plasma emission is the source of the soft X-ray radiation we find that we can model also the observed 1/4 keV excess emission using a plasma temperature of $T \simeq 10^{6.2-6.3}$ K.
Using the plasma temperature and the 3/4 keV excess emission intensity we derive an emission measure of $EM_{\rm HVC} \simeq 0.015\,{\rm cm}^{-6}\,{\rm pc}$. Assuming a distance to HVC 90.5+42.5-130 of roughly 4 kpc, we find an electron density of $n_e \simeq 0.02\,{\rm cm}^{-3}$. This is roughly a factor of ten larger than the Galactic halo density expected at this distance from the Galactic plane (Kalberla et al., this volume). This indicates that the X-ray emission close to the HVCs originates not from the heated neighbourhood of the HVC. The HVC matter itself is heated up.
Kerp et al. (1997) show that soft X-ray enhancements in the 1/4 keV and 3/4 keV energy regime are not only associated with the here presented HVC complex C. They found excess emission towards the HVC complexes D, GCN, A, and the Magellanic Stream. Moreover, they verified that the entire HVC complex C is associated with soft X-ray emission.
These findings are independently confirmed by Pietz et al. (1997) while focussing their view onto the intensity distribution of the 3/4 keV and 1/4 keV SXRB. Their difference maps, observed minus modelled SXRB intensity distribution, reveal excess emission just at the position of the HVC complexes mentioned above.

References

Barber C.R., Roberts T.P., & Warwick R.S. (1996), MNRAS 282, 157
Gendreau K.C. et al. (1995): PASJ 47, 5
Hartmann Dap, & Burton W.B. (1997): "Atlas of Galactic Neutral Hydrogen" Cambridge University Press
Hasinger G. et al. (1993): A&A 275, 1
Herbstmeier et al. (1995): A&A 298, 606
Kerp J. et al. (1997): A&A submitted
Kerp J., Lesch H., & Mack K.-H. (1994): A&A 286, L13
Kerp J., Lesch H., Mack K.-H., & Pietz J. (1995): Adv. in Space Res. Vol. 16, No. 3, 119
Kerp J. et al. (1996): A&A 312, 67
Kuntz K., & Danly L. (1996): ApJ 457, 703
Pietz J. et al. (1997): A&A, submitted
Snowden S.L., McCammon D., & Verter F. (1993): ApJ 409, L21
Snowden S.L. et al. (1997): ApJ 485, in press
Wakker B.P., & van Woerden H. (1997): ARA&A, in press

The High-Velocity Clouds: Galactic or Extragalactic?

Hugo van Woerden[1], Bart P. Wakker[2], Ulrich J. Schwarz[1], Reynier F. Peletier[1], and Peter M.W. Kalberla[3]

[1] Kapteyn Institute, Postbus 800, 9700 AV Groningen, The Netherlands
[2] Department of Astronomy, University of Wisconsin, Madison WI, USA
[3] Radioastronomisches Institut der Universität, Bonn, Germany

Abstract. We present firm evidence that one of the major high velocity clouds (HVCs), Complex A, lies in the Milky Way Halo, at a vertical distance $z = 3$ - 7 kpc from the Galactic plane. For clouds MII/MIII, Danly et al. and Keenan et al. had already found $z < 5$ kpc. We further report that the metallicity in the largest HVC, Complex C, is at least 0.1 solar. CaII/HI ratios in 6 HVCs, ranging from 0.002 to 0.07 times solar, set lower limits to their metallicities.

Blitz et al. have recently suggested that most of the HVCs are relatively unprocessed, extragalactic remnants of the gas which formed the Local Group of galaxies. However, the results mentioned above indicate that several major HVC complexes are neither primordial nor extragalactic. For the smaller HVCs, some of which have much higher velocities, a location in the Local Group remains a possibility.

1 Introduction

High velocity clouds (HVCs) are HI clouds with velocities inconsistent with simple differential galactic rotation. Their origins remain unclear; major candidates include a Galactic Fountain and accretion of intergalactic gas (see review by Wakker & Van Woerden 1997).

Blitz et al. (1997) claim the HVCs are 'remnants of Local Group formation', and 'most plausibly explained as members of the Local Group of galaxies'. While no details of their work have been published, the main arguments of Blitz et al. appear to be the following. 1) The velocity centroid of the HVC ensemble is the same as that of the Local Group. 2) If the HVCs are gravitationally bound and tidally stable, their distances must be similar to those of Local Group galaxies. 3) The angular sizes and LSR velocities of HVCs appear to be inversely correlated. These arguments are all of a statistical nature and hence, even if applicable in general, carry no weight for individual HVCs. Moreover, the conditions for the second argument may not be fulfilled. What is required is direct distance determinations for individual HVCs – a program we have pursued over the last several years.

The only direct method available is to search for interstellar absorption at the HVC's velocity in the spectra of stars at different distances. Schwarz et al. (1995) have discussed the method, and its pitfalls, in detail. In view of the uncertain abundance of metal ions in HVCs, and their strong small scale structure (Wakker & Schwarz 1991), we require a combination of high resolution optical (or ultraviolet) spectra of both stellar and extragalactic probes with HI column density measurements at high angular resolution.

The CaII/HI abundance ratios measured for 6 HVCs (Wakker & Van Woerden 1997) range between 0.002 and 0.07 times the *total* solar Calcium abundance. Given the unknown degree of ionization of Calcium, its depletion onto interstellar grains, and the possible destruction of grains in interstellar shocks, these CaII abundances give loose lower limits only to the total Calcium abundances in HVCs. Abundances of various other metal ions indicate metallicities of order 0.1 in several HVCs (Wakker & Van Woerden 1997).

2 Distance Limits for Three Major HVC Complexes

For *Complex M*, Danly et al. (1993) and Keenan et al. (1995) found a distance d <5 kpc (z <4.4 kpc above the plane) from absorptions in a stellar probe spectrum, and abundances > 0.15 solar for SiII and 0.02 solar for CaII. (Note that this distance limit applies to clouds MII/MIII only. Cloud MI has a different velocity, is not connected, and may be at a different distance.)

For *Complex H*, Wakker et al. (1997) have analyzed IUE spectra of OB stars at distances 1.3 – 5.0 kpc. The lack of MgII, CII and OI absorptions shows that Complex H has $d > 3.5$ kpc, and probably $d > 5$ kpc. (Because of its low latitude, the limits on z are insignificant.) Since no extragalactic probe spectrum was available, no abundances could be derived; however, if Complex H were not beyond the stellar probes. the non-detections of absorption would imply extremely low abundances (factors 10^2 – 10^5 below solar).

For the largest HVC, *Complex C*, two extragalactic probes have given CaII abundances of 0.008 and 0.013 times solar (Wakker et al. 1996b). Recent HST spectra of Mark 290 by Wakker suggest an SII abundance of order 0.10 solar. Several UV and CaII studies of stars at distances d up to 2.5 kpc have yielded $z > 1.7$ kpc (Wakker & Van Woerden 1997). Recent observations with WIYN by Wakker and with the WHT by Van Woerden & Peletier of stars at distances d up to 5 – 8 kpc even indicate $z > 3 - 4$ kpc, implying an HI mass exceeding 10^6 $M_\odot$ for this huge complex.

3 Complex A: the First Firm Distance Bracket

Complex A, the first HVC discovered, has often served as prototype for HVCs. HST spectra of the Seyfert galaxy Mark 106 show strong MgII absorption (Wakker et al. 1996a), indicating a MgII abundance > 0.03 solar. The lack of high velocity MgII absorption in a neighbouring star, at distance d =4 kpc, implies $z > 2.5$ kpc for the distance of this HVC above the plane, unless its MgII abundance would vary by more than a factor 20 over a few degrees.

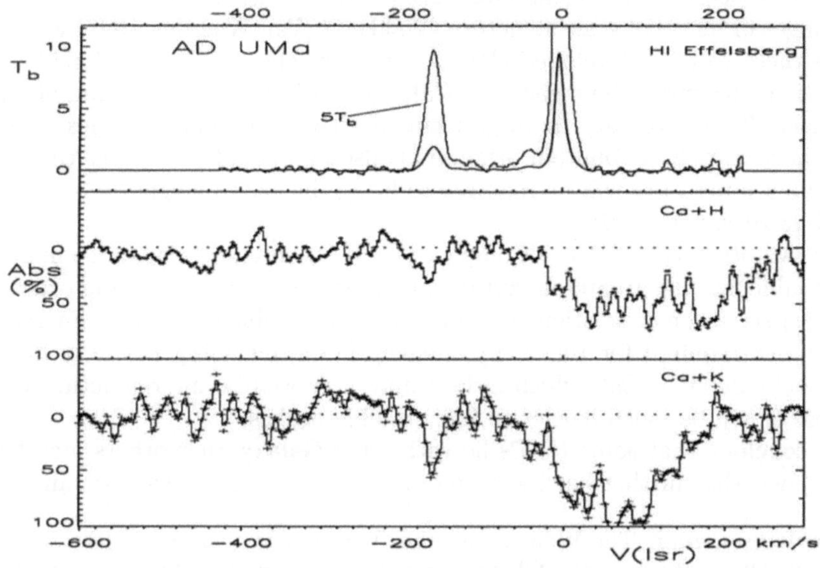

Fig. 1. HI emission and CaII absorption measured towards the RR Lyr variable AD UMa, which lies at 10 kpc distance, projected on HVC Complex A (Van Woerden et al., in preparation). Velocities are given relative to the local standard of rest. The Effelsberg 21 cm profile, shown on two intensity scales, different by a factor 5, has three components; that at – 160 km/s is due to the HVC. The CaII H and K lines, measured with the William Herschel Telescope at La Palma (10 km/s resolution) both show, in addition to a broad stellar component, and H-epsilon near the CaII H-line, the same three interstellar components as the 21 cm line. This proves that HVC Complex A, as well as the other two clouds, lies in front of the star.

Recent spectra of the RR Lyr star AD UMa (distance $d = 10$ kpc), which lies very close to Mark 106, display strong absorption by the HVC in both CaII lines (Fig. 1). Together with the previous result, this sets a firm bracket of $2.5 < z < 7$ kpc on the distance of Complex A from the plane. The corresponding HI mass lies between 1.6 and 10 x 10^5 $M_\odot$.

4 Conclusions and Prospects

Two major HVC complexes, A and M, have now been firmly found to lie within our Galaxy. Both these and several other HVCs have significant metal abundances, showing that they do not consist of primordial gas.

What does the location of Complex A, high in the Galactic halo, tell us about its origin? Its pronounced small scale structure (Wakker & Schwarz 1991) may be consistent with formation through instabilities in a Galactic Fountain. On the other hand, if Complex A originated in accretion of intergalactic gas, possibly of Magellanic debris, the time scales of 10^7 years derived for changes in its small scale structure (Wakker & Van Woerden 1997) would require these to have developed close to their present location, probably by interaction with high-z halo gas. The CaII and MgII abundances found allow a low-metallicity intergalactic origin; but in view of the uncertainties mentioned in Section 2, a Fountain origin may also be possible. In either case, a high halo temperature will be required to bring the gas to heights $z > 3$ kpc (Houck & Bregman 1990).

For two other large HVCs, Complex C and H, only lower distance limits are available, leaving greater uncertainty concerning their location and origin. Further spectra of more distant stars may solve the problem, perhaps. Further work is also required for the many smaller HVCs, including the very high velocity clouds (VHVCs), which indeed may very well be intergalactic. But for these it will be much harder to find suitable probes.

We conclude that some HVCs lie within the Galaxy, that others may lie outside, and that further hard work on distance determinations is required.

Acknowledgements. The William Herschel Telescope is operated by the Royal Greenwich Observatory, in the Observatorio del Roque de los Muchachos of the Instituto de Astrofisica de Canarias. We thank the NFRA Programme Committee, and in particular Huib Henrichs, for their support of our program. The Effelsberg Telescope belongs to the Max Planck Institute for Radio Astronomy in Bonn. Wakker was partly supported by NASA through grant GO-04536.01-92A from STScI, which is operated by AURA, Inc., under NASA contract NAS5-26555. Wakker also thanks Blair Savage for financial support and useful discussions. HvW thanks Dieter Breitschwerdt, the IAU and the Leids Kerkhoven-Bosscha Fonds for support of his participation in this meeting.

References

Blitz L., Spergel D.N., Teuben P.J., Hartmann L., & Burton W.B. (1997): BAAS 28, 1349.

Danly L., Albert C.E., & Kuntz K.D. (1993): ApJ 416, L29.

Houck J.C., & Bregman J.N. (1990): ApJ 352, 506.

Keenan F.P., Shaw C.R., Bates B., Dufton P.L., & Kemp S.N. (1995): MNRAS 272, 599.

Schwarz U.J., Wakker B.P., & van Woerden H. (1995): A&A 302, 364

Wakker B.P., & Schwarz U.J. (1991): A&A 250, 484.

Wakker B.P., & van Woerden H. (1997): ARA&A 35, in press.

Wakker B.P., Howk C., Schwarz U.J., van Woerden H., Beers T.C., Wilhelm R., Kalberla P.M.W., & Danly L. (1996a): ApJ 473, 834.

Wakker B.P., van Woerden H., Schwarz U.J., Peletier R.F., & Douglas N.G. (1996b): A&A 306, L25.

Wakker B.P., van Woerden H., de Boer K.S., & Kalberla P.M.W. (1997): ApJ, submitted.

The Galactic X-ray Halo

J. Pietz[1], J. Kerp[1], P.M.W. Kalberla[1], W.B. Burton[2], Dap Hartmann[2,3], and U. Mebold[1]

[1] Radioastronomisches Institut der Universität Bonn, Germany,
[2] Sterrewacht Leiden, The Netherlands
[3] Harvard-Smithsonian Center for Astrophysics, USA

Abstract. We analyzed the soft X-ray background (SXRB) source intensity distribution of the 3/4 keV and 1/4 keV photons by correlating the public *ROSAT* PSPC All-Sky Survey data with the new Leiden/Dwingeloo H I Survey. We found that the 3/4 keV and 1/4 keV distant background source intensity distribution can be modelled by an isothermal ($T = 1.5 \cdot 10^6$ K) disk-like (scale length $A_1 = 15$ kpc) X-ray halo in addition to the radiation of the extragalactic X-ray background.

1 Introduction

Early investigations of the soft X-ray background at photon energies below $E \leq 1$ keV suggested the existence of a hot thermal plasma component, absorbed by the neutral galactic interstellar matter (Nousek et al. 1982, Marshall & Clark 1984). Using the new Leiden/Dwingeloo H I Survey (Hartmann 1994, Hartmann & Burton 1997) as quantitative tracer for the photoelectric absorption we analyzed the 1/4 keV and 3/4 keV data of the *ROSAT* All-Sky Survey (Snowden et al. 1995) to construct an X-ray halo model for the distant hot plasma, explaining the observed X-ray background intensities.

2 Method

At photon energies above 1 keV the X-ray background is predominantly of extragalactic origin (Hasinger et al. 1993), whereas at lower energies additional soft X-ray emitting components become important (e.g. Chen et al. 1997). In our approach we made use of the different contributions of three individual spectral soft X-ray background (SXRB) components to the different PSPC energy bands (Kerp & Pietz, this volume):

- **3/4 keV:** comprising an extragalactic, approximated by a power-law $E^{-\Gamma}$ with a photon index of $\Gamma \sim 1.4$, plus a distant hot thermal component with a plasma temperature of $T \sim 1.5 \cdot 10^6$ K; both components are absorbed by the galactic interstellar matter.
- **1/4 keV:** consists of extragalactic plus distant thermal radiation, with both components being absorbed by the galactic interstellar matter, and

the emission of an unabsorbed foreground component with a temperature of $T \sim 0.9 \cdot 10^6$ keV.

Because the unabsorbed foreground component accounts only for an insignificant fraction of the 3/4 keV emission, the observed intensity distribution in the 3/4 keV energy band is caused by the combined effects of photoelectric absorption and "real" X-ray source intensity variations across the sky. Using the Leiden/Dwingeloo H I survey as a main tracer for the absorption by the galactic interstellar medium – neglecting the absorption by H_2 and HII – it is possible to disentangle both components.

3 Results

An analysis of the 3/4 keV energy band reveals that the photoelectric absorption cannot account for the observed intensity variations in contrast to the 1.5 keV energy band where we found no intensity variations. We found that a model for the distant thermal component has to fulfill the following conditions:

- The modelled halo X-ray count rate decreases smoothly from the galactic center towards the galactic anticenter.
- The halo X-ray count rate decreases smoothly from the galactic plane to the galactic poles.

In the following, we assume that the galactic X-ray halo is caused by an isothermal hot gas which is in hydrostatic equilibrium with the gravitational potential of the Galaxy. The pressure $p(z) = n(z)T(z)$ is assumed to balance the gravitational potential $\Phi(z)$:

$$n(R,z) = n_0 \cdot \frac{\mathrm{sech}^2(R/A_1)}{\mathrm{sech}^2(R_\odot/A_1)} \exp\left(-120 \cdot \Phi(z)/T)\right) \qquad (1)$$

According to Taylor & Cordes (1993) the scale length parameter A_1 defines a density gradient as function of the galactocentric radius, R, normalized to $R_\odot = 8.5\,\mathrm{kpc}$. We used the potential $\Phi(z)$ derived by Kuijken & Gilmore (1989). The normalization n_0 is determined by the emission measure assuming a collisional ionization equilibrium plasma model (Mewe-Kaastra). The temperature of the distant plasma is constrained by the count-rate ratios between the 1/4 keV and 3/4 keV energy bands (Pietz et al. 1997), confirming the used plasma temperatures of $T = 1.5 \cdot 10^6$ K. Such a model distribution with best-fit parameters of $n_0 \approx 1.4 \cdot 10^{-3}\,\mathrm{cm}^{-3}$ and $A_1 = 15$ kpc can explain simultaneously the background intensity variations of the 3/4 keV and 1/4 keV energy bands (see Fig. 1).

4 Comparison with a Galactic Fountain Model

Bregman (1980) constructed a galactic fountain model to explain the origin of high velocity clouds (HVCs). Note that the halo parameters of the isothermal

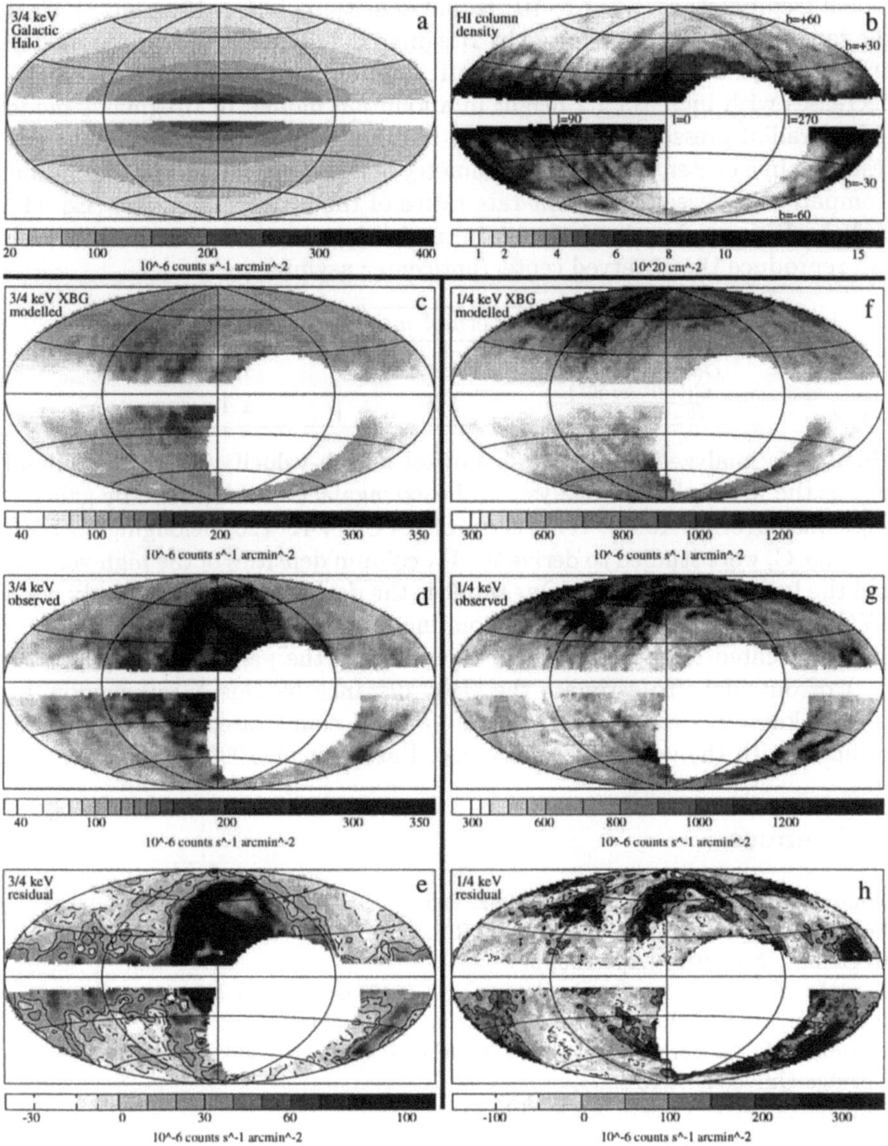

Fig. 1. Comparison of the modelled and observed 3/4 keV and 1/4 keV SXRB maps centered on the galactic center in Hammer-Aitoff projection. (a) proposed unabsorbed 3/4 keV intensity distribution of the distant galactic X-ray halo; (b) observed H I column density distribution derived from the Leiden/Dwingeloo survey; (c) modelled 3/4 keV SXRB intensity distribution; (d) observed 3/4 keV intensity distribution; (e) difference map representing observed minus modelled SXRB intensities; (f) modelled SXRB intensity distribution of the 1/4 keV energy range; (g) observed 1/4 keV intensity distribution; (h) the difference map between the observed X-ray intensity distribution and the model situation.

model (temperature $T \sim 1.5 \cdot 10^6$ K and density $n_0 = 1.4 \cdot 10^{-3}\,\mathrm{cm}^{-3}$) are in the range of the values preferred by Bregman (1980), although his parameters are only valid at the base of the galactic fountain flow, since the temperature decreases with increasing z height in a fountain model. In Bregman's model (1980) radial pressure gradients cause a radial outflow of the rising hot gas, resulting in a center/anticenter asymmetry of the X-ray intensity distribution. Comparison between the count-rate ratios of the isothermal model (Eq. (1)) and the Bregman's galactic fountain model demonstrates that both models can reproduce the observed center/anticenter asymmetry:

Count-rate ratios	isothermal model	galactic fountain
$\frac{C(l=0°,b=30°)}{C(l=180°,b=30°)}$	2.1	1.9
$\frac{C(l=0°,b=60°)}{C(l=180°,b=60°)}$	1.2	1.4

Finally, we analyzed the X-ray shadow of a high-velocity cloud in a similar way as the X-ray shadow analysis of the extragalactic background by galaxies (e.g. Barber et al. 1996). H I spectra of HVC87+41–120, belonging to HVC complex C, were studied to derive the H I column densities of the high velocity and the low velocity matter. Our result is the detection of an extremely weak HVC X-ray shadow in the PSPC pointing 200721p which suggests that this HVC is located close to the upper boundary of the galactic halo ($|z| > 2.5$ kpc) or outside[1]. To estimate the HVC distance by this X-ray shadow we have to know the exact extragalactic 1/4 keV count rate, which is still not available with the needed accuracy (e.g. Barber et al. 1996).

References

Almaini O., et al. (1996): MNRAS 282, 295
Barber, C.R., Roberts, T.P., Warwick, R.S. (1996): MNRAS 282, 157
Bregman, J.N. (1980): ApJ 236, 577
Chen, L.-W., Fabian, A.C., Gendreau, K.C. (1997): MNRAS 285, 449
Hartmann, D. (1994): Ph.D. thesis, University of Leiden
Hartmann, D. & Burton, W.B. (1997): Atlas of Galactic Neutral Hydrogen, Cambridge University Press
Hasinger G., et al. (1993): A&A 275, 1
Kuijken, K. & Gilmore, G. (1989): MNRAS 239, 605
Marshall, F.J. & Clark, G.W. (1984): ApJ 287, 633
Nousek, J.A., et al. (1982): ApJ 258, 83
Pietz, J., et al. (1997): A&A submitted
Taylor, J.H. & Cordes, J.M. (1993): ApJ 411, 674

[1] Note that if the HVC is located *outside* the galactic X-ray halo the observed X-ray shadow implies a value of $\Gamma = 1.5^{+0.25}_{-0.5}$ for the extragalactic power-law index, consistent with the results by Almaini et al. (1996).

The Galactic Halo in Hydrostatic Equilibrium

P.M.W. Kalberla, J. Pietz, and J. Kerp

Radioastronomisches Institut der Universität Bonn, Auf dem Hügel 71, 53121 Bonn, Germany

Abstract. The large scale distribution of gas, magnetic field and cosmic rays in the Galaxy is explored. We analyze recent all-sky surveys of H I gas (Leiden/Dwingeloo survey), soft X-ray radiation (*ROSAT*) and high energy gamma-rays (*EGRET* > 100 MeV) in combination with the 408 MHz survey. From these observations we derive a Galactic halo model consistent with hydrostatic equilibrium conditions.

The large scale equilibrium configuration consists of two separate domains with different properties and scale sizes:

1) The gaseous halo consists of collisionally ionized gas at T=$10^{6.2}$K as well as H I and has an exponential scale height $h_z \sim 4.4$ kpc. The radial distribution is characterized by the galactocentric scale length $A_1 \sim 15$ kpc. All components of the halo – gas, magnetic field and cosmic rays – contribute equally to the pressure. The magnetic field in the halo is oriented parallel to the Galactic plane.

2) The disk extends up to a scale height of $h_z \sim 1$ kpc and is associated with irregular magnetic fields. Here the pressure of the magnetic field is only 1/3 of the gas pressure. The cosmic rays are only weakly coupled to the disk.

1 Model Assumptions

We assume that the average properties of the Galactic halo can be described by a hydrostatic equilibrium situation as defined by Parker (1966) and more recently by Boulares & Cox (1990, B&C). We represent the gaseous disk-halo system by three basic components with their characteristic vertical exponential scale heights h_z: the neutral interstellar medium (ISM) with $h_z = 400$ pc as used by B&C, the diffuse ionized gas (DIG) with $h_z = 1$ kpc as described by Reynolds (1997), and a halo gas with $h_z = 4.4$ kpc.

New in our hydrostatic model are the parameters for the gaseous halo which are from Pietz et al. and Kalberla et al. (both, this volume). The halo is composed of an X-ray emitting gas of a temperature of $T = 10^{6.2}$ K and a highly turbulent neutral component. Both halo components, according to their best-fit distributions, have similar density distributions and an identical scale height of $h_z = 4.4$ kpc.

Fig. 1 displays the vertical density distributions of the gas phases (ISM, DIG and halo gas) in the solar vicinity. Fig. 2 displays our normalized radial density distribution g_1 at $z = 0$ kpc which was selected according to Taylor & Cordes (1993) after modifying the radial scale length to $A_1 = 15$ kpc.

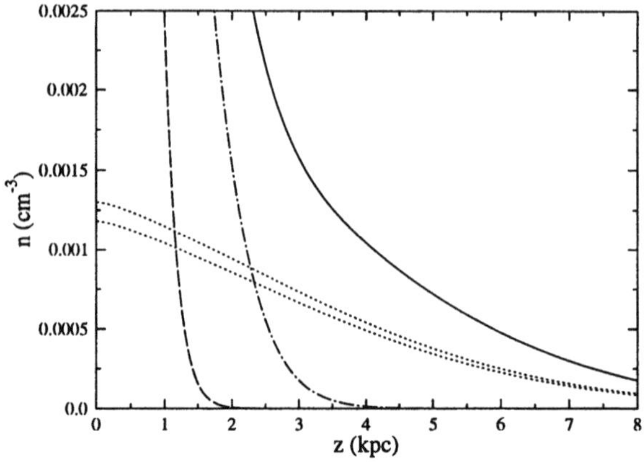

Fig. 1. Vertical density distribution of the gas components in the solar vicinity. The dotted lines mark the halo gas (lower: neutral gas, upper: ionized gas). The neutral ISM gas is indicated by the dashed line. The Galactic DIG layer is represented by the dot-dash line; the sum of all components is given by the solid line.

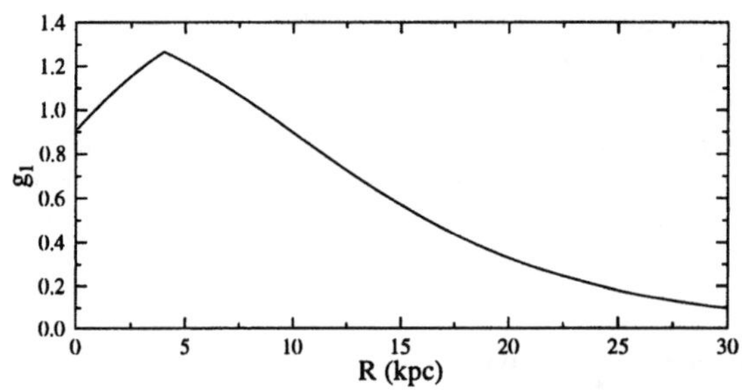

Fig. 2. Radial distribution for the gas density at $z = 0$ kpc normalized to the solar vicinity, $g_1(R_\odot)$=1. The scale length is $A_1 = 15$ kpc.

2 Gas, Magnetic Field and Cosmic Rays in Equilibrium

Following Parker's (1966) suggestion we studied whether gas, magnetic fields and cosmic rays in the Galactic halo may be in pressure equilibrium. Indeed, model calculations fit best the all-sky-averaged observations if one assumes such an equilibrium situation. To verify Parker's suggestion we have modelled the synchrotron emission at 408 MHz as observed by Haslam et al. (1982) as well as the γ-ray emission as observed with *EGRET* at energies > 100 MeV (Fichtel et al. 1994). A detailed description of the model calculations and a comparison with the observations are beyond the scope of this short contribution; details will be published elsewhere.

We derive a complete pressure equilibrium between gas, magnetic field and cosmic rays for the halo component. In the halo, the magnetic field is oriented parallel to the Galactic plane. Close to the disk ($|z| < 1$ kpc) the pressure of the magnetic field and the cosmic ray component is only 1/3 of the gas pressure. The orientation of the field is irregular.

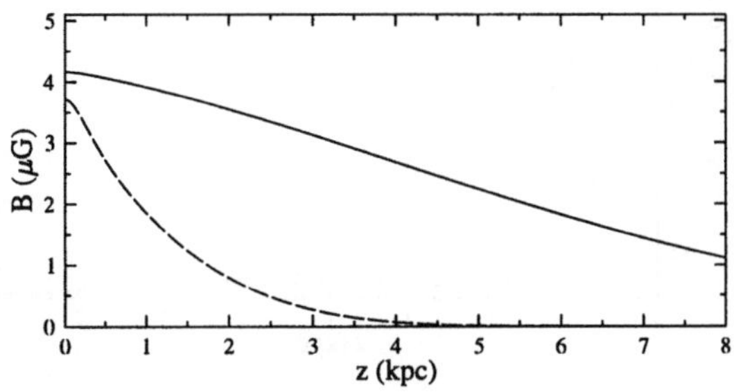

Fig. 3. Vertical distribution of the large scale magnetic-field components as expected from our model for the solar vicinity ($R=R_\odot$). The solid line represents the regular magnetic field component in the halo which is oriented parallel to the Galactic plane. The irregular component of the magnetic field which is associated with the disk is marked by the dashed line.

The strengths of regular and irregular magnetic field components derived by us as a function of vertical distance from the Galactic plane are plotted in Fig. 3. We can trace the γ-ray emission up to intermediate latitudes with good agreement between model and observations. In Fig. 4 we display the derived average radial distribution of the γ-ray emissivity for $z = 0$ kpc. Our radial distribution is in good agreement with the one derived by Webber et al. (1992) which is plotted for comparison. We note further that our vertical scale height is also consistent with Webber et al. (1992).

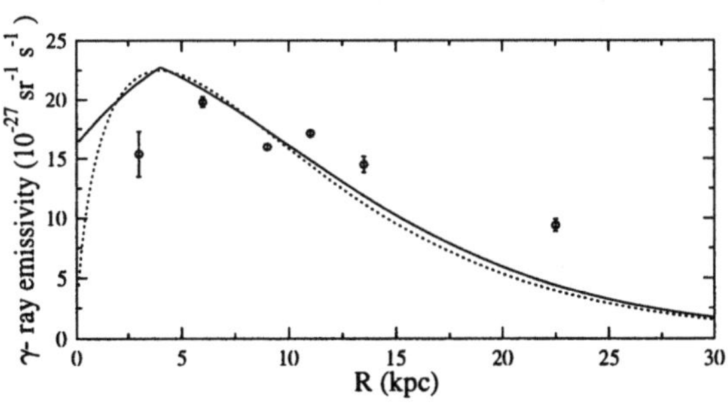

Fig. 4. Radial distribution of the γ-ray emissivity derived from our model (solid line). The dashed line represents the best fit model given by Webber et al. (1992, Fig. 2). The data are obtained from a gradient model analysis of *EGRET* diffuse Galactic γ-ray emission (Strong & Mattox 1995).

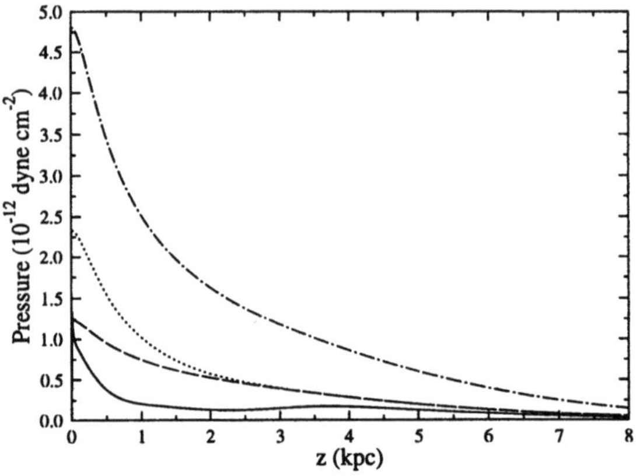

Fig. 5. Vertical pressure distribution in the solar vicinity for gas (dotted), magnetic field and cosmic rays (dashed). The solid line represents the minimum gas pressure required for stability. The dash-dotted line gives the total pressure. Stability for the disk-halo system is ensured for $|z| \lesssim 4.5$ kpc.

Since Parker's (1966) classical publication it was not possible to describe a halo which is in quantitative agreement with observations and which is stable at the same time. Applying stability criteria as derived by Lachièze-Rey et al. (1980, see B&C for discussion) we find that our modelled disk-halo system is stable. In Fig. 5 we display our vertical pressure distributions for gas, magnetic fields and cosmic rays. Stability is ensured if the gas pressure exceeds a minimum pressure given by the lower curve. Instabilities may be caused by small scale pressure fluctuations. However, according to Fig. 5 the formation of high velocity clouds due to such fluctuations appears to be restricted to $z \gtrsim 4.5$ kpc.

Remarkable is the high gas pressure which results from our model. In the local vicinity we expect $\mathrm{n}T/k \sim 17000$ K cm^{-3}.

References

Boulares A. & Cox D.P., (1990): ApJ 365, 544 (B&C).
Fichtel C.E. et al. (1994): ApJS 94, 551.
Haslam C.G., Stoffel H., Salter C.J. & Wilson W.E., (1982): A&AS 47, 1.
Lachiže-Rey M., Asséo E., Cesarsky C.J. & Pellat R., (1980): ApJ 238, 175.
Parker E.N., (1966): ApJ 145, 811.
Reynolds R.J. (1997): in Proc. of 156. WE-Heraeus-Seminar on 'The Physics of Galactic Halos', eds. H. Lesch, R.-J. Dettmar, U. Mebold, & R. Schlickeiser, Akademie Verlag, Berlin, 57.
Taylor J.H. & Cordes J.M., (1993): ApJ 411, 674.
Strong A.W. & Mattox J.R., (1995): A&A 308, L21.
Webber W.R., Lee M.A. & Gupta, M. (1992): ApJ 390, 96.

The Galactic HI Halo

P.M.W. Kalberla[1], G. Westphalen[1], U. Mebold[1], D. Hartmann[2], and W.B. Burton[3]

[1] Radioastronomisches Institut der Universität Bonn, 53121 Bonn, Germany
[2] Center for Astrophysics, 60 Garden Street, Cambridge, MA 02138, U.S.A.
[3] Sterrewacht Leiden, P.O. Box 9513, RA 2300 Leiden, The Netherlands

Abstract. We find indications for diffuse HI gas at substantial z heights in our Galaxy, with a velocity dispersion of 60 km s^{-1}and a vertical projected column density of $1.4{\cdot}10^{19}$ cm^{-2}. This pervasive component of the emission spectrum could be identified in the Leiden/Dwingeloo 21 cm Survey (LDS) after increasing the accuracy further by correcting the observations for reflections from ground. Assuming hydrostatic equilibrium an exponential scale height of 4.4 kpc for the observed diffuse high-dispersion HI component is deduced. This differs from the scale height of 1 kpc derived by Lockman & Gehman (1991), which corresponds to a velocity dispersion of 34 km s^{-1}, based on an analysis of the the Bell Laboratories HI Survey (BLS). A comparison of BLS and LDS data explains the differences in the derived model parameters in terms of baseline uncertainties at a level of $\approx$ 30 mK. We find additional indications for baseline uncertainties in the BLS data. Concerning the LDS we cannot, however, exclude that this survey may also be affected by baseline uncertainties. Receiver bandpass and stray-radiation effects need a more thorough analysis before drawing firm conclusions.

1 The Bell Laboratories HI Survey (BLS)

This survey (Stark et al. 1992) is based on observations with the 20-foot horn reflector at AT&T Bell Laboratories. This antenna has the advantage of very low sidelobe contamination due to its main beam efficiency of 92%. The width of the main beam is 2° × 2.°8 at FWHM. The average rms noise is 17 mK in 5.2 km s^{-1}wide channels. Using the BLS, Kulkarni & Fich (1985) found evidence for HI gas with dispersions of up to 35 km s^{-1}. Lockman & Gehman (1991, hereafter LG) proposed that such components may be interpreted as emission from neutral gas in hydrostatic equilibrium with the gravitational potential of our Galaxy. Due to its turbulent pressure such HI gas would extend to z distances of $\approx$ 1 kpc, thus contributing a neutral component to the gaseous halo.

2 The Leiden/Dwingeloo HI Survey (LDS)

The LDS (Hartmann 1994; Hartmann & Burton 1997) is the first large scale 21 cm line survey which has been corrected for stray-radiation from the side-

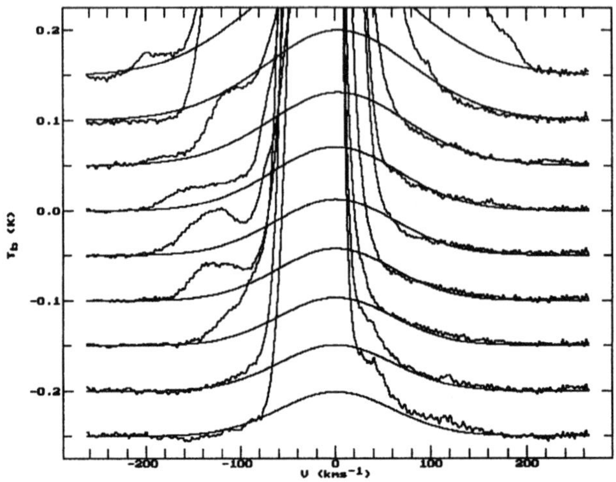

Fig. 1. Profiles averaged over all Galactic longitudes and 10° in latitude. The bottom profile is centered at latitude 85°, the top one at latitude 5°. The solid lines are not results of Gaussian decomposition, but follow from the model.

and backlobes of the antenna pattern (Hartmann et al. 1996). The width of the main beam is 0.°6 at FWHM; the average rms noise is 70 mK at a resolution of 1 $km\,s^{-1}$. The LDS covers velocities between -450 < V_{LSR} < 400 $km\,s^{-1}$. We searched the LDS for low-surface-brightness features. In our analysis we first averaged spectra over large areas. After Hanning smoothing we verified that the noise went down as expected. We then determined that ground reflections limit the accuracy of the averaged profiles. Based on 2700 spectra showing the typical signature of reflections, a proper correction for such reflections from the ground was calculated. This additional correction was applied to the entire LDS data set. To make sure that the extended wings are due to smooth emission and not to a superposition of some scattered high-velocity clouds or to interference, we carefully monitored the quality of the profiles used. While averaging over 10° × 10° we simultaneously calculated the variance of the line emission for each velocity channel. Multiple observations at the same position were compared. Averaged spectra at ~250 different positions were decomposed into Gaussian components. Only those channels of the averaged spectra which were found not to be contaminated by fluctuations (due to small scale structures like HVCs or interference) have been used for the fit. Thus the analysis was most sensitive to extended HI components. To verify that these lines are not due to residual stray radiation we applied a similar Gaussian analysis to the observed stray-radiation components. We found that the velocity dispersion of the stray components is low compared to the large velocity dispersion (LVD) component. We conclude that the outermost profile wings cannot be due to stray radiation. HI lines with LVD

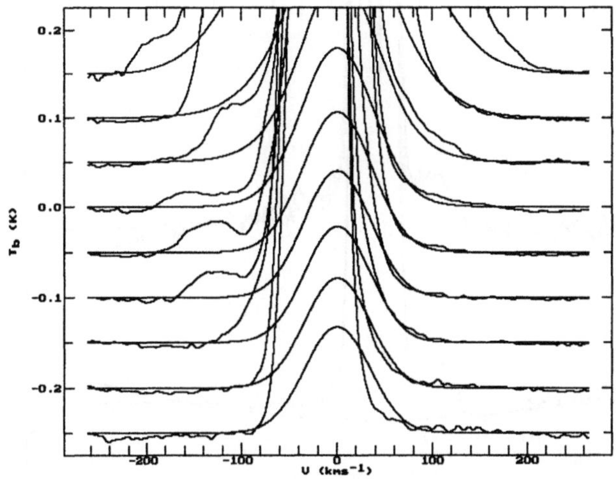

Fig. 2. Profiles averaged over all Galactic longitudes as in Fig. 1, but derived from the BLS. The solid lines represents the Lockman layer (LG model 2).

were found in all of the averaged spectra. The mean velocity dispersion of these lines is $\sigma = 60(\pm3)\mathrm{km\,s^{-1}}$ at the north galactic pole and increases toward lower latitudes, consistent with line broadening due to Kolmogoroff turbulence along the line of sight which increases toward lower Galactic latitude. Based on the hypothesis of hydrostatic equilibrium as proposed by LG, we modelled the HI emission for all latitudes $b > 0°$. Our model is based on the observed velocity dispersion of 60 $\mathrm{km\,s^{-1}}$ and a column density of $N_{\mathrm{HI}} = 1.4 \cdot 10^{19}$ $\mathrm{cm^{-2}}$ in the direction of the north galactic pole. Fig. 1 shows the LDS H I emission averaged over all longitudes and 10° in latitude. In Fig. 2 we show for comparison the BLS H I emission and the emission modelled according to LG. The basic parameters in this case are the velocity dispersion of 34 $\mathrm{km\,s^{-1}}$ and the column density of $N_{\mathrm{HI}} = 1.9 \cdot 10^{19}$ $\mathrm{cm^{-2}}$.

3 Discussion

From the comparison of Figs. 1 and 2 it is obvious that both data sets and the corresponding models differ with respect to the profile wings. Smoothly varying offsets suggesting baseline problems are visible. To study whether one of the surveys may be affected by systematic baseline problems we averaged, for both surveys, fields of $5° \times 5°$ in size. We found in the majority of cases significant baseline deviations between the averaged BLS and LDS profiles. Fig. 3 displays a typical case. The BLS profile is found to be consistent with the LDS data after applying a second-order baseline correction to the BLS profile. There is no indication for any residual stray radiation contamination. While

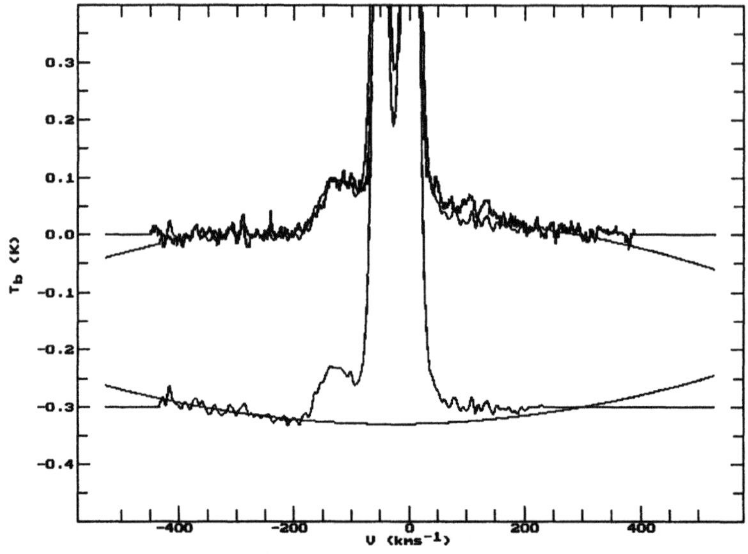

Fig. 3. Intercomparison of LDS and BLS profiles centered at $(l, b) = (142\overset{\circ}{.}5, 42\overset{\circ}{.}5)$. The lower panel was derived from the BLS. A second-order baseline readjustment for the BLS profile is indicated. The upper panel is an overlay of the corresponding LDS profile (thick line) and the readjusted BLS profile (thin line).

it is obvious from Fig. 3 that the BLS profile suffers from an ill-determined baseline we assume that the LDS data set is free from residual baseline errors. The differences in the baselines are up to 20 or 30 mK, about 50% of the amplitude of the LVD component which was determined by analyzing the LDS. It is essential to verify that the LVD components derived from the LDS are unaffected. Our attempts to improve the BLS baseline correction failed due to the fact that the number of channels which are safely assumed to be free from possible H I emission are usually limited. A reanalysis of the LDS with the aim to get an improved insight to the baseline accuracy is in progress.

References

Hartmann D. (1994): PhD thesis, University of Leiden

Hartmann D., & Burton W.B. (1997): Atlas of Galactic Neutral Hydrogen, Cambridge University Press

Hartmann D., Kalberla P., Burton W.B., & Mebold U. (1996): A&AS 119, 115

Kulkarni S.R., & Fich M. (1985): ApJ 289, 792

Lockman F.J., & Gehman C.S. (1991): ApJ 382, 182

Stark A.S. et al.(1992): ApJS 79, 77

The Galactic Fountain

F.D. Kahn

Department of Physics and Astronomy, The University of Manchester, Manchester M13 9PL, England

Abstract. The Galactic fountain is driven by the hot inter-cloud medium (ICM). The ICM itself is heated by successive supernova explosions. This process introduces sufficient irregularities to prevent the fountain flow from being smooth. Instead it leads to the formation of a frothy medium above the disk (Avillez, Berry & Kahn 1997). This paper describes the effect of successive supernova explosions, in a rather idealised case. The real case is still more complicated, but the present approach shows why it is so.

1 Introduction

Most of the Galactic interstellar matter is present in a disk, a relatively thin layer extending to some 100 pc on either side of the Galactic plane. This material is largely contained in cool clouds where hydrogen, the dominant constituent, is present either in atomic (H) or (H_2) molecular form. The atomic hydrogen or HI clouds have the kinematical properties of extreme Population I: they follow the rotation of the Galactic disk and have random speeds large enough to raise them to a maximum distance of 100 pc or so from the plane. The typical temperature in an HI region may be taken to be 100 K and the smoothed-out density 2×10^{-24} gm cm^{-3}. The adjective smoothed-out is important here: the cool gas in reality fills only a few per cent of the available volume.

The inter-cloud medium (ICM) fills the space between the clouds. The ICM is hot, at typically 10^6 K, in contrast to the HI clouds. Its density is typically 10^{-26} gm cm^{-3} so that its pressure is typically 10^{-12} dyne cm^{-2}. An important physical parameter is the sound speed in the ICM

$$c_s \equiv \left(\frac{\gamma P_\circ}{\rho_h}\right)^{1/2} \sim 1.3 \times 10^7 \text{ cm s}^{-1} \tag{1}$$

an order of magnitude larger than the typical random speeds of extreme Population I objects. (In this formula $P_\circ$ and ρ_h are respectively the pressure and the density in the ICM, and γ is the ratio of specific heats.) The gravitational field of the Galaxy confines Population I objects to a disk with a half-thickness of 100 pc; the hot ICM should consequently rise to a height c_s^2/g_z which is larger by two orders of magnitude. Here g_z is the component of the Galactic gravitational field perpendicular to the disk; it has a value

10^{-8}cm s^{-2} at points well away from the Galactic plane. Clearly then the ICM should extend into a volume much thicker than that of the disk.

The dynamical relaxation time will be

$$t_{\rm dyn} = \frac{c_s}{g_z} \tag{2}$$

in any equilibrium structure attained by the ICM; for comparison, the rate of radiative heat loss is given by the cooling law

$$\frac{D}{Dt}\kappa^{3/2} = -q \tag{3}$$

which is applicable in the range of temperatures $10^5 < T < 3 \times 10^7$ K (Kahn 1976). In this relation

$$\kappa = P/\rho^{5/3} \tag{4}$$

and P and ρ are respectively the pressure and the density of the ICM. The cooling rate is set by the parameter q and itself depends on the composition of the gaseous medium concerned. For solar abundances, that is no depletion to dust grains, its value is 4×10^{32}cm^6 gm^{-1} s^{-4}. The resulting cooling time, due to radiative loss, is

$$t_{\rm cool} = \frac{\kappa^{3/2}}{q} = \frac{P^{3/2}}{q\rho^{5/2}} \tag{5}$$

and the hot ICM can reach a hydrostatic equilibrium only if

$$t_{\rm cool} \gg t_{\rm dyn} \tag{6}$$

or

$$\frac{P}{\rho^2} \gg \frac{q}{g_z} \tag{7}$$

With the present values of the physical parameters, the left hand side in this inequality is smaller, by a factor 4, than the right hand side, with the result that the ICM can never attain equilibrium unless it is heavily depleted in metals. The hot gas consequently escapes from the thin Population I disk, but it eventually falls back because its thermal energy is too small for escape from the Galaxy.

In the most familiar models this flow is described as a fountain, or a fountain with some loss into a wind (Bregman 1980, Breitschwerdt, McKenzie & Völk, 1993, Kahn 1994). The medium is generally treated as though it were smooth, having no structure on small scales. The upward flow either remains subsonic throughout (in Bregman's version) or starts subsonic and then passes through a critical level above which it is supersonic (in Kahn's version). The continuing radiative heat loss creates conditions which allow the transit through the sonic level. In all cases the gas is cold on its return flow.

At first sight such models provide a very reasonable solution to the problem posed by the existence of a reservoir of hot gas in the Galactic disk with so much thermal energy that it cannot be held. The residence time of the ICM is only of order $Z_o/c_h \approx 10^6$ years, and here Z_o is the half-thickness of the disk. Supernovae of type II continually heat more interstellar gas to replenish the ICM. According to the usual estimates, there are three such events per century in the Galaxy, or 3×10^4 every million years. The Galactic disk has an area of some 3×10^8 pc^2, so that there is one explosion per million years for every 10^4 pc^2 of disk. In any period of a million years, that is within the characteristic escape time of the ICM gas, the typical distance between neighbouring explosions is therefore 100 pc, similar to the thickness of the disk.

Successive supernova explosions introduce a rather rugged structure into the interstellar medium. The outer parts of a SNR are shocked, compressed, and heated but not enough to stay hot for a million years. The gas on the outside cools off again, and then condenses into HI clouds with random speeds which are too small to smooth out the irregularities before the next explosion. The gas from the inner part of the SNR, by contrast, stays hot. If each remnant yields some 200 $M_\odot$ of hot gas, then there is an adequate supply to compensate for the outflow to the fountain. But the question remains whether this flow can be smooth.

Recently Avillez et al. (1997) have tackled this problem in a computer simulation. Their results, presented in Figures 1 and 2 (see description in Avillez et al. (1997), show what the Galactic disk and halo can be expected to look like after evolving some hundreds of million years.

The simulation starts with the interstellar gas distributed in a smooth thin disk. Successive explosions cause the formation of a frothy disk, with typical half-thickness 1 kpc. Once disrupted by the explosions, the disk never returns to its initial state. Instead it seems to approach a state where there is cool gas present predominantly in a thin but irregular layer around the Galactic plane. The ICM is shown to extend into a thicker irregular layer where it is mixed with some of the cool gas, and where the mixture is in some sort of turbulent equilibrium. A smoother fountain flow appears to exist at greater heights. There seem also to be chimneys present (in Figure 2 (a) located at $x = 200$ pc, $z = -150$ pc), where high pressure gas has forced its way out through relatively narrow channels (for a recent observation of a chimney, see Normandeau, Taylor & Dewdney 1996.). The froth in the thick disk consists of bubbles of hot gas separated by thin regions of cold gas.

One great problem at present with such a calculation is that the computational grid is still rather coarse. The resolution is limited to 10 pc. It is therefore impossible to resolve clearly any cool sheets present. Typically such sheets are expected to be no more than one or two parsecs thick. Clearly there are important parts of the physics which the computation can only handle in an approximate way. Nevertheless the model developed by Avillez constitutes

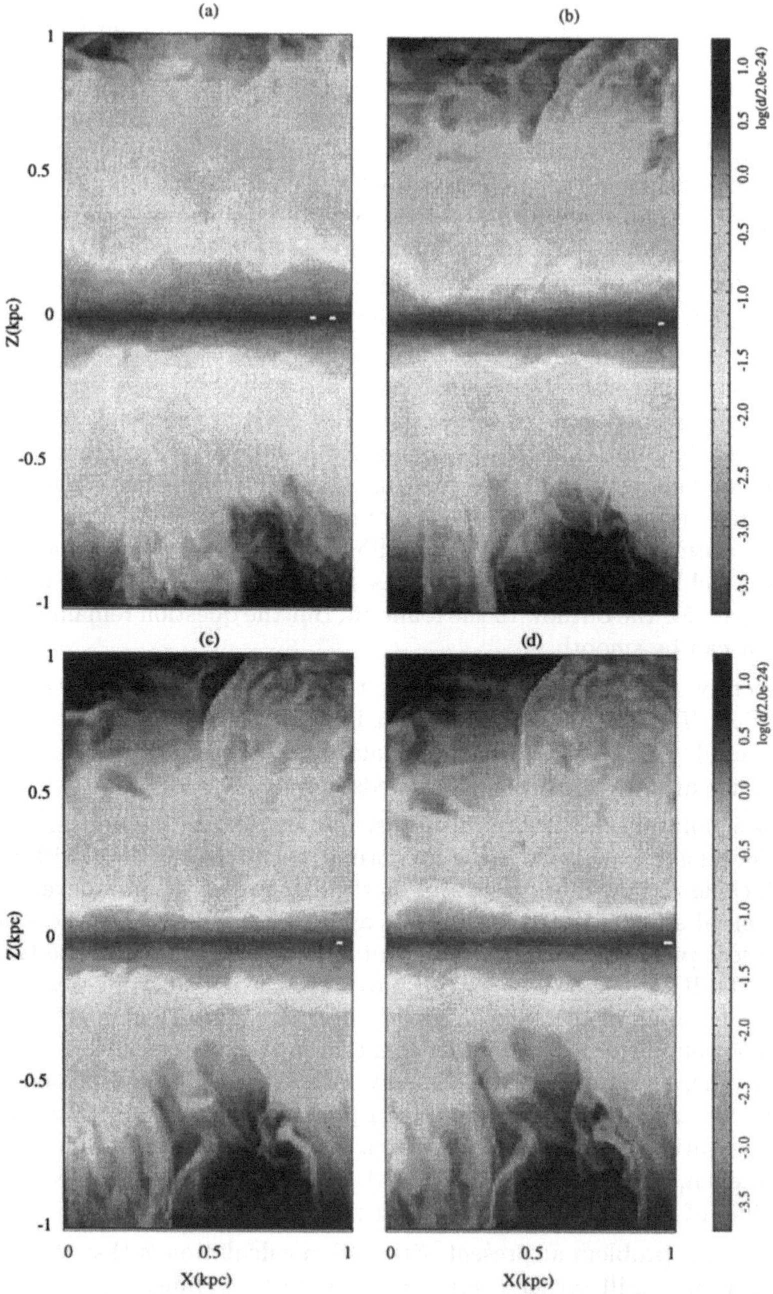

Fig. 1. Density distribution in a vertical section perpendicular to the Galactic plane for times 280, 283, 286, and 289 Myr after start of calculations; (see Plate 5).

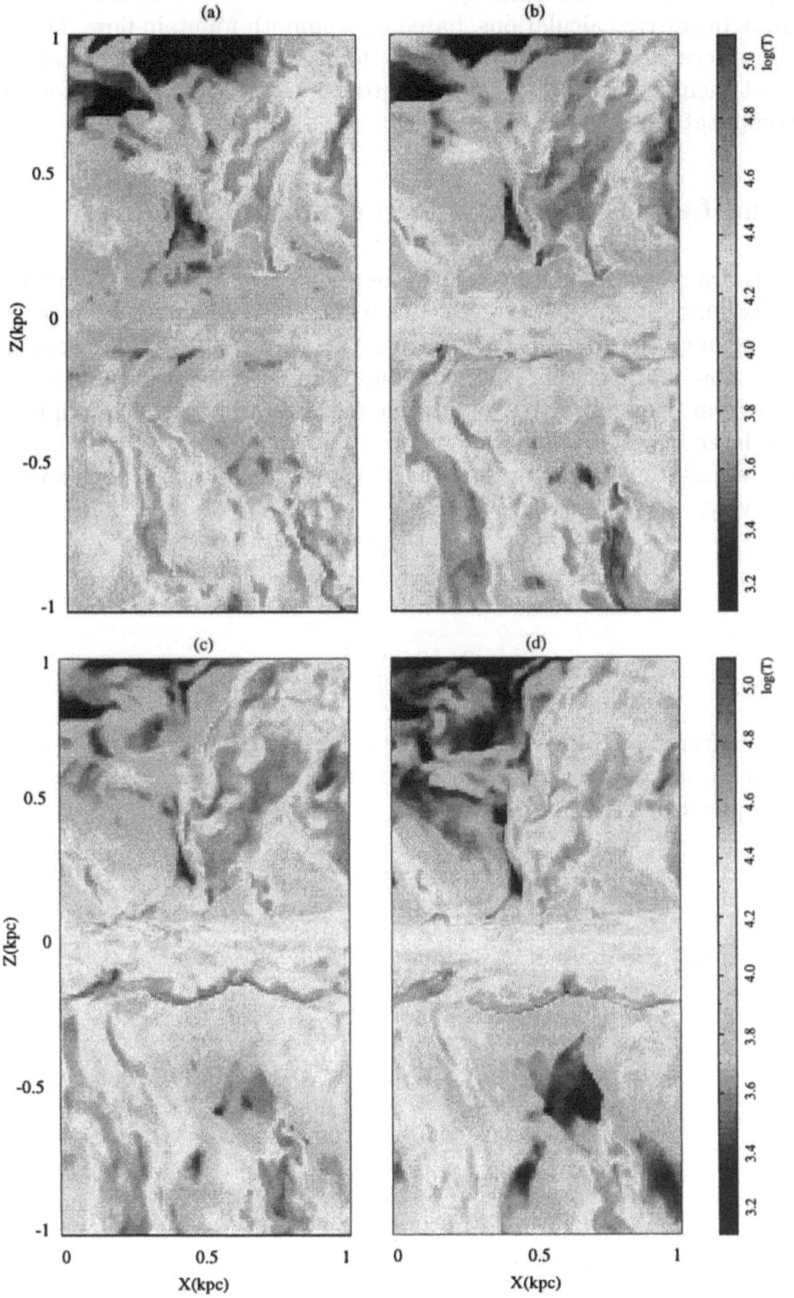

Fig. 2. Temperature distribution in a vertical section perpendicular to the Galactic plane for times 280, 283, 286, and 289 Myr after start of calculations; (see Plate 6).

an important step forward and agrees very much better with observation than the more primitive calculations, based on a smooth fountain flow. There may well be interesting magnetic effects in the flow, but the present calculation does not include them. Much better resolution would be needed for that in the computation.

2 The Late Evolution of Supernova Remnants

A supernova releases some 10^{53} erg of energy, mostly in the form of neutrinos; a smaller amount E, typically 3×10^{51}erg, is released in the form of kinetic energy of baryonic matter and destroys the parent star. It is important to discover how much of this energy is passed to the interstellar medium, and in what form. It is therefore worth while to follow the evolution of a remnant into its later stages.

The standard results are that the shock radius and the shock speed are, respectively,

$$r = \left(\frac{2E}{\rho_\circ}\right)^{1/5} \tau^{2/5} \tag{8}$$

and

$$\dot{r} = \frac{2}{5}\left(\frac{2E}{\rho_\circ}\right)^{1/5} \tau^{-3/5} \tag{9}$$

at time τ after the explosion; here ρ_o is the density of the interstellar medium, assumed to be homogeneous. The adiabatic parameter κ immediately post-shock is given by

$$\kappa_s^{3/2} = 0.02\frac{\dot{r}_s^3}{\rho_\circ} = 0.002\frac{E^{3/5}}{\rho_\circ^{8/5}}\tau^{-9/5}; \tag{10}$$

for comparison the mass of gas which has been shocked at that time is

$$\mathcal{M}_s = \frac{4\pi}{3}\rho_\circ\left(\frac{2E}{\rho_\circ}\right)^{3/5} \tau^{6/5} \tag{11}$$

so that there is a simple relation

$$\kappa_s \mathcal{M}_s = 0.10\frac{E}{\rho_\circ^{2/3}} \tag{12}$$

between the mass of gas engulfed by the shock, at any time, and the value of the adiabatic parameter immediately post-shock. The mass enclosed is conserved following any spherical shell, and so is the adiabatic parameter on the surface of that shell, as long as radiative heat loss has not yet become significant, that is during Phase II. It follows that relation (10) remains valid without the suffixes, until serious cooling sets in.

A parcel of gas that was shocked at time τ will have cooled completely at time

$$t = \tau + \frac{0.002E^{3/5}}{q\rho_\circ^{8/5}}\tau^{-9/5}; \tag{13}$$

the first element of gas to do so must have passed through the shock at time

$$\tau = \tau_\star = \frac{0.13E^{3/14}}{q^{5/14}\rho_\circ^{4/7}} \tag{14}$$

and will be completely cold when

$$t = t_{II} = \frac{0.20E^{3/14}}{q^{5/14}\rho_\circ^{4/7}}; \tag{15}$$

this time marks the end of Phase II. Relation (6), which is still reasonably accurate then, shows that the radius of the SNR is

$$r_{II} = \frac{0.60E^{2/7}}{q^{1/7}\rho_\circ^{3/7}} \tag{16}$$

at that time. With the parameters assumed here $r_{\rm II} = 9.7 \times 10^{19}$ cm and $t_{\rm II} = 1.7 \times 10^{12}$ s. Much later, well into Phase III, only the gas for which

$$\kappa_s^{3/2} > qt \tag{17}$$

will still be hot; according to eq. (12) it amounts to a mass

$$\mathcal{M}_h = \frac{0.10E}{\rho^{2/3}q^{2/3}}t^{-2/3} \tag{18}$$

At partial mass $\mathcal{M}$ within this distribution of hot gas, assumed to be spherically symmetrical, the adiabatic parameter is

$$\kappa = \frac{0.10E}{\rho_\circ^{2/3}\mathcal{M}} \tag{19}$$

and the specific volume of the gas is

$$v = \left(\frac{\kappa}{P}\right)^{3/5} = \frac{0.25E^{3/5}}{\rho_\circ^{2/5}\mathcal{M}^{3/5}}P^{-3/5} \tag{20}$$

under pressure P. The volume occupied by the hot gas is therefore, after some algebra,

$$V = \int_0^{\mathcal{M}_h} v d\mathcal{M} = \frac{0.25E}{\rho_\circ^{2/3}P^{3/5}q^{4/15}}t^{-4/15} \tag{21}$$

and so

$$P = \frac{0.0092E^{5/3}}{\rho_\circ^{10/9}q^{4/9}}r^{-5}t^{-4/9} \tag{22}$$

for a spherical volume of radius r at time t. It is convenient to introduce dimensionless radius R and time θ by setting

$$r = Rr_{II} \text{ and } t = \theta t_{II} \tag{23}$$

and then

$$P = p_\star R^{-5}\theta^{-4/9} \tag{24}$$

where

$$p_\star = 0.24E^{1/7}q^{3/7}\rho_\circ^{9/5} \tag{25}$$

The pressure of the hot gas assists the expansion of the remnant shell even after the end of Phase II. The equation of motion

$$\frac{\rho_\circ}{3}r\ddot{r} + \rho_\circ\dot{r}^2 = P \tag{26}$$

differs from the usual snowplough equation by the presence of the pressure term on the right. The equation makes no allowance for the mass of hot gas, and is really valid only long after time $t_{\rm II}$. It can be written

$$\frac{1}{3}\frac{d}{dt}\rho_\circ r^3\dot{r} = Pr^2 \tag{27}$$

in dimensionless variables and after some transformations becomes

$$S^{3/4}\frac{d^2S}{d\theta^2} = \frac{12p_\star t_{II}^2}{\rho_\circ r_{II}^2}\theta^{-4/9} = 0.32\theta^{-4/9} \tag{28}$$

with the definition

$$S \equiv R^4 \tag{29}$$

The term on the right in eq. (28) represents the effect of the pressure of the hot gas on the dynamics of the shell. Without this term the equation describes the motion in the snowplough approximation, and then S is linear in θ. Clearly the snowplough treatment is valid at late times, when θ is large. At early times, during Phase II, the motion of the shell is described by equation (8), or in dimensionless form

$$S = \theta^{8/5} \tag{30}$$

The solution, for late times, which matches smoothly onto that for Phase II, at time $\theta = 1$, is

$$S = 2.44\theta\left(1 - 0.43\theta^{-7/36} - 0.16\theta^{-1}...\right) \tag{31}$$

and so the dimensionless radius is

$$R = 1.25\theta^{1/4}\left(1 - 0.11\theta^{-7/36} - 0.041\theta^{-1} + ...\right) \tag{32}$$

The dimensionless volume of the remnant is

$$\mathcal{V} = \frac{4\pi}{3}R^3 = 8.18\theta^{3/4} \tag{33}$$

to adequate accuracy. Expressed in dimensionless terms the rate at which supernovae occur is

$$N \equiv \nu r_{II}^3 t_{II}; \tag{34}$$

in this formula

$$\nu = (30\text{years})^{-1}\left(6 \times 10^{10}\text{pc}^3\right)^{-1} = 6 \times 10^{-76}\ \text{cm}^{-3}\text{s}^{-1} \tag{35}$$

is the rate in physical variables, and it follows that $N = 0.00093$. A simple argument leads to an estimate for the length of time that passes before virtually the whole space in the Galactic disk must be filled by old SNR's. Suppose that to begin with the interstellar medium is homogeneous and that supernovae occur at rate N from time $\Theta = 0$ on. A remnant started by an explosion at time $\Theta - \theta$ will have grown to volume $\mathcal{V}$ at time Θ, and so the fractional volume occupied at time Θ is

$$f = 8.18N \int_0^{\Theta} \theta^{3/4} d\theta = 4.67N\Theta^{7/4} \tag{36}$$

The estimate is consequently that the remnants have filled all the available space when $f = 1$ and $\Theta = 0.41N^{-4/7}$ so that $\Theta = 22.1$, equivalent to a physical age 3.8×10^{13} s or 1.2×10^6 years. The radius of the remnant will then be $R = 2.17$, equivalent to 2.1×10^{20} cm or 70 pc.

These estimates are simple but have the great drawback that they are inappropriate. The simulations described by Avillez, Berry & Kahn (1997) show that the interstellar medium never returns to its initial state and that all supernova remnants evolve in an inhomogeneous medium.

Still, in the framework of the simple model, the thermal energy content of a remnant in Phase III is

$$\mathcal{E}_{\text{ther}} = 2\pi P r^3 = 0.21E\theta^{-17/18} \tag{37}$$

and always small when compared with E. Typically $\mathcal{E}_{\text{ther}} = 0.024\,\text{E}$ and $0.012\,E$ at dimensionless times $\theta = 10$ and 20, respectively. With the data adopted here the total energy released, per unit time, in dynamical form by supernovae amounts to some 3×10^{42} erg s^{-1} for the Galactic disk as a whole. Of this power about one or two per cent becomes available to drive the Galactic Fountain, enough for a mass loss rate of some 10^{-19} gm cm^{-2} s^{-1} with a speed of 100 km s^{-1} over an area of 6×10^8 pc^2. There seems to be satisfactory agreement.

3 A Supernova Exploding in a Low Density Region

A second supernova may explode inside an existing SNR, an event much more likely to occur during Phase III than Phase II, and later rather than sooner. The dynamical energy of the second explosion goes almost entirely to the remaining hot gas. Before being shocked again this gas has a density distribution of the form

$$\rho = \Omega r^n \tag{38}$$

and $n = 9/2$ in the present application. An expression for Ω is found from the condition that the mass of hot gas $\mathcal{M}_h$ (see relation (18)) is contained in a sphere of radius $1.25 r_{\rm II}\Theta^{1/4}$, so that, with some algebra,

$$\Omega = 0.148 \rho r_{II}^{-9/2} \Theta^{-61/24} \tag{39}$$

The effect of the second explosion can be found analytically, but only at the cost of some apparently drastic approximations. A similarity solution requires the second explosion to occur at the same place as the first, and the energy content of the remaining hot gas to be small compared with E. The second condition is easily met, but the first seems highly artificial. Nevertheless it makes little real difference if the second explosion takes place rather off centre, for the following reason. The shock from the second blast advances very fast through the interior, where the density is low, and slows down considerably later, as it runs into the higher density gas further out. An off-centre explosion creates a blast wave whose surface is elongated, to begin with, in the direction where the ambient density is lowest, but this period of asymmetrical expansion is only brief. The shock spends much more time moving through the bulk of the mass, which is near the boundary of the first remnant, and its surface becomes progressively more spherical.

Another difference is that a similarity solution strictly applies only when the explosion itself ejects a negligible mass. An actual supernova ejects a finite mass M_{ej}, say $5M_\odot$; the mass of hot gas in the remnant can be shown to be

$$\mathcal{M}_h = 0.32 \mathcal{M}_{II} \Theta^{-2/3} \tag{40}$$

$\mathcal{M}_{\rm II}$ being about 4000 $M_\odot$ here. So if the second explosion takes place say 500,000 years after the first, when $\Theta \approx 9$, then the similarity solution holds if

$$5 \ll 1280 f \times 9^{-2/3} \tag{41}$$

Here f is the fraction of the hot diffuse gas that has been overrun by the second shock; according to (41) it has to exceed about 2 per cent. In general the approximation holds.

The shock path is easily found from the virial theorem

$$\frac{1}{2}\ddot{I} = 2\left(E_{\rm ther} + E_{\rm kin}\right) \equiv 2E \tag{42}$$

where

$$I \equiv \int_0^{\mathcal{M}_\star} r^2 d\mathcal{M} \tag{43}$$

Here define $\xi \equiv r/r_s$ and $\mu \equiv \mathcal{M}/\mathcal{M}_s$ where r_s and $\mathcal{M}_s$ respectively are the shock radius and the mass that it encloses. Relation (42) holds provided no significant energy loss has occurred by radiation. The mass $\mathcal{M}_\star$ is fixed and larger than $\mathcal{M}_s$. The motion of the shock leads to a change ΔI in the value of I because material within radius r_s has been displaced outwards, and in the case of a general n

$$\Delta I = \mathcal{M}_s r_s^2 \int_0^1 \xi^2 d\mu - 4\pi\Omega \int_0^{r_s} r^{n+4} dr = \mathcal{M}_s r_s^2 \left[\int_0^1 \xi^2 d\mu - \frac{n+3}{n+5} \right] \tag{44}$$

The quantity $\mathcal{M}_s r_s^2$ varies like t^2 in a similarity solution; it follows that

$$2E = \frac{1}{2}\frac{d^2 I}{dt^2} = \frac{\mathcal{M}_s r_s^2}{t^2} \left[\int_0^1 \xi^2 d\mu - \frac{n+3}{n+5} \right] = \frac{4\pi\Omega r_s^{n+5}}{(n+3)\, t^2} \left[\int_0^1 \xi^2 d\mu - \frac{n+3}{n+5} \right] \tag{45}$$

The strong shock condition leads to the boundary condition

$$\frac{d\mu}{d\xi} = 4\,(n+3) \tag{46}$$

immediately behind the shock at $\xi = 1$. The integral in (45) can then be well approximated by assuming that μ increases exponentially with ξ as ξ approaches unity and then (45) leads to

$$2E = \frac{4\pi\Omega r_s^{n+5}}{(n+3)\, t^2} \left[\frac{2\,(n+3)}{2n+7} - \frac{n+3}{n+5} \right] \tag{47}$$

and so

$$r_s^{n+5} = \frac{(n+5)\,(2n+7)}{6\pi} \frac{Et^2}{\Omega} \tag{48}$$

and

$$\dot{r}_s^2 = \frac{2\,(2n+7)}{3\pi\,(n+5)} \frac{E}{\Omega} r_s^{-(n+3)} \tag{49}$$

since

$$\dot{r}_S = \frac{2r}{(n+5)\, t} \tag{50}$$

Here $n = 9/2$ and so

$$\dot{r}_s = 0.36 \frac{E}{\Omega} r_s^{-15/2} \tag{51}$$

The second shock reaches the edge of the original remnant when

$$r_s = 1.25 r_{II} \Theta^{1/4} \tag{52}$$

and then, from (38), (39) and (51),

$$\dot{r}_s^2 = 0.46 \frac{E}{\rho_o r_{II}^3} \Theta^{2/3} = 0.24 \left(\frac{r_{II}}{t_{II}} \right)^2 \Theta^{2/3} \tag{53}$$

and the shock speed is $280\Theta^{1/3}$ km s^{-1} there. The dynamical time, that is the time to reach the outer edge after the second explosion, is

$$t_{\rm dyn} = \frac{4}{19} \frac{r_s}{\dot{r}_s} = 0.43 t_{II} \Theta^{-1/12} \tag{54}$$

This time has to be compared with the cooling time $t_{\rm cool}$ of the newly shocked gas. When the blast waves reaches the boundary of the old remnant

$$t_{\rm cool} = \frac{\kappa_s^{3/2}}{q} = \frac{0.02 \dot{r}_s^3}{q \Omega r_s^{9/2}} = 0.78 t_{II} \Theta^{29/12} \tag{55}$$

and so

$$\frac{t_{\rm cool}}{t_{\rm dyn}} = 1.82 \Theta^{5/2} ; \tag{56}$$

with the shock speed found above, the post-shock temperature becomes

$$T_s = 10^6 \Theta^{2/3} \;\; \mathrm{K} \tag{57}$$

and can be quite high if Θ is large enough, in other words if the second supernova is delayed sufficiently long.

4 Conclusions

The analytical approach can hardly be carried any further, since great complications will arise from the combined effect of many supernovae. The dynamical consequence of any one explosion is sensitively dependent on the state of the surrounding interstellar medium. If an explosion takes place in a region of low density then the surrounding gas is heated, but takes a long time to cool. The energy therefore becomes available to drive further motions. By contrast an explosion in a high density region is followed by (relatively) rapid heat loss, and much less dynamical effect. Any realistic calculation must strive to give an accurate description of the interstellar medium on quite small length scales.

References

Avillez M.A., Berry D.L. & Kahn F.D., 1997, these proceedings.
Bregman J.N. 1980, ApJ 236, 577.
Breitschwerdt D., McKenzie J.F., Völk H.J., 1993, A&A, 269, 54.
Kahn F.D., 1994, ApSS, 216, 325.
Normandeau M., Taylor A.R. & Dewdney P.E., 1996, Nature 380, 687.

3D Hydrodynamical Simulations of a Galactic Fountain

M.A. de Avillez[1], D.L. Berry[1], and F.D. Kahn[2]

[1] Departamento de Física, Universidade de Évora, R. Romão Ramalho 59, 7000 Évora, Portugal
[2] Department of Physics and Astronomy, University of Manchester, Schuster Laboratory, Manchester M13 9PL, England

Abstract. Galactic fountains are thought to be responsible for the formation of the observed intermediate and high velocity clouds in the Galactic halo. Three-dimensional simulations have been carried out to determine the evolution of the disk gas as it enters the fountain, cools and returns to the Galactic disk. The descending cold gas, headed by a shock, sweeps up the ascending flow, triggering the formation of an unstable layer where Rayleigh-Taylor instabilities grow, which may provoke the formation of the observed clouds.

1 Introduction

Gas that escapes from the Galactic disk, rises into the halo, cools and falls back, constitutes a Galactic fountain. The term "Galactic fountain" should include the outflows that result from isolated as well as clustered supernovae scattered in the Galactic disk. These outflows have been classified as Galactic fountains (Shapiro & Field, 1976) and chimneys (Tomisaka & Ikeuchi, 1986), and have been studied independently. Up to now three-dimensional calculations with the combined effects of such outflows have not been carried out. The fountain flow considered here results from combined processes that limit the dimensions of the chimneys and where turbulence becomes a very important factor in the formation of the observed HI clouds. Three-dimensional modelling of such a Galactic fountain will be presented here.

2 Computational Modelling

A three-dimensional hydrodynamical scheme described in Avillez (1997) with the adapted mesh refinement algorithm of Berger & Collela (1989) is used to follow the evolution of the Galactic fountain. Thermal conduction is not included in the scheme but, cooling is allowed by approximating the cooling curve of Raymond et al. (1976) except in the range of temperatures 5×10^5-5×10^7 K where Kahn's (1976) approximation is used. A section of the Galaxy centered on the Galactic plane with an area of 1 kpc^2 and a vertical extent from -4 kpc to 4 kpc is used with the finest grid resolution being 10 pc. The

Galactic disk is populated with randomly distributed population I stars and OB associations. Population I stars have a vertical density distribution, $\rho_\star$, given by (Avillez, 1997)

$$\rho_\star = \rho_{\star,o}\mathrm{sech}^2\left[(2\pi G\beta_\star\rho_\star)^{1/2}\, z\right] \tag{1}$$

where $\rho_{\star,o} = 3.0 \times 10^{-24}$ g cm^{-3} is the mass density near the Galactic plane (Allen, 1991) and the constant $\beta_\star = 1.9 \times 10^{-13}$ cm^{-2} s^2. This mass distribution generates a local gravitational potential with the form

$$\Phi = -\frac{2}{\beta_\star}\ln\cosh\left[(2\pi G\rho_\star\beta_\star)^{1/2}\, z\right] . \tag{2}$$

Supernovae occur at a rate of three per century in the Galaxy, which corresponds to a rate of one per 1.1×10^4 years in the disk volume under study. The supernova remnants enter these calculations at the beginning of phase II, with initial radii depending on the density of the surrounding medium.

3 Evolution of the Disk Gas and Halo

The simulations have been carried out for a period of 300 Myr. The first 20 Myr are characterized by radiative cooling excess over heating because of the small number of supernovae during this period. Thus, the gas originally located in the lower halo starts cooling and moves ballistically towards the disk, colliding there with gas falling from the opposite side of the Galactic plane. In consequence, a very thin disk of gas is formed around the Galactic plane as can be seen in Figure 1 (a) which shows HI emission at 150 Myr after the start of calculations. The largest column density observed in the image is caused by the low scale height of the cold gas. Figure 1 (b) and (c) show the subsequent evolution of the gas at 250, and 275 Myr, respectively. Figure 1 (b) shows the presence of low column density regions that persist for long periods. These areas are surrounded by cool shells resulting from clustered supernovae that have blown holes in the disk. In consequence hot gas escapes into the halo in a chimney-like structure. In Figure 1 (c) this region has cooled with the corresponding increase in the column density.

Colour maps of the density distribution in the galactic disk and halo obtained in these calculations are presented in Figures 1 and 2 in Kahn (1997). The density maps (Figure 1 in Kahn, 1997) show that the disk gas extends up into the halo with a stratified distribution. Low density gas located within 20 pc of the Galactic plane, having temperatures varying between about 10^5 K and 10^6 K, is shown in Figure 1 (in Kahn, 1997). The figure reveals the presence of cold filaments and clouds with temperatures about 10^3 K distributed in the disk up to a height of 200 pc. A warm medium with temperatures of $10^4 - 10^5$ K is distributed up to heights of 800 – 900 pc. This medium is mixed with cool and cold gas that descends toward the Galactic

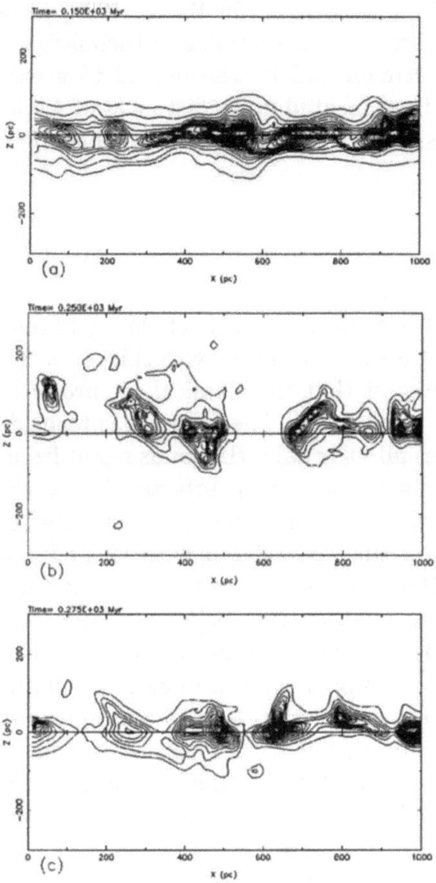

Fig. 1. Position maps of HI emission for the gas in a vertical section through the Galactic plane at (a) 150 Myr, (b) 250 Myr and (c) 275 Myr after start of calculations. The 20 equally spaced contours represent values of column density between $\log(N_{HI}) = 18$ and $\log(N_{HI}) = 19$, where N_{HI} is in cm^{-2}.

disk. The turbulent warm medium supports the cool gas that sits on top of it in a kind of temporary equilibrium.

Cold HI clouds falling towards the Galactic disk are observed in the southern hemisphere (dark blue spots in Figures 2 (b-d) in Kahn, 1997), and make part of a complex with a size about 200 pc. These clouds with sizes of 20 pc typically have LSR velocities of -190 km s^{-1}. Most of the cold gas detected in the northern hemisphere shows LSR velocities of -20 km s^{-1} to -90 km s^{-1}. There are two possible fates for these structures: (1) they break up into cloudlets due to Rayleigh-Taylor instabilities whenever they interact with the ascending gas from the galactic disk (Berry et al., 1997) or (2) they are disrupted by the stream of hot gas that escapes from the Galactic disk in chimneys (in Figure 2 (a) in Kahn, 1997; located at $x = 200$ pc, $z = -150$ pc). In either case convection and turbulence occur. The disrupted gas enters the flow stream and is transported to greater heights becoming part of a diffuse medium located up to distances of 4 kpc above the disk. Gas with temperatures smaller than 10^4 K has not been detected there in these calculations.

4 Summary

The results reported here have a drawback, the grid resolution. Therefore, it is impossible to resolve any cool sheets with thickness of some 2 pc. Nevertheless, these results suggest that the classical picture of the Galactic fountain and chimneys must be changed. The fountain is formed by a turbulent convective flow that rises into the halo. HI clouds result from (a) cool dense shells on the outside of old supernova remnants and (b) Rayleigh-Taylor instabilities at the interface of the cooling structures that interact with the ascending flow. These structures may be related with the HI complexes observed in the halo (Wakker, 1991 and references therein). As they interact with more energetic fountain gas coming from below and specially from chimneys the clouds break up and the cool gas is swept along by the hot gas stream that eventually flows into the upper halo. Chimneys are limited to sizes no bigger than 100 – 120 pc in the disk and lower halo in agreement with the observations of Normandeau et al. (1996). A stratified density distribution is found in the Galaxy, with cool and diffuse media residing in equilibrium with the layers below; this result is in agreement with observations by Kalberla et al. (1997).

Acknowledgements. This work has been sponsored by a joint grant from JNICT/ ESO (Project number PESO/C/PRO/1022/94). M.A. de A. would like to thank the LOC of this colloquium for a travel grant.

References

Allen, C.W., 1991, *Astrophysical Quantities*, The Athlone Press, London.
Avillez M.A., 1997, Ph.D Thesis, University of Évora.
Berry D.L., Avillez M.A. & Kahn F.D., 1997, these proceedings.
Berger M.J. & Colella P., 1989, J.Comput.Phys. 82, 64.
Kahn F.D., 1976, A&A 50, 145.
Kahn F.D., 1997, these proceedings.
Kalberla P.M.W., Pietz J. & Kerp J., 1997, these proceedings.
Normandeau M., Taylor A.R. & Dewdney P.E., 1996, Nature 380, 687.
Raymond J.C., Cox D.P. & Smith B.W., 1976, ApJ 204, 290.
Shapiro, P.R., Field G.B., 1976, ApJ 205, 762.
Tomisaka, K. & Ikeuchi S., 1986, Publ. Astron. Soc. Japan 38, 697.

Cloud Formation in a Galactic Fountain Resulting from Rayleigh-Taylor Instabilities

D.L. Berry[1], M.A. de Avillez[1], and F.D. Kahn[2]

[1] Departamento de Física, Universidade de Évora, R. Romão Ramalho 59, 7000 Évora, Portugal

[2] Department of Physics and Astronomy, University of Manchester, Schuster Laboratory, Manchester M13 9PL, England

Abstract. Intermediate and high velocity clouds are produced in a Galactic fountain as a result of Rayleigh-Taylor instabilities that grow at the interface separating the cool descending gas and the hot ascending gas within the fountain flow. It is proposed here that there is no need to introduce any initial perturbation to the system in order to trigger such instabilities. They will arise naturally as a result of the difference in the adiabatic parameters above and below the interface. A three-dimensional study of the unstable layer and the resulting cloud formation is presented.

1 Introduction

We have carried out three-dimensional hydrodynamical simulations of a Galactic fountain in a section of the Galaxy, having an area of 1 kpc^2 and a vertical extent ranging from -4 kpc to 4 kpc, using a grid resolution of 10 pc and have presented them in these proceedings (Avillez et al., 1997). The simulations show that as the hot intercloud medium enters the fountain it starts cooling and after some 200 Myr the lower halo is populated with intermediate velocity structures having sizes ranging from $\sim$ 20 to 100 pc. The descending gas, headed by a shock, sweeps up the ascending flow thereby heating it. In consequence the interface separating the descending fountain structure from the ascending flow becomes Rayleigh-Taylor unstable and breaks up into clouds. Several small clouds are observed which have previously been detached from the larger structures; however, as a result of grid resolution cloudlets with sizes smaller than 10 pc are not resolved. Here a three-dimensional hydrodynamical scheme is applied to a typical fountain structure in order to simulate the formation of the cloudlets; this scheme has been described in Avillez (1997) and uses an adapted mesh refinement algorithm of Berger & Collela (1989). Thermal conduction is not included but, cooling is allowed by approximating the cooling curve of Raymond, Cox & Smith (1976) but with Kahn's (1976) approximation in the range of temperatures $5 \times 10^5 - 5 \times 10^7$ K. Because magnetic fields are important, a minimum temperature in the gas is arranged so that the effects of magnetic pressure support are simulated.

2 Numerical Modelling

Rayleigh-Taylor instabilities that occur at the interface of two states in perfect but unstable equilibrium, with a sharp variation at the interface, have been carried out among others by Hachisu et al. (1990), and Jun et al. (1995). In their papers initial perturbations introduced at the interface are necessary to trigger the instabilities. Such an initial set-up introduces spurious effects that lead to non-linear growth of the instabilities. In consequence non-linear modes of the perturbation start dominating the flow prematurely. For the present study this picture must be modified because (a) there is a smooth variation at the interface between the two gases and (b) the refined grid itself introduces a type of smoothing at the interface. The modes of perturbation are grid resolution dependent; that is, the smallest wavelengths are of the order of the cell width, and as a result there is a maximum temperature for which the instability will occur in the descending cool sheet. No initial perturbations are introduced explicitly but numerical inhomogeneities in the medium may be responsible for triggering the instabilities.

2.1 Threshold Conditions for Rayleigh-Taylor Growth

Initially, the gas moves with velocities very small compared with the gas sound speed at the interface and so the effects of gas compressibility are negligible (Jun, Norman & Stone, 1995). Thus, the linear growth rate of the instability, σ_{RT}, is expected to be similar to that of an incompressible fluid, given by (Chandrasekhar, 1961)

$$\sigma_{\mathrm{RT}} = \left(\frac{2\pi g}{\lambda}\right)^{1/2} \left(\frac{\rho_+ - \rho_-}{\rho_+ + \rho_-}\right)^{1/2} \tag{1}$$

where λ is the wavelength of the perturbation at the interface between the two fluids, g is the gravitational acceleration, and ρ_+/ρ_- are the gas densities above/below the interface. For a numerical scheme with a grid resolution Δx, instabilities only develop if the growth time, τ_{RT} $(= 1/\sigma_{\mathrm{RT}})$ is shorter than the hydrodynamical time scale given by the Courant-Friedrich-Lewy condition (Richtmyer & Morton, 1967). This leads to an upper limit for the initial wavelengths given by

$$\lambda_{\mathrm{in}} \leq 2\pi g \left(\frac{\rho_+ - \rho_-}{\rho_+ + \rho_-}\right) \left(\frac{\Delta x}{\max(c_s)}\right)^2 . \tag{2}$$

For $\rho_+ \gg \rho_-$, this inequality becomes

$$\lambda_{\mathrm{in}} \leq 2\pi g \left(\frac{\Delta x}{\max(c_s)}\right)^2 . \tag{3}$$

The minimum wavelength must cover at least 4 grid cells ($\lambda \sim 4\Delta x$) and so the local sound speed at the interface must satisfy

$$\max(c_s) \leq \left(\frac{\pi}{2} g \Delta x\right)^{1/2}. \tag{4}$$

So, for the cool gas, the sound speed must be smaller than 22 km s^{-1} for the resolution considered in these calculations, which corresponds to a temperature of 3.5×10^4 K. Gas with temperatures above this value would not develop any instabilities.

2.2 Evolution of the Unstable Interface

The results presented in this section refer to an HI cloud having a LSR speed of -25 km s^{-1}, and a density of 10^{-24} g cm^{-3}. The falling sheet headed by a shock sweeps up the ascending flow which rises with a speed $\sim$ 60 km s^{-1}. The shocked gas is heated to temperatures of $\sim 10^5$ K. Figure 1 (a-d) shows the evolution of a section of the descending sheet with a width of 100 pc and with a grid resolution of 0.25 pc. Both the shocked and cooled layers are shown in Figure 1-a. Small amplitude perturbations at the interface separating both fluids are observed. The width of the shocked layer is approximately 10 pc - a value in accordance with the predictions of Kahn (1981). The other gray scale images show the evolution of the interface until some clouds start detaching from the descending layer. As the instabilities set in, smaller structures appear first, because the growth rate is larger for smaller wavelengths. At later stages fingers expand and interact with their neighbouring fingers introducing turbulence in the flow (Figure 1-b). As a consequence, non-linear effects are originated by these interactions which, in addition to larger wavelengths dominating the flow at later stages, give the appearance shown in Figures1 (c-d). The cool gas, as it expands in finger-like structures, leaves behind empty spaces that are filled with the hot ascending flow. No mixing occurs between the gases apart from numerical mixing. As the fingers stretch and break up, filaments and cloudlets of cool gas with sizes of several parsecs detach from them as shown in Figure (1-d).

3 Summary

It has been shown here that hydrodynamic simulations of the Rayleigh-Taylor unstable layer in a fountain do not require any initial perturbation to trigger the growth of the instabilities. Furthermore, refinement recalculations show the formation of clouds having sizes of the order of several parsecs; these clouds were missing in the previous simulations carried out using coarser grids.

Acknowledgements. This work has been sponsored by a joint grant from JNICT/ ESO (Project number PESO/S/PRO/1053/95). M.A. de A. would like to thank the LOC for the travel grant.

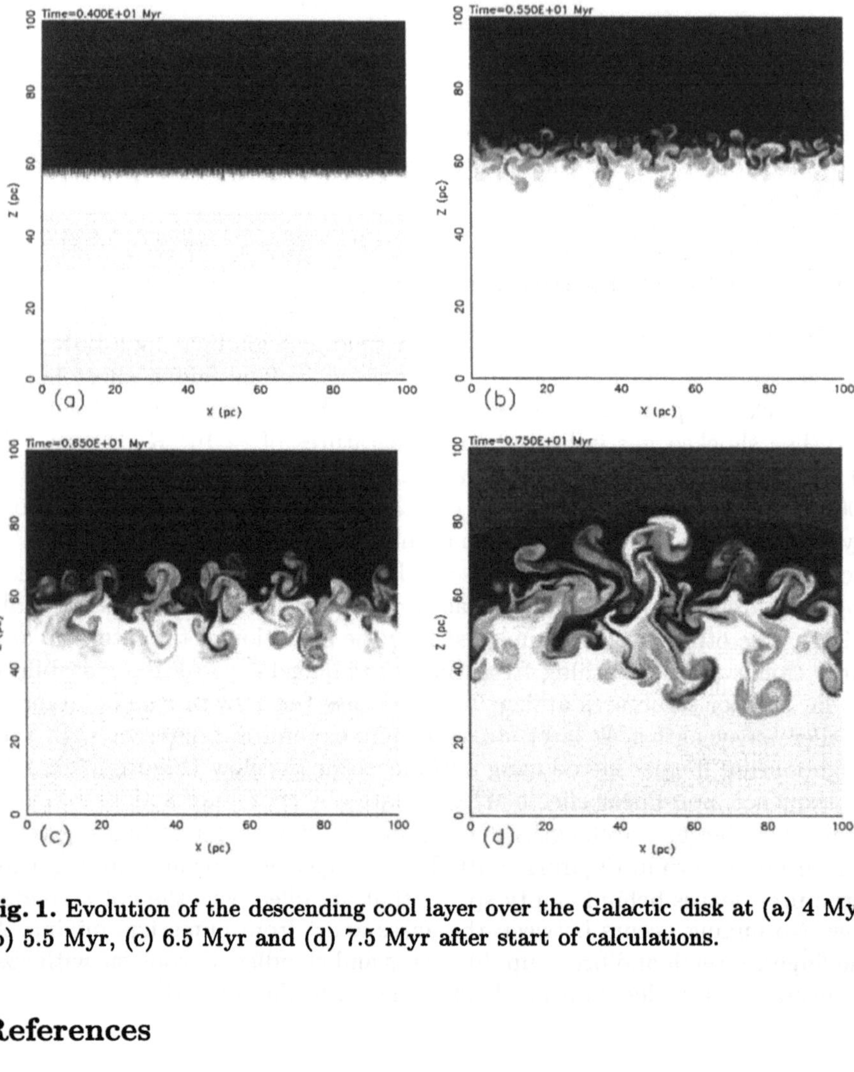

Fig. 1. Evolution of the descending cool layer over the Galactic disk at (a) 4 Myr, (b) 5.5 Myr, (c) 6.5 Myr and (d) 7.5 Myr after start of calculations.

References

Avillez M.A., 1997, Ph.D Thesis, University of Évora.

Avillez M.A., Berry D.L. & Kahn F.D., 1997, these proceedings.

Berger M.J. & Colella P., 1989, J.Comput.Phys. 82, 64.

Chandrasekhar S., 1961, in *Hydrodynamic and Hydromagnetic Stability*, Clarendon Press.

Jun B., Norman M.L. & Stone J.M., 1995, ApJ 453, 332.

Kahn F.D., 1976, A&A 50, 145

Kahn F.D., 1991, in *Investigating the Universe*, D. Reidel Publ. Co., Dordrecht.

Raymond J.C., Cox D.P. & Smith B.W., 1976, ApJ 204, 290.

Richtmyer R.D. & Morton K.W., 1967, *Difference Methods for Initial-Value Problems*, 2ed, Wiley Interscience.

The Hot Galactic Corona and the Soft X-ray Background

Q. Daniel Wang

Dearborn Observatory, Northwestern University 2131 Sheridan Road, Evanston, IL 60208-2900, USA

Abstract. I characterize the *global* distribution of the 3/4 keV band background with a simple model of the hot Galactic corona, plus an isotropic extragalactic background. The corona is assumed to be approximately polytropic (index = 5/3) and hydrostatic in the gravitational potential of the Galaxy. The model accounts for X-ray absorption, and is constrained iteratively with the *ROSAT* all-sky X-ray survey data. Regions where the data deviate significantly from the model represent predominantly the Galactic disk and individual nearby hot superbubbles. The global distribution of the background, outside these regions, is well characterized by the model; the 1σ relative dispersion of the data from the model is $\sim 15\%$. The electron density and temperature of the corona near the Sun are $\sim 1.1 \times 10^{-3}$ cm^{-3} and $\sim 1.7 \times 10^6$ K. The same model also explains well the 1.5 keV band background. The model prediction in the 1/4 keV band, though largely uncertain, qualitatively shows large intensity and spectral variations of the corona contribution across the sky.

1 Introduction

Although Spitzer (1956) speculated the presence of a hot ($\sim 10^6$ K) Galactic corona around the Milky Way more than 40 years ago, direct observational evidence of it comes only recently. The very existence of hot gas far away from the Sun is shown convincingly first by the the detection of the shadow cast by the Draco cloud ($D \gtrsim 300$ pc; $z \gtrsim 200$ pc) against the 1/4 keV background (Burrows & Mendenhall 1991; Snowden et al. 1991). Based on the X-ray shadowing of the Magellanic Bridge ($D \sim 60$ kpc), Wang & Ye (1996) further find that $\sim 30\%$ of the background observed at ~ 0.7 keV is Galactic in origin. At this energy, however, the contribution from the Local Hot Bubble around the Sun appears negligible (Snowden, McCammon, & Verter 1993; Kuntz, Snowden, & Verter 1997), although the Bubble is responsible for most of the background in the 1/4 keV band (e.g., Snowden et al. 1997a). Moreover, the *ROSAT* all-sky survey (Snowden et al. 1995; 1997b) clearly shows an overall intensity enhancement in the 0.5-2 keV range over the Galactic center hemisphere. Snowden et al. have suggested that this enhancement may represent a bulge of hot gas around the Galactic center.

I have tested how a simple, physically self-consistent model of the hot Galactic corona may account for the soft X-ray background, especially in the 3/4 keV band. The radiation in this band is sensitive to gas at temperatures $\gtrsim 1.5 \times 10^6$ K and is not as heavily attenuated by the X-ray-absorbing interstellar medium as in the 1/4 keV band. Furthermore, the effect of the small-scale ($\lesssim 1°$) clumpiness of the medium may be negligible over a unit X-ray absorption depth of $\sim 2 \times 10^{21}$ cm^{-2}. It is clear, though, that the observed background is strongly contaminated by various discrete X-ray sources in regions near the Galactic plane and by a few high Galactic latitude features such as Loop I and Eridanus superbubbles. I have therefore devised data reduction and analysis algorithms to minimize the effects of these contaminations. The results are encouraging. The model provides a frame work for both characterizing the global hot gas distribution of the Galactic corona and defining discrete X-ray emitting features.

In the following, I first talk about the data analysis and modeling procedures, and then present some preliminary results. Finally, I make some comparisons of the results with other independent measurements and extend the results into the 1/4 keV and 1.5 keV bands.

2 Data Analysis

The data used in this analysis come from the all-sky surveys made with *ROSAT* (e.g., Snowden et al. 1995) and *IRAS* (e.g., Boulanger & Perault 1988). The *ROSAT* data, in three energy bands ($\sim$ 1/4 keV, 3/4 keV, 1.5 keV), were down-loaded from MPE (http://www.rosat.mpe-garching.mpg.de/), and the *IRAS* 100μm data from HEASARC (http://skview.gsfc.nasa.gov/). The data were presented as surface brightness intensity maps (480$\times$240 pixels), Aitoff-projected in the Galactic coordinates. These maps, though not reflecting the intrinsic spatial resolutions (a few arcminutes) of the surveys, provide a convenient database for investigating the global distributions of both the X-ray background and the X-ray-absorbing medium.

For my data modeling purpose, I processed the data in three major steps. First, to approximately correct for the zero intensity level of the *IRAS* data, I shifted the 100μm map to match the estimated total absorption 9×10^{19} cm^{-2} in the Lockman Hole region (Snowden et al. 1994), using a conversion of 1.14 $\times 10^{20}$ cm^{-2}/(MJy sr^{-1}) (Boulanger & Perault 1988; Snowden et al. 1997a). Second, I removed strips of pixels that were not, or very poorly, covered in the surveys. Third, I compressed the maps by taking a median summary in each *non-overlapping* square of 3$\times$3 pixels. I discarded any square that contained one or more removed pixels (including those outside the sky projection). The median summary, as a resistant statistic, effectively removed the effects of outstanding small-scale ($\lesssim$ 1 deg) features (e.g., nearby galaxies and AGNs). The top panels of Fig. 1 present the processed data.

3 Modeling

To capture the global distribution of the X-ray background, I have considered a model that consists of two contributions: an isotropic extragalactic background (EGB) and an axisymmetric Galactic corona. The corona is assumed to be quasi-hydrostatic in the gravitational potential of the Galaxy (Wolfire et al. 1995; Johnston, Spergel, & Hernquist 1995). The potential is a sum of three components: a Miyamoto-Nagai disk, a spheroid bulge, and a logarithmic halo. These components together provide a nearly flat rotation curve from 1 to 30 kpc with a circular velocity of 225 km s^{-1} at the Galacto-centric radius 8.5 kpc of the Sun. It is the disk component, however, that is chiefly responsible for the corona structure visible in the *ROSAT* band. For ease of modeling, I neglect the angular momentum of the corona. The momentum should be small if the corona is fueled primarily by hot gas from the Galactic central region. Observations of nearby disk galaxies do indicate that hot gas outflows happen primarily in galactic center regions (e.g., Wang et al. 1995; Pietsch, Supper, & Vogler 1995). Furthermore, because the cooling of the hot gas at temperatures $\gtrsim 10^6$ K is most likely adiabatic, a polytropic equation of state of index 5/3 may be a reasonably good description of the hot gas. The shape and normalization of this model corona are determined by two adjustable parameters, chosen here to be the electron density (n_o) and temperature (T_o) of the corona near the Sun.

I calculate a model X-ray background, accounting for both the X-ray absorption and the *ROSAT*/PSPC spectral response. The calculation makes the following assumptions: (1) The EGB has a spectrum characterized by a power law of energy slope equal to 1 (e.g., Hasinger et al. 1993); (2) The hot gas of the corona is in a collisional ionization equilibrium (Raymond & Smith 1977); (3) Both the hot gas and the X-ray-absorbing medium are of solar metal abundances; (4) The absorption is foreground, which should be reasonably good except for regions close to the Galactic plane ($|b| \lesssim 10°$). X-ray surface brightness intensities of a unit volume are calculated first in a grid as a function of hot gas temperature and X-ray absorbing medium column density. Based on this grid, an intensity integration along a line of sight can be carried out efficiently. The integrated intensity of the corona, together with a partially absorbed EGB contribution, constitutes the model intensity m_i, where i represents a line of sight, or a pixel in a background map (Fig. 1).

The intrinsic intensity I_e of the EGB as well as the corona parameters, n_o and T_o, are constrained by a model fit to the *ROSAT* survey data d_i in the 3/4 keV band. Specifically, the fit minimizes the statistic $\chi^2 = \sum[(d_i - m_i)w_i]^2$, where the summation is over the adopted map pixels. The weight w_i is inversely proportional to the pixel-to-pixel intensity dispersion, $\sim 15\% \, m_i$, estimated in regions away from local X-ray enhancements (see also §4). In comparison, the counting statistical uncertainty is considerably smaller ($\lesssim 5\%$) in a ~ 4 deg^2 pixel with a typical exposure ~ 500 s of the

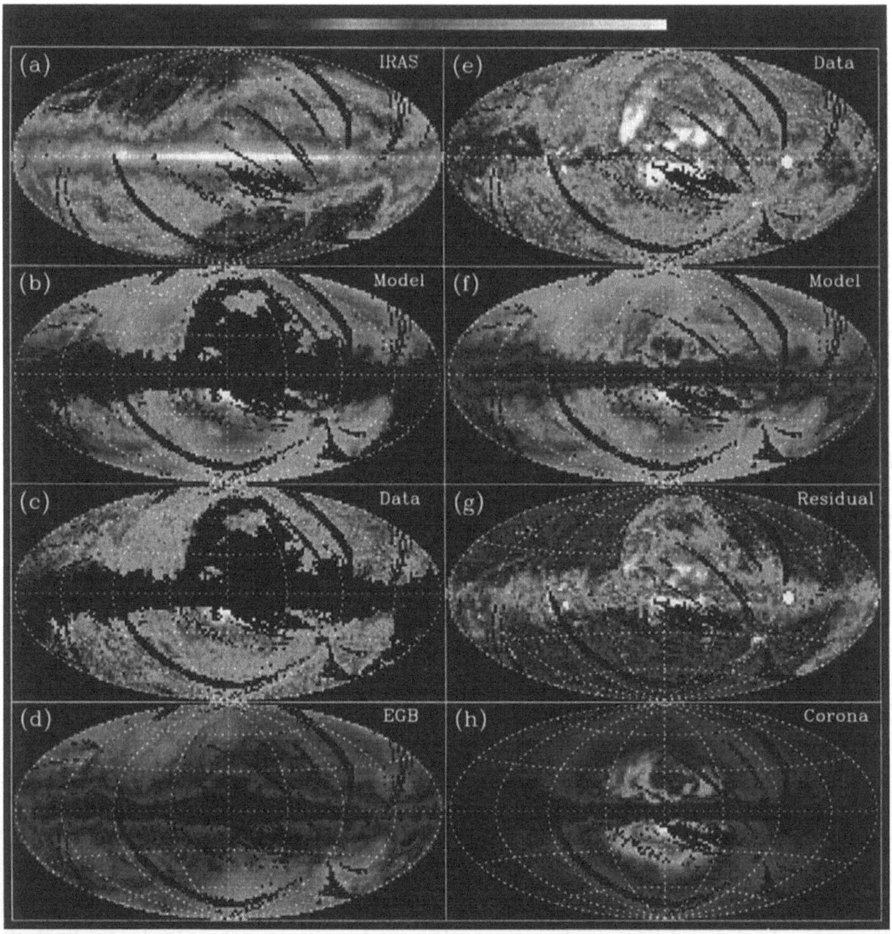

Fig. 1. Surface brightness intensity maps in the zero-centered Galactic coordinates with Aitoff equal-area projections. Both X-ray data and model are in the 3/4 keV band. From the top panel to the bottom one in the left column are the zero-level corrected *IRAS* 100μm survey (a); Model map covering only pixels used in the fit at the threshold 0.4 (b); *ROSAT* data in the same region (c); and the extragalactic background component of the model (d). The panels in the right column, arranged for an easy comparison with Fig. 4, are the *ROSAT* data (e); the model including regions not used in the fit (f); the residual of the *ROSAT* survey minus the model (g); the corona component of the model (h). The false color range is between 1 and 440 MJy sr^{-1} logarithmically for (a), 0 to 5 $\times 10^{-4}$ counts s^{-1} arcmin^{-2} linearly for (b), (c), (d), (e), (f), and (h), and -0.7 to 5 $\times 10^{-4}$ counts s^{-1} arcmin^{-2} for (g); the black represents the lowest intensity while the yellow is the highest; (see Plate 7).

ROSAT survey. Adding this uncertainty in the weight would cause only minor ($\lesssim 5\%$) changes in the results. The fit is conducted iteratively to remove outliers — pixels where the data deviate significantly from the model. Starting with w_i = constant, the fit gets an initial estimate of the model m_i. w_i is then recalculated. After removing those pixels with relative deviations $(d_i - m_i)w_i > 2$, an arbitrarily chosen initial threshold, the fit proceeds to minimize χ^2 again and so on, until no more pixel has $(d_i - m_i)w_i > 2$. Next, the threshold is reduced by 10%, an arbitrarily chosen step, and a new round of fitting and pixel removing starts. Fig. 2 shows how the model parameter fits evolve with the deviation threshold for pixel-removing. When the threshold is large, the outliers strongly influence the fit of the parameters, especially n_o and I_e. The parameter values also fluctuate until the threshold is reduced to ~ 0.2, when nearly half of the map pixels are removed. The results are not particularly sensitive to either the initial threshold or the step.

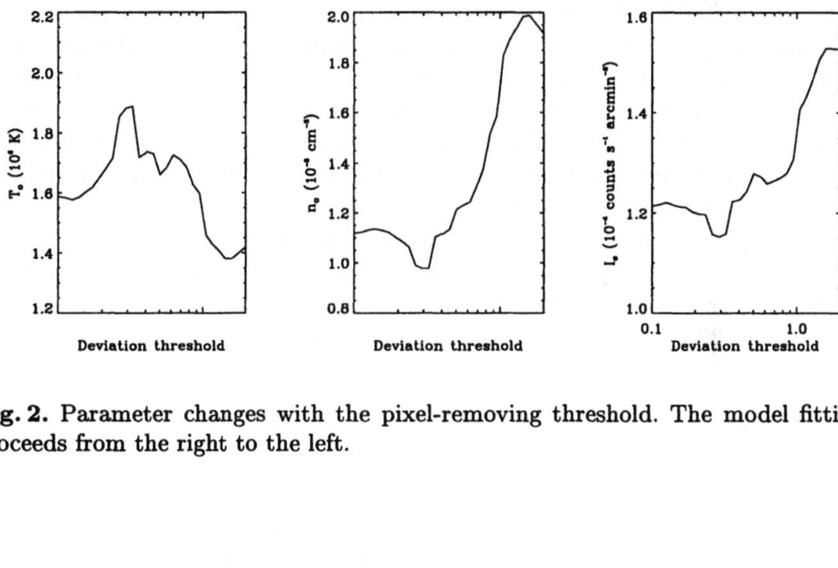

Fig. 2. Parameter changes with the pixel-removing threshold. The model fitting proceeds from the right to the left.

4 Results

Fig 1 includes a comparison of the data with the model background fitted at the threshold equal to 0.4, at which the parameter values are close to the converging values at $\lesssim 0.2$ and only a quarter of the pixels are removed. While the model is *intrinsically* symmetric relative to the Galaxy rotation axis, the asymmetric appearance of the model background in Fig. 1 (e.g., Panel f) is due to the foreground X-ray absorption (a). The dispersion of the data from the model is $\sim 0.15\%$, averaged over the pixels remained in the fit (Panels b and c in Fig. 1). Pixels excluded from the fit are almost entirely at the Galactic plane and in high Galactic latitude regions that are contaminated by Loop I (including the Sco-Cen starforming region) and Eridanus superbubbles as well as the Large Magellanic Cloud near the south Ecliptical pole (see Snowden et al. 1995 for a graphic illustration). These features and the disk component stand out in the residual map, because of the removal of global emission and absorption effects of the corona and EGB contributions. Part of the residual emission near the Galactic plane, however, may result from an incomplete removal of the corona contribution. The foreground assumption of the X-ray absorbing medium leads to a slight overestimation of the absorption near the Galactic plane, where some of the medium is located behind the corona emission. While the corona contribution (h) is strongly concentrated in the Galactic center hemisphere, the EGB (d) is more uniformly distributed in the sky, except for regions close to the Galactic plane.

Since the model describes the global distribution of the 3/4 keV background reasonably well, it is tempting to use the model fit as a characterization of the hot Galactic corona. From Fig. 2, I estimate the local hot gas electron density and temperature of the corona as $\sim 1.1 \times 10^{-3}$ cm^{-3} and $\sim 1.7 \times 10^{6}$ K. The uncertainties in these two parameters are $\sim 10\%$, corresponding to the fluctuations of the parameter values in the threshold range of $\lesssim 0.5$. Fig. 3 shows the spatial distribution of hot gas temperature in the corona. The density and temperature at the Galactic center are $\sim 1.2 \times 10^{-2}$ cm^{-3} and $\sim 8.5 \times 10^{6}$ K. The integrated mass, thermal energy, and bolometric luminosity of gas at temperatures $> 10^{6}$ K are $3 \times 10^{7} M_{\odot}$, $\sim 2 \times 10^{56}$ ergs, and $\sim 2 \times 10^{40}$ ergs s^{-1} (only 15% in the 0.5-2 keV range), respectively. The inferred mean radiative cooling time scale of the gas is then $\sim 3 \times 10^{8}$ years. The radiative cooling accounts for about 2% of the total mechanical energy input from supernovae in the Galaxy (assuming one supernova per 30 years and 10^{51} ergs per supernova).

5 Discussion

The above corona model, constrained by the spatial distribution of the 3/4 keV band background, makes specific predictions that can be compared with various independent measurements. First, the model predicts a thermal pressure of the corona near the Sun as $P/k \sim 3.6 \times 10^{3}$ K cm^{-3}. This prediction is within the uncertainty range of $\sim 2 \times 10^{3} - 4 \times 10^{3}$ K cm^{-3} inferred from CIV emission/absorption lines (Martin & Bowyer 1990; Shull & Slavin 1994) and from the two-phase structure of high-velocity clouds (Wolfire et al. 1995). Second, Fig. 3 shows that the spectral characteristics of coronal gas varies strongly in the sky. Based on broad-band spectral properties of the soft X-ray background, various attempts have been made to estimate the average temperature of hot gas beyond the Local Bubble (e.g., Garmire et al. 1992; Wang & McCray 1993; Kerp 1994; Sidher et al. 1996; Snowden et al. 1997b). Such an estimate depends on an assumption about the EGB spectrum, which remains poorly constrained in the 0.1-2 keV range. Nev-

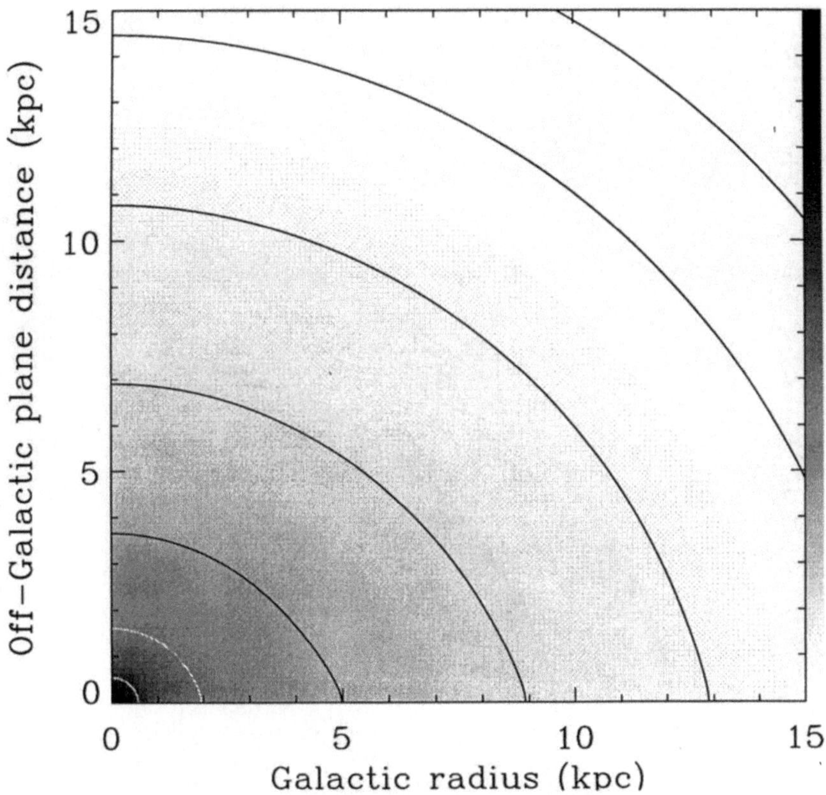

Fig. 3. Model gas temperature distribution of the corona in the Galactic cylindrical coordinates. The gray-scale is in the range of 0.18 to 8.4 $\times 10^6$ K, while the contours are at 0.5, 0.75, 1.1, 1.7, 2.5, 3.8, and 5.7 $\times 10^6$ K.

ertheless, the estimated hot gas temperature falls within a range between $\sim 10^{6.0}$ K and $\sim 10^{6.4}$ K. This range can in principle be reproduced with the model, depending on specific lines of sight (Fig. 3). Third, the model predicts a corona contribution of $\sim 3 \times 10^{-5}$ counts s^{-1} $arcmin^{-2}$ in the PSPC R4 band, which is centered around 0.7 keV (Snowden et al. 1997b), toward the X-ray shadowing cloud ($l, b = 295°, -42°$) in the Magellanic Bridge. This contribution is consistent with the measured *Galactic* component of $2.7(1.5 - 3.7) \times 10^{-5}$ counts s^{-1} $arcmin^{-2}$ (90% confidence interval; Wang & Ye 1996). Thus the corona model is useful for a uniform explanation of various observations of distant hot gas at high Galactic latitudes.

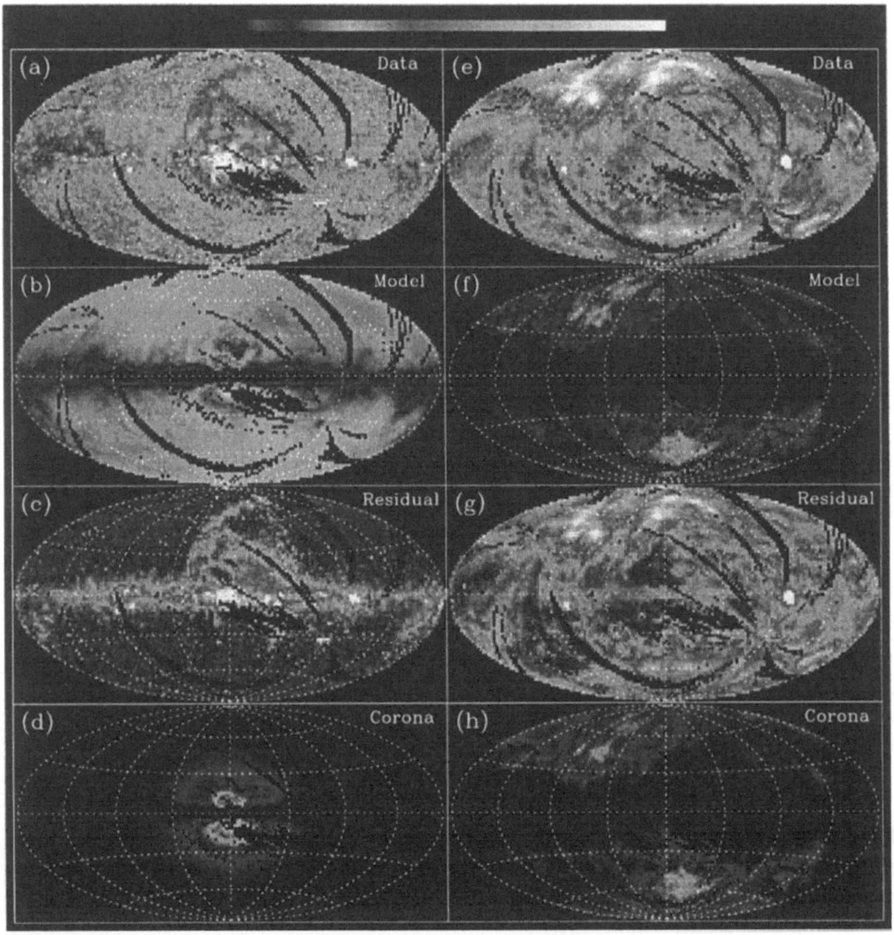

Fig. 4. Data and model predictions in the 1.5 keV band (left column) and in the 1/4 keV band (right column). The projections and sky coverage are the same as those in Fig. 1. From the top row to the bottom one are the *ROSAT* survey data, the model predictions of the corona plus the extragalactic background, the residual maps of the survey data minus the model, and the corona components of the model in the two bands. The false color range in the left column spans between 0 to 5 $\times 10^{-4}$ counts s^{-1} $arcmin^{-2}$ for the panels (a), (b) and (d) and -0.7 to 5 $\times 10^{-4}$ counts s^{-1} $arcmin^{-2}$ for (c); the range in the right column is all between 0 to 15 $\times 10^{-4}$ counts s^{-1} $arcmin^{-2}$; (see Plate 8).

Fig. 4 further compares the model *predictions* in the 1.5 keV and 1/4 keV bands with the corresponding *ROSAT* survey data. The data in these bands are processed in the same way as in the 3/4 keV band (§2). The intrinsic EGB intensity is fixed as 1.4×10^{-4} counts s^{-1} $arcmin^{-2}$ in the 1.5 keV band, accounting for the total background observed in high Galactic latitude, anti-Galactic center hemisphere, and as 4.2×10^{-4} counts s^{-1} $arcmin^{-2}$ in the 1/4 keV band (Barber, Roberts, & Warwick 1996). The model explains well the global distribution of the background in 1.5 keV band. Residual features (Panel c) morphologically mimic those in the 3/4 keV band. The corona contribution is confined more into the central region of the Galaxy than in the 3/4 keV band.

In the 1/4 keV band, the model demonstrates that the corona contribution may vary significantly across the sky in both intensity and spectrum. The model predicts that up to $\sim 45\%$ of the observed background intensity may arise in the corona, depending on individual lines of sight. In the direction of the Draco cloud ($l, b \sim 90°, 39°$), for example, the corona contribution is about 30%, which is significantly less than the distant component ($\sim 60\%$) inferred from the X-ray shadowing measurement (e.g., Burrows & Mendenhall 1991). The extragalactic component can account for an additional $\sim 10\%$. The rest of the distant component may arise in a disk component that is not included in the model. In addition, the model predication may be an underestimation of the true corona contribution in the 1/4 keV band. The assumed ionization equilibrium may break down for gas of $< 10^6$ K, because the radiative cooling time scale becomes shorter than the relevant recombination time scales (e.g., Breitschwerdt & Schmutzler 1994). Furthermore, the unit absorption depth ($\sim 1 \times 10^{20}$ cm^{-2}) in the band is substantially shorter than in the higher energy bands. Thus the background intensity and distribution, even at high Galactic latitudes, is sensitive to both the uncertainty in the *IRAS* map as an X-ray-absorbing medium tracer and the clumpiness of the medium. Accounting for these effects would tend to increase the corona contribution in the 1/4 keV band.

In conclusion, the simple model presented here, though admittedly simplistic, describes reasonably well the global X-ray background distribution at high Galactic latitudes ($\gtrsim 20°$) in the 0.5-2 keV range, and appears to be consistent with independent measurements of hot gas beyond the Local Bubble. The model can be improved when better data products (e.g., source-removed high resolution *ROSAT* maps and *COBE* DIRBE calibrated *IRAS* maps — Snowden et al. 1997a) become available.

Acknowledgements. The author is grateful to S. L. Snowden for preprints, W. T. Reach for information about the *IRAS* and *COBE* DIRBE data, and the conference organizers for the invitation to give this talk. This work is supported partly by NASA under the grant NAG 5-2716.

References

Barber, C. R., Roberts, T. P., & Warwick, R. C. (1996): MNRAS 282, 157
Boulanger, F., & Perault, M. (1988): ApJ 330, 964
Breitschwerdt, D., & Schmutzler, T. (1994): Nature 371, 774
Burrows, D. N., & Mendenhall, J. A. (1991): Nature, 351, 629
Garmire, G. P., et al. (1992): ApJ 399, 694
Hasinger, G., et al. (1993): A&A 275, 1
Johnston, K. V., Spergel, D. N., & Hernquist, L. (1995): ApJ 451, 598
Kerp, J. (1994): A&A 289, 597
Kuntz, K. D., Snowden, S. L., & Verter, F. 1997, ApJ, submitted
Martin, C., & Bowyer, S. (1990): ApJ 350, 242
McCammon, D., & Sanders, W. T. (1990): ARA&A 28, 657
Pietsch, W., Supper, R., & Vogler, A. (1995): in The Interplay between Massive Star Formation, The ISM, and Galaxy Evolution, p179
Raymond, J. C, & Smith, B. W. (1977): ApJS 35, 419, and updated by Raymond, J. C. and installed in the XSPEC software package
Sidher, S. D., Sumner, T. J., Quenby, J. J., & Gambhir, M. (1996): A&A 305, 308
Shull, J. M., & Slavin, J. (1994): ApJ 427, 784
Snowden, S. L., et al. (1991): Science 252, 1529
Snowden, S. L., McCammon, D., & Verter, F. (1993): ApJL 409, 21
Snowden, S. L., et al. (1994): ApJ 430, 601
Snowden, S. L., et al., (1995): ApJ 454, 643
Snowden, S. L., Egger, R., Finkbeiner, D., Freyberg, M. J., & Plucinsky, P. P. (1997a): ApJ, submitted
Snowden, S. L., et al. (1997b): ApJ, in press
Spitzer, L. (1956): ApJ 124, 20
Wang, Q. D., & McCray, R. (1993): ApJL 409, 37
Wang, Q. D. et al. (1995): ApJ 453, 783
Wang, Q. D., & Ye, T. (1996): New Astronomy 1, 245
Wolfire, K. G., McKee, C. F., Hollenbach, D., & Tielens, A. G. G. M. (1995): ApJ 453, 673

A Soft X-ray Halo of the LMC?

M.J. Blondiau

Radioastronomisches Institut der Universität Bonn, D-53121 Bonn, Germany

Abstract. *ROSAT* X-ray observations of the LMC show several X-ray shadows in the 3/4 keV and 1.5 keV bands. The 1/4 keV band is dominated by emission of the galactic foreground. It is uncertain whether there is additional soft X-ray contribution from the LMC. The so-called HI-L-component is located in front of the HI-D-component and therefore the matter of the L-component accounts for most of the observed X-ray shadows. The 1/4 keV *ROSAT* image of the LMC reveals a slightly fainter X-ray emission region at the position of the eastern part of the HI-L-component. This suggests the existence of a soft X-ray halo in the LMC. Investigations of the X-ray shadows show that also the lower energy end of the 3/4 keV band originates partly from this soft X-ray plasma which is located in front of the absorption features. This additional soft X-ray emission, in the lower energy end of the 3/4 keV band, is brighter than the typical diffuse galactic emission at the same galactic latitude.

1 Introduction

The investigation of X-ray shadows is a powerful tool to disentangle the 3-D structure of the interstellar medium on large angular scales. Wang et al. (1991) verified the existence of a widely distributed X-ray gas in the LMC. These authors also discovered the first X-ray shadow in the LMC. The largest X-ray shadow in the LMC is located south of the giant HII region 30 Doradus. In the *ROSAT* X-ray images, different spatial parts of this shadow show up in different energy bands. The dark western part shows up in the *ROSAT* 3/4 keV and 1.5 keV bands (Blondiau et al. 1997), while the easternmost part is detected in the softer *ROSAT* 1/4 keV band. In the *ROSAT* 3/4 keV and 1.5 keV bands the structure of a $T = 10^{6.7}$ K hot plasma in the LMC is visible (Blondiau 1996). The question is whether there is a contribution from a softer diffuse X-ray emitting plasma of the LMC also in the 1/4 keV band.

2 Data

I have analyzed pointed X-ray data of the LMC extracted from the *ROSAT* public archive. The whole area of Fig. 2 is a mosaic of 172 individual pointed PSPC observations. The minimum integration time is 1 ksec while the maximum is about 204 ksec and is reached in the region of the supernova 1987A. The data are corrected for the various effects associated with the so-called non-cosmic X-ray backgrounds (for details see Blondiau et al. 1997).

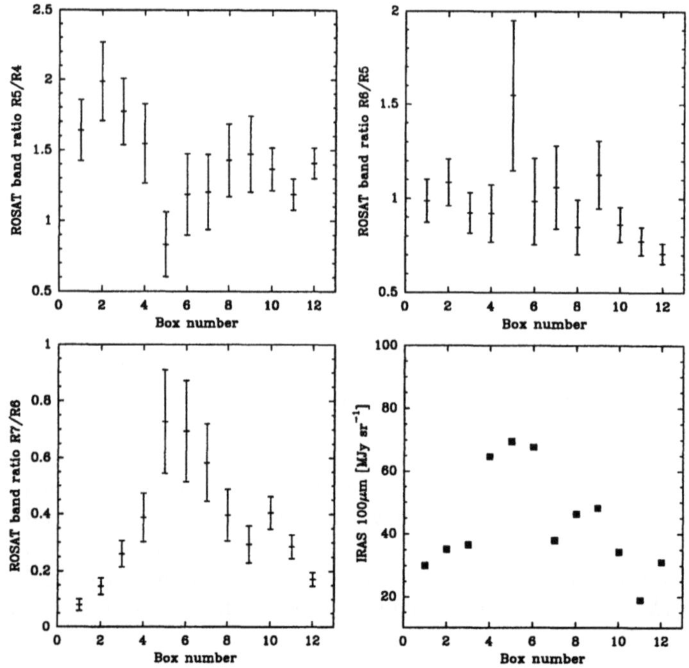

Fig. 1. This figure shows the *ROSAT* energy band ratios $\frac{\mathrm{R5}}{\mathrm{R4}}, \frac{\mathrm{R6}}{\mathrm{R5}}$ and $\frac{\mathrm{R7}}{\mathrm{R6}}$ across the area of the X-ray shadow described in Blondiau et al. (1997). The boxes 5 to 9 are located at the shadow while the other boxes are located in X-ray bright "OFF-shadow-positions". In the bottom right-hand panel the *IRAS* 100μm intensity which serves as a tracer for the total X-ray absorbing column density is plotted versus the same box positions. The $\frac{\mathrm{R7}}{\mathrm{R6}}$ energy band ratio shows the expected behaviour, because the X-ray spectrum is hardened after absorption. The $\frac{\mathrm{R5}}{\mathrm{R4}}$ energy band ratio shows the opposite behaviour, which can be explained if one assumes the existence of a foreground diffuse soft X-ray emitting gas.

3 Energy Band Ratios and Radiation Transfer

First hints to disclose X-ray shadows are given by a positional anticorrelation of X-ray intensity versus HI column density or *IRAS* 100μm intensity. The latter serves as a tracer for the total X-ray absorbing column density of neutral and molecular gas. The behaviour of *ROSAT* band ratios in the regions of X-ray shadows in the LMC was investigated in my thesis (Blondiau 1996). It turned out that in every case the behaviour of the R5/R4 band ratio across the X-ray shadow is in contrast to the behaviour in the R7/R6 and R6/R5 band ratios. For a definition of the *ROSAT* energy bands see Snowden et al. (1994). The best explanation for this is the existence of a cooler X-ray plasma in front of the absorbing clouds which gives a significant contribution to the R4 band intensity. Of course, this contribution can be produced by the

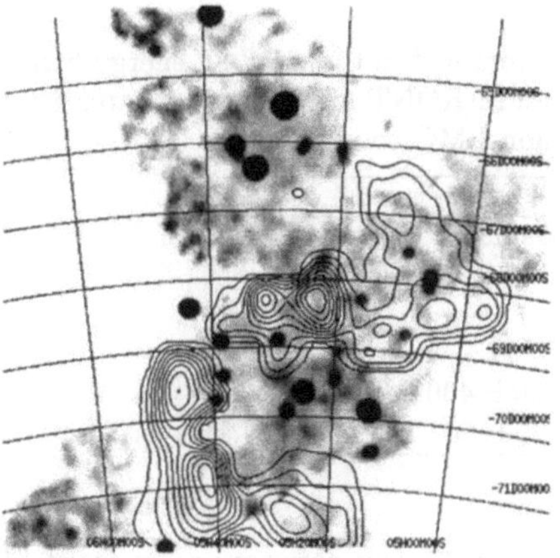

Fig. 2. This image shows a map of the X-ray emission in the 1/4 keV band in the region of the LMC. It consists of 172 individual *ROSAT* pointed observations and is centered at RA(2000)=5^h30^m, DEC(2000)=$-68°$. The map covers a sky area of 4° radius and its angular resolution is about 8′. The linear intensity scale ranges from 5 to $10 \cdot 10^{-4}$ cts s^{-1} arcmin^{-2}. The contour plot shows the distribution of HI column density in the so-called HI-L-component (Luks & Rohlfs 1992). The contour lines are in steps of 10% to 100% of the maximum value. The maximum column density is about $1.8 \cdot 10^{21}$ cm^{-2}. The angular resolution of the HI-data is 15′.

galactic halo emission of our own galaxy. Fig. 2 shows the distribution of the 1/4 keV diffuse X-ray emission across the whole LMC region which is rather uniform in comparison to the emission in the R4 to R7 *ROSAT* energy bands (Blondiau et al. 1996). This is what one would expect if the galactic X-ray emission and the galactic absorption do not much change on angular scales of Fig. 2 (8′). Therefore one has to solve a radiation transfer equation to decide whether there is an additional LMC contribution in the *ROSAT* R4 band or not. In case of the largest X-ray shadow I presented a radiation transfer calculation in the *ROSAT* R4 – R7 bands. I used typical values for galactic halo X-ray intensities (Snowden et al. 1993) and for the cosmological background (Hasinger et al. 1993) intensities and subtracted them from the observed values. Then one has to solve the following equation to determine whether there is LMC X-ray intensity originating in the foreground (FRONT) of the absorbing cloud or not, $I_X = I_{FRONT} + I_{BACK}\, e^{-\sigma_X N_X}$. Here X denotes both, the ON-shadow and OFF-shadow-position. The absorbing column densities are derived from the *IRAS* 100μm emission. Converting these emission values leads to an estimation of the total X-ray absorbing column density on the lines of sight. At the OFF-position one gets $N_{OFF} = (5.2 \pm 0.5) \cdot 10^{21}$ cm^{-2} and on the ON-position $N_{ON} = (11.1 \pm 0.5) \cdot 10^{21}$ cm^{-2}. As expected Tab. 1 shows that there is an additional LMC X-ray foreground only in the R4 band present, which accounts for the correct solution of the radiation transfer. So,

Table 1. Count rates in units of 10^{-6} cts s^{-1} $arcmin^{-2}$ in the *ROSAT* energy bands R4 to R7 ON and OFF the shadow. BACK and FRONT give the count rates from the absorption and after subtraction of the non-LMC components.

I_X	R4	R5	R6	R7
OFF-shadow	363± 5	627± 7	586± 7	144± 4
ON-shadow	107± 2	127± 2	139± 2	76± 2
Galactic halo	37	49	8	0
Cosm. background	33	26	72	45
BACK	3500± 700	4600± 900	2800± 400	260± 20
FRONT	70± 20	20± 40	-20± 50	-10± 20

there should be a second softer LMC intrinsic X-ray component in front of the absorbing cloud which also should show up in the *ROSAT* 1/4 keV band.

4 Interpretation

As a result of my thesis, it turned out that the HI-L-component is located in front of the D-component, because most X-ray shadows are associated with absorbing HI-clouds of low radial velocities, therefore belonging to the HI-L-component. In the last section I showed that there is some evidence for a second soft X-ray plasma that should show up below the energy of the *ROSAT* R4 band, that means in the 1/4 keV band. Fig. 2 shows the HI column density distribution of the L-component superimposed on the X-ray intensity distribution in the *ROSAT* 1/4 keV band in the LMC area. Right at the eastern wing of the L-component the X-ray intensity is relatively low. In this area the L-component seems to absorb the LMC contribution of the 1/4 keV band radiation. So, there are hints for a soft X-ray plasma ("I_{FRONT}") situated in front of the X-ray absorption clouds which weaken the $T = 10^{6.7}$ K "I_{BACK}" X-ray plasma. It is only absorbed by the eastern wing of the HI-L-component, which is the absorbing gas layer of the LMC that is located also in front of the softer X-ray plasma denoted as I_{FRONT}.

References

Blondiau, M.J. (1996): Ph.D. thesis, University of Bonn
Blondiau, M.J., Kerp, J., Mebold, U., Klein, U. (1996): MPE Report, 263, 321
Blondiau, M.J., Kerp, J., Mebold, U., Klein, U. (1997): A&A 323, 585
Hasinger, G., et al. (1993): A&A 275, 1
Luks, Th., Rohlfs, K. (1992): A&A 263, 41
Snowden, S.L., McCammon, D., Verter, F. (1993): ApJL 409, L21
Snowden, S.L., et al. (1994): ApJ 424, 714
Wang, Q., Hamilton, T., Helfand, D.J., Wu, X. (1991): ApJ 374, 475

The ORFEUS Far Ultraviolet Spectrum of the LMC Binary Star HDE 269546

H. Widmann[1], G. Krämer[1], I. Appenzeller[2], J. Barnstedt[1], A. Fromm[1], M. Gölz[1], M. Grewing[3], W. Gringel[1], C. Haas[1], W. Hopfensitz[1], N. Kappelmann[1], J. Krautter[2], A. Lindenberger[1], and H. Mandel[2]

[1] Institut für Astronomie und Astrophysik, D-72076 Tübingen, Germany
[2] Landessternwarte Heidelberg, Königsstuhl, D-69117 Heidelberg, Germany
[3] Institut de Radio Astronomie Millimétrique, F-38406 St. Martin d'Hères, France

Abstract. Far UV high resolution spectra of 3 LMC and SMC stars were obtained with the Echelle spectrograph during the second *ORFEUS* mission in Dec. 1996. We present the first results from observations of the LMC star HDE 269546. We find definitely components of very hot gas identified as OVI and SVI absorption in the galactic halo of the Milky Way and in the LMC. Additionally, more than 30 ions of the most abundant elements in different stages of ionization can be identified in both our galaxy and the LMC. For the first time we can identify a significant absorption component of molecular hydrogen in the *ORFEUS* II Echelle spectrum with a redshift of 200 km s^{-1}, doubtlessly to be attributed to the LMC.

1 Introduction

With the *Copernicus* satellite the first detection of OVI (Rogerson 1973) showed the presence of a widely spread hot ISM component in the Milky Way. *IUE* and more recently *HST* spectra from distant, high galactic latitude stars and from many extragalactic objects, including stars in the LMC and SMC, gave insight into the structure of the ISM in the disc and halo of our galaxy (Savage 1987, Spitzer 1990, Danly 1991) and the existence of a hot corona surrounding the Magellanic Clouds. *Copernicus* observations of the Lyman and Werner bands of molecular hydrogen gave a first insight into the distribution of this most abundant molecule in the ISM (Spitzer & Jenkins 1975, Shull & Beckwith 1982). *ORFEUS* offered the first opportunity to study the very hot OVI, and the very cool H_2 ISM component in greater distances and with sufficient spectral resolution for ISM studies. Results from *ORFEUS* I observations of the hot halo gas were published by Hurwitz et al (1995) and Hurwitz & Bowyer (1996).

2 Instrumentation, Observations, and Data Reduction

The *ORFEUS* 1m-telescope is equipped with two alternatively operating spectrometers. The details about the telescope and the Echelle spectrometer

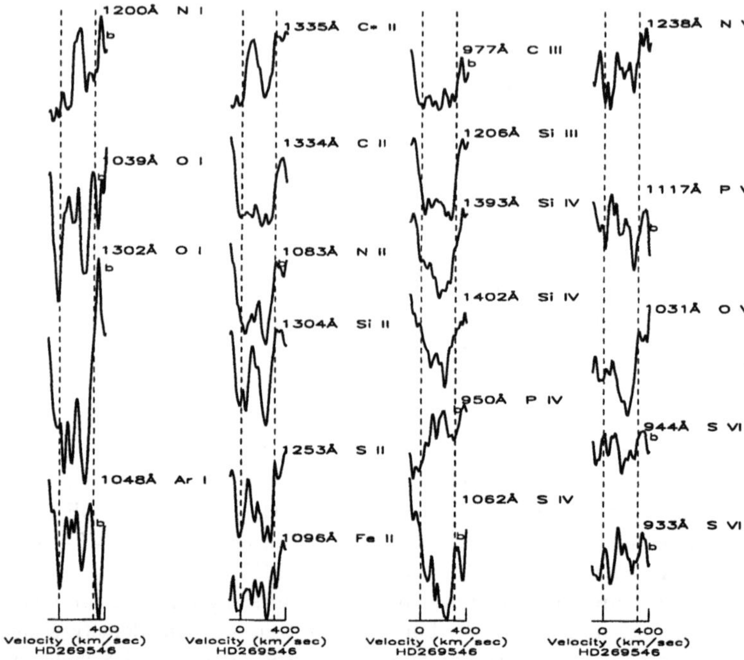

Fig. 1. A compilation of several absorption lines from neutral, singly and multiply ionized elements, including the most important O VI doublet. (b: blends)

Table 1.

Ions	center velocities km s^{-1}				
N I 1200 Å; O I 1039 Å; O I 1302 Å Ar I 1048 Å; C* II 1335 Å; C II 1334 Å N II 1083 Å; Si II 1304 Å; S II 1253 Å Fe II 1096 Å	-50	0	100	200-250	
C III 977 Å; Si III 1206 Å; Si IV 1393 Å Si IV 1402 Å; P IV 950 ÅS IV 1062 Å	-30	0	70	220	270
N V 1238 Å; P V 1117 Å; O VI 1031 Å S VI 944 Å; S VI 933 Å		0(?)	150	200-300	

are discussed in Krämer et al (1990), the Berkeley spectrometer is described in Hurwitz & Boywer (1995). The main properties of the Echelle spectrometer used for the measurements discussed herein are: spectral range: 912 Å to 1410 Å; spectral resolution: $\frac{\lambda}{\Delta\lambda} > 10,000$; effective area: 1 cm^2. *ORFEUS* has been flown in orbit twice with the free flying carrier *ASTRO-SPAS*, launched and retrieved onboard the Shuttles STS 51, Discovery in Sept. 93 , and STS 80, Columbia in Nov.-Dec.96. During the *ORFEUS* II Mission several background stars in the LMC and SMC have been observed, three of them with

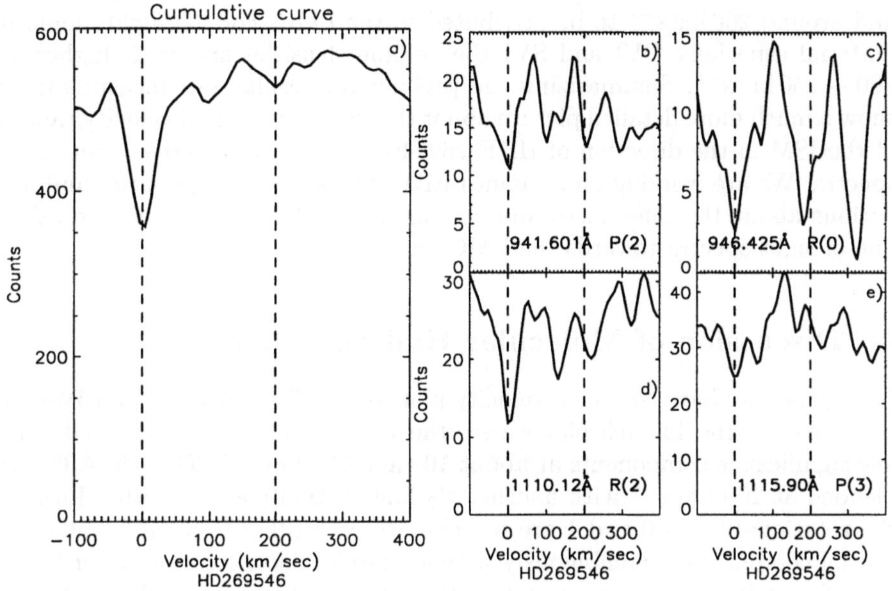

Fig. 2. a: Coaddition of 25 Lyman and Werner lines. b–e: Velocity plot of 4 different Lyman or Werner lines of molecular hydrogen.

the Echelle spectrometer. Here we present the first results from the observation of the background source HDE 269546, a B3Ia-WN3 binary that belongs to the superbubble Hen 144 (Grewing & Schulz-Lüpertz (1980)). The total integration time of this target was 110 min. We used a preliminary data reduction package: The intrinsic wavelengths scale is correct to about 0.1 Å. The velocity scale is not yet finally corrected, according to *IUE* spectra the scale should be shifted by about $-30\,\mathrm{km\,s^{-1}}$ (Chu et al (1994)). The flux scale is arbitrary.

3 Discussion of the ion spectra

We find in our spectrum more than 50 interstellar lines. Some of these are blended mostly with molecular hydrogen lines, but there are more than 30 lines of many important elements in different ionization stages, most of them not accessible with *HST*. Figure 1 shows a velocity plot of some of these lines, the spectra are smoothed with a 3 pixel boxcar, the pixel size is 0.03 Å. We identify between 4 and 5 significantly different components at each ionization stage with slightly different center velocities (Tab. 1). There is also strong indication that at least at low and intermediate stages there is some material in all velocity ranges. Qualitatively we can state that the highest column densities for the neutral and weakly ionized material are found at 0 and around 200 km s^{-1} to be attributed to the LMC. For the highly ionized material especially OVI and SVI the column densities are much higher at 200 – 250 km s^{-1}. Summarizing the preliminary results, our measurements show a much more detailed picture about the temperature and velocity ranges of the ISM in the direction of HDE 269546 than could be derived from *IUE* spectra. We are working on a quantitative analysis of this spectrum and are curious about the differences and similarities in the results from the other line of sight spectra towards the LMC and SMC.

4 Discussion of Molecular Hydrogen Lines

In Fig. 2a-e we have compiled velocity plots of 4 different Lyman and Werner band lines of H_2. In each plot we see the v=0 km s^{-1} component and with low significance components at about 100 and 200 km s^{-1}. The 946 Å line is blended at 200 km s^{-1} with another H_2 line. Actually also all other lines of the rotational levels 0 and 1 are too close to each other to identify different spatial components unambigously without careful deconvolution. In order to identify significant H_2 components with nonzero velocities we selected 25 H_2 lines from higher rotational levels which are not contaminated by any other known line within a range from -200 to +400 km s^{-1} and coadded them. The result, seen in Fig. 2a, shows definitely an H_2 component at about 200 km s^{-1} and, less significantly another component at about 90 km s^{-1}. At least the higher velocity material belongs to the LMC, consistent with the neutral gas discussed in Sect. 3.

Acknowledgements. *ORFEUS* could only be realized with the support of all our German and American colleagues and collaborators. The *ORFEUS* program was supported by DARA grants WE3 OS 8501 and WE2-QV 9304and NASA grant NAG5-696.

References

Chu, Y.-H., Wakker, B., Mac Low, M.-M.Garcia-Segura, G. (1994), AJ 108, 1696
Danly, L.(1991), in *IAU Symp. 144*, ed H.Bloemen (Kluwer Acad. Publishers), 53
Grewing, M. et al (1980), in *Proc. of the Second IUE Conference*, ESA SP-157, 357
Hurwitz, M., Bowyer, S. (1995), ApJ 446, 812
Hurwitz, M., Bowyer, S. (1996), ApJ 465, 296
Hurwitz, M., Bowyer, S., Kudritzki, R.-P., Lennon, D.J. (1995), ApJ 450, 149
Krämer et al. (1990) in *Observatories in Earth Orbit and Beyond*, ed. Y. Kondo (Kluwer Academic Publishers), 177
Rogerson, J.B. et al (1973), ApJ, 181, L110
Savage, B.D., Massa, D. (1987), ApJ 314, 380
Shull, J.M., Beckwith, S. (1982), ARA&A 20, 163
Spitzer, L., Jenkins, E.B. (1975), ARA&A 13, 133
Spitzer, L. (1990), ARA&A 28, 71

The HI Supergiant Shells in the Large Magellanic Cloud

S. Kim[1], L. Staveley-Smith[2], R.J. Sault[2], M.J. Kesteven[2], D. McConnell[2], M.A. Dopita[1], and M. Bessell[1]

[1] Mount Stromlo and Siding Spring Observatories, Weston Creek PO. ACT 2611, Australia
[2] Australia Telescope National Facility, CSIRO, 76 Epping, NSW 2121, Australia

Abstract. We present the result of an HI aperture synthesis mosaic of the Large Magellanic cloud (LMC), made recently with the Australia Telescope Compact Array (ATCA). The resolution of the mosaiced images is $1\rlap{.}'0$ (15 pc, using a distance to the LMC of 50 kpc). In contrast to its appearance at other wavelengths, the LMC is remarkably symmetrical in HI on the largest scales, with the bulk of the HI residing in a disk of diameter $8\rlap{.}^\circ4$ (7.3 kpc). Outer spiral structure is clearly seen, though the features appear to be due to differential rotation, therefore transient in nature. On small to medium scales, the combined action of numerous shells and supershells dominate the structures and motions of the HI gas in the LMC. A good correlation is seen between supershells previously identified in Hα (e.g. Meaburn 1980) and HI structures. We compare the results with a new wide-field Hα image.

1 Introduction

We completed the high resolution HI survey of the Large Magellanic Cloud with the Australia Telescope Compact Array (ATCA). The LMC has been studied in neutral hydrogen gas with Parkes Telescope by McGee and Milton (1966), Rohlfs et al. (1984) and Luks and Rohlfs (1992), but the spatial resolution of these previous surveys was 14.9' (220 pc). The new survey with the ATCA has much higher spatial resolution, 1' (15 pc). We compare the HI distribution in the LMC to the distribution of Hα emission, taken with a camera lens mounted on the 16 inch at Mount Stromlo and Siding Spring Observatories. The Hα images cover the same area of HI Mosaic survey with a spatial resolution of 20". These two surveys offer a unique probe of the detailed relationship between the ionized phase and atomic phase in the gas of the LMC.

2 Observations and Data Reduction

The detailed observing strategy and the data reduction, which involves the mosaicing together of 1344 fields are described in Kim et al. (1997). The central observing frequency was 1.419 GHz, corresponding to a central heliocentric velocity of 297 km s^{-1} , a velocity coverage of -33 to 627 km s^{-1} with a velocity resolution of 1.65 km s^{-1} .

For the wide-field survey of Hα emission in the LMC, a Nikon 30.72 mm *f*/5.0 camera lens was mounted on the 16 inch Telescope with a filter assembly and a cooled (2048 × 2148) CCD. Each 15 μm pixel corresponds to 20."63 on the sky, giving a total field size of 11.7° square. The Hα filter was centered at 6569 Å with a FWHM of 15 Å. Continuum subtraction was done with another image centered at 6620 Å. The exposure time for both Hα and continuum images was 4 × 900 s.

3 Results

Figure 1 shows the HI brightness temperature map of the LMC, constructed from the brightest HI component at each spatial position. The coordinate scale is a J2000 tangent plane centered on $\alpha = 05^h23^m$, δ=-69°44'(J2000). The heliocentric velocity range in which we find HI emission is 190 - 370 km s^{-1} . On the large scales, the HI distribution in the LMC is nearly axisymmetric, in contrast with Hα emission from the result of a new Hα survey (Fig. 2). The geometric inclination obtained from the axial ratio (a/b) of the apparent ellipse is 22°.

In the channel maps, we see the southern spiral pattern which is close to the optical bar and extends out from the 'B3' stub, tentatively identified by de Vaucouleurs and Freeman (1972). This pattern appears in the velocity range, 215 - 258 km s^{-1} . In the velocity range, 202 - 289 km s^{-1} , the HI gas in the LMC has stretched toward the HI bridge between the LMC and the SMC.

On the smaller scales, the peak brightness temperature map of the LMC shows filamentary structures, combined with HI holes and shells. In our preliminary analysis we have defined 32 HI supergiant shells (Fig. 2). The selection criteria we applied are 1) The HI shells (edge-brightened ring-like structure), larger than 600 pc are visible in at least three integrated channel maps (the line width of each integrated channel map is 5 km s^{-1}), and 2) The approaching and receding hemispheres (or one of the hemispheres) are visible in position-velocity cuts (*P*-*V*) diagrams.

Many of the shells are difficult to identify as expanding shells in the *P*-*V* diagrams due to relatively low sensitivity resulting from the lack of HI inside shells and confusion of interlocking shells.

If we compare the distribution of HI Supergiant shells in each of the four quadrants of the LMC, we clearly see that the largest number of shells are located in the south-eastern quadrant, near the 30 Doradus nebula (Fig. 2). Fig. 2 also shows the age of each shell, t_s (in units of Myr), calculated from $t_s = (3/5)(R_s/V_s)$ given by the standard theory of wind-driven bubbles

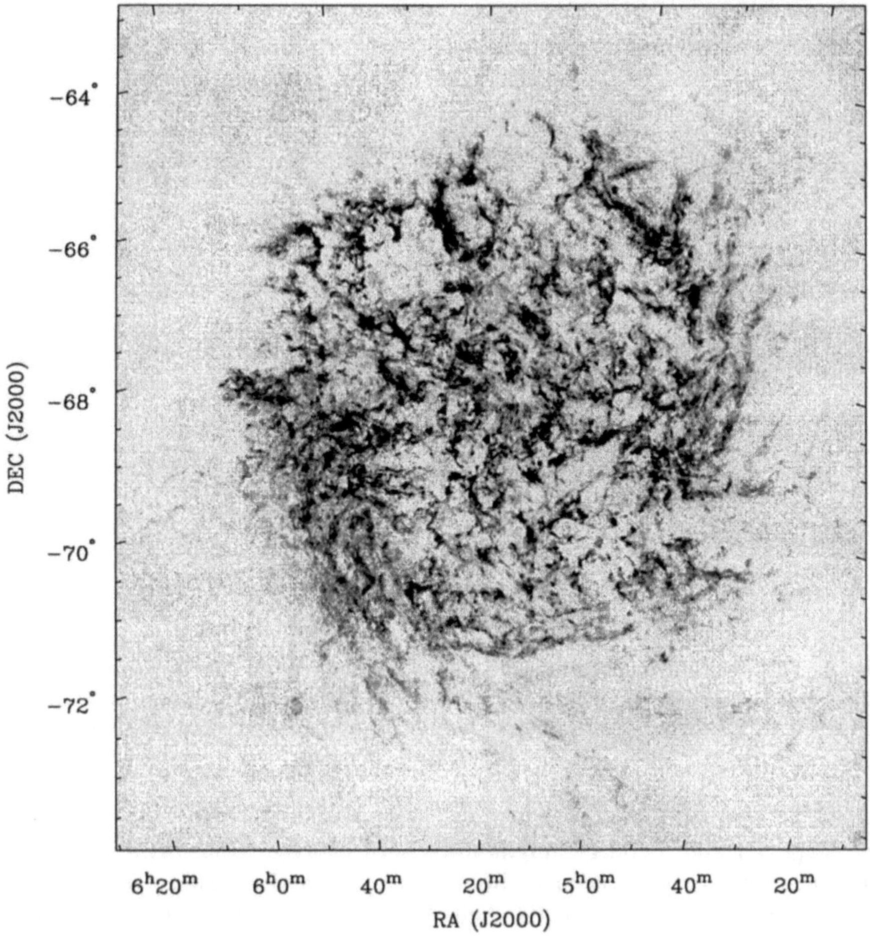

Fig. 1. The peak 21-cm brightness temperature map (measured at each channel) from the ATCA survey of HI in the LMC.

(Weaver et al. 1977). R_s is the shell radius in units of pc and V_s is the expansion velocity of the shell in units of km s^{-1} .

The measured ages of these expanding supergiant shells are distributed between 3 Myr and 25 Myr. The bubbles in Fig. 2 may also be classified morphologically by comparing the HI and Hα structure. We define five categories for the geometrical correlation between HI and Hα emission. This should be approximately an age sequence. In Type I shells, the HI shells are filled with ionized gas or else include discrete HII regions inside the HI shell. In Type II shells the ionized gas forms a thin shell and has been trapped by the HI shell. Type III represents HI shells which have discrete HII regions on the wall of the shell. Type IV is an HI shell which is not associated with ionized

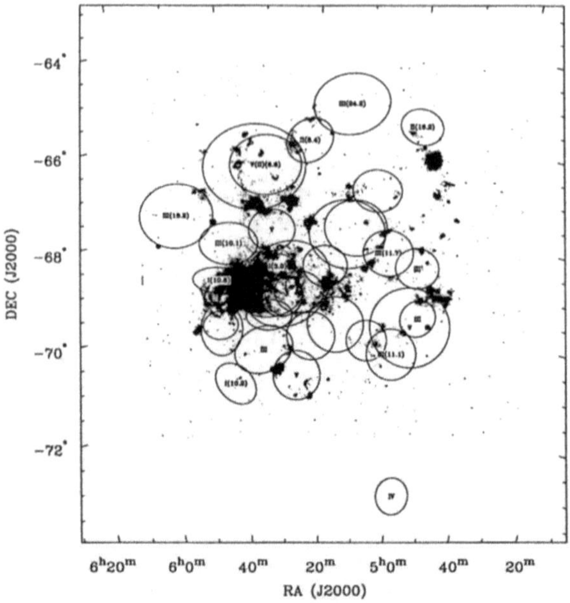

Fig. 2. The position of preliminarily defined HI supergiant shells are overlaid as ellipses on the Hα image from this survey. The classification, based on the comparison with HI shells and Hα is marked and the numbers indicate the age of the shells in Myrs.

gas. Type III and Type IV might have both evolved from Type II and the difference of two types is probably dependent on the density of the interstellar medium. Type V are HI shells which have morphologically complex structure, which might be caused by the triggered star formation due to the expanding shell. Shells near 30 Dor and some other regions are very confused and have not been classified. No age in some of the shells marked in Fig. 2 is due to these objects having very uncertain or unmeasurable expansion velocities.

References

de Vaucouleurs, G. and Freeman, K.C. (1972): Vistas in Astronomy 14, 163
Kim, S., Staveley-Smith, L., Sault, R.J., Kesteven, M. J., Mcconnell, D.(1997): A&A, to be submitted
Luks, T., & Rohlfs, K. (1992): A&A 263, 41
McGee, R.X., & Milton, J.A. (1966): Aust.J.Phys. 19, 343
Meaburn, J. (1980): MNRAS 192, 365
Rohlfs, K. et al.(1984): A&A 137, 343
Weaver, R., McCray, R., Castor, J., Shapiro, P., Moore, R. (1977): ApJ 218, 377

Part IX

External Galaxies

After-dinner speaker Jay Lockman attracts full attention ...

... with funny anecdotes about astronomers.

Diffuse Ionized Gas in Halos of Spiral Galaxies

Ralf-Jürgen Dettmar

Astronomisches Institut, Ruhr-Universität Bochum, D-44780 Bochum, Germany

Abstract. Over the last couple of years Diffuse Ionized Gas (DIG) has been identified as an important constituent of the interstellar medium (ISM) in the halos of spiral galaxies. Imaging in and spectroscopy of optical emission lines allow us to study the distribution and excitation of this gas with a spatial resolution not achievable for other phases of the ISM in external galaxies. Its origin and ionization is under debate and give important constraints for models of the ISM in general and on the large scale exchange of matter between disk and halo in particular. This review summarizes more recent observational results and compares them with model predictions. The data available now demonstrate that the presence of DIG in the disk-halo interface of spiral galaxies is related to star formation processes in the underlying disk. While photoionization by OB stars in the disk seems a viable source for the power required to ionize the DIG, additional processes are needed to explain some of the spectral features. The observed correlation with properties of the non-thermal radio continuum indicate that magnetic fields and cosmic rays could play a rôle for the physics of this medium.

1 Introduction

The detection of X-ray emitting gas in halos of late type spiral galaxies by ROSAT (Bregman and Pildis 1994, Wang et al. 1995, Vogler et al. 1996) provides important evidence for the long proposed scenario of a large scale exchange of matter between the interstellar medium (ISM) in galactic disks and their halos. This scenario is closely related to dynamical models of the ISM which also try to explain the formation of bubbles as, e.g., discussed in the contribution by Ikeuchi (1997). While the existence of X-ray halos supported by such processes was long expected, the presence of Diffuse Ionized Gas (DIG) or WIM (Warm Ionized Medium) as an important constituent of the ISM in the disk-halo interface has been recognized only more recently (Kulkarni and Heiles 1988, Reynolds 1990).

For the understanding of the complex interplay of the various phases of the ISM and star formation in galaxies, DIG plays an important rôle not only as the latest addition to the list of principal constituents of the ISM with filling factors of ~0.2 and up to 30% of the total atomic hydrogen content, but in particular also as a valuable diagnostic tool. Its existence and energy balance is closely related to the UV radiation field and radiative transfer in galactic disks, e.g., the question how leaky H II regions and how porous the ISM are. This specific question was addressed in several recent papers (Lehnert and Heckman 1994, Ferguson et al. 1996a, Hoopes at al. 1996) which all conclude that DIG contributes 25% – 50% of the Hα flux in spiral galaxies. In addition, DIG is an important tracer for halo components of the ISM in other galaxies since its emission line spectrum is rather easily accessible by optical imaging and spectroscopy. Most other tracers, such as radio continuum from cosmic rays or X-rays from hot plasma, either can not be observed with comparable angular resolution or sufficient sensitivity.

In the following, the known detections of extraplanar DIG in halos of galaxies are summarized, the current evidence for a correlation with star formation in the underlying disks and other components of gaseous halos is compiled, and the ionization problem for DIG is discussed. More detailed reviews of this topic can be found in, e. g., Dettmar (1992), Rand (1995), or Dahlem (1997).

2 Hα Imaging and Morphology of Extraplanar DIG

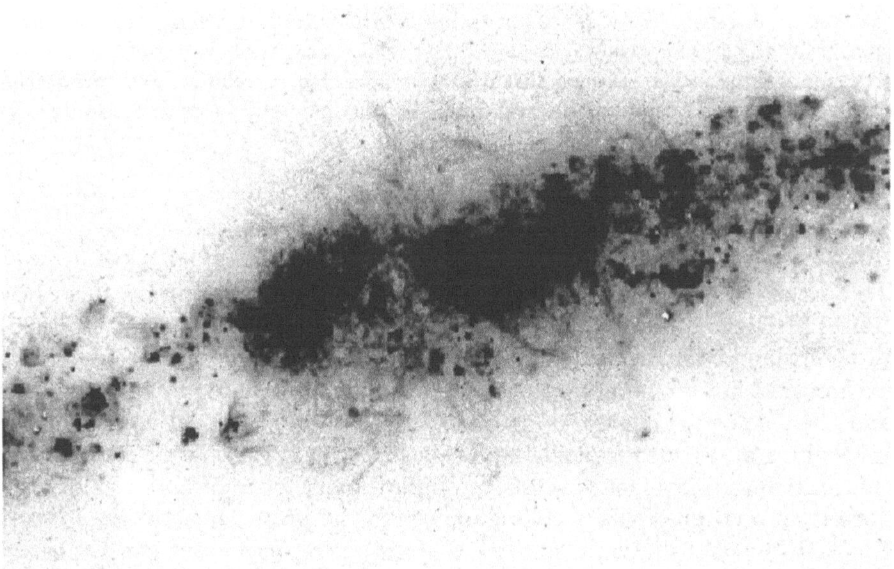

Fig. 1. Central part of the Sculptor group late type edge–on galaxy NGC 55 in Hα from Ferguson et al. (1996b)

The detection of a thick layer of DIG in NGC 891 which was found to be similar to the Reynolds layer of the Milky Way (Rand et al. 1990, Dettmar 1990) was followed by several Hα imaging searches. About two dozen 'normal' (i.e., excluding nuclear starbursts) edge-on galaxies have published results (Dettmar 1992, Rand et al. 1992, Pildis et al. 1994b, Rand 1996). Only few of them are showing evidence for a wide spread DIG in the halo comparable to NGC 891. In NGC 891 the DIG is distributed in long filaments (Rand et al. 1990, Dettmar 1990) and bubbles (Pildis et al. 1994a) of ionized gas embedded in a smooth background. In those objects showing Hα emission from the halo the spatial distribution is found to be very different, from thick layers with filaments and bubbles (e.g., NGC 4631, NGC 5775) to individual filaments and isolated plumes (e.g., UGC 12281). To demonstrate the complex morphologies of DIG two extreme cases are presented here.

NGC 55 is a late-type Sculptor Group galaxy with a very modest level of star formation. However, the energy released by starforming regions creates a spectacular distribution of the ionized gas (Fig. 1) with loops, bubbles, and long filaments extending far into the halo. The Hα image shows a clear difference in DIG morphology if regions of high star formation rate (SFR) are compared to quiet parts of the galaxy. The 'active' part shows (up to several hundred pc) long filaments as well as a smooth background of DIG. Some of the filaments can be traced back to individual HII regions in the disk. In the more quiet parts of the galaxy the DIG is confined to bubble like structures.

In the case of the prototype radio continuum- and X-ray halo galaxy NGC 4631 (Hummel and Dettmar 1990, Wang et al. 1995, Rand et al. 1992) the total far infrared (FIR) luminosity – 15 times higher than in NGC 55 (Rice et al. 1988) – indicates a high SFR. For this object Donahue et al. (1995) report the detection of a 30 kpc diameter halo of DIG down to an emission measure of $0.3\,cm^{-6}pc$. If confirmed, this very extended halo component of the ISM will be of interest for the discussion of gas halos in the early Universe.

With the latter example of NGC 4631 the key question with regard to the physics of DIG becomes obvious: what kind of source can supply the power required to keep the gas ionized. For the Milky Way DIG it was shown that only the UV radiation by OB stars meets the power requirements and that 10% – 20% of the UV photons have to leak out of HII regions (Reynolds 1990, 1997), a reasonable number if compared to the above mentioned results for other galaxies. However, even if the UV photons escape the HII regions, a special morphology of the ISM is still required to allow for the large free pathlengths that are needed to ionize gas several kpc out of the disk. With typical HI column densities of $N_{HI} \sim$several$\times 10^{20}\,cm^{-2}$ and an absorption cross-section of $\sigma=6\times10^{-18}\,cm^{-2}$ near the Lyman edge, an ionizing photon could travel only very short distances if the extended HI layer would be distributed homogeneously. We will come back to the ionization problem in sect. 5 where the emission line spectra will be discussed. In the following we first address the relation of DIG to other phases of halo ISM.

3 Correlation with Other Gas Phases in the Halo ...

The best studied galaxies with respect to their halo ISM – and in particular their halo component of DIG – are the objects NGC 891 and NGC 4631.

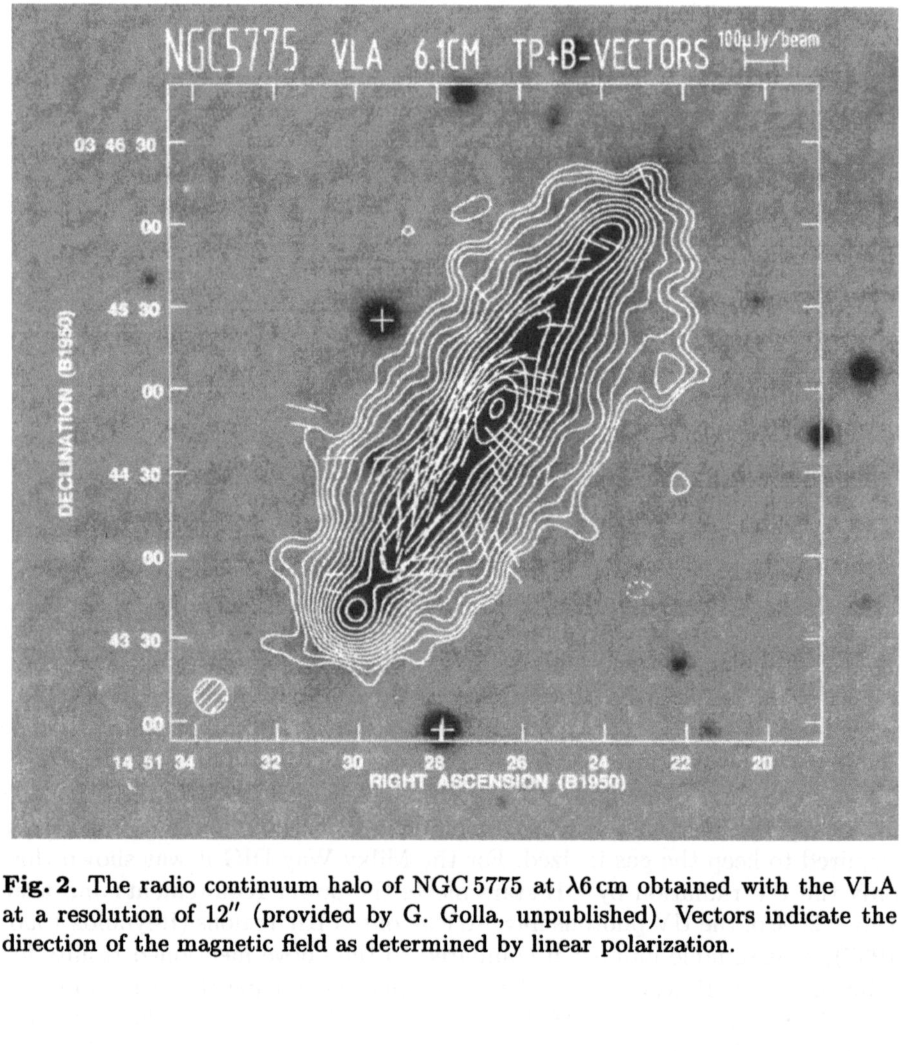

Fig. 2. The radio continuum halo of NGC 5775 at $\lambda 6$ cm obtained with the VLA at a resolution of 12″ (provided by G. Golla, unpublished). Vectors indicate the direction of the magnetic field as determined by linear polarization.

Both of them not only show prominent thick layers of DIG, they also possess extended radio continuum, H I, and X-ray halos. The spatial correlation of radio continuum emission, indicative of cosmic rays in a magnetic field found in a thick disk, and extraplanar DIG has been discussed for NGC 891 in detail (Dettmar 1992, Dahlem et al. 1994). It is shown there that the scale height for the ionized gas correlates with the scale height of cosmic rays. A particular active region 2′ NE of the nucleus not only shows individual filaments of ionized gas that can be traced out to 4–5 kpc above the disk, this region also shows H I emission high above the plane (see sect. 7). This H I gas could be comparable to Galactic High-Velocity Clouds. In addition, interferometer CO observations indicate a molecular spur reaching 520 pc above the plane in this region (Handa et al. 1992).

A search for more radio continuum halo or thick disk candidates by Hummel et al. (1991a) could identify only a few more suitable candidates for further studies of this suggestive correlation. One example is NGC 5775 (Fig. 2) which indeed also has an extended Hα halo. Again, radio continuum spurs are spatially correlated with Hα filaments (Dettmar 1992) and in the case of NGC 253 a possible correlation of such localized Hα filaments with radio continuum and X-ray spurs (Carilli et al. 1992, Pietsch 1994) is suggested by the comparison of these maps with a recent Hα image (Ehle, priv. comm.).

The radio continuum thick disks in some objects show rather high degrees of linear polarization (for a review see Beck 1997). In the case of NGC 891 (Hummel et al. 1991b) the highest degree of polarization is again observed above the most active part in the disk, the region 2′ NE of the nucleus (Hummel et al. 1991b, Dahlem et al. 1994). This is very surprising as one would expect the polarization here to be small due to Faraday depolarization by the thermal gas and part of the observed distribution of the polarized emission in NGC 891 must be influenced by Faraday depolarization along the different lines of sight. Although the halo of NGC 891 is missing a well ordered field on larger scales, a substantial ordered field has to be present locally and similar degrees of polarization (10%) are found in other radio continuum thick disks, too. Therefore magnetic field could be important for the physics of DIG.

Only for a few galaxies high resolution studies at various wavelengths are available to analyze such localized correlations. In particular we are still lacking FIR data that could be used for such comparisons; in this respect progress can soon be expected from ISO results. Although the examples above show that the presence of gas in galactic halos is in some cases a more localized phenomenon, evidence for the hypotheses that star formation activity in the disk is driving it can be found by comparing global properties.

4 ... and Star Formation in the Disk

To demonstrate this, we have compiled some properties of halo gas from the literature for all objects that have been searched for halo DIG. This includes a recently found new object, namely NGC 4634, with a very prominent and thick DIG layer with a FWHM of almost 500 pc (Fig. 3). Evidence for radio continuum or X-ray halo components is given together with a description of the DIG morphology. The thermal emission of dust in galaxies observable in the FIR is known to be a good tracer of star formation (e.g., Thronson and Telesco 1986) and therefore the SFR in the disk is characterized by the thermal IRAS FIR luminosity normalized to the optical diameter. Given the lack of angular resolution this results in a first order estimate for the energy input into the ISM per unit area.

FIR fluxes and the normalized FIR luminosities of these galaxies differ by almost two orders of magnitude and from this small sample the earlier

Tab. 1: DIG properties, SFR, and halo ISM

Object(Ref.)	$L_{FIR}/D^2_{25} \times 10^{40(a)}$ (erg/s/kpc^2)	DIG distrib.&morphol.	radio halo	X-ray$^{(b)}$ halo
NGC 4666 (1)	14.5	bright, diffuse filaments	X	X
NGC 3079 (2,3)	8.9	bright, diffuse filaments	X	
NGC 5775 (4,5)	8.1	bright, diffuse+ filament	X	
NGC 253 (6-8)	8.1	bright, diffuse+ filamen	X	X
NGC 4634 (24)	5.9	very bright, extended		
NGC 3044 (5,9)	4.0	bright, diffuse	X	
NGC 4402 (26)	3.8	faint diffuse		
NGC 891 (10-13)	3.3	bright, diffuse+ filamen	X	X
Galaxy (14-16)	3.0	diffuse	X	X
NGC 4013 (5,9)	2.6	faint diffuse	?	
NGC 4631 (17-19)	1.8	bright diffuse	X	X
NGC 4302 (5,10,20)	<2.3	faint diffuse	?	
NGC 3432 (25)	1.6	?	X	
NGC 973 (5,20)	1.0	plumes	–	
NGC 55 (28,29)	0.9	giant filaments		?
UGC 3326 (5,20)	0.8	–	–	
NGC 5907 (9, 27, 30)	0.8	–	–	
UGC 12281 (20)	0.5	plumes		
UGC 2092 (5,20)	0.4	plumes	–	
UGC 10288 (5,9)	0.4	3 or 4 plumes	–	
NGC 4565 (10,21,30)	0.3	–	–	X
NGC 5746 (9)	0.2	–		
NGC 4244 (22,23)	<0.04	–	–	
NGC 5023 (9)	<0.09	1 or 2 plumes		
NGC 4217 (5,9)	<0.12	2 faint patches	X	
UGC 4278 (9)	<0.04	–		
NGC 4762 (9)	<0.15	–		

Notes: *(a)* distances from Tully 1988; FIR fluxes from Rice et al. 1988 or Fullmer and Lonsdale 1989; FIR for Galaxy from Cox and Mezger 1989. *(b)* only ROSAT detections are considered here.

(1) Dahlem et al. 1997 (2) Veilleux et al. 1995 (3) Duric et al. 1983 (4) Dettmar 1992 (5) Hummel et al. 1991a (6) Carilli et al. 1992 (7) Pietsch 1994 (8) Ehle (priv. comm.) (9) Rand 1996 (10) Rand et al. 1990 (11) Dettmar 1990 (12) Dahlem et al. 1994 (13) Bregman and Pildis 1994 (14) Reynolds 1990 (15) Beuermann et al. 1985 (16) Freyberg (1997) (17) Rand et al. 1992 (18) Hummel and Dettmar 1990 (19) Wang et al. 1995 (20) Pildis et al. 1994a (21) Sukumar and Allen 1991 (22) Walterbos 1991 (23) Schlickeiser et al. 1986 (24) this paper (25) English and Irwin 1997 (26) Dettmar 1995 (27) Dumke et al. 1995 (28) Ferguson et al. 1996b (29) Schlegel et al. 1997 (30) Vogler et al. 1996

– : non-detection

? : possible candidates that still lack confirmation or high resolution observations

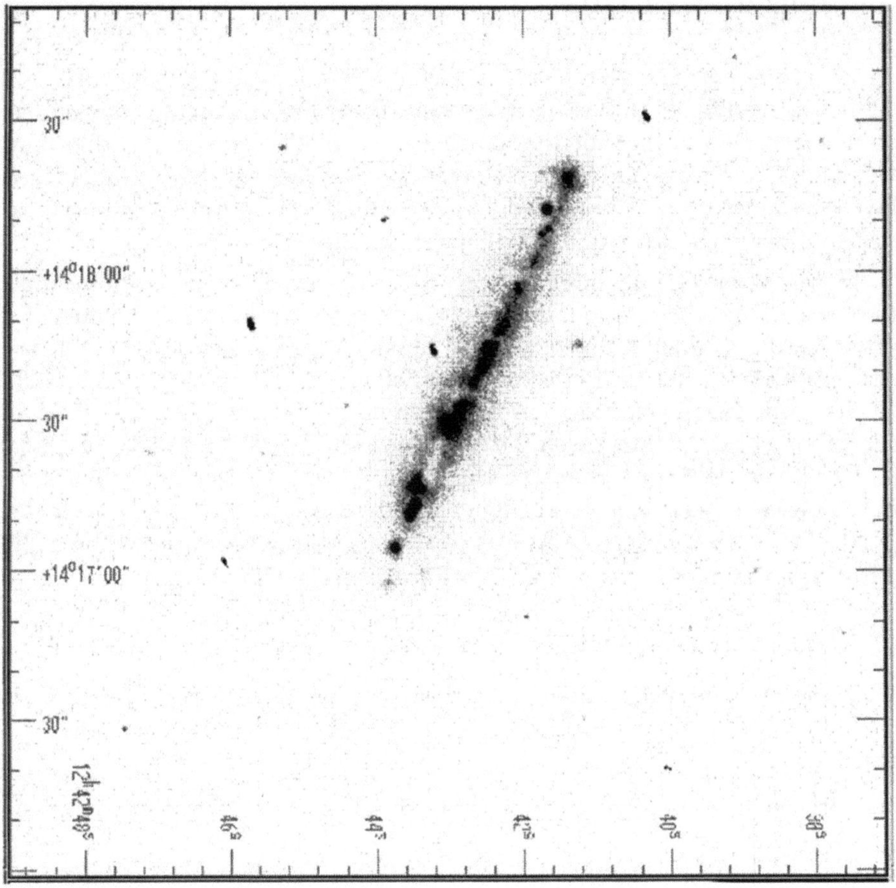

Fig. 3. Hα image of NGC 4634 obtained with the ESO/MPI 2.2 m telescope on La Silla using EFOSCII.

impression, that extraplanar DIG is related to star formation, emerges again. Bright and extended, widespread DIG layers as well as thick disks of radio continuum emission are found only in those objects with high SFR per unit area. In less active galaxies only individual plumes of DIG are found, which still could be caused by individual starforming regions. Only the observed X-ray halo in NGC 4565 and the radio data for NGC 4217 do not fit into a scenario that relates extraplanar DIG, radio, and X-ray halos to the SFR in the disk. It is worth mentioning that the Milky Way actually fits into this picture, details are described in Dettmar (1992). One should keep in mind that the compilation in Tab. 1 is still based on very inhomogeneous data, in particular with regard to sample selection and detection limits, and that it is therefore difficult to perform a more quantitative analyses.

5 Ionization of DIG

As it was mentioned in Sect. 2, the most important question for our understanding of DIG is that for the ionizing process. The papers by Miller and Cox (1993) and Dove and Shull (1994) show, that the thickness of the ionized layer can indeed be attributed to the UV radiation field caused by a typical distribution of OB stars in disks of spiral galaxies.

However, the detailed information on the ionization and excitation conditions from long-slit emission line spectroscopy is not that easily explained by photoionization models. Spectroscopic observations of extraplanar DIG are now available for NGC 891 (Dettmar and Schulz 1992, Rand 1997), NGC 4631 (Golla et al. 1996), NGC 2188 (Domgörgen and Dettmar 1997), and NGC 55 (Ferguson et al. 1996b). Generally the observed line ratios of [NII]/Hα and [SII]/Hα vary with distance from the plane. Both, [NII] and [SII] lines, are becoming relatively stronger at higher z and ratios up to [SII]/H$\alpha = 0.6$ and [NII]/H$\alpha = 1.4$ are observed. As a typical example the line ratios along a slit parallel to the minor axis of NGC 4631 are given in Fig. 4.

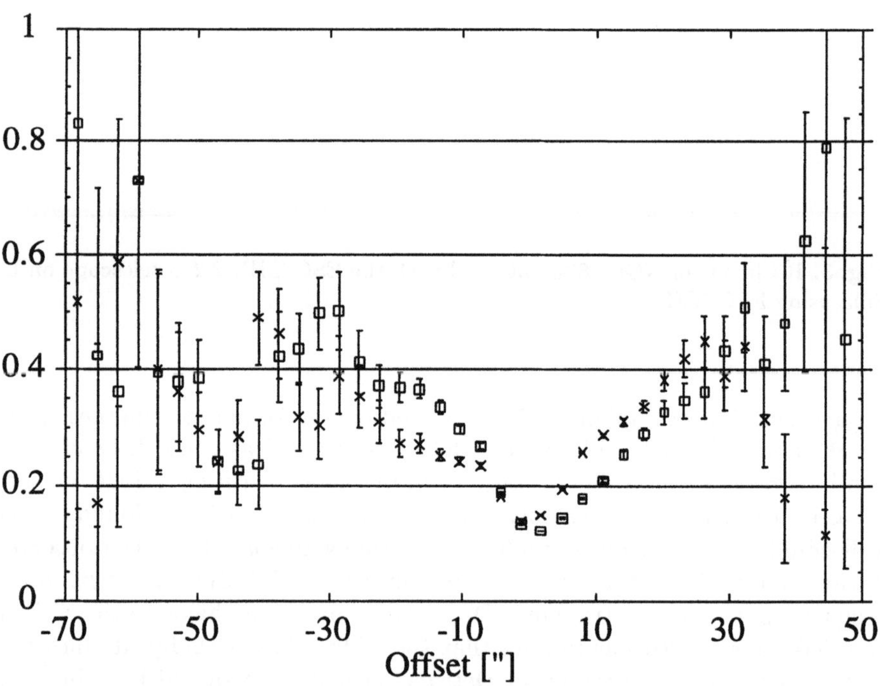

Fig. 4. Variation of emission line ratios [NII]/Hα (□) and [SII]/Hα (x) as a function of distance above the galactic plane in NGC 4631 (from Golla et al. 1996)

To explain the relatively strong [NII] and [SII] lines together with the weak [OI]λ6300 Å and [OIII]λ5007 Å lines observed in the Reynolds-layer of the Galaxy the radiation transfer for photoionization models in a low density regime (Mathis 1986) was modified to account for these observations (Domgörgen and Mathis 1994). Since these models require a certain hardness of the radiation field a prediction was made for the expected fraction of ionized He. The reported upper limit for the HeI λ5876 Å line along two lines of sight in the Reynolds-layer (Reynolds and Tufte 1995), however, is inconsistent with the model predictions. The upper limit corresponds to an ionized fraction of helium relative to hydrogen $\chi_{\rm He}/\chi_{\rm H} \lesssim 0.25$ and such a small fraction of ionized He in the Milky Way is supported by radio-recombination-line observations of Galactic DIG filaments ("worms") by Heiles et al. (1996) with even stricter limits on $\chi_{\rm He}/\chi_{\rm H}$.

With regard to He^+ the extragalactic situation is slightly different since Rand (1997) presents a very deep spectrum of NGC 891 with the detection of HeI λ5876 Å in the DIG. From this observation one can conclude that the ionizing radiation field is significantly harder than in the Reynolds-layer. However, in NGC 891 the line ratio for [NII]/Hα reaches 1.4, a value that is still not explainable with the radiation field deduced from the HeI/Hα ratio. To account for this enhanced, low-ionization emission there is still additional heating required.

It is possible that shocks may contribute at higher z and some indication for such effects are given by Dahlem et al. (1997). Additional contributions could come from photoelectric heating by dust (e.g., Reynolds and Cox 1992), and a lack of coolants due to metal depletion, or, of course, a combination of these effects. The Decaying Dark Matter theory, in which the decay of a 27eV neutrino produces ionizing UV photons, was for example applied to the DIG in NGC 891, too (Sciama and Salucci 1991). A completely different approach to the ionization problem is attempted by models that try to explain the DIG as the cooling phase of a galactic fountain phenomenon. In this context mixing, heat conduction, and non-equilibrium processes might be important (e.g., Breitschwerdt 1997, Slavin et al. 1993, Shapiro and Benjamin 1993). This discussion makes clear that the heating and ionization of DIG is still not well understood and several contributions at this conference address the question of additional sources of energy such as magnetic reconnection (Birk and Lesch 1997) or by turbulence (Minter and Balser 1997).

6 Dust (Mixed in)

All the various phases of the ISM come with some "contamination" by dust and from theoretical arguments dust is expected to be present in the halo too, driven out either by coupling to the gas "blow out" or by radiation pressure (Ferrara 1997). Indeed, in many edge-on galaxies filaments of dust perpendicular to the disk or giant loops of dust reaching into the halo can be identified (Keppel et al. 1990, Sofue et al. 1995). Sofue et al. suggest, that the dusty loop structures and filaments observed in NGC 253 are influenced by magnetic fields. This dust can be important for the physics of halo gas, e.g., since it may contribute to the energy balance by photoelectric heating (Reynolds and Cox 1992) or radiative losses due to fast spinning small grains (Ferrara and Dettmar 1994). It could also influence the emission line diagnostics of the disk since its scattering properties will change the observed line ratios (Ferrara et al. 1996).

7 Velocity Information

If DIG (and other components of the ISM) in the halo are due to dynamical processes, important information on its origin and ionization could come from kinematical studies. This source of information has not been much used yet, mainly for sensitivity reasons. In the case of NGC 891, however, there is evidence for peculiar velocities of DIG (Pildis et al. 1994a, Rand 1997) and HI (Swaters et al. 1996). For both components a deviation from corotation is observed on scales of 2 kpc above the disk in the sense that the halo gas is rotating slower as it is expected in "fountain flow" models in which uprising gas is moving outwards (Bregman 1980).

8 Conclusions

DIG is an important tracer for gas in halos of galaxies and its emission line spectrum offers several important diagnostic tools, such as kinematic information, whose possibilities have not yet been fully explored. The evidence for a correlation of extraplanar DIG with other phases of gas in galactic halos and with the SFR in the underlying disk is supporting models that describe a large scale circulation of matter between actively starforming regions in the disk and the halo such as *galactic fountains* (Shapiro and Field 1976, Kahn 1997), *chimneys* and *superbubble blow outs* (Norman and Ikeuchi 1989, Mac Low et al 1989., Mac Low and Ferrara 1997), or *winds* (Habe and Ikeuchi 1980). However, details of the processes and the importance of specific models remain still unclear and the uncertainties with regard to the ionization and excitation of DIG reflect this situation.

Acknowledgements. It is a pleasure to thank H. Domgörgen, A. Ferguson, V. Bunzel, G. Golla, and J. Rossa for their help preparing this manuscript. Financial support for research in this field was provided by DFG and MPIfR.

References

Beck, R. (1997) in *The Physics of Galactic Halos*, eds.: H. Lesch, R.-J. Dettmar, U. Mebold, R. Schlickeiser, Akademie Verlag, p. 135

Beuermann, K., Kanbach, G., Berkhuijsen, E. M. (1985): A&A 153, 17

Birk, G. T., Lesch, H. (1997): these proceedings

Bregman, J. N. (1980): ApJ 236, 577

Breitschwerdt, D. (1997): in *The Physics of Galactic Halos*, eds.: H. Lesch, R.-J. Dettmar, U. Mebold, R. Schlickeiser, Akademie Verlag, p.237

Bregman, J. N., Pildis, R. A. (1994): ApJ 420, 570

Carilli, C. L., Holdaway, M. A., Ho, P. T. P., de Pree, C. G. (1992): ApJ 399, L59

Cox, P., Mezger, P. G. (1989): A&AR 1, 49

Dahlem, M. (1997): PASP, in press (ESA preprint)

Dahlem, M., Dettmar, R.-J., Hummel, E. (1994): A&A 290, 284

Dahlem, M., Petr, M. G., Lehnert, M. D., Heckman, T. M., Ehle, M. (1997): A&A 320, 731

Dettmar, R.-J. (1990): A&A 232, L15

Dettmar, R.-J. (1992): Fund. Cosmic Phys. 15, 148

Dettmar, R.-J. (1995): in *The Physics of the Interstellar Medium and Intergalactic Medium*, eds. A. Ferrara, C. F. McKee, C. Heiles, P. R. Shapiro, ASP conf. Ser. Vol. 80, p. 398

Dettmar, R.-J., Schulz, H. (1992): A&A 254, L25

Domgörgen, H., Dettmar, R.-J. (1997): A&A 322, 391

Domgörgen, H., Mathis, J. S. (1994): ApJ 428, 647

Donahue, M., Aldering, G., Stocke, J. T. (1995): ApJ 450, L45

Dove, J. B., Shull, J. M. (1994): ApJ 430, 222

Dumke, M., Krause, M., Wielebinski, R., Klein, U. (1995): A&A 302, 691

Duric, N., Seaquist, E. R., Crane, P. C., Bignell, R. C., Davies, L. E. (1983): ApJ 273, L11

English, J., Irwin, J. A. (1997): AJ 113, 2006

Ferguson, A. M. N., Wyse, R. F. G., Gallagher, J. S., Hunter, D. A. (1996a): AJ 111, 2265

Ferguson, A. M. N., Wyse, R. F. G., Gallagher, J. S. (1996b): AJ 112, 256

Ferrara, A. (1997): in *The Physics of Galactic Halos*, eds.: H. Lesch, R.-J. Dettmar, U. Mebold, R. Schlickeiser, Akademie Verlag, p. 189

Ferrara, A., Dettmar, R.-J. (1994): ApJ 427, 155

Ferrara, A., Bianchi, S., Dettmar, R.-J., Giovanardi, C. (1996): ApJ 467, L69

Freyberg, M. (1997): these proceedings

Fullmer, L., Lonsdale, C. (1989): Cataloged Galaxies and Quasars observed in the IRAS Survey, JPL

Golla, G., Dettmar, R.-J., Domgörgen, H. (1996): A&A 313, 439

Habe, A., Ikeuchi, S. (1980): Prog. Theor. Phys., 64, 1995

Handa, T., Sofue, Y., Ikeuchi, S., Kawabe, R., Ishizuki, S. (1992): PASJ 44, L227

Heiles, C., Koo, B.-C., Levenson, N. A., Reach, W. T. (1996): ApJ 462, 326

Hoopes, C. G., Walterbos, R. A. M., Greenawalt, B. E. (1996): AJ 112, 1429

Hummel, E., Dettmar, R.-J. (1990): A&A 236, 33

Hummel, E., Beck, R., Dettmar, R.-J. (1991a): A&AS 87, 309

Hummel, E., Beck, R., Dahlem, M. (1991b): A&A 248, 23

Ikeuchi, S. (1997): these proceedings
Kahn, F. D. (1997): these proceedings
Keppel, J. W., Dettmar, R.-J., Gallagher, J. S., Roberts, M. (1990): ApJ 374, 507
Kulkarni, S. R., Heiles, C. (1988): in *Galactic and Extragalactic Radioastronomy, 2nd edition*, eds. G. L. Verschuur, K. I. Kellermann, Springer, p. 95
Lehnert, M., D., Heckman, T. M. (1994): ApJ 426, L27
Mac Low, M.-M., McCray, R., Norman, M. L. (1989): ApJ 337, 141
Mac Low, M.-M., Ferrara, A. (1997): these proceedings
Mathis, J. S. (1986): ApJ 301, 423
Miller, W. W., Cox, D. P. (1993): ApJ 417, 579
Minter, A. H., Balser, D. S. (1997): ApJ 484, L133
Norman, C. A., Ikeuchi, S. (1989): ApJ 345, 372
Pietsch, W. (1994): in *Panchromatic Views of Galaxies*, eds.: Hensler, Theis, Gallagher, p. 134
Pildis, R. A., Bregman, J. N., Schombert, J. M. (1994a): ApJ 423, 190
Pildis, R. A., Bregman, J. N., Schombert, J. M. (1994b): ApJ 427, 160
Rand, R. J. (1995): in *Gas Disks in Galaxies* ed. J. M. van der Hulst, Kluwer, in press (Univ. of Maryland preprint)
Rand, R. J. (1996): ApJ 462 712
Rand, R. J. (1997): ApJ 474, 129
Rand, R. J., Kulkarni, S. R., Hester, J. J. (1990): ApJ 352, L1 (Err. 362, L35)
Rand, R. J., Kulkarni, S. R., Hester, J. J. (1992): ApJ 396, 97
Raymond, J. C. (1992): ApJ 384, 502
Reynolds, R. J. (1990): in *IAU Symp. 144 The interstellar disk-halo connection in galaxies* ed. H. Bloemen, Kluwer, p.
Reynolds, R. J. (1997): in *The Physics of Galactic Halos*, eds. H. Lesch, R.-J. Dettmar, U. Mebold, R. Schlickeiser, Akademie Verlag Berlin 1997, p. 57
Reynolds, R. J., Cox, D. P. (1992): ApJ 400, L33
Reynolds, R. J., Tufte, S. L. (1995): ApJ 439, L17
Rice, W., Lonsdale, C. J., Soifer, B. T. et al. (1988): ApJS 68, 91
Schlickeiser, R., Werner, W., Wielebinski, R. (1984): A&A 140, 277
Sciama, D. W., Salucci, P. (1991): MNRAS 247, 506
Shapiro, P. R., Benjamin, R. A. (1993): in *Starforming Galaxies and their Interstellar Medium* eds. J. J. Franco and F. Ferrini, Cambridge, p. 273
Shapiro, P. R., Field, G. B. (1976): ApJ 205, 762
Slavin, J. D., Shull, J. M., Begelman, M. C.: (1993): ApJ 407, 83
Sofue, Y., Wakamatsu, K.-I., Malin, D. F. (1994): AJ 108, 2102
Sukumar, S., Allen, R. J. (1991): ApJ 382, 100
Swaters, R. A., Sancisi, R., van der Hulst, J. M. (1996): in *IAU Symp. 171 New Light on Galaxy Evolution* eds. R. Bender and R. L. Davies, Kluwer, p. 450
Thronson, H. A., Telesco, C. M. (1986): ApJ 311, 98
Tully, R. B. (1988): Nearby Galaxies Catalog, Cambridge
Veilleux, S., Cecil, G., Bland-Hawthorn, J. (1995): ApJ 445,112
Vogler, A., Pietsch, W., Kahabka, P. (1996): A&A 305, 74
Walterbos, R. A. M. (1991): in *IAU Symp. 144 The interstellar disk-halo connection in galaxies* ed. H. Bloemen, Kluwer, p. 223
Wang, D., Walterbos, R., Steakley, M., Norman, C. N., Braun, R. (1995): ApJ 439, 176

Magnetic Reconnection and Extraplanar Diffuse Ionized Gas

Guido T. Birk[1], Harald Lesch[1] and Thomas Neukirch[2]

[1] Institut für Astronomie und Astrophysik, LMU München, Scheinerstr. 1, 81679 München, Germany
[2] School of Mathematical and Computational Sciences, University of St. Andrews, St. Andrews, Fife KY16 9SS, Scotland, United Kingdom

Abstract. Many observations indicate the occurrence of ionized gas in the distant halos of galaxies (among them our own). Since photoionization by stars (mainly O stars, young or evolved low mass stars depending on the kind of galaxy) does not seem to be exclusively responsible for the ionization of the hydrogen filaments that should otherwise cool and recombine quickly, the question arises which energy source may result in the quasistationary ionization. We propose that localized magnetic field dissipation in current filaments may contribute to the ionization of the extraplanar halo gas. The dissipated magnetic energy is converted into heat, which then is transferred to the neutrals, resulting in re-ionization on a time scale shorter than the recombination time scale. Quasistatic solutions are characterized by plasma temperatures and densities that agree nicely with the observed values for the diffuse ionized gas component of the Galactic interstellar medium.

1 Introduction

During the last decade it has become obvious that the existence of a diffuse ionized gas (DIG) component is of importance for our understanding of the interstellar medium. This component is widespread in the Galactic Halo, in the disk halo interface of spiral galaxies as well as in irregular, active and early type galaxies (recent reviews on the detection of DIG are given by Walterbos (1991) and Dettmar (1992)). The detection of the DIG component of the interstellar medium gives rise to the ionization problem: What are the energy sources that keep the extraplanar DIG ionized?

In this contribution we consider a fundamental plasma process that may help to solve the ionization problem: magnetic reconnection. Reconnection has proved to be an efficient heating process in quite different cosmical plasma regimes. Recently, it was shown, e.g. that this process can explain the X-ray emission of boundary regions of high-velocity clouds hitting the Galactic disk (Zimmer et al. 1997) and that it contributes to the small-scale solar coronal heating in bright points (Priest et al. 1993; Birk et al. 1997).

The train of thought is as follows: Extraplanar gas far away from O-stars cools and recombines effectively. In gas associations that are located in regions with inhomogeneous magnetic fields and localized electric current filaments, the recombination results in a lower thermal plasma pressure that counteracts the magnetic pressure in the central current regions and in a reduced electric conductivity mainly due to electron-neutral collisions. In this situation magnetic reconnection is likely to occur (e.g. Parker 1994). During this process, due to the local violation of ideal Ohm's law, magnetic field lines are reconnected and magnetic flux is dissipated. The free magnetic energy is mainly converted into heat via Ohmic dissipation and, consequently, the gas is re-heated and can be re-ionized.

2 Quasistatic Stationary Heating and Ionization by Magnetic Dissipation

Consider a mainly hydrogen quasi-neutral partially ionized plasma in ionization equilibrium in a quasistatic steady state. The relevant equations are the stationary ionization and energy balance equations

$$\alpha n_p^2 = \iota n_n \tag{1}$$

$$\mathbf{j}^2 = \sigma(L^{\mathrm{rad}}_{\mathrm{brems}} + L^{\mathrm{rad}}_{\mathrm{recom}}) \tag{2}$$

where n_p, n_n, $\mathbf{j}$, α, ι, σ, $L^{\mathrm{rad}}_{\mathrm{brems}}$ and $L^{\mathrm{rad}}_{\mathrm{recom}}$ denote the plasma (HII) and neutral gas (HI) particle densities, the current density, the recombination coefficient, the ionization frequency, the collisional conductivity due to electron-neutral collisions and the radiative loss functions due to recombination radiation as well as bremsstrahlung (cf. Huba 1994 for the detailed expressions for the transport coefficients). The electric current filaments are assumed to be of the Harris sheet type (Harris 1962) for simplicity (we assume the magnetic field to vary in the y-direction):

$$\mathbf{j} = 7.7 \cdot 10^{-16} \left(\frac{B}{\mu\mathrm{G}}\right) \left(\frac{\delta}{\mathrm{pc}}\right)^{-1} \mathrm{sech}^2\left(\frac{y}{\delta}\right) \mathbf{e}_z \ \mathrm{statamp\ cm}^{-2} \tag{3}$$

where δ is the length scale of the inhomogeneity of the magnetic field, i.e. the half thickness of the current layer.

These equations can be rewritten as a set of non-linear algebraic equations for the plasma density n_p and the electron temperature T_e (measured in eV):

$$n_p^3 + 2.7 \cdot 10^{-23} \frac{B^2(n_p - n_t)}{L^2 \cosh^4(y/\delta)(1 + E_\infty T_e^{-1/4})} = 0 \tag{4}$$

$$0.4 + 0.5 \ln\left(\frac{E_\infty}{T_e}\right) + 0.5\left(\frac{T_e}{E_\infty}\right)^{1/3} = \frac{2 \cdot 10^8 \, T_e(n_t - n_p)e^{-E_\infty/T_e}}{E_\infty^{5/2}(6 + T_e E_\infty^{-1/4})n_p} \tag{5}$$

The ionization energy for hydrogen is $E_\infty = 13.6$ eV and the lower estimate for the magnetic field strength is about 5μG from radio data (Beck et al.

1996). The total particle density n_t and δ are the only free input parameters of this idealized model. Fig. 1 shows for example the spatial profiles of the electron temperature T_e, plasma density n_p and neutral gas density n_n for $n_t = 1\,\mathrm{cm}^{-3}$ and $\delta = 10^{-11}$ pc. The electron temperature is close to the observed value of the HII temperature. The electron and ion temperatures should not differ significantly since electron-ion collisions lead to a relatively fast (as compared to the recombination time scale) thermalization. The plasma density has its maximum in the central reconnection region (i.e. the central region of the current filament) and decreases with increasing distance from the current sheet. The ionization rate is 60% which agrees quite well with observed values in some Galactic DIG regions. We note that

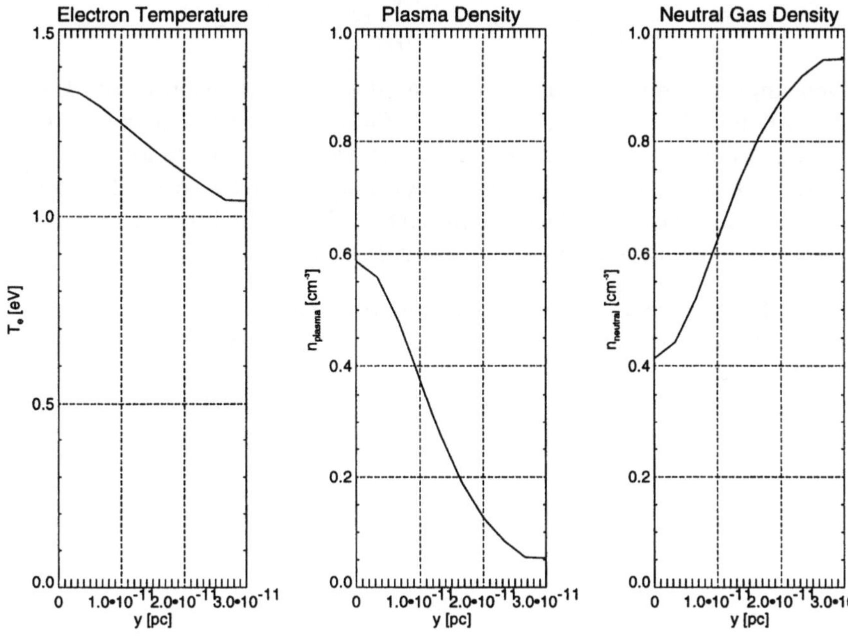

Fig. 1. Spatial profiles of the electron temperature, the plasma and the neutral gas density for a total densities of $n_t = 1\mathrm{cm}^{-3}$

at least for the relatively wide parameter range studied, $n_t = 0.15..20\mathrm{cm}^{-3}$ and $\delta = 10^{-11}..10^{-8}$ pc, we find reasonable values for the HII densities and temperatures.

3 Stationary Heating by Magnetic Reconnection

Dropping the assumption of a static configuration we are dealing with stationary magnetic reconnection (e.g. Parker 1994). In the framework of this description the magnetic diffusivity gives rise to dynamical dissipation of the inflowing magnetic field as rapidly as magnetic flux is convected in. The time scale of the dissipation process depends on the width of the current filament Δ, and can be estimated (cf. Parker 1994) as $\tau_{\rm rec} = 5 \cdot 10^4$ s for Δ comparable to the thickness of the filament δ ($\delta = 10^9$ cm obtained from the condition of stationarity) up to $\tau_{\rm rec} = 10^{12}$ s for the slowest case ($\Delta = \delta S$ where S is the magnetic Reynolds number) which is still faster than the recombination time scale $t_{\rm rec} = 10^{13}$ s.

The rate of conversion of magnetic energy caused by magnetic reconnection at maximum has to satisfy (cf. Dettmar 1992):

$$\frac{d}{dt}\int \frac{B^2}{8\pi} d^3r \approx 10^{-12} \frac{V}{\tau_{\rm rec}} {\rm G}^2 \approx 10^{42} {\rm erg s}^{-1} \qquad (6)$$

which implies a quite small (as compared to the extension of the Galactic halo) volume $V \approx 10\,{\rm pc}^3$ in which a sufficient amount of magnetic energy can be dissipated in many individual reconnection processes for $\Delta \approx \delta$.

We note that the refreshment rate of magnetic fields (caused by galactic fountains, supernovae, Galactic dynamo etc.) has to be taken into account in order to put quantitative restrictions on our model.

Acknowledgements. This work was supported by the DFG through the grant ME 745/18-1 and by the United Kingdom Particle Physics and Astronomy Research Council.

References

Beck, R., Brandenburg, A., Moss, D. et al. (1996): ARAA 34, 155–206

Birk, G.T., Dreher, J., Neukirch, T. (1997): *Magnetic Reconnection in the Solar Atmosphere*, eds. R. D. Bentley and J. T. Mariska, ASP Conf. Ser. 111, 89–94

Dettmar, R.-J. (1992): Fund. Cosm. Phys. 15, 143–208

Harris, E.G. (1962): Nuovo Cim. 23, 115

Huba, J.D. (1994): *NRL Plasma Formulary* (Naval Research Laboratory, Washington), 39–57

Parker, E.N. (1994): *Spontaneous Current Sheets in Magnetic Fields* (Oxford University Press), 291–300

Priest, E.R., Parnell, C.E., Martin, S.F. (1993): ApJ 427, 459–474

Walterbos, R.A.M. (1991): *The Interstellar Disk-Halo Connection in Galaxies*, ed. H. Bloemen, IAU Symp. 144, 223–236

Zimmer, F., Lesch, H., Birk, G.T. (1997): A&A 320, 746–756

Turbulent Heating in the Galactic Diffuse Ionized Gas

Anthony Minter and Dana S. Balser

National Radio Astronomy Observatory**, Green Bank, West Virginia, U.S.A. 24944

Abstract. The observed properties of the diffuse ionized gas (DIG) in our Galaxy are not easily reconcilable with simple photoionization models. This suggests that there are different or additional physical processes at work in the DIG. We have developed a model of the DIG whereby it is ionized by a relatively soft ionizing spectrum ($T_{\text{eff}} \leq 32,000\,\text{K}$) and is also heated by an additional thermal mechanism: the dissipation of turbulence. This model predicts the same electron temperature, [N II] λ 6583/Hα ratio, [S II] λ 6716/Hα ratio and He I λ 5876/Hα ratio as observed in the DIG. Without the turbulent thermal heating term, this model will not reproduce the observed properties of the DIG. The dissipation of turbulence may also be important in other phases of the ISM.

1 Modeling the Diffuse Ionized Gas

The diffuse ionized gas (DIG), also referred to as the Reynolds layer, is a nearly fully ionized gas with a mean midplane density of $0.025\,\text{cm}^{-3}$ and a scale height of $\sim 1\,\text{kpc}$. The energy required to keep the DIG ionized is $\geq 10^{-4}\,\text{ergs}\,\text{cm}^{-3}\,\text{s}^{-1}$, corresponding to 5×10^6 hydrogen ionizing photons per cm^{-2} per sec in the Galactic disk (Reynolds 1992). The He I λ 5876/Hα ratio indicates that the ionizing spectrum incident on the DIG is relatively soft and corresponds to a helium to hydrogen ionization factor of $\chi_{\text{He}}/\chi_{\text{H}} \lesssim 0.5$ (Tufte 1997). The observed ranges of [N II] and [S II] relative to Hα require a much harder ionizing spectrum if photoionization is to explain the line ratios (Reynolds 1985a). Previous models (e.g., Domgörgen & Mathis 1994) have attempted to explain the conditions in the DIG but have not been entirely successful (Reynolds & Tufte 1995; Rand 1997).

In classical turbulence theory, the energy contained in the large scale "eddies" cascades to smaller scale eddies. This cascade of energy to smaller scales occurs on all spatial scales down to a dissipative scale, at which point the turbulent energy is converted into thermal energy via some dissipation mechanism. The dissipation rate of the turbulence is proportional to the energy contained in the turbulence (Minter & Spangler 1996). Minter & Spangler

** The National Radio Astronomy Observatory is a facility of the National Science Foundation operated under cooperative agreement by Associated Universities, Inc.

Table 1. Observed Properties of the DIG

Quantity	Observed Range	Reference
T_e	$8,000 \pm 2,000$ K	Reynolds 1989, Reynolds 1985a
He I λ 5876/Hα	$\lesssim 0.02$	Reynolds & Tufte 1995, Tufte 1997
[N II] λ 6583/Hα	$0.2 - 0.5$	Reynolds 1985a
[S II] λ 6716/Hα	$0.2 - 0.5$	Ogen & Reynolds 1985, Reynolds et al. 1977
[O I] λ 6300/Hα	≤ 0.04	Reynolds 1990, Tufte 1997
χ_H	$90 - 97\%$	Tufte 1997

(1996) determined the turbulent energy density in the DIG. From this information, Minter & Spangler (1997) determined that the turbulent dissipation rate, and thus the turbulent heating rate, is likely to be in the range $10^{-25} - 10^{-27}$ ergs cm^{-3} s^{-1} for the DIG, assuming $\chi_H > 0.9$. This value is approximately equal to the cooling rate of the DIG, indicating that the heating of the DIG via the dissipation of turbulence could be an important mechanism. The turbulent heating rate in H II regions is of the order of $10^{-21} - 10^{-24}$ ergs cm^{-3} s^{-1}which is at least two orders of magnitude less than the cooling rate (Minter & Balser 1997). Therefore, heating via the dissipation of turbulence is not a significant process in determining the observed line emissions of H II regions whereas it may be for the DIG and other phases of the ISM.

The numerical program CLOUDY (version 90.02) is used to model the diffuse ionized gas (DIG) near the Galactic midplane (see Ferland 1996 for a complete description of CLOUDY). The interstellar medium is composed of several different phases. We use CLOUDY to model only the diffuse ionized component which we assume is physically distinct, *i.e.* the DIG is composed of nearly fully ionized hydrogen. Observations of the DIG are used to constrain the physical parameters defined in CLOUDY (see Minter & Balser 1997 for a detailed discussion of how CLOUDY was used).

2 Results and Conclusions

Figure 1 summarizes the models for a range of effective stellar temperatures ($T_{\rm eff}$). Notice that none of the models with *only* photoionization heating (dashed lines) satisfy the observational constraints. When the turbulent heating term is included (solid lines) there are many effective temperatures which satisfy all of the observational constraints. The He observations constrain the radiation field to be softer than $Q({\rm He})/Q({\rm H}) \sim 5.2 \times 10^{-2}$ ($T_{\rm eff} \leq 32,000$ K) for all models. The [N II] λ 6583/Hα and [S II] λ 6716/Hα ratios predicted

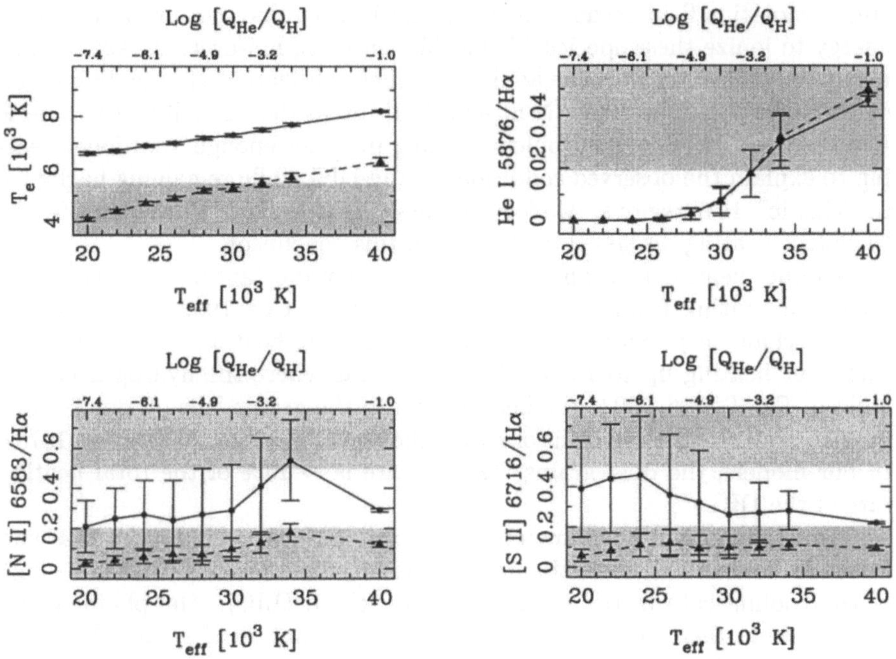

Fig. 1. Model results of the diffuse ionized gas (DIG). Plotted are the average values, over approximately the first 100 pc out of the plane, of the electron temperature (T_e), the He I λ 5876/Hα ratio, the [N II] λ 6583/Hα ratio, and the [S II] λ 6716/Hα ratio as a function of the stellar effective temperature ($T_{\rm eff}$). The hardness of the ionizing spectrum is also described by Q(He)/Q(H) which is shown at the top of each figure. N.B., the effective temperature only determines the *shape* of the incident continuum and not the overall *intensity*. Each panel has two curves: the solid line (filled circles) includes the additional turbulent heating term, while the dashed line (filled triangles) excludes this term. The vertical lines correspond to the 1σ variation in the mean values. Regions which are excluded observationally have been shaded gray (see Table 1).

by the models are consistent with the observations in Table 1 only when the turbulent heating is included.

All models predict that the [O III] λ 5007/Hα ratio is < 0.005, while Reynolds (1985b) measured an [O III] λ 5007/Hα ratio of 0.06 in the diffuse ISM. The origin of this [O III] emission is not known. Our models suggest that the observed [O III] emission does not originate from the oxygen in the DIG. Our models also predict that the [O II] λ 3729/Hα line ratio ranges from 0.15 at $T_e = 20,000$ K to 0.85 at $T_e = 40,000$ K. This ratio is at least five times greater than the values predicted by the models without turbulent heating.

A simple physical picture of the DIG evolves naturally from these results. The Hα and He I recombination lines are produced almost entirely by the stellar radiation field because the turbulent heating does not produce enough energy to ionize these species. The forbidden lines, however, are excited primarily by collisions; thus the addition of thermal heating via the dissipation of turbulence increases the [N II] and [S II] intensities as well as the kinetic temperature. Therefore, turbulent heating provides enough additional heating to explain the observed conditions in the DIG while remaining negligible in "classic" H II regions. Models of classic H II regions which include the turbulent heating discussed in §2 confirm this conclusion.

Another heating mechanism which could be important in the DIG is photoelectric heating from dust grains (Reynolds & Cox 1992, Draine 1978). This mechanism provides approximately the same heating rate as does the turbulent heating up to $T_e \sim 6,000\,\mathrm{K}$ in a gas where the hydrogen is fully ionized. For $T_e \geq 6,000\,\mathrm{K}$ the heating due to the grains falls off quickly due to cooling of the gas via collisions with the cooler dust grains (Draine 1978). In our models, the dust grains' heating rate is $\sim 15\%$ of the total heating rate of the DIG.

From the measured parameters of the local ISM (Wood & Linsky 1997 and Frisch & Slavin 1996), we estimate that the turbulent heating rate is $10-50\%$ of the cooling rate in the local interstellar cloud (LIC). The photoelectric heating from dust grains should also be important in the LIC (Frisch & Slavin 1996). Thus the dissipation of turbulence should be important in the local ISM and may also be important in other phases of the ISM.

References

Domgörgen, H. & Mathis, J.S., 1994, ApJ428, 647

Draine, B.T., 1978, ApJS36, 595

Ferland, G. J. 1996, *Hazy, A Brief Introduction to Cloudy*, University of Kentucky Department of Physics and Astronomy Internal Report

Frisch, P.C. & Slavin, J.D., 1996, Space Science Reviews 78, 223 (Kluwer Academic Publishers)

Minter, A.H. & Balser, D.S., 1997, ApJ484, in press

Minter, A.H. & Spangler, S.R., 1996, ApJ458, 194

Minter, A.H. & Spangler, S.R., 1997, ApJ485, in press

Ogden, P. M., & Reynolds, R.J. 1985, ApJ290, 238

Rand, R.J., 1997, ApJ474, 129

Reynolds, R.J., 1985a, ApJ294, 256

Reynolds, R.J., 1985b, ApJL298, L27

Reynolds, R.J., 1989, ApJ345, 811

Reynolds, R. J. 1990, ApJ 349, L17

Reynolds, R.J., 1992, ApJL392, L95

Reynolds, R.J. & Cox, D.P., 1992, ApJL400, L33

Reynolds, R.J., Roesler, F. L., & Scherb, F. 1977, ApJ211, 115

Reynolds, R.J. & Tufte, S.L., 1995, ApJL439, L17

Tufte, S.L., Ph.D. thesis, University of Wisconsin-Madison

Wood, B. E. & Linsky, J. L. 1997,ApJL474, L39

ROSAT Observations of NGC 4258

Andreas Vogler and Wolfgang Pietsch

Max-Planck-Institut für extraterrestrische Physik, Giessenbachstr.,
D-85740 Garching, Germany

Abstract. We performed deep follow up observations of the active spiral galaxy NGC 4258 with the ROSAT HRI and PSPC to study the different emission components. The overall luminosity of NGC 4258 is $L_x \sim 2 \times 10^{40}$ erg s^{-1} in the 0.1–2.4 keV ROSAT band. Besides a weak nuclear point source ($L_x \lesssim 10^{38}$ erg s^{-1}) fourteen point sources are detected in the NGC 4258 disk (integral $L_x \sim 3 \times 10^{39}$ erg s^{-1}). The bulk of the X-ray emission is not resolved. The main contributions are explained by hot interstellar medium along the anomalous spiral arms of NGC 4258 ($L_x \sim 1 \times 10^{40}$ erg s^{-1}, $T \sim 0.4$ keV) and by interstellar medium escaping from NGC 4258 disk into the approaching halo hemisphere ($L_x \sim 4 \times 10^{39}$ erg s^{-1}, $T \sim 0.2$ keV)

Surface brightness, temperature and absorption profiles have been taken for the X-rays in the regions of the anomalous arms. The highest and lowest absorption was received for the NW and SE tips of the arms, respectively. Our X-ray findings are compared to a model of the anomalous arms which suggests that the bar of the galaxy causes shocks in the interstellar medium and might heat a small part of it to X-ray temperatures.

1 The Active Galaxy NGC 4258

The highly inclined barred spiral galaxy NGC 4258 ($i = 72°$, $d = 6.4$ Mpc) hosts a Seyfert 1.9 nucleus (Courtès et al. 1993 and references therein). Courtès & Cruvellier (1961) discovered a pair of anomalous spiral arms (cf. Fig. 2d). While the arms have been interpreted as twisted jets ejected from the nucleus and tunneling through the disk (cf. Cecil et al. 1995 and references therein), CO and Hα measurements performed by Cox & Downes (1996) suggest that the arms are due to interstellar medium excited by bar shocks. The total length of the bar is $\sim 3'$ (5 kpc). Water maser emission arising inside the obscuring torus of the AGN has been measured by Miyoshi et al. (1995) indicating the existence of a massive black hole in the center of the galaxy.

We reported on a 7.8 ks observation with the ROSAT PSPC in Pietsch et al. (1994). The low Galactic foreground ($N_H = 1.2 \times 10^{20}$ cm^{-2}) allows a sensitive search for X-ray emission in the ROSAT 0.1–2.4 keV band. The emission was separated into three components: Seven point like sources in the spiral arms (total $L_x = 2.5 \times 10^{39}$ erg s^{-1}), extended emission from hot ($T \sim 4 \times 10^6$ K) interstellar medium above the NGC 4258 disk to the East and structured diffuse emission connected to the anomalous spiral arms (altogether $L_x = 2.1 \times 10^{40}$ erg s^{-1}). ROSAT HRI observations (27 ks, Cecil et al. 1995) mostly reveal X-ray emission from the anomalous spiral arms. The authors suggested that the X-rays emerge from hot interstellar medium, which is heated by interactions of the jets with molecular clouds (the jets in this case are believed to be material expelled by the AGN). ASCA revealed hard X-rays from the highly absorbed ($N_H \sim 10^{23}$ cm^{-2}) AGN (Makashima 1994).

2 Deep ROSAT Follow-Up Observations

We performed deep follow-up observations of NGC 4258 with the ROSAT PSPC and HRI to better characterize the different emission components. The total PSPC and HRI observation times have been 32.9 ks and 55.8 ks, respectively. The (smoothed) images are shown in Fig. 1. Point sources are marked with squares. The central source was detected with a luminosity[1] of $L_x \sim 1.2 \times 10^{39}$ erg s^{-1} The main luminosity of the central source is due to extended emission (cf. Fig 2b). For a possible point source on top of the extended emission we receive $L_x \lesssim 10^{38}$ erg s^{-1}. This value can be viewed as an upper limit of contributions of the AGN in the ROSAT band. Fourteen point sources besides the X17 are located within the disk of NGC 4258, and they have luminosities in the range $(1-10) \times 10^{38}$ erg s^{-1}. Three sources (X6, X14, and X18) coincide with HII regions reported by Courtès et al. (1993).

HRI images of the central region of NGC 4258 have been produced using different filtering (Fig. 2a-c). A comparison of the diffuse X-ray emission features with the radio map (1480 MHz; Albada & van der Hulst 1982; our Fig. 2d) suggests a relationship of the diffuse emission with the anomalous radio arms. In both wavelengths the arms have a position angle of 34° and bend off at radial distances between 1!5 and 2′ (3 – 4 kpc) from the nucleus. The HRI images shown in Figs. 2a, 2b trace the anomalous arms out to distances of $\sim$ 2!7 (5 kpc) from the center of the galaxy, the radio observations (2d) out to $\sim$ 3!5 (6 kpc). The adaptively filtered HRI image (2c) might indicate that also the diffuse X-ray emission has such an extent.

Temperature, absorption, and brightness profiles along the anomalous arms (9 boxes from the SE tip of the arms to the NW tip) were calculated from the PSPC observations. We measure low and high absorption for the southeast and northwest tips of the anomalous arms, respectively. This might indicate that hot gas along the jets first is confined within the NGC 4258 disk and then escapes into the halo. The gas in the SW flows into the approaching halo hemisphere and is only absorbed by Galactic foreground. A diffuse emission region with a diameter of $r \sim 4' - 6'$ is visible in the PSPC soft (0.1–0.4 keV) band with $L_x \sim 4 \times 10^{39}$ erg s^{-1} and $T \sim 0.2$ keV. The emis-

[1] All luminosities are corrected only for Galactic foreground.

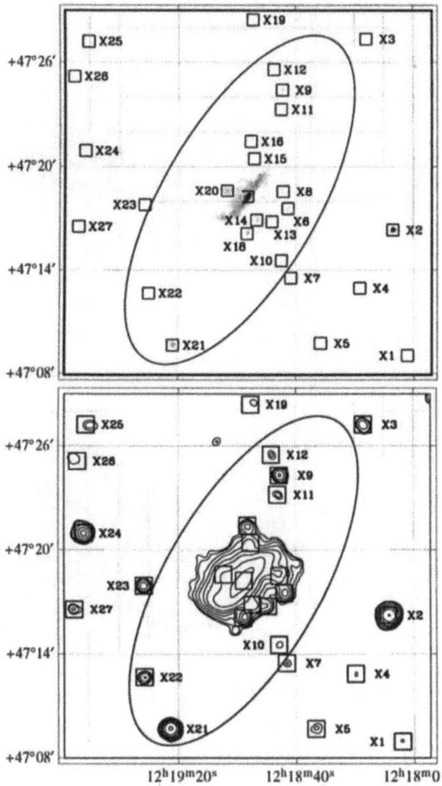

Fig. 1. ROSAT–Observations of NGC 4258 with the HRI (top) and PSPC (bottom). Point sources are marked with squares and enumerated (the central source (X17) has not been enumarated)

sion is displaced from the center towards the SE halo hemisphere and seems to suffer no absorption from cold gas in the NGC 4258 disk. It could be hot interstellar medium ejected from the southwestern arm into the approaching halo hemisphere. The gas in the NW flows into the reverse hemisphere, and the soft X-rays are absorbed by cold gas in the intervening NGC 4258 disk.

Our X-ray findings are compatible with the model of the anomalous arms as suggested by Cox & Downes (1996). They interpret the anomalous arms as interstellar medium excited by bar shocks. Their CO observations indicate that the inner parts of the anomalous arms are confined to the galactic disk (there the bar shocks take place), and the gas flows out of the disk in the regions of the radio plateaus at radial distances of $\sim 1\rlap{.}'5$ from the nucleus.

We measure an X-ray luminosity from the SE tip to NW tip of the arms of $L_{\rm x} \sim 9 \times 10^{39}$ erg s^{-1} (corrected for Galactic foreground). If we correct for the absorption according to our absorption profiles we obtain an intrinsic

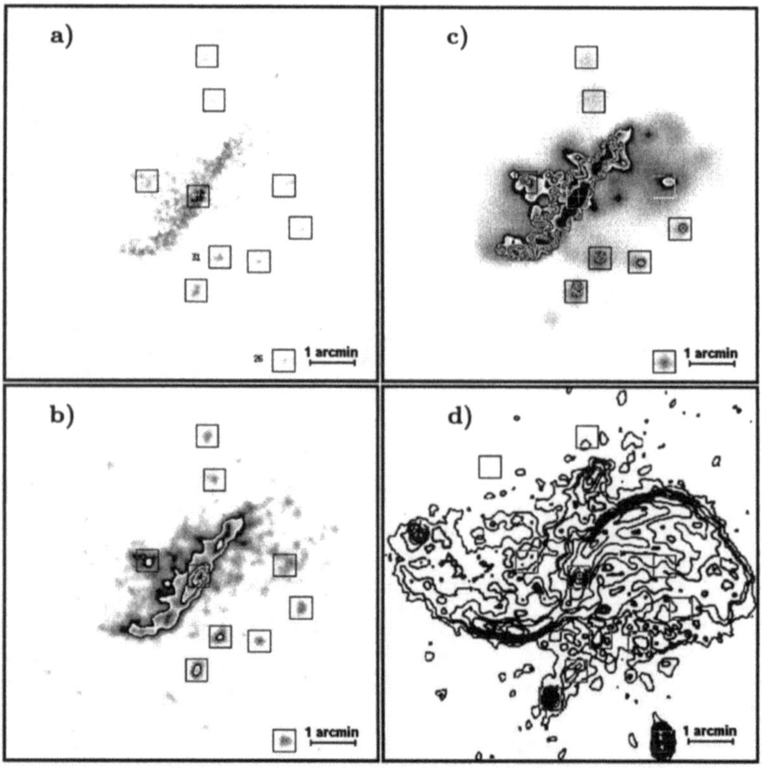

Fig. 2. The inner region of NGC 4258. **a)** ROSAT HRI image at full resolution (smoothed with a gaussian function of 5″ FWHM) **b)** at 10″ resolution **c)** after applying an adaptive filtering technique **d)**1480 Mhz radio observation (Albada & van der Hulst 1982)

luminosity that is higher by a factor of five. The X-ray data allow to estimate the density and mass of the hot gas along the arms. For a filling factor of unity we calculate $n_e \sim 3 \times 10^{-2}$ cm^{-3} and $m \sim 7 \times 10^6$ M$_\odot$. The values are of the same order of magnitude as predicted by Cox & Downes (1996).

References

Albada G.D., van der Hulst J.M., 1982, A&A 115, 263
Cecil G., Wilson A.S., de Pree C., 1995, Apj 440, 181
Courtès G., Cruvellier P., 1961, Compt.Rend.Acad.Sci.Paris, 253, 218
Courtès G., Petit H., Hua C.T., et al., 1993, A&A 268, 419
Cox P., Downes D., 1996, ApJ 473, 219
Miyoshi M., Moran J., Herrnstein J., et al., 1995, Nature vol. 373, p. 127
Pietsch W., Vogler A., Kahabka P., et al., 1994, A&A 284, 386

A Ring of X-rays from the Cartwheel Galaxy

Ginevra Trinchieri[1,2], Anna Wolter[2], and Angela Iovino[2]

[1] Max-Planck-Institut für extraterrestrische Physik, D–85748 Garching, Germany
[2] Osservatorio Astronomico di Brera, via Brera 28, I–20121 Milano, Italy

Abstract. ROSAT HRI observations of the Cartwheel galaxy indicate that there is X-ray emission from the outer ring in connection with the strongest HII regions in the southern portion of the ring. No significant emission is detected from the nucleus or from the inner ring. This implies a strong connection between star formation activity and X-ray emission in this ring galaxy.

1 The Cartwheel: a Spectacular Ring Galaxy

Nearly head-on collisions between a compact galaxy and a gaseous disk are thought to be the origin of ring galaxies (Lynds & Toomre 1976; Theys & Spiegel 1977; Toomre 1978). The density wave produced in the collision propagates outwards from the nucleus, and sweeps up material to form a relatively symmetrical ring, where high densities can be reached and star formation phenomena induced. Evidence of high levels of star formation in rings is based on their blue optical colors, large far infrared luminosities and large luminous HII regions (Appleton & Struck-Marcell 1987; Fosbury & Hawarden 1977, Higdon 1995).

One of the more spectacular and famous examples of a recently formed ring galaxy is the A0035-324 system, named "Cartwheel" because of its optical appearance with a bright, sharp outer ring linked by spiral spokes to the nucleus. An inner ring close to the core is also well studied recently with HST (Borne et al 1996; Struck et al. 1996). Many detailed observations of this object are available, from radio line and continuum, to near and far infrared, optical and Hα images and line spectroscopy. All confirm the presence of a recent starburst in the outer ring, without corresponding activity in the inner ring, nucleus or spokes, believed to be relatively devoid of gas. Most of the activity is in fact in the S-SW quadrant of the ring, where massive and luminous HII regions characterized by large Hα luminosities and equivalent widths are found (Higdon 1995). Both dynamical considerations and stellar evolution models suggest an age of 2-4 $\times 10^8$ yr for the starburst; the evidence of very low metallicity, with large deficiency factors in O, N and Ne relative to the solar value, and the estimated supernova rate, almost two orders of magnitude higher than in normal galaxies, also support the view that star formation in the ring is a recent phenomenon and that gas currently forming stars was nearly primordial at the time of the impact (Fosbury & Hawarden 1977; Higdon 1995; Marcum et al 1992). The Cartwheel is located in a tight

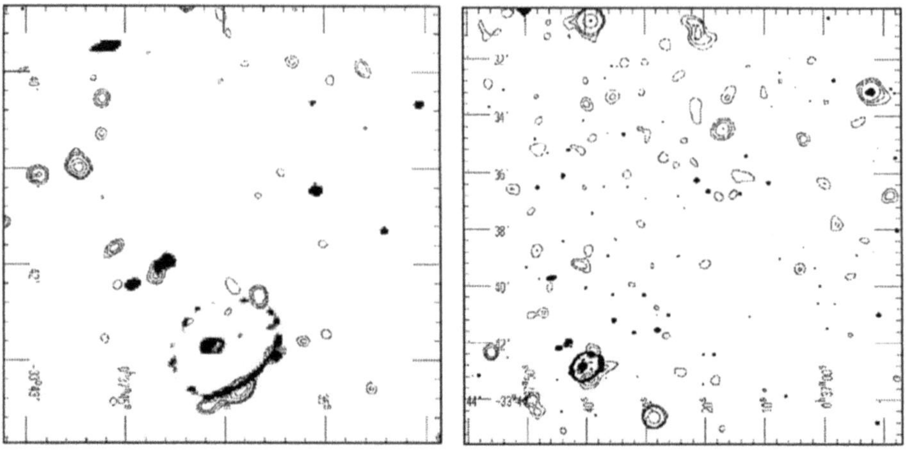

Fig. 1. Left: X-ray emission from the Cartwheel and its 3 companions superposed on the Digitized Sky Survey. The HRI data are smoothed with a $\sigma = 4''$ Gaussian function. **Right**: In this larger field several optical counterparts coincident with X-ray sources are visible. A $\sigma = 10''$ smoothing is applied to the X-ray data.

group ($\sim$ 0.2 Mpc size) that includes three nearby companions of comparable size and magnitude, with a redshift within 400 km s^{-1} from one another.

All of the above indicates that X-ray observations could show both the galaxy and the group as as a source of emission, as a consequence of the star formation activity and in analogy with similar systems. As the Cartwheel is relatively large in size ($80''$ along the major axis, corresponding to 65 kpc at the distance of 180 Mpc), the outer ring can be well separated from the inner core, and the emission from the Cartwheel from that of other components of the group. We present here some preliminary results from a $\sim$ 60 ks observation we obtained with the ROSAT HRI.

2 The Cartwheel in X-rays

ROSAT data (Fig. 1) indicate clearly that the emission comes from the outer ring, and in particular from the Southern quadrant, while the nucleus, inner ring and spokes do not emit. The coincidence of several other X-ray sources in the field with optical counterparts (see Fig. 1, right panel) ensures that the emission is really coming from the ring, and that no errors in the pointing or in the aspect solution could have determined this coincidence.

We have detected $\sim$ 80 net counts above an average field background of $\sim$ 0.07 cts $arcsec^{-2}$ from the Cartwheel. For an assumed *Raymond*-type spectrum with kT= 1 keV and 20% cosmic abundance, and low energy absorption corresponding to the line-of-sight column density $N_H = 1.8 \times 10^{20}$

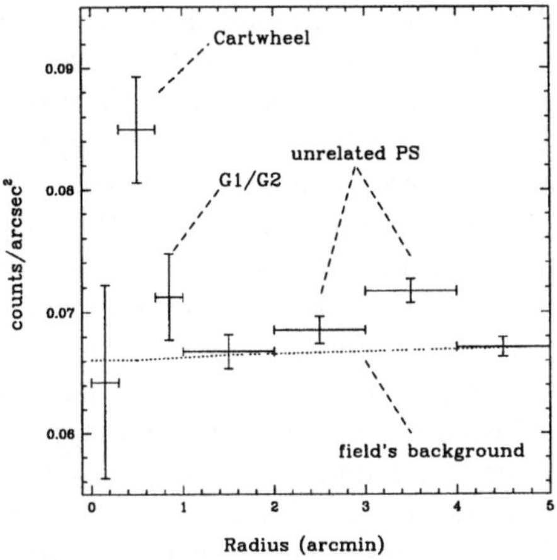

Fig. 2. Radial distribution of the total counts detected in the HRI observation (+) and of the field's background, both centered on the Cartwheel.

cm^{-2} (Stark et al. 1992), this corresponds to a flux $f_x \sim 5.5 \times 10^{-14}$ erg cm^{-2} s^{-1} and luminosity $L_x \sim 2.0 \times 10^{41}$ erg s^{-1}. The luminosity could increase to $L_x \sim 4.0 \times 10^{41}$ erg s^{-1} if an intrinsic absorption of $N_H = 2 \times 10^{21}$ cm^{-2} (corresponding to the reddening observed in the HII regions), is assumed. Note that the result is virtually insensitive to the choice of kT and abundance.

As shown by Fig. 2, all of the emission is in an annulus at $\sim 0.3' - 1'$ from the nucleus, and can be attributed to the outer ring of the Cartwheel (the majority) and to the two nearmost companions G1-G2 (which are probably a separate interacting pair, Higdon 1996). No emission is detected interior to the ring, or associated with the group or with the companion G3, which is now believed to be the intruder responsible for the impact with the Cartwheel (Higdon 1996). A small fraction ($\sim 1/10$) of the flux attributed to the Cartwheel might be associated with a star superimposed on the outer ring, well visible in the HST image (NASA Press Release PRC96-36a, Nov. 26, 1996) and already identified as an object with the colors of a G star (Marcum et al. 1992). The X-ray to optical flux ratio (estimated at 5×10^{-15} erg cm^{-2} s^{-1} for a 17-18 mag. object) is consistent with values of stars in this class (see nomogram in Maccacaro et al. 1988).

3 Why X-rays on the Ring?

We see a remarkable coincidence between X-ray and Hα emission, indicative of a close association with the star formation phenomenon. X-ray emission is higher in connection with the brightest HII regions while no emission is observed from the nucleus or inner ring, where no star formation enhancement is inferred from optical, radio or IR data.

In spite of the relatively large number of young stars, coronal emission from normal stars is unlikely to account for the observed L_X. Optical and infrared data infer the presence of a few $\times 10^5$ O stars in the whole ring (Marcum et al. 1992). This implies $\lesssim 10^7$ O-B type stars, which would produce < 10% of the observed luminosity at most, even in the generous assumption that all emit at the highest level of $\sim 10^{33}$ erg s^{-1} detected from stars in our galaxy (Vaiana et al. 1981; Chlebowski et al. 1989). Galactic scale strong shocks are excluded from radio or optical observations (Higdon 1996).

It is therefore likely that the evolved stellar population, in the form of binaries, supernovae and supernova remnants, is responsible for the observed emission. Giant HII regions and complex structures, typically coincident with peaks of Hα emission, are also found in actively star forming objects like the interacting system "The Antennae" (Read et al. 1995; Fabbiano et al. 1997) as extremely bright X-ray sources, with intrinsic luminosities up to several $\times 10^{40}$ erg s^{-1} and could be analogous to sources in the Cartwheel's ring.

Emission from the intergroup gas, if any, is very low, at least a factor of 10 less than other groups. The lack of emission from diffuse hot gas could be consistent with the suggestion that the this component is less dense, colder or simply less in late-type dominated groups (see Ponman et al. 1996; Mulchaey et al. 1996) or could be a further indication that most of the emission detected in these systems is due to singular phenomena (like for example shocks between interacting members like in Stephan's Quintet; Pietsch et al. 1997).

References

Appleton P.N. and Struck-Marcell C. (1987): ApJ 318, 103
Borne, K.D. et al. (1996): Science with the Hubble Telescope II, P. Benvenuti, F.D. Macchetto, E. Schreier eds. p. 239.
Chlebowski, T., Harnden, F.R., and Sciortino, S. (1989): ApJ 341, 427.
Fabbiano, G., Schweizer, F., and Mackie, G. (1977); ApJ
Fosbury, R.A. and Hawarden, T.G. (1977): MNRAS 178, 473
Higdon, J.L. (1995): ApJ 455, 524.
Higdon, J.L. (1996): ApJ 467, 241
Lynds, R. and Toomre, A. (1976): ApJ 209, 382
Marcum, P.M., Appleton, P.N. and Higdon, J. (1992): ApJ 399, 57.
Maccacaro, T. et al. (1988): ApJ 326, 680
Mulchaey J.S. et al. (1996): ApJ 456, 80
Pietsch, W. Trinchieri, G., Arp, H., and Sulentic J.W. (1997): A&A 322, 89
Ponman, T.J., et al., (1996): MNRAS 283, 690
Read A.M., Ponman T.J. and Wolstencroft R.D. (1995): MNRAS 277, 397
Stark, A.A, et al. (1992): ApJS 79, 77
Struck, C., Appleton, P.N., Borne, K.D. and Lucas, R.A. (1996): AJ 112, 1868
Theys, J.C. and Spiegel, A.E. (1977): ApJ 212, 616
Toomre, A. (1978): IAU Symp. 79, p. 109
Vaiana, G.S. et al. (1981): ApJ 245, 163

Radio and Polarization Properties in the Disk and Halo of Edge-On Spirals

M. Dumke and M. Krause

Max–Planck–Institut für Radioastronomie, Auf dem Hügel 69, D-53121 Bonn, Germany

Abstract. We have observed a small sample of edge-on galaxies in total and linearly polarized radio continuum emission at $\lambda\,6.2\,\text{cm}$, using the Effelsberg 100 m telescope. The resulting maps were combined with available VLA data in order to obtain high-sensitivity radio images which have a sufficient angular resolution and do not suffer from the so-called missing-spacing problem.

From an analysis of these images (together with 2.8 cm and 20 cm data) we get the following results: 1. The radio emission perpendicular to the plane can be described by a two-component exponential distribution with scale-heights of $\sim 300\,\text{pc}$ and $\sim 1.8\,\text{kpc}$, independently of the star forming activity and interaction state. 2. In the disk of the non-interacting galaxies of our sample the thermal fraction of the emission seems to be higher than usually deduced from observations of face-on galaxies. 3. The fractional polarization p increases with increasing galactic height, reaches a maximum at $h \sim 1 - 3\,\text{kpc}$, and decreases again. In general p is lower in actively star forming galaxies.

1 Introduction

The distribution of relativistic electrons as well as the morphology of uniform and random magnetic fields in galactic halos are strongly affected by dynamical processes in the underlying disk. During the evolution of a galaxy, ongoing star formation leads to subsequent supernova explosions and the creation of superbubbles. These processes may cause the transport of high-energetic particles and small-scale turbulences away from the star forming regions into the disk-halo interface and to high galactic latitudes. As a consequence, the magnetic field properties as well as the spectral distribution of the emission is expected to be different in galaxies of differing star forming activity.

Edge-on galaxies are ideal candidates for observing these effects using radio continuum and polarization measurements at cm-wavelengths, where the emission is mainly due to relativistic electrons gyrating around magnetic field lines. Especially the combination of observations at different wavelengths provide useful information necessary to understand the various relationships between star-formation in the disk and its consequences on the interstellar medium.

2 Combination of Interferometer and Single-Dish Observations

From February to August 1996 we have observed a small sample of edge-on galaxies, consisting of NGC 891, NGC 3628, and NGC 4565, at λ 6.2 cm using the Effelsberg 100 m-telescope. The data were reduced in a standard manner using the NOD2 software package available at the MPIfR Bonn. The resulting maps were cleaned from the sidelobes of the telescope pattern, and the instrumental polarization was removed down to a level of 0.3 %.

The spatial resolution of large single-dish telescopes is limited. On the other hand, data from interferometers like the VLA suffer from the so called "missing-spacing problem": because of the lack of visibilities at short baselines they are not sensitive to extended emission. In order to get final maps with both the high resolution of the interferometer data and the short-spacing information provided by the single-dish observations, we combined the maps from the Effelsberg telescope with VLA maps at this wavelength from Sukumar & Allen (1991), Dumke (1994), and Dumke & Krause (in prep.). This was done in the Fourier plane using the SDE task IMERG (which is available from the NRAO) for the total power as well as the polarization data (Holdaway 1992). Remaining interferences in the combined maps around strong sources, which limited the dynamic range, could successfully be removed using a Fourier filter technique.

The combined maps including the short-spacing information differ significantly from the VLA maps. It is obvious that the usage of interferometer data alone may lead – and have probably led in many investigations in the past – to unreliable results concerning the quantitative and qualitative behaviour of radio and polarization properties. Our results also show that even at λ 20 cm the absence of short spacing information in the interferometer data may be a serious problem. Hence the single-dish data from the Effelsberg telescope are absolutely necessary to investigate the emission of extended radio sources at cm-wavelengths.

3 Results and Discussion

Fig. 1 shows the distributions of the radio emission perpendicular to the major axis in the observed galaxies. For all three objects the distribution could be well fitted by a two-component exponential function, convolved with the beam size. Considering also inclination effects, the scale-heights of both components are similar in all galaxies: $h_{\mathrm{thin\,disk}} \sim 300$ pc and $h_{\mathrm{thick\,disk}} \sim 1.8$ kpc, independently of the star forming activity in the underlying disk and the interaction state. Since all three galaxies exhibit a large-scale magnetic field predominantly parallel to the disk (Dumke et al. 1995), the field configuration could possibly be the determinant factor for the emission scale-heights.

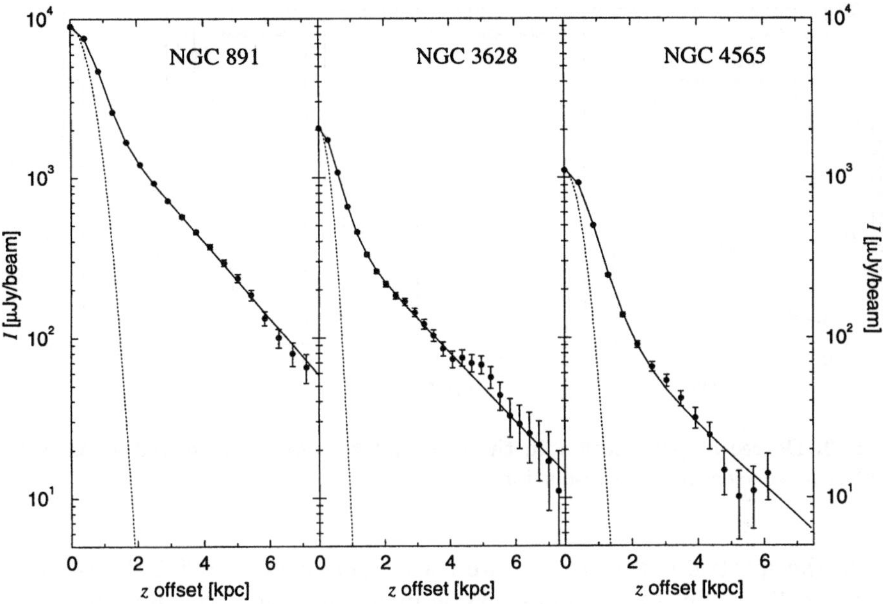

Fig. 1. Distribution of the radio continuum emission perpendicular to the major axis of the galaxies. The dotted lines show the telescope beam, the solid lines the fitted function for a two-component distribution model

Differing results in previous investigations may be caused by a lack of sensitivity/resolution and the lack of large-scale structure information in interferometric data. Strong star forming activity, however, increases the absolute intensities of both components and may be responsible for a relatively high intensity of the thick disk component compared to the thin disk, as it is the case in NGC 891.

The spectral index distribution perpendicular to the plane shows a similar behaviour in all three galaxies: an increase with galactic height z. We see (with the available resolution of $2-3$ kpc) no evidence for different transport mechanisms for relativistic electrons. The non-interacting objects NGC 891 and NGC 4565, however, have flatter spectra in the disk and in the halo. Especially the spectra in the disk of these two objects can only be explained with a very high thermal fraction of the total emission, about 50 % at λ 6.2 cm. This is also suggested by a spatial correlation of flat spectra and low fractional polarizations in localized regions in the disk. The spectra observed in face-on galaxies are usually steeper, because those objects must have significant halo emission (which has a steeper spectrum) to yield an observable surface brightness.

The two galaxies NGC 891 and NGC 4565 seem to be very similar, and the first one appears to be just an upscaled version of the latter. On the other

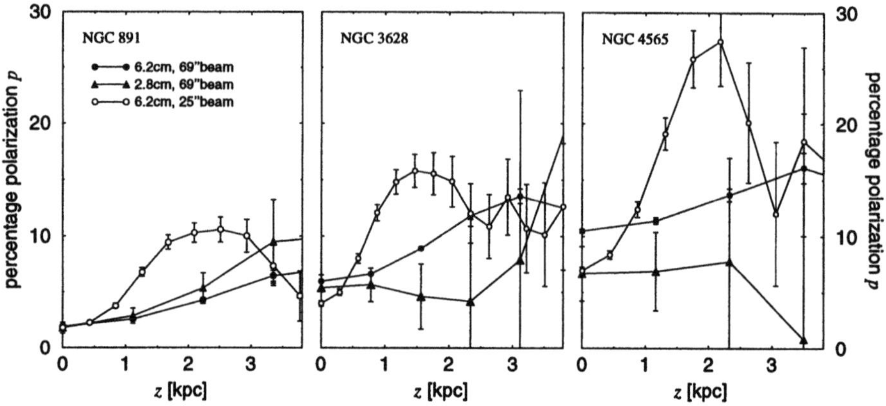

Fig. 2. Degree of polarization in the disk and halo of the observed galaxies for different wavelengths and resolutions

hand the spectral indices in the interacting galaxy NGC 3628 indicate just a very small amount of thermal emission in the galactic disk, and strong star formation seems to be restricted to the central region. Hence the relatively strong thick disk component cannot be due to recent star formation in the thin disk. Possible explanations may be the gravitational interaction with other galaxies and/or that the observed thick disk is a remnant of higher star forming activity in earlier times.

Fig. 2 shows the behaviour of the fractional polarization with increasing distance from the galactic plane, as a function of wavelength and beam size. From the high-resolution 6.2 cm data (white circles) we find that p increases over the first $\sim$ 2 kpc, but then decreases again. This suggests strong depolarization effects in the disk and in the upper halo of the galaxies. In general, the degree of polarization is highest in NGC 4565, the galaxy with lowest star forming activity, and smallest in NGC 891. This is expected, because star formation leads to enhanced turbulence and more efficient depolarization. At λ 20 cm (which is not shown in Fig. 2) the polarized intensity is always very weak, due to Faraday depolarization. In case of strong star formation, as in the central region of NGC 3628 and over the whole disk of NGC 891, there is already strong Faraday depolarization at λ 6.2 cm.

References

Dumke M. (1994): Diploma Thesis, University of Bonn

Dumke M., Krause M., Wielebinski R., Klein U. (1995): A&A 302, 691

Holdaway M.A. (1992): Mosaicing with Even Higher Dynamic Range, Millimeter Array Memo No. 73. NRAO, Socorro

Sukumar S., Allen R.J. (1991): ApJ 382, 100

Superbubbles in Dwarf Galaxies: Blown Out or Blown Away?

Mordecai-Mark Mac Low[1] and Andrea Ferrara[2]

[1] MPI für Astronomie, Königstuhl 17, D-69117 Heidelberg, Germany
[2] Osservatorio Astrofisico di Arcetri, Largo Enrico Fermi 5, I-50125 Firenze, Italy

Abstract. We numerically model the impact of superbubbles produced by starbursts on the gas in dwarf galaxies, using realistic gravitational potentials including the contributions from dark matter haloes, and galaxy radii based on empirical laws. We find that the metal-rich gas is very efficiently ejected in many cases, contradicting previous results; the ambient gas is far more difficult to eject. X-ray emitting plumes can last more than 10^8 yr after the end of the active starburst.

1 Introduction

One of the most extreme examples of bubble dynamics occurs in dwarf galaxies undergoing starbursts. The stellar winds and supernovae from massive stars produce huge shock waves that can blow out of the dwarf's interstellar medium, and might even be able to blow away all the gas from the disk (e.g. De Young & Heckman 1995), raising its radial velocity above the local escape velocity from the gravitational potential. The rotation curves of dwarf galaxies suggest that their gravitational potentials are dominated by dark matter (Persic, Salucci & Stel 1996). There is now observational evidence in support of the existence of outflows from dwarf galaxies including NGC 1705 (Meurer et al.1992), I Zwicky 18 (Martin 1996), NGC 1569 (Della Ceca et al.1996) and a number of others in a survey by Marlowe et al.(1995). In this paper, we numerically model the effect of nuclear starbursts on dwarf galaxies, including the effect of dark matter for the first time.

2 Models

We model a dwarf galaxy as a system made of a gaseous and stellar (i.e. visible) disk of mass M_d and a dark halo of mass M_h. We assume that the gaseous disk has constant surface density, motivated by observations of our Galaxy. Persic, Salucci & Stel (1996) find that the dark-to-visible mass ratio $\phi = M_h/M_d$ is a function of the galactic mass, $\phi \simeq 34.7 M_{d,7}^{-0.29}$, where $M_{d,7} = M_d/10^7\ M_{\odot}$. We assume that the halo can be approximated by a softened isothermal sphere, with scale radius and central density dependent on total mass (Burkert 1995). From these assumptions, we can derive the gravitational potential of the halo, and compute the exact, hydrostatic equilibrium of the

Table 1. Efficiencies

	Mass Eject. Eff.			Metal Ret. Eff.		
Lum	Mass ($M_\odot$)			Mass ($M_\odot$)		
(10^{38} erg s^{-1})	10^7	10^8	10^9	10^7	10^8	10^9
0.1	4.6 (-3)	0.0	0.0	4.1 (-1)	1.0	1.0
1.0	1.4 (-2)	6.5 (-4)	1.0 (-6)	2.7 (-3)	4.0 (-1)	9.7 (-1)
10.0	6.8 (-2)	4.8 (-3)	1.3 (-4)	4.9 (-3)	1.7 (-2)	4.5 (-1)

gas in this potential. From an empirical fit to a sample of dwarf galaxies Ferrara & Tolstoy (in preparation) find that the radius-HI mass relation is well approximated by the law $R \simeq (3\ \mathrm{kpc}) M_{g,7}^{0.338}$, so we cut off our gas distributions at this radius.

For our numerical models, we use ZEUS-3D[1], a second-order, Eulerian, astrophysical gas dynamics code (Stone & Norman 1992), including equilibrium radiative cooling (MacDonald & Bailey 1981), and a tracer field in order to maintain a sharp interface at the edge of the hot bubble (using the Yabe-Xiao transform on the tracer function [Yabe & Xiao 1993]).

We set up the gas at a constant sound speed of 10 km/s, with adiabatic index $\gamma = 5/3$, and minimum density $n_{\mathrm{IGM}} = 6.4 \times 10^{-8}\ \mathrm{cm}^{-3}$. We take the central energy source to be a constant luminosity wind driven by a thermal energy source. We use 2D, azimuthally-symmetric grids centered on the center of the galaxies, with reflecting boundary conditions along the symmetry axis and along the galaxy mid-plane, and outflow boundary conditions on the other two axes, and having a resolution of 20 zones per scale height, with the outer regions covered by a grid ratiod at 1.03 per zone.

3 Results

In Figure 1 we show the density distribution of every model in the parameter study. We see the holes in the plane of the galaxy recollapsing under the influence of gravity and the pressure of the disk gas. At the same time, the hot bubbles rising above the disks expand out into the surrounding intergalactic medium, leaving most of the swept-up shell material behind.

We can directly compute the efficiency of mass ejection and retention of metal-enriched gas from our models. In Table 1 we show the fraction of mass accelerated beyond escape velocity. Even in our most extreme model, with a supernova every 30,000 years in a galaxy of mass $10^7\ M_\odot$, less than 10% of the mass is ejected, and the efficiency drops rapidly from there.

[1] Developed by the Laboratory for Computational Astrophysics at the NCSA, and available for community use by registration at lca@ncsa.uiuc.edu

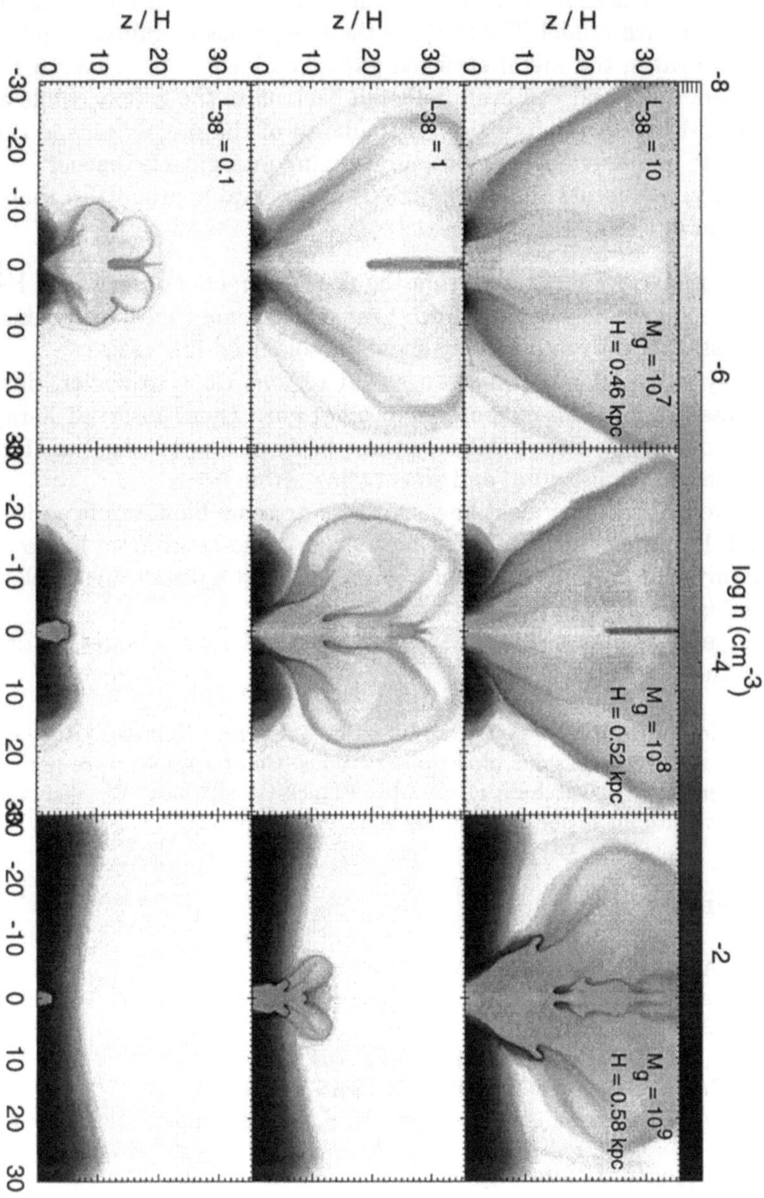

Fig. 1. Density distributions of models of galaxies with the given masses in solar masses, containing starbursts of the given mechanical luminosities in units of 10^{38} erg s^{-1} at a time of 100 Myr, some 50 Myr after the end of energy input from the starbursts, with the values of density given by the colorbar at the top of the figure. Lengths are given in terms of the best-fitting exponential scale height for each model, with values quoted in the figure. Luminosities are equivalent to supernovae every 30,000 to 3 million yr.

Using our tracer field, we can also find the fraction of metal-enriched gas *retained* for each model (Table 1). In the more massive galaxies and at lower supernova rates, the metal enriched gas is retained in the gravitational well of the dark halo, and will eventually fall back on to the galaxy, while at lower masses and higher luminosities, virtually all of the metals escape the grasp of the halo and travel freely into surrounding intergalactic space.

Finally, we discuss some predictions of observable properties that we can draw from our models:

- Hot gas is well separated from the dense filaments observed in Hα.
- These filaments are not shock-heated; they may be photoionized from OB stars below, and so may show photoionized line ratios.
- X-ray emission may be strongest in regions close to cooler, dense filaments, where mass evaporates into hot gas. Thus, observed X-ray emission may come from relatively thin sheets of evaporating gas, far out of ionization equilibrium, and so emitting strongly.
- Blowouts can carry field lines, so more extreme blowouts from low-mass, high-luminosity galaxies will have larger radio-continuum haloes.
- Highly-enriched gas far from a galaxy might be detectable in absorption against background QSO's.
- These ideas may also apply to starbursts in larger galaxies, as described by Suchkov et al. (1994).

Acknowledgements. We are grateful to E. Tolstoy, D. Bomans, R.-J. Dettmar, and G. Golla for discussions of the observations. Computations were performed at the Rechenzentrum Garching of the Max-Planck-Gesellschaft. We each thank the other's institute for hospitality during work on this paper.

References

Burkert, A. (1995): ApJ 447, L25

Della Ceca, R., Griffiths, R. E., Heckman, T. M., MacKenty, J. W. (1996): ApJ 469, 662

De Young, D. S., Heckman, T. M. (1994): ApJ 431, 598

MacDonald, J., Bailey, M. E. (1981): MNRAS 197, 995

Marlowe, A. T., Heckman, T. M., Wyse, R. F. G., Schommer, R. (1995): ApJ 438, 563

Martin, C. (1996): ApJ 465, 680

Meurer, G. R., Freeman, K. C., Dopita, M. A., Cacciari, C. (1992): AJ 103, 60

Persic, M., Salucci, P., Stel, F. (1996): MNRAS 281, 27

Stone, J. M., Norman, M. L. (1992): ApJS 80, 753

Suchkov, A. A., Balsara, D. S., Heckman, T. M., Leitherer, C. (1994): ApJ 430, 511

Yabe, T., Xiao, F. (1993): J. Phys. Soc. Japan 62, 2537

Part X

Conference Summary

Relaxed atmosphere during the conference dinner.

Joachim Trümper closes the conference: what have we learnt?

The Local Bubble and Beyond: Summary

Christopher F. McKee

Departments of Physics and Astronomy, University of California, Berkeley CA 94720, USA

1 Introduction

Lyman Spitzer, Jr, the founder of modern studies of the interstellar medium (ISM), passed away March 31, 1997. This conference occurred shortly thereafter and is dedicated to his memory. While many of his contributions underlie the work that was discussed at this meeting, one paper stands out in particular: his theoretical "discovery" of the hot gaseous halo of the Galaxy based on the need to confine the clouds of H I observed above the Galactic plane (Spitzer 1956). We now know that much of the ISM within about 100 pc of the Sun is largely filled by very low density gas, which is generally inferred to be hot, and as a result this region is termed the Local Bubble (Cox and Reynolds 1987). This conference was convened to establish the current state of our knowledge of the Local Bubble, both observational and theoretical, and its relation to the rest of the ISM. Because it is nearby, the Local Bubble is a laboratory for interstellar astrophysics, making the dedication to Spitzer's memory particularly appropriate.

Everyone I have talked to agrees that this conference was a great success. That success is due to two individuals. One is an experimentalist who has realized his dream of making a high resolution map of the entire X-ray sky. After enormous effort, the *ROSAT* All-Sky Survey is nearing completion. When it is released, it will become an invaluable resource for all astronomers, just as the *IRAS* infrared survey did over a decade ago. This person is Joachim Trümper. The second is a theorist who has developed a highly original model for the Local Bubble. Don Cox and I both refereed the paper describing this model for *Nature*. Although we rarely agree (I like to think of our approaches as lying along the real and imaginary axes–I'll let you decide which is which), we both felt that the paper was so original that it should be published there. This person is Dieter Breitschwerdt. Dieter had not only the vision to create the conference, but also the energy to ensure that everything worked perfectly. We owe both Joachim and Dieter a debt of gratitude.

It is impossible to adequately summarize such a large conference with so many threads of thought. The community is attempting to weave these threads into a coherent tapestry, but our work is far from done and the final image unclear. To bring some order to my discussion, I shall largely follow the outline suggested by Cox and Reynolds (1987) in their review:

1. What is the local interstellar medium (LISM) like?
2. Why is it like that?
3. Is the LISM typical of the ISM as a whole?

Additional comprehensive reviews of the Local Bubble are to be found in Bochkarev (1987) and Frisch (1995).

2 What Is the LISM Like?

2.1 The Very Local ISM–Interaction with the Heliosphere

It is traditional to distinguish space physics, which relies on in situ observations, from astrophysics, which relies on remote observation. The two fields converge in the study of the very local ISM that is streaming through the solar system. The exciting recent developments in this field were described by Lallement*.[1] Theoretically, one expects the solar wind to expand until its ram pressure becomes comparable to the pressure of the ambient medium, at which point the solar wind undergoes a termination shock. Because of the motion of the Sun relative to the surrounding medium, the shocked solar wind is swept backwards. The interstellar medium also undergoes a shock (the bow shock) when it interacts with the solar wind. The interface between the shocked solar wind and shocked ISM is termed the heliopause. The bow shock occurs primarily in the ionized component of the ISM. Interstellar H^0 interacts with the ionized gas through charge exchange, creating a "wall" of hot hydrogen. On the other hand, the charge exchange cross section for He^0 is small; interstellar He^0 readily penetrates the heliosphere, enabling direct measurement of its properties by the *Ulysses* spacecraft (Witte et al., 1992). Lallement* reported that the Sun's velocity relative to the local ISM is found to be 26 $\mathrm{km\,s^{-1}}$, the densities are in the range $n(\mathrm{H}^0) = 0.1 - 0.15\ \mathrm{cm}^{-3}$ and $n_e = 0.15 - 0.05\ \mathrm{cm}^{-3}$, the temperature is $T \sim 7000$ K, and the thermal pressure is in the range $P/k = 2700 - 1700\ \mathrm{K\ cm}^{-3}$. The sun is thus embedded in a cloud of warm gas (the Local Cloud, sometimes termed the "Local Fluff"), which in turn is embedded in the Local Bubble.

The *Voyager* spacecraft, which have yet to reach the termination shock, have detected electromagnetic radiation with a lower cutoff at 1.8 kHz (Kurth et al., 1984). According to Lallement*, this radiation is now believed to originate beyond the heliopause, which suggests that the interstellar electron density is $n_e \simeq 0.04\ \mathrm{cm}^{-3}$, just below the lower limit inferred from *Ulysses* data. The termination shock is inferred to be more than 75 AU from the Sun, and this limits the total pressure of the incoming ISM, including the magnetic pressure. Knowing the density and the velocity of the local ISM, it is then possible to place an upper limit on $B_\perp$, the component of the interstellar magnetic field that is normal to the relative velocity: $B_\perp < 4.4\ \mu$G

[1] References to talks and posters at this conference are denoted by an asterisk.

(Lallement*). Once the *Voyager* spacecraft penetrate the heliopause, it will be possible to make a direct measurement of the shocked interstellar field.

The most surprising recent result on the very local ISM is described in a poster by Landgraf*: the dust size distribution measured by the *Ulysses* and *Galileo* spacecraft is inconsistent with that inferred from analysis of interstellar extinction (e.g., Kim, Martin, and Hendry 1994; Kim and Martin 1996). Whereas the mass distribution of interstellar grains is inferred to peak at about 0.2 μm, the in situ observations imply a mass distribution that peaks at about 0.4 μm. Although large grains are less efficient at causing extinction than small grains, it does not appear possible that the extinction observations have simply missed the large grains: virtually all the silicon in the ISM is needed to make the grain size distribution inferred by Kim et al. (1994), so there is none left to make the large grains observed by *Ulysses* and *Galileo*. Furthermore, the astrophysical model for interstellar grains has passed a major test with *ISO* observations, which show that the infrared emission spectrum of interstellar grains is just what was predicted (Reach and Boulanger*). The convergence of space physics and astrophysics appears to have produced a real puzzle in this case.

2.2 Hot Gas in the Local Bubble

There have been two recent major discoveries about the hot gas in the Local Bubble. First, as discussed by Snowden*, the observation of X-ray shadows by *ROSAT* has demonstrated that a significant fraction of the observed emission comes from the halo of the Galaxy, not the Local Bubble. He infers that the hot gas in the Local Bubble has a size of 40 – 130 pc, a temperature $T \simeq 10^{6.07}$ K and a pressure $P/k \simeq 1.4 \times 10^4$ K cm^{-3}. The hot gas in the halo is not uniform; on average, its emission measure is comparable to that in the Local Bubble, and its temperature is about the same ($T \simeq 10^{6.02}$ K). After allowing for the halo emission, it is still possible to describe the hot gas in the Local Bubble by a modified displacement model in which the X-ray brightness is proportional to the radius of the Local Bubble. There is little, if any, hard X-ray (M-band) emission from the Local Bubble. The hot gas in the Local Bubble appears to be approximately isobaric: analysis of the emission toward Eridanus gives $P/k \simeq 1.2 \times 10^4$ K cm^{-3} (Burrows and Guo*), similar to the value inferred by Snowden* for the entire Local Bubble. Historically, Bochkarev (1987) appears to have been the first to use X-ray shadows to establish the existence of Galactic X-ray emission from beyond the Local Bubble, although the data available at that time did not permit him to derive any quantitative conclusions. It should be noted that there is also spectroscopic evidence for two components of thermal X-ray emission, a local one with no evidence for absorption, and a distant one with evidence for absorption by the gaseous disk of the Galaxy that is consistent with a halo origin (Garmire et al., 1992).

The second major discovery, albeit an expected one, is the demonstration that the soft X-ray background is dominated by line emission, as expected for emission from a hot plasma (Sanders*). These results, which were obtained with the DXS experiment, were initially very puzzling, since at first thermal emission models did not fit the data. However, inclusion of better atomic physics in the models and allowance for depletion of Si, S, and Fe results in a significantly better fit for collisional ionization equilibrium (CIE) models, albeit with a χ^2 greater than 2. In fact, gas at a temperature of about 10^6 K is almost certainly not in ionization equilibrium (Böhringer*), so the data should prove to be consistent with thermal emission models. Significantly greater spectral resolution will be needed in order to determine just how the ionization deviates from that expected in equilibrium.

EUV observations reported at the conference also shed light on the nature and distribution of the hot gas in the Local Bubble. In what is possibly the longest single observation in the history of astronomy, 2×10^7 s of data from the Extreme Ultraviolet Explorer *(EUVE)* have been searched for line emission without success (Vallerga and Slavin*). *EUVE* was not designed for observations of diffuse emission, however, so this negative result is consistent with thermal emission from hot gas if iron is depleted.

EUV shadowing observations reported by Berghöfer et al* show that the pressure of the hot gas in the Local Bubble is consistent with that inferred from X-ray observations and is the same along several different lines of sight. X-ray data have given a consistent value of the pressure of the local hot gas over the years: McKee and Ostriker (1977) inferred $P/k \simeq 1.0 \times 10^4$ K cm^{-3} from the data available 20 years ago, whereas the current estimate from *ROSAT* is 1.4×10^4 K cm^{-3} (Snowden*). After correcting for the background, *EUVE* data give a value of 1.3×10^4 K cm^{-3}. Both the X-ray and EUV estimates are based on the assumption of CIE and use the atomic physics in the Raymond and Smith code; if the gas is far from ionization equilibrium, as in the model of Breitschwerdt and Schmutzler (1994), then the pressure of the hot gas could be much different. Provided the CIE estimates are even approximately correct, however, there is a major conundrum: why is the thermal pressure of the hot gas so much greater than that of the Local Cloud? It is not that the Local Cloud is anomalous, since its thermal pressure is similar to that of another cloud in the Local Bubble (Jenkins*).

It is also possible to study hot gas through UV absorption lines of highly ionized species. Gry and Dupin* presented a tantalizing result, a 2σ detection of C IV absorption at the velocity of the Local Cloud. Confirmation of this possible detection is important. C IV and Si IV absorption lines have also been detected in the spectra of stars located in Loop I (Freire Ferrero*).

2.3 Warm and Cold Gas in the Local Bubble

Although most of the volume in the Local Bubble appears to be filled with hot gas, most of the mass is either warm and partially ionized or cold and neutral. The ionization of helium in the warm gas seems comparable to that of hydrogen: Bowyer* reported values of the column density ratios for helium and hydrogen as $N(\mathrm{He}^+)/N(\mathrm{He}^0) \simeq 0.3 - 0.4$ and $N(\mathrm{H}^+)/N(\mathrm{H}^0) \lesssim 0.25$; Barstow et al* found values of 0.27 and 0.35 for these ratios. The relatively high ionization of helium is quite different from that in most of the ISM: optical (Reynolds and Tufte 1995) and radio (Heiles et al., 1996) observations of gas beyond the Local Bubble show that He II recombination lines are quite weak, indicating a low level of ionization compared to that of hydrogen. However, Slavin and Frisch* have shown that emission from evaporating clouds has an ionizing spectrum that can account for the observed ionization.

21 cm observations show that the Local Bubble corresponds to a cavity in the distribution of H I (Mebold*). For a disk of uniform column density $N_\perp$, one expects the column density observed at Galactic latitude b to satisfy the relation $N \sin b = N_\perp$ =const. Instead, observations show that $N \sin b$ increases away from the poles, reaching a constant value only for $b < 45°$. The column density at negative latitudes is somewhat greater than that at positive latitudes, as expected for the Sun's position above the Galactic plane.

Mapping the distribution of the warm and cold gas in and around the Local Bubble is a major undertaking, and several groups reported progress on this problem. Fluctuations in the X-ray background at high latitudes are consistent with absorption by clouds of column density $2 - 3 \times 10^{19}$ cm^{-2} (Warwick et al*), similar to those seen by Spitzer and Fitzpatrick (1993) toward a star in the halo. EUV observations can give a coarse picture of the distribution of gas with no velocity information; observations of over 200 white dwarfs and late-type stars in the EUV showed that the wall of absorbing gas toward Loop I is very nearby (Hutchinson et al*). Optical absorption line observations can provide much higher spatial resolution and offer velocity resolution as well. Beckman et al* have measured optical absorption lines toward a number of stars within 200 pc in order to constrain the three dimensional distribution of the interstellar gas, and they developed a technique to combine these data with 21 cm observations to extract more information about the absorbing clouds. They have found cold atomic clouds, cold clouds with warm ionized envelopes, and warm clouds without any cold cores. The cold atomic clouds have masses up to 2000 $M_\odot$, and the warm clouds have masses up to several hundred $M_\odot$. Some of the clouds appear to be roughly spherical, whereas others are sheet-like. Ultra-high resolution spectroscopy of such clouds indicates a complex structure (Welty*): even at a resolution of 0.3 km s^{-1} , only about half the velocity components appear to have been resolved. The complexity of the velocity structure is apparent even for very nearby stars, with distances less than 30 pc (Welsh et al*). Finally, interstellar scintillation studies are suggestive of turbulent structures in the ionized component of the LISM (Bhat et al*; Bochkarev and Ryabov*).

Spectral line observations also allow the determination of the overall kinematics of the gas. The Local Cloud is moving away from the Scorpius-

Centaurus Association (Frisch*); indeed, the Local Cloud velocity is approximately parallel to the velocity of the H I associated with the center of the North Polar Spur (Heiles*). On the other hand, somewhat more distant gas, which appears to be outside the Local Bubble, is moving in the opposite direction, toward Sco-Cen (Genova et al*).

There is very little molecular gas in the Local Bubble. Studies of the molecular gas at the boundaries of the Local Bubble are quite valuable because of the high spatial resolution that is possible. Observations of CO emission at very high velocity resolution ($\lambda/\Delta\lambda \simeq 2 \times 10^7$) suggest that the gas is in structures only 200 AU in size (Falgarone*). Direct observations of these structures with interferometry would be highly desirable.

2.4 Magnetic Field and Cosmic Rays

Although it is widely recognized that magnetic fields and cosmic rays play a central role in the dynamics of the ISM (e.g., Parker 1969), they are often slighted because it is so difficult to observe them. Heiles*, who has devoted a non-negligible fraction of his scientific life to measuring interstellar magnetic fields, gave a graphic demonstration of this problem: acknowledging that there are very few data on the magnetic field in the Local Bubble (except for the inferred upper limit discussed in §2.1 and the observation of optical polarization in some nearby stars by Tinbergen 1982), he presented an analysis of the structure of the field in radio Loop I, which lies in the North Polar Spur. He argued that the surface brightness of the synchrotron emission is inconsistent with a limb–brightened shell; instead, the field must be concentrated in flux ropes that have been displaced into a spherical shell. Heiles's results suggest that the field inside Loop I is very weak (Kahn*). It appears that interchange instabilities have not acted to bring the flux ropes back into the interior. Furthermore, the lack of limb brightening suggests that the relativistic electrons can escape from the compressed flux ropes, at least in some cases.

Cosmic rays outside the solar system can be observed only indirectly, through the gamma ray emission and, for cosmic ray electrons, the synchrotron emission they produce. It is therefore of some interest that they can also be observed back through time. As discussed by Morfill*, the intensity of the cosmic rays has been remarkably constant over the last 10^3, 10^6, and 10^9 yr. (However, fluctuations in the ^{10}Be concentration in Antarctic ice cores have been interpreted as being due to brief fluctuations in the cosmic ray intensity—Sonett et al., 1987.) The constancy of the mean cosmic ray intensity is somewhat surprising, since the radius of the heliopause varies substantially as the Sun moves into different regions of the Galaxy, leading to variations in the amount of solar modulation (Frisch*).

2.5 Relation to the Sco-Cen Association

The Scorpius-Centaurus Association is the nearest OB association to the Sun, at a distance of about 170 pc. Berkhuijsen et al. (1971) identified the origin of the Loop I radio source with a "super-supernova" in Sco-Cen. Subsequently, Weaver (1979) pointed out that the stellar winds from the OB stars in Sco-Cen would create a large bubble surrounded by an H I shell. He identified the H I filaments associated with the Loop I radio source as the shell produced by Sco-Cen, and pointed out that the near side of the bubble must be very close to the Sun since the filaments subtend an angle of 170°. At about that time, Heiles (1979) showed that H I shells and supershells are quite common in the Galaxy. Weaver identified Radio Loop I as a supernova remnant that exploded in the Sco-Cen Bubble relatively recently. Weaver's picture of the interstellar structures associated with Sco-Cen is quite close to the one we have today, except that it is now known that the supernovae dominate stellar winds after several million years (Bruhweiler et al., 1980).

Egger* presented evidence for the interaction of the Sco-Cen Bubble with the Local Bubble. He identified an elliptical ring of H I in the Sco-Cen shell with the ring of dense gas formed by the collision of two bubbles (Yoshioka and Ikeuchi 1990). Absorption line data indicate that this H I ring is about 70 pc away. Soft X-rays are absorbed by this gas. However, there is also a lack of hard X-ray emission on the western side of Loop I, and he attributed this to cooling of the X-ray emitting gas by the H I ring. The pressure of the X-ray emitting gas within the Sco-Cen Bubble was estimated to be $P/k \simeq 2 \times 10^4$ K cm^{-3}, somewhat greater than that of the Local Bubble.

As remarked above, the very local ISM is moving away from Sco-Cen, just as if it were part of the Sco-Cen shell. How can this be consistent with the geometry inferred by Weaver and by Egger*, in which the shell has yet to reach us? In the discussion following Egger's talk, Kahn* pointed out that the shell of shocked gas between the Local Bubble and the Sco-Cen Bubble would be accelerated by the greater pressure in the latter, leading to a Rayleigh-Taylor instability. In addition, the colliding shells would be subject to a bending mode instability (Vishniac 1994). Hydrodynamic instabilities such as these could account for the complex structure of the H I in the LISM.

Frisch's* view of the Local Bubble is somewhat different from Egger's: she associates the creation of a separate bubble and shell with each of the three subassociations in Sco-Cen. The first two shells have already swept past us; we are now in the third shell. The low density of the Local Bubble is due in part to our location in an interarm region; it was not created by local supernova explosions. The question yet to be quantitatively addressed by this model is how the bubbles and shells created by the successive subassociations in Sco-Cen could avoid merging into a single large superbubble and supershell.

3 Why Is the Local Bubble Like That?

3.1 Models of the Local Bubble

In the beginning was Cox and Smith (1974). They were the first to show that the supernova rate in the Galaxy is large enough that supernova remnants (SNRs) should occupy a significant fraction of the volume of the ISM. They envisioned old SNRs forming a network of tunnels in the ISM. To determine the importance of the interaction among old SNRs, they introduced the porosity

$$Q = S \int V(t)\, dt, \tag{1}$$

where S is the SN rate per unit volume and $V(t)$ is the volume of an SNR of age t. If the porosity Q is small, then it is just the filling factor of SNRs; for large values of Q, interactions among SNRs will be important. Q is proportional to the four-volume of the SNR in space-time (McKee 1990), and it gives the filling factor of the SNRs in space when the ISM is in a steady state.

Next, McKee and Ostriker (1977) showed that simple estimates of the porosity suggest that it was large, and that this required a revision of our picture of the ISM. Instead of H I filling most of space, with isolated H II regions and SNRs, it is the hot gas in old SNRs that fills space. H I is in the form of clouds embedded in this hot gas; ionizing radiation propagating through the hot gas forms ionized envelopes around the clouds (the Warm Ionized Medium). The evolution of SNRs is dominated by the evaporation of embedded clouds. They conjectured that the Local Bubble was heated by a supernova remnant that is several hundred thousand years old. Such a remnant, with a pressure $P/k \simeq 1.0 \times 10^4$ K cm^{-3}, could account for the soft X-ray data then available. They did not attempt to explain the low mean density of the Local Bubble; however, in contrast to most other models, this model naturally accounts for the presence of clouds in the Local Bubble.

The first detailed model for the Local Bubble was developed by Cox and Anderson (1982). They considered the evolution of an SNR in a low density ($n \simeq 0.004$ cm^{-3}) ambient medium, which is similar to assuming that the remnant evolves in the hot phase of the McKee-Ostriker model without interacting with the embedded clouds. They included non-equilibrium ionization and thermal conduction, and showed that they could reproduce the soft X-ray count rates for a remnant with an age of about 10^5 yr.

Innes and Hartquist (1984) attempted to model the Local Bubble as an old superbubble. By making several assumptions that they did not attempt to justify—treating the evolution of a superbubble as being the same as that of a blast wave with an energy of 10 SNe, assuming that the ambient pressure of the interstellar medium is very large (7×10^4 K cm^{-3}), and neglecting thermal conduction—they were able to achieve consistency with the X-ray data for a bubble that is 4×10^6 yr old. Related models were considered by Edgar and Cox (1993); they attributed the high external pressure ($3.5 - 6.5 \times 10^4$ K cm^{-3} in their models) to a strong ambient magnetic field.

Three models of the Local Bubble dominated the discussion at this conference. The most innovative is that of Breitschwerdt and Schmutzler (1994), who assumed that a superbubble created by a local OB association evolved in a dense ($n \sim 10^4$ cm^{-3}) cloud, and that the superbubble broke out of this dense gas about 4×10^6 yr ago. Expanding in a steep density gradient, the hot gas in the bubble underwent catastrophic adiabatic expansion, cooling to $T \sim 6 \times 10^4$ K while remaining highly ionized. The advantage of this model is that the thermal pressure ($P/k \simeq 2600$ K cm^{-3}) is comparable to that observed in the Local Cloud, and the electron density ($n_e \simeq 0.024$ cm^{-3}) is comparable to the mean electron density to a nearby pulsar. However, precisely because the model is so innovative, it faces a number of challenges: according to Snowden*, it predicts more M-band emission than is observed; it may not produce the line emission observed in the DXS experiment; the confinement of the superbubble by its natal molecular cloud is problematic in view of the destructive effects of photoionization (Whitworth 1979); the low thermal pressure of the gas is achieved at the expense of a large ram pressure, which was not accounted for; and there is no evidence for the other stars in the association, nor the enormous mass of gas ($M \sim 10^6$ $M_\odot$ for a molecular cloud of radius 10 pc and density 10^4 cm^{-3}). This model is readily testable by the next generation of soft X-ray experiments.

Cox* and Smith and Cox* presented the most sophisticated models yet developed for the Local Bubble. These models include thermal conduction, non-equilibrium ionization calculated with current plasma codes, and the effects of magnetic pressure. Agreement with the data is obtained for a model with 2 or 3 supernova explosions in the Local Bubble, with the last one having occurred several million years ago. An advantage of this model is that it accounts for the low mean density of the Local Bubble. Since the last SN occurred so long ago, the model is also consistent with the presence of relatively quiescent H I in the Bubble. However, the model rests on a crucial assumption: even though thermal conduction both within the bubble and to the walls of the bubble is allowed, it is assumed that no heat is conducted from the hot gas to the embedded clouds (i.e., there is no cloud evaporation).

Finally, Slavin* presented models that allowed for cloud evaporation. His models are incomplete in that they do not attempt to explain the origin of the hot gas–presumably, the Local Bubble was reheated by a supernova several hundred thousand years ago. He accounted for the Wisconsin Be band, B band, and C band intensities with models in which the hot gas far from the clouds has a temperature $T \simeq 10^{6.0-6.2}$ K. Slavin and Frisch* also showed that it is possible to account for the relatively high ionization of helium observed in the LISM, a problem not addressed by the other two models.

3.2 Evolution of Superbubbles

The early work of Cox and Smith (1974) and McKee and Ostriker (1977) treated supernovae as occurring randomly in space and time. Beginning with the work of Bruhweiler et al. (1980), however, it was realized that many supernovae occur in OB associations, and that this would profoundly affect the interaction of supernovae with the ISM. In particular, the larger amounts of energy available meant that gas in the disk could be lofted into the halo, revitalizing the fountain model of Shapiro and Field (1976). Ikeuchi* and his collaborators were major contributors to these studies. His work culminated in the "chimney model" (Ikeuchi 1987; Norman and Ikeuchi 1989), in which superbubbles break out of the Galactic disk, spewing hot gas and nucleosynthesis products into the halo and then onto the disk below.

Several new results on superbubble evolution were presented: Chernin* showed how small superbubbles can be created by the merger of two SNRs, and described the complex hydrodynamics that results. Bisnovati-Kogan and Silich* showed how the three-dimensional evolution of superbubbles can be treated in the thin-shell approximation. Theis et al* and Ehlerova et al* discussed fragmentation and star formation in expanding supershells.

Radiative losses can affect the evolution of interstellar bubbles and superbubbles more than is generally realized. Although the solutions for momentum-conserving bubbles (Avedisova 1971; Steigman, Strittmatter, and Williams 1975) and adiabatic bubbles (Castor, McCray, and Weaver 1975) were developed at about the same time, it has generally been assumed that actual bubbles are described by the adiabatic solution for most of their lives. However, photoevaporation of embedded clouds (including the molecular cloud out of which the association is born) can "poison" the bubble by injecting enough material into the bubble interior so that it can cool, thereby making the expansion intermediate between the momentum-conserving and adiabatic cases (McKee, Van Buren, and Lazareff 1984). Evidence for cloud photoevaporation in the Cepheus Bubble was presented by Patel et al*.

The Sco-Cen Bubble appears to show evidence for the effects of radiative cooling. Egger* presented evidence that the hot gas in the western part of the bubble is cooling (see above). The oldest subassociation is 14-15 Myr old (Frisch 1995), so that the bubble would have begun expanding about 12 Myr ago. The effective wind luminosity of the association is now about 6×10^{37} erg s^{-1} and the ambient density is about 0.6 cm^{-3} (Egger*), so that the predicted bubble radius is $R_b \simeq 130$ pc for the momentum-conserving case and 290 pc for the adiabatic case. Such a model is not very realistic because it does not allow for the lower wind luminosity early in the life of the association, nor for the fact that the density is larger near the association than it is near the Sun (Frisch 1995). We allow for these effects by assuming a wind luminosity $L_w = \mathcal{L}t$ and an ambient density $\rho_0 = \rho_{01} r^{-1}$, where $\mathcal{L}$ and ρ_{01} are constants. These assumptions lead to a bubble radius that increases linearly with time: For the momentum-conserving bubble $R_b = (\mathcal{L}/2\pi\rho_{01} v_{\rm in})^{1/3} t \simeq 70$ pc, where

$v_{\rm in} \simeq 2000$ km s^{-1} is the average velocity of the gas injected by the stellar winds and supernovae; for the adiabatic bubble, $R_b = (\mathcal{L}/8\pi\rho_{01})^{1/4}t \simeq 210$ pc (Koo and McKee 1992). The observed radius of about 150 pc is intermediate between these two cases, suggesting that energy loss has been important in the evolution of the Sco-Cen Bubble. Our approximate attempt to include the effects of the evolution of the Sco-Cen association and the density structure of the ISM is not fully consistent, however: for $R \propto t$, the shell velocity should be $v = R/t = 12.5$ km s^{-1}, less than the observed 19 km s^{-1} (Sofue et al., 1974). If the superbubble is as old as suggested by the ages of the OB stars, then the shell must have undergone acceleration, at least for part of its expansion; this would have been unstable, contributing to the injection of matter into the hot bubble and adding to the radiative losses.

3.3 The LISM as an Astrophysical Laboratory

Study of the LISM enables us to address two fundamental astrophysical questions: How does hot gas cool? How is pressure balance achieved in the ISM? The nature of the cooling of hot gas is important not only for the interstellar medium of galaxies, but also for hot gas in clusters of galaxies and in active galactic nuclei. The best-studied cooling mechanism is radiation by the gas due to collisional excitation and to recombination (Böhringer*). The emitted spectrum depends on the nature of the ionization. Broadly speaking, there are three general cases: the gas can be overionized due to cooling outpacing recombination for $T \lesssim 10^6$ K or to adiabatic expansion; it can be in collisional ionization equilibrium; or it can be underionized, as in the case of evaporation. For $T \gtrsim 10^6$ K, another cooling mechanism can dominate, emission by embedded dust grains (Ostriker and Silk 1973). This is self-limiting because the dust is destroyed by sputtering as it cools the gas (Draine 1981; Dwek et al., 1996). In the Local Bubble, the sputtering time $t_{\rm sp}$ for the large grains that contain most of the mass is about 10^8 yr (Ferrara*); for the smaller grains that dominate the cooling, $t_{\rm sp} \propto$ (grain radius) is less. Thermal conduction can also cool the gas, both by conducting the heat to cooler gas, which radiates it, and by evaporating cooler gas, thereby sharing the heat among more particles. Finally, the gas can cool by adiabatic expansion, as in the model of Breitschwerdt and Schmutzler (1994); however, the radiative cooling time is unaffected by adiabatic expansion for $T \gtrsim 10^5$ K (Kahn 1976). Each of these cooling mechanisms depends on the history of the hot gas, and the Local Bubble provides an ideal laboratory for studying how history can determine the physical conditions and spectrum of hot gas.

What is the nature of pressure balance in the ISM? On the average, the pressure must equal the weight of the material above it (Parker 1969). The ISM is heavy: the weight is of order 3×10^4 K cm^{-3} (Boulares and Cox 1990), far greater than the typical thermal pressure in the ISM, 4000 K cm^{-3} (Jenkins et al., 1983). Thermal pressure $P_{\rm th}$, turbulent pressure $P_{\rm turb} = \rho\delta v^2$, magnetic pressure P_B, and cosmic ray pressure $P_{\rm cr}$ all contribute to the pressure support. Kalberla et al* presented a model for the hydrostatic equilibrium of the ISM based on the inclusion of an H I component with a velocity dispersion of 60 km s^{-1}; such a component is not seen in the Bell Labs data (Blitz, private communication), however, and requires confirmation.

On small scales, cosmic rays cannot sustain pressure gradients in a static medium because they are diffusive. If the nonthermal pressure is truly turbulent, then it depends on the size scale ℓ since $\delta v \propto \ell^p$, where $p \sim \frac{1}{3} - \frac{1}{2}$. (In molecular clouds, this is just the line width–size relation [Larson 1981]). Thus, turbulent pressure is also ineffective on small scales. Recent Faraday rotation observations indicate an rms fluctuating field of only about 1 μG on scales $\lesssim$ 4 pc, confirming the weakness of turbulent fluctuations on small scales in the diffuse ISM (Minter and Spangler 1996). Locally, then, $P_{\rm th} + P_B = {\rm const}$ in equilibrium. The rms interstellar magnetic field is about 4.2 μG (Heiles 1996), corresponding to a pressure of 5100 K cm^{-3}. This is only slightly greater than the typical interstellar thermal pressure, so we expect thermal pressure to be significant on small scales. Why then does the thermal pressure in the Local Cloud ($P_{\rm th} \sim$ 2000 K cm^{-3}) appear to be much less than that of the hot gas in the Local Bubble ($P_{\rm th} \sim$ 14000 K cm^{-3} for CIE models)? A strong magnetic field in the Local Cloud has been suggested as one possibility (Cox and Reynolds 1987, McKee 1996), but the field at the solar bow shock (Lallement*) is inadequate. Understanding the pressure balance in the LISM has broad implications for interstellar gas dynamics, and the LISM provides the laboratory in which we can hope to find the answer.

4 Is the LISM Typical of the ISM as a Whole?

4.1 Observation

The Local Bubble is too small to be readily detected from a vantage point elsewhere in the Galaxy. However, the Sco-Cen Bubble appears to be large enough that it could be detectable. Observations of superbubbles in Eridanus (Burrows and Guo*), Sagittarius (Miville-Deschênes et al*), and Cepheus (Patel et al*) were presented at the conference. No one addressed the important issue of what fraction of the ISM is filled with these structures. Such an estimate should be more straightforward for the LMC, since we are not embedded within it. The remarkable 21 cm map of the LMC obtained by Kim and Stavely-Smith* with the Australia Telescope Compact Array shows large numbers of superbubbles and provides the basis for such an estimate.

How typical are the properties of the gas observed locally? Dettmar* showed that Warm Ionized Medium of the type seen locally in the Galaxy can be observed in other galaxies as well; the source of ionization of this gas has not been conclusively identified. Dixon et al* showed that the distribution of H I and H_2 beyond the Local Bubble does not vary much over several long sightlines at the solar circle. As for the hot gas, the *ROSAT* shadowing experiments discussed above clearly show that there is a significant amount of emission from the Galactic halo. It is very difficult to unambiguously separate halo emission from that due to the Local Bubble, the disk, and extragalactic sources (Freyberg*). Several groups attempted such a separation nonetheless: Sumner et al* inferred the existence of a halo component by spectral modeling of *ROSAT* PSPC data. Kerp and Pietz* modeled the radiative transfer of X-rays through the ISM and estimated the temperatures of the Local Bubble and halo components. Pietz et al* developed an isothermal model ($T = 1.6 \times 10^6$ K) and estimated a total Galactic X-ray luminosity of 7×10^{39} erg s^{-1}. Finally, in order to model the 0.5-1 keV emission, Wang* generalized the isothermal model of Wolfire et al. (1995) to a polytrope ($P \propto \rho^{5/3}$) and developed a clever iterative technique to fit the parameters for the emitting gas in regions that are not contaminated by disk emission. Locally, he found that the pressure of the hot gas is about 3000 K cm^{-3}, comparable to that of the H I. Kerp et al* showed evidence for interaction of high-velocity clouds with the X-ray background. This supports the interpretation of an X-ray halo for the Galaxy, since two of the larger HVCs are now known to be in the halo: Complex A is in the range $2.5 < z < 7$ kpc (Van Woerden et al*) and Complex M is in the range $z < 4.4$ kpc (Danly et al., 1993). There may also be an X-ray halo around the LMC (Blondiau*).

4.2 Theory

Several theories for the origin of hot gas in the halo were discussed. The fountain model was reviewed by Kahn* but the basic premises of the model were not confirmed in 3D hydrodynamic simulations presented by Avillez et al*. Supernovae in the halo can directly heat the gas there, but Shelton et al* showed that such supernovae could not account for the observed X-ray emission; on the other hand, the remnants of halo supernovae could survive a long time and possibly account for the highly ionized gas seen in absorption in the halo. Simulations of magnetic reconnection, which has been suggested as a heating mechanism for halo gas by Parker (1990) and by Raymond (1992), were presented by Lesch and Birk*. They suggested that reconnection could account for Kerp et al's* observations of excess soft X-ray emission near several high-velocity clouds. It is a challenge to understand how a microscopic process like reconnection, acting on ordered magnetic fields, can account for the large volume emission measure of hot gas inferred by Kerp et al. Finally, there is the cosmic-ray driven wind model of Breitschwerdt and Schmutzler (1994), which produces an X-ray spectrum dominated by adiabatic expansion. This model is confronted with the difficulty of explaining why the H I in the solar vicinity is observed to be falling toward the plane; indeed, there are *no* observed cases in which outflowing gas is seen in UV absorption (deBoer*).

Before closing, I would like to briefly discuss the status of the theory of the ISM that Ostriker and I developed twenty years ago (McKee and Ostriker 1977). The role of this theory in the development of our understanding of the ISM is discussed by Ikeuchi*. The theory is based on the premise that the porosity of hot gas in the ISM (eq. 1) is large. This premise has been criticized in a detailed study by Slavin and Cox (1993), but we have seen ample evidence at this conference that it is valid locally. Frisch's* map of the local ISM shows that hot (or at least very low density) gas is seen along virtually every line of sight from the Sun. Once a line of sight intersects a cloud, however, no further hot gas appears on the map (with the exception of the Sco-Cen Bubble). Is this because the Sun is in a unique place in the ISM, or is it because of the difficulty in detecting the presence of hot gas under typical interstellar conditions?

Several results of the theory have been confirmed by subsequent observation: Large pressure fluctuations were predicted and have been observed (Jenkins et al., 1983); the properties of the Warm Ionized Medium predicted for the disk of the Galaxy are consistent with observations (Reynolds 1993); and the weakness of X-ray emission from other disk galaxies (e.g., McCammon and Sanders 1984) is also consistent with the model, since evaporative cooling is expected to reduce the temperature of the hot gas to $\sim 5 \times 10^5$ K. On the other hand, the weakness of O VI absorption lines (Shelton and Cox 1994) remains a puzzle. Direct evidence for cloud evaporation in the ISM is weak, although Spitzer (1996) concluded that conductive interfaces provide the best explanation for the presence of highly ionized atoms in the disk. The possible detection of a conductive interface in the Local Bubble (Gry and Dupin*), together with the development of an evaporative model for the X-ray spectrum (Slavin*) and the ionizing spectrum (Slavin and Frisch*) of the Local Bubble, suggest that further study of the Local Bubble should prove fruitful in determining the role of thermal conduction in the ISM.

5 Prospects for the Future

The Holy Grail for studies of the Local Bubble is the spectrum of the hot gas, which contains the key to the origin of the gas. A number of missions that are aimed at achieving this goal were described at the conference. On the first day of the conference we heard the dramatic announcement of the successful launch of the EUV spectrograph *EURD*, an international collaboration between Spain (PI: Morales*) and the US (PI: Bowyer*). As this is being written, the spectrograph is successfully gathering data. Other missions that will attempt to measure the spectrum of the hot gas are *CUBIC* (Burrows), *UCB* (Edelstein et al*), *GRADES* (Hurwitz et al*), and *FUSE*. McLean et al* described a rocket-borne spectrograph that will carry out spectroscopy of the C IV and O VI lines. Planned X-ray missions, including *ASTRO-E* and *ABRIXAS*, may also shed light on the spectrum.

It is remarkable that our extragalactic colleagues have made far more progress in mapping the three dimensional distribution of galaxies many megaparsecs away than we have made in mapping the distribution of local interstellar clouds within a hundred parsecs. What is needed is an organized effort to map the LISM, a Local Interstellar Absorption Line Survey. Steps in this direction were described by Beckman*, Gry and Dupin*, Genova et al*, Welsh et al*, and Welty*. The advent of accurate stellar distances from *Hipparcos* should improve the maps considerably.

The Local Bubble poses a number of theoretical problems as well. As emphasized by Breitschwerdt*, the cooling of hot plasma reflects its history; how can we learn to read that history? How can one reconcile the low thermal pressure observed in the Local Cloud with the substantially greater thermal pressure inferred in the Local Bubble? How do SNRs and superbubbles evolve in a magnetized, multiphase ISM, including photoevaporation of clouds by EUV radiation from OB stars? What is the efficiency of thermal conduction in magnetized, turbulent media? How do magnetic fields reconnect in the ISM? We can hope that when the next Local Bubble conference is organized, some of these observational and theoretical questions will have been answered.

Acknowledgements. I wish to thank Dieter Breitschwerdt and Joachim Trümper for the opportunity to participate in this conference. Comments by Dieter, Don Cox, Pris Frisch, Wilt Sanders, and Jon Slavin were invaluable. My research is supported in part by NSF grant AST95-30480.

References

Avedisova, F.S. (1971) AZh 48, 894 (Sov Astron 15, 708 (1972))
Berkhuijsen, E.M., Haslam, C.G.T., Salter, C.J. (1971) A&A 14, 252
Bochkarev, N. (1987) Ap&SS 138, 229
Boulares, A., Cox, D.P. (1990) ApJ 365, 544
Breitschwerdt, D., Schmutzler, T. (1994) Nature 371, 774
Bruhweiler, F.C., Gull, T.R., Kafatos, M., Sofia, S. (1980) ApJ 238, L27
Castor, J., McCray, R., Weaver, R. (1975) ApJ 200, L107
Cox, D.P., Anderson, P.R. (1982) ApJ 253, 268
Cox, D.P., Reynolds, R. (1987): ARA&A 25, 303
Cox, D.P., Smith, B. (1974) ApJ 189, L105
Danly, L., Albert, C.E., Kuntz, K.D. (1993) ApJ 416, L29
Draine, B.T. (1981) ApJ 245, 880
Dwek, E., Foster, S.M., Vancura, O. (1996) ApJ 457, 244
Edgar, R.J., Cox, D.P. (1993) ApJ 413, 190
Frisch, P.C. (1995) Space Sci. Rev. 72, 499
Garmire, G.P., Nousek, J.A., Apparao, K.M.V., Burrows, D.N., Fink, R.L., Kraft, R.P. (1992) ApJ 399, 694
Heiles, C. (1979) ApJ 229, 533
Heiles, C. (1990) ApJ 354, 483
Heiles, C. (1996) in *Polarimetry of the Interstellar Medium*, ed. W.G. Roberge and D.C.B. Whittet (ASP, San Francisco), p. 457
Heiles, C., Koo, B.-C., Levenson, N.A., Reach, W.T. (1996) ApJ 462, 326
Ikeuchi, S. (1987) in *Starbursts and Galaxy Evolution*, ed. T. Thuan, T. Montmerle, J. Tran Thanh Van (Gif-sur-Yvette, Editions Frontieres), p. 27

Innes, E.E., Hartquist, T.W. (1984) MNRAS 209, 7
Jenkins, E.B., Jura, M., Loewenstein, M. (1983) ApJ 270, 88
Kahn, F. (1976) A&A 50, 145
Kim, S.-H., Martin, P.G. (1996) ApJ 444, 293
Kim, S.-H., Martin, P.G., Hendry, P.D. (1994) ApJ 422, 164
Koo, B.-C., McKee, C.F. (1992) ApJ 388, 103
Kurth, W.S., Gurnett, D.A., Scarf, F.L., Poynter, R.L. (1984) Nature 312, 27
Larson, R.B. (1981) MNRAS 194, 809
McCammon, D., Sanders, W.T. (1984) ApJ 287, 167
McKee, C.F. (1990) in *The Evolution of the Interstellar Medium*, ed. L. Blitz (ASP, San Francisco), p. 3
McKee, C.F. (1996) in *The Interplay between Massive Star Formation, the ISM, and Galaxy Evolution*, ed. D. Kunth et al (Gif-sur-Yvette, Editions Frontieres), p. 223
McKee, C.F., Ostriker, J.P. (1977) ApJ 218, 148
McKee, C.F., Van Buren, D., Lazareff, B. (1984) ApJ 278, L115
Minter, A.H., Spangler, S.R. (1996) ApJ 458, 194
Norman, C., Ikeuchi, S. (1989) ApJ 345, 372
Ostriker, J.P., Silk, J.I. (1973) ApJ 184, L113
Parker, E.N. (1969) Space Sci. Rev. 9, 651
Parker, E.N. (1990) in *Galactic and Intergalactic Magnetic Fields*, ed. R. Beck, R.P. Kronberg, and R. Wielebinski (Dordrecht, Kluwer), p. 169
Raymond, J. (1992) ApJ 384, 502
Reynolds, R. (1993) in *Massive Stars: Their Lives in the Interstellar Medium*, ed. J.P. Cassinelli and E.B. Churchwell (San Francisco, ASP), p. 338
Reynolds, R., Tufte, S. (1995) ApJ 439, L17
Shapiro, P.R., Field, G.B. (1976) ApJ 205, 762
Slavin, J.D., Cox, D.P. (1993) ApJ 417, 187
Sofue, Y., Hamajima, K., Fujimoto, M. (1974) PASJ 26, 399
Sonett, C.P., Morfill, G.E., Jokipii, J.R. (1987) Nature 330, 458
Spitzer, L. Jr (1956) ApJ 124, 20
Spitzer, L. Jr (1996) ApJ 458, L29
Spitzer, L. Jr, Fitzpatrick, E.L. (1993) ApJ 409, 299
Steigman, G., Strittmatter, P.A., Williams, R.E. (1975) ApJ 198, 575
Tinbergen, J. (1982) A&A 105, 53
Vishniac, E.T. (1994) ApJ 426, 186
Weaver, H. (1979) in *The Large-Scale Characteristics of the Galaxy*, ed. W.B. Burton (Reidel, Dordrecht), p. 295-300
Whitworth, A. (1979) MNRAS 186, 59
Witte, M., Rosenbauer, H., Banaszkiewicz, M., Fahr, H.J. (1992) Adv. Space Res. No. 6, 13, 121
Wolfire, M.G., McKee, C.F., Hollenbach, D.J., Tielens, A.G.G.M. (1995) ApJ 453, 673
Yoshioka, S., Ikeuchi, S. (1990) ApJ 360, 352

Part XI

Workshop

The Local Bubble is a breath-taking subject ...

... as a spontaneously organized workshop on Friday afternoon demonstrated.

Workshop for Continued Discussion of the Local Bubble

Wilton T. Sanders[1] and Priscilla C. Frisch[2]

[1] Department of Physics, University of Wisconsin–Madison, 1150 University Avenue, Madison, WI 53706, USA
[2] Department of Astronomy and Astrophysics, University of Chicago, 5640 South Ellis Avenue, Chicago, IL 60637, USA

Abstract. At the end of the regularly scheduled conference, the organizers made available a large sunny conference room for the remaining participants to continue discussion. The main topics were the structure of the Local Bubble and the properties of the Galactic X-ray halo. Here we try to summarize the main points of those discussions.

1 Introduction

The above topics, and others that participants brought to the workshop, generated three more hours of creative and lively discussion. Remarkably, the discussion was able to stay focused on one particular topic for quite a long time, although excursions into related areas were common. The broad-ranging fast-paced nature of the discussions and the at-times-dizzying shifts in topic make it impossible to give a detailed summary of what was said or who said it. Over half of the conference attendees participated in the workshop, and we apologize for mentioning only a few names here.

2 The Structure of the Local Bubble

By far the most time was spent trying to find agreement on what might naively appear to be easy at a conference on the Local Bubble: what is the gross structure of the Local Bubble? and how did it get that way? It was apparent that each person had a slightly different picture in mind. Two competing pictures came out in the presentations by Egger and by Frisch and their contributions to this volume should be read for more details. An important part of each of these pictures is the relationship between the Local Bubble and disturbances in the interstellar medium caused by the hot stars of the nearby Scorpius-Centaurus OB association.

In the Egger picture, the interaction of the Local Bubble shell with the Loop I superbubble shell generated by stellar winds or supernovae from the Sco-Cen stars results in a circular ring of neutral hydrogen at the location of the interaction of these two shells. Inside this ring is a sheet of neutral hydrogen, a relatively dense "interaction wall," separating the two bubbles with the distance from the Sun to the wall set at 40 ± 25 pc by optical absorption lines. In this picture, the Local Bubble shell and the soft X-ray background were formed by a separate supernova explosion near the Sun.

The arguments in support of this view come from the ROSAT XRT/PSPC observations of the soft X-ray sky within ~ 70 degrees of the center of the Sco-Cen association. These show large portions of a ring in absorption against the soft X-ray brightening from the interior of the Sco-Cen bubble. There are neutral hydrogen ridges or filaments coincident with this ring of increased X-ray absorption, and optical absorption measurements put their distance at roughly 70 pc. Earlier, these filaments were viewed as lying on the stellar wind bubble created by the Sco-Cen stars (Weaver 1979).

The Frisch picture begins with the view that there is not a separate supernova remnant surrounding the Sun, but that the Sun is located in an interarm region of our Galaxy where superbubble expansion would have taken place in an inhomogeneous medium. Frisch argues that superbubble shells formed by earlier star formation episodes in Sco-Cen will have swept past the Sun into the low density interarm region, and that the Sun is now embedded in a shell of a ~ 4 Myr old superbubble or supernova from Sco-Cen. In this picture, the evolution of stars in Sco-Cen drives the observed outflow of interstellar gas, and the Local Bubble cavity is an artifact of the interaction of the initial Sco-Cen superbubble with the interarm region. Frisch claims that the general flow of interstellar matter from the Sco-Cen region, in which virtually all interstellar absorption lines seen in stars within 50 pc of the Sun have velocities between 0 and -20 km s^{-1} in the local standard of rest, is difficult to interpret if the Sun were immersed in a supernova remnant distributed symmetrically about the Sun, but is naturally explained in her picture.

In the Frisch picture the first and second of the Sco-Cen shells have passed the solar neighborhood and are responsible for the local hot plasma that we call the Local Bubble, but the third shell is only now reaching the Sun. A variation on this picture was suggested in which the third shell has not yet reached the Sun and is responsible for the hydrogen feature that Egger calls the wall, and possibly the ring also. The clouds seen in optical absorption lines and moving through the solar neighborhood from the direction of Sco-Cen at ~ 15 km s^{-1} could be stragglers from the passing of the second shell, or they could be early arrivals associated with the third shell.

Heiles indicated that the 21 cm data show neutral material flowing towards the Sun from the general direction of Sco-Cen, but the direction of origin of the flow is not centered on Egger's ring, nor is it particularly associated with any of the Sco-Cen sub-associations.

Egger and Snowden argued that the lack of a general soft X-ray enhancement at energies less than 0.28 keV towards Sco-Cen required an N_{H} "wall" of at least 10^{20} cm^{-2} between the Sun and Sco-Cen. Frisch countered that a uniformly distributed neutral wall separating the Local Bubble and the

Sco-Cen superbubble does not exist because there are stars well beyond the distance of the "wall" that have an $N_{\rm H}$ of 1 - 3 $\times 10^{19}$ cm^{-2} (λ Sco and β Cen), so if there is a wall, it must be patchy or partially ionized.

Other data were considered to see if they could help decide between these two pictures. ROSAT WFC observations of white dwarfs and late-type stars are consistent with the Sun being in a Local Bubble with a wall towards the Galactic Center direction, but at a distance of 30 pc. The pressure of the Local Bubble was discussed, but did not appear to favor either picture. The velocities of the walls of the Local Bubble in various directions could be used to check for consistency with either picture, but there was not enough data in the hands of the workshop participants to come to a conclusion. If the wall velocities were more nearly in radial expansion in all directions, it would favor the Egger picture, while if there were an asymmetric flow in the wall velocities aligned with an axis away from Sco-Cen, it would favor the Frisch picture. Further analysis of the X-ray spectrum of the soft X-ray background (see presentation by Sanders), or additional measurements to look for spectral differences in different directions, may be able to distinguish among different plasma emission mechanisms and therefore models for the origin of the hot plasma.

A problem that was brought out in the workshop about the Frisch hypothesis is that it relies on a simple model of superbubble formation and neglects the radiative cooling of the bubble interior that occurs between star formation epochs.

At some point in the discussion, we mostly agreed that there was a ring of neutral hydrogen between the Sun and the Sco-Cen bubble and that its distance was of order 70 pc, but the cause of the ring was still being debated. The existence of a dense "wall" interior to the ring was accepted by the adherents to the Egger picture, but denied by the Frisch camp. And that's how we left it.

3 The Properties of the Galactic X-ray Halo

Two quite different pictures of the Galactic X-ray halo were presented by Snowden and by Wang, and their contributions to this volume should also be read for more details. The X-ray halo described by Snowden is very patchy and has a temperature, T $\lesssim 10^6$ K, while the corona described by Wang varies smoothly over the sky and covers a range of temperatures, from T $< 0.5 \times 10^6$ K in the outer Galaxy to T $> 5 \times 10^6$ K towards the center of the Galaxy. The easiest way to reconcile these two pictures is for the Snowden halo and the Wang corona to have different origins and to be located at different heights above the plane of the Galaxy.

Wang assumed that the corona is quasi-hydrostatic in the gravitational potential of the Galaxy, using the potential described by Wolfire et al. (1995), and assumed a polytropic equation of state for the gas with index 5/3. The

Wang corona extends more than 10 kpc above the galactic plane and more than 15 kpc radially from the galactic center, and could be interpreted as a hot outflow from the center of the Galaxy. The Snowden halo needs only to be above the neutral hydrogen gas layer of the galaxy, less than 1 kpc from the plane, and to have characteristic scale sizes of hundreds of pc.

The origin of the Snowden halo is not clear. Galactic fountains and winds, the Breitschwerdt and Schmutzler model, and supernovae in the halo were among the many mechanisms touched upon in the discussion. Again, understanding the spectra of the soft X-ray background was mentioned as a way to distinguish among the possible mechanisms for the Snowden halo, and as a check on the Wang corona.

A tangent to this discussion led to high latitude clouds and it was noted that there was still much information about the halo contained in observations of high latitude clouds that was not yet fully exploited. Of particular interest were the velocity distributions of neutral clouds and the wings of lines from low ionization state ions. But the difficulty of separating clouds from the intercloud gas was also noted as part of the reason for the lack of progress.

4 Conclusions

The origin of the soft X-ray background, both the halo component and the Local Bubble component, will not be understood until the spectrum is adequately modeled.

The resolution of the issue of whether or not the Local Bubble is the result of a separate supernova remnant will rest on both the correct interpretation of the soft X-ray data and an understanding of the sequence of propagating star formation in Sco-Cen which has evidently taken place, and how this star formation relates to the bright X-ray emission associated with Sco-Cen.

In the 3/4 keV band, Galactic center emission must be separated from the Sco-Cen contribution to make sense of either emission component.

Acknowledgements. This work was supported in part by NASA grants NAG5-3080, NAG5-3524, NAGW-5061, and NAG5-5142.

References

Weaver, H. F. (1979): in *Large Scale Characteristics of the Galaxy*, IAU Symposium No. 84, ed. W. B. Burton (Reidel:Dordrecht), 295

Wolfire, M. G., McKee, C. F., Hollenbach, D., & Tielens, A. G. G. M. (1995): ApJ 453, 673

Part XII

Appendix

Plates

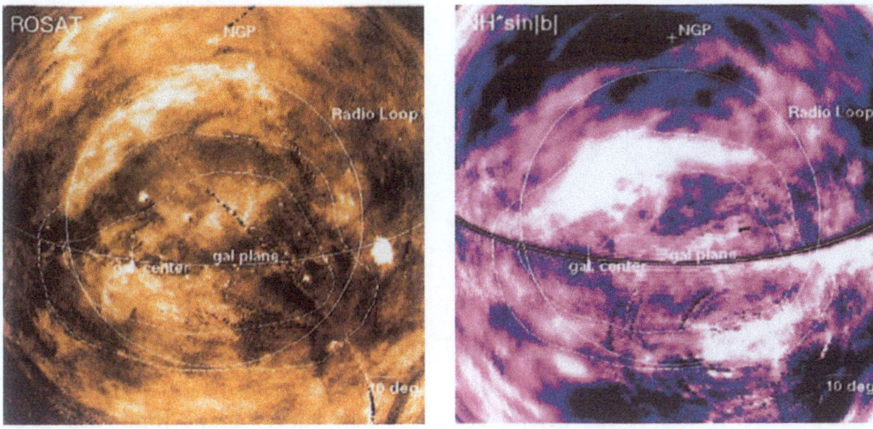

Plate 1. *Left:* ROSAT Survey map centred on Loop I in the energy range 0.1–2.0 keV. The circle of 58° radius follows the radio continuum loop. The dashed lines outline the contours of the annular shadow. *Right:* H I map divided by cosec$|b_{II}|$; (see article by *Egger*).

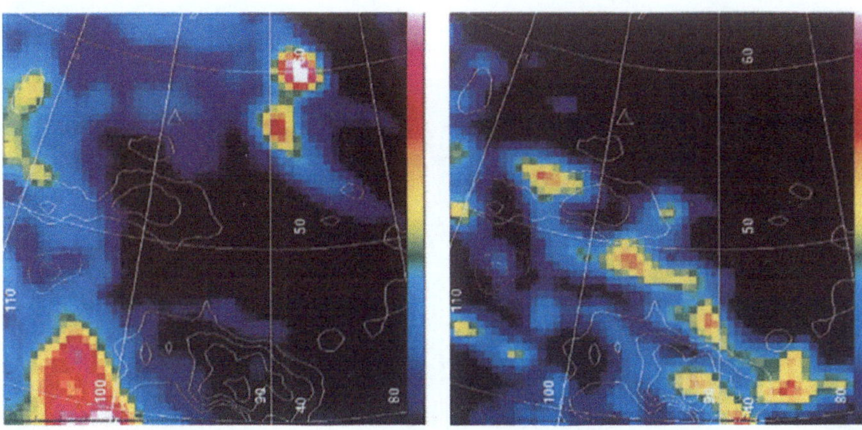

Plate 2. *Left:* H I column density distribution (black: low, white: high) of the IVCs across the field of interest, $N_{\rm HI}(\min) = 0.2 \cdot 10^{20}\,{\rm cm}^{-2}$ and $N_{\rm HI}(\max) = 2.0 \cdot 10^{20}\,{\rm cm}^{-2}$ ($v_{\rm LSR} \in [-90; -25]\,{\rm km\,s}^{-1}$). Contour lines mark the 1/4 keV excess emission (solid lines) starting at 4σ with steps of 2σ. *Right:* The HVC column density distribution ($v_{\rm LSR} \in [-450; -90]\,{\rm km\,s}^{-1}$) with $N_{\rm HI}(\min) = 0.1 \cdot 10^{20}\,{\rm cm}^{-2}$ and $N_{\rm HI}(\max) = 1.0 \cdot 10^{20}\,{\rm cm}^{-2}$; (see article by *Kerp et al.*).

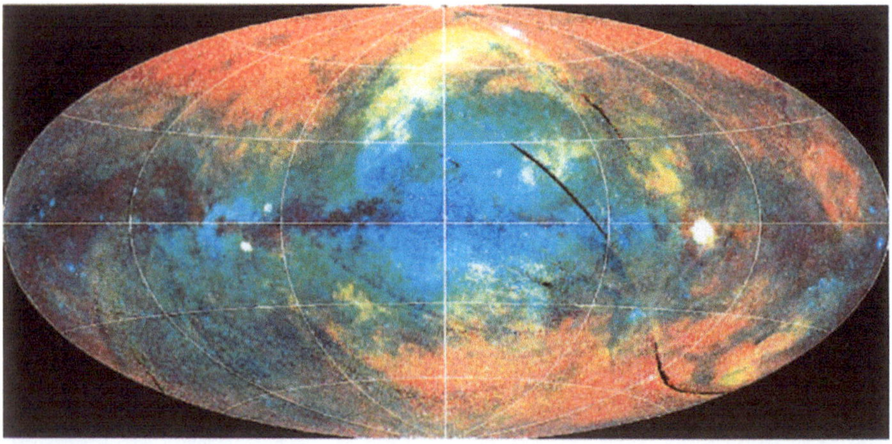

Plate 3. RGB image of the SXRB using the new survey maps (Snowden et al. 1997). The three energy bands, 1/4, 3/4, and 1.5 keV, are colour-coded with red, green, and blue, respectively. As in all our maps we use galactic coordinates with increasing longitudes to the left and $l = 0°$ in the center; (see article by *Freyberg & Schmitt*).

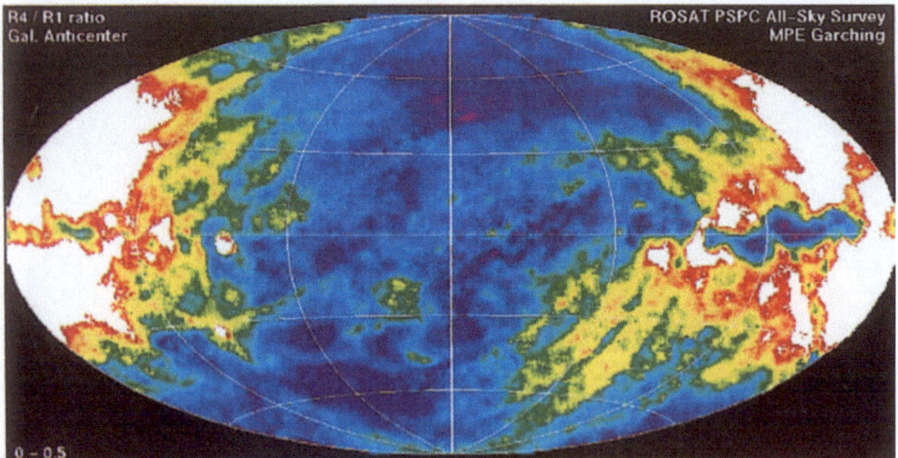

Plate 4. R4/R1 ratio centered at $l = 180°$, the colour bar ranges from 0.0 to 0.5; (see article by *Freyberg & Schmitt*).

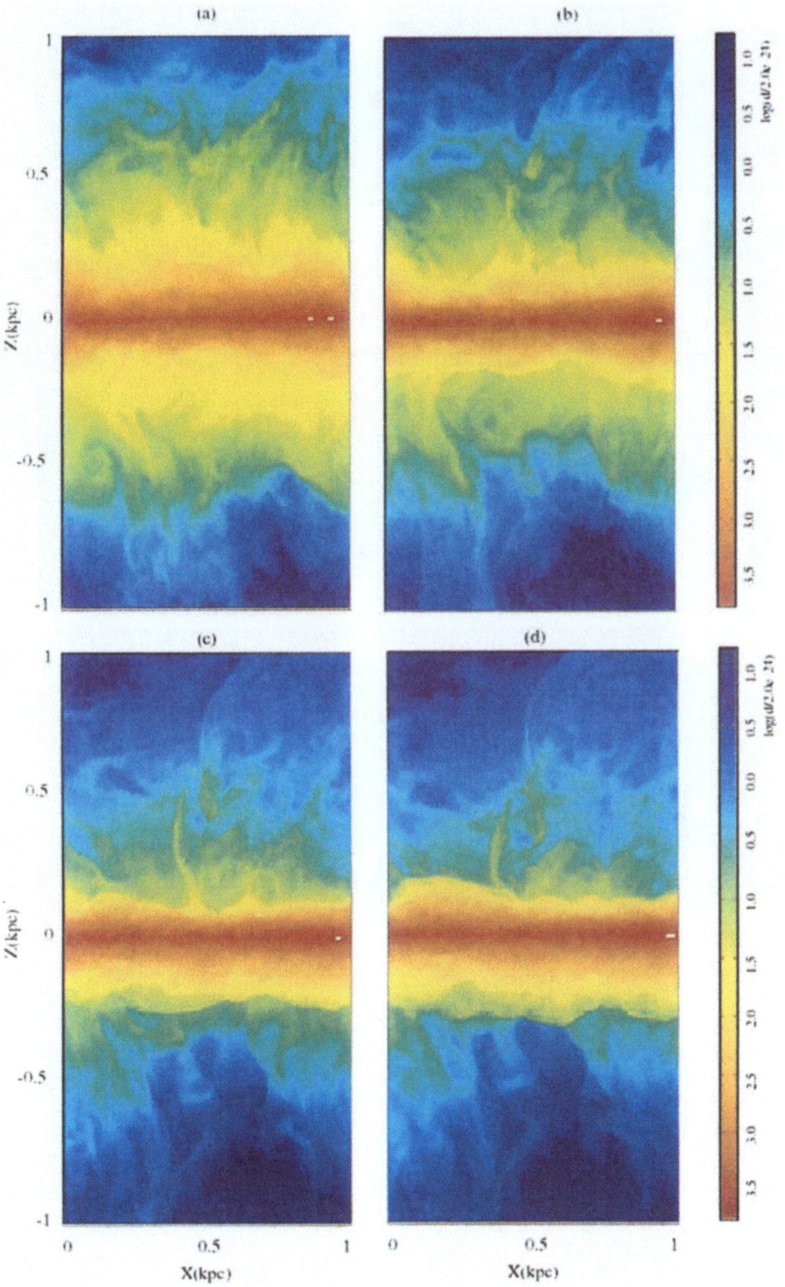

Plate 5. Density distribution in a section perpendicular to the Galactic plane for times 280, 283, 286, and 289 Myr after start of calculations; (see article by *Kahn*).

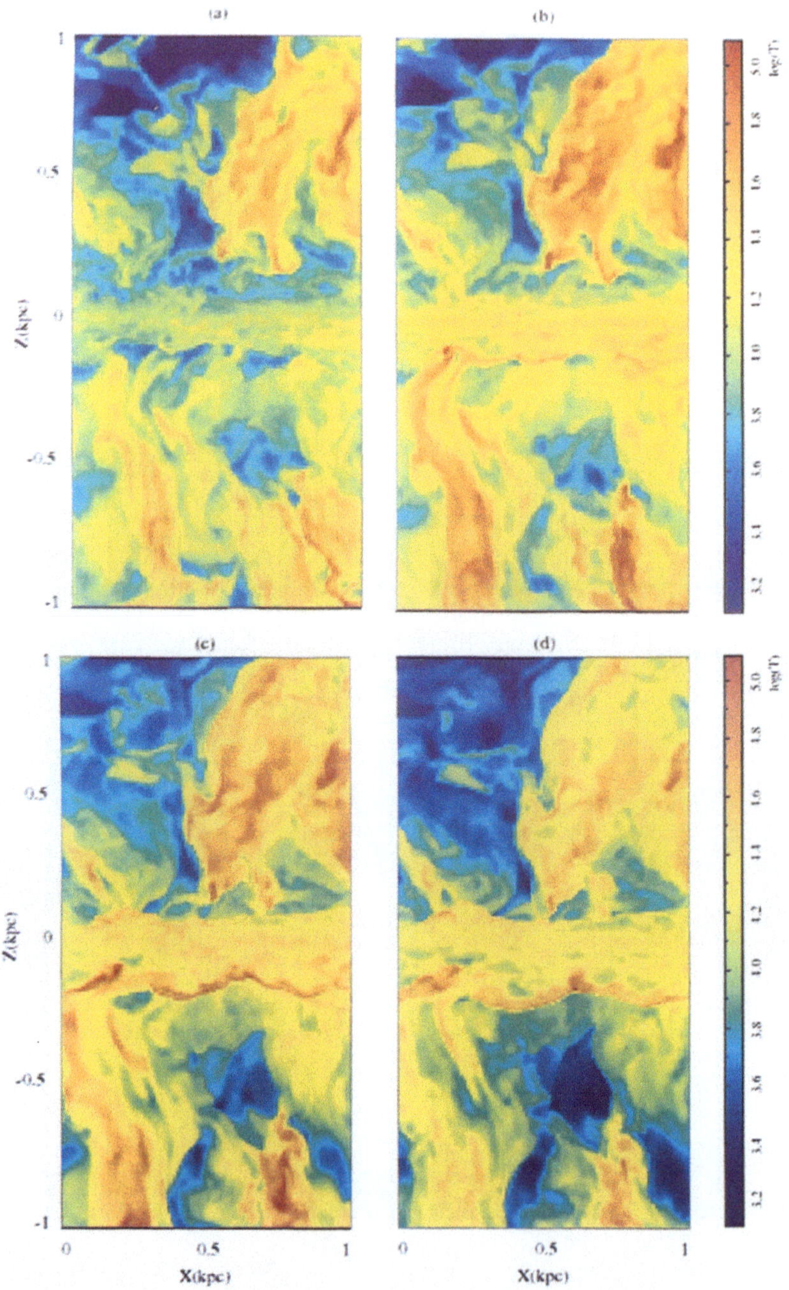

Plate 6. Temperature distribution perpendicular to the Galactic plane for times 280, 283, 286, and 289 Myr after start of calculations; (see article by *Kahn*).

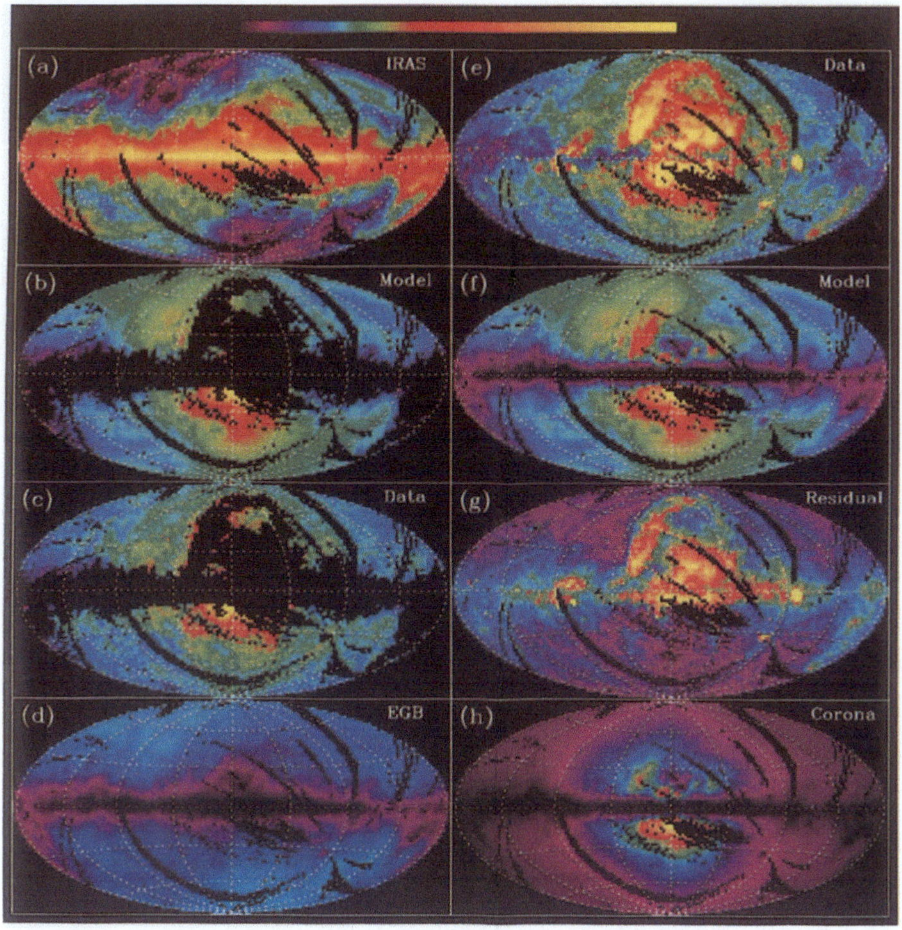

Plate 7. Surface brightness intensity maps in the zero-centered Galactic coordinates with Aitoff equal-area projections. Both X-ray data and model are in the 3/4 keV band. From the top panel to the bottom one in the left column are the zero-level corrected *IRAS* 100μm survey (a); Model map covering only pixels used in the fit at the threshold 0.4 (b); *ROSAT* data in the same region (c); and the extragalactic background component of the model (d). The panels in the right column, arranged for an easy comparison with Fig. 4, are the *ROSAT* data (e); the model including regions not used in the fit (f); the residual of the *ROSAT* survey minus the model (g); the corona component of the model (h). The false color range is between 1 and 440 MJy sr^{-1} logarithmically for (a), 0 to 5 $\times 10^{-4}$ counts s^{-1} arcmin^{-2} linearly for (b), (c), (d), (e), (f), and (h), and -0.7 to 5 $\times 10^{-4}$ counts s^{-1} arcmin^{-2} for (g); the black represents the lowest intensity while the yellow is the highest; (see article by *Wang*).

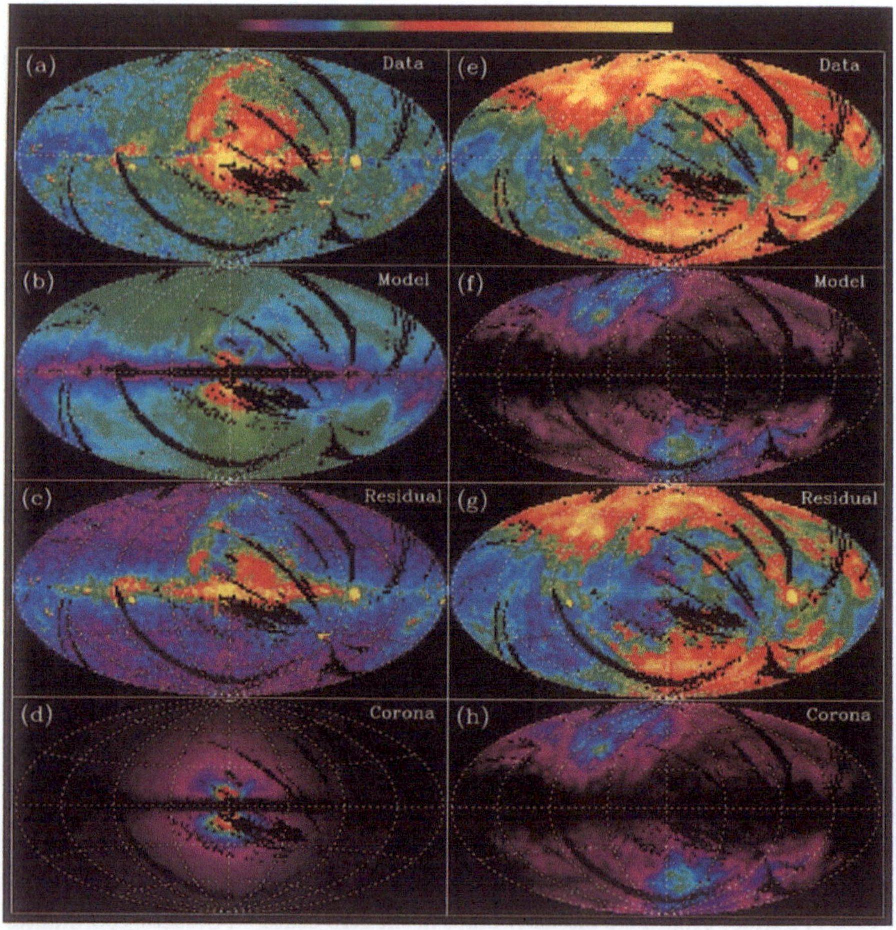

Plate 8. Data and model predictions in the 1.5 keV band (left column) and in the 1/4 keV band (right column). The projections and sky coverage are the same as those in Fig. 1. From the top row to the bottom one are the *ROSAT* survey data, the model predictions of the corona plus the extragalactic background, the residual maps of the survey data minus the model, and the corona components of the model in the two bands. The false color range in the left column spans between 0 to 5 $\times 10^{-4}$ counts s^{-1} $arcmin^{-2}$ for the panels (a), (b) and (d) and -0.7 to 5 $\times 10^{-4}$ counts s^{-1} $arcmin^{-2}$ for (c); the range in the right column is all between 0 to 15 $\times 10^{-4}$ counts s^{-1} $arcmin^{-2}$; (see article by *Wang*).

Subject and Object Index

Author Index

FSC
www.fsc.org
MIX
Papier aus verantwortungsvollen Quellen
Paper from responsible sources
FSC® C105338

If you have any concerns about our products,
you can contact us on
ProductSafety@springernature.com

In case Publisher is established outside the EU,
the EU authorized representative is:
Springer Nature Customer Service Center GmbH
Europaplatz 3, 69115 Heidelberg, Germany

Printed by Libri Plureos GmbH
in Hamburg, Germany